HIGHLIGHTS OF ASTRONOMY
Volume 14

COVER ILLUSTRATION

THE ASTRONOMICAL TOWER OF THE KLEMENTINUM COMPLEX IN PRAGUE'S OLD TOWN, CZECH REPUBLIC

Klementinum is the oldest Jesuit college in Czech Lands founded immediately after Jesuits came to Prague in 1556. In 1622 Klementinum merged its activities and libraries with the Charles University, Prague (founded 1348) and formed Charles-Ferdinand University in 1654. After the Jesuit order was abolished in 1773, Klementinum continued to house the University and focused on philosophical and mathematical studies, astronomy and theology.

The Astronomical Tower was erected together with the completion of the Baroque library and of the Mirror Chapel in 1722. The tower is crowned with a statue of Atlas holding the celestial sphere on his shoulders.

Josef Stepling (1716-'78), a study director at the Philosophical Faculty, equipped the tower with astronomical instruments and thus transformed it into an astronomical observatory for both scientists and students. The computation and designs necessary to construct these instruments were provided by an excellent mechanician, Jan Klein (1684-1762). The instruments he produced were among the best of what his epoch could offer.

Antonín Strnad (1746-'99) was appointed the observatory adjunct in 1774. He improved significantly systematic meteorological observations performed in Klementinum since 1752. This series of daily temperature measurements of a high quality, later enriched by other meteorological data and continuing without any interruption up to now, is used as a benchmark for Central European climatologic research.

The Astronomical Tower was used by astronomers to keep time in Prague (Tempus Pragense). From 1842, a man waving a flag from the top of the tower signaled mean Prague noon. This signal was followed by a cannon shot from 1891. This tradition was maintained till 1926. This all came to an end in 1928, when the astronomical observations were moved to the new Ondřejov Observatory and all scientific research, except the meteorological recording, came to an end in 1939.

IAU XXVI GENERAL ASSEMBLY

HIGHLIGHTS OF ASTRONOMY

EDITORIAL BOARD

Editor

Chief co-editors

INTERNATIONAL ASTRONOMICAL UNION
UNION ASTRONOMIQUE INTERNATIONALE

HIGHLIGHTS OF ASTRONOMY

VOLUME 14

AS PRESENTED AT THE IAU XXVI GENERAL ASSEMBLY, 2006

Edited by

KARÉL A. VAN DER HUCHT
General Secretary of the Union

CAMBRIDGE UNIVERSITY PRESS
The Edinburgh Building, Cambridge CB2 8RU, United Kingdom
32 Avenue of the Americas, New York, NY 10013-2473, USA
477 Williamstown Road, Port Melbourne, VIC 3207, Australia
Ruiz de Alarcón 13, 28014 Madrid, spain
Dock House, The Waterfront, Cape Town 8001, South Africa

First published 2007

Printed in the United Kingdom at the University Press, Cambridge

Typeset in System LaTeX 2ε

A catalogue record for this book is available from the British Library

Library of Congress Cataloguing in Publication data

ISBN 9780521896832 hardback
ISSN 1743-9213

Table of Contents

JD 5. Calibrating the Top of the Stellar M-L Relation

Editors: Claus Leitherer, Anthony F.J. Moffat & Joachim Puls

JD 6. Neutron Stars and Black Holes in Star Clusters

Editor: Frederic A. Rasio

JD 7. The Universe at $z > 6$

Editors: Daniel Schaerer & Andrea Ferrara

JD 8. Solar and Stellar Activity Cycles

Editors: Klaus G. Strassmeier & Alexander Kosovichev

JD 10. Progress in Planetary Exploration Missions
Editor: Guy J. Consolmagno

JD 15. New Cosmology Results from the Spitzer Space Telescope
Editors: George Helou & David T. Frayer

JD 16. Nomenclature, Precession and New Models in Fundamental Astronomy
Editors: Nicole Capitaine, Jan Vondrák & James L. Hilton

III. SPECIAL SESSIONS

SpS 1. Large Astronomical Facilities of the Next Decade

Editors: Gerard F. Gilmore & Richard T. Schilizzi

SpS 2. Innovation in Teaching/Learning Astronomy Methods

Editors: Rosa M. Ros & Jay M. Pasachoff

SpS 3. The Virtual Observatory in Action: New Science, New Technology, and Next Generation Facilities

Editors: Nicholas A. Walton, Andrew Lawrence & Roy Williams

SpS 5. Astronomy for the Developing World

Editors: John B. Hearnshaw & Peter Martinez

SpS 6. Astronomical Data Management

Editor: Raymond P. Norris

SpS 7. Astronomy in Antarctica

Editor: Michael G. Burton

Preface

The first volume of the IAU *Highlights of Astronomy* (Perek 1968) recorded the proceedings of *Invited Discourses*, covering broad fields of astronomy, *Joint Discussions* and *Special Meetings*, covering specialised fields of astronomy, as presented at the IAU XIII General Assembly held in Prague, August 1967, and started a tradition. Ever since the IAU XXII General Assembly in Den Haag, 1994, also six IAU Symposia have been held during IAU General Assemblies, the proceedings of which have appeared in the regular IAU Symposium Proceedings series.

In August 2006, the International Astronomical Union returned to Prague, Czech Republic, for the IAU XXVI General Assembly which hosted, next to the business meetings of its 12 Divisions, 40 Commissions and 75 Working Groups, a comprehensive scientific program.

The proceedings of the six IAU GA *Symposia*: No. 235 (Combes & Palouš 2007), No. 236 (Milani, Valsecchi & Vokrouhlický 2007), No. 237 (Elmegreen & Palouš 2007), No. 238 (Karas & Matt 2007), No. 239 (Kupka, Roxburgh & Chan 2007), and No. 240 (Hartkopf, Guinan & Harmanec 2007) have all been published in the regular IAU Symposium Proceedings series.

The present *Highlights of Astronomy* Volume 14 comprises the proceedings of five *Invited Discourses*, including the discourse presented during the Inaugural Ceremony on 'Astronomy in Prague, from the past to the present' by Alena Adravova, of seventeen *Joint Discussions* and of six *Special Sessions*. Proceedings of Special Session No. 4 on Hot Topics are not available, but many papers presented therein have been presented also elsewhere during the GA.

The organizers-editors of JD 4 (Gómez de Castro & Barstow 2007) and JD 13 (Corbally, Bailer-Jones, Giridhar & Lloyd Evans 2006) and SpS 5 (Hearnshaw & Martinez 2007) have also published complete proceedings of their scientific meetings.

Numerous scientific sessions of smaller size have been held during the GA Business Meetings of the IAU Commissions and Working Groups, also providing a wealth of highly important scientific results. Reports of some of those sessions are available on the individual web sites of those IAU Commissions and Working Groups.

It is my pleasure to thank the organizers-editors of the scientific meetings' proceedings for their prompt and courteous response to my requests and for timely submitting their draft manuscripts resulting in the present volume, in spite of certain delays caused by substantial changes at the IAU Secretariat that came about immediately after the Prague 2006 General Assembly. Each co-editor used the room available in the offered style file in a different way, giving some local flavour to each of the chapters in this volume.

Last but not least, I wish to acknowledge the formidable efforts of my predecessor Oddbjørn Engvold, IAU Executive Assistant Mme. Monique Léger-Orine, and the Prague NOC/LOC chaired by Jan Palouš, Jan Vondák and Cyril Ron, in organizing the IAU XXVI General Assembly. Together they have dealt in a most outstanding and commendable way with the enormous task of organizing a plethora of large and small scientific and business meetings during the two weeks of the GA, entertaining 2412 participating astronomers from 73 countries, involving 650 oral papers and over 1550 poster papers.

The present volume is faithful to its title: its provides a substantial part of the Highlights of Astronomy in 2006.

Karel A. van der Hucht
IAU General Secretary
Paris, Utrecht, September 2007

References

Combes, F., & Palouš, J. (eds.), 2007, *Galaxy Evolution across the Hubble Time*, Proc. IAU Symp. No. 235 (Cambridge: CUP)

Corbally, C., Bailer-Jones, C., Giridhar, S., & Lloyd Evans, T. (eds.), 2006, *Exploiting Large Surveys for Galactic Astronomy*, Proc. IAU XXVI GA Joint Discussion No. 13, *Mem. S.A.It.* 77, no. 4, 1026 - 1190

Elmegreen, B. G., & Palouš, J. (eds.), 2007, *Triggered Star Formation in a Turbulent ISM*, Proc. IAU Symp. No. 237 (Cambridge: CUP)

Gómez de Castro, A. I., & Barstow, M. A. (eds.), 2007, *The Ultraviolet Universe: Stars from Birth to Death*, Proc. XXVI GA Joint Discussion No. 4 (Madrid: Editorial Complutense)

Hartkopf, W. I., Guinan, E. F., & Harmanec, P. (eds.), 2007, *Binary Stars as Critical Tools and Tests in Contemporary Astrophysics*, Proc. IAU Symp. No. 240 (Cambridge: CUP)

Hearnshaw, J. B., & Martinez, P. (eds.), 2007, *Astronomy for the Developing World*, Proc. IAU XXVI GA Special Session No. 5 (Cambridge: CUP)

Karas, V., & Matt, G. (eds.), 2007, *Black Holes: from Stars to Galaxies – across the Range of Masses*, Proc. IAU Symp. No. 238 (Cambridge: CUP)

Kupka, F., Roxburgh, I. W., & Chan, K. L. (eds.), 2007, *Convection in Astrophysics*, Proc. IAU Symp. No. 239 (Cambridge: CUP)

Milani, A., Valsecchi, G. B., & Vokrouhlický, D. (eds.), 2007, *Near Earth Objects, our Celestial Neighbors: Opportunity and Risk*, Proc. IAU Symp. No. 236 (Cambridge: CUP)

Perek, L. (ed.) 1968, *Highlights of Astronomy*, Proc. IAU XIII General Assembly (Dordrecht: Reidel)

I. Invited Discourses

Highlights of Astronomy, Volume 14
IAU XXVI General Assembly, 14-25 August 2006
Karel A. van der Hucht, ed.

doi:10.1017/S1743921307009817

Astronomy in Prague: from the past to the present

Alena Hadravová[1] and Petr Hadrava[2]

[1]Research Center for the History of Science and Humanities, Academy of Sciences, Prague, Czech Republic
[2]Astronomical Institute, Academy of Sciences, Ondřejov, Czech Republic
email: had@pleione.asu.cas.cz

Welcome to Prague. Welcome to this Congress Centre built in a close neighbourhood of the ancient seat of the first Czech dukes (Fig. 1). Its name Vyšehrad means the Upper Town. According to the oldest Czech chronicles, it was here where the legendary princess Libuše ordered her people to found the city of Prague and where she envisaged its glory touching the stars (Fig. 2). It was also here where the canon of Vyšehrad recorded in the first half of the 12th century into his chronicle some observed astronomical and meteorological phenomena.

Centuries later, Johannes Kepler, whose results achieved in Prague have really reached a starry fame, reminded Libuše in his treatise *Somnium seu De astronomia Lunari*, dealing with Lunar astronomy (Fig. 3, cf. Rosen 1967, p. 11): "Stimulated by the widespread public interest, I turned my attention to reading about Bohemia, and came upon the story of the heroine Libussa, renowned for her skill in magic." Kepler's own work is a nice example of links between astronomy and history, which he also studied seriously. He

Figure 1.

Figure 2.

Figure 3.

was searching for historical records of astronomical events and he also judged the impact of astronomical knowledge on the human mind.

Astronomy is not only a science, which is inseparable from other natural sciences and which yields different particular results of practical use, but it is also an important part of culture.

Figure 4.

It is thus natural that – like in many other countries – astronomy played here this role from prehistoric times. It was cultivated both by the church in monasteries and schools, as well as at courts of nobility in the medieval Czech Kingdom.

Let me show several examples: On the picture we can see the celestial globe from the possession of the Czech King Přemysl Otakar II or Wenceslas II (Fig. 4). It is supposed to be a gift from their relative, the King of Castille Alfonso X the Wise.

Richly illuminated Ptolemaic *Catalogue of Fixed Stars* written in Northern Italy in the fourteenth century belonged to the collection of astronomical and astrological manuscripts of King Wenceslas IV (Fig. 5). This manuscript follows the tradition of the 9th century Arabian scholar Al-Sufi. For each constellation a table with description of its stars, their coordinates and magnitudes, is accompanied by a figure with the positions of the stars marked.

In another codex from the library of Wenceslas IV we can find a illumination with an astronomer named Těříško observing the sky using a quadrant (Fig. 6). This is the first preserved portrait of some Czech astronomer, unfortunately otherwise unknown.

An illustration from the *Book of Mandeville's Travels* shows astronomers observing stars using quadrants and astrolabes, the most frequently used medieval astronomical instruments (Fig. 7). This Czech work from the early 15th century is nowadays in British Library.

The study of astronomy in Bohemia improved significantly when in the year 1348 the Prague University was established by the Emperor Charles IV as the first one in the Central Europe (Fig. 8). Astronomy reached a high level at the beginning of the 15th century. Among other outstanding scholars of this period was also Master Cristannus de Prachaticz, influential by his works in astronomy, mathematics, medicine and theology. In 1407 he wrote his *Composition* and *Use of the astrolabe* as the basis of his university lectures. The astrolabe, a universal astronomical and geodesic instrument, was widely used from Antiquity up to the early modern times. In the Middle Ages, study of the

Figure 5.

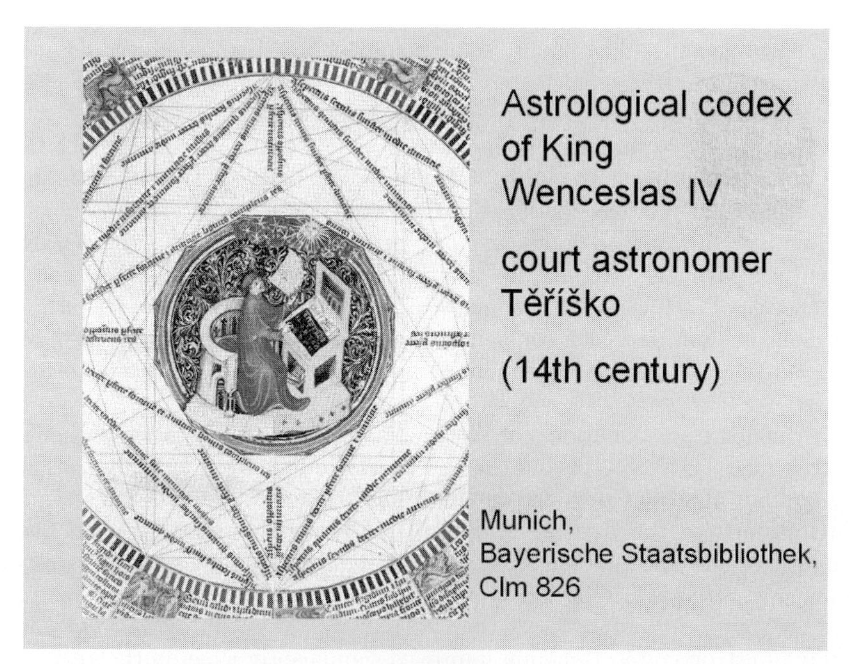

Figure 6.

astrolabe was a fundamental part of the astronomy curriculum in European universities. World literature contains many texts about this instrument, based on Ptolemy's Greek-language treatise *Planisphaerium* (2nd century A.D.). Cristannus's treatises circulated throughout Europe in many copies and later they were also the first printed texts on the astrolabe (the Perugia incunabulum 1477–1479). By tradition, both works were wrongly attributed to Robertus Anglicus or Prosdocimo de Beldomandi until their Prague origin

Figure 7.

Figure 8.

was proved by the critical edition of the Latin text (cf. Hadravová & Hadrava 2001, 2007). In the 1420s or 1430s, a version of Cristannus' treatises on the astrolabe was written by Master of Viennese University Johannes von Gmunden (the predecessor of Georg von Peuerbach and Regiomontanus at the same university), who borrowed it in its entirety and developed some its passages.

Figure 9.

The importance of Prague as a stage of astronomy culminated in the Rudolphine period at the turn of the 16th and 17th century. Prague was a manifold European cultural center under the rule of the Emperor Rudolph II (Fig. 9).

A prominent Czech scholar Thaddaeus Hagecius observed here the 'new star' of 1572 in the constellation of Cassiopeia. In his treatise *Dialexis* he proved from the measurement of its parallax the supralunar nature of this object. Hagecius was in friendly relation with Tycho Brahe and being a physician at Emperors court, he used his influence to reach an offer of the position of Imperial Mathematician to the Danish astronomer Tycho Brahe. When Tycho Brahe came to Prague, he invited also Johannes Kepler to work here with him. This promising collaboration of the best observer of that time with the bright theoretician finished soon by Tycho's death in 1601, but Johannes Kepler stayed in Prague for twelve years, which were the most fruitful in his life (Fig. 10). In this period he wrote his main works as *Astronomia nova, Optica* and others (cf. Christianson *et al.* 2002). Here in Prague Johannes Kepler also formulated the first two of his laws of planetary motion.

Instruments made by famous mechanicians on the Prague Imperial court like Erasmus Habermel or Joost Bürgi are preserved in National Technical Museum as well as in many museums all over the world (Fig. 11, cf. Horský & Škopová 1968).

In the epoch of re-catholicization after the Thirty Years War (1618–1648) the education in Bohemia was dominated by the Jesuit order established in 1540. Jesuits also cultivated science, including the astronomy (cf. Kašparová & Mačák 2002; Voit 1990). They came to Prague already in 1556 and in the middle of the 17th century they started to build the Clementinum college – a monumental complex of buildings. Owing to a frequent migration of Jesuits from one college to another or to missions, several skilled astronomers and mathematicians were professors in Clementinum or got their education here and worked later abroad. Among others, let us name Roderigo de Ariaga, Theodore Moretus, Valentin Stansel, François Noël, Karel Slavíček, Johann Klein, Joseph Stepling,

Figure 10.

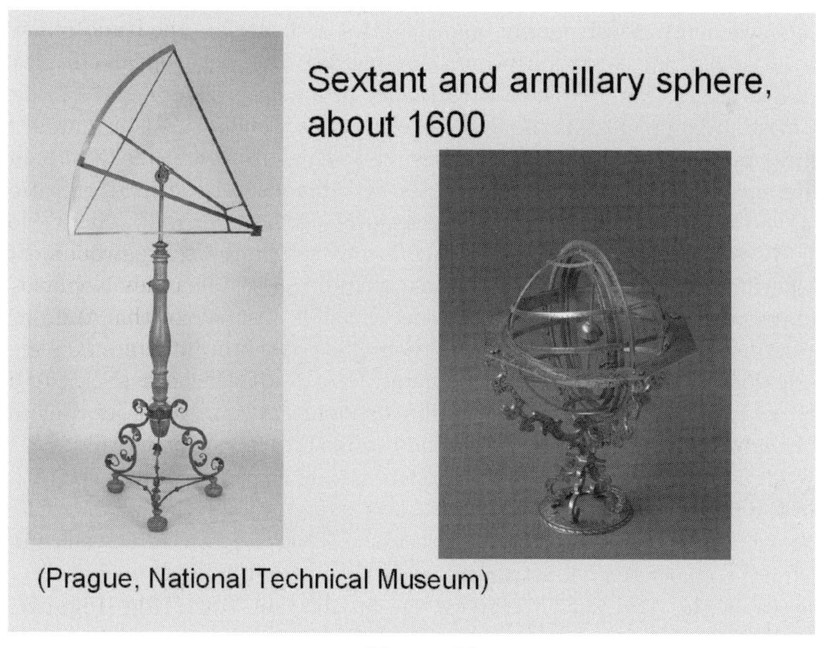

Figure 11.

Antonín Strnad or the premonstratensian Alois David. In spite of their overridingly ideological mission, Jesuits in Prague were relatively tolerant with respect to heliocentrism and related opinions, what is testified by some their writings as well as by frescoes in Clementinum, which show as comparable theories the Ptolemy's, Copernicus', Brahe's and Riccioli's model of the Solar System, or even the Giordano Bruno's multiplicity of the worlds.

Figure 12.

Several achievements in astronomy, mathematics and physics were also reached here later on at the time of the Austrian Empire. For instance the mathematician and philosopher Bernard Bolzano (Fig. 12) dealt with theory of functions, mathematical logics and analysis. Christian Doppler gave the first lecture on his principle in Karolinum, the historical central building of Charles University, in 1842 and he published his attempt to explain differences in colours of some binaries by difference of their radial velocities in Proceedings of the Royal Czech Learned Society (cf. Štoll *et al.* 1992). Ernst Mach was a professor of experimental physics at Prague university for three decades. He made here his experiments in acoustics and optics, including a study of shock waves generated by supersonic motion (cf. Janta & Niederle 2005). Astronomy was also one of the sciences promoted by Czech patriots during their effort at cultural, economical and political emancipation of Czech nation – let me name Josef Jan Frič, the founder of Ondřejov observatory, as example. The work of Slovak astronomer Milan Rastislav Štefánik, who studied in Prague, turned out to be crucial in establishing of Czechoslovakia.

A review of history of astronomy in Bohemia as well as of its present status can also be found in the book by Hadrava (2006).

Prague is said to be a city of a hundred towers. Allow me to remind you of three of them, which are connected with astronomy:

The first one is the tower of Charles Bridge on the Old Town side (Fig. 13). It was founded by the Emperor Charles IV as his triumphal arc at a unique event – in 1357 on the 9th of July at 5:31 in the place, from where the summer solstice sunset is seen behind the tomb of St. Vitus in the cathedral at Prague Castle. The importance of this structure is also emphasized by medieval cosmological model encrypted in its decorations – individual layers correspond to spheres of the Earth, Moon, Sun and stars (cf. Horský 1979).

The second tower at the Town hall on the Old Town Square is equipped by horologium – the famous Astronomical clock of Prague (Fig. 14). It was constructed in 1410 by clock-

Figure 13.

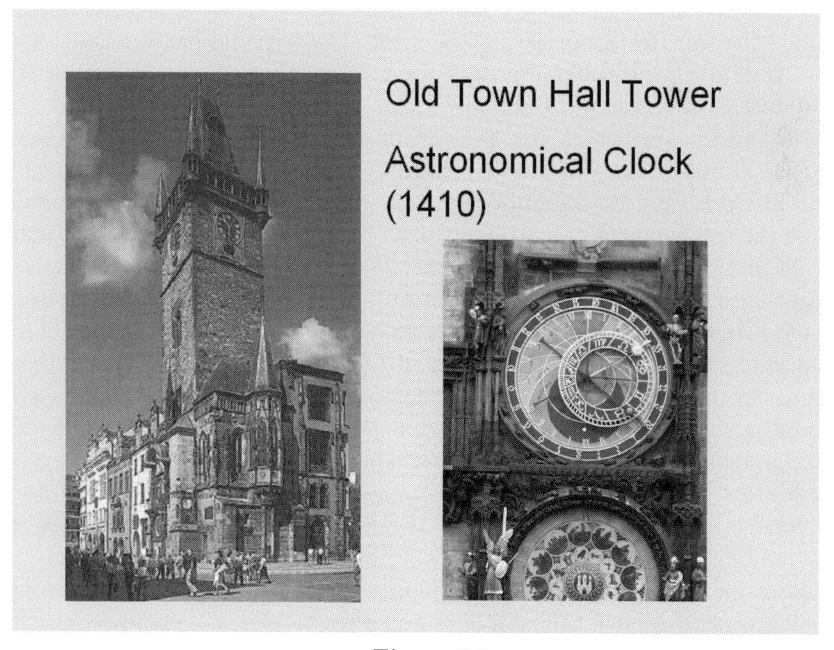

Figure 14.

maker Mikuláš of Kadaň according to recommendations of the Master of Prague university Iohannes Šindel, to use the then astronomical knowledge for practical purposes of time-keeping. Its dial is an astrolabe, which shows the position of Sun and Moon on the ecliptic, as well as the position of the ecliptic with respect to the horizon of Prague. It enables to read not only the nowadays commonly used equal hours, but also the unequal

Figure 15.

hours dividing the time from sunrise to sunset into twelve equal parts, as well as the old Czech time measured from the sunset and the sidereal time. The horologium still runs with the original gothic clockwork (Horský 1988).

Finally, the third tower, which you know from the poster of this General Assembly, is the Astronomical observatory of Jesuit college Clementinum erected in 1722 (Fig. 15). Its purpose was not only to measure the time with increasing precision, but also to contribute to the development of astronomy itself (cf. Šíma 2006). When the Jesuit Order was dissolved in 1773, the Astronomical tower became the Czech state observatory. Its tradition continues to the present day, and is developed by its follower, the Astronomical Institute of the Academy of Sciences of the Czech Republic, which gradually moved into its new seat in the astrophysical observatory at Ondřejov.

These three examples illustrate different historical roles of astronomy: as a symbol in culture and arts, a tool in daily practice, and an advanced science challenging the human mind and promoting its development. The present day sciences tend to diverge into many narrow fields, which are nearly inaccessible to non-specialists. However, an important feedback to the general culture persists. This is again remarkably seen in astronomy.

For instance, one of subjects of this meeting is the physics of black holes, the historical roots of which lead to Albert Einstein acting in Prague nearly a century ago (Fig. 16). Mathematical details of this subject are similarly difficult for layman as the systems of epicycles centuries ago. Nevertheless, basic notions and results have already became a part of common knowledge.

At the end of my talk, I would like to wish you for your coming meeting a great success in revealing new secrets of the Universe. And at the same time, I wish to the public to get from you a new enlargement of the horizons of mind. Thank you for your attention.

Figure 16.

References

Christianson, J. R., Hadravová, A., Hadrava P., & Šolc, M. (eds.) 2002, *Tycho Brahe and Prague: Crossroads of European Science, Acta Historica Astronomiae*, vol. 16 (Frankfurt am Main: Harri Deutsch Verlag)

Hadrava, P. 2006, *The European Southern Observatory and Czech Astronomy* (Prague: Academia)

Hadravová, A. & Hadrava, P. 2001, *Křišťan z Prachatic, Stavba a Užití astrolábu* (Prague: Filosofia)

Hadravová, A. & Hadrava, P. 2007 – in press, *Prachaticz, Cristannus de.* In: Dictionary of Scientific Biography (New York: Charles Scribner's Sons)

Horský, Z. 1979, *Založení Karlova mostu a kosmologická symbolika Staroměstské mostecké věže.* In: Staletá Praha (ed. Z. Buřival), vol. 9 (Prague: Panorama), pp. 197–212

Horský, Z. 1988, *Pražský orloj* (Prague: Panorama)

Horský, Z. & Škopová, O. 1968, *Astronomy Gnomonics. A Catalogue of Instruments of the 15th to the 19th Centuries in the Collections of the National Technical Museum* (Prague: Panorama)

Janta, J. & Niederle, J. (eds.) 2005, *Physics and Prague* (Prague: Academia)

Kašparová, J. & Mačák, K. 2002, *Utilitas matheseos. Jesuit Mathematics in the Clementinum (1602-1773)* (Prague: National Library)

Rosen, E. 1967, *Kepler's Somnium. The Dream or Posthumous Work on Lunar Astronomy* (Madison, London: The University of Wisconsin Pess)

Šíma, Z. 2006, *Astronomy and Clementinum* (Prague: National Library)

Štoll, I., Seidlerová, I., Schwippel, J., Pöss, O. & Šolc, M. 2006, *The Phenomenon of Doppler* (Prague: The Czech Technical University)

Voit, P. 1990, *Pražské Klementinum* (Prague: Národní knihovna)

[Extended version of lecture read by Alena Hadravová at the Inaugural Ceremony of the IAU XXVI General Assembly, Prague, 15 August 2006.]

Highlights of Astronomy, Volume 14
IAU XXVI General Assembly, 14-25 August 2006
Karel A. van der Hucht, ed.

doi:10.1017/S1743921307009829

The evolution of life in the Universe: are we alone?

Jill C. Tarter

SETI Institute, 515 N Whisman Road, Mountain View, CA 94043, USA
email: tarter@seti.org

Abstract. In his book *Plurality of Worlds*, Steven J. Dick (1984) has chronicled the millennia of discourse about other inhabited worlds, based upon deeply held religious or philosophical belief systems. The popularity of the idea of extraterrestrial life has waxed and waned and, at its nadir, put proponents at mortal risk. The several generations of scientists now attending this General Assembly of the International Astronomical Union at the beginning of the 21st century have a marvelous opportunity to shed light on this old question of habitable worlds through observation, experimentation, and interpretation, without recourse to belief systems and without risking their lives (though some may experience rather bumpy career paths). The newly-named and funded, multi-disciplinary field of astrobiology is extremely broad in its scope and is encouraging IAU members to learn and speak the languages of previously disparate disciplines in an attempt to answer the big picture questions: *'Where did we come from?'*, *'Where are we going?'*, and *'Are we alone?'* These are questions that the general public understand and support, and these are questions that are attracting students of all ages to science and engineering programs. These questions also push the limits of modern instrumentation to explore the cosmos remotely across space and time, as well as to examine samples of interplanetary space returned to the laboratory and samples of distant time teased out of our own Earth.

Within my personal event horizon, the other planetary systems long-predicted by theorists have been uncovered, along with many whose structures were not predicted. The 'just-so' conditions requisite for the comfort of astronomers have been understood to be only a very narrow subset of the conditions that nurture extremophilic, microbial life. Thus the potentially habitable real estate beyond Earth has been greatly expanded and within the next few decades it may be possible to detect the biosignatures or technosignatures of inhabitants on distant worlds, should there be any.

---

1. Introduction

Since we are here, it must be the case that this Universe, at least, is biofriendly. I leave the question of whether or not there are other universes to those of you in the audience far more qualified than I am. I will content myself with wondering just how biofriendly this Universe is, and whether it is possible to find scientific evidence of the existence of life elsewhere.

Humans have been asking the *'Where did we come from?'* question throughout oral and recorded histories. Every culture has its own creation myth, sometimes more than one. The scientific culture is no exception; we have our own scientific creation myth called cosmology. Because our myth is bolstered by exquisite observational evidence and is capable of predictive precision, we correctly view it in a different light than other creation myths; though some fundamental cosmological questions remain unanswered, and we do not know whether they ever can be (Greene 2000).

The first widely published science fiction work (Shelley 1818) in the early 19th century is a good marker for human interest in the *'Where are we going?'* question. In this audience, we study the fate of our Universe as a result of its dark-energy-driven, accelerated expansion. Modern cultures, experiencing accelerated change in their lives due to technological innovation, have begun speculating about, and manipulating, the future course of the human race. Today none of us can say whether there will yet be many long and happy chapters to the story of humanity, or whether this history will terminate abruptly, or perhaps purposefully evolve into something other than human. But contrary to our ancestors, we expect that the future will be different and we struggle to anticipate the change.

Since my graduate student days, two phenomena have been given more and more attention: extremophiles and extrasolar planets. Both are relevant to the question of other habitable worlds and inhabitants.

2. Astrobiology

Astrobiology is the science that deals with the origin, evolution, distribution, and future of life in the Universe. It has been successful in bringing together scientific specialists from many different disciplines to tackle these big-picture questions. Many of my colleagues at the SETI Institute are astrobiologists, studying organisms living in extreme conditions (by human standards) in an attempt to better understand the origin of life on Earth and the potential habitable real estate for life beyond Earth. In the past few decades we have expanded the range of conditions recognized as suitable for life. Life is no longer confined between the boiling and freezing points of water. Hyperthermophiles live at high temperatures (and sometimes also high pressures), the current record holder being archean microbial *Strain 121* that thrives at 121°C metabolizing iron, but it can survive up to 130°C (Kashefi & Lovley 2003). At the other extreme, the psychrophilic bacterium *Psychromonas ingrahamii* survives and reproduces (very slowly) at −12°C in the ice off Point Barrow, Alaska (Breezee *et al.* 2004). Macroscopic ice worms occupy and move through the Alaskan glaciers as well as the methane ice seeps on the floor of the Gulf of Mexico using natural antifreeze to protect their cellular structures (Fisher *et al.* 2000). Sunlight, once argued to be the source of energy for all life, is completely absent miles beneath the surface of the ocean, around the deep hydrothermal vents, where a rich and diverse community of organisms thrives in the dark, at enormous pressures. Small blind shrimp there have developed IR-sensing eye spots to navigate the vent environs or to travel from one vent to another using their thermal signatures (Pelli & Chamerlain 1989). Some chemical process (or processes) within the vents also produces minute quantities of visible light that are harvested by green sulfur bacteria for photosynthesis even though many hours go by between photons (Beatty *et al.* 2005). Humans increasingly protect themselves from exposure to UV radiation as the protective layer of atmospheric ozone thins, since our DNA lacks sufficient repair mechanisms to survive intense radiation environments. Yet organisms inhabiting the highest freshwater lakes on Earth, in the caldera of the Lincancabur volcano overlooking the Atacama desert, have adapted to the huge UV load their altitude and evaporating environment present (Cabrol *et al.* 2005). Colleagues from the SETI Institute free-dive in these lakes each austral spring to catalog these organisms and study the DNA repair mechanisms they have elaborated. If life once occupied Mars, similar mechanisms might have been employed by organisms seeking to survive the dual stresses of increased radiation and desiccation due to loss of planetary atmosphere. Even more spectacular, *Deinococcus radiodurans* can withstand millions of rads of hard radiation because its DNA repair mechanisms are so effective. This skill probably evolved, not because it encountered naturally high radiation environments, but because desiccation causes the same sorts of breaks in DNA linkage; a sufficiently robust repair mechanism can endow survival independent of the damage source (Mattimore & Battista 1996). As a result, this microbe is the focus of many bioremediation programs to deal with radioactive materials, and is being sought in the Atacama to demarcate areas just too dry for life. Neutral pH was once thought essential for life, but acidophiles are plentiful; cyanobacteria and fish can survive at $pH \simeq 4$, but the red alga *Cyanidium caldarium* and the green alga *Dunaliella acidophila* can live at $pH < 1$ (Rothschild & Mancinelli 2001). In the ground waters of industrial slag heaps, extremely alkaline-tolerant microbes have been found thriving at a pH of 12.8 (Roadcap *et al.* 2006). While salt has been used historically to preserve food from decay due to bacterial action, halophilic archean microbes have been found living within pure $NaCl$ crystals (Rothschild *et al.* 1994). Astronomers may find it extremely unpleasant to live beyond the 'just right' bounds of our current terrestrial environment, but clearly life has a greater tolerance and no lack of innovative ways of making a living. Within the past few years, we have begun to accept the concept of the 'deep hot biosphere', and to acknowledge that perhaps ten times as much biomass is resident in the crust beneath our feet, as compared to the surface biomass with which we have long been familiar (Gold 1998). All these biological adaptations must inform our searches for life elsewhere in the Universe.

Although it appears that life populated the Earth in a geologic eye blink, following the appearance of liquid water on its surface, and although we have astronomical evidence of ubiquitous organic chemistry together with observations of uniformity in the laws of physics and chemistry through vast stretches of space and time, and although there is much reason to sympathize with Christian De Duve who sees life as a cosmic imperative (De Duve 1996), we cannot yet conclude that life of any form actually exists elsewhere in our Universe. Astrobiologists are engaged in a large number of different efforts to look for the proof of existence.

While no fundamental definition of 'life' exists, nor for that matter of 'planet' (until this particular IAU General Assembly) it is still a valid claim that life-as-we-know-it is a planetary phenomenon. That is, all examples of life show a remarkable genetic connection, suggesting that life originated and evolved on this planet. As it did so, it was profoundly affected by the planetary environment and life, in turn, profoundly affected the planet Earth. Thus a search for life elsewhere necessarily starts with a search of the other planets in our solar system, and a search for exoplanets beyond. Mindful of the adaptability of life to extreme environments, within our own solar system (where we may have some hope of systematic *in-situ* sampling) we should expand our inspection to any bodies capable of providing raw materials and energy sources that might be exploited by biology of any sort. Thus, in addition to the terrestrial planets (which were all biologically connected during an earlier epoch of planet building and bombardment), we should consider the large icy satellites of Jupiter and Saturn, namely Europa, Ganymede, Callisto, and Enceladus where liquid, briny, water oceans are thought to exist beneath their icy outer crusts. Titan might be a world hosting biology without liquid water as a solvent, though the slow reaction rates there challenge the notions of reproducibility and heritable evolutionary change that often appear in attempts at the definition of life. Perhaps results from the ESA's *Venus Express* mission will shed light on whether the chemical disequilibria previously noted in the Venusean atmosphere require explanations involving Grinspoon's revival of the Sagan and Salpeter 'sinkers, floaters, hunters and scavengers' (Schulze-Makuch *et al.* 2004). For the foreseeable future NASA will continue a 'follow the water' strategy, returning to Mars with robots and humans to look for signs of extinct life from a wetter, warmer epoch or even for subsurface extant life. In addition they will try to understand whether a biotic mechanism must be invoked to explain the claimed, current source of an inhomogeneous distribution of atmospheric methane (Mumma *et al.* 2005). Subsequent missions may venture to Europa to verify the existence of a massive water ocean and to examine the tantalizing discolorations near the surface cracks that could be the end products of organic molecule irradiation (Dalton *et al.* 2003), and then, later still, return to make a sterile penetration of the ice and search for life in the water below. No missions are yet planned for the other bodies of interest, but from the point of view of astrobiology, all such missions would be warranted.

3. Exoplanets

Moving out beyond the solar system, the focus will be on exoplanets. The current prize being the detection of, and susequent imaging of, a terrestrial-mass planet within the so-called 'habitable zone' of its host star. While traditionally referring to the region of space where a planet might be expected to have surficial liquid water, the habitable zone should now be enlarged to include giant planets with potentially habitable moons, and other environments that might host extremophiles.

Today, exoplanets are primarily detected by indirect techniques: astrometry, radial velocity studies, transits, and gravitational micro-lensing. Most of the planets discovered are giants; a $5.5\,\mathrm{M}_{\mathrm{Earth}}$ planet from a micro-lensing event is the lowest mass to date (Beaulieu *et al.* 2005). 'Hot Jupiters' in short-period orbits and an abundance of high-eccentricity orbits were surprising discoveries, though we are beginning to have reasonable explanations based on interactions with viscous protoplanetary disks and a version of cosmic billiards. Initially it was assumed that the presence of hot Jupiters would doom any terrestrial planets within the habitable zone, but recent studies by Raymond *et al.* (2006) argue that it is possible to have wet, terrestrial planets, even though a Jupiter-mass planet has migrated through their orbital radii on its way towards the star. We believe that the absence to date of a true Jupiter analog (in a circular orbit at 5 AU distance from a solar-type star) is the result of an observational bias that will be overcome once sufficiently long datasets from high-precision, radial-velocity instruments have been collected. Radial-velocity instruments with precisions of 1 m/s are currently in operation at Lick, Keck, and La Silla Observatories, and the first purpose-built observatory, the Automated Planet Finder is, nearing completion. Precise photometric measurements of transit events may soon yield terrestrial mass planets from the *CoRot* spacecraft to be launched later this year, or the more capable *Kepler* mission to launch in 2008. Of note: *Kepler* is sized such that a failure to find any Earth-sized planets will be a significant null result, drastically changing our ideas about the processes that form planetary systems. The *Space Interferometry Mission* (*SIM*) will achieve the astrometric precision necessary to detect planets of a few Earth-masses orbiting the closest stars. Recent progress has been made in networking together a significant number of smaller, optical observatories. An alert system makes possible rapid follow-up to detect the brief microlensing events caused by a propitiously placed planet orbiting a foreground star during the gravitational lensing of a background star. This should greatly enhance the detection rate of exoplanets, though further observations of these particular planetary bodies will not be possible. The near-term future holds promise for the detection of other Earths, large moons of gas-giant planets, and other potentially habitable cosmic real estate. The next obvious question is; will they be inhabited?

4. Exolife

Just how exactly like the Earth does another environment have to be in order to host life? Ward & Brownlee (2001) have argued that an exact duplicate of the terrestrial environment, its history, large moon, and giant-planet shields are required for anything beyond microbial life. However, Darling (2001) reviews the arguments and concludes that other astronomers might exist on many worlds. In fact the evidence is consistent with life, including intelligent life, existing on many worlds or exclusively only on Earth; there is as yet no evidence. We have an example of 'one', we cannot know from tracing the detailed history of that single example what the branching ratio might be for the experiment of life; how many other ways might things have gone but didn't, at what rate, with what end result? The number "two" will be all important in answering this question – as in the second example of an independent origin of life. In 1996 McKay *et al.* (1996) claimed that the best explanation for features found in the Allen Hills 84001 meteorite (actually an ancient piece of Mars) was that the features represented a fossilized form of life. This claim stirred up great controversy and subsequently, very innovative experimentation. Additionally it reminded us that the detection of life elsewhere in the solar system must necessarily raise the further questions of biological infection *vs.* second genesis. While both answers are intriguing, only the latter has significant implications for life beyond the solar system.

Within the solar system, *in-situ* measurements and sample return missions perhaps have the opportunity of detecting life by knowing-it-when-we-see-it. Observations searching for life elsewhere must rely on predetermined biosignatures or technosignatures; if the wrong filter is assumed and applied, it may miss valid evidence that is there. Similarly, in non-terrestrial environments, unappreciated, abiotic causes may be explanation for the results of any observations, thus leading to the potential for false positives.

It seems impossible to avoid this particular trap of being 21st century humans. In seeking life, or its technological by-products, we cannot search for what we cannot conceive, and it is also impossible to guarantee that we will correctly interpret what we find. This conundrum has been shared by all past explorers. Those who were successful pushed ahead, with the tools at their disposal, or tools they invented. As my colleague Seth Shostak is fond of saying "Columbus didn't wait for a 747 to cross the Atlantic". Neither should astrobiologists.

5. Biosignatures

The *Viking* exobiology experiments on Mars in 1976 were the first life-detection attempts on another world. While most scientists agree that *Viking* detected evidence of an unexpected superoxide soil chemistry, a few still believe that the results are consistent with a biological source (Levin 1997). The *Viking* laboratory-in-a-box experimental suite attempted to find life by looking for it with cameras (nobody looked back), looking for a changes in gas composition after soil samples were moistened and then 'fed' nutrients, providing soil samples with labeled nutrients and water then looking for metabolism in the form of $C^{14}O_2$ release (this is the contested result), and using pyrolized soil samples as input into a gas chromatograph (no mass fragments corresponding to organic compounds were detected). Whether or not the *Viking* results were indeed due to unexpected soil chemistry, it is clear from our improved understanding of extremophiles, that *Viking* would have failed to detect many different me tabolic pathways based on, e.g., Fe, S, or H that were not considered three decades ago. The current strategy for the Martian rovers (*Sojourner*, then *Spirit* and *Opportunity*) is one of 'following the water'. Having verified that early Mars did have flowing surface water, the current topic of debate is how much, for how long, and where is it now (Squyres *et al.* 2006)? The *Martian Science Laboratory*† will be the next on the scene, trying to detect extant biology and the by-products of biological activities on the surface and as far below the surface as the next generation of drilling technologies can achieve. Experiments capable of detecting metabolic strategies of life-as-we-know-it are being tested in the remotest and least habitable areas on Earth, but *Viking* reminds us that there is no guarantee these particular signatures will be all-inclusive or indicative of any Martian life-as-we-don't-yet-know-it.

Moving out beyond the Solar System, the next step will be to attempt to conduct a chemical assay of the atmosphere of any terrestrial planet imaged in orbit around nearby stars. Transiting hot Jupiters have already permitted the first analysis of chemical constituents of exoplanet atmospheres. Observations of HD 209458b reveal sodium, hydrogen, oxygen and carbon in an extended atmosphere and/or escaping from the planet (Vidal-Madjar *et al.* 2004). The *Terrestrial Planet Finder* (*TPF*) being studied by NASA and the *Darwin* constellation under development by ESA will eventually launch during the first quarter of the 21st century, perhaps combined as an international mission. Telescopes on these spacecraft are intended to suppress the light from a central star, using either an occulting coronagraph at visible wavelengths or interferometric nulling in the IR, thereby achieving a point spread function capable of spatially resolving and directly detecting reflected starlight from any terrestrial exoplanets orbiting within the stellar habitable zone: contrast ratios as great as 10^{-10} are required‡¶. Once an image has been formed, additional observations will attempt to collect sufficient photons to reveal absorption lines in the exoplanet spectrum due to trace atmospheric constituents that might be clues to the presence of biology on the planetary surface.

As difficult a technical challenge as implementing these spacecraft will be, perhaps an even greater challenge will be deciding what spectral signature(s) does or does not constitute a reliable biomarker. Using the present Earth as an example, chemical disequilibrium is one very promising sign. The coexistence of molecular oxygen and methane (more than 30 orders of magnitude out of equilibrium) in our current atmosphere is the direct result of photosynthetic cyanobacteria and plants as well as the fermenting bacteria within termites, ruminants, and rice paddies (Lovelock 1965). But the paleoearth would have presented a very different picture during the billions of years when life was present, but had not yet participated in the elevation of atmospheric O_2 levels; for that world, we must ask about the undeniable biosignatures of methanogens? Additionally, one must ask if these biosignatures are not only necessary, but sufficient – are there no abiotic processes that can yield the same result? For exoplanets, the harsh realities of the remote observational circumstances are further challenges. The available spectral coverage on any single mission will be limited; it does not now seem feasible to simultaneously observe the visible bands of O_2 and the thermal IR signature of CH_4 in one instrument. All observations of atmospheric chemistry will be photon limited, and therefore only the broadest and deepest absorption features will be usable. Detectable amounts of O_3 (at $\lambda 9.3\,\mu$m) in the atmosphere of an exoplanet believed to have liquid water on its surface (indicative of an active carbonate-silicate cycle that would quickly drawdown any initial reservoir of ozone) is currently the favored biosignature for complex life-as-we-know-it when the primary is a Sun-like star (Leger *et al.* 1999). When the primary is an M-type dwarf, other molecules such as N_2O, and CH_3Cl may be detectable along

† http://mars.jpl.nasa.gov/msl/overview/
‡ http://planetquest.jpl.nasa.gov/TPF/tpf_index.cfm
¶ http://www.darwin.rl.ac.uk/overview.htm

with ozone (Segura *et al.* 2005). While there are abiotic sources of oxygen, and therefore ozone, an ongoing biological source appears to be required for substantial atmospheric ozone concentrations on a geologically active planet. Any future announcements of atmospheric ozone detection from an exoplanet and a potential linkage to biota are likely to be accompanied by a long list of caveats.

Photon collection from the atmosphere of a terrestrial exoplanet will take a very long time and spatial resolution will be limited to one, or a few, pixels. Therefore, the spectra will be averaged over the full exoplanet disk, and time-averaged by planetary rotation and orbital phase. Clouds, if present, will further dilute any evidence of vegetation signatures such as the 'red edge' of chlorophyll (a sharp increase in leaf reflectance around 700 nm) and the differences in albedo or specular reflection caused by land masses and water oceans (Montanes-Rodriguez *et al.* 2005). To prepare for these future missions, to optimize the onboard spectroscopic instrumentation, and to plan for interpreting realistic data returns, the globally averaged signature of the Earth is being measured in earthshine from the moon (Woolf *et al.* 2002). In addition, the surface biota and atmosphere of the Earth through time are being used as input to numerical models of exoplanets in orbit around stars of different spectral type (Tinetti *et al.* 2006). Investigations into the metabolic strategies of microbial extremophiles can also help prepare for the interpretation of possible future spectral biosignatures from life-as-we-don't-yet-know-it. However, spectral absorption features alone are unlikely to distinguish between a future detection of alien microbes or mathematicians.

6. Technosignatures

'Are we alone?' is really a loaded question. When humans pose it, they are generally not asking whether another world is teeming with bacterial analogs. The detection of life on another world (extant or extinct) is a real possibility within the lifetimes of many of us at this General Assembly. That is a thrilling possibility. Detection of irrefutable biosignatures will provide the pivotal 'number two' in the record of life in the Universe, but at a very deep level, humans want to know whether other intelligent creatures also view the cosmos and wonder how they came to be. Like the term 'life', there is really no acceptable definition of 'intelligence'. Nevertheless we may be able to remotely deduce its existence over interstellar distances. If we can find technosignatures – evidence of some technology that modifies its environment in ways that are detectable – then we will be permitted to infer the existence, at least at some time, of intelligent technologists. As with biosignatures, it is not possible to enumerate all the potential technosignatures of technology-as-we-don't-yet-know-it, but we can define systematic search strategies for equivalents of some 21st century terrestrial technologies. Recognizing that we are a very young technology in a ten billion year old galaxy, we should continue to explore our Universe utilizing every wavelength and every information-bearing particle, with all of the observational tools we can construct, remaining vigilant for anomalies that might be new 'Class A' astrophysical phenomena (Harwit 1981) or might just be the detectable consequences of someone else's astro-engineering project or catastrophe. That is, we should pursue astronomy with vigor and in the spirit of Jocelyn Bell Burnell (1977), letting no bit of 'scruff' go unheeded.

SETI (the Search for ExtraTerrestrial Intelligence) is that subspecialty of astrobiology that currently conducts systematic explorations for other technology-as-we-know-it; primarily searches for electromagnetic radiation. SETI predates the current field of astrobiology, beginning life as a valid field of exploratory science in 1959 with the publication of the first paper on this subject in a refereed journal (Cocconi & Morrison 1959). This first paper advocated a search for radio signals, but the suggestion of searching for pulsed optical laser signals followed shortly thereafter in 1961 (Townes & Schwartz 1961). SETI provides another plausible avenue for discovering habitable worlds by attempting to detect the actions of technological inhabitants. More than a hundred SETI searches can be found in the literature (Tarter 2001) beginning with Project Ozma in 1960 (Drake 1960), the first search for radio signals from nearby Sun-like stars. While this may seem like a large effort, the sum total of all these investigations has covered only a minute fraction of the search-parameter space.

Information can be encoded with photons for the benefit of the transmitting technology or to deliberately attract the attention of another civilization. While it may be possible for us to detect unintentional leakage radiation from another technology, deliberate signals, transmitted to be detectable, are the most likely to be detected. Furthermore, any detectable signals will have originated from a technology far older than our own. There may be no other detectable technological civilizations in the Milky Way or beyond, but if there are others, we can be confident that we are youngest. On Earth, we've had at least one viable technology for interstellar communication for about a hundred years, prior to that our technology was undetectable at a distance . Today we are in a period of rapid exponential improvements in information processing and communication technologies. Whether that exponential will continue indefinitely into a future dominated by machine rather than biological intelligence (the 'singularity' of Kurzweil and other futurists (Kurzweil 2005), or whether the exponential will saturate and sustain due to some resource limitation is currently unknown. There is of course another alternative; that our technology will cease, due to its destruction or deliberate disuse. If technology tends to be a long-lived phenomenon among Galactic civilizations, then statistics favor the detection of signals from a technology during its old age. If technology, in general, is a short-lived phenomenon, then it will be undetectable because the chance that two technological civilizations would be not only co-located, but co-temporal during the 10 billion year history of the Milky Way is vanishingly small. For this reason, Philip Morrison has called SETI the archeology of the future. The finite speed of light guarantees (for current understanding of physics and available technologies) that any detected signal will tell us about the transmitter's past, but the detection of any signal tells us that it is possible for us to have a technological future (Mallove 1990). The potential for deliberate disuse of any given technology is another argument against the likelihood of detecting leakage radiation; any transmission mechanism for internal use is likely to be superseded by a more efficient technology within a short time. However, with respect to deliberately transmitted signals, a long-term strategy is implicit (the signal must be there whenever the intended receiver evolves to listen, if transmission is to be successful). Furthermore, an older technological civilization might choose a transmission technology that favors detection by emergent technologies (such as ourselves) rather than reflecting its own capabilities. Or an older technological civilization might choose to do the opposite, that is to adopt a transmission strategy that is only detectable by another technological civilization with some minimum level of competence, not by us. *Ab initio*, we cannot unravel alien motivation, we can only choose to search, or not to search, using the technologies within our grasp. Gott (1993) has used a statistical argument based on the Copernican Principle to suggest that because we find ourselves to be a very young technology in a very old galaxy, it is unlikely that we will become an old technology. Yet if, to the contrary, a long technological existence is indeed in our future, it would be impossible to avoid passing through our current emergent state, at which point we are capable of making Gott's calculation. Considerations of the evolutionary history of galactic supernovae activity, star formation, and metallicity have led to the idea of the galactic habitable zone, in which 75 % of the stars are older than the Sun (Lineweaver *et al.* 2004). Thus, technological civilizations older than our own are indeed a possibility, and we should plan any search strategies with our asymmetric position (we are the youngest detectable technology) in mind.

7. Exploring the cosmic haystack

As previously noted, most SETI projects rely on photon detection. Even restricting ourselves to electromagnetic radiation, the cosmic haystack we wish to explore for a signal (by analogy the difficult-to-find needle) is at least nine-dimensional: three dimensions of space and one dimension of time, two senses of orthogonal polarization, one unknown photon frequency, some form of modulation scheme for information encoding, and finally an undetermined minimum sensitivity (because the transmitter power and distance are unknown). Recently it has been suggested that a photon's quantized orbital angular momentum could encode information so that should become another dimension of the cosmic haystack (Harwit 2003). Since there is no near-term, practical implementation of that search strategy for SETI, it will for the moment be counted as part of the dimension dealing with modulation schemes. Contrary to popular belief, since Project Ozma forty-six years ago, we have only managed to pull a few straws from the cosmic haystack for close examination. Most of the time, most SETI searches are off the air. Given the nine-dimensional haystack, searches to date have been unable to set meaningful limits on the existence (or non-existence) of other technological civilizations. Explanations for 'the great silence' are numerous, but in fact they are not yet required. Today the situation is changing as new instruments intended for dedicated SETI use are commissioned and beginning to look at the sky. This is an exciting time because our tools may finally be getting to be commensurate with the magnitude of our task.

As above, we should concentrate on detecting deliberately broadcast signals and try to devise systematic search strategies of sufficient scope that null results will be significant, if positive results continue to elude us. Deliberate signals might be modulated in one of two ways: they could appear to be 'almost astrophysical', or they could appear to be 'obviously technological'. The distinct benefit of the former scheme is that such signals are very likely to be captured as an emergent technology begins to deploy multiple sensors to study the Universe around it. Detectors will be optimized to record signals with just these sorts of characteristics, and they will be entered into databases as pulsars, GRBs, SNs, eclipsing binaries, etc., etc. Detailed study of the individual entries in any such databases will eventually discover that a particular 'pulsar' had a spindown rate of exactly zero (not just momentarily zero, or zero to instrumental accuracy) or that 'pulsar' might glitch back and forth between two precise rotation periods, or the GRBs might end up being aligned in 4-dimensional space time because they were the acceleration events of some matter-anti-matter annihilation rocket traveling through space, or the light curve of an 'eclipsing' binary star turned out to be caused by an occulter(s) with non-spherical cross-section, or widely separated SNs might have synchronized light curves, or .... As an emergent technology, our rapidly improving astronomical observing capability and our curiosity about the cosmos should insure that we eventually discover any such 'almost astrophysical' signals. Almost by definition, the detection of 'obviously technological' signals will require construction of specific instrumentation not available from astronomical observing programs, because the characteristics of the signals are precisely those that we expect nature to be unable to produce. Within this class of signals, SETI has historically concentrated on those signals whose time-bandwidth product approaches the minimum value permitted by the uncertainty principle. At radio frequencies, signals exhibiting extreme frequency compression (CW or narrowband pulses) with bandwidths less than the $\sim 300\,\mathrm{Hz}$ exhibited by line-narrowed, satura ted OH masers (Cohen *et al.* 1987) are being sought. Because the natural background from Galactic synchrotron emission rises rapidly below 1 GHz, while the noise from atmospheric water vapor and oxygen contributes above 10 GHz, radio SETI searches have a goal of systematically exploring the naturally quiet Terrestrial Microwave Window from 1 - 10 GHz. While there is no natural background at these frequencies except the 3 K CMB, radio searches must expend significant computing resources to mitigate the effects of human-caused RFI, and giving up pieces of some spectral bands altogether, even at the quietest sites. At optical frequencies, signals exhibiting extreme time compression (broadband laser pulses) are searched for with photon counters having nanosecond rise times, a regime with no known sources of astrophysical background (Howard & Horowitz 2001). Because interstellar dust begins to absorb optical pulses over distances beyond $\sim 1000\,\mathrm{ly}$, it is desirable to extend the optical SETI search into the infrared so that more of the Galaxy becomes accessible. This will happen when, and if, the requisite fast photon counters become available and affordable in the IR. Other modulation schemes employed by current terrestrial communications technologies can produce signals whose statistical properties differ from the Gaussian noise of astrophysical emitters. Detection of this class of signals is far more compute-intensive than detection of the simple $B\tau \simeq 1$ signals. As Moore's Law delivers more affordable computing, efforts are beginning to accommodate searches for more complex signals at radio frequencies.

At any frequency, there are two basic search strategies that can be implemented: move quickly across the sky (or a portion thereof) to cover as much of the spatial dimension of the cosmic haystack as possible, or select individual directions deemed to have a higher *a priori* probability of harboring a technological civilization and make targeted observations in those directions for longer periods of time. The former strategy minimizes the assumptions about the source of the signal, and is better suited to the case where the location of the transmitter might no longer be associated with the point of origin of the technological civilization. But, in general, the surveys will achieve poorer sensitivity as the result the smaller telescopes and shorter dwell times typical of this strategy, and they are more likely to be out of phase with transient or periodic emitters. By developing a list of plausible targets, it is possible to achieve significantly better sensitivity through integration and signal processing gain for a wider range of signal types. However, if the target list is constructed under the wrong assumptions, detection probabilities are lowered rather than improved. In both cases it is desirable, but seldom affordable, to accomplish the task of signal detection and recognition in real-time, or near-real-time, so that immediate follow up of candidate signals is enabled before they can vary with time, and opportunities for distinguishing between terrestrial and extraterrestrial technologies can be exploited. Discrimination against terrestrial signals can also be improved by the simultaneous use of multiple telescopes at widely separated sites; lack of coincidence or, in the case of narrowband signals, differential Doppler signatures are powerful tools. Since Project Ozma, plausible arguments about 'magic' frequencies, places, and times have been used to constrain the scope of individual searches. After more than four decades there is still divergence among humans, rather than convergence, with regard to the proper magical parameters. With the exception of a timing scheme based on novae and supernovae (Lemarchand 1994), there is little opportunity to synchronize our magical thinking with the extraterrestrial transmitter.

In 2004, the SETI Institute completed Project Phoenix (Backus (2006)), and this summer UC Berkeley's SERENDIP IV system at Arecibo concluded its data taking (Cobb *et al.* 2000). Phoenix was the privately-funded continuation of a targeted search of 800 nearby solar-type stars from 1.2 to 3 GHz begun as the NASA High Resolution Microwave Survey in 1992, and continued by the SETI Institute following congressional termination of NASA SETI funding in 1993. SERENDIP IV was a commensal, random sky survey conducted with a feed and receiver system attached to the end of the Arecibo carriage house arm not being used by the primary observer. A few percent of the data SERENDIP IV collected was subsequently analysed on the pioneering SETI@home distributed computing platform. Until searches begin with the Allen Telescope Array (ATA, see below) in 2007, there remain only a few organized radio SETI observing programs on the air; the new SERENDIP V recorders attached to the ALFA multibeam feed at Arecibo†, Project Argus an amateur effort coordinated by the SETI League‡, and a SETI Italia project¶. None of the current projects are real-time searches, none can follow-up immediately on candidate signals. Over a 5-year period, Harvard's optical SETI targeted search ran on the Wyeth Telescope at the Oak Ridge Observatory completing observations of more than 6000 stars, many of them synchronously with the Fitz-Randolph Observatory at Princeton University as a successful method to suppress the false-positive event rate (Howard *et al.* 2004). Both sytems are now shut down, but the Harvard group is commissioning a new OSETI observatory for a survey of the sky visible from Oak Ridge (see below). Targeted OSETI systems are continuing to observe at UCSC Lick Observatory (remotely operated from the SETI Institute)‖, at UCB Leuschner Observatory††, and at the Campbelltown Rotary Observatory as part of an astrobiology instructional program at the University of Western Sydney‡‡. At this precise moment, observational progress is slow, as new dedicated SETI observing facilities are brought on line.

8. Observatories for tomorrow

The SETI Institute and the University of California Berkeley Radio Astronomy Lab have partnered to build the Allen Telescope Array (ATA) in Northern California at the Hat Creek Radio Observatory for the purpose of simultaneously surveying the radio sky for signals of astrophysical and technological origin (DeBoer *et al.* 2004). Ultimately the array will consist of 350 antennas, each 6.1 m in diameter, extending over a maximum baseline of 900 m. The ATA will do for the radio sky what the Sloan Digital Sky Survey has recently done for the optical sky. And it will do it so rapidly that it will also provide the first systematic look at the transient radio universe. The ATA provides simultaneous access to any frequency between 500 MHz and 11.2 GHz, a system temperature $\sim$50 K, four separately tunable intermediate frequency channels feeding a suite of signal processing backends. The backend instrumentation can produce wide-angle radio images of the sky with $\sim$20,000 resolution pixels and 1024 spectral channels per pixel, and at the same time, study up to 32 point sources of interest within its large field of view using phased up beams at up to four different frequencies. With 61,075 baselines feeding two spectral-imaging correlators, the ATA is a superb radio snapshot camera, yielding maps with exceptional image fidelity. This new approach to commensally sharing the sky allows SETI and traditional radio astronomical science to both utilize the telescope nearly full time. For efficient sky 'sharing' a large list of SETI target stars is essential. Two initial catalogs of stars capable of hosting habitable planets, or 'hab stars', have been created

† seti.berkeley.edu/casper/papers/2004-07-10_Iceland_SV_poster.pdf
‡ http://www.setileague.org/
¶ http://www.seti-italia.cnr.it/
‖ http://seti.ucolick.org/optical/
†† http://seti.ssl.berkeley.edu/opticalseti/
‡‡ http://www.atnf.csiro.au/pasa/17_2/bhathal/

from data provided in the *Hipparcos* and *Tycho* catalogs (Turnbull & Tarter 2003). The $\sim$250,000 stars in the current habstar catalogs will provide a few stellar targets in every array field of view at lower frequencies. Future results from the *Gaia* mission will permit expansion of these catalogs by a factor of 4 or more in order to maintain observing efficiency at the higher frequencies with smaller primary fields of view. The current generation of SETI signal processing backends (Prelude) are based on PCs, each with a special accelerator card, and provide over 200 millions channels of 1 Hz resolution channels that can be spread over as many as three dual-polarization phased array beams. The output data from every ATA beam is delivered in a standard IP format. Development is now ongoing to convert to the next generation of SETI signal processing equipment (SETI on the ATA or SonATA) as 'software only' implementations within generic computing clusters. Since the full 11 GHz bandwidth analog signal is transported from the antennas to a central processing lab next to the array, future improvements in the cost performance ratios of digitizers and other Moore's Law devices promise improved throughputs and observing speeds. Over the next observational decade, it should be possible to observe about a million SETI target stars through the terrestrial microwave window from 1 to 10 GHz. Today there are 42 antennas in the array, and they are spaced fairly close together for ease of commissioning. Therefore, the first SETI search that will be conducted with the ATA-42 will be a survey of 20 square degrees along the Galactic plane in the vicinity of the center and will cover the 1420 to 1720 MHz 'waterhole'. Along these lines of sight, the survey will interrogate $\sim 10^{10}$ stars, most of them distant, searching for strong technological signals. At the distance of the Galactic Center this survey would detect a transmitter with a power equivalent to 20,000 Arecibo planetary radars. A targeted search of a million nearby stars with the full 350 element array should have the sensitivity to detect transmitters with powers ranging from that of current terrestrial ionospheric backscatter radars up to the Arecibo planetary radar, depending on the stellar distance. The first radio science to be undertaken with the ATA-42 will be a 5 GHz survey of 10^4 square degrees of the sky (overlapping with the SDSS) using the spectral imaging correlator to search for transients. Selected sky patches of 100, and 10 square degrees, will be revisited weekly and daily. In this way, it should be possible to characterize the temporal variability of the radio sky on a range of timescales and provide finding lists of variable sources that will inform a number of different scientific studies in the community. The standard IP protocol data output for the phased array beams invites external users to construct their own backend processors for use on the ATA to enlarge the amount of commensal observing accomplished with the array.

The ATA is the first attempt to manufacture a radio telescope by taking advantage of cost discounts from economies of scale, consumer off-the-shelf components (primarily from the telcom industry), and inexpensive commercial manufacturing technologies. The least expensive way to build such an array is all at once. Paul Allen, Nathan Myhrvold, and the USNO funded the array-technology development and initial construction phase to implement the ATA-42. Completion of the ATA-350 requires additional private and federal partners. Absent a funding projection timeline, it is difficult to estimate the actual cost of the ATA-350, but our best estimate is that it will be somewhere between 1/6 and 1/5 the cost of building a traditional antenna of this size for cm λ astronomy. This is encouraging as we look to the future of SETI and centimeter-decimeter radio astronomy and to the Square Kilometre Array, a project to build an international observatory with 100 times the collecting area of the ATA. More work is needed to bring down the unit cost of $A_{\rm eff}/T_{\rm sys}$ but greater opportunities of scale, the direct involvement of industrial partners, and the lessons learned from actually building a Large N - Small D array of small telescopes like the ATA are suggesting several potentially successful pathways forward. We need to get through today and finish the ATA-350, before we can enjoy the promises of tomorrow. So at this General Assembly, I am hoping that 42 may soon turn out to be the answer to life-the-Universe-and-SETI! But just in case it doesn't, I am reduced to holding out my hands and quoting Jodie Foster – *we know what to do, now all we need is the money.*

Moore's Law improvements have helped OSETI programs every bit as much as they have enabled the ATA. At Harvard, Paul Horowitz and his students are just beginning to collect data with a new, dedicated, OSETI telescope for a survey of the 60 % of the northern sky visible from the Oak Ridge Observatory in Massachusetts (Horowitz 2006). The telescope is housed in a building enthusiastically constructed with student labor, featuring a roll-back roof and a removable section in the south-facing wall that accommodates drift scans with only a single axis of rotation. The 72-inch primary and 36-inch secondary mirrors have been manufactured inexpensively by fusing glass over a spherical form and then polishing, because the system does not require image quality optics. The detection system is based on eight pairs of 64-pixel Hamamatsu fast photo-diodes, and custom electronics to permit real-time pixel by pixel intercomparison for coincidence detection to eliminate false positives. This new telescope will search for powerful transmitters from a large collection of stars by conducting meridian transit scans of the sky in $1.^\circ6 \times 0^\circ.2$ strips (with a dwell time, due to the Earth's rotation, of about one minute). The sky visible from that site can be scanned in approximately 150 clear nights. The survey sensitivity should be adequate to detect laser pulses from the analog of a Helios-class laser being transmitted through a 10 m telescope up to a distance of 1000 ly. To improve the sensitivity of OSETI searches, larger aperture antennas are needed. To make these searches affordable, it is important to consider commensal observations with 10 m class optical light buckets being built for other purposes. Gamma-ray, Cherenkov-detecting telescopes are possibilities. SETI signals can be distinguished from the extended shower events in single dishes, such as the Whipple 10 m telescope, by means of the individual image shapes (Holder *et al.* 2005), or in the case of array like VERITAS, if a fast trigger could be developed, then single pixel events originating at the same point in the sky could be recovered as candidate SETI signals rather than being discarded as noise. Opportunities for commensal OSETI observations should also be considered for future terrestrial planet imaging missions as well as the overwhelmingly-large class of new groundbased telescopes being proposed for the future.

9. Broadcasting

This presentation has considered only one side of the communication equation – that of the receiver. However, if everybody listens and nobody transmits, then communication will not occur. Referring back to our asymmetric position *vis á vis* any other technological civilization in the Milky Way – we are the youngest – it is appropriate that as we listen first. Transmission is a harder job than receiving. It is more expensive to build and power transmitters, and it is also more expensive in the cultural currencies of diplomacy, ethics, and politics. Today we lack any form of global governance, we have no mechanism with which to reach concensus on whether we should transmit (either *ab initio* or in response to a detected signal) and what we should say, and who should say it on behalf of the human species on planet Earth. And we are impatient. While a listening strategy could yield results immediately. If transmission is required to elicit a reply, then the times involved are, at a minimum, the roundtrip light travel times between us and the intended target. As a species, we are also not very good at long term commitments. We focus on 2, 5, and 10 year plans, are beginning to come to grips with the environmental consequences hundreds of years in the future of actions we take today, but we are ill-equipped to commence and continue a transmission strategy whose duration needs to be measured in Galaxy-crossing times at a minimum, and potentially much longer if we wish to include technologies that may emerge after us. Listening comes first – any other technologies will have done that when they were young. Now that they are older, and have successfully grappled with the questions above, they may have assumed the burden of transmission, and so should we when we become older.

10. Finding our replacements

Kids really do like dinosaurs, ghosts, creepy-crawly things, and ET. This gives us a wonderful hook on which to hang a science education. Astrobiology is a particularly appealing story to tell in the classroom. It allows students to put themselves into the big picture and encourages them to reach across discipline boundaries to embrace whatever body of knowledge and tools that can help solve a problem. At the SETI Institute, where my colleagues and I have embarked on an exploration of the cosmic haystack that could well take far more than our lives to systematically conclude, we think a lot about how to keep the searching going after we retire or pass away. We think about it in financial terms – how do we establish an endowment that will be viable into the indefinite future. And we think about it in human terms – where will we find the next generation of SETI researchers. Though it is now far closer to being a mainstream concept than when I first got hooked on SETI, it is still distinct enough that we cannot assume that this exploration will endure through the traditional route of generation after generation of university students being educated and then educating more students within the larger academic structure. So we try to improve the odds a bit by developing formal curriculum materials for young students, to try to counter Herman Bondi's observation when told that all young children were fearlessly curious – *we have a cure for that, it's called education!* Our latest attempt is a year-long integrated science curriculum for ninth graders, called *Voyages Through Time*, and it has a subtitle *Everything Evolves*. So we're not standing on the sidelines, but trying to get involved where it counts to promote future science literacy for the next generation.

SETI might succeed in my lifetime, or in my granddaughter's, or never. There is no satisfactory way to make an estimate. The wisest summary still remains the last sentence in the original 1959 *Nature* journal article: *The probability of success is difficult to estimate, but if we never search the chance of success is zero.*

So I invite you to stay tuned, because some of us are determined to keep searching!

References

Backus, P.R. 2006, *in preparation*

Beatty, J. T., Overmann, J., Lince, M. T., Manske, A. K., Lang, A. S., Blankenship, R. E., Van Dover, C. L., Martinson, T. A., & Plumley, F. G. 2005, *Proc. Nat. Acad. Sci.*, 102, 9306

Beaulieu, J.-P., Bennett, D. P., Fouque, P., Williams, A., & Dominik, M. 2005, *Nature*, 439, 437

Bell Burnell, J. S. 1977, *Annals NY Acad. Sci.*, 302, 685

Breezee, J., Cady, N., & Staley, J.T. 2004, *Microb Ecol.*, 300, 4

Cabrol, N. A., Grin, E. A., Prufert-Bebout, L., Rothschild, L, & Hock, A. N. 2005, *Astrobiology*, 5, 305

Cobb, J., Lebofsky, M., Werthimer, D., Bowyer, S., & Lampton, M. 2000, in: G.A. Lemarchand & K.J. Meech (eds.), *Bioastronomy '99: A New Era in Bioastronomy*, *ASP-CS*, 213, 485

Cocconi, G., & Morrison, P. 1959, *Nature*, 184, 844

Cohen, R. J., Downs, G., Emmerson, R., Grimm, M., Gulkis, S., Stevens, G., & Tarter, J. C. 1987, *MNRAS*, 225, 491

Dalton, J. B., Mogul, R., Kagawa, H. K., Chan, S. L., & Jamieson, C. S. 2003, *Astrobiology*, 3, 505

Darling, D. 2001, *Life Everywhere: The Maverick Science of Astrobiology* (New York: Basic Books)

DeBoer, D., Welch, W. J., Dreher, J. D., Tarter, J. C., Blitz, L. Davis, M. D., Fleming, M., Bock, D., Bower, G., Lugten, J., Girmay-Keleta, G., D'Addario, L., Harp, G., Ackermann, R., Weinreb, S., Engargiola, G., Thornton, D., & Wadefalk, N. 2004, *Proc. SPIE*, 5489, 1021

De Duve, C. 1996, *Vital Dust: Life As a Cosmic Imperative* (New York: Basic Books)

Dick, S. J. 1984, *Plurality of Worlds: The Extraterrestrial Life Debate from Democritus to Kant* (Cambridge: CUP)

Drake, F. D. 1960, *Sky & Telescope*, 39, 140

Fisher, C. R., MacDonald, I. R., Sassen, R., Young, C. M., Macko, S., Hourdez, S., Carney, R., Joy, S., & McMullin, E. 2000, *Naturwissenschaften*, 87, 184

Gold, T. 1998, *The Deep Hot Biosphere* (Berlin: Springer)

Gott, J. R. 1993, *Nature*, 363, 315

Greene, B. 2000, *The Elegant Universe: Superstrings, Hidden Dimensions, and the Quest for the Ultimate Theory* (Vintage Press)

Harwit, M. 1981, *Cosmic Discovery: The Search, Scope, and Heritage of Astronomy* (New York: Basic Books)

Harwit, M. 2003, *ApJ*, 597, 1266

Holder, J., Ashworth, P., LeBohec, S., Rose, H.J., & Weekes, T. C. 2005, *Proc. 29th Int. Cosmic Ray Conf.*, 5, 387

Horowitz, P., Coldwell, C., Howard, A., Latham, D., Stefanik, B., Wolff, J., & Zajac, J. 2006, preprint <http://seti.harvard.edu/oseti/oseti.pdf>)

Howard, A. B., & Horowitz, P. 2001, *Proc. SPIE*, 4273, 153

Howard, A. W., Horowitz, P., Wilkinson, D. P., Coldwell, C. M., Groth, E. J., Jarosik, N., Latham, D. W., Stefanik, R. P., Willman, A. J. Jr., Wolff, J., & Zajac, J. M. 2004, *ApJ*, 613, 1270

Kashefi, K., & Lovley, D. R. 2003, *Science*, 301, 394

Kurzweil, R. 2005, *The Singularity Is Near: When Humans Transcend Biology* (Viking Press)

Leger, A., Ollivier, M., Altwegg, K., & Woolf, N. J. 1999, *A&A*, 341, 304

Lemarchand, G. A. 1994, *Ap. Space Sci.*, 214, 209

Levin, G. V. 1997, *SPIE Proc.*, 3111, 146

Lineweaver, C., Fenner, Y., & Gibson, B. 2004, *Science*, 303, 59

Lovelock, J. 1965, *Nature*, 207, 568

Mallove, E. F. *MIT Tech Talk*, Nov. 7

Mattimore, V., & Battista, J. R. 1996, *J. Bacteriology*, 178, 633

McKay, D. S., Gibson, E. K. Jr., Thomas-Keprta, K. L., Vali, H., Romanek, C. S., Clemett, S. J., Chillier, S. E. F., Maechling, C. R., & Zare, R. N. 1996, *Science*, 273, 924

Montanes-Rodriguez, P., Palle, E., Goode, P. R., Hickey, J., & Koonin, S. E. 2005, *ApJ*, 629, 1175

Mumma, M. J., Novak, R. E., Hewagama, T., Villanueva, G. L., Bonev, B. P., DiSanti, M. A., Smith, M. D., & Dello Russo, N. *BAAS*, 37, 669

Pelli, D. G., & Chamerlain, S. C. 1989, *Nature*, 337, 460

Raymond, S. N., Mandell, A. M., & Sigurdsson, S. 2006, *Science*, 313, 1413

Roadcap, G. S., Sanford, R. A., Qusheng, J., Pardinas, J. R., & Bethke, C. M. 2006, *Ground Water*, 44, 511

Rothschild, L. J., & Mancinelli, R. L. 2001, *Nature*, 409, 1092

Rothschild, L. J., Giver, L. J., White, M. R., & Mancinelli, R. L. 1994, *J. Phycol.*, 30, 431

Schulze-Makuch, D., Grinspoon, D. H., Abbas, O., Irwin, L. N., & Bullock, M. A. 2004, *Astrobiology*, 4, 11

Segura, A., Kasting, J. F., Meadows, V., Cohen, M., Scalo, J., Crisp, D., Butler, R. A. H., & Tinetti, G. 2005, *Astrobiology*, 5, 706

Shelley, M. 1818, *Frankenstein, or, The Modern Prometheus* (London: Lackington, Hughes, Harding, Mavor & Jones)

Squyres, S. W., Knoll, A. H., Arvidson, R. E., Clark, B. C., & Grotzinger, J. P. 2006, *Science*, 313, 1403

Tarter, J. C. 2001, *ARAA*, 39, 511

Tinetti, G., Meadows, V. S., Crisp, D., Fong, W., Fishbein, E., Turnbull, M., & Bibring, J.-P. 2006, *Astrobiology*, 6, 34

Townes, C. H., & Schwartz, R. N. 1961, *Nature*, 192, 348

Turnbull, M. C., & Tarter, J. C. 2003, *ApJS*, 145, 181

Vidal-Madjar, A., Desert, J.-M., Levavelier des Etangs, A., Hebrard, G., Ballester, G. E., Ehrenreich, D., Ferlet, R., McConnell, J. C., Mayor, M., & Parkinson, D. C. 2004, *ApJ* (Letters), 604, L69

Ward, P. & Brownlee, D. 2001, *Why Complex Life is Uncommon in the Univere* (Berlin: Springer)

Woolf, N. J., Smith, P. S., Traub, A., & Jucks, K. W. 2002, *ApJ*, 574, 430

Highlights of Astronomy, Volume 14
IAU XXVI General Assembly, 14-25 August 2006
Karel A. van der Hucht, ed.

doi:10.1017/S1743921307009830

The magnetic field and its effects on the solar atmosphere in high resolution

Alan M. Title

Lockheed Martin ATC, O/ADBS, B. 252, 3251 Hanover St., Palo Alto, CA, 94304, USA
email: title@lmsal.com

Abstract. The Sun's magnetic field is produced throughout the solar interior; it emerges and is dispersed by surface and subsurface flows, and then expands above the surface to dominate the structure of the corona. To resolve the effects of the magnetic field it is necessary to image the interior and measure its rotation and flow systems; track the responses of the magnetic fields to flows in the surface; and to follow the evolution of structures in the corona. Because the Sun is dynamic both high spatial and temporal resolution are essential. Because the Sun's magnetic field effects encompass the entire spherical exterior, the entire surface and outer atmosphere must be mapped. And because the magnetic field is cyclic high-resolution observations must be maintained over multiple cycles.

Keywords. Sun, magnetic fields, high-resolution

---

1. Introduction

The last 15 years has been a revolutionary period in Solar Physics because of new observatories on the ground and in space, advances in numerical simulations, and theoretical developments. The new instrumentation and analysis techniques have produced data with a greatly enhanced resolution in time, space, and spectral span. In parallel advances in theory, numerical simulation, and computer technology have allowed significant insights into the complex physical processes that occur in both in the solar atmosphere and throughout the heliosphere. The almost universal commitment of observatories and Principal Investigators to open data policies coupled with fast internet connections has enabled the international community of solar physicists, astrophysicists, plasma physicists, and fluid dynamicists to participate in the analysis process.

The Sun's magnetic field is most probably produced at a range of depths in the solar interior; it emerges and is dispersed by surface and subsurface flows, and then expands above the surface to dominate the structure of the corona and heliosphere. In order to resolve the effects of the magnetic field it is necessary to image the interior and measure its rotation and flow systems, track the responses of the magnetic fields to flows in the surface, and to follow the evolution of structures in the corona. Because the Sun is dynamic, high spatial resolution also implies high temporal resolution. Because of the wide range in temperatures, a broad spectral range is necessary to follow the heating processes above the photosphere. Because the Sun's magnetic field effects encompass the entire spherical exterior, the entire surface and outer atmosphere must be mapped. And because the magnetic field is cyclic, high-resolution observations must be maintained over multiple cycles.

Key to the advances of the last 15 years were the development of active and adaptive optics systems at a number of ground-based observatories and the launch and operation of the *YOHKOH*, *SOHO*, *TRACE*, and *RHESSI* scientific spacecraft. This year, 2006, has already seen the launch of the *Solar X-ray Imager (SXI)* on the *GOES-N* spacecraft and will soon see the launches of the *STEREO* and *SOLAR B* missions. In 2007 new observatories at Big Bear in California and Tenerife in the Canary Islands will start operation. In 2008 the *Solar Dynamics Observatory (SDO)* will be launched. In the next decade a 4-meter Advanced Technology Solar Telescope and a Frequency Agile Solar Radio telescope should come into operation. These new telescope facilities combined with expected developments in computer technology will lead to an acceleration in our understanding of both interior hydrodynamics and exterior magnetic field dominated processes.

2. Below the surface

The science of Helioseismology has allowed mapping of the local and global flows in the solar interior and is just beginning to image the flow systems in and around sunspots. Although Heiloseismology has produced many interesting results, the origin of the Sun's magnetic fields is still a mystery. It is still not established whether the field generation process is limited to the bottom of the convection zone or that fields are generated at multiple depths throughout the convection zone. At present it is difficult to distinguish observationally whether field is pulled to the surface by convection from stored fields in the interior or whether some of the fields are generated by local dynamo processes driven by convection. It is possible, and even probable, that multiple processes are responsible for the surface appearance of new magnetic fields.

The surface distribution of magnetic field is well reproduced by kinematic models that depend on meridional and convective flow systems to distribute the emerged fields over the surface. Helioseismic measurements of the subsurface layers have the potential for providing clues to the dispersal mechanisms and perhaps local and global dynamo processes.

The new high-resolution full disk helioseismograph and vector magnetograph, the *Helioseismic and Magnetic Imager (HMI)*, on the *SDO* spacecraft holds promise of imaging the top 10,000 km of the Sun with sufficient resolution to answer many questions about the origins, emergence, and dispersal of active region magnetic fields. *HMI* will provide imaging of the magnetic, temperature, and flow structures below spots. In addition it will provide moderate resolution (one arc second) vector magnetograms.

High-resolution observations of sunspots are causing a serious review of the nature of the sunspots and in particular the nature and role of penumbra. It is now clear that sunspots are just part of the plage/moat/sunspot phenomena. The *HMI* data combined with the high-resolution imagery now possible from the ground (see Fig. 1) should provide the essential data for understanding both the temperature and magnetic structure of sunspots. The continuous coverage of *HMI* should enable the understanding of the spot emergence and decay processes. The next generation of computers should allow numerical simulation of sunspot phenomena.

3. The visible surface

3.1. *Quiet regions*

The measurement of the line of sight magnetic field with *Magnetic Doppler Imager (MDI)* on *SOHO* has provided a nearly unbroken record of full disk magnetograms for a decade. This has allowed the accurate tracking of the evolution of the surface magnetic field. The *SOHO* mission was designed to follow individual magnetic features from their origin in active regions and sunspots to their eventual disappearance by merging with magnetic

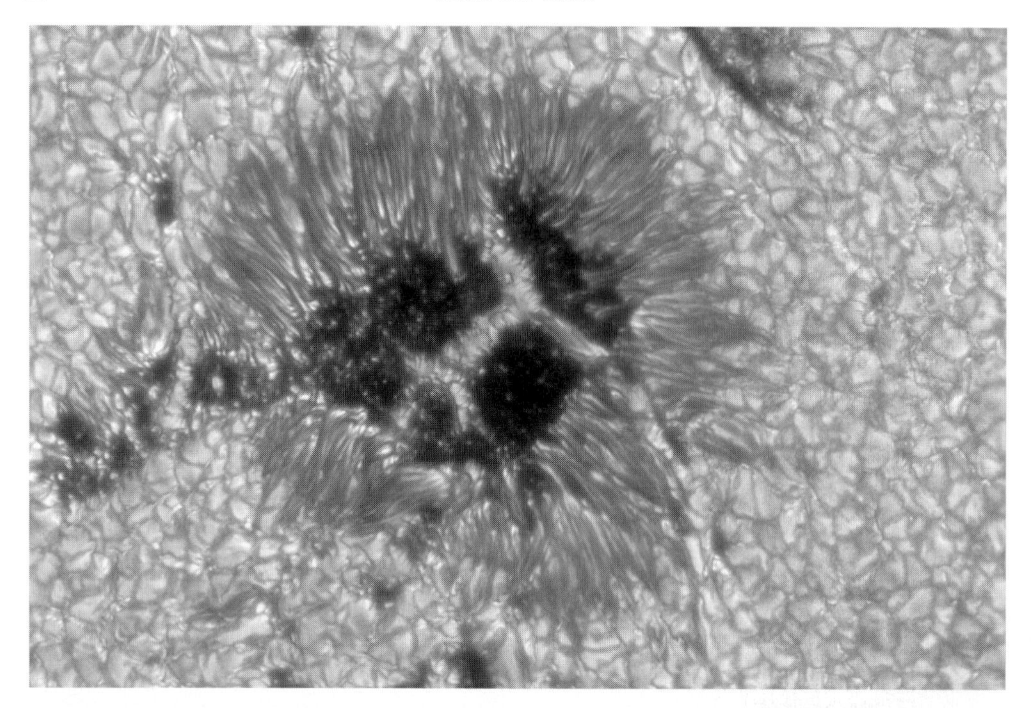

Figure 1. Image of sunspot in G-band taken on 7 June 2006 by T. Berger at the Swedish Solar Telescope, La Palma, Canary Islands, Spain.

flux elements with opposite polarity. This concept was based on the premise that virtually all the magnetic flux important for surface magnetic evolution emerged in active regions. Interestingly, it was discovered that flux emerged everywhere on the surface at a rate such that any magnetic field outside of active regions was replaced in about a day and even flux in active regions lived less than a month.

The *MDI* data showed that the mixed polarity fields (see Fig. 2) that are seen all over the solar surface, the "Magnetic Carpet", were the result of the constant emergence of small bipoles with a mean flux of $\sim 10^{-18}$ Mx rather than the result of a surface spreading process acting upon the active region field. The magnetic carpet component of field is not just a kind of magnetic noise, rather it has a number of significant effects. When observed at a resolution of 4 arc seconds the average of the absolute value of the field in quiet Sun was observed to be between 4 and 10 gauss. When observed with the resolution of $3'$ the mean flux in quiet Sun usually ranges between $\pm$ 0.1 gauss. As a result the majority of field lines emerging in quiet Sun, even in the network, are connected locally to the sea of magnetic carpet bipoles.

MDI magnetogram data from solar minimum to maximum demonstrated that the amount of flux in the magnetic carpet increases rather than decreases in solar minimum. This suggests that either there is local generation of magnetic fields or that field is stored in the convection zone throughout the cycle and is continuously pulled to the surface by the convective flows. Models of granular convection with rigid lower boundary conditions can create local dynamo action. However, more physically realistic models with open lower boundary conditions do not allow dynamo action because upflow and downflow regions do not mix significantly.

The *MDI* measurements also have shown that magnetic flux dispersal is not independent of the amount of flux in a local flux concentration. This had been assumed in

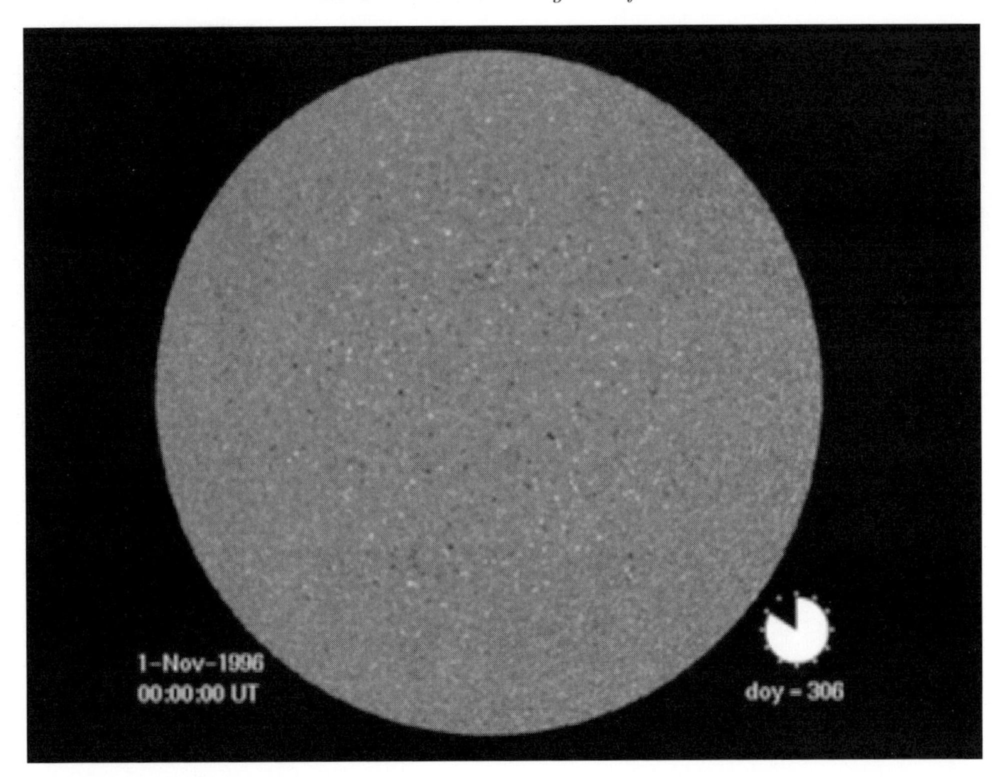

Figure 2. *MDI* magnetogram at Solar Minimum showing mixed polarity covering the solar diskmagnetic fields. The "Magnetic Carpet", over the entire solar surface.

earlier kinematic flux dispersal models. But larger flux concentrations last longer because it takes more time for the fragmentation and cancellation processes to act. Before this was recognized there was a problem with the kinematic models because they required a diffusion coefficient that was a factor of three larger than measured in order to fit the observed surface flux distributions. The *MDI* observation that larger concentrations moved more slowly resolved this problem. To measure the diffusion caused by supergranulation, magnetic fields must be tracked over several supergranule turnover times. Only the larger and hence longer-lived magnetic concentrations lasted long enough to measure their displacements. As a result many measurements of the magnetic diffusion coefficient, which results from supergranulation, were under estimates.

Even at *MDI* deep magnetogram resolution ($\sim 4''$) the total absolute value of the flux is usually an order of magnitude greater than the net flux averaged over arc minute regions. Observations at higher resolution are suggesting that the total flux may be one or even two orders of magnitude greater than the net flux on arc minute scales. The *Stokes Polarimeter* of the *Solar B Solar Optical Telescope (SOT)* in combination with ground based measurements will greatly aid in understanding just how much small scale flux resides on the Sun and how it evolves. The tasks for the new magnetic field measuring instruments on *Solar B* have been sharpened by the remarkable achievements in adaptive optics and post processing, in particular the fast image capture and adaptive optics operations at the Swedish Vacuum Tower Telescope in La Palma, Canary Islands, Spain and the post processing software developed by the teams at the observatory and the Institute of Theoretical Astrophysics at Oslo University. The ground observations made at the new German, Swedish, and American telescopes, especially chromosphere magnetograms

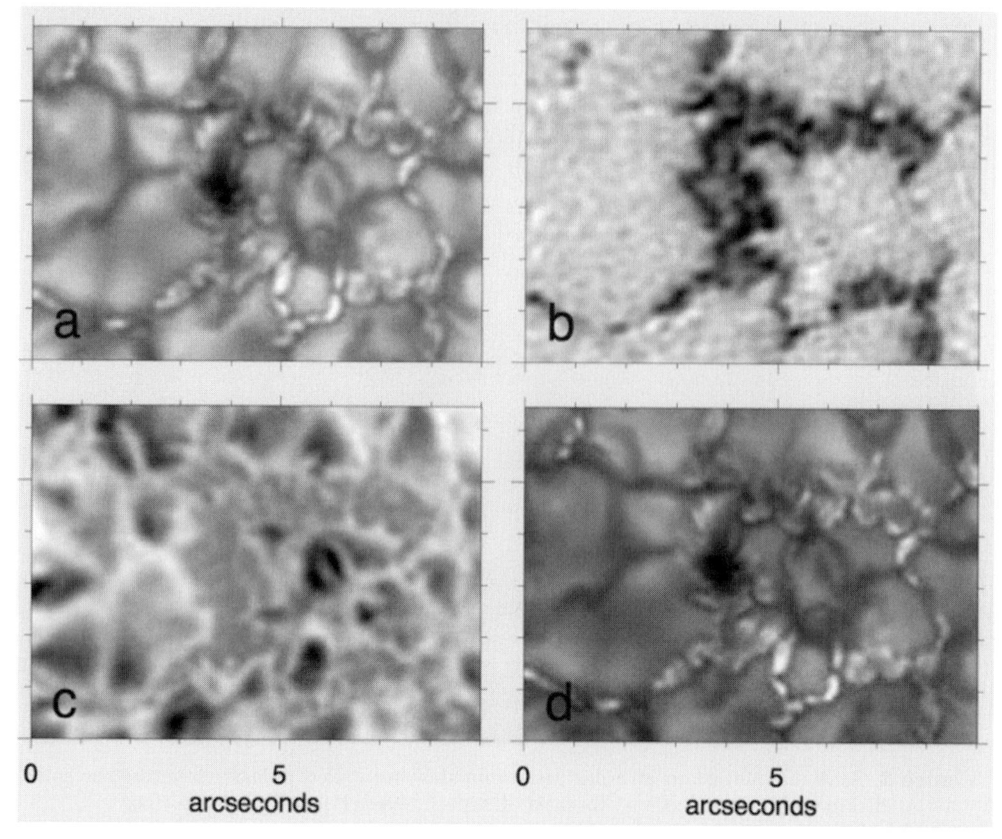

Figure 3. Image in white light (a), magnetogram (Fe 6302) (b), dopplergram (Fe6302) (c), and G-band (d). Images from Swedish Vacuum Tower telescope taken by T. Berger.

made in the IR should allow a new understanding of how the flow dominated vector fields in the photosphere evolve into the force free fields in the corona.

Until last few years properties of the smallest magnetic structures have been inferred from observations of the effects of the field on the atmosphere observed in the CN and CH bandheads. The presence of the field causes line weakening, which lowers the $\tau = 1$ surface. As a result magnetic structures appear as bright features in filter images, ~ 8 Å wide, centered on the bandheads. All CN and CH bright points are contained in the granule boundaries. With adaptive optics and post processing it has been possible to make magnetograms with resolutions comparable to the filter images. These show that the magnetic fields exist in the bandhead bright points and also in the adjacent intergranular lanes (see Fig. 3). Using a combination of narrow band Fabry-Perot filter images, ~ 0.05 Å, and spectral polarimetry in the visible and IR, it now seems that approximately equal amounts of the flux in the intergranular lanes has a strength of greater than a kilogauss and less than 500 gauss.

Magnetogram movies together with spectra and CH bandhead movies indicate that the bandhead bright points appear when the magnetic field strength exceeds a kilogauss and disappear when the field strength drops below that level. The picture that is emerging is that the granular flows alternately compress and allow the magnetic field to relax toward the equipartion strength of about 300 gauss. As a result the kilogauss regions move about in the intergranular lanes appear, disappear, and merge in a continuous dance modulated by the granular flows.

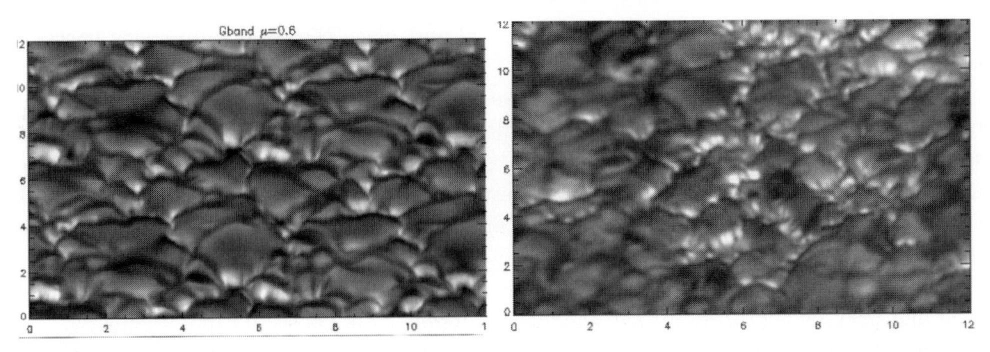

Figure 4. Numerical Simulation of granulation at $\mu = 0.6$ (left panel) and observation (right panel). Courtesy Mats Carlsson.

New observations now are showing the nature of the emergence process on very small scales. Numerical simulations indicate that there should be differences between the appearance of small-scale fields created by local dynamo processes and flux loops pushed onto the surface by convective plumes. In the case of dynamo action the magnetic field appears in swirls, "cinnamon buns", whereas convection drives the emergence of small bipoles. The new observations on the ground will provide the best discriminating data because of the larger apertures available. Combining the highest resolution ground observations with precision spectropolarimetry from space will provide critical data for determining the physics of the small magnetic features.

It is well known that the granular pattern lifetime is on the order of ten minutes. This means that the magnetic field concentrations change significantly on that same time scale. This is important because the loops seen in the corona and the structures in the chromosphere and transition region that are attached in to quiet Sun must be evolving on the same time scale. However, this is an upper limit to the time scale because all of the fields are changing position locally. The magnetic structure at any point is changing because of the strength and location of the magnetic fields both locally, in their vicinity, and globally.

A variety of arguments have been made for heating of the corona by twisting, cancellation, merging, and reconnection of small flux tubes. It is now straightforward to create models that can generate enough energy to heat the chromosphere, transition region, and corona, but at present it is not possible to distinguish these different scenarios observationally.

3.2. *Granulation*

Models of surface convection have for more than a decade created simulated images of granulation that are in excellent agreement with the appearance of granulation observed in white light. Recently observations of granulation at the limb with the new one-meter Swedish telescope revealed for the first time the three-dimensional structure of granulation. Initially, the model calculations did not reproduce these observations. This occurred because not enough ray angles were included in the numerical simulations to properly model the granule edges as seen near the limb. This has been corrected and observations and models are in remarkable agreement (see Fig. 4). These detailed numerical simulations validate the models made nearly three decades ago that showed that hot walls would occur because the lower density gas in the magnetic flux tubes allowed seeing deeper in to the sides of the granules. As a result the solar luminosity variations caused by magnetic fields are much better understood and place the solar luminosity variations seen at the Earth on a firm physical basis.

In regions of dense plage the granulation has been described as "abnormal". Abnormal granulation was characterized by smaller size and lower contrast. This was always somewhat of a mystery because the surface brightness in plage was not different from normal quiet Sun at sensitivity levels of less than 0.1%. The new observations indicate that the granules in plage appear different at lower resolution because magnetic field fills more of the lanes. As a result the lower resolution images show brighter lanes in plage than relatively field-free regions but, in the 100 km resolution images, contrast between lane and granule center is not significantly different from those in quieter Sun.

4. The upper atmosphere

Above the photosphere the temperature of the solar atmosphere initially drops, but about 500 km above the visible surface the temperature rises from ~4,000 K at the temperature minimum to 3,000,000 K in the corona. The processes responsible for the rise are essentially non-thermal and dynamic. Fig. 5 illustrates the complex mixture of cool and hot gas in the chromosphere and corona. In the region of the above temperature minimum the magnetic pressure dominates the gas pressure, while in the photosphere below the gas dynamics control the magnetic field. The problem of heating the chromosphere, transition region, and corona is the problem of understanding how the flow systems in the surface transfer energy to the magnetic fields and then how that energy is dissipated in the atmosphere above.

Observations using the *YOHKOH* spacecraft, which imaged the Sun in the soft X-ray region, ~2-3,000,000 K, forced a recognition that magnetic reconnection occurred both often and rapidly in spite of the very high conductivity of the coronal gasses. *SOHO* which observes in the EUV ~600,000 to 2,000,000 K, also showed that magnetic reconnection was a common process. Both *YOHKOH* and *SOHO* observed the entire Sun, but because of their orbits and the state of technology at the time of their design, they were somewhat limited in both temporal and spatial resolution. The *TRACE* satellite, which observes in the EUV ~600,000 to 2,000,000 K, sacrificed full field of view for higher spatial resolution and higher temporal cadence and observes 24/7 for nine months of the year. Together *SOHO* and *TRACE* provide complementary observations and normally operate as a single data collection system.

Besides rapid reconnection events, data from the *YOHKOH*, *SOHO*, and *TRACE* missions show that the corona is multi-thermal with no line of sight that can be characterized by a temperature and density (see Fig. 6). A line of sight may and most probably will contain gas in the range of temperatures from 6,000 to 3,000,000 K even in the absence of flares. Part of the temperature and density mixing may be local. That is, loops may consist of multiple threads that have different temperatures and densities. Some is non-local in that there maybe hotter or cooler material both above and below any particular structure under study. Further, loops fluctuate continuously with a time scale of a few minutes and it is difficult to identify an isolated loop for much more than twenty minutes.

A primary goal of the upper atmosphere missions is to collect images and movies in a set of spectral lines that span the temperature range from about 6,000 K to 3,000,000 K with a cadence that allows the entire temperature range to be sampled in a time short compared to the rate at which the solar structures are evolving. While the spectral imaging is valuable it suffers from the problem that as the temperature increases or decreases the structure disappears from the image because the line formed in a narrow temperature band. (However, features seldom disappear because of velocity shifts because

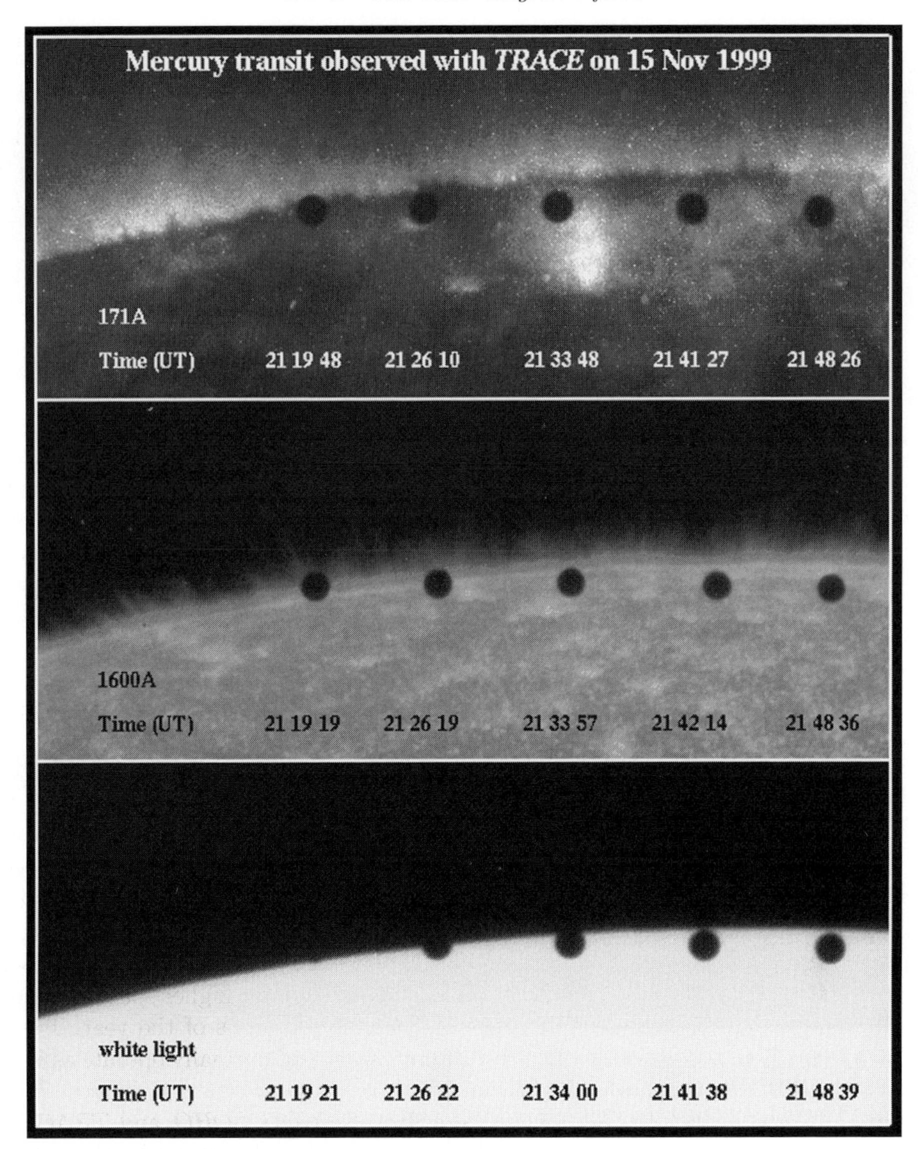

Figure 5. Each panel of this figure is a composite of images of a Mercury transit across the limb of the Sun captured by *TRACE*. In the vertical direction each set of Mercury images is taken within 30 seconds of time. The upper panel is taken in Fe IX-X 171 Å (1 MK), the middle in the 1600 Å band which contains C IV (60,000 K), and the lower image in the continuum (6,000 K). The diameter of Mercury is ~5 Mm. It is clear from the images that the transition region and corona overlap and the transition region extends at least 6 Mm above the visible surface.

the filter bands allow velocities of ± 5000 km/sec.) It is then difficult to establish whether the changes seen in a narrow band are due to temperature or density effects.

To sample the set of four EUV *TRACE* bands requires about 90 seconds. Most phenomena on the Sun change considerably in 20 to 40 seconds. As a result most *TRACE* observations have been in a single line or several lines with occasional sample of the full spectral range. Another short coming of both *TRACE* and *EIT* (*Extreme ultraviolet Imaging Telescope* on *SOHO*) is that the 2.5 to 5 million degree corona is not well sampled. The *Atmospheric Image Assembly (AIA)* on *SDO* will take full disk observations in

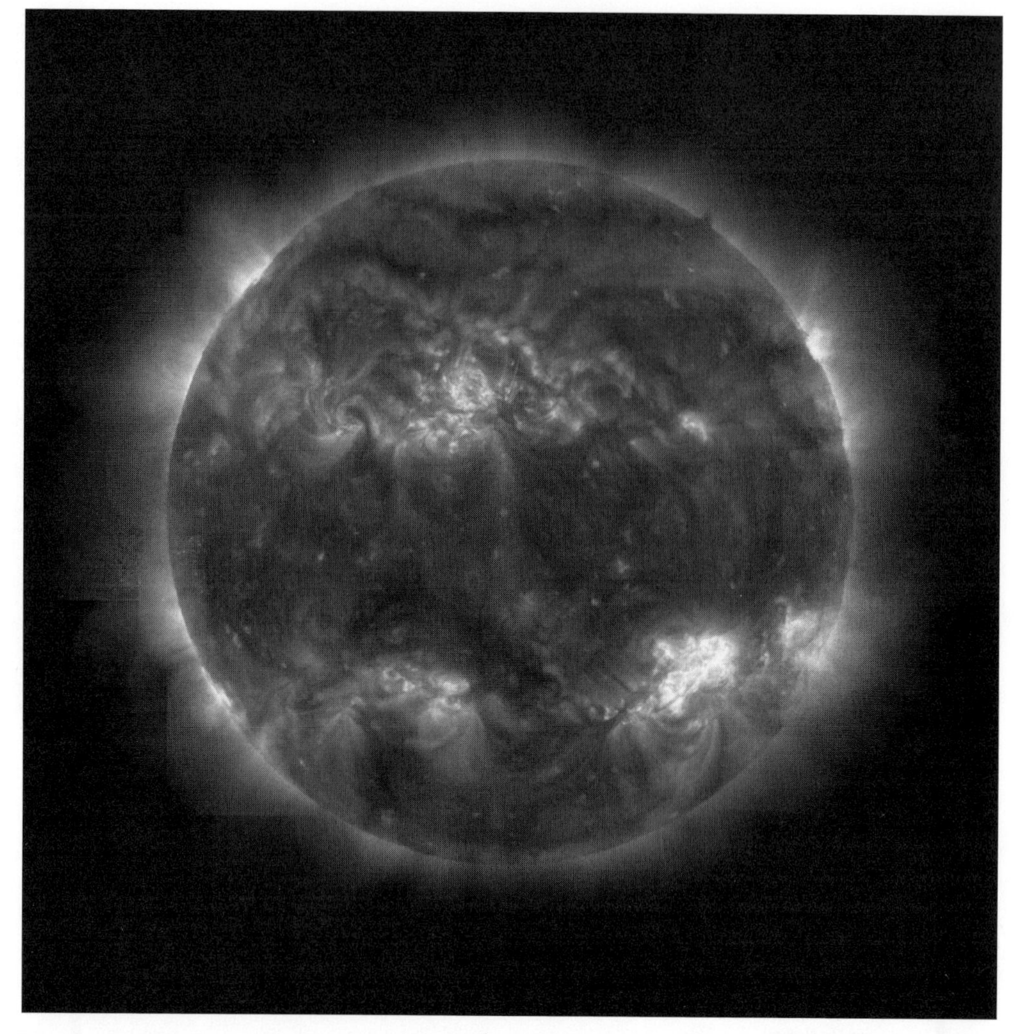

Figure 6. This is a composite of the 171 Å (1 MK-blue), 195 Å (1.5 MK-green), and 284 Å (2 MK-red)) *TRACE* images. Each *TRACE* full disk image is a composite of 37 different *TRACE* pointings. The entire composite consists of 111 separate images and it took 1.5 hr to complete. AIA will take 8 full disk images with nearly the same spatial resolution in 10 sec.

8 spectral bands in 10 seconds and with the spatial resolution of *TRACE*. The *SDO* data combined with the full disk the *X-Ray Telescope (XRT)* on *Solar B* and the new *Solar X-ray Imager* on *GOES-N* will for the first time acquire images spanning the temperature range from 6,000 K to 30,000,000 K over the entire Sun. The 3-D EUV images from *STEREO* will contribute greatly to removing some of the ambiguities of the structure and evolution of coronal features. The new *Extreme ultraviolet Imaging Spectrograph (EIS)* on *Solar B* will provide extremely valuable data velocity and line profile data, but more is needed.

Because the solar atmosphere is optically thin neither spectral images nor spectra can completely resolve the temperature and density of individual loop structures. The inversion problem is also intrinsically ambiguous. In order to confirm our understanding of loop heating mechanisms forward modeling is required. By using the surface and chromospheric magnetic fields measurements estimates of the magnetic structure in the corona

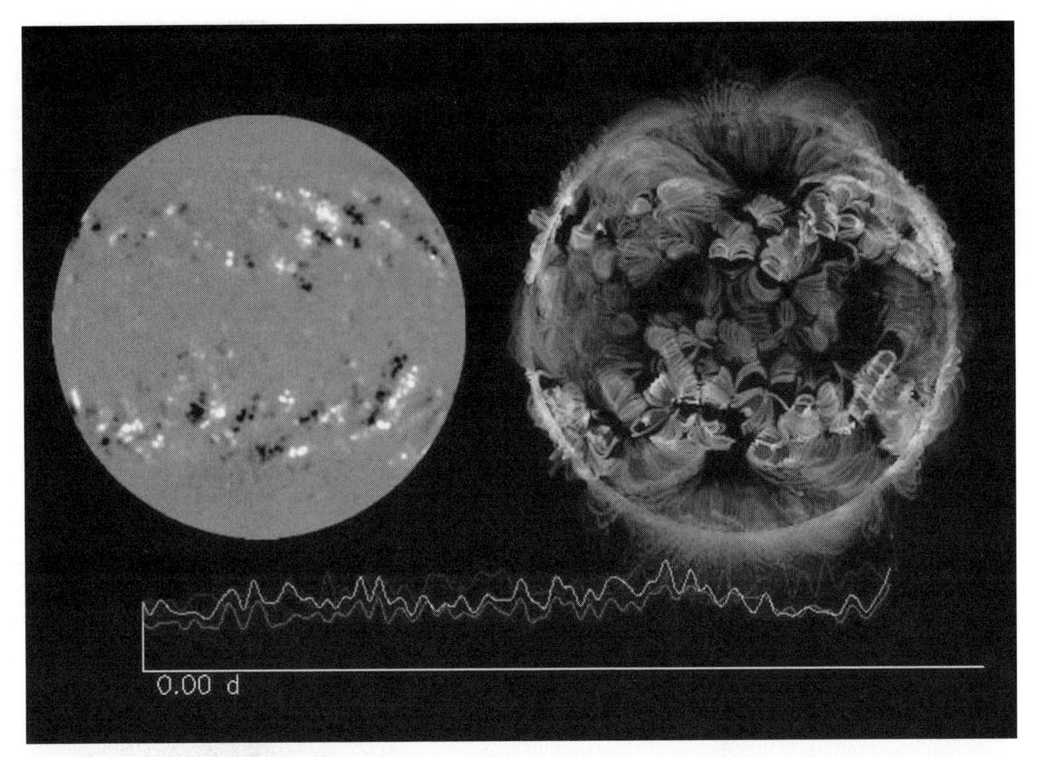

Figure 7. The left hand panel shows a magnetogram and the right hand panel shows the estimated loop structures. The loops are color by temperature 1 MK (blue), 1.5 MK (green), and > 2 MK (red). The plot below show the relative amount of gas in the temperature bands over a 27-day period. The magnetic field lines are a potential field calculation.

can be made. The surface flows can be measured. Then using models of loop heating a model corona can be constructed, which can then be compared with the observations. An example of the synthetic Sun concept is shown in Fig. 7. By continuous refinement of the physics in the models, it should be possible to define the basic physics of at least some of the coronal heating processes. The advances in both computer power and physics based numerical simulations have allowed real progress in understanding some of the major issues in coronal heating.

Another complication to understanding the corona is the lack of knowledge of the transition between the photosphere and the corona. The chromosphere and the transition region are hard to model and are intrinsically dynamic. Observations of the transition region are difficult because properties of materials limit the reflectivity of mirrors in the spectral region that has the most contribution, 50 to 100 nm. But there has been recent work that indicates normal incidence optics can be built to operate efficiently in this spectral region. If this work is successful, then more data on the transition region should become available. Meanwhile, because the coronal observations have demonstrated the importance of these regions significant effort in modeling has been expended and new results are exciting.

The paragraphs above on the corona have emphasized what we do not know. The amazing observations of the past 15 yr have provided many and varied wonderful surprises. We were surprised by how rapid and how pervasive were the magnetic reconnection processes, by how many loops were not in hydrostatic equilibrium, by the observations that most structures less that $\sim$2,000,000 K were not resolved with one arc second resolution,

Figure 8. *TRACE* image in 171 Å (1 MK) showing long loops at the limb that do not increase substantially in cross section with distance from the surface.

and by the observation that the width of most loops did not expand with distance from the solar surface (see Fig. 8). Completely unanticipated were large-scale loop oscillations. Once these oscillations were observed it was surprising that the oscillations damped so rapidly. The discovery of these coronal oscillations has created a new science of "Coronal Seismology".

The topology of the magnetic fields is now recognized an important factor in both the "steady" heating of the corona, the rapid release of energy in solar flares, and the large scale coronal mass ejections. The pervasive emergence of small-scale fields results in continuous reconnection with loops on larger scales. There is a continuous inverse cascade of energy released and transported by the evolution and interactions of smaller and larger magnetic structures. Additionally emergence of new active regions affects adjoining active regions both in the same hemisphere and across the equator. The emergence of flux locally causes rearrangements that can impact the entire Sun.

Surprising observations are the reason science is so exciting. Each new surprise drives new approaches, insights, and understanding. It is not discouraging that many complex interactions are occurring in the Sun; it is exciting, challenging, and important to all of physics and astrophysics.

Highlights of Astronomy, Volume 14
IAU XXVI General Assembly, 14-25 August 2006
Karel A. van der Hucht, ed.

doi:10.1017/S1743921307009842

Similar phenomena at different scales: black holes, the Sun, γ-ray bursts, supernovae, galaxies and galaxy clusters

Shuang Nan Zhang[1,2]

[1]Physics Department and Center for Astrophysics, Tsinghua University, Beijing, 100084, China
email: zhangsn@tsinghua.edu.cn

[2]Key Laboratory of Particle Astrophysics, Institute of High Energy Physics, Chinese Academy of Sciences, Beijing, China

Abstract. Many similar phenomena occur in astrophysical systems with spatial and mass scales different by many orders of magnitudes. For examples, collimated outflows are produced from the Sun, proto-stellar systems, gamma-ray bursts, neutron star and black hole X-ray binaries, and supermassive black holes; various kinds of flares occur from the Sun, stellar coronae, X-ray binaries and active galactic nuclei; shocks and particle acceleration exist in supernova remnants, gamma-ray bursts, clusters of galaxies, etc. In this report I summarize briefly these phenomena and possible physical mechanisms responsible for them. I emphasize the importance of using the Sun as an astrophysical laboratory in studying these physical processes, especially the roles magnetic fields play in them; it is quite likely that magnetic activities dominate the fundamental physical processes in all of these systems.

As a case study, I show that X-ray lightcurves from solar flares, black hole binaries and gamma-ray bursts exhibit a common scaling law of non-linear dynamical properties, over a dynamical range of several orders of magnitudes in intensities, implying that many basic X-ray emission nodes or elements are inter-connected over multi-scales. A future high timing and imaging resolution solar X-ray instrument, aimed at isolating and resolving the fundamental elements of solar X-ray lightcurves, may shed new lights onto the fundamental physical mechanisms, which are common in astrophysical systems with vastly different mass and spatial scales. Using the Sun as an astrophysical laboratory, "Applied Solar Astrophysics" will deepen our understanding of many important astrophysical problems.

Keywords. Sun: flares, Sun: corona, Sun: X-rays, X-rays: binaries: individual (Cygnus X-1), accretion disks, magnetic fields, gamma rays: bursts: individual (GRB 940217), supernovae: individual (SN 1987A), galaxies: active: individual (M 87), black hole physics

---

1. Introduction

To start with, I first categorize research methods in astronomy into four classes:

• Objects oriented: focus on studying particular types of objects, such as stars, galaxies, or clusters of galaxies, etc. Astronomers doing this type of research are normally called observers or observational astronomers;

• Physical processes oriented: apply known physical processes in order to explain all observed phenomena. Those doing such research are called theorists or theoretical astrophysicists.

• Data analysis/mining: apply new or existing data analysis methods to large quantities of data, in order to discover new phenomena. We call them data analysts.

• "Similar phenomena at different scales": look for similar phenomena from astrophysical systems with very different scales, in order to find common physical mechanisms operating in them. Perhaps they can be called "Observational Physicists".

The last class is the focus of this review article. This method may be appreciated from another perspective. According to the well-known astronomy ladder of J.J. Drake, the "fun" of astronomical subjects going down in this order: cosmology/black holes, quasars/AGN, planet hunting, galaxies, Milky Way/CVs, stars and the Sun. On the other hand, the amount or details of information astronomers can collect go almost in the opposite order. Clearly as the Sun is a star in our backyard, it is one of the most accessible astronomical objects. The Sun is the only object we can collect rather direct information on its core, radiative zone, convective zone, photosphere, chromosphere, corona and its winds. Interestingly, it has been gradually realized that many astrophysical phenomena occurring in the Sun also take place in many other astrophysical objects with enormously different scales. It is thus very likely that similar physical processes are operating in all these different systems. We therefore can use the Sun as our closest astrophysical laboratory for probing many physical processes and then apply the knowledge to understanding other astrophysical systems. Perhaps we can call this research discipline "Applied Solar Astrophysics".

The remaining part of this report is divided into four sections. In section 2, I will show some similar phenomena between the atmospheres of the Sun and that in accreting black hole systems, between jets produced in all kinds of different astrophysical objects with vastly different scales, and between the triple-ring structures in a young supernova remnant and in a galaxy cluster centered at an active galaxy. Magnetic activities seem to dominate or at least greatly influence all these phenomena. In section 3, I will focus on a particular kind of similar phenomenon, i.e., the non-linear dynamics of X-ray variations between that of the Sun, of a stellar mass black hole binary and of a gamma-ray burst. Their X-ray variations show remarkably consistent properties which can be modelled by multiplicative non-linear dynamical processes, suggesting that their X-ray emission regions are made of many inter-connected nodes or elements; multi-scale magnetic field topology most likely plays a key role in the non-linear dynamics observed in all of them. In section 4, I will make some concluding remarks on using the Sun as an astrophysical laboratory to understand these similar phenomena at different scales, and suggest a future X-ray instrument required to solve some existing open issues, in order to make further progress in "Applied Solar Astrophysics".

2. Common magnetic activities dominating processes

In this section, I review briefly three types of similar astrophysical phenomena over a huge range of astrophysical scales. Observational and theoretical studies indicate that these phenomena are all related to magnetic processes in which magnetic topology and energy release play fundamental roles.

2.1. *Similar atmospheric structures between the Sun and black hole systems*

The temperature inside the Sun decreases from about 15 million degrees in its nucleus, where thermal nuclear fusion produces about 99% of the Sun's radiation energy, to about six thousand degrees on its photosphere. However starting from its photosphere its atmospheric temperature increases to about ten thousand degrees in its chromosphere and to above one million degrees in its corona. Because the Sun's interior remains as the only energy source, this temperature inversion means that the solar atmosphere is not in an equilibrium with its interior thermal energy source. Such a temperature inversion, i.e.,

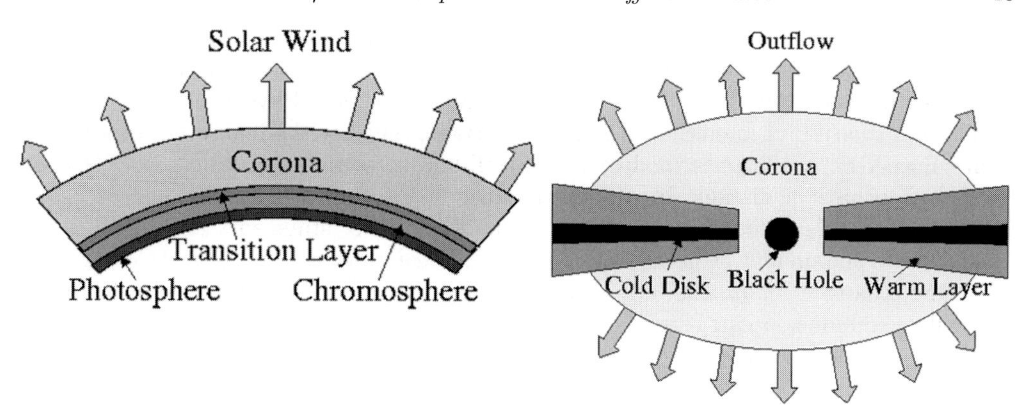

Figure 1. Schematic diagrams of the solar atmosphere and accretion disk structure. The temperatures in the solar atmosphere are approximately: 6×10^3 K (photosphere), 3×10^4 K (chromosphere), and 2×10^6 K (corona), respectively. For the black hole disk atmosphere, the corresponding temperatures are approximately 500 times higher: 3×10^6 K (cold disk), 1.5×10^7 K (warm layer), and 1×10^9 K (corona), respectively. (Re-produced from Zhang *et al.* 2000).

the solar corona heating problem still remains one of the most important problems in astrophysics. It is widely believed that solar flares driven by magnetic reconnections may be able to provide the required energy source for heating the solar corona, if sufficiently frequent 'nano'-flares (about 10^{24} erg energy per event) are produced. This is the so-called Parker's conjecture (Parker 1983, Parker 1988, Parker 1991). However up to now searches for these speculated 'nano'-flares still remain inconclusive (Walsh 2003). Theoretically it also remains uncertain if it is physically possible to produce enough 'nano'-flares (Klimchuk 2006), even within the most hopeful avalanche models (Charbonneau *et al.* 2001). On the other hand, alternative models of solar corona heating may involve ion cyclotron waves and turbulence (see, e.g., Markovskii & Hollweg 2004, Cranmer & van Ballegooijen 2005).

The accretion disk around a black hole in an X-ray binary system has a similar structure to solar atmosphere (Liang & Price 1977) and in particular the temperature inversion from its disk surface to its corona resembles that of the Sun, as shown in Fig. 1 (Zhang *et al.* 2000). Strikingly, the temperatures of the three regions in an accretion disk around a black hole are higher by approximately a factor of 500 than the corresponding regions in the Sun. This supports the notion that magnetic activity is responsible for powering the upper atmosphere in both cases, giving $T \propto E^{1/4} \propto B^{1/2}$ and thus $T_{\rm disk}/T_{\rm sun} \approx (B_{\rm disk}/B_{\rm sun})^{1/2} \approx (10^8 {\rm G}/500{\rm G})^{1/2} \approx 500$, where the typical magnetic field strengths of $\sim 10^8$ G for inner regions of black hole accretion disk (assuming equipartition) and ~ 500 G for the active regions of the Sun are used.

Since in an ionized accretion disk the angular momentum transport as well as energy dissipation are likely dominated by the so-called magneto-rotational instability mechanism (Hawley *et al.* 1999), X-ray flares or 'shots', driven by magnetic reconnection, may be responsible for the observed X-ray variability in systems ranging from accreting neutron star and black hole binaries, as well as active galactic nuclei (AGN) harboring supermassive black holes. Such a 'flare' model for accretion disks around black holes thus makes two generic predictions: (1) The X-ray emission region, including the corona in both stellar mass and supermassive black hole systems should have a 'disk-like' geometry, rather than a spherical-like geometry, which has been widely assumed; (2) Naturally their X-ray light curves should be made of individual flares or 'shots' if sufficiently good sensitivity and time resolution are available.

Kubota *et al.* 2001 have identified a 'disk-like' configuration for the hot corona producing a hard X-ray power-law component through inverse Comptonization process, in the stellar mass black hole binary GRO J1655−40, which was discovered by Zhang *et al.* 1994 and in fact the second microquasar (X-ray binaries with superluminal jets) in the Milky Way (Tingay *et al.* 1995, Harmon *et al.* 1995. The X-ray emitting coronae in AGNs are suggested to have a 'disk-like' geometry, in order to explain the decrease of fraction of Type II AGNs as a function of the observed apparent X-ray luminosity (e.g., Ueda *et al.* 2003), due to the smaller projected area of the 'disk-like' corona for Type II AGNs (Zhang 2005, Liu, Zhang & Zhang 2007). Flares or 'shots' are commonly observed from accreting systems of different sizes (Terrell 1972, Doi 1978, Miyamoto *et al.* 1988, Haardt *et al.* 1997, Nayakshin & Melia 1997, di Matteo *et al.* 1999, Poutanen & Fabian 1999, Negoro *et al.* 2001, Wang *et al.* 2004, Liu & Li 2004). Modelling of periodic or quasi-periodic X-ray and infrared flares from Sgr A*, the supermassive black hole in the center of the Milky Way, suggests that these flares are of accretion disk origin (Aschenbach *et al.* 2004). It is thus quite possible that these flares are also produced by processes driven by magnetic reconnections.

We note that a hard X-ray power-law component up to at least 50-100 keV have also been seen from from active galactic nuclei (Bassani *et al.* 1996) and weakly magnetized (with surface magnetic field strength of around $\sim 10^8$ G) neutron star X-ray binaries (e.g., Zhang *et al.* 1996, Zhang 1997, Chen, Zhang & Ding 2006), which often exhibit similar spectral state transitions as black hole binaries (Zhang 1997, Zhang *et al.* 1997a). It is well understood that the standard optically thick and geometrically thin accretion disk models (Shakura & Sunyaev 1973) cannot produce this power-law component. However the orbital kinetic energies of protons in the inner accretion disk regions in weakly magnetized neutron star X-ray binaries, stellar mass black hole binaries and supermassive black hole systems should be approximately the same. Therefore, if the energy source of the above mentioned 'flare' model for accretion disks is dominated by particle's kinetic energies, the similar hard X-ray power-law component may be generated naturally. We comment in passing that the solar magnetic fields are also believed to be generated by dynamo mechanisms due to differential rotation in the Sun.

2.2. *Astrophysical jets at different scales*

Collimated outflows or jets are common astrophysical phenomena, now found in the Sun, proto-stellar systems, isolated neutron stars, neutron star and black hole binaries, gamma-ray bursts and supermassive black holes. MHD simulations have shown that differential rotation and twisted magnetic fields are two generic ingredients for generating collimated outflows (Meier *et al.* 2001), somewhat similar to magnetic reconnection processes in which twist and shear of magnetic flux tubes play very important roles for solar flares and perhaps also coronal mass ejection events (see, e.g., Wang *et al.* 2001, Pevtsov 2002, Yamamoto *et al.* 2005). With these two generic ingredients, their different scales and observational appearances (such as Lorentz factor and degree of collimation) of different astrophysical jets may reflect their different astrophysical environments (such as gravitational potential, differential rotation energy, magnetic field strength and topology, spin of the central object, etc), where jets are produced, accelerated, collimated and transported. Therefore, astrophysical jets can be used as powerful astrophysical probes for many different types of objects. For example, prolonged optical and X-ray afterglow emissions from gamma-ray bursts have been used to probe the properties of interstellar media around the gamma-ray burst progenitors and consequently the nature of their progenitors, because gamma-ray burst afterglow emissions are believed to be produced from external shocks when their jets eventually plow into their surrounding media, in a similar

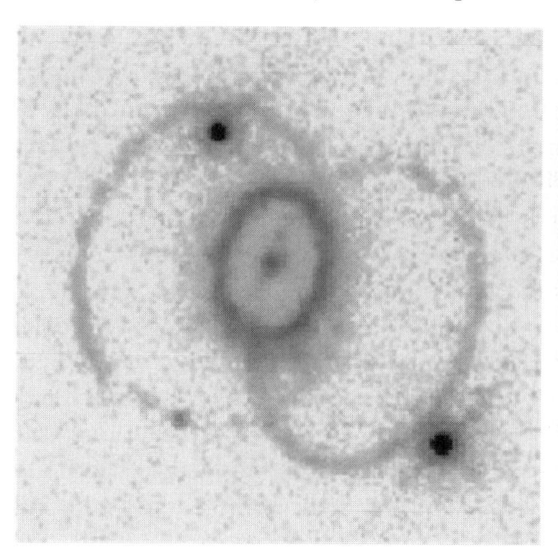

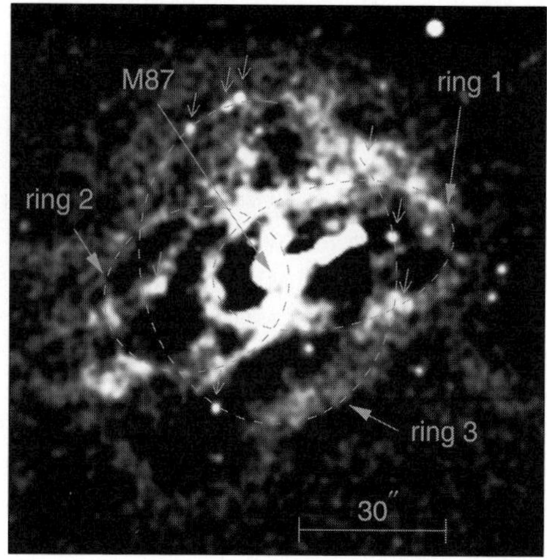

Figure 2. Similar triple-ring structures between the supernova remnant of SN 1987A and Virgo galaxy cluster centered at the active galaxy M 87. *Left*: The optical triple-ring structure of SN 1987A, as observed with *Hubble Space Telescope* (Burrows 1995). *Right*: The X-ray triple-ring structure of Virgo galaxy cluster as observed with the *Chandra* X-ray observatory (Feng *et al.* 2004).

way to an expanding supernova shell plowing into its surrounding medium. Interestingly, the same external shock model has been used to explain the decelerated relativistic jet from a black hole binary system (Wang, Dai & Lu 2003), implying a dense surrounding medium and thus a massive star as the progenitor of the stellar mass black hole, similar to the progenitors to many long duration gamma-ray bursts (see, e.g., Meszaros 2001 for a review and references therein).

In many cases jets from black hole binaries, gamma-ray bursts and supermassive black holes are relativistic and may share the same production mechanisms, and it is very likely that black hole's spin energy is extracted to power their relativistic jets (see, e.g., Zhang *et al.* 1997c, Mirabel & Rodriguez 1998, Mirabel & Rodriguez 2002 for reviews and references there in). Using the X-ray continuum spectra, Zhang, Cui & Chen (1997) first determined the spin of the black holes in several black hole binaries with relativistic jets and established the connection between black hole spin and jet observationally. This method of determining black hole's spin has since been further refined and widely applied to many more black hole binaries (for a recent review, see, e.g., Remillard & McClintock 2006); the existence of extremally spinning black hole is now well-established.

2.3. *Similar triple-ring structures between a young supernova remnant and a galaxy cluster*

Another intriguing pattern of similarity is between the supernova remnant of SN 1987A and M87, an active galaxy producing a large scale jet in the center of the Virgo Cluster (Fig. 2). The triple-ring structure of SN 1987A observed in visible band with the *Hubble Space Telescope* is perhaps the most beautiful astronomical picture taken so far (see Fig. 2 (left); Burrows 1995). A MHD simulation by Tanaka & Washimi (2002) showed that if its progenitor has similar large-scale magnetic field structure like the Sun, then its winds, similar to solar winds, will follow the twisted magnetic field topology and are thus preferentially located along its spin axis or in the equatorial plane. Suppose its progenitor has experienced a slow-wind phase (red-giant) and then a fast-wind phase (blue-giant), then the fast winds will eventually catch the slow winds to form the triple-ring like structure. It is possible that a triple-ring structure might be generic in a rotating system involving magnetized faster winds catching up slower winds. Meanwhile a central source of radiation is required for the rings to shine. For SN 1987A, the strong UV radiation from the supernova provides the necessary illumination.

In Fig. 2 (right) a triple-ring like structure in the X-ray band similar to that of SN1998A was shown, observed with the *Chandra* X-ray observatory (Feng *et al.* 2004). The ring sizes in the Virgo cluster are several thousand times larger than those of SN 1987A. It is possible that a catastrophic merging event around the M 87 nucleus is responsible for the triple-ring structure revealed here. In this scenario, a "slower wind" was present before the merging begins, e.g., a mixture of galactic winds from two merging galaxies. A "faster wind" was then driven during the merging process. The final merging of the two supermassive black holes releases a huge amount of high energy radiation to shine the triple rings in X-ray bands. Presumably, the resulting supermassive black hole spins rapidly to power the highly collimated M 87 jets.

3. Non-linear dynamical processes at different scales

In an extensive study of X-ray variability from a prototype black hole X-ray binary Cygnus-1 (Margon *et al.* 1973), Uttley *et al.* 2005 have shown that its X-ray flux fluctuations, characterized by rms (root-mean-squares) of its flux, is proportional to its mean flux. Such a flux-rms linearity cannot be modelled by additive or deterministic processes. Instead, Uttley *et al.* 2005 have demonstrated that the dynamical process for the X-ray production should be a non-linear 'multiplicative' process, which can be modelled by the following non-linear time-series model,

$$X_i = 1 + \sum_{j=0}^{\infty} G_j u_{i-j} + \sum_{j=0}^{\infty}\sum_{k=0}^{\infty} G_{jk} u_{i-j} u_{i-k} + \sum_{j=0}^{\infty}\sum_{k=0}^{\infty}\sum_{l=0}^{\infty} G_{jkl} u_{i-j} u_{i-k} u_{i-l} + \cdots \tag{3.1}$$

where the coefficients $G_j, G_{jk}, G_{jkl} \dots$ and the higher-order co-efficients are non-zero. Therefore the flux-rms linearity is a clear indication of non-linear dynamical process. Combining this and the fact that the flux distribution of Cygnus X-1 follows a log-normal distribution, but the peak flux distribution of solar flares follows approximately a power-law distribution (which is predicted by the self-organized criticality model), Uttley *et al.* 2005 thus claimed to have rejected essentially all previous models for X-ray variability from X-ray binaries and active galactic nuclei, including shot noise, self-organized criticality, or dynamical chaos models (Uttley *et al.* 2005).

Since the characteristic X-ray variability seen in Cygnus X-1 has also been seen from other black hole X-ray binaries (Uttley & McHardy 2001), as well as neutron star (Uttley 2004) and supermassive black hole accreting systems (Edelson *et al.* 2002, Vaughan *et al.* 2003a, Vaughan *et al.* 2003b), it has been suggested by Uttley *et al.* (2005) that the X-ray variability in these systems should be driven by fluctuating accretion flow (Kotov 2001, Lyubarskii 1997). Since accretion flow does not exist in the Sun, their conclusion rules out common physical mechanism for X-ray emission from the Sun, thus has far-reaching impacts to our understanding of some fundamental processes in a very broad range of astrophysical systems.

Motivated by this and in light of the outstanding solar corona heating problem, we examine the X-ray variability in the Sun's X-ray lightcurves collected by the *Reuven Ramaty High Energy Solar Spectroscopic Imager* (*RHESSI*) (Lin *et al.* 2002)†, which has been in productive operation since launch on 5 February 2002 (Dennis *et al.* 2005). We also re-examine the non-linear dynamical properties of Cygnus X-1 and a gamma-ray burst. We demonstrate that a simple model can describe the non-linear dynamical properties of these three kinds of systems of very different astrophysical objects at very different scales.

3.1. *Flux-rms relation and flux distributions for individual solar X-ray flares*

In Fig. 3 the lightcurve, rms evolution and flux-rms correlation for *RHESSI* 2022619 solar flare are shown. The flux-rms relationship shown in the inset of Fig. 3 can be described by a simple model (plotted as solid line),

$$\mathrm{rms}^2 = (\alpha C)^2 + C, \tag{3.2}$$

where $\alpha = 0.27$ is a constant, C is the average counts within each 1 s interval. Clearly, the total rms observed is made of two terms, a term linearly proportional to the average flux of the solar flux at a given time (plotted as dashed line), and a Poisson fluctuation term (plotted as dotted line); for a random Poisson process, $\mathrm{rms} = \sqrt{C}$. It can be seen from Fig. 3 that the linear term starts to dominate at flux above 20 counts per 0.1 s.

In Fig. 4 and Fig. 5, eight *RHESSI* solar flares with different peak fluxes, durations and lightcurve morphologies are shown to illustrate their common properties. Interestingly, not only can the same simple model in equation (3.2) describes their flux-rms relation perfectly, but also $\alpha = 0.27$ for all of them. In the insets the flux distribution of each solar flare is also plotted, which can be approximated by one or several log-normal components, i.e., the logarithm of the flux follows normal distributions, in comparison for the single log-normal peak of flux distribution for the black hole X-ray binary Cygnus X-1 (Uttley *et al.* 2005), and see the inset of Fig. 8 in section 3.3.

3.2. *Flux-rms relation for continuous solar X-ray light curves*

In Figs. 6 & 7 several continuous segments of characteristic *RHESSI* solar X-ray lightcurves are shown in 3-6 keV and 6–12 keV bands with 1 s time resolution. These figures reveal X-ray variability at all time scales with a large range of amplitudes over about five orders of magnitudes. Despite of its apparently very different lightcurves observed at well separated times, the solar X-ray variability exhibits remarkably consistent and simple relationship between the rms and its mean flux as shown in Fig. 8, which are again modelled by equation (3.2) with the same value of $\alpha = 0.27$. We note that the flux-rms linearity also exists in a very broad frequency range if calculated from the FFT power-spectra of its lightcurves (Tang & Zhang 2007), again similar to that of Cygnus X-1 (Uttley *et al.* 2005). It should be noted that a continuous solar X-ray lightcurve is collected over the whole surface of the Sun, not just from a single active region of the solar surface as for the case of a single solar flare. According to the mathematical description of equation (3.1) for the linear flux-rms relationship, this implies that many active regions in the solar surface are not isolated, but somehow connected and react to common triggers or ignitions. This is because the flux fluctuations from many isolated or independent regions should be proportional to the square-root of the total combined flux. Such inferred inter-connection between different active regions may offer deep insight for the solar activity mechanisms.

† http://hesperia.gsfc.nasa.gov/hessi/

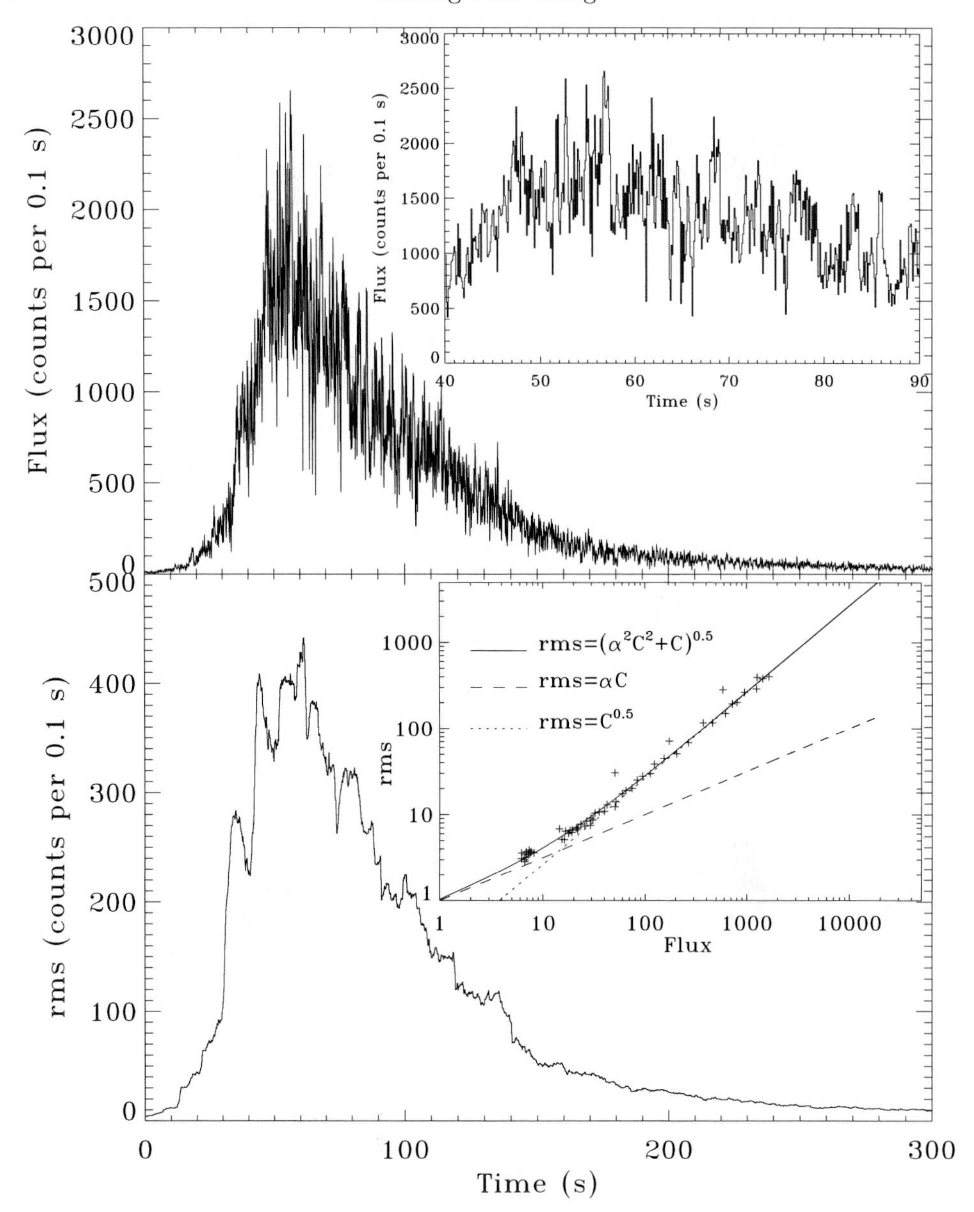

Figure 3. *RHESS* 2022619 solar X-ray flare between 3–50 keV. The upper panel and inset show its lightcurve at 0.1 second resolution; variability beyond Poisson counting fluctuations is clearly visible. The lower panel shows the rms (root-mean-squares), in units of counts in 0.1 sec, for each 1 sec internal as a function of time and the average flux, in units of counts per 0.1 sec, during each 1 sec interval. For the model lines shown in the inset, $\alpha = 0.27$ is the best fit result with negligible error.

3.3. *Cygnus X-1 and a γ-ray burst*

In Fig. 9 the flux-rms correlation for the canonical stellar mass black hole binary Cygnus X-1, observed with the PCA instrument onboard the *Rossi X-ray Timing Explorer*†, is shown (see Uttley *et al.* 2005 for comprehensive results on the nonlinear behaviours of

† http://astrophysics.gsfc.nasa.gov/xrays/programs/rxte/pca/

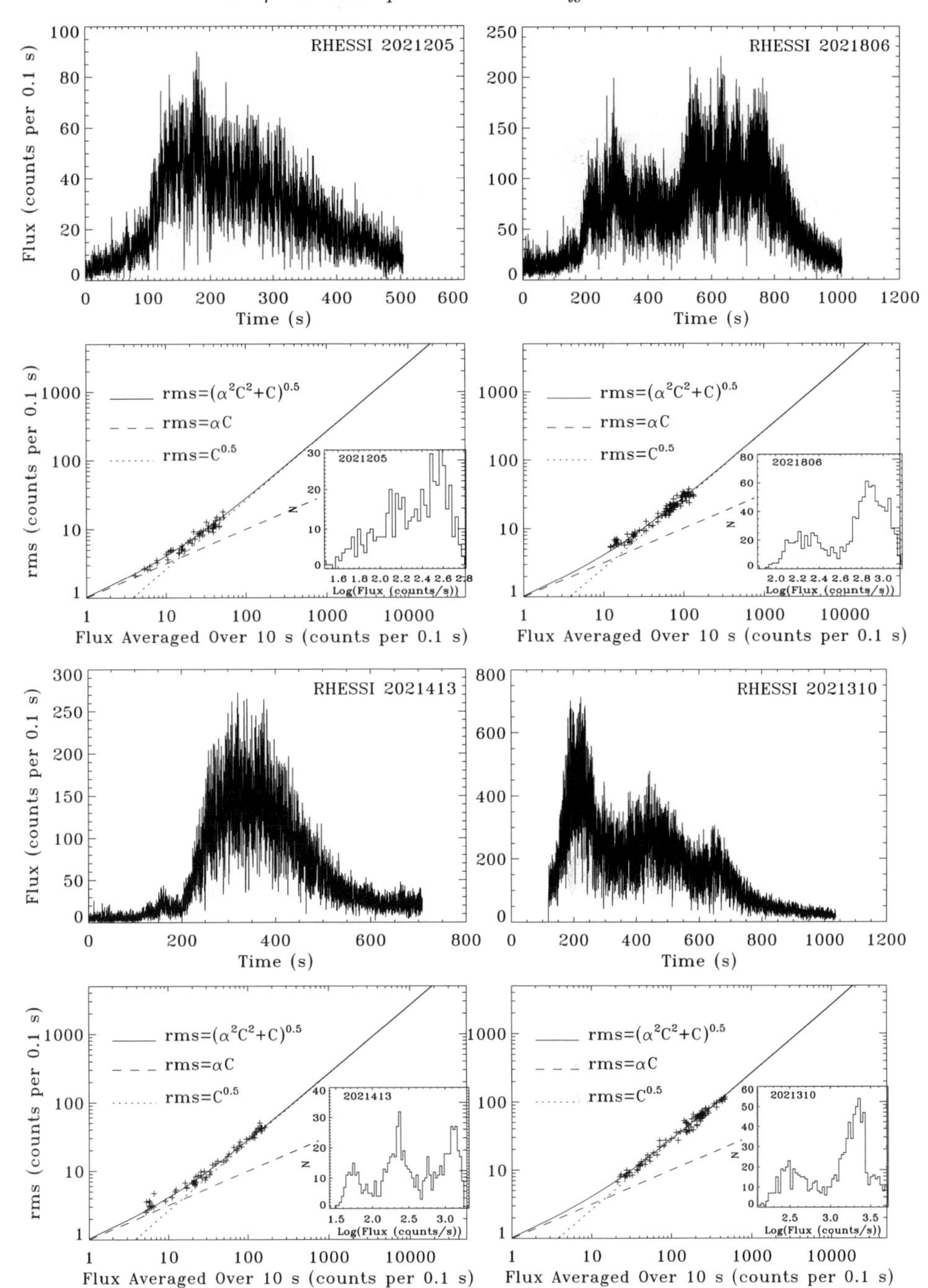

Figure 4. The lightcurves, flux-rms correlation and distribution of the logarithm of the flux for four individual solar X-ray flares. It should be noted $\alpha = 0.27$ for all of them, and that their distributions of the logarithm of the flux are made of one or several normal distributions.

Cygnus X-1 with more data). The crosses are for the average values of rms for each flux interval divided uniformly in linear scale due to the narrow flux range. Compared to the flux-rms relation for *RHESSI* solar X-ray flares, the model describing the correlation is

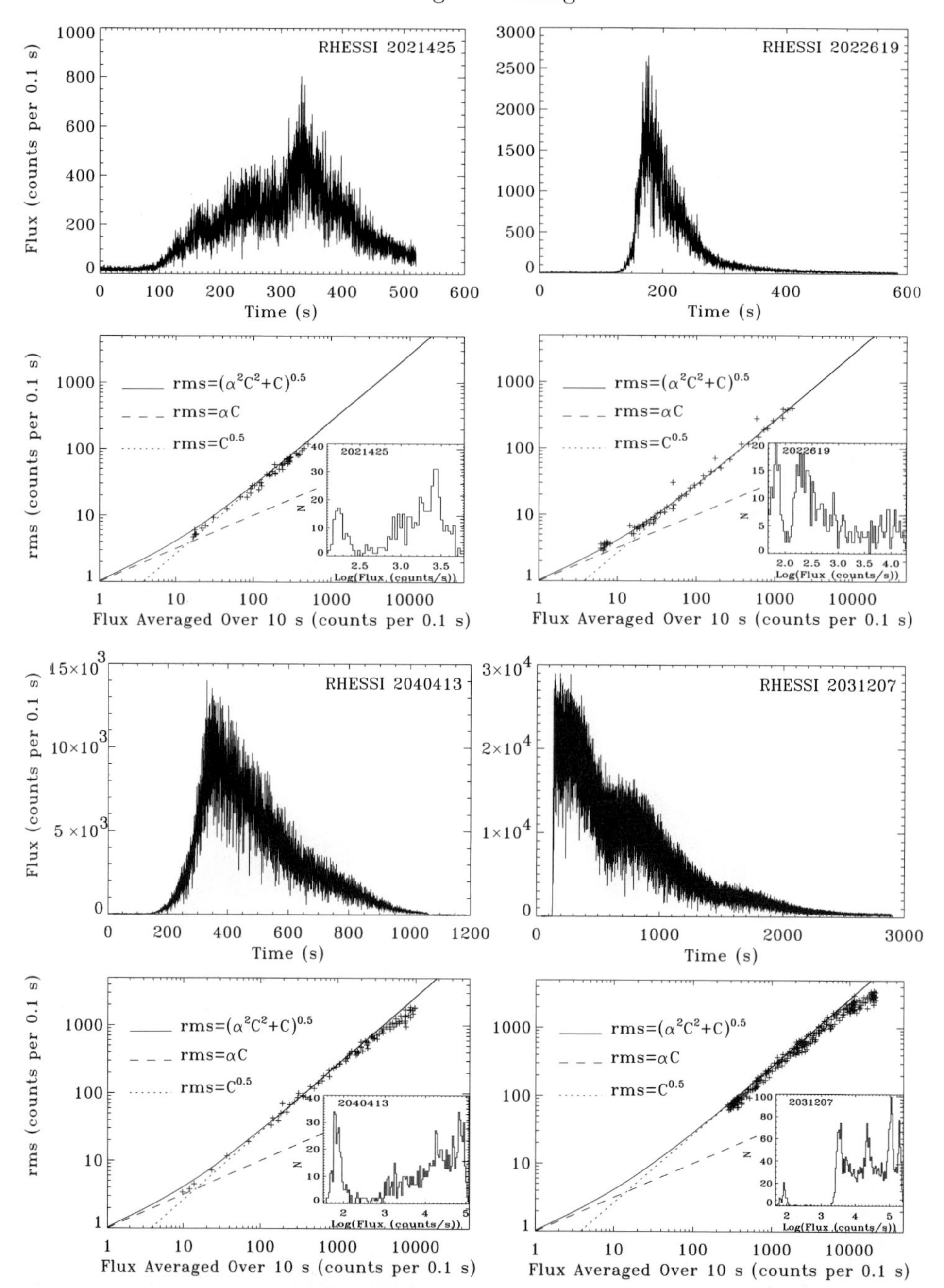

Figure 5. Continuation of Fig. 4.

only slightly more complicated,

$$\mathrm{rms}^2 = (\alpha F)^2 + C, \ F = C - C_0, \tag{3.3}$$

where the total count C consists of a constant term $C_0 = 57$ and a variable flux with rms $= 0.38 \times F$ over the Poisson counting fluctuations. Note that $C_0 = 57$ is well above the PCA background level, and is thus an intrinsic component from Cygnus X-1. The inset in Fig. 9 shows the log-normal flux distribution of Cygnus X-1. It has been known that

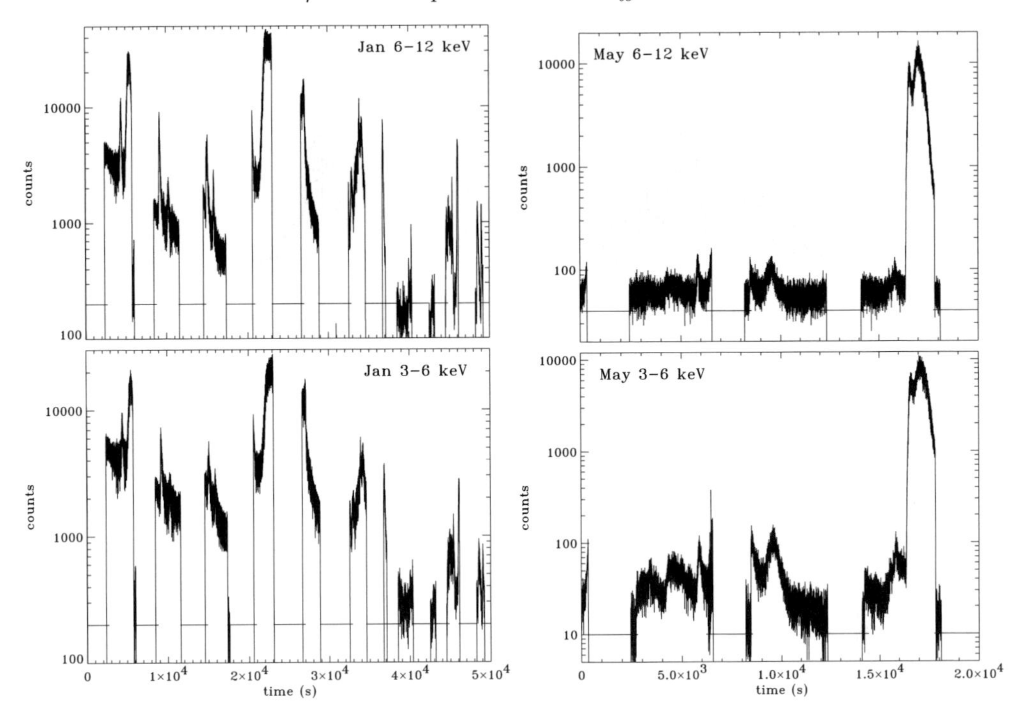

Figure 6. Solar X-ray lightcurves in two energy bands with 1 s resolution, on 2004-Jan-1.0 (left panels) for 50,000 consecutive seconds and on 2004-May-1.0 (right panels) for 20,000 consecutive seconds. Horizontal straight bars near the bottom in each panel mark data gaps, i.e., 'night' times for the *RHESSI* satellite.

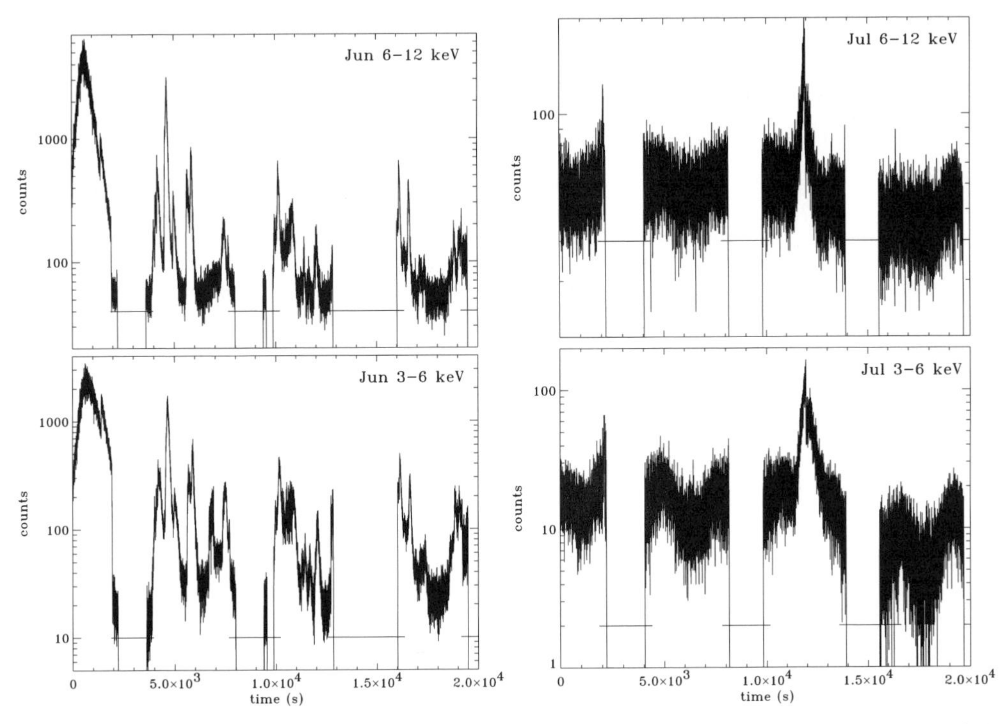

Figure 7. Solar X-ray lightcurves in two energy bands with 1 s resolution, on 2004-Jun-1.0(left panels) for 20,000 consecutive seconds and on 2004-Jul-1.0 (right panels) for 20,000 consecutive seconds. Horizontal straight bars near the bottom in each panel mark data gaps, i.e., 'night' times for the *RHESSI* satellite.

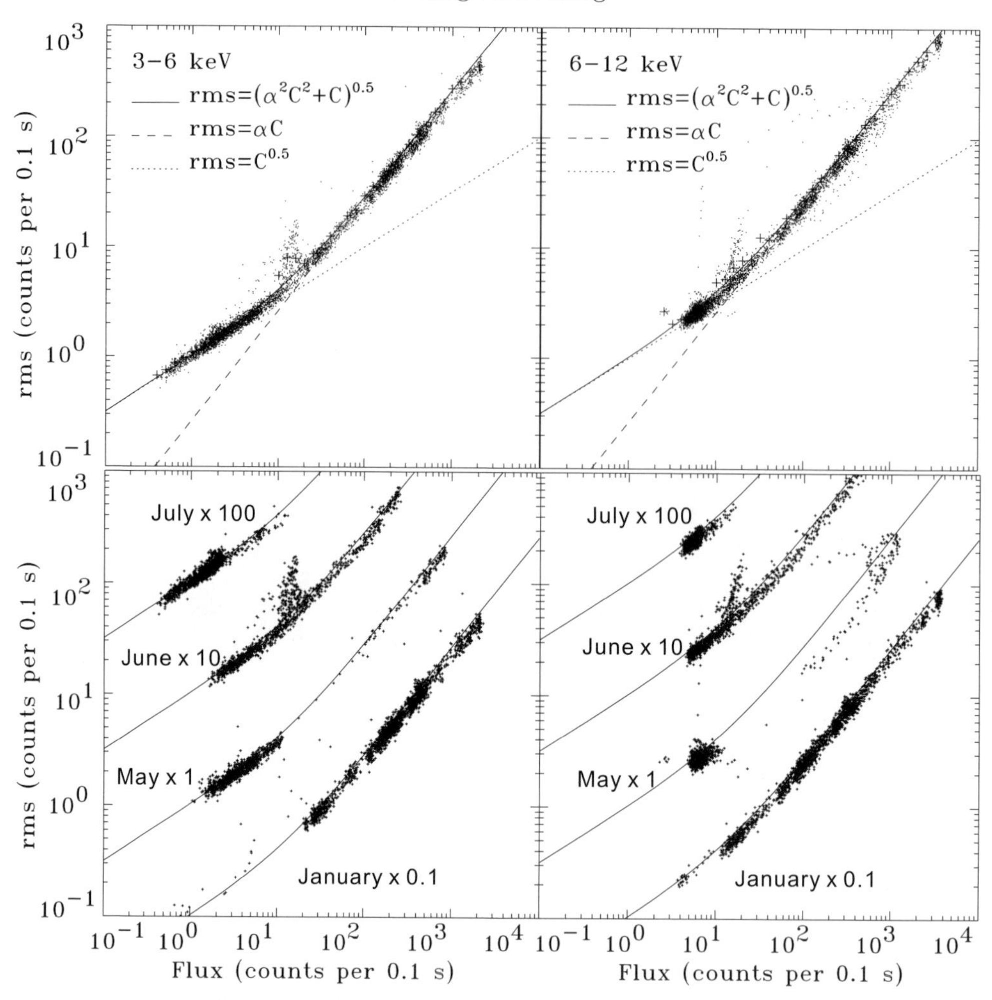

Figure 8. *Upper panels*: The relationship between flux and rms for all four segments of lightcurves in two energy bands. Each point in the scatter plots is for a 10 s interval of data with 0.1 s resolution. *Lower panels*: The flux-rms relations for the four lightcurves are shown separately; the data and model lines for January, June and July are shifted vertically for clarity. Clearly all four lightcurves have consistent flux-rms relations.

similar linear flux-rms relation exists for neutron star (Uttley 2004) and black hole binaries (Uttley & McHardy 2001), as well as for supermassive black hole binaries (Edelson *et al.* 2002, Vaughan *et al.* 2003a, Vaughan *et al.* 2003b).

In Figs. 10 & 11 the lightcurve and flux-rms relation for GRB 940217, detected with the BATSE instrument onboard the *Compton Gamma-ray Observatory*†, are shown for comparison. Equation (3.3) provides a consistent description for the flux-rms relation with $\alpha = 0.25$ and $C_0 = 600$. Compared to the flux-rms relation for Cygnus X-1, $C_0 = 600$ is entirely due to the background level of the *BATSE* instrument. Therefore, the flux-rms relation for GRB 940217 is described by exactly the same model for *RHESSI* solar X-ray flares, if the different instrumental background levels are taken into account.

† http://cossc.gsfc.nasa.gov/docs/cgro/batse/

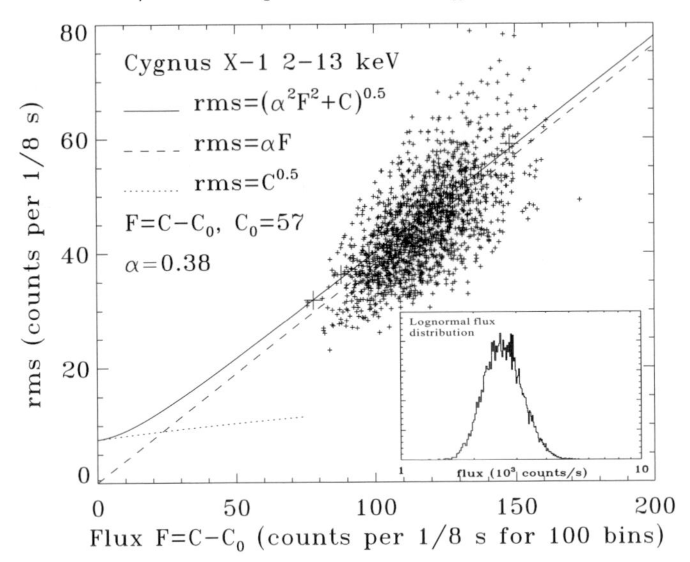

Figure 9. The flux-rms relation for the canonical stellar mass black hole binary Cygnus X-1 in the 2-13 keV, with data collected with the *PCA* instrument onboard the *Rossi X-ray Timing Explorer* from 1996-10-23-18:30:24 to 1996-10-24-02:30:26. The crosses for the average values of rms for each flux interval divided uniformly in linear scale due to the narrow flux range. Note counts are for per 1/8 s time bin; each point in the scatter plots is for a 10 s interval of data with 1/8 s resolution. The inset shows its log-normal flux distribution.

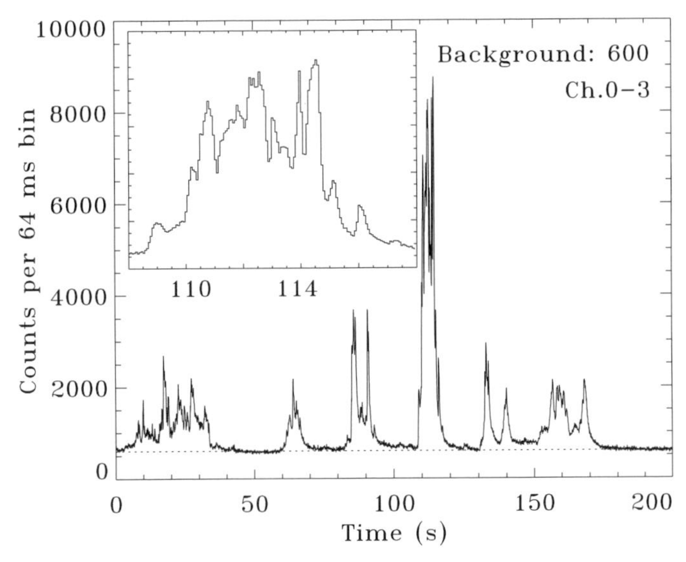

Figure 10. The lightcurve of GRB 940217 (trigger No. 2831) in channels 0–3 for the 16 channel data, i.e., between 13 to 54 keV, collected with the *BATSE* instrument onboard the *Compton Gamma-ray Observatory*. The inset shows clear variability above Poisson fluctuations. Note the background level is about 600 per 64 ms bin.

We chose to use GRB 940217 as an example of γ-ray bursts, because of its large dynamical range of flux and very long duration (thus many photons are available for a statistically meaningful correlation for a single γ-ray burst), as well as its very complex light curve morphology, for illustrating the robustness of the correlation. We comment in passing that GRB 940217 is a very important gamma-ray burst in its own right,

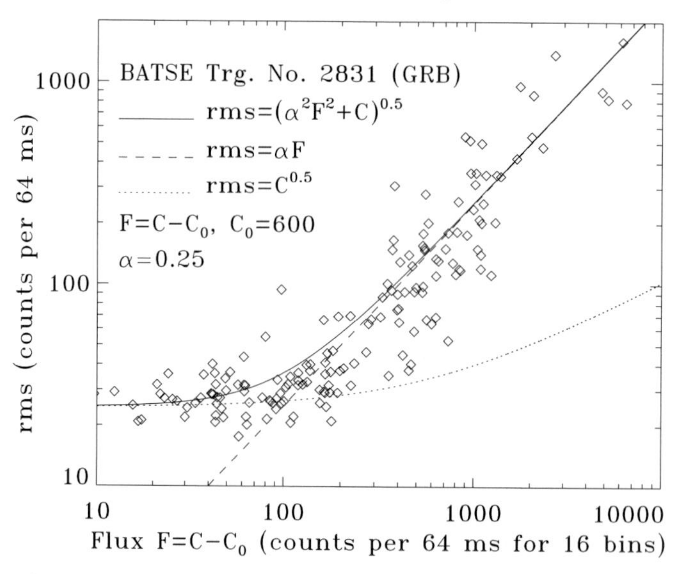

Figure 11. The flux-rms relation for the lightcurve of GRB 940217. The model is the same as equation (3.3).

because of its long-lasting high energy afterglow emission detected; >30 MeV photons were recorded for about 5400 s, including an 18 GeV photon about 4500 s after the low energy γ-ray emission had ended in the *BATSE* band as shown in Fig. 10 (Hurley *et al.* 1994). However, despite of the similar rms-flux correlation between GRB 940217 and solar flares, it is not meaningful to include multiple γ-ray bursts in one correlation plot, because different gamma-ray bursts are expected to have quite different characteristic time scales, since it is now known that long duration gamma-ray bursts have a large range of relativistic beaming factors and are produced in the whole Universe up to at least the redshift of 10 (see, e.g., Lin, Zhang & Li 2004, Zhang 2007).

3.4. *Further comments on the flux-rms correlations*

For the X-ray flux-rms correlations from the Sun, Cygnus X-1, and GRB 940217, about the only qualitative differences are: (*a*) There is a non-negligible constant flux from the total X-ray flux of Cygnus X-1 but not from the Sun or GRB 940217; (*b*) The range of rms variations from the Sun and GRB 940217 covers over several orders of magnitudes, but only about a factor of two for Cygnus X-1; (*c*) The flux distribution for each solar flare is made of one or more log-normal peaks, in comparison for only a single log-normal peak for that of Cygnus X-1. For the Sun and GRB 940217, even the portion of Poisson noise domination is seen clearly. It is commonly known that solar X-rays are produced from flare regions on solar surface, prompt X/γ-rays from γ-ray bursts (at least for long duration gamma-ray bursts such as GRB 940217) originate from the colliding relativistic jets (the so-called internal shock model, see, e.g., Piran 2004 for a recent review and references therein), and X-rays from black hole binaries are emitted from the accretion disks around their central black holes. The remarkable common X-ray flux variations between the apparently very different systems at different scales suggest that there should be a common and dominating mechanism operating in all of them, although we cannot exclude the possibility that different physical mechanisms may generate very similar X-ray flux variations. Because the Sun is our closest astrophysical laboratory from which we can obtain a wealth of information, in particular the direct measurements of its magnetic field topology and activities, understanding the solar X-ray production mechanism may be a key to revealing the black hole accretion disks and gamma-ray burst jets.

The success of this simple model with only one parameter in describing the very complex solar X-ray flares and continuous lightcurves, observed at well separated times in two energy bands, implies that the linear flux-rms relationship is very helpful for our understanding of X-ray emission from the Sun, and perhaps also the corona heating problem. Previously EUV and X-ray variability in the pixels of focal plane detectors of *TRACE* and *YOHKOH* have been analysed in order to probe the contribution of the speculated 'nano'-flares to the solar corona heating (Shimizu & Tsuneta 1997, Krucker & Benz 1998, Aschwanden *et al.* 2000a, Aschwanden *et al.* 2000b, Parnell & Jupp 2000, Katsukawa & Tsuneta 2001, Benz & Krucker 2002, Katsukawa 2003). However it is still not settled if the amount of 'nano'-flares can fulfill the requirement of Parker's conjecture (Walsh 2003). Nevertheless, these studies have indicated that probing X-ray variability is probably so far the best way for studying this outstanding problem in astrophysics. However due to the very limited counting statistics in those previous studies, no individual counting excess above noise fluctuation can be identified as real flaring events. It is thus uncertain if these variability studies have revealed unambiguously 'nano-'flares as extrapolated smaller events from observed solar flares.

It is interesting to note that the detected X-ray variability, with the *Yohkoh Soft X-ray Telescope*, is related to its mean flux as (Katsukawa & Tsuneta 2001), $\mathrm{rms} \propto F^{0.93\pm0.10}$, in good agreement with the linear relationship shown here with *RHESSI* data for X-ray variability including all flaring events detected in the X-ray lightcurves. This should be compared to the case of Cygnus X-1 in which the X-ray variability is also mostly composed of unidentifiable flaring events, despite that occasionally very strong flares are detectable (Gierlinski& Zdziarski 2003), which can be considered as analogy of solar flares in active regions of the Sun. Therefore, the results shown in Fig. 8 for the flux-rms linear relationship over several orders of magnitudes of X-ray flux variations from the Sun can be regarded as a missing link between the extremely violent flaring events from the Sun, and X-ray flux variations with small relative amplitudes from the Sun and from accreting neutron star and black hole binaries systems.

These results therefore demonstrate that common dynamical physical processes may dominate the X-ray emission from the Sun, γ-ray bursts, accreting neutron star and black hole binaries, as well as in accreting supermassive black hole systems (Uttley *et al.* 2005). Since accretion flow does not exist in the Sun, this may rule out immediately the fluctuating accretion flow model (Lyubarskii 1997, Kotov 2001) for this type of dynamical properties occurring in neutron star and black hole binaries with accretion disks. Because essentially all other models for X-ray emission from neutron star and black hole systems have already been ruled out by such dynamical behaviours (Uttley *et al.* 2005), new physical models for them may be required, which should have common mechanisms to X-ray emission from the Sun. We, therefore, probably should return to the solar flare-like models for accreting neutron star and black hole systems, as well as to γ-ray bursts.

4. Common physical mechanisms: self-organized criticality, percolation and dynamical driving

Then what are the possible common physical mechanisms? Magnetic energy releases, perhaps due to magnetic reconnections, should be the engines in all these systems (Nayakshin & Melia 1997, di Matteo *et al.* 1999, Poutanen & Fabian 1999, Zhang *et al.* 2000, Liu *et al.* 2003, Wang *et al.* 2004, Liu & Li 2004, Priest & Forbes 2002). For the Sun, an emerging flux and reconnection model has been proposed for the triggering of the coronal mass ejections (Chen & Shibata 2000). In addition, some small scale activities, such as Ellerman bombs, and type II white-light flares are possibly the results of magnetic reconnection in the solar lower atmosphere (Chen, Fang & Ding 2001). The avalanche mechanisms (Charbonneau *et al.* 2001) driven by self-organized criticality (Bak *et al.* 1988, Christensen *et al.* 1992) have been widely invoked to explain previously observed distributions of solar flare peak flux and event interval, as well as timing properties in accreting black hole systems (Mineshige *et al.* 1994, Takeuchi *et al.* 1995). The multiplicative nature of the generated events in the self-organized criticality model can produce the observed flux-rms linear relationship. However the simple self-organized criticality model also predicts power-law distributions of flux, which is not found in accreting neutron star and black hole systems, and in fact also not verified satisfactorily to the whole range of amplitudes of solar flares; the power-law flux distribution of solar flares is the basic idea of Parker's conjecture (Parker 1983, Parker 1988, Parker 1991). As shown in Figs. 4 & 5, the flux distributions of individual solar flares consist of multiple log-normal peaks, rather than a single power-law.

This mismatch between observed and predicted flux distributions may be due to oversimplified self-organized criticality models which only describe events occurring locally. However X-ray photons from accreting neutron star and black hole systems cannot come from only a localized self-organized criticality spot, yet events occurred are still interrelated as indicated by the multiplicative property. Therefore multiple spots or a major part of the whole inner accretion disk region responsible for X-ray emission must be inter-organized together somehow; this is further supported by the fact that only about 30% or less of its total X-ray flux does not follow the flux-rms relationship in Cygnus X-1 (see Fig. 9), i.e., only this 30% flux is the superposition of many randomly occurring and independent emission events. The *RHESSI* solar X-ray lightcurves we analysed here are collected from the whole Sun exposed to us, and its non-varying X-ray flux is very low, suggesting that essentially all detectable X-ray emission in a time series over the whole Sun is produced from many inter-related X-ray emission regions. In fact, it has already been observed that many active regions well separated and spread nearly over the whole solar surface show networked activities; these regions are now known to be connected by large scale magnetic field lines (Zhang *et al.* 2006), as shown in Fig. 12.

In a simulation study of particle acceleration during solar flares, Vlahos *et al.* 2004 have found that multi-scale magnetic fields are involved, which may be described mathematically by the percolation model of Albert & Barabasi 2002, as illustrated in Fig. 13. In the percolation model, two nodes are connected by an edge with a certain probability. If the probability is below the percolation threshold, the connected nodes form small and local clusters. In this case, most nodes will act independently, and thus their total contributions will be additive, rather than multiplicative. On the other hand, if the connection threshold is above the percolation threshold, global and large clusters are formed, and thus most nodes will act inter-connectively. In this case their total contributions will be multiplicative, because the action of each node will influence many other nodes, even if they are well separated spatially. Such networked activities or fluctuations can be studied by identifying the dominating driving mechanisms, i.e., internal or external driving. If the connection probability between two neighboring nodes is below the percolation threshold, the dynamical system will be dominated by internal driving, or otherwise dominated by external driving.

Following Argollo de Menezes & Barabasi (2004), for a dynamical system, we can record the time dependent activity of N components, expressed by a time series $\{f_i(t)\}$, $t = 1, \ldots, T$ and $i = 1, \ldots, N$. As each time series reflects the joint contribution from the system's internal dynamics and external fluctuations, we assume that we can separate

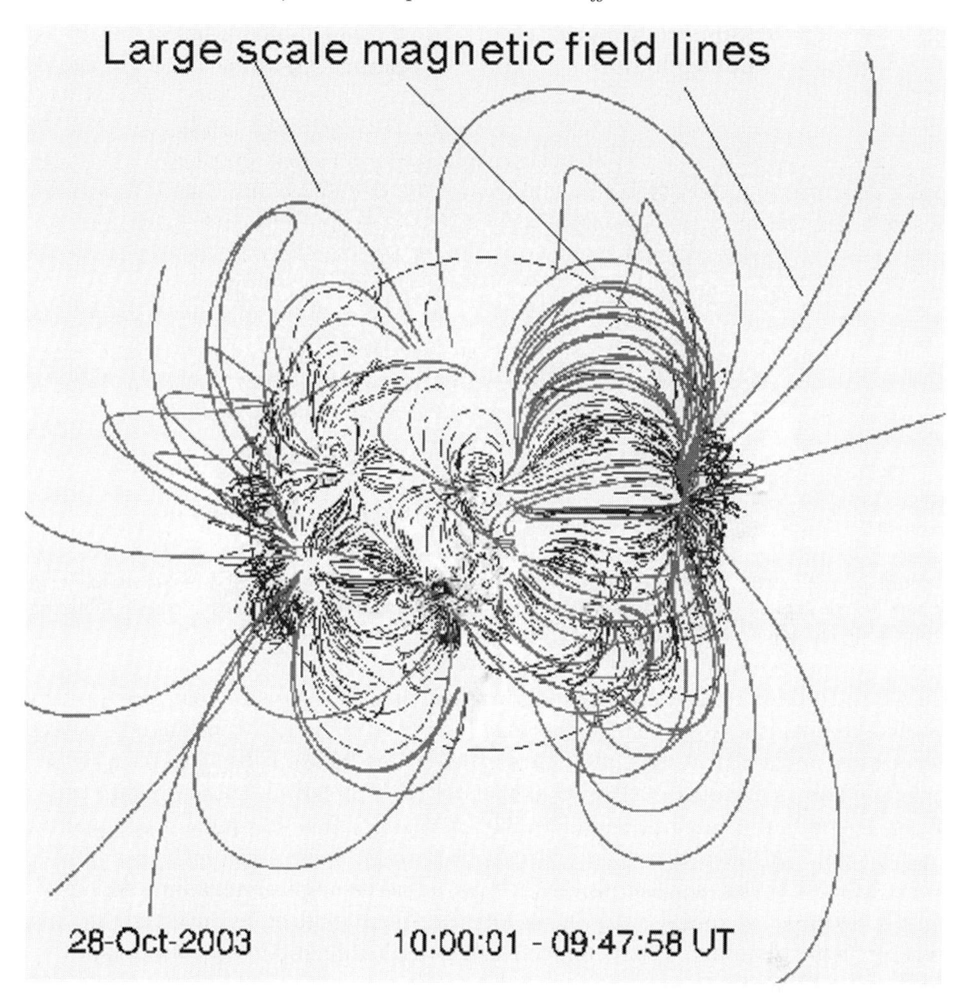

Figure 12. Solar magnetic field topology during the major solar activities on October 28th, 2003. Many active regions are connected by complex magnetic field lines. (This figure is adapted from Zhang *et al.* 2006.)

the two contributions by writing

$$f_i(t) = f_i^{int}(t) + f_i^{ext}(t). \tag{4.1}$$

It has been shown by Argollo de Menezes & Barabasi (2004) that the fluctuation or rms is linearly proportional to f if the dynamical system is externally driving, or the rms is proportional to $\sqrt{f}$ if the dynamical system is internally driving.

Since the dominating component of the X-ray flux fluctuation or rms is proportional to flux, for solar X-ray flares and continuous light curves, the X/γ-ray lightcurve of GRB 940217, accreting neutron star and black hole systems, we can infer that their X-ray production mechanisms are all dominated by external driving dynamics. We thus suggest that perhaps similar large scale magnetic field topology also exists in the accretion disks around neutron stars and black holes, and also maybe in γ-ray bursters. We comment in passing that no current γ-ray burst model involves inter-connected emission nodes. However, contrary to the widely accepted internal shock model of prompt γ-ray emissions for γ-ray bursts, an alternative model, based on electrodynamic accretion mod-

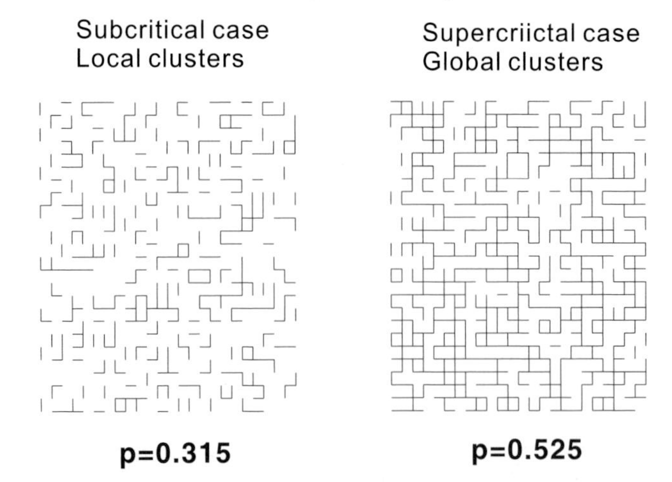

Figure 13. Illustration of the percolation model of Albert & Barabasi 2002. Two nodes are connected by an edge with probability p. For $p = 0.315$ (*left*), which is below the percolation threshold $p_c = 0.5$, the connected nodes form small and local clusters. For $p = 0.525$ (*right*), which is above the percolation threshold, the global and large cluster are formed. (This figure is adapted from Albert & Barabasi 2002.)

els in which the electromagnetic process turns rotational energy into particle energy in a pulsarlike mechanism (Katz 1997), has been used to unite jets in γ-ray bursts, accreting supermassive and stellar mass black hole systems (Katz 2006). It has also been suggested a magnetic field dominated outflow model for GRBs is preferred, based on data from the dedicated γ-ray burst satellite *Swift* (Kumar *et al.* 2007). These results are in qualitative agreement with our common energy release picture for these different types of objects at very different scales, suggesting that it may be necessary to re-examine seriously the currently accepted standard γ-ray burst internal shock models, by investigating energy release processes dominated and connected by complex magnetic field topology.

Although a component of rms proportional to the square-root of total counts is identified in all of them, especially when the total counts are small, this component is not due to internal driving, because this component can be completely described as Poisson counting fluctuations. This means that currently no internal driving dynamics has been detected in any of them. It is interesting to note that the individual solar flares show the same flux-rms linearity as for the total continuous lightcurves of all those systems, indicating that each individual solar X-ray flare is produced from many different inter-connected emission nodes or elements, if the percolation model is also applicable here. It is thus reasonable to assume that regardless what triggers the initial onset of a solar X-ray flare, the event immediately spreads out to many nodes or elements. Imaging observations of X-ray flares have identified magnetic loops and their foot-points as X-ray and hard X-ray production sites. Therefore it will be important to find the relationship between the above inferred inter-connected X-ray emission nodes or elements and observed magnetic loops and their foot-points.

5. Concluding remarks

In section 1, I briefly categorized different research methods in astronomy and emphasized the importance of using the Sun as an astrophysical laboratory to study many astrophysical phenomena across vastly different scales, but appearing to be similar to what occurring in the Sun. I call this research discipline "Applied Solar Astrophysics". A beautiful example of the progress in this discipline is the use of magnetic reconnection model by Dai *et al.* 2006 to explain the late X-ray flux re-brightening in the X-ray afterglow of a short γ-ray burst, which is believed to have produced a spinning young neutron star after two neutron stars merged together (this merging event triggered the initial onset of the short γ-ray burst).

In section 2, I briefly summarized several phenomena: (1) the atmospheric structures around accreting black holes which are similar to solar atmosphere; (2) astrophysical jets from many different kinds of astrophysical systems, whose production mechanisms involve differential rotation and twisted magnetic fields similar to magnetic reconnections in the Sun for solar flares and coronal mass ejection; (3) similar triple-ring structure in SN 1987A and Virgo cluster which are understood to be related to rotating large scale magnetic field topology and winds, again similar to the solar magnetic fields and solar winds. Clearly our understanding of related solar phenomena are important for studying these similar phenomena at different scales.

In section 3, I focused on the similar non-linear dynamical properties of X-ray variations of solar flares, a stellar mass black hole binary Cygnus X-1 and a γ-ray burst GRB 940217. The remarkable consistency of the underlying non-linear dynamics model for these three kinds of apparently very different astrophysical systems at different scales suggests that many X-ray emission nodes or elements are somehow inter-connected and act in response to common triggering. Again using the Sun as an astrophysical laboratory, we suggest that complicated magnetic field topology across many different scales in each system plays an important role in connecting many X-ray emission nodes or elements together. It is possible that these individual nodes produce 'nano'-flares required for heating the solar corona.

However previous and current solar X-ray instruments do not have the required capability to resolve individual X-ray flares or the quiet Sun X-ray lightcurves into many 'nano'-flares. We suggest that a future *Solar X-ray Timing and Imaging)* (*SXTI*) instrument with a large effective area and direct high-resolution imaging capability is needed to further resolve the fine timing structures of X-ray flares, i.e., inter-connected individual 'nano'-flares, to study the development of solar flares in fine details, and overcome the Poisson counting fluctuations with a high counting rate during the quiet Sun, in order to identify the internal driving component, which may be made of isolated 'nano'-flares.

Although it is beyond the scope of this report for proposing a conceptual design for the *SXTI* telescope, here we outline only the basic requirements: (1) an effective area of $>100\,\mathrm{cm}^2$ at a few keV to be compared with the $30\,\mathrm{cm}^2$ effective area at 10 keV for *RHESSI*, however *RHESSI* relies on rotation modulation imaging and thus cannot offer simultaneous imaging and high timing resolution capability; (2) direct imaging angular resolution of < 1 arcsec, to be compared with *HINODE* (Solar-B)'s X-ray telescope† which has 2 arcsec direct imaging resolution, but with only $1\,\mathrm{cm}^2$ effective area at 0.523 keV and 2 s timing resolution; and (3) timing resolution < 10 ms which requires advanced X-ray CCDs, such as that in current developments (e.g., Zhang *et al.* 2007).

For comparison the currently operating *Chandra* X-ray telescope‡ has a peak effective area of about $800\,\mathrm{cm}^2$ at around 2 keV, with a direct imaging resolution of 0.5 arcsec and each X-ray CCD's full-frame readout time of around 3.3 s. Although telescopes pointed at the Sun involve significantly more complex technical difficulties, it is possible that a solar X-ray observatory with the above proposed *SXTI* will be technically feasible within

† http://solar-b.nao.ac.jp/xrt_e/fact_sheet_e.shtml
‡ http://chandra.harvard.edu/about/science_instruments.html

the near future. With such an instrument, our understanding of solar X-ray production mechanism, and subsequently many currently open issues on solar magnetic fields, particle acceleration and coronal heating, etc., will be advanced significantly. Because of many astrophysical phenomena at different scales should have similar underlying physical processes to what occurring in the Sun, we should also make significant progresses towards understanding many important astrophysical problems. Therefore "Applied Solar Astrophysics" has a bright future.

Acknowledgements

I shall thank Cheng Fang of Nanjing University for nominating me, the IAU Executive Committee for inviting me, as one of the four discourse speakers at the XXVIth IAU GA in Prague, Czech Republic, August 2006, and Ding-Qiang Su of Nanjing University, who was then the President of the Chinese Astronomical Society, for encouragements and advises during my preparation of this invited discourse. I also appreciate discussions with Bob Lin of UC Berkeley and Don Melrose of University of Sydney during the 26th IAU GA on *RHESSI* data analysis and solar particle accelerations, respectively. The many kind compliments and stimulating discussions from the audience in the discourse are the best rewards for my discourse, which has indeed taken me a lot of time and efforts to prepare for.

I also thank Shui Wang of University of Science and Technology of China for discussions on magnetic reconnections, Jun-Han You of Shanghai Jiao-Tong University for discussions on radiation mechanisms, Wei-Qun Gan of Purple Mountain Observatory for advises on *RHESSI* data analysis and conversations on solar flares, Jing-Xiu Wang of National Astronomical Observatories of China and Louise Harra of Mullard Space Science Laboratory of University College London for exchanges on solar physics, Jian-Min Wang of Institute of High Energy Physics for many discussions on accretion physics, and many of my previous collaborators and formal students for the privilege and fun of working with them and for their important contributions to our works discussed and/or referenced in this report. My student Shi-Cao Tang of Tsinghua University is greatly acknowledged for assisting me in most of the data analysis work shown in Figs. 3–11 of this report. Finally Cheng Fang, Zi-Gao Dai & Peng-Fei Chen of Nanjing University, Tan Lu of Purple Mountain Observatory, Jun Lin of Yunnan Observatory, Jing-Xiu Wang of National Astronomical Observatories of China, Phil Uttley of Southampton University and Simon Vaughan of Leicester University are thanked for comments and suggestions on the manuscript of this report.

This work is supported in part by the Ministry of Education of China, Directional Research Project of the Chinese Academy of Sciences and by the National Natural Science Foundation of China under project no. 10521001, 10327301, 10233010 and 10233030.

References

Albert, R., & Barabasi, A.L. 2002, *Rev. Mod. Phys.*, 74, 47

Argollo de Menezes, M., & Barabasi, A.L. 2004, *Phys. Rev. Lett.*, 93, 068701

Aschenbach, B., Grosso, N., Porquet, D., & Predehl, P. 2004, *A&A*, 417,71

Aschwanden, M.J., Nightingale, R.W., Tarbell, T.D., & Wolfson, C.J. 2000a, *ApJ*, 535, 1027

Aschwanden, M.J., Tarbell, T.D., Nightingale, R.W., *et al.* 2000b, *ApJ*, 535, 1047

Bak, P., Tang, C., & Wiesenfeld, K. 1988, *Phys. Rev. A*, 38, 364

Bassani, L., Malaguti, G., Paciesas, W.S., Palumbo, G.G.C., & Zhang, S.N. 1996, *A&AS*, 120C, 559

Burrows, C.J. 1995, *ApJ*, 452, 680

Benz, A.O., & Krucker, S. 2002, *ApJ*, 568, 413

Charbonneau, P., McIntosh, S.W., Liu, H.-L., & Bogdan, T.J. 2001, *Sol. Phys.*, 203, 321
Chen, P.F., & Shibata, K. 2000, *ApJ*, 545, 524
Chen, P.F., Fang, C., & Ding, M.D. 2001, *ChJAA*, 1, 176
Chen, X., Zhang, S.N., & Ding, G.Q. 2006, *ApJ*, 650, 299
Christensen, K., Olami, Z., & Bak, P. 1992, *Phys. Rev. Lett.*, 68, 2417
Cranmer, S.R., & van Ballegooijen, A.A. 2005, *ApJS*, 156, 265
Dai, Z.G., Wang, X.Y., Wu, X.F., & Zhang, B. 2006, *Science*, 311, 1127
Dennis, B.R., Hudson, H.S., & Krucker, S. 2005, in: Proc. CESRA Workshop 2004, Isle of Skye, Scotland Review of selected RHESSI solar results
di Matteo, T., Celotti, A., & Fabian, A.C. 1999, *MNRAS*, 304, 809
Doi, K. 1978, *Nature*, 275, 197
Edelson, R., Turner, T.J., Pounds, K.A., *et al.* 2002, *ApJ*, 568, 610
Feng, H., Zhang, S., Lou, Y., & Li, T. 2004, *ApJ* (Letters), 607, L95
Gierlinski, M., & Zdziarski, A.A. 2003, *MNRAS* (Letters) 343, L84
Haardt, F., Maraschi, L., & Ghisellini, G. 1997, *ApJ*, 476, 620
Harmon, B.A, Wilson, C.A., Zhang, S.N., Paciesas, W.S., Fishman, G.J., Hjellming, R.M., Rupen, M.P., Scott, D.M., Briggs, M.S., & Rubin, B.C. 1995, *Nature*, 374, 703
Hawley, J.F., Balbus, S.A., & Winters, W.F. 1999, *ApJ*, 518, 394
Hurley, K., Dingus, B.L., Mukherjee, R., *et al.* 1994, *Nature*, 372, 652
Katsukawa, Y. 2003, *PASJ*, 55, 1025
Katsukawa, Y., & Tsuneta, S. 2001, *ApJ*, 557, 343
Katz, J.I. 1997, *ApJ*, 490, 633
Katz, J.I. 2006, [astro-ph/0603772]
Klimchuk, J.A. 2006, *Sol. Phys.*, 234, 41
Kotov, O., Churazov, E., & Gilfanov, M. 2001, *MNRAS*, 327, 799
Krucker, S., & Benz, A.O. 1998, *ApJ*, 501, L213
Kubota, A., Makishima, K., & Ebisawa, K. 2001, *ApJ* (Letters), 560, L147
Kumar, P., McMahon, E., Panaitescu, A., *et al.* 2007, *MNRAS* (Letters) 376, L57
Liang, E.P.T., & Price, R.H. 1977, *ApJ*, 218, 247
Lin, R.P., Dennis, B.R., Hurford, G.J., *et al.* 2002, *Sol. Phys.*, 210, 3
Lin, J.R., Zhang, S.N., & Li, T.P. 2004, *ApJ*, 605, 819
Liu, C.Z., & Li, T.P. 2004, *ApJ*, 611, 1084
Liu, B.F., Mineshige, S., & Ohsuga, K. 2003, *ApJ*, 587, 571
Liu, Y., Zhang, S.N., & Zhang, X.L. 2007, *PASJ*, 59, 185
Lyubarskii, Y.E. 1997, *MNRAS*, 292, 679
Margon, B., Bowyer, S., & Stone, R.P.S. 1973, *ApJ*, 185, L113
Meier, D.L., Koide, K., & Uchida, Y. 2001, *Science*, 291, 84
Markovskii, S.A., & Hollweg, J.V. 2004, *ApJ*, 609, 1112
Mineshige, S., Ouchi, N.B., & Nishimori, H. 1994, *PASJ*, 46, 97
Meszaros, P. 2001, *Science*, 291, 5
Mirabel, I.F., & Rodriguez, L.F. 1998, *Nature*, 392, 673
Mirabel, I.F., & Rodriguez, L.F. 2002, *Sky & Telescope*, May 2002, 32
Miyamoto, S., Kitamoto, S., Mitsuda, K., & Dotani, T. 1988, *Nature*, 336, 450
Nayakshin, S., & Melia, F. 1997, *ApJ* (Letters), 490, L13
Negoro, H., Kitamoto, S., & Mineshige, S. 2001, *ApJ*, 554, 528
Parker, E.N. 1983, *ApJ*, 264, 642
Parker, E.N. 1988, *ApJ*, 330, 474
Parker, E.N. 1991, *ApJ*, 376, 355
Parnell, C.E., & Jupp, P.E. 2000, *ApJ*, 529, 554
Pevtsov, A.A. 2002, *Sol. Phys.*, 207, 111
Piran, T. 2004, *Rev. Mod. Phys.*, 76, 1143
Poutanen, J., & Fabian, A.C. 1999, *MNRAS* (Letters), 306, L31
Priest, E.R., & Forbes, T.G. 2002, *A&A Rev.*, 10, 313
Remillard, R.A., & McClintock, J.E. 2006, *ARA&A*, 44, 49

Tang, S.C., & Zhang, S.N. 2007, in preparation
Shakura, N.I., & Sunyaev, R.A. 1973, *A&A*, 24, 337
Shimizu, T., & Tsuneta, S. 1997, *ApJ*, 486, 1045
Takeuchi, M., Mineshige, S., & Negoro, H. 1995, *PASJ*, 47, 617
Tanaka, T., & Washimi, H. 2002, *Science*, 296, 321
Terrell, N.J.J. 1972, *ApJ* (Letters), 174, L35
Tingay, S.J., Jauncey, D.L., Preston, R.A., *et al.* 1995, *Nature*, 374, 141
Ueda, Y., Akiyama, M., Ohta, K., & Miyaji, T. 2003, *ApJ*, 598, 886
Uttley, P. 2004, *MNRAS* (Letters), 347, L61
Uttley, P., & McHardy, I.M. 2001, *MNRAS* (Letters), 323, L26
Uttley, P., McHardy, I.M., & Vaughan, S. 2005, *MNRAS*, 359, 345
Vaughan, S., Fabian, A.C., & Nandra, K. 2003a, *MNRAS*, 339, 1237
Vaughan, S., Edelson, R., Warwick, R.S., & Uttley, P. 2003b, *MNRAS*, 345, 1271
Vlahos, L., Isliker, H., & Lepreti, F. 2004, *ApJ*, 608, 504
Walsh, R.W., & Ireland J. 2003, *A&A Rev.*, 12, 1
Wang, J.-M., Watarai, K.-Y., & Mineshige, S. 2004, *ApJ* (Letters), 607, L107
Wang, J., Zhang, J., Wang, T., Zhang, C., Liu, Y., Nitta, N., & Slater, G.L. 2001, in: P. Brekke, B. Fleck & J.B. Gurman (eds.), *Recent Insights into the Physics of the Sun and Heliosphere: Highlights from SOHO and Other Space Missions*, Proc. IAU Symp. No. 203 (San Francisco: ASP), p. 331
Wang, X.Y., Dai, Z.G., & Lu, T. 2003, *ApJ*, 592, 347
Yamamoto, T.T., Kusano, K., Maeshiro, T., Yokoyama, T., & Sakurai, T. 2005, *ApJ*, 624, 1072
Zhang, B. 2007, *ChJAA*, 7, 1
Zhang, C., Lechner, P., Lutz, G., Porro, M., Richter, R., Treis, J., Struder, L., & Zhang, S.N. 2006, *Nucl. Instr. & Meth. A*, 568, 207
Zhang, S.N., Wilson, C.A., Harmon, B.A., *et al.* 1994, *IAUC* 6046
Zhang, S.N., Harmon, B.A., Paciesas, W.S., Fishman, G.J., Grindlay, J.E., Barret, D., Tavani, M., Kaaret, P., Bloser, P., Ford, E., & Titarchuk, L. 1996, *A&AS*, 120C, 279
Zhang, S.N. 1997, in: D.T. Wickramasinghe, G.V. Bicknell & L. Ferrario (eds.), *Accretion Phenomena and Related Outflows*, Proc. IAU Coll. No. 163, *ASP-CS*, 121, 41
Zhang, S.N., Cui, W., Harmon, B.A., Paciesas, W.S., Remillard, R.E., & van Paradijs, J. 1997, *ApJ* (Letters), 477, L95
Zhang, S.N., Cui, W., & Chen, W. 1997, *ApJ* (Letters), 482, L155
Zhang, S.N., Mirabel, I.F., Harmon, B.A., Kroeger, R.A., Rodriguez, L.F., Hjellming, R.M., & Rupen, M.P. 1997, in: C.D. Dermer, M.S. Strickman & J.D. Kurfess (eds.), Proc. Fourth Compton Symposium, *AIP-CP*, 410, 141
Zhang, S.N., Cui, W., Chen, W., Yao, Y., Zhang, X., Sun, X., Wu, X.-B., & Xu, H. 2000, *Science*, 287, 1239
Zhang, S.N. 2005, *ApJ* (Letters), 618, L79
Zhang, Y.-Z., Wang, J.-X., Attrill, G., & Harra, L.K. 2006, in: V. Bothmer & A.A. Hady (eds.), *Solar Activity and its Magnetic Origin*, Proc. IAU Symposium No. 233 (Cambridge: CUP), p. 357

Highlights of Astronomy, Volume 14
IAU XXVI General Assembly, 14-25 August 2006
Karel A. van der Hucht, ed.

doi:10.1017/S1743921307009854

The power of new experimental techniques in astronomy: zooming in on the black hole in the Center of the Milky Way

Reinhard Genzel

Max-Planck Institut für extraterrestrische Physik (MPE), Garching, FRG
and
Department of Physics, University of California, Berkeley, USA
email: genzel@mpe.mpg.de

Abstract. It is becoming ever easier to obtain first class astronomical data by working solely from one's computer terminal, the modern equivalent of the armchair. This wonderful development may tempt us to forget that astronomical discoveries and findings are first and foremost driven by hard-won progress in observational and experimental capabilities. The present article is meant to demonstrate this assertion by reviewing recent developments and future possibilities in studying the Center of the Milky Way, our best case for the existence of a (massive) black hole and a superb laboratory for studying the physical processes in the immediate vicinity of such an enigmatic object.

Keywords. Galaxy: center, infrared: stars, stars: individual (Sgr A*)

1. Black Holes

Rev. John Michell was in 1784 the first to note that a sufficiently compact star may have a surface escape velocity exceeding the speed of light. He argued that an object of the mass of the Sun (or larger) but with a radius of 3 km would thus be invisible. A proper mathematical treatment of this problem then had to await Albert Einstein's General Relativity (Einstein 1916). Karl Schwarzschild's solution of the vacuum field equations in spherical symmetry (Schwarzschild (1916)) demonstrated the existence of a characteristic event horizon, the Schwarzschild radius $R_s = 2GM/c^2$, within which no communication is possible with external observers. Kerr (1963) generalized this solution to spinning black holes. The mathematical concept of a black hole was established (although the term itself was coined only later by John Wheeler (1968). But are these theoretical objects of General Relativity realized in Nature?

Astronomical evidence for the existence of black holes started to emerge in the 1960s with the discovery of distant luminous quasars (QSOs: Schmidt 1963) and variable X-ray emitting binaries in the Milky Way (Giacconi *et al.* 1962). Early on it became clear from energetic arguments that the enormous luminosities and energy densities of QSOs can most plausibly be explained by accretion of matter onto massive black holes (Lynden-Bell 1969; Rees 1984). We now know from high resolution imaging that QSOs are located (without exception) at the nuclei of large, massive galaxies. QSOs just represent the most extreme and spectacular among the general nuclear activity of most galaxies. This includes variable X-ray and γ-ray emission and highly collimated, relativistic radio jets, all of which cannot be plausibly accounted for by stellar activity. It is thus tempting to conclude that most galactic nuclei may harbor massive black holes.

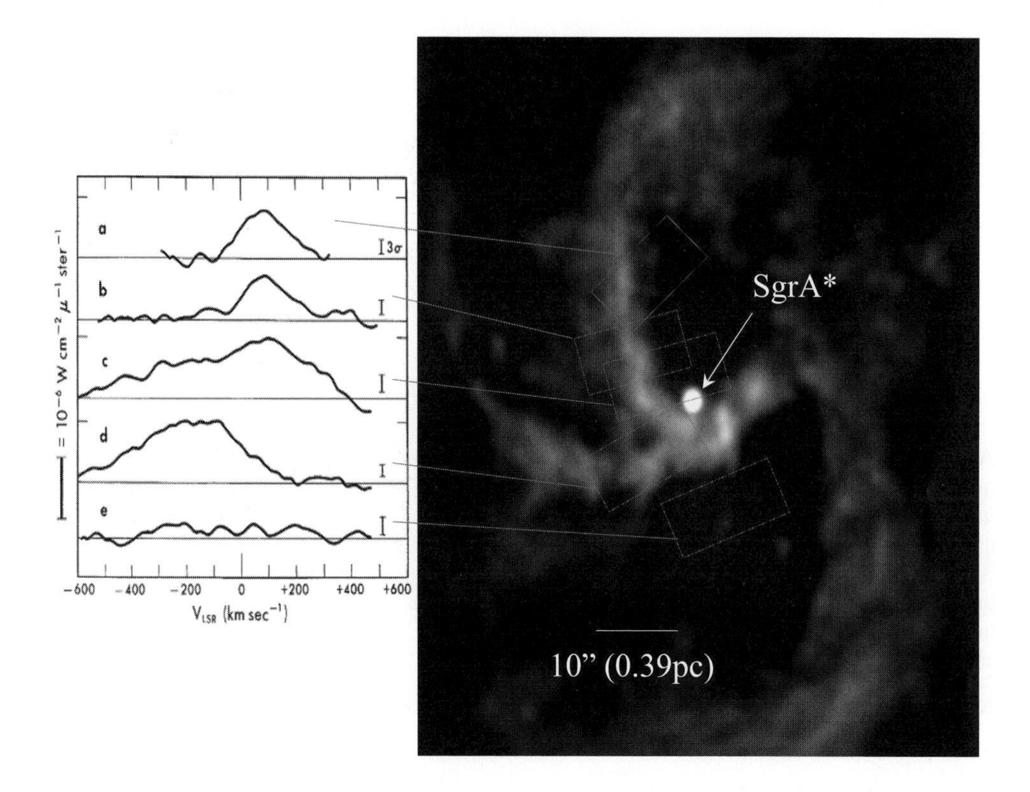

Figure 1. *Right:* 3.6 cm VLA radio continuum map of the central parsec (Roberts & Goss 1993). *Left:* 12.8μm [Ne II] line profiles (Wollman *et al.* 1977) for the apertures indicated on the radio map. The radio emission delineates ionized gas streamers (the 'mini-spiral') orbiting the compact radio source Sgr A*. The Wollman *et al.* observations provided the first dynamic evidence from large gas velocities that there might be a hidden mass of 3-4 million $M_\odot$ located near Sgr A*.

In the case of Galactic X-ray stellar binary systems (see Giacconi, 2003, for a historic account) dynamical mass determinations from Doppler spectroscopy of the visible primary star have established in a number of cases that the mass of the X-ray emitting secondary is significantly larger than the maximum stable neutron star mass of about $3\,M_\odot$ (McClintock & Remillard 2006). These X-ray sources thus are excellent candidates for stellar black holes. To apply similar, direct dynamical mass determinations to massive black hole candidates it is necessary to determine the motions of test particles (interstellar gas or stars) in close orbit around the nucleus. The aim is to show from measurements at different separations from the center that the gravitational potential is dominated by a compact, non-stellar mass that cannot be in any other form than that of a massive black hole. Such measurements are not possible (yet) in distant QSOs but have become feasible in nearby galaxy nuclei, where solid evidence for central dark mass concentrations in about 30 galaxies has emerged over the past two decades (Kormendy 2004), from *Hubble Space Telescope* imaging and spectroscopy and from very long baseline radio interferometry (VLBI).

The first really convincing case that these dark mass concentrations cannot just be dense conglomerates of white dwarfs, neutron stars and perhaps stellar black holes, emerged in the mid 1990s from spectacular VLBI observations of the nucleus of NGC 4258,

Figure 2. Some the key facilities that have contributed to Galactic Center research over the past two decades. *Lower left:* the TeV-Gamma Ray telescope HESS. (High Energy Stereoscopic System) in Namibia. *Top left:* the X-ray telescopes *XMM-Newton* (ESA) and *Chandra* (NASA). *Center:* large optical/near-infrared telescopes, such as the Keck telescopes and the ESO VLT. *Top right:* the Very Long Baseline Array (NRAO). *Bottom right:* The Submillimeter Array. The main institutions/groups involved in Galactic Center research with these facilities are listed.

a mildly active galaxy at a distance of 7 Mpc (Miyoshi *et al.* 1995). The VLBI observations showed that the galaxy nucleus contains a thin, slightly warped disk of H_2O masers in Keplerian rotation around a point mass of 40 million $M_\odot$. The inferred effective density of this mass exceeds a few $10^9\ M_\odot pc^{-3}$ and thus cannot be a long-lived cluster of 'dark' astrophysical objects of the type mentioned above.

Over the past decade, measurements of stellar orbits in the nearest galaxy nucleus, namely the Center of the Milky Way, have now shown that the central dark mass concentration in the Galactic Center must indeed be a massive black hole, beyond any reasonable doubt.

2. Sagittarius A*

The central light years of our Galaxy contain a dense and luminous star cluster, as well as several components of neutral, ionized and extremely hot gas (Genzel *et al.* 1994). In the late 1970s and early 1980s, Professor Charles Hard Townes, Nobel Laureate for the invention of the maser and laser, with his outstanding group of students and postdocs at the University of California, made the discovery, with the then emerging technique of mid-infrared spectroscopy, that the nucleus of our Milky Way contains a non-stellar, central mass concentration of 3-4 million $M_\odot$ (Fig. 1). The Berkeley group concluded that this mass concentration might be a massive black hole (Wollman *et al.* 1977; Townes *et al.* 1982). The Galactic Center also contains a very compact radio source, Sgr A* (Figs. 1 and 3; Balick & Brown 1974), which is located at the center of the nuclear star cluster and ionized gas. Short-wavelength centimeter and millimeter VLBI observations

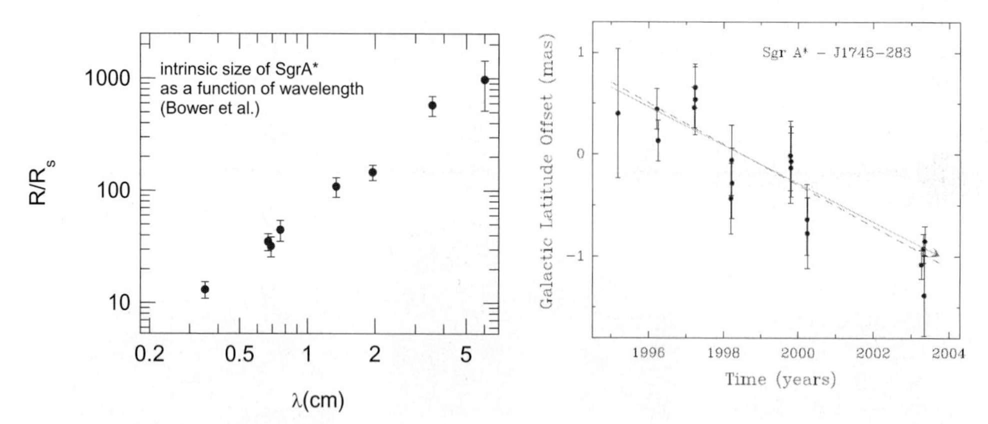

Figure 3. Radio characteristics of Sgr A*. *Left:* Inferred intrinsic major axis size of the radio source as a function of wavelength. The intrinsic sizes are in units of the Schwarzschild radius of a 3-4 million $M_\odot$ black hole (Bower *et al.* 2004; Bower, private communication). *Right:* Galactic latitude position residuals of Sgr A* relative to the QSO J1745−283. The dashed line is the variance-weighted best-fit proper motion of -0.202 ± 0.019 milli-arcseconds/yr, or $-7.6 \pm 0.7\,\mathrm{km\,s^{-1}}$ for $R_o = 8.0$ kpc. The solid arrow indicates the apparent motion of Sgr A* expected for the $7.17\,\mathrm{km\,s^{-1}}$ motion of the Sun perpendicular to the plane of the Galaxy. The difference translates into a $2\,\sigma$ upper limit of the residual motion of Sgr A* in the Galactic Center reference frame of less than $18.8\,\mathrm{km\,s^{-1}}$ (from Reid & Brunthaler 2004).

have established that its intrinsic radio size is a mere 10 light minutes (Fig. 3; Bower *et al.* 2004; Shen *et al.* 2005). Sgr A* is also an X-ray emission source, albeit of only modest luminosity (e.g., Baganoff *et al.* 2001). Most recently, Aharonian *et al.* (2004) have discovered a source of TeV γ-ray emission within 10″ of Sgr A*.

It is not yet clear whether these most energetic γ-rays come from Sgr A* itself, or whether they are associated with the nearby supernova remnant Sgr A East. Sgr A* thus may be a supermassive black hole analogous to QSOs, albeit of much lower mass and luminosity. Because of its proximity – the distance to the Galactic Center is about 10^5 times closer than the nearest quasars – high-resolution observations of the Milky Way nucleus offer the unique opportunity of stringently testing the black hole paradigm and of studying stars and gas in the immediate vicinity of a black hole, at a level of detail that will not be accessible in any other galactic nucleus in the foreseeable future. Since the Center of the Milky Way is highly obscured by interstellar dust particles in the plane of the Galactic disk, observations in the visible are not possible. Investigations require measurements at longer wavelengths, in the infrared and microwave bands, or at shorter wavelengths, at hard X-ray and γ-rays, where the veil of dust is transparent. The dramatic progress in our knowledge of the Galactic Center over the past two decades is a direct consequence of the development of novel facilities, instruments and techniques across the whole range of the electromagnetic spectrum (Fig. 2).

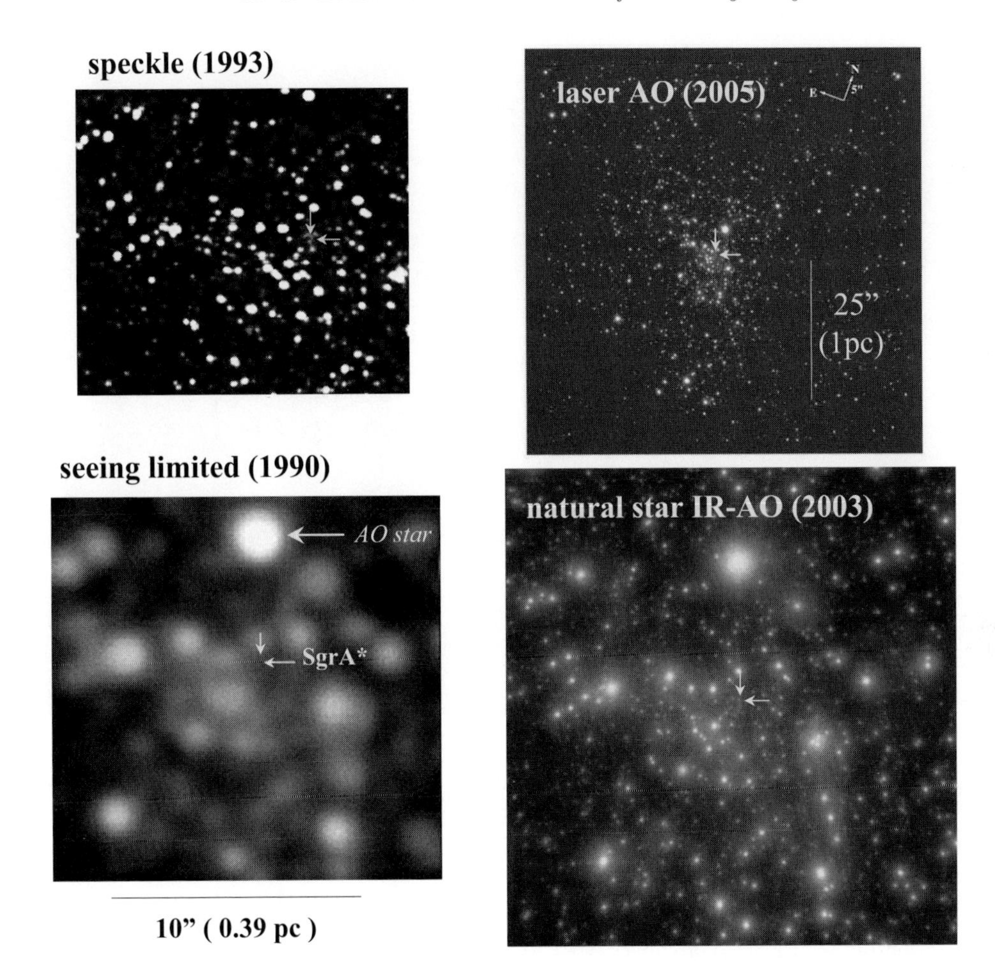

Figure 4. Progress in high-resolution near-infrared imaging of the central few light years of our Milky Way. *Lower left:* a 'atmospheric seeing' limited ($\sim 0.5''$ resolution) image in three colors (1.6/2.2/3.5 μm: DePoy & Sharp (1991)). *Top left:* a 'cleaned' shift-and-add speckle 2.2 μm image at $0.15''$ resolution (ESO NTT) from the work of (Eckart *et al.* 1993, 1995). *Bottom right:* the same region (at 1.6/2.2/3.7 μm) at the diffraction limit of the 8 m ESO VLT ($\sim 0.05\,\mu$), taken with the NACO AO-camera and an infrared wavefront sensor (Genzel *et al.* 2003b). All these three images are on the same spatial scale. *Top right:* a Keck diffraction limited image of a ~ 3 times larger field, this time with a sodium laser guide star for AO correction (Ghez *et al.* 2005b). The diffraction limited images are much sharper and also much deeper than the seeing limited images. In addition to increasing the sky coverage the laser allows easy mosaicing of a large field of view. The arrows denote the position of the compact radio source Sgr A* and the location of the AO-star IRS 7. NACO is a collaboration between ONERA (Paris), Observatoire Paris, Observatoire Grenoble, MPE (Garching) and MPIA (Heidelberg) (Lenzen *et al.* 1998; Rousset *et al.* 1998). The Keck AO camera NIRC2 was developed at Caltech (Matthews *et al.* 2007) and the Keck laser system was developed by a consortium of Lawrence Livermore National Laboratory and the Keck Observatory (Wizinovich *et al.* 2007).

3. High angular resolution astronomy

The key obviously lies in very high angular resolution measurements. The Schwarzschild radius of a 3.6 million solar mass black hole at the Galactic Center subtends a mere 10^{-5} arcseconds. For high-resolution imaging from the ground it is necessary to correct for the

Figure 5. Image of a Na 589 nm laser beacon (PARSEC) projected from UT4 of the ESO VLT in Chile. The laser beam is focused on a layer of atomic sodium in the upper atmosphere at $\sim$90 km altitude where it creates an artifical laser 'star' by resonant backscattering. The wavefront of the laser star traveling back from the sky to the telescope can then be used for adaptive optics correction. In the background one can see the band of the southern Milky Way, as well as the Large Magellanic Cloud (close to the ground near UT4). The VLT laser guide star facility is a collaboration between ESO (Garching), MPE (Garching) and MPIA (Heidelberg).

distortions of an incoming electromagnetic wave by the refractive and dynamic Earth atmosphere. VLBI overcomes this hurdle by phase-referencing to nearby QSOs; sub-milli-arcsecond resolution can now be routinely achieved. In the optical/near-infrared waveband the atmosphere smears out long-exposure images to a diameter at least ten times greater than the diffraction limited resolution of large ground-based telescopes (Fig. 4). From the early 1990s onward initially 'speckle imaging' (recording short exposure images, which are subsequently processed and co-added to retrieve the diffraction limited resolution) and then later 'adaptive optics' (AO: correcting the wave distortions on-line) became available. With these techniques it is possible to achieve diffraction limited resolution on large ground-based telescopes. In the case of AO (Beckers 1993) the incoming wavefront of a bright star near the source of interest is analyzed, the necessary corrections for undoing the aberrations of the atmosphere are computed (on time scales shorter than the atmospheric coherence time of a few milli-seconds) and these corrections are then applied to a deformable optical element (e.g., a mirror) in the light path. The requirements on the brightness of the AO star and on the maximum allowable separation between star and source are quite stringent, resulting in a very small sky coverage of 'natural star' AO. Fortunately, in the Galactic Center there is a bright infrared star only 6″ away from Sgr A*, such that good AO correction can be achieved with an infrared wavefront sensor system (see lower left inset of Fig. 4). Artificial laser beacons can

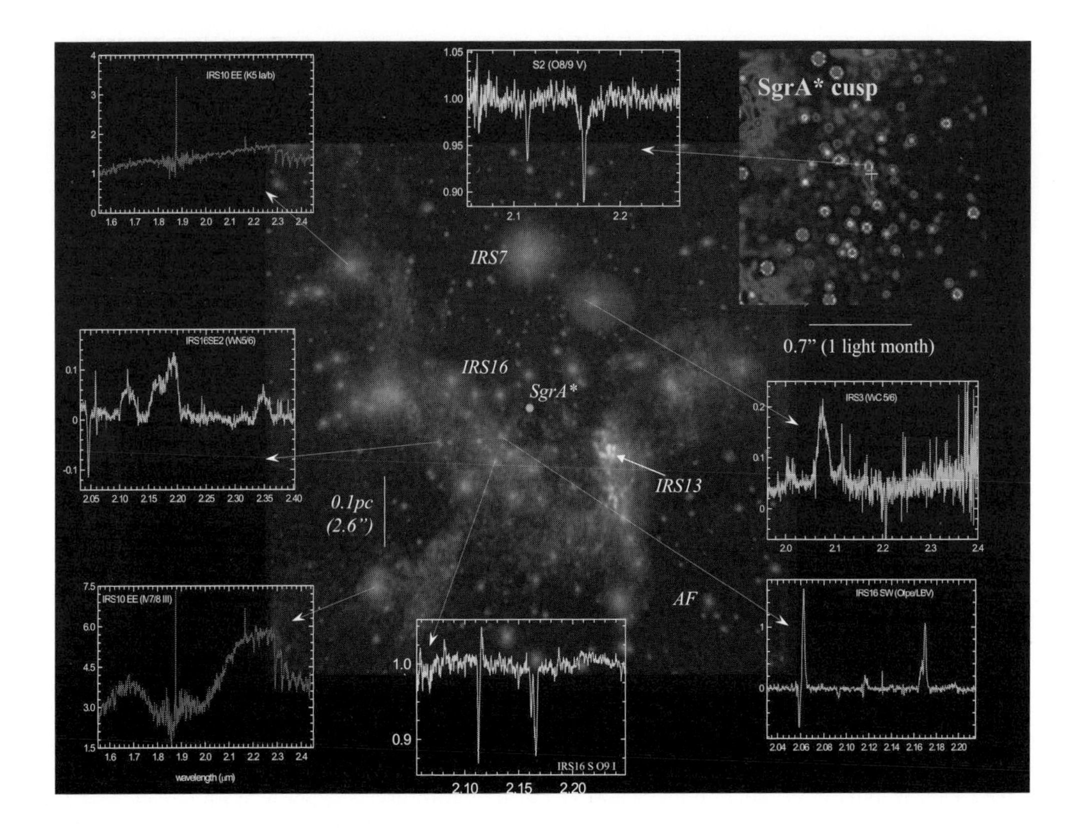

Figure 6. *Center:* composite of 2.2 μm (= K: blue), 3.7 μm (L': red) and 1.3 cm (green) continuum emission (NACO/VLT & VLA), probing stars of different types, hot dust, ionized gas and the compact radio source Sgr A*. *Top right:* 40 milli-arcsecond 1.6 μm image (VLT) of the dense central stellar cusp around Sgr A* (cross). *Surrounding:* selected 2 μm spectra of late-type (red) and early-type (light blue) stars in the nuclear cluster taken with the AO assisted integral field spectrometer SINFONI on the VLT (Eisenhauer *et al.* 2003b; Bonnet *et al.* 2004).

overcome the sky coverage problem to a considerable extent. For this purpose, a laser beam is projected from the telescope into the upper atmosphere and the backscattered laser light can then be used for AO correction (Fig. 5). The Keck telescope team has already begun successfully exploiting the new laser guide star technique for Galactic Center research (Ghez *et al.* 2005b).

After AO correction, the images are an order of magnitude sharper and also much deeper than in conventional seeing limited measurements (Figs. 4, 6). The combination of AO techniques with advanced imaging and spectroscopic instruments (e.g., 'integral field' imaging spectroscopy) have resulted in a major breakthrough in high-resolution studies of the Galactic Center.

4. Nuclear star cluster and 'paradox of youth'

Looking at the radio/IR composite image in Fig. 6 one of the immediate surprises is the fairly large number of bright stars, a number of which were already apparent on the discovery infrared images of Becklin & Neugebauer (1975): IRS 7, 13, 16. High-resolution

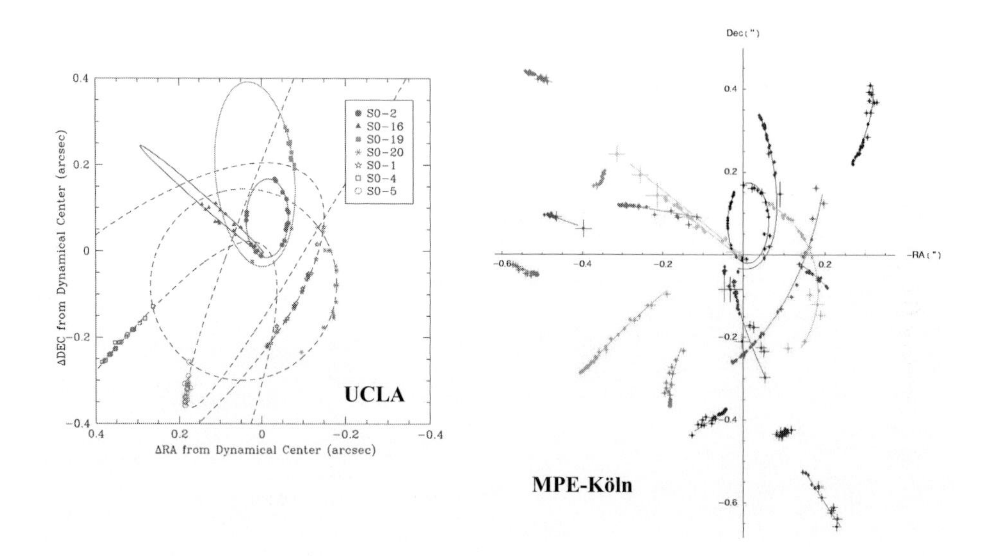

Figure 7. Positions on the sky as a function of time of the central stars ('S'-stars) orbiting the compact radio source Sgr A*. *Left inset:* the data from the UCLA group working with the Keck telescope (Ghez *et al.* 2005a). *Right inset:* the data from the MPE-Cologne group at the ESO-VLT (Schödel *et al.* 2003; Gillessen *et al.*, in preparation).

infrared spectroscopy reveals that many of these bright stars are somewhat older, late-type supergiants and AGB stars. Starting with the discovery of the 'AF'-star (Fig. 6 lower right; Allen *et al.* 1990; Forrest *et al.* 1987), however, an ever increasing number of these bright stars has been identified as being young, massive and early-type. The most recent count from deep SINFONI integral field spectroscopy yields about 100 OB-type stars, including various luminous blue supergiants and Wolf-Rayet stars, but also normal main-sequence OB-type stars (Paumard *et al.* 2006). The nuclear star cluster is one of the richest concentrations of young massive stars in the Milky Way.

The deep adaptive optics images also trace the surface density distribution of the fainter stars, to about $K \simeq 17$-18 mag, corresponding to late B or early A stars (mass 3-6 $M_\odot$), which are a better probe of the density distribution of the overall mass density of the star cluster. While the surface brightness distribution of the star cluster (dominated by the bright stars) is not centered on Sgr A*, the surface density distribution is. There is clearly a cusp of stars centered on the compact radio source (Fig. 6, upper right inset; Genzel *et al.* 2003b; Schödel *et al.* 2007). The inferred volume density of the (observable) cusp is a power-law scaling with $R^{-1.4\pm0.1}$. This is quite consistent with the expectation for a stellar cusp around a massive black hole (Alexander 2005).

If there is indeed a central black hole associated with Sgr A* the presence of so many young stars in its immediate vicinity constitutes a significant puzzle (Allen & Sanders 1986; Morris 1993; Alexander 2005). For gravitational collapse to occur in the presence of the tidal shear from the central mass, gas clouds have to be denser than $\sim 10^9\,(R/(10''))^{-3}$ hydrogen atoms per cm^3. This 'Roche' limit exceeds the density of any gas currently observed in the central region. Recent near-diffraction limited AO

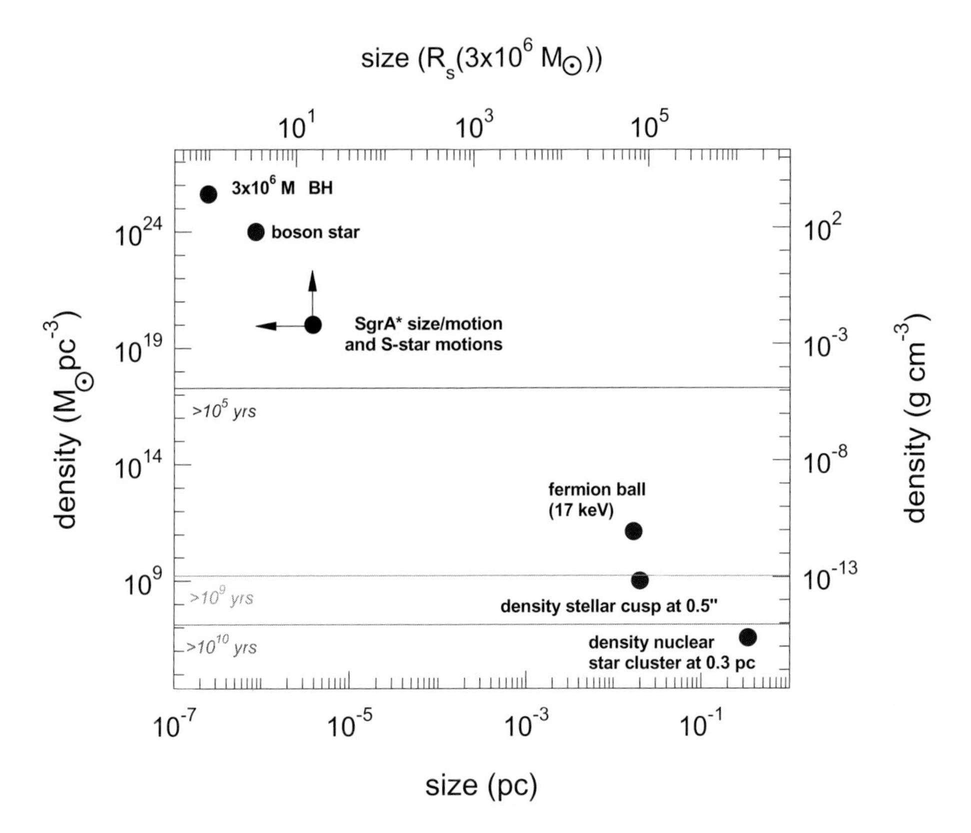

Figure 8. Constraints on the size (horizontal axis) and density (vertical axis) of Sgr A*. 1 pc corresponds to 3.26 light years. The circle with arrows denotes the observational constraints based on stellar orbits, radio size and the upper limit to Sgr A*'s proper motion. Filled circles denote various possible configurations of the mass. A black hole of 3 million $M_\odot$ is located in the upper left corner, for which 'density' refers to the mass smeared out over the volume spanned by the event horizon. Top, middle and bottom horizontal lines mark the location of hypothetical dark astrophysical clusters (of neutron stars, white dwarfs, stellar black holes) of life-times 10^5, 10^9 and 10^{10} years, respectively.

spectroscopy with both the Keck and VLT shows that almost all of the cusp stars brighter than $K \simeq 16$ mag appear to be normal, main sequence B-type stars (Fig. 6, top central spectrum of S2; Ghez *et al.* 2003; Eisenhauer *et al.* 2005b). If these stars formed *in situ* also, the required cloud densities approach conditions in outer stellar atmospheres. Several scenarios have been proposed to account for this 'paradox of youth'. The most prominent are *in situ* formation in a dense gas accretion disk that can overcome the tidal limits, rapid in-spiral of a compact, massive star cluster that formed outside the central region, and re-juvenation of older stars by collisions or stripping (see Alexander 2005 for a detailed discussion and references). Based on the fact that most of the massive stars live in one of two fairly thin, rotating and coeval (~ 6 Myr) disks with a total mass of not much exceeding 10^4 $M_\odot$, Paumard *et al.* (2006) conclude that the *in situ* star formation scenario is the likely answer for stars outside the central arcsecond, at least for the most prominent, so-called 'clockwise' disk. For B-type stars in the cusp immediately around Sgr A* the riddle remains unsolved, although most experts suspect some sort of efficient

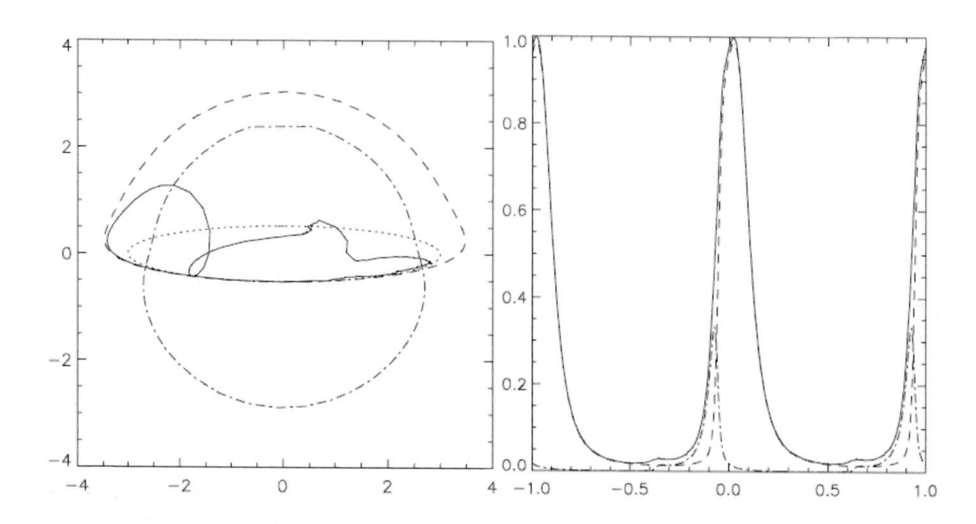

Figure 9. Photo-center wobbling (*left*) and light curve (*right*) of a hot spot on the innermost stable orbit around a Schwarzschild black hole, at an inclination of 80° as derived from ray tracing in a Schwarzschild metric. *Dotted curve:* true path of the hot spot; *dashed curves:* apparent path and light curve of the primary image; *dash-dotted curves:* same for secondary image; *solid curves:* path of centroid and integrated light curve. Axes of left panel are in Schwarzschild radii of a 3 million $M_\odot$ black hole, roughly equal to the astrometric accuracy of 10 μ-arcseconds. Abscissa axis of right panel is in cycles. The loop in the centroid's track is due to the secondary image, which is strongly sensitive to the space time. The overall motion can be detected at good significance at the anticipated accuracy of GRAVITY. Details can be obtained by analyzing several flares simultaneously (Gillessen *et al.* 2006b; Paumard *et al.* 2005).

transport, such as resonant relaxation and three-body collisions (Hopman & Alexander 2006), with stripping of late-type stars as a possible dark horse (Davies & King 2005).

5. Compelling evidence for a central massive black hole

With diffraction limited imagery starting in 1991 on the 3.5 m ESO New Technology Telescope and continuing since 2002 on the VLT, a group at MPE was able to determine proper motions of stars as close as $\sim 0.1''$ from Sgr A* (Eckart & Genzel 1996, 1997). In 1995 a group at the University of California, Los Angeles started a similar program with the 10 m diameter Keck telescope (Ghez *et al.* 1998). Both groups independently found that the stellar velocities follow a 'Kepler' law ($v \propto R^{-1/2}$) as a function of distance from Sgr A* and reach $\geqslant 10^3$ km s^{-1} within the central light month. This implies that the 3–4 million $M_\odot$ found earlier by Prof. Townes and his group must be concentrated within this volume, making any configuration other than a massive black hole fairly unlikely.

Only a few years later both groups achieved the next and crucial step. They were able to determine individual stellar orbits for several of the stars very close to the compact radio source (Fig. 7; Schödel *et al.* 2002, 2003; Ghez *et al.* 2003, 2005a; Eisenhauer *et al.* 2005a). In addition to the astrometric imaging, they obtained near-diffraction limited

Doppler spectroscopy of the same stars (Ghez *et al.* 2003; Eisenhauer *et al.* 2003a, 2005a), yielding precision measurements of the three dimensional structure of several orbits, as well as the distance to the Galactic Center (see below). At the time of writing, orbits have been determined for about a dozen stars in the central light month (Fig. 7). The central mass and most stellar orbital parameters derived by the two teams agree mostly very well. These orbits show that the gravitational potential indeed is that of a point mass centered on Sgr A* within the relative astrometric uncertainties of $\sim$10 mas. The stars orbit the position of this dark mass like planets around the Sun. However, most orbits are fairly elliptical and their orientations appear to be random. Most of the mass must be concentrated well within the peri-approaches of the innermost stars, $\sim$10–20 light hours, or 70 times the Earth orbit radius and about 1000 times the event horizon of a 3.6 million $M_{\odot}$ black hole. There is presently no indication for an extended mass greater than about 5 % of the point mass.

Simulations indicate that current measurement accuracies already are sufficient to detect the first and second order effects of Special and General Relativity (to $(v/c)^2$) in a few years time (Zucker *et al.* 2006). Observations with future 30 m+ diameter telescopes should be able to measure the mass and distance to the Galactic Center to $\sim$0.1% precision, detect radial precession in stellar orbits due to General Relativity and constrain the extended mass to $< 10^{-3}$ of the massive black hole (Weinberg *et al.* 2005). At that level a positive detection of a halo of stellar remnants (stellar black holes and neutron stars) and perhaps dark matter would appear to be likely. Future interferometric techniques will push capabilities yet further (see below).

As mentioned above the simultaneous measurement of positions/proper motions and radial velocities for a number of stars yields a direct ('primary') determination of the Sun-Galactic Center distance, $R_{\circ}$. Eisenhauer *et al.* (2003a, 2005b), Eisenhouwer (in preparation), and Ghez *et al.* (in preparation) find $R_{\circ} \simeq 7.6$ kpc, in good agreement with the values determined from other techniques (Reid 1993). While the statistical uncertainty of the inferred distance is only a few percent, superior to other methods, substantial systematic uncertainties presently dominate the error budget: *a priori* assumptions about the motion of Sgr A* itself, the weight given to the data sets near peri-approach (where flares from Sgr A* have the greatest impact on the stellar positions) and the selection of stars included in the fitting procedure.

Long term VLBA-observations have set 2σ upper limits of about 20 km s^{-1} and 2 km s^{-1} (or 50 micro-arcseconds per year!) to the motion of Sgr A* itself, along and perpendicular to the plane of the Milky Way, respectively (Fig. 3; Reid & Brunthaler 2004; see also Backer & Sramek 1999). This precision measurement demonstrates very clearly that the radio source itself must indeed be massive, with simulations indicating a lower limit to the mass of Sgr A* of $\sim 10^5$ $M_{\odot}$. The intrinsic size of the radio source at millimeter wavelengths is less than 5 to 20 times the event horizon diameter (Fig. 3; Bower *et al.* 2004; Shen *et al.* 2005). Combining radio size and proper motion limit of Sgr A* with the dynamical measurements of the nearby orbiting stars leads to the conclusion that Sgr A* can only be a massive black hole, beyond any reasonable doubt (Fig. 8). An astrophysical dark cluster fulfilling the observational constraints would have a life-time less than a few 10^4 years and thus can be safely rejected, as can be a possible 'fermion' ball of hypothetical heavy neutrinos. All configurations but a massive black hole and a hypothetical 'boson' star (which, however, is not stable when accreting baryons as in the Galactic Center) can be excluded by the available measurements (Schödel *et al.* 2003; Ghez *et al.* 2005a). Under the assumption of the validity of General Relativity, the Galactic Center is now the best quantitative evidence astrophysics has that (massive) black holes do indeed exist.

6. Zooming in on the event horizon

Recent millimeter, infrared and X-ray observations have detected irregular, and sometimes intense outbursts of emission from Sgr A* lasting anywhere between 30 minutes and a number of hours and occurring at least once per day (Baganoff *et al.* 2001; Genzel *et al.* 2003a; Marrone *et al.* 2006). These flares originate from within a few milli-arcseconds of the radio position of Sgr A*. They probably occur when relativistic electrons in the innermost accretion zone of the black hole are significantly accelerated (from $\gamma \simeq 10$ to $\geqslant 10^3$) so that they are able to produce infrared synchrotron emission and X-ray synchrotron or inverse Compton radiation (Markoff *et al.* 2001; Yuan *et al.* 2003; Liu *et al.* 2005). This interpretation is also supported by the detection of significant polarization of the infrared flares (Eckart *et al.* 2006b), by the simultaneous occurrence of X- and IR-flaring activity (Eckart *et al.* 2006a; Yusef-Zadeh *et al.* 2006) and by variability in the infrared spectral properties (Ghez *et al.* 2005a; Gillessen *et al.* 2006a; Krabbe *et al.* 2006). There are indications for quasi-periodicities in the light curves of some of these flares, perhaps due to orbital motion of hot gas spots near the last circular orbit around the event horizon (Genzel *et al.* 2003a; Aschenbach *et al.* 2004; Belanger *et al.* 2006).

These and a number other recent pan-chromatic observations of Sgr A*, in conjunction with increasingly detailed theoretical simulations, are also beginning to elucidate why Sgr A* is so underluminous, $\sim 10^{-5}$ to 10^{-8} times the Eddington luminosity of a 3.6 million $M_\odot$ black hole. It appears that the lack of radiation is due to a combination of several factors (Melia & Falcke 2001). Only a small fraction ($\sim 0.1\%$, Marrone *et al.* 2006) of the gas apparently on its way to the black hole at the Bondi radius of $\sim 1''$ ($\sim 10^{-5}\,M_\odot\,yr^{-1}$, Quataert 2004) actually penetrates to the event horizon. The rest is probably ejected back out because of inefficient angular momentum transport and a strong outflow from the inner accretion zone (Blandford & Begelman 1999; De Villier *et al.* 2005). In addition, the accretion flow is probably radiatively inefficient, due to poor thermal coupling between the electrons and the ions (Narayan *et al.* 1996; Yuan *et al.* 2003). Finally, the accretion rate and luminosity of Sgr A* may be strongly time-variable on time scales of a few hundred years (Revnivtsev *et al.* 2004).

The infrared flares as well as the steady microwave emission from Sgr A* may be important probes of the dynamics and space time around the black hole (Broderick & Loeb 2006). Future long-baseline interferometry at short millimeter or submillimeter wavelengths may be able to map out the strong light bending ('shadow') around the photon orbit of the black hole (Falcke *et al.* 2000). Eisenhauer *et al.* (2005b) are developing 'GRAVITY' for the VLT-Interferometer, which is intended to yield dual-beam, 10 microarcsecond precision infrared astrometric imaging of faint sources. GRAVITY may be able to map out the motion on the sky of hot spots during flares with a high enough resolution and precision to determine the size of the emission region and possibly detect the imprint of multiple gravitational images (Fig. 9; Paumard *et al.* 2005; Gillessen *et al.* 2006b). In addition to studies of the flares, GRAVITY will also be able to image the orbits of stars very close to the black hole, which should then show radial and Lense-Thirring orbital precessions due to General Relativity. Both the microwave 'shadows' as well as the infrared hot spots are sensitive to the space time and metric in the strong gravity regime (Broderick & Loeb 2006). As such, these ambitious future experiments can potentially test the validity of the black hole model near the event horizon and perhaps even the validity of General Relativity in the strong field limit.

References

Aharonian, F., Akhperjanian, A.G., Aye, K.-M., *et al.* 2004, *A&A* (Letters), 425, L13

Alexander, T. 2005, *Phys. Rep.*, 419, 65
Allen, D.A., & Sanders, R. 1986, *Nature*, 319, 191
Allen, D.A., Hyland, A.R., & Hillier, D.J. 1990, *MNRAS*, 244, 706
Aschenbach, B., Grosso, N., Porquet, D., & Predehl, P. 2004, *A&A*, 417, 71
Backer, D.C., & Sramek, R.A. 1999, *ApJ*, 524, 805
Baganoff, F., Bautz, M.W., Brandt, W.N., *et al.* 2001, *Nature*, 413, 45
Balick, B., & Brown, R. 1974, *ApJ*, 194, 265
Beckers, J.M. 1993, *Ann. Rev. A&A*, 31, 13
Becklin, E.E., & Neugebauer, G. 1975, *ApJ*, 200, L71
Becklin, E.E., Matthews, K., Neugebauer, G., & Willner, S.P. 1978, *ApJ*, 219, 121
Belanger, G., Terrier, R., de Jager, O.C., Goldwurm, A., & Melia, F. 2006, *J.Phys.CS*, 54, 420
Blandford, R., & Begelman, M.C. 1999, *MNRAS* (Letters), 303, L1
Bonnet, H., Abuter, R., Baker, A., *et al.* 2004, *ESO Messenger*, 117, 17
Bower, G.C., Falcke, H., Herrnstein, R.M., *et al.* 2004, *Science*, 304, 704
Broderick, A., & Loeb, A. 2006, *J.Phys.CS*, 54, 448
Davies, M.B., & King, A. 2005, *ApJ* (Letters), 624, L25
DePoy, D.L., & Sharp, N.A. 1991, *AJ*, 101, 1324
De Villier, J.P. Hawley, J.F., Krolik, J.H., & Hirose, S. 2005, *ApJ*, 620, 878
Eckart, A., Genzel, R., Hofmann, R., Sams, B.J., & Tacconi-Garman, L.E. 1993, *ApJ* (Letters), 407, L77
Eckart, A., Genzel, R., Hofmann, R., Sams, B.J., & Tacconi-Garman, L.E. 1995, *ApJ* (Letters), 445, L23
Eckart, A., & Genzel, R. 1996, *Nature*, 383, 415
Eckart, A., & Genzel, R. 1997, *MNRAS*, 284, 576
Eckart, A., Baganoff, F.K., Schödel, R., *et al.* 2006a, *A&A*, 450, 535
Eckart, A., Schödel, R., Meyer, L., *et al.* 2006b, *A&A*, 455, 1
Einstein, A. 1916, *Ann. Phys.*, 49, 50
Eisenhauer, F., Schödel, R., Genzel, R., *et al.* 2003a, *ApJ* (Letters), 597, L121
Eisenhauer, F., Abuter, R., Bickert, K., *et al.* 2003b, *Proc. SPIE*, 4841, 1548
Eisenhauer, F., Genzel, R., Alexander, T., *et al.* 2005a, *ApJ*, 628, 246
Eisenhauer, F., Perrin, G., Rabien, S., Eckart, A., Lena, P., Genzel, R., Abuter, R., & Paumard, T. 2005b, *Astron. Nachr.*, 326, 561
Falcke, H., Melia, F., & Algol, E. 2000, *ApJ* (Letters), 528, L13
Ferrarese, L., & Merritt, D. 2000, *ApJ* (Letters), 539, L9
Forrest, W.J., Shure, M.A., Pipher, J.L., & Woodward, C.A. 1987, in: D.C. Backer (ed.), *The Galactic Center*, Proc. Symp. honoring C.H. Townes, Berkeley, CA, USA, 1986, *AIP-CP*, 155, 153
Genzel, R., Hollenbach, D., & Townes, C.H. 1994, *Rep. Prog. Phys.*, 57, 417
Genzel, R., Schödel, R., Ott, T., *et al.* 2003a, *Nature*, 425, 934
Genzel, R., Schödel, R., Ott, T., *et al.* 2003b, *ApJ*, 594, 812
Ghez, A.M., Klein, B.L., Morris, M., & Becklin, E.E. 1998, *ApJ*, 509, 678
Ghez, A.M., Duchêne, G., Matthews, K., *et al.* 2003, *ApJ* (Letters), 586, L127
Ghez, A.M., Salim, S., Hornstein, S.D., *et al.* 2005a, *ApJ*, 620, 744
Ghez, A.M., Hornstein, S.D., Lu, J.R., *et al.* 2005b, *ApJ*, 635, 1087
Giacconi, R., Gursky, H., Paolini, F., & Rossi, B.B. 1962, *Phys. Rev. Lett.*, 9, 439
Giacconi, R. 2003, *Rev. Mod. Phys.*, 75, 995
Gillessen, S., Eisenhauer, F., Quataert, E., *et al.* 2006a, *ApJ* (Letters), 640, L163
Gillessen, S., Perrin, G., Brandner, W., *et al.* 2006b, in: J.D. Monnier, M. Schöller & W.C. Danchi (eds.), *Advances in Stellar Interferometry*, *SPIE*, 6268, 33
Hopman, C., & Alexander, T. 2006, *ApJ*, 645, 1152
Kerr, R. 1963, *Phys. Rev. Lett.*, 11, 237
Kormendy, J. 2004, in: L.C. Ho (ed.), *Coevolution of Black Holes and Galaxies*, Proc. *Carnegie Observatories Astrophysics Series* (Cambridge: CUP), p. 1
Krabbe, A., Iserlohe, C., Larkin, J.E., *et al.* 2006, *ApJ* (Letters), 642, L145

Lenzen, R., Hofmann, R., Bizenberger, P., & Tusche, A. 1998, in: A.M. Fowler (ed.), *Infrared Astronomical Instrumentation*, *Proc. SPIE*, 3354, 606
Liu, S., Melia, F., & Petrosian, V. 2005, *ApJ*, 636, 798
Lynden-Bell, D. 1969, *Nature*, 223, 690
Markoff, S., Falcke, H., Yuan, F., & Biermann, P.L. 2001, *A&A* (Letters), 379, L13
Marrone, D., Moran, J.M., Zhao, J.-H., & Rao, R. 2006, *ApJ*, 640, 308
Matthews, K., *et al.* 2007, in preparation
Melia, F., & Falcke, H. 2001, *Ann. Rev. A&A*, 39, 309
McClintock, J.E., & Remillard, R.A. 2006, in: W. Lewin & M. van der Klis (eds.), *Compact Stellar X-ray Sources*, *Cambridge Astrophysics Series* No. 39. (Cambridge: CUP), p. 157.
Miyoshi, M., Moran, J., Herrnstein, J., *et al.* 1995, *Nature*, 373, 127
Morris, M. 1993, *ApJ*, 408, 496
Narayan, R., Yi, I., & Mahadevan, R. 1996, *Nature*, 374, 623
Paumard, T., Perrin, G., Eckart, A., Genzel, R., Lena, P., Schödel, R., Eisenhauer, F., Müller, T., & Gillessen, S. 2005, *Astron. Nachr.*, 326, 568
Paumard, T., Genzel, R., Martins, F., *et al.* 2006, *ApJ*, 643, 1011
Quataert, E. 2004, *ApJ*, 613, 322
Rees, M. 1984, *Ann. Rev A&A*, 22, 471
Reid, M.J. 1993, *Ann. Rev. A&A*, 31, 345
Reid, M.J., & Brunthaler, A. 2004, *ApJ*, 616, 872
Revnivtsev, M.G., Churazov, E.M., Sazonov, S.Yu., *et al.* 2004, *A&A* (Letters), 425, L49
Rivera, E.J., Lissauer, J.J., Butler, R.P., *et al.* 2005, *ApJ*, 634, 625
Roberts, D.A., & Goss, W.M. 1993, *ApJS*, 86, 133
Rousset, G., Lacombe, F., Puget, P., *et al.* 1998, in: D. Bonaccini & R.K. Tyson (eds.), *Adaptive Optical System Technologies*, *Proc. SPIE*, 3255, 508
Schödel, R., Ott, T., Genzel, R., *et al.* 2002, *Nature*, 419, 694
Schödel, R., Ott, T., Genzel, R., *et al.* 2003, *ApJ*, 596, 1015
Schödel, R., Eckart, A., Muzic, K., *et al.* 2007, *A&A* (Letters), 462, L1
Schmidt, M. 1963, *Nature*, 197, 1040
Schwarzschild, K. 1916, *Sitzungsber. Preuss. Akad. Wiss.*, 424
Shen, Z.Q., Lo, K.Y., Liang, M.C., Ho, P.T.P., & Zhao, J.H. 2005, *Nature*, 438, 62
Townes, C.H., Lacy, J.H., Geballe, T.R., & Hollenbach, D.J. 1982, *Nature*, 301, 661
Weinberg, N.N., Milosavljevic, M., & Ghez, A.M. 2005, *ApJ*, 622, 878
Wheeler, J.A. 1968, *American Scientific*, 56, 1
Wizinovich, P., *et al.* 2007, in preparation
Wollman, E.R., Geballe, T.R., Lacy, J.H., Townes, C.H., & Rank, D.M. 1977, *ApJ* (Letters), 218, L103
Yuan, F., Quataert, E., & Narayan, R. 2003, *ApJ*, 598, 301
Yusef-Zadeh, F., Bushouse, H., Dowell, C.D., *et al.* 2006, *ApJ*, 644, 198
Zucker, S., Alexander, T., Gillessen, S., Eisenhauer, F., & Genzel, R. 2006, *ApJ* (Letters), 639, L21

II. Joint Discussions

Highlights of Astronomy, Volume 14
IAU XXVI General Assembly, 14-25 August 2006
Karel A. van der Hucht, ed.

doi:10.1017/S1743921307009866

Joint Discussion 1 Particle acceleration – from solar system to AGN

Marian Karlický[1] and John C. Brown[2] (eds.)

[1]Astronomical Institute of the Academy of Sciences of the Czech Republic, CZ-25165 Ondřejov, Czech Republic
email: karlicky@asu.cas.cz

[2]Department of Physics & Astronomy, University of Glasgow, G12 8QW, Scotland, UK
email: john@astro.gla.ac.uk

Abstract. The scene is set for IAU JD01 on Cosmic Particle Acceleration: from Solar System to AGNs

Keywords. cosmic particle acceleration

Preface

The first suggestion of this meeting came from Marian Karlický and others at the June 2004 meeting of CESRA at Sabhal Mhor Ostaig (Gaelic College), Isle of Skye. Aptly, this is only about 100 km from the Observatory on Ben Nevis, where C.T.R. Wilson carried out his pioneering cosmic ray (CR) work and whose Brocken Spectre inspired Wilson's invention (`<www.gla.ac.uk/adulteducation/Personnel/alec/wilson/index.html>`) of the cloud chamber.

The meeting was also very timely in view of the fact that, since the last IAU meeting dedicated to accelerated particles (IAU Colloquium No. 142 in 1993, Maryland – *Particle Acceleration Phenomena in Astrophysical Plasmas*), as well as theoretical progress, there have been major observational advances in all high-energy wavebands. The field of very-high-energy γ-ray astronomy has finally achieved maturity, with confirmed detections of both point and diffuse sources of galactic and extra-galactic origin by experiments such as VERITAS, CANGAROO, HEGRA, CAT and HESS, using imaging Cerenkov telescopes. All of these high-energy advances have been further enhanced by coordinated multi-wavelength observations using both established and new facilities, especially radio. These and other facilities have expanded our knowledge of particle acceleration on all cosmic scales, including:

- Planetary and interplanetary particles – *CLUSTER, Ulysses, Cassini, WIND*
- Solar flare particles – *Yohkoh, Compton Observatory, RHESSI, WAVE, Granat, Koronas*
- Stellar fast particles in colliding hot star winds and cool star flares - *Chandra, XMM, Compton GRO*
- Pulsars, plerions, shell-type SNRs and GRBs – CANGAROO, HESS, *Integral, RXTE, Beppo-SAX*
- Extra galactic jets and AGNs – *Chandra, XMM, Integral, Compton GRO*, VERITAS, HEGRA, HESS
- Ultra-high-energy cosmic rays – Auger, Fly's eye, AGASA

While reports on these major observational developments are included, the main thrust of this meeting was toward improving our knowledge and understanding of the physical processes involved – especially plasma/particle kinetic and wave phenomena and intense beam electrodynamics, as opposed to MHD. Mean free paths of cosmic energetic particles are generally very large compared to length scales of primary energy conversion (current sheet thickness, shock scale, mirroring length, etc.). Given that fast particles constitute much of the total energy, it is clear that particle kinetics is a vital aspect that must be integrated with MHD approaches in order to achieve credible theories of the overall energy release and particle acceleration process in high-energy sources.

Particle acceleration is a ubiquitous phenomenon on all cosmic scales and one of the remaining great enigmas of astrophysics. In the big picture, within the realms of 'visible' matter, energetic particles (cosmic rays or CRs) rank with galaxy rotation and magnetic fields in the league table of cosmic energy densities, falling behind only the mean densities of stellar thermal and gravitational energies. In many, if not all, energetic sources , from solar flares to GRBs, the total fast particle numbers and energy can be a large fraction of the total system content. This poses efficiency problems of energy and particle supply in addition to those of achieving high energies of individual particles, again found on all scales from terrestrial lightning through aurorae and solar/stellar flares to supernovae (SN), pulsars, GRBs and AGNs. Even in our own daily lives, accelerated cosmic particles play roles ranging over the disruption of power and data transmission by flares and CMEs, the solar cycle linked influence of CRs on cloud formation and climate, and the potentially lethal effect of CRs from a nearby SN.

Thus, the observed properties of, and acceleration theory for, energetic particles lie at the heart of a number of key problems on all cosmic scales, as well as being important in both hot fusion and cool lab plasmas. Despite this, particle acceleration processes remain among the greatest unsolved problems of plasma (astro)physics. These facts were the original driver in our proposal of this Joint Discussion. The aim of the meeting was thus to bring together experts from across the whole range of IAU disciplines concerned with accelerated particles and from all pertinent domains of observed energy, diagnostic modelling, and theory, to share and enhance understanding of all acceleration mechanisms and their relevance in each cosmic regime. As the following pages show, this goal was amply achieved via the excellent invited and contributed talks, and the posters, which ranged both observationally and theoretically from impulsive solar electrons above 10 keV to ultra-relativistic radiation dominated plasmas.

The aim of the meeting was thus to bring together experts from across the whole range of IAU disciplines concerned with accelerated particles and from all pertinent domains of observed energy, diagnostic modelling, and theory, to share and enhance understanding of all acceleration mechanisms and their relevance in each cosmic regime.

We are grateful to the IAU for supporting our proposal and to all the participants who made it a great success. We look forward to further discussions of this exciting field at a session in the Rio GA 2009. In the same spirit as that in which cosmologists have recently re-discovered issues of star formation under the name of *cosmic re-ionization*, we should very possibly hold that session with a re-badging of our field as *cosmic re-acceleration*. Particles of the early universe declined rapidly in energy as the Big Bang proceeded, but some of them are now being returned to very high energies by plasma electric fields, having their ultimate origins in the same gravity driven formation of hot condensed plasmas (stars and galaxies) as is responsible for radiative cosmic re-ionization. May progress and funding of acceleration studies itself accelerate as quickly as possible.

Scientific Organizing Committee

Jonathan Aarons (USA), Roger D. Blandford (USA), John C. Brown (UK, co-chair), Mary K. Hudson (USA), Marian Karlicky (Czech Republic, co-chair), John Kirk (BRD, co-chair), Robert P. Lin (USA), Donald B. Melrose (Australia), Kazunari Shibata (Japan), Nicole Vilmer (France), and Alan M. Watson (UK).

John C. Brown, co-chair
Glasgow, September 2006

Highlights of Astronomy, Volume 14
IAU XXVI General Assembly, 14-25 August 2006
Karel A. van der Hucht, ed.

doi:10.1017/S1743921307009878

Acceleration: problems and progress

Donald B. Melrose
School of Physics, University of Sydney, NSW 2006, Australia
email: melrose@physics.usyd.edu.au

Abstract. Older ideas on acceleration are linked to current ideas to identify both the major successes in the field, and long-standing problems and difficulties that remain unresolved.

Keywords. acceleration of particles, shock waves, plasmas

---

1. Introduction

The recent literature on acceleration separates into four broad applications: to space plasmas, for which we have *in situ data*; to solar particles, for which we have both signatures of the particles in the solar atmosphere and direct measurements of particles that escape: to Galactic cosmic rays (CR); and to synchrotron sources, AGN, bursters, etc., for which we have only the electromagnetic spectrum. Rather than attempt to review progress in all these areas, I take a look back to early ideas on acceleration, consider the progress we have made, and the major problems that have remained unsolved for decades.

2. Brief historical review

Swann (1933) suggested that solar CR are accelerated by a changing magnetic field (betratron effect), and he estimated the available potential to be $\Phi \approx 10^{10}$ V. Although Swann's specific model is no longer taken seriously, there are two important ideas that remain relevant today: effective acceleration is due to inductive electric fields, and the maximum potential available can be estimated from simple arguments, e.g., $\Phi \approx BLv$ for magnetically connected conducting regions separated by L in relative motion at speed v or, for a unipolar inductor, $\Phi \approx (Z_0 P)^{1/2}$, where $Z_0 \simeq 300\,\Omega$ is the impedance of free space and P is the power. Fermi (1949) suggested that CR are accelerated by reflections off moving interstellar clouds, and this became the prototype for stochastic acceleration. It is now recognized that the acceleration corresponds to an isotropic diffusion of particles in momentum space (Tverskoi 1967) that can be attributed to the damping of compressive MHD waves (Achterberg 1981). Fermi (1954) proposed first-order acceleration between 'closing jaws' and this idea underpins diffusive shock acceleration (DSA), proposed independently by various authors in 1997-8 (e.g., the review by Malkov & Drury 2001) – a major success in this field. An essential ingredient in both stochastic acceleration and DSA is the efficient scattering of fast particles. Another major success was the recognition in the mid 1950s that resonant scattering by waves causes pitch-angle diffusion and spatial diffusion along field lines; moreover, anisotropic particles cause their resonant waves to grow (e.g., Melrose 1980), but this is ineffective at high energy.

3. Outstanding problems

In simple models for DSA, effective diffusion at high energy and across field lines are simply postulated, without adequate justification. Ideas on how these problems might be addressed were suggested by Bell (2004) and Achterberg & Ball (1994), respectively.

The Galactic Cosmic Ray (GCR) CR spectrum is separated into (Axford 1994) GCRI at $< 10^{15}$ eV, GCRII and EGCR. GCRI is attributed to DSA by shocks from SN explosions that permeate the ISM. GCRI has nearly normal cosmic abundance implying that the injection into DSA is insensitive to charge (Z) and mass (A)– maybe super-Alfvénic jets of fluid due to magnetic reconnection. This contrasts with anomalous abundances in other contexts, notably overabundance of ^{3}He, notably in 3 Cent. A (Sargent & Jugaku 1961) and in some solar particle events. Sensitivity to Z, A in pre-acceleration by cyclotron waves (Fisk 1978) remains the favored explanation (Liu, Petrosian & Mason 2006). The acceleration of the GCRII and its relation to GCRI is still a topic of debate, and the acceleration of the presumably extragalactic EGCR is a major unsolved problem.

Flat synchrotron spectra correspond to $f(p) \propto p^{-3}$, while DSA at a single nonrelativistic shock implies $f(p) \propto p^{-b}$ with $b > 4$. The 'cosmic conspiracy' model for flat spectra involves broadening a self-absorbed peak due to a special geometry (Cotton *et al.* 1980). Flat spectra can be produced by DSA at multiple shocks (White 1985): energy gains, due to DSA at each shock, and losses, due to decompression between shocks, constitute a stochastic acceleration mechanism. Synchrotron pile-up occurs for $b < 4$, and DSA at multiple shocks with synchrotron losses is an alternative model for flat spectra (Melrose & Crouch 1997). The emission is dominated by the compressed regions around the shocks.

A major outstanding problem is 'bulk energization' of electrons to 10–100 keV in solar flares. Magnetic energy is favored, but this must involve a statistically large number of tiny dissipation regions. How these are coupled together and how all the electrons can be processed through a tiny net volume is unclear. Network theory (e.g., Vlahos, Isliker & Lepreti 2004) might provide a useful framework.

4. Conclusions

Progress has been slow, with resonant scattering, stochastic acceleration and DSA being major successes. Many problems remain inadequately understood, and many more could be added to my list.

References

Achterberg, A. 1981, *A&A*, 97, 259
Achterberg, A., & Ball, L. 1994 *A&A*, 285, 687
Axford, W.I. 1994, *ApJS*, 90, 937
Bell, A.R. 2004, *MNRAS*, 353, 550
Cotton, W.D., Wittels, J.J., Shapiro, I.I., *et al.* 1980, *ApJ*, 238, L123
Fermi, E. 1949, *Phys. Rev.*, 75, 1169
Fermi, E. 1954, *ApJ*, 119, 1
Fisk, L.A. 1978, *ApJ*, 224, 1048
Liu, S., Petrosian, V., & Mason, G.M. 2006, *ApJ*, 636, 462
Malkov, M.A., & Drury, L.O.'C. 2001, *Rep. Prog. Phys.*, 64, 429
Melrose, D.B. 1980, *Plasma Astrophysics II* (New York: Gordon & Breach), p. 12
Melrose, D., & Crouch, A. 1997, *Pub. Astron. Soc. Aust.*, 14, 251
Sargent, W.L.W., & Jugaku, J. 1961, *ApJ*, 134, 777
Swann, W.F.G. 1933, *Phys. Rev.*, 43, 217
Tverskoi, B.A. 1967, *Sov. Phys. JETP*, 25, 317
Vlahos, L., Isliker, H., & Lepreti, F. 2004, *ApJ*, 608, 540
White, R.L. 1985, *ApJ*, 289, 698

Highlights of Astronomy, Volume 14
IAU XXVI General Assembly, 14-25 August 2006
Karel A. van der Hucht, ed.

doi:10.1017/S174392130700988X

The solar system: a laboratory for the study of the physics of particle acceleration

Robert P. Lin

Physics Department and Space Sciences Laboratory,
University of California, Berkeley CA 94720-7450, USA
email: rlin@ssl.berkeley.edu

Abstract. A remarkable variety of particle acceleration occurs in the solar system, from lightning-related acceleration of electrons to tens of MeV energy in less than a millisecond in planetary atmospheres; to acceleration of auroral and radiation belt particles in planetary magnetospheres; to acceleration at planetary bow shocks, co-rotating interplanetary region shocks, shocks driven by fast coronal mass ejections, and possibly at the heliospheric termination shock; to acceleration in magnetic reconnection regions in solar flares and at planetary magnetopause and magnetotail current sheets. These acceleration processes often occur in conjunction with transient energy releases, and some are very efficient. Unlike acceleration processes outside the solar system, the accelerated particles and the physical conditions in the acceleration region can be studied through direct in situ measurements, and/or through detailed imaging and spectroscopy. Here I review recent observations of tens of MeV electron acceleration in the Earth's atmosphere and in the Earth's radiation belts, electron and ion acceleration related to magnetic reconnection in solar flares, electron acceleration to $\geqslant 300$ keV in magnetic reconnection regions in the Earth's deep magnetotail, and acceleration of solar energetic particles (SEPs) by shocks driven by fast coronal mass ejections (CMEs).

Keywords. acceleration of particles in solar system

1. Terrestrial gamma-ray flashes (TGFs)

Bursts of ~ 30 to > 300 keV γ-rays from the Earth lasting ~ 1 ms, were first detected by the *BATSE* instrument (designed to detect cosmic γ-ray bursts, CGRBs) on the *Compton Gamma-Ray Observatory* (Fishman *et al.* 1994). 75 TGFs were detected over nine years, and it was shown that they occurred simultaneous with lightning. Recent observations from the *RHESSI* mission showed that TGFs typically extend up to > 10 MeV in energy; the spectrum obtained by averaging over many bursts is consistent with *bremsstrahlung* of a minimum of $\sim 10^{15}$ electrons of ~ 30 MeV energy (Smith *et al.* 2005). The *RHESSI* observations show no energy dispersion down to ~ 0.1 ms time scale, indicating that the acceleration is extremely fast. About 10–15 TGFs are detected a month; the much higher rate is likely due to the fact that *RHESSI* records every photon while *BATSE* depends on a trigger algorithm that is optimized for CGRBs. Since *RHESSI* can only detect TGFs within line of sight, this implies an order of a thousand TGFs per day over the latitude band from -38 to +38 degrees. The TGFs are definitely associated with lightning, and appear to be preferentially observed in the tropics. The present hypothesis for TGFs is that they are due to runaway of cosmic ray secondary relativistic electrons in the very large potential (~ 30 megavolts) above the cloud set up by the lightning stroke. Electrons at energies of a few MeV have the lowest energy loss to Coulomb collisions and therefore require the smallest electric field to run away, but many questions remain. It is remarkable that the Earth can be a high energy accelerator and gamma-ray source, and it should be noted that lightning orders of magnitude more intense has been detected (through the associated radio emission) from Jupiter and Saturn.

2. Radiation belt MeV electron acceleration

In recent years the importance of transients in the acceleration of radiation belt particles has become more apparent. Spectrogram plots of the MeV electron fluxes *vs.* equatorial distance as a function of time are dominated by sudden ($\leqslant$1 day) increases followed by decay over months, with more intense fluxes around solar minimum when high speed streams are present. Thus, the transient response to high-speed solar wind streams appears to accelerate much of the radiation belt electrons through a combination of injection of particles from the magnetotail followed by inward radial diffusion driven by ULF waves. Much rarer are situations where a strong shock from a large CME impinging on the magnetosphere produces an intense transient electric field which is able to accelerate electrons up to $>$ 10–15 MeV in an electron drift period (such as occurred on March 24, 1991). These events, however, produce the highest energy electron belts, and those can last for years.

3. Solar flare electron and ion acceleration associated with magnetic reconnection

It has become clear from *RHESSI* imaging spectroscopy measurements of solar flares that the energy release process for solar flares, e.g., magnetic reconnection, is able to efficiently accelerate ions and electrons to high energies. The closely similar temporal variations of pairs of hard X-ray (HXR) footpoints seen in large γ-ray flares such as the 23 July 2002 provide strong evidence for reconnection and the formation of new closed loops (Lin *et al.* 2003). The temporal variation of the HXR intensity also shows a rough correlation with the rate of reconnection, as inferred from the footpoint motions, suggesting that the energy released goes into the acceleration of the HXR-producing electrons. For the first time, the accelerated ions were located through *RHESSI* imaging of the 2.223 MeV neutron-capture γ-ray line. Surprisingly, the line emission comes from a different location, $\sim$ 15,000 km away from the HXR emission, indicating that acceleration and/or propagation of the electrons and ions are different. In the large flare of 28 October 2003 two footpoint sources were observed in the gamma-ray line image, but they were again separated from the HXR footpoint sources by $\sim$ 10,000 km (Hurford *et al.* 2006). How electrons and ions are efficiently accelerated to high energies in the reconnection process is presently unknown.

4. Electron acceleration in the magnetotail reconnection region

The *Wind* spacecraft has provided in situ measurements in traversing the ion diffusion region (where the ions decouple from the magnetic field because of their large gyroradii) of a magnetic reconnection event observed deep ($\sim$ 60 Earth radii) in the Earth's magnetotail. Electrons up to $\geqslant$ 300 keV were detected peaking in the ion diffusion region, with fluxes decreasing and the spectrum softening with distance away (Oieroset *et al.* 2002). No acceleration of ions was detected, however, and no unusual plasma wave activity was detected. Again the acceleration mechanism is presently unknown.

5. Solar energetic particle (SEP) acceleration by shocks

SEPs are hypothesized to be accelerated by shocks driven by fast ($\sim$1-3 thousand km/s) Coronal Mass Ejections (CMEs). In large SEP events, the efficiency, defined as the percentage of the total energy in the CME (dominated by the kinetic energy of the CME) that is in SEPs, is typically of order $\sim 10\%$ (Emslie *et al.* 2004). This efficiency is also what is needed for supernova shocks to accelerate galactic cosmic rays. Fast CMEs and shocks are observed at 1 AU on occasion, sometimes with coincident increases in SEP ions up to $\sim$10 MeV. Above $\sim$20 MeV, however, SEP increases related to interplanetary shocks near 1 AU are almost never observed, implying that acceleration to those energies does not occur for some reason. The time profiles for SEP fluxes at high energies indicate that those particles are accelerated close to the Sun. Some very fast ($>$2000 km/s) and wide ($\geqslant$100 degrees) CMEs are unaccompanied by SEP events, and for a given CME speed, the peak SEP fluxes can vary by several orders of magnitude. Clearly, our understanding of the shock acceleration process is, at best, qualitative. NASA's planned *Solar Sentinels* mission is intended specifically to make progress on this fundamental problem of shock acceleration to high energies, with in situ measurements of pristine, freshly accelerated SEPs and of the unevolved shock and upstream waves that are presumably key to the acceleration process, and close (0.25 AU) to the near-Sun acceleration region to avoid the blurring effects of scattering and diffusion of the SEPs in reaching 1 AU.

References

Emslie, A.G., Kucharek, H., Dennis, B.R., *et al.* 2004, *J. Geophys. Res.*, 109, A10104
Fishman, G.J., Bhat, P.N., Mallozzi, R., *et al.* 1994, *Science*, 264, 1313
Hurford, G., Krucker, S., Lin, R.P., *et al.* 2006, *ApJ* (Letters), 644, L93
Lin, R.P., Krucker, S., Hurford, G.J., *et al.* 2003, *ApJ* (Letters), 595, L69
Oieroset, M., Lin, R.P., Phan, T.D., *et al.* 2002, *Phys. Rev. Lett.*, 89, 195001
Smith, D.M., Lopez, L.I., Lin, R.P., & Barrington-Leigh, C.P. 2005, *Science*, 307, 1085

Highlights of Astronomy, Volume 14
IAU XXVI General Assembly, 14-25 August 2006
Karel A. van der Hucht, ed.

doi:10.1017/S1743921307009891

Evidence for solar shock production of heliospheric near-relativistic and relativistic electron events

Stephen W. Kahler

Air Force Research Laboratory/VSBXS, Hanscom AFB, MA 01731, USA
email: stephen.kahler@hanscom.af.mil

Abstract. Properties of near-relativistic ($E \gtrsim 30$ keV, NR) and relativistic ($E \gtrsim 0.3$ MeV) electron events produced near the Sun and observed within 1 AU are reviewed. Observations suggest the CME-driven shocks are the sources of many events, but flares are often sources for NR events.

Keywords. acceleration of particles, shock waves, Sun: particle emission, flares, CMEs

---

Transient electron events extend from thermal solar-wind to relativistic energies. Two candidate acceleration processes for these events are MHD shocks driven by fast ($\gtrsim 1000$ km/s) coronal mass ejections (CMEs) and electric field or turbulent wave acceleration during magnetic reconnection in flares. Which process is responsible for observed events?

NR electrons have inferred solar injection onsets delayed by ~ 10 min from the flare impulsive phases. The delays, if real, could indicate either a shock or later flare acceleration. Peak electron intensities correlate with CME speeds but also equally well with peak soft and hard X-ray flare fluxes, and most events are not associated with type II burst shocks. Electron spectral variations are ordered by flare soft X-ray flare durations and by CME speeds, but not by the presence or absence of associated CMEs needed to produce shocks. Impulsive ^{3}He-rich ion events produced in flares are well associated with electron events, but ion injections may be delayed. Simultaneous ion and electron injections occurred in the big 20 January 2005 event, but both were near the flare impulsive phase and during a very fast CME. Events observed on both *Ulysses* and *ACE* during their wide angular separations provide strong evidence for shock acceleration, but most *ACE* events are not seen at *Ulysses*, consistent with flare injections. An impulsive component was followed 11 minutes later by a gradual component with a harder spectrum in the 28 October 2003 electron event, suggesting separate flare and shock accelerations and injections. In summary, correlations with CME speeds and type II bursts, associations with gradual SEP events, and large angular distributions suggest that some NR electron events are shock accelerated, but events from impulsive flares are also common.

Relativistic electron injections coincided with flare impulsive phases in *Helios* observations at $r < 0.5$ AU, but some events consisted of second injections with harder energy spectra. Event onset delays and rise times increase with longitudinal separation from flare sites, and both event rise times to maxima and spectral hardness correlate with CME speeds, as expected for shock acceleration, but event peak intensities correlate with flare X-ray peak fluxes. Two electron events were associated with eruptions with no active region flares. The electrons share the spectral invariance relative to shock locations exhibited by the ions. Similar time-intensity profiles and common e/p ratios in many events suggest that SEPs and electrons are produced together in shocks. Event size distributions match better those of SEPs than of flares, consistent with shocks.

Highlights of Astronomy, Volume 14
IAU XXVI General Assembly, 14-25 August 2006
Karel A. van der Hucht, ed.

doi:10.1017/S1743921307009908

Electron acceleration in solar flares: observations *versus* numerical simulations

Arnold O. Benz, Paolo C. Grigis and Marco Battaglia
Institute of Astronomy, ETH Zurich, 8092-Zurich, Switzerland
email: benz@astro.phys.ethz.ch

Abstract. We use *RHESSI* hard X-ray observations to constrain acceleration of solar flare electrons, generally considered to be a primary recipient of the released energy.

Coronal X-ray sources have been previously discovered and tentatively associated with bremsstrahlung emission near the acceleration site. Now, *RHESSI* imaging spectroscopy (Lin *et al.* 2002) can temporally resolve the non-thermal spectrum of the coronal source for the first time (Battaglia & Benz 2007). We compare the time behaviour with the predictions of stochastic acceleration, as described by transit-time damping of MHD turbulence excited by reconnection. The results in five limb events indicate Soft-Hard-Soft (SHS) behaviour of the coronal source emission in the course of an X-ray peak (the more flux, the harder the spectrum, Grigis & Benz 2004; Battaglia *et al.* 2005). The SHS behaviour thus constitutes a conspicuous property of the acceleration process. The temporal behaviour of the spectrum can be quantitatively described by a pivot point at a photon energy of about 20 keV at which the flux remains constant in time.

We solve a diffusion equation for the interaction of waves and particles including trapping, escape and particle replenishment (Grigis & Benz 2006). The solution yields a spectrum that is approximately a power-law in the observed range of energies. However, the theoretically derived pivot point is generally at energies lower than observed. For this reason we include transport effects, such as produced by an electric potential, or scattering in the coronal source (collisional trapping) to bring the pivot energy up to the observed value. Escaping particles propagate to the base of the loop in the dense chromosphere. These precipitating particles are identified as origin of the observed hard X-ray footpoints.

Observations and simulations show that solar flare electron acceleration cannot be modeled without transport effects. The observations are consistent with stochastic acceleration in a relatively dense medium (up to 10^{11} cm^{-3}) high wave energy densities (up to 0.001 magnetic) and a return current. The observations also constrains the global flare geometry, requiring coupling between the coronal source and the chromospheric footpoints.

Acknowledgments. The analysis of *RHESSI* data at ETH Zurich is partially supported by the Swiss National Science Foundation (grant nr. 20-67995.02).

References

Battaglia, M., & Benz, A.O. 2007, *A&A*, 466, 713
Battaglia, M., Grigis, P.C., & Benz, A.O. 2005, *A&A* 439, 737
Grigis, P.C., & Benz, A.O. 2004, *A&A* 426, 1093
Grigis, P.C., & Benz, A.O. 2006, *A&A*, 458, 641
Lin, R.P., Dennis, B.R., Hurford, G.J., *et al.* 2002, *Solar Phys.*, 210, 3

Highlights of Astronomy, Volume 14
IAU XXVI General Assembly, 14-25 August 2006
Karel A. van der Hucht, ed.

doi:10.1017/S174392130700991X

Radio and X-ray diagnostics of electrons accelerated in solar flares

Miroslav Bárta and Marian Karlický
Astronomical Institute of the Academy of Sciences of the Czech Republic,
CZ-25165 Ondřejov, Czech Republic
email: barta,karlicky@asu.cas.cz

Abstract. Starting from 2.5D MHD modelling of solar flares on a global scale we calculate (using the PIC and test-particle simulations) the radio and X-ray emissions generated in solar flare reconnection. Our results – the radio and X-ray spectra and brightness distributions, and their dynamics – are directly comparable with observations providing thus a test of particle acceleration models as well as of the 'standard' global flare scenario.

Keywords. Solar flares, acceleration of particles, radio and X-ray emissions

---

It is a well known fact that solar flares are efficient particle accelerators – even several concurrent acceleration processes can take place in a single flaring volume. Using MHD and particle simulations we study some of them – namely direct acceleration in a current sheet torn during magnetic reconnection and pinch-effects acting on electrons trapped in non-equilibrium plasmoids newly created by the reconnection process (see also Kliem *et al.* 2000), and acceleration in collapsing magnetic traps (Karlický & Bárta 2006; Karlický & Kosugi 2004) formed by shrinking reconnected magnetic field lines inside cusp structures.

In order to relate our models more directly to the real world, we not only calculate the dynamics of electron distribution functions, but we extend our results to forms comparable with observations. As accelerated particles in the solar atmosphere manifest themselves most remarkably by the radio and X-ray emissions, the final outputs of our modelling are radio and X-ray spectra and their dynamics as well as spatial structures of X-ray and radio sources. The approach used provides not only a test of particle acceleration mechanisms, but also checks the validity of the global flare model in the framework of which the acceleration processes have been studied. Comparing simulated X-ray and radio data with observations, we found that Drifting Pulsating Structures (DPS, see also Karlický *et al.* 2005) can be interpreted as the radio emission of accelerated electrons trapped in ejected plasmoids, and electrons accelerated in the collapsing magnetic trap can account for the brightness distributions and spectra of X-ray flare loop-top sources.

The results obtained are sensitive to the parameters of the flare model used. Thus, fitting the observed and the modelled data provides us with a diagnostic tool for investigation of conditions under which the acceleration in the solar flares proceeds.

References

Karlický, M., & Bárta, M. 2006, *ApJ*, 647, 1472
Karlický, M., & Kosugi, T. 2004, *A&A*, 419, 1159
Karlický, M., Bárta, M., Meszárosová, H., & Zlobec, P. 2005, *A&A*, 432, 705
Kliem, B., Karlický, M., & Benz, A.O. 2000, *A&A*, 360, 715

Highlights of Astronomy, Volume 14
IAU XXVI General Assembly, 14-25 August 2006
Karel A. van der Hucht, ed.

doi:10.1017/S1743921307009921

Solar flare particle heating via low-β reconnection

Dietmar Krauss-Varban and Brian T. Welsch

Space Sciences Laboratory, University of California, Berkeley, CA 94720-7450, USA

Abstract. Observations give tight constraints on the temporal and spatial scales of particle heating in solar flares, and on the required efficiency. Electrons are accelerated into a quasi-thermal population of a few tens of keV. X- and γ-rays imply tails in electron and ion distributions reaching tens of MeV and above. Simple estimates indicate that all available electrons are accelerated at least once to moderate energies, pointing to an initial process resembling bulk heating rather than acceleration of a small or localized population. In the absence of effective collisions, wave-particle interactions are the prime candidate. Here we address the outstanding questions, *(i)* what process can heat the entire reconnecting plasma to the above energies, and *(ii)*, what provides the free energy for wave-particle interactions? We propose a process in which initially the ions are heated and provide the free energy for electron heating and tail formation.

Keywords. Solar flares, acceleration of particles

Our hybrid simulations (kinetic ions, fluid electrons) for the far magnetotail (Krauss-Varban & Omidi 1995) indicate that in low-β Petschek-type reconnection, the ions are accelerated at the attached discontinuities into counter-streaming beams with close to the Alfvén speed and do not properly thermalize for vast extents of the outflow region – as recently observed in the solar wind (Gosling *et al.* 2005). Our new simulations show $\sim$ half of the available reconnection energy goes into heating, implying the ions to get energized by a factor of $\sim 0.5\,\beta$ (Fig. 1a). The final ion energy is that observed for flare electrons (Fig. 1b) and within *ms* a seed particle pool for MeV ions is generated. Linear theory for the ion beams predicts growth of bi-directional fast/magnetosonic waves to exceed that of e.s. A/IC waves (Fig. 1c). We suggest that these waves, i.e., the free energy contained in the beams causes the efficient coupling to the electrons via transit-time damping.

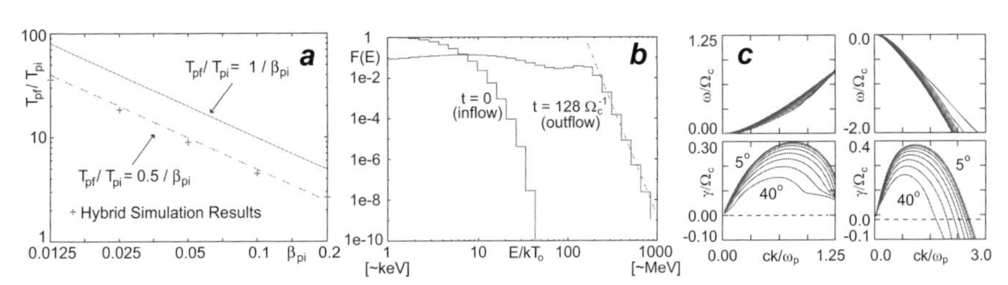

Figure 1. *(a)* Scaling with β, *(b)* ion energy, and *(c)* bi-directional F/MS waves.

This work was supported by NSF CISM and by NASA grant NNG04GH38G.

References

Krauss-Varban, D., & Omidi, N. 1995, *Geophys. Res. Lett.*, 22, 3271
Gosling, J.T., Skoug, R.M., McComas, D.J., & Smith, C.W. 2005, *J. Geophys. Res.*, 110, A01107

Highlights of Astronomy, Volume 14
IAU XXVI General Assembly, 14-25 August 2006
Karel A. van der Hucht, ed.

doi:10.1017/S1743921307009933

Physical conditions of the acceleration region of a solar flare with an unusually narrow gyrosynchrotron spectrum

Guillermo G. Giménez de Castro, Joaquim E. Rezende Costa and Adriana Válio Roque da Silva

Centro de Rádio Astronomia e Astrofísica Mackenzie
R. da Consolação 896, 01302-907, São Paulo, Brazil
email: guigue@craam.mackenzie.br

During the rising phase of the August 30, 2002 X1.5 flare a short pulse with a total duration of 8 seconds was observed. Its background-subtracted radio spectrum ranges only from 5 to 12 GHz with a maximum flux density of approximately 900 s.f.u. at 7 GHz and a steep optically thin spectral index $\alpha \simeq 8$. Maximum degree of polarization at 7 GHz is around 5%. The hard X-ray pulse emission above the background in the range of 30–150 keV observed by *RHESSI* is coincident in time with the microwave observation. Hard X-ray images reveal very compact ($\sim 10''$) footpoint sources. Below 30 keV, a thermal source is observed.

In order to explain the microwave spectrum we analyzed different alternatives. A thermal source would need extremely high temperature to produce the photons of 150 keV observed with *RHESSI*. Harmonic or maser emission generate spectra which are narrower than ours, and are strongly polarized. We used instead a distribution of accelerated electrons represented by a double power law, with $\delta_{E\ <\ 250\ \mathrm{keV}} = 5.3$ and $\delta_{E\ \geqslant\ 250\ \mathrm{keV}} = 13$, a 35 keV low energy cutoff, and computed the gyrosynchrotron and thick target *bremsstrahlung* fluxes of a homogeneous source. A thermal source with $EM = 10^{48}$ cm^{-3} and $T = 3.2\,10^7$ K was added to fit the X-ray spectrum below 35 keV.

From our fitting we determine a rate of $\dot{N} = 7.5\ 10^{34}\ \mathrm{s}^{-1}$ accelerated electrons. The power deposition of the nonthermal electrons is $5.5\ 10^{28} \leqslant Q \leqslant 5.5\ 10^{29}$ ergs s^{-1}, depending on the plasma density assumed: $10^{10} \leqslant n \leqslant 10^{11}$ cm^{-3}. The ratio $\dot{N}/Q$ can be related to the accelerating DC electric field. Using the relation established by Holman *et al.* (1989), we may deduce that $17 \leqslant E_d/E \leqslant 24$ here with E the electric field, and E_d the Dreicer field. The critical energy above which the electrons runaway can also be determined, yielding $24 \leqslant W_c \leqslant 33$ keV. Finally, the total number of current sheets is $7.5\ 10^4 \leqslant s \leqslant 10^5$.

This uncommon event, short in time and intense in flux, is explained by means of a beam of accelerated electrons with energies mainly between 35 and 250 keV, because the electron index above the break energy is so high that it acts as a high-energy cutoff. Assuming that a DC field accelerated the electrons, we concluded that the electric field should be sub-Dreicer. The critical energy obtained, around 30 keV, is compatible with the 35 keV low-energy cutoff of our model. The large number of current sheets needed to accelerate the electrons could be overestimated if the current sheet density is higher or if the resistivity is non classical.

Reference

Holman, G. D., Kundu, M. R., & Kane, S. R. 1989, *ApJ*, 345, 1050

Highlights of Astronomy, Volume 14
IAU XXVI General Assembly, 14-25 August 2006
Karel A. van der Hucht, ed.

doi:10.1017/S1743921307009945

Galactic acceleration phenomena

Yves A. Gallant

Laboratoire de Physique Théorique et Astroparticules, UMR 5027, CNRS/IN2P3,
Université Montpellier II, 34095 Montpellier Cedex 5, France
email: gallant@lpta.in2p3.fr

Abstract. I review the current status of our observational knowledge of prominent classes of particle accelerators in the Galaxy, namely shell-type supernova remnants (SNRs) and pulsar wind nebulae. I highlight in particular the contribution of the recent improvement in sensitivity of very-high-energy (VHE) γ-ray observations, which are currently the most direct probe of particle acceleration in the Galaxy up to energies of several hundreds of TeV.

Shell-type SNRs have long been proposed as sources of the Galactic cosmic rays. In recent years, X-ray observations have revealed very thin, non-thermal rims in many young SNRs, and I discuss the implications of these observations for magnetic field amplification and the maximum particle energy attainable by acceleration at the blast wave. I then review the current status of the evidence for accelerated nuclei in these objects, and summarise current uncertainties.

The most numerous class of identified Galactic VHE gamma-ray sources is currently that of pulsar wind nebulae (PWNe). The emission from these objects is generally assumed to be predominantly leptonic, and I outline the new information provided by VHE gamma-ray observations beyond what could be inferred from observations of synchrotron emission.

Keywords. acceleration of particles, cosmic rays, gamma rays: observations, supernova remnants, shock waves, magnetic fields, pulsars: general, radiation mechanisms: nonthermal

---

1. Galactic cosmic rays and VHE γ-rays

Galactic cosmic rays (GCRs) were known to fill the Galaxy from *EGRET* observations of diffuse high-energy γ-ray emission. The recent discovery of diffuse very-high-energy (VHE) γ-ray emission (with energy $E_\gamma > 100\,\mathrm{GeV}$) from the Galactic Centre ridge by the *HESS* experiment (Aharonian *et al.* 2006a) illustrates the potential of this higher energy regime to close in on the sources of the GCRs. The spectrum of this VHE emission, thought to be hadronic in nature, is harder than the generic GCR spectrum, betraying the presence of recently accelerated cosmic rays.

2. Shell-type supernova remnants

Shell-type supernova remnants (SNRs) are widely believed to be the sources of the GCRs, due to energetics and composition arguments, and because of observational evidence for particle acceleration at their outer blast waves. In particular, the non-thermal X-ray emission seen in most young SNRs is generally thought to be synchrotron emission from accelerated electrons. Recent X-ray observations have revealed this emission to originate in narrow rims in many young SNRs (see, e.g. Warren *et al.* 2005), which has been interpreted as evidence for turbulent magnetic field amplification at the shock. Such amplification would increase the maximum energy attainable by ions accelerated at this shock, but this still appears insufficient to reach the 'knee' of the GCR spectrum (Parizot *et al.* 2006). Indirect evidence for the efficient acceleration of ions is provided by SNR hydrodynamics, in particular the relative radii of the blast wave and contact discontinuity in some young SNRs (Decourchelle, Ellison & Ballet 2000).

VHE γ-ray observations have the potential to yield direct evidence for GCR acceleration in SNRs. Two shell-type SNRs have been detected and studied in great morphological and spectral detail by *HESS*, namely G 347.3−0.5 (or RX J1713.7−3946, Aharonian *et al.* 2006b) and G 266.2−1.2 (or RX J0852.0−4622). Both remnants emit dominantly non-thermal X-rays, with which the γ-ray emission shows a high degree of morphological correlation, and have relatively hard γ-ray spectra, with power-law indices $\Gamma \approx 2.25$. It is nonetheless still unclear whether this γ-ray emission is hadronic, from π^0 decay, or leptonic, via the inverse Compton (IC) process. The spectrum would tend to favour the former hypothesis, but the latter more naturally explains the morphological correspondence with X-rays.

At least four of the new VHE sources discovered in the *HESS* survey of the Galactic plane (Aharonian *et al.* 2006c) show convincing positional coincidences with SNRs, namely HESS J1640−465 (with G 338.3−0.0), J1713−381 (CTB 37B), J1834−087 (W41), and J1813−178 (G 12.8−0.0). In some of these SNRs, however, the VHE emission could originate in a yet unconfirmed central plerion (see below) rather than in the shell. More detailed, multi-wavelength studies should help clarify the nature of the VHE γ-ray emission in these SNRs.

3. Pulsar wind nebulae

In pulsar wind nebulae (PWNe), also known as plerionic SNRs, accelerated particles are thought to originate in the relativistic wind of a central pulsar. The observed nonthermal emission is usually assumed to be predominantly leptonic, consisting of synchrotron emission from radio waves to X-rays and beyond, and IC scattering in VHE γ-rays. As the target photon density is in general almost uniform on the scale of the PWN, VHE γ-rays then directly trace the spatial distribution of energetic electrons, unlike synchrotron emission which may also reflect magnetic field variations. The electron spectrum can also be directly inferred from the observed IC emission, provided the target photon density is known (Aharonian *et al.* 2006d; Porter, Moskalenko & Strong 2006).

VHE γ-ray emission has been detected by *HESS* from seven established PWNe: the Crab Nebula, MSH 15-5*2*, G 0.9 + 0.1, Vela X, two sources in the Kookaburra, and the nebula of PSR B1823−13. In this last case, γ-ray spectral steepening away from the pulsar supports the identification of HESS J1825−137 as the PWN (Aharonian *et al.* 2007). In several of these sources, the centroid of the PWN appears significantly offset from the pulsar, possibly due to the passage of an asymmetric reverse shock. Several other sources discovered in the *HESS* survey of the Galactic plane may also be associated with energetic pulsars (Aharonian *et al.* 2006c). VHE γ-rays are thus opening a new and promising observational window on particle acceleration in PWNe.

References

Aharonian, F., Akhperjanian, A. G., Bazer-Bachi, A. R., *et al.* 2006a, *Nature*, 439, 695
Aharonian, F., Akhperjanian, A. G., Bazer-Bachi, A. R., *et al.* 2006b, *A&A*, 449, 223
Aharonian, F., Akhperjanian, A. G., Bazer-Bachi, A. R., *et al.* 2006c, *ApJ*, 636, 777
Aharonian, F., Akhperjanian, A. G., Bazer-Bachi, A. R., *et al.* 2006d, *A&A* (Letters), 448, L43
Aharonian, F., Akhperjanian, A. G., Bazer-Bachi, A. R., *et al.* 2007, *A&A*, 661, 236
Decourchelle, A., Ellison, D. C., & Ballet, J. 2000, *ApJ* (Letters), 543, L57
Parizot, E., Marcowith, A., Ballet, J., & Gallant, Y. A. 2006, *A&A*, 453, 387
Porter, T. A., Moskalenko, I. V., & Strong, A. W. 2006, *ApJ* (Letters), 648, L29
Warren, J. S., Hughes, J. P, Badenes, C., Ghavamian, P., McKee, C. F., Moffett, D., Plucinsky, P. P., Rakowski, C., Reynoso, E., & Slane, P. 2005, *ApJ*, 634, 376

Highlights of Astronomy, Volume 14
IAU XXVI General Assembly, 14-25 August 2006
Karel A. van der Hucht, ed.

doi:10.1017/S1743921307009957

Particle acceleration via converter mechanism

Evgeny V. Derishev[1,2], Vitaly V. Kocharovsky[1,3], Vladimir V. Kocharovsky[1] and Felix A. Aharonian[2]

[1]Institute of Applied Physics RAS, Nizhny Novgorod, Russia
email: kochar@appl.sci-nnov.ru

[2]Max-Planck-Institut für Kernphysik, Heidelberg, Germany

[3]Texas A&M University, College Station, TX, USA

Abstract. We review the problem of particle acceleration in relativistic shocks or shear flows. We propose a converter mechanism, which operates via continuous conversion of accelerated particles from charged into neutral state and back, and show that it is capable of producing the highest energy cosmic rays.

Keywords. Relativistic jets and shock waves, acceleration of particles, high-energy cosmic rays

---

We compare different acceleration mechanisms and show that the converter mechanism, suggested recently (see Derishev *et al.* 2003), is the least sensitive to the geometry of the magnetic field in accelerators and can routinely operate up to cosmic-ray energies close to the fundamental limit. The converter mechanism utilizes multiple conversions of charged particles into neutral ones (protons to neutrons and electrons/positrons to photons) and back by means of photon-induced reactions or inelastic nucleon-nucleon collisions. It works most efficiently in relativistic shocks or shear flows under the conditions typical for Active Galactic Nuclei, Gamma-Ray Bursts, and microquasars, where it outperforms the standard diffusive shock acceleration. The main advantages of the converter mechanism in such environments are that it greatly diminishes particle losses downstream and avoids the reduction in the energy gain factor, which normally takes place due to highly collimated distribution of accelerated particles.

Paradoxically, interactions with photons, which have been always treated as dissipative processes leading to degradation of particle energy, in this scenario play a positive role: they allow (through the charge-changing particle conversion, e.g., via photopionic reactions) the maximum energy gain to remain up to the largest particle energies. Indeed, a particle in the neutral state can freely cross the magnetic field lines, which allows it to avoid both particle losses downstream and reduction in the energy gain factor, which normally takes place due to highly collimated distribution of accelerated particles.

We also analyze the properties of gamma-ray radiation, which accompanies acceleration of particles via the converter mechanism and can provide evidence for the latter. In particular, we point out the fact that the opening angle of the radiation beam-pattern is different at different photon energies, which is relevant to the observability of the cosmic-ray sources as well as to their timing properties. Various consequences for observations are discussed.

Reference

Derishev, E. V., Aharonian, F. A., Kocharovsky, V. V., & Kocharovsky, Vl. V. 2003, *Phys. Rev. D*, 2003, 68, 043003

Highlights of Astronomy, Volume 14
IAU XXVI General Assembly, 14-25 August 2006
Karel A. van der Hucht, ed.

doi:10.1017/S1743921307009969

Diffusive shock acceleration and radio emission from shell-type SNRs

A. I. Asvarov

Institute of Physics, Baku AZ1143, Azerbaijan; email: asvarov@physics.ab.az

Abstract. Diffusive shock acceleration (DSA) in test-particle approximation is used for explaining the radio emission from shell-type SNRs.

Keywords. supernova remnants, radiation mechanisms: nonthermal, acceleration of particles

---

Almost all known SNRs are sources of synchrotron radio emission. Naturally, much richer observational information on the SNRs is available in the radio range that can be used for explaining the origin of high energy electrons and magnetic fields responsible for the synchrotron emission. DSA is the most suitable mechanism for explaining the origin of radio-emitting electrons. But in the theory of DSA some principal aspects remain unclear. Modeling of the SNR radio evolution by using the most reliable and common theoretical results of DSA and comparison with observational data can be used for resolving these questions. A model of radio emission of shell-type SNRs using test-particle DSA has been developed in a recent paper by the author (Asvarov 2006). It is based on the assumptions that electrons are injected into the mechanism directly from the high energy tail of the downstream Maxwellian distribution function, and the magnetic field is compressed at the typical interstellar field shock .

The model predicts that the radio surface brightness ($\Sigma_{\rm R}$) evolves with diameter as $\propto D^{-(0.3\div0.5)}$ while the bounding shock wave is strong (Mach number $\mathcal{M} \geqslant 10$) followed by steep decrease (steeper than $\propto D^{-4.5}$) for $\mathcal{M} < 10$. This result is in good agreement with the conclusion of Berkhuijsen (1986) about the $\Sigma_{\rm R}$ -D relations. The dependence of the nearly constant $\Sigma_{\rm R}$ on the environmental and initial parameters is expressed as $\Sigma_{\rm R} \propto n_{\rm e0}^{2/3}\ B_0^{3/2}\ E_{\rm SN}^{1/2}\ M_{\rm ej}^{-1/6}$. The shape of the dependence of $\Sigma_{\rm R}$ on D and other initial and environmental parameters greatly reduce the usefulness of $\Sigma_{\rm R}$ - D relations as a tool for determining the distances to SNRs. Our model easily explains both very large diameter radio sources such as Galactic Loops and candidates for Hypernova radio remnants and small size radio sources as the remnant of Nova Persei 1901. The model predicts no radio emission from the radiative SNRs and the existence of radio quiet but relatively active SNRs is possible.

From the comparison of the model results with the statistics of evolved shell-type SNRs we were able to estimate the fraction of electrons accelerated from the thermal pool as in the range $(3 \div 11) \times 10^{-4}$. If acceleration takes place from a Maxwellian distribution function then the corresponding injection momentum is estimated as $p_{\rm inj} \simeq (2.7 \div 3) \cdot p_{\rm th}$. If the Galactic Loops are actually SNRs then from our models it follows that DSA still acts at very weak shocks (say, with $\mathcal{M} \leqslant 3$).

References

Asvarov, A. I. 2006, *A&A*, 459, 519
Berkhuijsen, E. M. 1986, *A&A* 166, 257

Highlights of Astronomy, Volume 14
IAU XXVI General Assembly, 14-25 August 2006
Karel A. van der Hucht, ed.

doi:10.1017/S1743921307009970

Extragalactic shocks as cosmic accelerators

Mikhail V. Medvedev

Department of Physics and Astronomy, University of Kansas, Lawrence, KS 66045, USA
email: medvedev@ku.edu

Abstract. It is quite well established that shocks accelerate particles via the Fermi mechanism. We discuss common features of various extragalactic sources, ranging from Gamma-Ray Bursts and jets of Active Galactic Nuclei to Large-Scale Structure shocks and address how they affect particle acceleration. In particular, we address constraints on the maximum energy of ultra-high-energy cosmic rays. Interestingly, some recent studies indicate that Fermi acceleration in relativistic shocks (and GRBs, in particular) faces severe difficulties. We will address this issue and demonstrate that the 'observed' shock acceleration of electrons may have nothing to do with Fermi acceleration, but may rather be associated with micro-physics of collisionless shocks.

Keywords. cosmic rays, acceleration of particles, shock waves, plasmas, gamma rays: bursts, quasars: general, galaxies: jets, galaxies: clusters: general

---

1. General constraints on electromagnetic CR acceleration

The maximum energy of an accelerated particle (in the absence of any losses) is determined from the confinement argument: the gyro-radius of the particle should not exceed the size of the system, R. This sets the maximum energy of the accelerated particle (Hillas 1984), $E_{\rm acc} = Ze\,B\,R \simeq 9.3\times10^{23}\,Z\,B\,R_{\rm kpc}$ eV where Ze is the charge of a particle, B is the characteristic magnetic field strength in the acceleration region (in gauss), and $R_{\rm kpc}$ is the size in kiloparsecs. Except for a special case (a rather unnatural one for cosmic ray (CR) sources, such as shocks, jets, supernova, etc.) of the accelerating electric field being parallel to the magnetic field, the confining magnetic field induces a radiation friction force. A particle of very large energy, $E\to\infty$, traversing a region of size R filled with B-field of strength $B(x)$, will leave the region with energy not exceeding $E_{\rm cr} = (3/2)(Am_pc^2/Ze)^4\left(\int_0^R B^2(x)\,dx\right)^{-1} \simeq 2.9\times10^{16}\,(A/Z)^4B^{-2}R_{\rm kpc}^{-1}$ eV where we assumed that $B\simeq$ constant. We now consider a general case, when a particle is accelerated by an electric field $E_{\rm ind}\simeq|\mathbf{v}\times\mathbf{B}|/c$ which, in the case of a relativistic system, is thus $\simeq B$. The energy evolution equation reads: $dE/dx \simeq Ze\,B - (2/3)\left(Ze/(Am_pc^2)\right)^4\,B^2\,E^2$. For a small initial energy of an accelerated particle, the solution of this equation takes a simple and elegant form (Medvedev 2003):

$$E = \sqrt{E_{\rm acc}E_{\rm cr}}\ \tanh\sqrt{E_{\rm acc}/E_{\rm cr}}, \tag{1.1}$$

where E is the terminal energy of the particle. This solution is plotted in Fig. 1, as a generalized Hillas plot (Medvedev 2003).

2. Non-Fermi acceleration in GRBs

Magnetic fields are generated at shocks by the Weibel instability. The fields are associated with current filaments made of protons, whereas the electrons, being much lighter than the protons, are quickly isotropized in the random fields and form a uniform background. Because of uncompensated charge of the filaments, they are also sources of

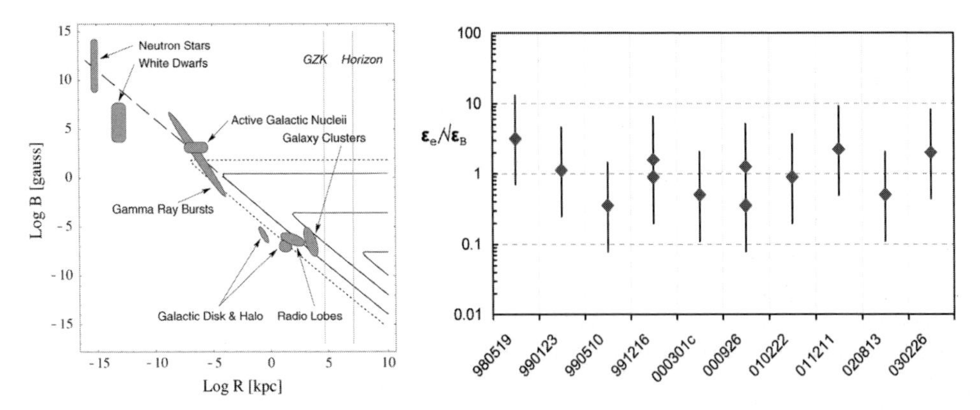

Figure 1. The modified Hillas diagram for UHECR sources. The long-dashed line is the original Hillas relation for a proton of energy 3×10^{20} eV. The dotted and solid lines represent the boundaries of the allowed parameter regions from the solution given by Eq. (1.1) for 3×10^{20} eV iron nuclei (*dotted line*) and for protons of energies 3×10^{20}, 10^{22}, and 3×10^{23} eV, respectively. Only those astronomical objects which fall inside the 'wedges' are, in principle, capable of accelerating the particles to such energies. The gray vertical lines mark two characteristic scales: the GZK attenuation distance (~ 20 Mpc) and the Hubble horizon size (~ 4 Gpc).

Figure 2. The $\epsilon_e/\sqrt{\epsilon_B}$ ratio for ten GRB afterglows analysed by Panaitescu 2005. The parameters of the best fit model (with the lowest χ^2/dof) are used. Clustering of data points around unity is evident.

electrostatic fields. The fields are related to each other as $B=\beta E$, where $\beta=\sqrt{1-\Gamma^{-1}}\simeq 1$. An electron, moving toward a filament gains energy $u_e\simeq elE\simeq elB$. The typical radial distance the electron travels is comparable to the filament size, which in turn, is of order the relativistic skin depth, $c/\omega_{pp,rel}$ (here $\omega_{pp,rel}=(4\pi e^2 n/m_p\gamma_p)^{1/2}$ is the relativistic proton plasma frequency) – the only characteristic scale of the system. Thus $l\simeq\lambda(c/\omega_{pp,rel})$, where $\lambda\simeq 1$ is the dimensionless parameter, which accounts for geometric effects, electrostatic shielding in plasmas, etc. All these effects can introduce a factor of few uncertainty. Finally, the electron energy density behind the shock front is $U_e=nu_e\simeq\lambda eBnc/\omega_{pp,rel}$. Once expressed in terms of the equipartition parameters $\epsilon_B=B^2/4\pi/(m_pc^2n\Gamma)$ and $\epsilon_e=U_e/(m_pc^2n\Gamma)$ measuring the amount of the total energy of the shock that goes into magnetic fields and the electron heating, the equation for U_e takes a simple and elegant form (Medvedev 2007):

$$\epsilon_e \simeq \lambda\sqrt{\epsilon_B}, \tag{2.1}$$

with $\lambda\sim 1$. The value of $\lambda=\epsilon_e/\sqrt{\epsilon_B}$ derived from spectral fits to GRB afterglows (Panaitescu 2005) is indeed consistent with unity, as our theory predicts, see Fig. 2.

Acknowledgements

The work is supported by grants DE-FG02-04ER54790 and NNG-04GM41G.

References

Hillas, A. M. 1984, *ARAA*, 22, 425
Medvedev, M. V. 2003, *PRE*, 67, 045401
Medvedev, M. V. 2007, *ApJ* (Letters), 662, L11
Panaitescu, A. 2005, *MNRAS*, 366, 1357

Highlights of Astronomy, Volume 14
IAU XXVI General Assembly, 14-25 August 2006
Karel A. van der Hucht, ed.

doi:10.1017/S1743921307009982

Stochastic re-acceleration in the ICM

Gianfranco Brunetti[1] and Alex Lazarian[2]

[1]INAF – Istituto di Radioastronomia, via P. Gobetti 101, I–40129, Bologna, Italy
[2]Astronomy Dep., Univ. of Wisconsin, 475 North Charter Street, Madison, WI 53706, USA

Abstract. Here we suggest that efficient stochastic particle re–acceleration in galaxy clusters may be driven by compressible modes. The damping of these modes is severely dominated by the TTD–resonance with thermal electrons and protons in the ICM. However, a small energy flux of these modes may be channelled into particle re-acceleration and this gives re-acceleration times of the order of $\sim 10^8$ yrs, sufficient to mantain GeV radiating electrons in the ICM.

Keywords. acceleration of particles, turbulence, galaxies: clusters: general

---

The Mpc-scale radio non-thermal emission detected in a growing number of galaxy clusters (GC) proves the presence of GeV radiating electrons (e.g., Feretti 2005). In addition, relativistic hadrons should accumulate in GC and direct measurements of the hadronic content may come from future γ-ray observations (e.g., Blasi 2004). The origin of the radio emitting electrons in GC is still a matter of debate – they may either be re-accelerated or secondary particles originating from hadronic collisions (e.g., Brunetti 2004; Sarazin 2004). Present observations seem to favour the first scenario, but the role of future radio and γ-ray observations remains crucial. Detailed calculations of particle-turbulent mode coupling in the framework of this scenario were restricted to the case of the Alfvén modes, since they channel the bulk of their energy flux into acceleration of relativistic particles (Brunetti *et al.* 2004; Brunetti & Blasi 2005). On the other hand, unless these modes are injected at small scales (e.g., Lazarian & Beresnyak 2006), they develop an anisotropic cascade and this makes the acceleration process less efficient (Yan & Lazarian 2004). An additional possibility comes from the action of compressible modes injected in GC during cluster-cluster mergers. Here turbulence is sub-sonic but strongly super-Alfvénic so that the bending of the magnetic field lines actually limits the particle mean free path. This implies that the dissipation of turbulence happens via collisionless dampings preferentially on thermal electrons and protons. However, fast modes and magnetosonic waves also couple with relativistic particles via TTD-resonance and non-resonant turbulent compression. In the case of hot GC, we find that the acceleration time is $\sim 10^8 (\frac{V_o}{c_s}/0.5)^3/(L_o/300\,\mathrm{kpc})$ yr, V_o being the rms velocity of magnetosonic modes at the max scale L_o, and c_s the sound speed. Thus, if the bulk of turbulence in GC is in the form of compressible modes, stochastic particle re-acceleration may maintain the GeV radiating particles.

References

Blasi, P. 2004, *J. Korean Astron. Soc.*, 37, 483
Brunetti, G. 2004, *J. Korean Astron. Soc.*, 37, 493
Brunetti, G., Blasi, P., Cassano, R., & Gabici, S. 2004, *MNRAS*, 350, 1174
Brunetti, G., & Blasi, P. 2005, *MNRAS*, 363, 1173
Feretti, L. 2005, *Adv. Sp. Res.*, 36, 729
Lazarian, A., & Beresnyak, A. 2006, *MNRAS*, 373, 1195
Sarazin, C. 2004, *J. Korean Astron. Soc.*, 37, 433
Yan, H., & Lazarian, A. 2004, *ApJ*, 614, 757

Highlights of Astronomy, Volume 14
IAU XXVI General Assembly, 14-25 August 2006
Karel A. van der Hucht, ed.

doi:10.1017/S1743921307009994

Jets of energetic particles generated by magnetic reconnection at a three-dimensional magnetic null

Silvia Dalla and Philippa K. Browning

School of Physics and Astronomy, University of Manchester,
PO Box 88, Manchester, M60 1QD, U.K.
email: s.dalla@manchester.ac.uk

Abstract. Magnetic reconnection is a candidate mechanism for particle acceleration in a variety of astrophysical contexts. It is now widely accepted that reconnection plays a key role in solar flares, and reconstructions of coronal magnetic fields indicate that three-dimensional (3D) magnetic null points can be present during flares. We investigate particle acceleration during spine reconnection at a 3D magnetic null point, using a test particle numerical code. We observe efficient particle acceleration and find that two energetic populations are produced: a trapped population of particles that remain in the vicinity of the null, and an escaping population, which leave the configuration in two symmetric jets along field lines near the spine. While the parameters used in our simulation aim to represent solar coronal plasma conditions of relevance for acceleration in flares, the fact that the 3D spine reconnection configuration naturally results in energetic particle jets may be of importance in other astrophysical situations. We also compare the results obtained for the spine reconnection regime with those for the other possible mode of 3D reconnection, fan reconnection. We find that in the latter case energetic particle jets are not produced, though acceleration is observed.

Keywords. acceleration of particles, Sun: flares

1. Methodology and results

We developed a numerical code that integrates the trajectories of a population of test particles in the magnetic and electric fields characteristic of 3D spine reconnection (Priest & Titov 1996). Particles were injected at random locations on a sphere centered in the magnetic null, with initial energy distribution given by a Maxwellian. The parameters used in the simulation, aiming to represent flare-like conditions in the solar corona, are described in Dalla & Browning (2006). Single particle trajectories in the configuration were previously studied by Dalla & Browning (2005).

We find that spine reconnection at a 3D null can efficiently accelerate particles, and produces two energetic populations: a trapped one and an escaping one. Escaping particles leave the magnetic configuration along the spine line, in two symmetric jets. The average energy of the escaping population is lower than that of the trapped one. The spectrum produced can be approximated by a power law of spectral index 0.92. The acceleration time scales of $\sim 60\,\mathrm{ms}$ obtained from our simulation, are broadly consistent with solar flare observations.

References

Dalla, S., & Browning, P. K. 2005, *A&A*, 436, 1103
Dalla, S., & Browning, P. K. 2006, *ApJ* (Letters), 640, L99
Priest, E. R., & Titov, V. S. 1996, *Phil. Trans. R. Soc. Lond. A*, 354, 2951

Highlights of Astronomy, Volume 14
IAU XXVI General Assembly, 14-25 August 2006
Karel A. van der Hucht, ed.

doi:10.1017/S1743921307010009

Particle acceleration theory

Roger D. Blandford

Kavli Institute for Particle Astrophysics and Cosmology,
Stanford University, Menlo Park, CA 94025, USA
email: rdb3@stanford.edu

Abstract. Astrophysical particle acceleration involves the efficient conversion of bulk energy to individual charge particle energy through work done by electric field. The ways in which this happens are quite varied but when considered from a physics perspective, commonalities can found between acceleration in quite different sites.

1. Direct acceleration

The energy gained by a high energy particle is $-q\int d\vec{r}\cdot\vec{E}$. Sometimes,these fields are created through a charge deficiency which is neutralized through breakdown of a unionized region or pair production from the vacuum, where there is a large scale electrostatic field as in a pulsar 'gap' or a double layer in an aurora. The potential difference is then given by that required for breakdown. This can be as large as $\sim 1\,\mathrm{TV}$ in pulsars. Some forms of magnetic reconnection create large current density which induce a large local 'anomalous' resistance and consequently a large potential difference in the background thermal plasma which can be traversed, collisionlessly, by high energy particles.

One generic way to create a potential difference is through unipolar induction. Here a rotating magnetized conductor creates an induced EMF, V, of order the product of the magnetic flux and the angular frequency. In the case of the Crab pulsar, $V\simeq 30\,\mathrm{PV}$ and the associated power is of order $V^2/Z_0\simeq 10^{31}\,\mathrm{W}$, where Z_0 is the impedance of free space. The induced currents may close near the neutron star and drive a pair dominated wind. Alternatively, they may close within the nebula so that the energy transport is primarily electromagnetic and the observed X-ray jets and equatorial emission can be interpreted as ohmic dissipation. Massive spinning black holes in active galactic nuclei also act as conductors and induce ZV potentials adequate to accelerate UHE cosmic rays provided that the currents close well outside the galactic nucleus. (The jets are also interpreted as ohmic dissipation.)

In source where the electromagnetic energy dominates, it is instructive to use relativistic force-free electrodynamics which is the limit of relativistic MHD when the inertia of the plasma can be ignored. Maxwell's equations are supplemented by the constitutive relation, $\rho\vec{E}+\vec{j}\times\vec{B}=0$. These lead to an evolutionary set of equations that appear to allow the Lorentz invariant E^2-B^2 to become positive. The conditions that are necessary for this to happen are not understood. However, if it does happen, the inertia of the plasma particles must be included. and there should be a catastrophic acceleration of electrons and positrons at a rate probably limited by radiation reaction. The end result is likely to be a prodigious burst of gamma rays. Another way in which this can happen less abruptly is if there is a cascade of electromagnetic energy down to smaller length scales so that the spectral energy density of shear Alfvén modes $\mathcal{E}_k$ falls off with k slower than k^{-3}. A microscale will then develop where there is insufficient plasma to compensate the divergence of the electric field and there will volumetric dissipation of the electromagnetic energy. These mechanisms may be relevant to pulsars, gamma ray bursts and blazars.

2. Stochastic acceleration

There are many circumstances where particle acceleration is likely to be stochastic and independent of the history, as in the traditional Fermi approach. The scattering is often associated with wave modes that produce a diffusion in momentum space. The creation and evolution of the wave spectrum must be considered alongside the Fokker-Planck evolution of the particles. One example that has been discussed recently is particle acceleration by sound waves created by radio sources in clusters of galaxies. Another example may be provided by a force-free electromagnetic wave spectrum as discussed above.

3. Non-Markovian acceleration

There is a third, hybrid, possibility where the steps are small but non-Markovian. A good example is provided by diffusive shock acceleration where, in the simplest description, the energy gains in the the de Hoffmann-Teller frame is due to the electric field associated with magnetic fluctuations that scatter in pitch angle. (There is an impediment to scattering through 90°, where additional acceleration may occur.) The non-Markovian character arises because the scattering waves are likely to have been created by the diffusing particles. The simple theory of diffusive shock acceleration can account for the inferred injection spectrum of primary cosmic rays. However, it does not easily explain injection, shock mediation, the level of scattering, and the observation that cosmic ray protons appear to be accelerated up to almost PeV energies – the so-called knee in the cosmic ray spectrum. (Heavier particles are accelerated to even higher energy.) This requires that magnetic field strengths are increased to values far higher than the microgauss strengths associated with the ambient interstellar medium. Indeed, there is accumulating evidence that magnetic field is strongly amplified at shock fronts. These observations suggest a new model of diffusive shock acceleration.

Consider a shock that accelerates particles up to some maximum energy. These particles are likely to have the largest diffusion coefficient (proportional to the energy under the Bohm assumption). They will therefore stream furthest ahead of the shock front and will make the first contact with the undisturbed interstellar medium. If these particles are still being scattered at this point, then they will return to the shock for further acceleration; if they are streaming forward faster than the shock then they will escape. It is conjectured that the highest energy streaming particles have a pressure along the magnetic field in excess of the magnetic pressure and create a relativistic firehose instability which quickly grows to nonlinear strength as it is convected towards the shock front. The fluctuations in the field will be on length scales in excess of the gyro radii of these highest energy particles. As these large amplitude magnetic fluctuations approach the shock front, they encounter progressively lower energy particles which they scatter through the generation of Alfven modes in an effectively uniform local field. If this 'magnetic bootstrap' can actually occur, the maximum field strength that can actually be generated ahead of a shock will be fixed by the cosmic ray pressure, a fraction of the rhermal pressure downstream. This leads to near milligauss fields at young supernova remnants and the acceleration of near PeV protons as observed.

Overall, the prospects for developing a general physical description of particle acceleration in astrophysical and space-physical sites is good. Numerical simulations are transforming our understanding of the plasma physics and high energy density experiments are likely to have a large impact on the field. Observationally, the biggest new impact is likely to come from *GLAST*.

Highlights of Astronomy, Volume 14
IAU XXVI General Assembly, 14-25 August 2006
Karel A. van der Hucht, ed.

doi:10.1017/S1743921307010010

Kinetic approaches to non-linear particle acceleration at shock fronts

Stefano Gabici[1], Elena Amato[2], Pasquale Blasi[2] and Giulia Vannoni[1]

[1]Max-Planck-Institut für Kernphysik, Heidelberg, Germany

[2]INAF/Osservatorio Astrofisico di Arcetri, Firenze, Italy

Abstract. We review some recent progresses in semi–analytic kinetic approaches to the problem of non linear particle acceleration at non relativistic shock waves.

Keywords. acceleration of particles

Diffusive shock acceleration is thought to be responsible for acceleration of cosmic rays in several astrophysical environments. Despite the success of this theory, some issues are still subjects of much debate, for the theoretical and phenomenological implications that they may have. One of the most important of these is the reaction of the accelerated particles on the shock: the violation of the *test particle approximation* occurs when the acceleration process becomes sufficiently efficient that the pressure of the accelerated particles is comparable with the incoming gas kinetic pressure. Both the spectrum of the particles and the structure of the shock are changed by this phenomenon, which is therefore intrinsically non-linear (Malkov & Drury 2001). Non-linear effects in shock acceleration result in the appearance of multiple solutions in certain regions of the parameter space. This phenomenon is very general and was found in both the two-fluid and kinetic models. Using a semi-analytic model developed by Blasi (2002), we showed that the appearance of multiple solutions is dramatically reduced if a self consistent model for injection, so called *thermal leakage*, is adopted (Blasi *et al.* 2005). The model has been further developed in Amato & Blasi (2005), Amato & Blasi (2006), and Amato & Blasi (2006). We find that the phenomenology of particle acceleration at modified shocks is characterized by three main features:

1) The modification of the shock increases with the Mach number of the fluid. For low Mach numbers the quasi-linear solution is recovered, but departures from it are evident already at relatively low Mach numbers. The modification of the spectra manifests itself with a hardening at high momenta and a softening at low momenta. The $p^4 f_0(p)$ shows a characteristic dip at intermediate momenta, typically around $p/mc \simeq 1-100$.

2) The total efficiency for particle acceleration saturates at large Mach numbers at a number of order unity. However, the largest fraction of the energy is not advected downstream but rather escapes from upstream infinity at the maximum momentum.

3) The high efficiency for particle acceleration reflects in a reduced role of cosmic ray modified shocks in the heating of the background plasma.

References

Malkov, M. A., & Drury, L. O'C. 2001, *Rep. Prog. Phys.*, 64, 429
Blasi, P. 2002, *Astropart. Phys.*, 16, 429
Blasi, P., Gabici, S., & Vannoni, G. 2005, *MNRAS*, 361, 907
Amato, E., & Blasi, P. 2005, *MNRAS*, 364, L76
Amato, E., & Blasi, P. 2006, *MNRAS*, 371, 1251
Amato, E., Blasi, P., & Gabici, S. 2007, arXiv:0705.3723v1 [astro-ph]

Highlights of Astronomy, Volume 14
IAU XXVI General Assembly, 14-25 August 2006
Karel A. van der Hucht, ed.

doi:10.1017/S1743921307010022

Particle acceleration by relativistic expansion of magnetic arcades

Hiroyuki Takahashi[1], **Eiji Asano**[1] **and Ryoji Matsumoto**[2]

[1]Graduate School of Science and Technology, Chiba University,
1-33 Yayoi-cho, Inage-ku, Chiba 263-8522, Japan

[2]Department of Physics, Faculty of Science, Chiba University,
1-33 Yayoi-cho, Inage-ku, Chiba 263-8522, Japan
email: takahasi,asano,matumoto@astro.s.chiba-u.ac.jp

Abstract. We carried out relativistic force free simulations and Particle In Cell (PIC) simulations of twist injection into the magnetic arcades emerging on the surface of a magnetar. As the magnetic energy is accumulated in the arcades, they expand self-similarly. In the arcades, a current sheet is formed and magnetic reconnection takes place. We also carried out 2-dimensional PIC simulations for the study of particle acceleration through magnetic reconnection. As a result, the energy spectrum of particles can be fitted by a power-law.

Keywords. acceleration of particles, magnetic field, plasmas, flare, neutron star

---

As a model of bursts of Soft Gamma-ray Repeaters (SGRs), Thompson & Duncan (1995) proposed magnetic flares on the surface of a magnetar (see also Lyutikov 2006). To study the dynamical evolution of magnetic arcades, we carried out axisymmetric relativistic force free simulations. At the footpoints of a force-free magnetic arcade, shear motion is imposed with speed 10% of the light speed. After twist injection, fast waves propagate isotropically. Alfven waves propagating along the magnetic field lines transport magnetic energy from the the footpoints. As the magnetic energy is accumulated, they expand self-similarly. The Lorentz factor is about $3 \sim 4$ but locally exceeds 10. Anti-parallel magnetic field, and thus current sheets, are created inside the arcade. Magnetic reconnection can take place in this region. In front of the arcades, a fast shock is formed and particles can be accelerated by shock acceleration.

Next, we concentrated on the the reconnection region and carried out 2-dimensional relativistic PIC simulations to study particle acceleration. We assumed an electron-positron plasma. We also assumed that the plasma is collisionless and the plasma density is larger than the Goldreich-Julian density. By the dissipation of magnetic energy, particles are accelerated up to almost the potential energy difference given by shear motions. The energy spectrum of particles can be fitted by a power-law.

Acknowledgements

We would like to thank Dr. M. Hoshino and S. Shibata for useful discussions.

References

Lyutikov, M. 2006, *MNRAS*, 267, 1594
Thompson, C., & Duncan, R. C. 1995, *MNRAS*, 275, 300

Highlights of Astronomy, Volume 14
IAU XXVI General Assembly, 14-25 August 2006
Karel A. van der Hucht, ed.

doi:10.1017/S1743921307010034

Cosmic particle acceleration in astrophysical shear flows

Frank M. Rieger and Peter Duffy

UCD School of Mathematical Sciences, University College Dublin, Dublin 4, Ireland
email: frank.rieger,peter.duffy@ucd.ie

Abstract. Shear flows are ubiquitous phenomena in astrophysical environments. In the present contribution the stochastic acceleration of energetic charged particles by Fermi-type processes in relativistic shear flows is considered. We briefly summarize recent theoretical progress in the field of viscous shear acceleration and indicate its significance for particle energization in relativistic flows of AGNs and GRBs.

Keywords. acceleration of particles, galaxies: active, gamma rays: bursts

---

Relativistic shear flows are a natural outcome of density and velocity gradients in the extreme astrophysical environments around AGNs and GRBs and, at least in the case of AGNs, observationally well established (see, e.g., Rieger & Duffy 2004 for review). We have shown elsewhere (see references) that scattering off magnetic turbulence structures embedded in such (collisionless) flows can lead to efficient Fermi-type particle acceleration and thus allow conversion of a non-negligible part of the bulk kinetic energy of the flow into nonthermal particles and radiation. For a momentum-dependent particle mean free path $\lambda(p) \propto p^{\alpha}$, acceleration of energetic particles in non-relativistic gradual shear flows is known to lead to (local) steady-state power-law particle distributions $n(p) \propto p^{-(1+\alpha)}$ above injection p_0 for $\alpha > 0$. Hence, for a particle mean free path $\lambda \propto$ gyro-radius $r_g \propto p$, one has $n(p) \propto p^{-2}$. As particles sample a higher shear with increasing λ, i.e., experience a higher scattering impact, the characteristic acceleration timescale is inversely related to λ, i.e., one finds $t_{\rm acc} \propto 1/\lambda$ in marked contrast to shock-type acceleration processes (Rieger & Duffy 2006). In particular, for $\lambda \simeq r_g$ (Bohm case), acceleration $t_{\rm acc}$ and radiative (synchrotron) cooling timescale $t_{\rm cool}$ have the same scaling with γ, so that radiative losses are no longer able to balance acceleration once it has started to work efficiently. The results can be generalized to the relativistic case with a similar outcome (e.g., Rieger & Duffy 2004). As particle energization appears possible as long as the shear persists, shear acceleration represents a natural candidate for distributed acceleration processes thought to be at work in large-scale relativistic AGN jets beyond the usual $2^{\rm nd}$ order Fermi mechanism (Rieger *et al.* 2007). Moreover, efficient shear acceleration of protons up to UHE cosmic ray energies appears possible within the jets of AGNs and GRBs, in the latter case probably more efficiently than via internal shock-type processes (Rieger & Duffy 2005). Finally, in μQuasar jets, shear acceleration may allow electron energies to reach those required to produce inverse Compton TeV emission (Rieger *et al.* 2007).

References

Rieger, F. M., & Duffy, P. 2004, *ApJ*, 617, 155
Rieger, F. M., & Duffy, P. 2005, *ApJ* (Letters), 632, L21
Rieger, F. M., & Duffy, P. 2006, *ApJ*, 652, 1044
Rieger, F. M., Bosch-Ramon, V., & Duffy, P. 2007, *Ap&SS*, 309, 119

Highlights of Astronomy, Volume 14
IAU XXVI General Assembly, 14-25 August 2006
Karel A. van der Hucht, ed.

doi:10.1017/S1743921307010046

Summary of Joint Discussion 1

Loukas Vlahos

Department of Physics, University of Thessaloniki, Thessaloniki, GR-54124, Greece
email: vlahos@astro.auth.gr

Abstract. We review the main ideas discussed during the meeting and propose methods for a new generation of space accelerators

Keywords. particle acceleration

Overall theories of particle acceleration were divided into three broad classes: (1) DC E-fields (double layers, thin current sheets), (2) stochastic acceleration in turbulent electromagnetic and electrostatic fields (Fermi acceleration or wave-particle interaction), and (3) shock waves. Blandford called these three processes (1) Direct or coherent, (2) stochastic (Markovian), and (3) systematic (Non-Markovian), while Melrose argued that ultimately all must involve induced rather than electrostatic E-fields. As we will show, all acceleration process discussed up to now fall into one of these classes. Let us call these three fundamental mechanisms, cosmic accelerators (CA).

The next important aspect in analysis of cosmic particle acceleration is the relation of the CA(s) to the host environment in which they operate, e.g., solar active regions host transient explosions, like Coronal Mass Ejections (CMEs) and flares); shell-type SNR host spherical shocks; extragalactic jets and AGNs host turbulent relativistic flows and shocks; and clusters of galaxies host MHD turbulence.

Finally we have to adopt an analytic or a computational method for analysis of the interaction of the CA with the particles. It is well known (e.g., Melrose) that at least two spatio-temporal scales are involved : (1) that of the macroscopic evolution of the environment hosting the CA; and (2) that of the microscopic interaction of particles with the CA. We usually try to split the analysis of the problem in two steps: (1) use the MHD equations to derive the electric and magnetic fields structure/spectrum of MHD waves, shock(s,) or current sheets; and (2) use these 'slowly' changing fields to study the fast evolution of particles by solving the diffusion equation, or following numerically test particles, in prescribed fields. Both methods lack self-consistency and, since efficient acceleration means fast transfer of energy from the CA to the particles, this approach will eventually break down.

During the meeting we had a series of interesting presentations on the topics mentioned above. Lin presented a full account of energetic particles in the heliosphere and touched on many unsolved issues. He stressed that the Earth is a strong γ-ray source, referring to terrestrial γ-ray flashes which are closely correlated with lightning strokes. The acceleration of particles is extremely fast and reaches very high energies (a few MeV). Is this a clear example of Direct acceleration? Other places where particles are efficiently accelerated are : (1) terrestrial radiation belts; (2) solar flares and CMEs; and (3) planetary magnetospheres/tails, and interplanetary space. A key point in Lin's talk was that heliospheric physics still has many unsolved problems so cannot be seen as the paradigm for the rest of CAs. These problem areas includes shock wave acceleration which is so popular in astrophysics. Kahler reported a search for evidence of non-relativistic and relativistic particles during CMEs, hoping to discover whether these particles are accelerated in the vicinity of an associated flare or are clearly associated with the CME shock. He had a hard time deciding from the available data the relative contributions of these two accelerators. The best he can say at the moment is that relativistic SEP particles seem to show better correlation with the CME than do the non-relativistic particles. In my view, the truth of the matter is that shock acceleration has not been unanimously accepted as an efficient accelerator for very high energy particles in the heliosphere.

Benz *et al.* analyzed the hard X-ray spectral evolution of coronal sources and tried to explain the often observed Soft-Hard-Soft spectral behavior of the radiation in the course of each Hard X-ray peak. They invoked the turbulence excited by reconnection as their CA mechanism and found that they cannot model the observations without additional transport effects (return current, high wave density, etc). The main issue for turbulent acceleration of particles during solar flares is the fact that the correlation of magnetic reconnection with the MHD waves assumed in this study remains conjectural.

Magnetic reconnection was analyzed in three separate contributions presented by Kraus-Verban and Welsch, by Bárta and Karlický, and by Dalla and Browning. Using a hybrid code (fluid electrons and kinetic ions) Kraus-Verban and Welsch proved that half of the available magnetic energy in low-β Petschek reconnection will go to ion heating. Within msecs a beam of MeV ions is created which can drive fast magnetosonic waves and, as Benz et all suggested, these waves may accelerate the particles. All these multi-level mechanisms (reconnection-MHD waves-particles) are of low efficiency. Observations, on the other hand, suggest very efficient transfer of magnetic energy to energetic particles (Lin) during solar flares.

Bárta and Karlický, using a 2.5D MHD code, created a flaring environment and then followed test particles to study its influence on the particles. They studied the combined action of pinch effects (betatron action) on electrons trapped in non-equilibrium plasmoids and collapsing magnetic traps. Dalla and Browning set up a 3D magnetic structure (magnetic null) which is known to host a magnetic reconnection environment, and followed test particles to study the efficiency of energy dissipation. The test particles were injected in random places inside the 3D structure. In 60 msecs the initial Maxwellian distribution has absorbed a considerable fraction of the available magnetic energy in the simulation box and developed a tail with spectral index -1. All such test particle codes are again lacking in self consistency in the sense of particle feedback on the CA mechanism. For efficient particle accelerators, as data indicate for e.g. magnetic reconnection, this is not a fully adequate method for analysis. A full 3D particle code is needed.

Medvedev presented extensive evidence suggesting that many well known applications of Fermi acceleration in relativistic shocks are problematic. His suggestion is that shock acceleration of electrons may be the effect of micro-physics on collisionless processes. 3D kinetic aspects of shock acceleration are currently a very active research topic (Gabici). However, shock acceleration is still the main candidate for acceleration of Galactic cosmic rays in shell-type SNRs. Gallant presented evidence for particle acceleration in blast waves (see also Asvarov).

Stochastic acceleration in shear flows was proposed by Rieger and Duffy as the acceleration mechanism in AGNs and GRBs. Brunetti and Lazarian used stochastic acceleration to explain the non-thermal emission in galaxy clusters. Obviously in both studies the MHD waves are put there by hand, rather than arising from basic physics, but the results agree with the observations using reasonable spectra and amplitudes for the MHD waves.

Based on these many fascinating contributons, I would like to close the summary with a few personal thoughts.

(*a*) So far 'turbulence' has meant 'particle scattering by a spectrum of **low amplitude** waves'. However 'Strong MHD turbulence' is a much wider ranging topic which may unite many of the CAs mentioned above. Increasing the amplitude of MHD waves to reach $\delta B/B \sim 1$ leads shocks and magnetic reconnection seemingly to co-exist and provide a new meaning for the concept of stochastic acceleration.

(*b*) The sharp separation of the three classes of CA, presented at the beginning of this summary, loses its meaning inside a major cosmic explosion. We may start with a large 3D current sheet which soon collapses to a "turbulent structure with many short lived current sheets, flows and shocks" or a large scale 3D shock may lose its character inside a strong turbulent flow. The mixing of acceleration mechanisms is a new concept and has not yet been discussed in depth.

(*c*) So far MHD codes have been used extensively for providing the environment in which kinetic aspects of particle accretion were investigated. This approach is possible when the non-linear structures are isolated in a single shock (CME, Shell-type SNRs) or one large scale 3D current sheet and stable for a long time. Our studies so far show that, inside the non linear structures (current sheets, shocks), acceleration is extremely efficient and requires a fully kinetic treatment. The problem becomes almost impossible to handle when the environment follows the ideal MHD equations and, at specific isolated short lived points, randomly appearing inside the large scale structure, the energy is dissipated kinetically. The communication between the large scale MHD code and short lived nonlinear dissipation structures remain a major unsolved problem.

Acknowledgements

I would like to thank John C. Brown for his help and suggestions in preparing this summary.

Highlights of Astronomy, Volume 14
IAU XXVI General Assembly, 14-25 August 2006
Karel A. van der Hucht, ed.

doi:10.1017/S1743921307010058

Posters presented at Joint Discussion 1

Costa, J.E.R., & Simoes, P.J.A.
Possible thermal emission by relativistic electrons at submillimeter range.

Dasso, S., Matthaeus, W.H., Smith, C.W., & Milano, L.J.
Observations of solar wind fluctuations: anisotropy properties.

Dieckmann, M.E., Drury, L.O.C., & Shukla, P.K.
On the ultrarelativistic two-stream instability, electrostatic turbulence and Brownian motion.

Dieckmann, M.E., Shukla, P.K., Eliasson, B., Sircombe, N.J., & Dendy, R.O.
Phase speed of electrostatic waves: The critical parameter for efficient electron surfing acceleration.

Dorman, L.I.
Great solar energetic particle events: monitoring and forecasting of radiation hazard by using on-line one-min neutron monitor data.

Elyiv, A., Petruk, O., & Hnatyk, B.
Spectrum of turbulence of the extragalactic magnetic field and IRAS PSCz catalogue.

Gabici, S., & Blasi, P.
Non-linear shock acceleration and high-energy γ-rays from clusters of galaxies and filaments.

Gan, W.Q., & Li, Y.P.
Acceleration region revealed from *RHESSI* X-ray observations.

Giovannini, G., Giroletti, M., & Taylor, G.B.
The superluminal giant radio galaxy 1144+35.

Gopalswamy, N., Yashiro, S., Kaiser, M.L., & Bougeret, J.-L.
Solar energetic particles and CME-driven shocks.

Hashimoto, T.H., Iye, M.I., & Aoki, K.A.
Investigation of ionization mechanism of extended narrow line region.

Jones, T.W., & Kang, H.
Time evolution of cosmic-ray modified MHD shocks

Kang, H., & Ryu, D.
Particle acceleration at cosmological shock waves.

Karlický, M., & Bárta, M.
Collapsing magnetic trap as accelerator of electrons in solar flares.

Kashapova, L.K., Zharkova, V.V., Chornogo, S.N., & Kotrč, P.
Signatures of high energy particle beams in the chromospheric events before the 25 July 2004 flare onset.

Kašparová, J., Kontar, E.P., & Brown, J.C.
RHESSI survey of photospheric albedo and directivity of solar flare hard X-ray spectra.

Kellogg, E.M., Anderson, C., DePasquale, J., Korreck, K., Nichols, J., Pedelty, J., & Sokoloski, J.
Hot gas at 10^7 K, Fe K-α reflection, ongoing non-thermal outbursts, jet formation and 10^6 K shocks in the symbiotic binary R Aqr.

Kotrč, P., & Kashapova, L.K.
On possible manifestation of non-thermal electrons in the $H\alpha/H\beta$ line profile ratio in the June 26, 1999 flare.

Kryvdyk, V.
Acceleration of cosmic rays by stellar collapse.

Kurtanidze, O.M., Fan, J., & Nikolashvili, M.G.
Photometric observations of γ-ray loud blazars.

Kurtanidze, O.M., & Kapanadze, B.Z.
Optical variability of X-ray selected blazars.

Lee, T.-L., & Hwang, C.-Y.
A new character to distinguish different types of galaxy clusters.

Lopez, E.D.
The Faraday rotation effect in quasar jets.

Marcowith, A.
Turbulence in collisionless shocks.

Mel'nik, V.N., & Rutkevych, B.P.
Propagation of relativistic electron beams in the plasma of the solar corona.

Nakamura, K., & Shigeyama, T.
Type Ic supernovae as sources of cosmic rays.

Peterson, B.A., & Schmidt, B.
Time resolved spectra of the $z = 3.7$ GRB 060605 from $+37.1\,\text{min} < t < +93.1\,\text{min}$ covering 320 nm to 950 nm

Petri, J.A., & Kirk, J.G.
Polarisation of high-energy emission in a pulsar striped wind.

Silva, A.V.R., Costa, J.E.R., Castro, C.G.G., Raulin, J.P., Kaufmann, P., Share, G.H., & Murphy, R.J.
The increasing submillimeter spectral component of intense solar flares.

Štěpán, J., Heinzel, P., Kašparová, J., & Sahal-Bréchot, S.
Polarization diagnostics of proton beams in solar flares.

Stepanov, A.V., & Tsap, Yu.T.
Wave-particle interaction and peculiarities of propagation and emission of accelerated particles in solar flares.

Tang, Y., Li, Ch., & Dai, Y.
The direct electric field acceleration in reconnection current sheet in three GLE events.

Topinka, M., & Karlický, M.
Simulation of magnetic field dissipation in astrophysical jets.

ud-Doula, A., Townsend, R.H.D., & Owocki, S.P.
Centrifugal breakout of magnetically confined line-driven stellar winds.

Vandas, M., & Karlický, M.
Electron acceleration in a wavy shock front.

Varady, M., Karlický, M., & Kašparová, J.
Return current and the energy deposit in flares.

Zharkova, V.V.
Hard X-ray spectral indices in solar flares as probes of the variations of magnetic field topology during a solar cycle.

Highlights of Astronomy, Volume 14
IAU XXVI General Assembly, 14-25 August 2006
Karel A. van der Hucht, ed.

doi:10.1017/S174392130701006X

Joint Discussion 2
On the present and future of pulsar astronomy

Werner Becker[1], Janusz A. Gil[2] and Bronislaw Rudak[3] (eds.)

[1]Max-Planck-Institut für extraterrestrische Physik,
Giessenbachstrasse 1, D-85741 Garching, Germany
email: web@mpe.mpg.de

[2]Johannes Kepler Institute of Astronomy, University of Zielona Gora, Poland
email: jag@astro.ia.uz.zgora.pl

[3]Centrum Astronomiczne im. M. Kopernica, Rabianska 8, 87-100 Torun, Poland
email: bronek@ncac.torun.pl

1. Introduction

Neutron stars are formed in supernova explosions. They manifest themselves in many different ways, for example, as pulsars, anomalous X-ray pulsars (AXPs) and soft γ-ray repeaters (SGRs) and the so-called 'radio-quiet neutron stars'. These objects are made visible by high-energy processes occurring on their surface or in the surrounding region. In most of these objects, ultra-strong magnetic fields are a crucial element in the radio, optical, X-ray and gamma-ray emission processes which dominate the observed spectrum.

The physics of pulsars spans a wide range of disciplines, from nuclear and condensed matter physics of very dense matter in neutron star interiors, to plasma physics and electrodynamics of the magnetospheres, to relativistic magnetohydrodynamics of electron-positron pulsar winds interacting with the surrounding ambient medium. Not to forget the test bed they provide for general relativity theories as well as being sources of gravitational waves.

Observationally, pulsar research is advancing rapidly. A great array of space instruments (the *Hubble Space Telescope*, *ROSAT*, *ASCA*, *BeppoSAX*, *RXTE* and the *Compton Gamma-Ray Observatory*), launched in the last decade of the previous century, have opened new windows in pulsar research with high quality data in energy bands from optical to the gamma-rays. With the more recently launched satellite X-ray observatories *Chandra* and *XMM-Newton*, upgraded radio observatories and ground based optical telescopes like Keck, GEMINI, Subaru and the VLT lots of the questions remained unanswered from the previous generation of observatories could be addressed and resulted in new and exciting results which changed the previous picture of neutron star evolution substantially (e.g., making Crab-like pulsars the exception not the rule for the appearance of a young neutron star). The X-ray Observatory *SUZAKU* (formerly *ASTRO-E2*) launched in 2005, and the γ-ray observatories *AGILE* and *GLAST*, which are supposed to be launched end of 2007 or beginning of 2008, will complement the pulsar studies and will enlarge the class of pulsars detected at their energy bands.

However, even in view of these great observational capabilities, the physical processes responsible for the pulsars' broad-band emission, observed from the infrared to the γ-ray band are still poorly known. Although no uniquely accepted theory exists so far, a notable progress was made very recently in this respect. New developments include caustic slot gaps as well as modified outer gaps. Last but not least: polarization-dependent treatment of high-energy radiative processes is under way, since X-ray and γ-ray polarimetry – a new powerful tool to discriminate between competing models – is around the corner.

To face recent observational results obtained in multi-wavelength studies from neutron stars and pulsars with the various theoretical models and to discuss on future perspectives on neutron star astronomy we organized this Joined Discussion (JD02) during the IAU XXVI General Assembly which took place in 2006 August in Prague. More than 150 scientists took actively part in this Joint Discussion. Fourteen invited review talks were presented to view the present and future of pulsar astronomy. Fifty three poster contributions displayed new and exciting results. PDF-files of all review talks and most of the posters displayed during the meeting are available on the meeting website < http://www.mpe.mpg.de/IAU_JD02 >.

The following sections give an overview of the invited review talks and contributed posters. The review talks are subject of review articles which will be published elsewhere. More information on this will be available at the URL given above.

2. Invited Review papers

2.1. *Radio emission properties of pulsars*

Richard N. Manchester: Currently, 1765 pulsars are known. 170 of them are millisecond pulsars. 131 pulsars are in binary systems. 129 pulsars are detected in 24 globular clusters. Recent observational results on the radio emission properties of pulsars were reviewed.

2.2. *New results on rotating radio transients*

Maura A. McLaughlin: A new population of radio-bursting neutron stars discovered in a large scale search for transient radio sources was discussed. Unlike normal radio pulsars, these objects, which are called Rotating RAdio Transient (or RRATs), cannot be detected through their time-averaged emission and are radio sources for typically less than 1 second per day. The spin periods of these objects range from 0.4 to 7 seconds, with period derivatives indicating that at least one RRAT has a magnetar-strength magnetic field. Recent developments were detailed, including X-ray observations and observations with more sensitive radio telescopes, and it was discussed how these objects are related to other neutron star populations. Also, the implications of this new source class for neutron star population estimates was described.

2.3. *Isolated neutron stars in optical and X-rays: room for discovery*

Patricia A. Caraveo: The multi-wavelength behavior of isolated neutron stars evolves as they age. In particular, the X-ray and optical emissions allow us to follow the shift from the non-thermal regime, typical of young objects, to a mostly thermal one, typical of older specimen. New observations unveil tale-telling details both on young and old objects, reminding us that a lot remains to be discovered in the complex INS family tree.

2.4. *Gamma-ray and TeV emission properties of pulsars and pulsar wind nebulae*

Ocker C. De Jager: Although more than 1,600 radio pulsars have been discovered, only a few have been detected in the γ-ray band. This is not because they are intrinsically faint, but because the pulsed component seems to cut off below about 30 GeV (the *EGRET* range), where the sensitivity was severely limited. However, ground-based atmospheric Cerenkov telescopes operating above 100 GeV (the Very High Energy or

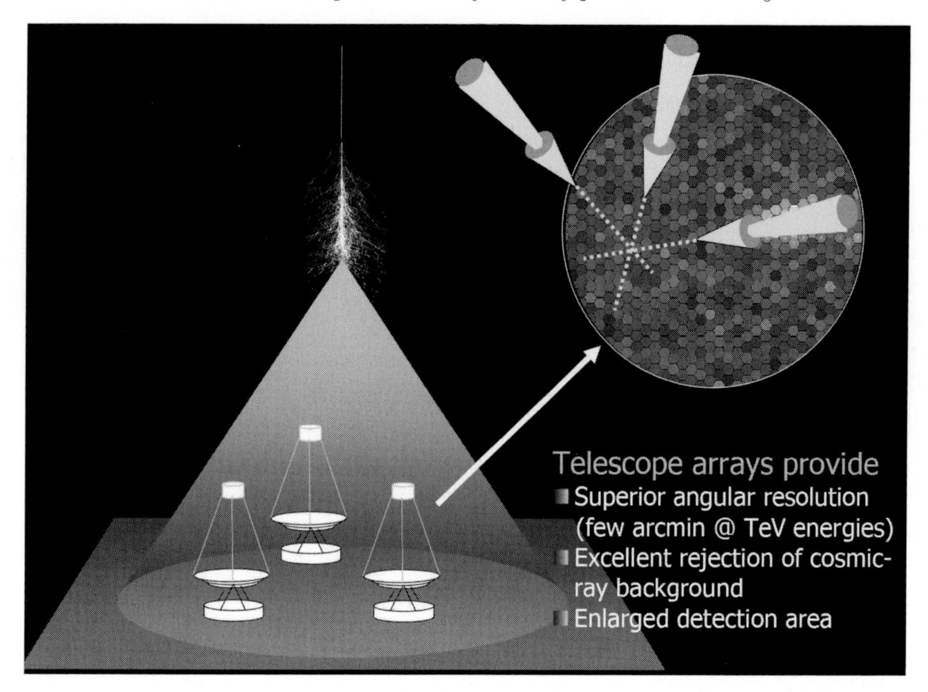

Figure 1. The detection of high energy gamma rays with, e.g., the H.E.S.S. telescopes is based on the imaging air Cherenkov technique. An incident high-energy γ-ray interacts high up in the atmosphere and generates an air shower of secondary particles. The number of shower particles reaches a maximum at about 10 km height, and the shower dies out deeper in the atmosphere. Since the shower particles move at essentially the speed of light, they emit the so-called Cherenkov light, a faint blue light. The Cherenkov light is beamed around the direction of the incident primary particle and illuminates on the ground an area of about 250 m diameter. Chart from the talk (2.4) presented by O.C. De Jager.

VHE domain), have both good sensitivity and good angular resolution to resolve several pulsar wind nebulae (PWN) in the VHE γ-ray domain. This review talk summarized the progress to date on pulsar and pulsar wind nebula observations and theory. Since γ-ray observations below 30 GeV have been limited by poor sensitivity, an instrument like *GLAST* should be able to resolve the pulsed component of a significant fraction of radio pulsars. This talk showed how the discovery potential of *GLAST* would be limited for fainter sources in the absence of contemporary radio pulsar parameters. This calls for the introduction of wide field-of-view radio pulsar monitors like KAT to resolve this problem. Most progress on PWN in the γ-ray domain is made by the HESS telescope system in Namibia. In this case we progressed to the level where VHE Gamma-ray Astronomy is taking the lead at all wavelengths (radio, IR, optical, X-ray and γ-ray) in the identification and understanding of new PWN. It was shown how the spin history of the PWN is more relevant to such VHE observations rather than X-rays, although the latter probe the more recent history of PWN evolution. It was then shown how these complementary wavebands can be combined to obtain new information about aspects such as the birth periods of pulsars and conversion efficiency of spin-down power to injected ultra-relativistic electrons.

2.5. *Pulsar timing and its future perspective*

David J. Nice: The present state of radio pulsar timing and prospects for future progress was surveyed. In recent years, pulsar timing experiments have grown rapidly in both

Masses of Neutron Stars in Neutron Star-Neutron Star Binaries

Pulsar	Recycled Pulsar	Companion Star
PSR B1913+16	1.4408±0.0003	1.3873±0.0003
PSR B2127+11C	1.349 ±0.040	1.363 ±0.040
PSR B1534+12	1.3332±0.0010	1.3452±0.0010
PSR J0737–3039	1.337 ±0.005	1.250 ±0.005
PSR J1756–2251	1.40 ±0.03	1.18 ±0.03
PSR J1518+4904	average mass=1.352±0.003	
PSR J1811–1376	average mass=1.300±0.450	
PSR J1829+2456	average mass=1.250±0.010	
PSR J1906+0746*	1.31 ±0.05	1.31 ±0.05

*Preliminary

Figure 2. Masses of neutron stars in NS-NS binaries. Chart from the talk (2.5) of D. Nice.

quantity and quality, yielding new tests of gravitation, new constraints on the structure of neutron stars, new insight into neutron star kicks and the dynamics of supernovae, and better understanding of the evolution of compact objects in the Galaxy.

The tremendous success of recent pulsar surveys, especially those with the Parkes multi-beam receiver, have vastly increased the number of known pulsars. Ongoing surveys at Arecibo and elsewhere promise to continue yielding new and exciting pulsars. At the same time, advances in instrumentation – including the development of wide-band receivers and spectrometers and the routine use of software coherent de-dispersion data acquisition systems – are increasing the precision attainable in timing experiments. State of the art timing precision is now around 100 nanoseconds. Limitations have been discussed on present timing experiments and prospects for future improvements.

2.6. *Radio emission theories of pulsars*

Vladimir Usov: Pulsar magnetospheres contain a multi-component, strongly magnetized, relativistic plasma. The present review was mainly concerned with generation and propagation of coherent radio emission in this plasma, emphasizing reasons why up to now there is no commonly-accepted model of the radio emission of pulsars. Possible progress in our knowledge about the mechanism of the pulsar radio emission was discussed.

2.7. *Theory of high-energy emission from pulsars*

Kwong Sang Cheng: In this talk various models of high energy emission from pulsars were briefly reviewed. In particular it was pointed out that the light curves can provide important constraints in the radiation emission regions and the location of the accelerators (gaps). Furthermore, the energy dependent light curves and phase-dependent spectrum could not be explained in terms of simple two dimensional models, three dimensional models must be used to explain the full detail of the observed data. A three dimensional outer gap model was presented to study the magnetospheric geometry, the light curve and the phase-resolved spectra of the Crab pulsar. Using a synchrotron self-Compton mechanism, the phase-resolved spectra with the energy range from 100 eV to

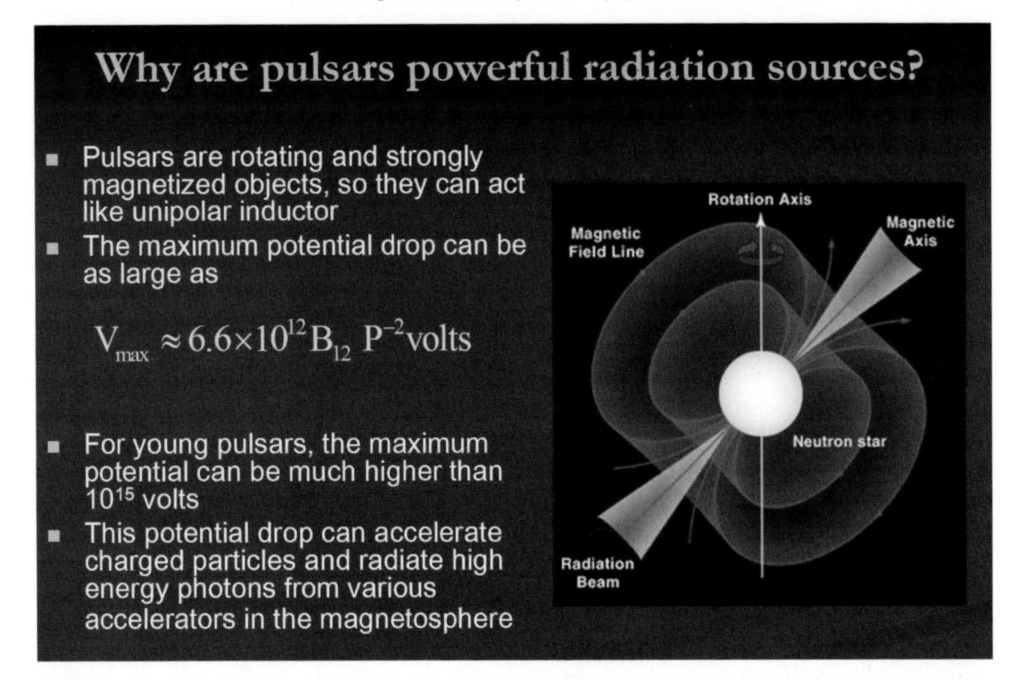

Figure 3. Why pulsars are powerful emitters of high-energy radiation (K.S. Cheng, 2.7).

10 GeV of the Crab pulsar can also be explained. The observed polarization angle swing of optical photons was also used to determine the viewing angle.

2.8. *On the theory of radio and high-energy emission from pulsars: where to go?*

John Arons: Recent progress in the theory of the energy loss from pulsars was discussed, focusing on the advances in force free models of the magnetosphere and on dissipation in the wind, with implications of the latter for unusually models of pulsed emission. Attention was also drawn to the implications the new global magnetosphere models may have for the generation of parallel electric fields, and the implications of these for high-energy emission, radio emission and the global mass loss rate through pair creation. Parallel electric field formation and consequences for photon emission was discussed in scenarios with self-consistent currents, which are rather different from the standard gap models with their starvation electric fields. Finally, remarks were presented on the importance of radiation transfer effects in unraveling the continuing mystery of pulsar radio emission.

2.9. *Cooling neutron stars: the present and the future*

Sachiko Tsuruta: Recent years have seen some significant progress in theoretical studies of physics of dense matter. Combined with the observational data now available from the successful launch of *Chandra* and *XMM-Newton* X-ray space missions as well as various lower-energy band observations, these developments now offer the hope for distinguishing various competing neutron star thermal evolution models. For instance, the latest theoretical and observational developments may already exclude both nucleon and kaon direct Urca cooling. In this way we can now have a realistic hope for determining various important properties, such as the composition, superfluidity, the equation of state and stellar radius. These developments should help us obtain deeper insight into the properties of dense matter.

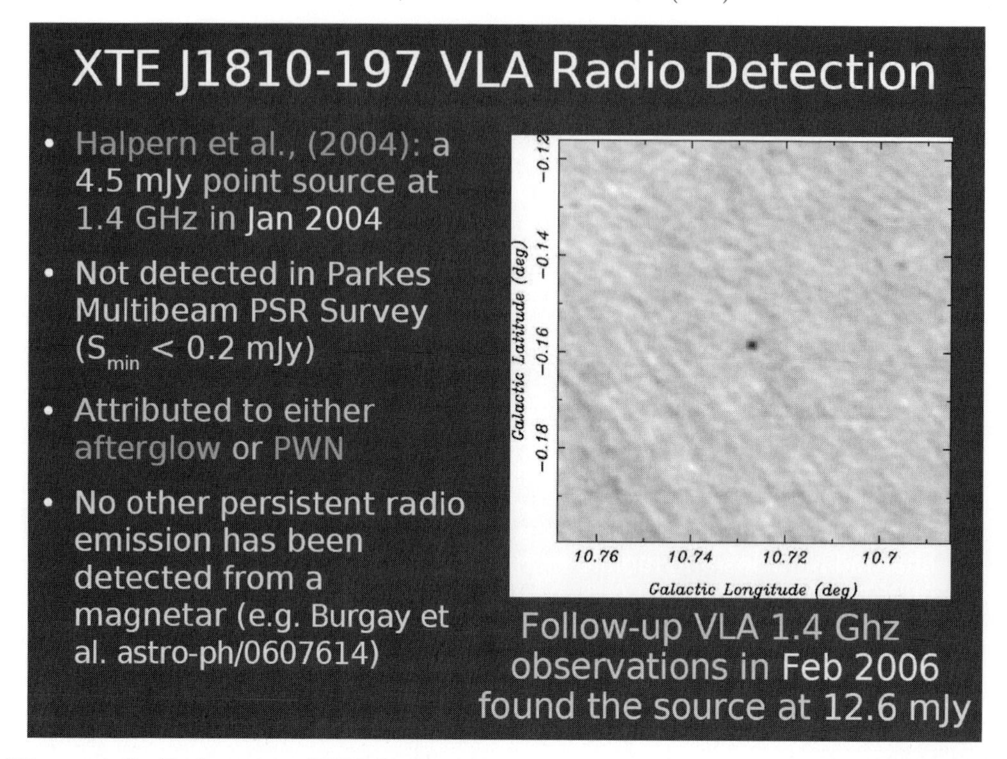

Figure 4. Radio detection of XTE J1810−197, as presented in the talk of S.M. Ransom (2.10).

2.10. *A decade of surprises from the anomalous X-ray pulsars*

Scott M. Ransom: A decade ago, the defining characteristics of the Anomalous X-ray Pulsars (AXPs) included slow spin periods (5–9 s), relatively soft but constant X-ray luminosities in the range of 10^{35}–10^{36} erg/s, and steady spin-down rates. The X-ray luminosities are too large to be powered by pulsar spin-down, and given the lack of evidence for accretion, are thought to be caused by the decay of $\sim 10^{14}$–10^{15} G magnetic fields (i.e., the magnetar theory as proposed by Thompson and Duncan). Within the past decade, though, detailed X-ray monitoring observations have shown that these sources are anything but constant and steady. Timing noise, glitches, X-ray bursts, and pulse profile and pulsed flux variations are now known to be relatively common in these sources. In addition, at least one recently discovered AXP, XTE J1810−197, is a full-fledged transient object. Detections in the optical and infra-red (including pulsations) and recent hard X-ray observations have complicated our views of their emission mechanisms. Finally, very recent detections of magnetar-like radio pulsars, as well as strong (and transient) radio pulsations from XTE J1810−197, show that these sources are linked (at least in some way) with the much more common radio pulsars.

2.11. *Pulsars and gravity*

Ingrid H. Stairs: Radio pulsars are superb tools for testing the predictions of strong-field gravitational theories. The currently achieved tests of equivalence principles and of predictions for relativistic binary orbital parameters were described, as well as limits on a gravitational-wave background, and future prospects in each of these areas were discussed.

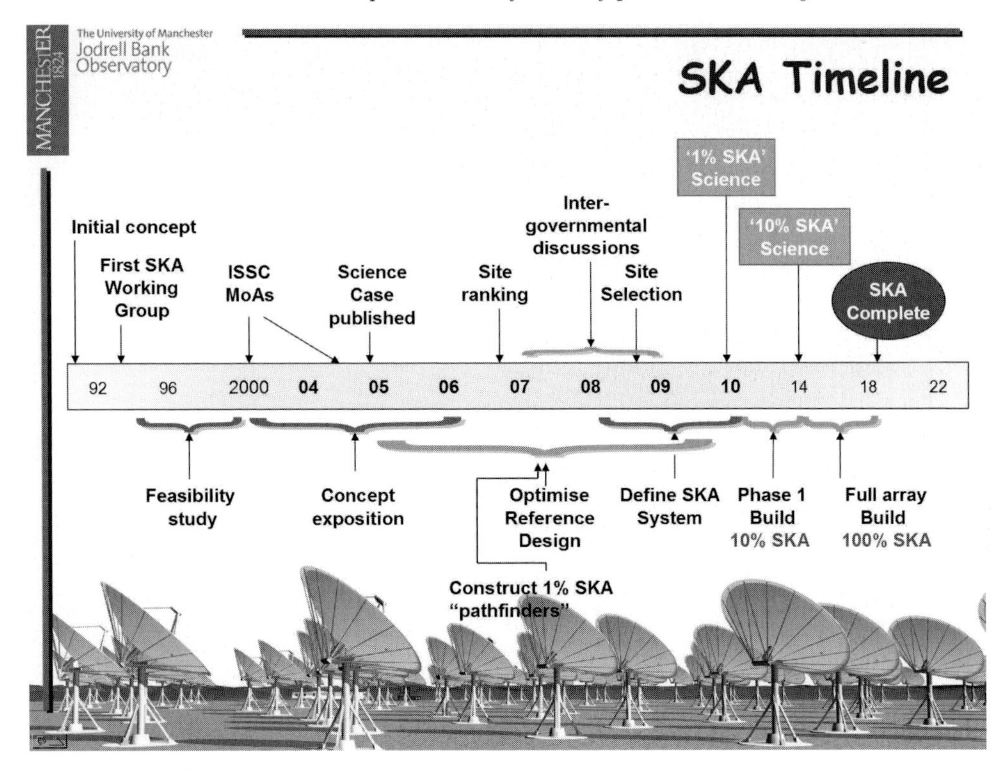

Figure 5. Time line of the SKA project, as presented by M. Kramer (2.12).

2.12. *Future radio observatories for pulsar studies*

Michael Kramer: Over the next decade, radio astronomers will have new, exciting instruments available to answer fundamental questions in physics and astrophysics. Without doubts, new discoveries will be made, revealing new objects and phenomena. We can expect pulsar astronomers to receive their fair share. In many respects, the field of pulsar astrophysics will change, as the science will not simply be a continuation of what has been done so successfully over the past 40 years. Instead, the huge number of pulsars to be discovered and studied with an unprecedented sensitivity will provide a step forward into exciting times. Instruments like the Low Frequency Array (LOFAR), the Square Kilometer Array (SKA) and their powerful pathfinder telescopes will enable a complete census of Galactic pulsars, ultimate tests of gravitational physics, unique studies of the emission process and much more. This talk demonstrated the potential of the future instruments by presenting highlights of the science to be conducted.

2.13. *Future optical and X-ray observatories for pulsar studies*

Werner Becker: Optical and X-ray astronomy has made great progress in the past several years thanks to telescopes with larger effective areas and greatly improved spatial, temporal and spectral resolutions. The next generation instruments like *XEUS*, *Constellation-X*, *Simbol-X*, *eROSITA*, the *James Webb Space Telescope* and the ESOs Extremely Large optical Telescope are supposed to bring again a major improvement in sensitivity. The purpose of this talk was to summarize the future plans for X-ray and optical telescopes with the emphasis of their application for pulsar and neutron star astronomy.

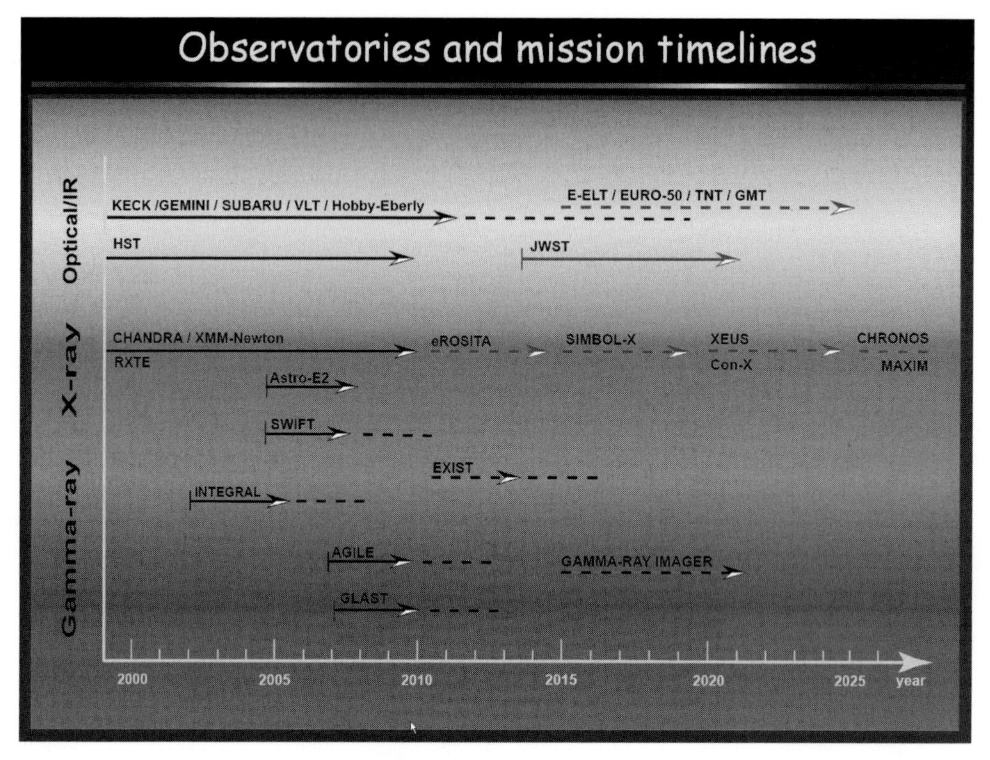

Figure 6. Time line of future high-energy missions, as presented by W. Becker (2.13).

2.14. *Future γ-ray and TeV observatories for pulsar searches*

David A. Smith: GeV measurements of pulsar lightcurves *versus* energy provide information on beam geometry, and spectral cut-offs give insights into the particle acceleration region(s) around the neutron star. The two together can help build a better picture of neutron star populations in the Galaxy. We are on the verge of instrumental break-through that promise to increase the sample of measured objects by ~ 10. This talk described instruments able to detect multi-GeV pulsations for the next few years. Developments of ground-based atmospheric Cherenkov detectors were covered, AGILEs prospects were reviewed, and then the talk focused on the *LAT (Large Area Telescope)* on *GLAST*, including sensitivity estimations. Accurate radio ephemerides can greatly enhance the γ-ray pulsar science. Efforts to build a large ephemerides database were described.

3. Poster papers

3.1. *Effects of core magnetic fields in evolution of binary neutron stars*

Taghi Mirtorabi, Atefeh J. Khasraghi, and Shohre Abdolrahimi: The standard scenario for evolution in a close binary system in which the neutron star pass through four evolutionary phases (isolated star, propeller, wind accretion and Roche-lobe accretion) was employed to calculate transport of orbital angular momentum to the neutron star or loss of angular momentum from the whole system. The evolution of core magnetic field of neutron stars in close binary systems with a low mass main sequence companion was explored. Assuming the core as a type II superconductor so the magnetic flux can be transported as quantized fluxoids, calculation have been performed to determine magnetic filed decay and its interaction with the mater accreted from the companion. The evolution of semi major axis of the binary in a time scale of 10^9 years comparable with the main sequence life time of the low mass companion was also investigated.

3.2. *External electromagnetic fields of slowly rotating relativistic magnetized NUT stars*

Bobomurat J. Ahmedov and A.V. Khugaev: Analytic general relativistic expressions for the electromagnetic fields external to a slowly-rotating magnetized NUT star with non-vanishing gravitomagnetic charge have been presented. Solutions for the electric and magnetic fields have been found after separating the Maxwell equations in the external background spacetime of a slowly rotating NUT star into angular and radial parts in the lowest order approximation in specific angular momentum and NUT parameter. The relativistic star was considered isolated and in vacuum, with different models for stellar magnetic field: (i) mono-polar magnetic field, and (ii) dipolar magnetic field aligned with the axis of rotation.

It was shown that the general relativistic corrections due to the dragging of reference frames and gravitomagnetic charge were not present in the form of the magnetic fields but emerge only in the form of the electric fields. In particular, it was argued that the frame-dragging and gravitomagnetic charge provide an additional induced electric field which is analogous to the one introduced by the rotation of the star in the flat spacetime limit.

3.3. *PSR J0538+2817 as the remnant of the first supernova explosion in a massive binary*

Vasilii V. Gvaramadze: It is generally accepted that the radio pulsar PSR J0538+2817 is associated with the supernova remnant (SNR) S 147. The only problem for the association is the obvious discrepancy (Kramer *et al.* 2003) between the kinematic age of the system of $\sim$30 kyr (estimated from the angular offset of the pulsar from the geometric center of the SNR and pulsar's proper motion) and the characteristic age of the pulsar of $\sim$600 kyr. To reconcile these ages one can assume that the pulsar was born with a spin period close to the present one (Kramer *et al.* 2003; Romani & Ng 2003).

An alternative explanation of the age discrepancy based on the fact that J0538+2817 could be the stellar remnant of the first supernova explosion in a massive binary system and therefore could be as old as indicated by its characteristic age was proposed. The proposal implied that S 147 is the diffuse remnant of the second supernova explosion (that disrupted the binary system) and that a much younger second neutron star (not necessarily manifesting itself as a radio pulsar) should be associated with S 147. The existing observational data on the system PSR J0538+2817/SNR S 147 were used to suggest that the progenitor of the supernova that formed S 147 was a Wolf-Rayet star (so that the supernova explosion occurred within a wind bubble surrounded by a massive shell) and to constrain the parameters of the binary system. The magnitude and direction of the kick velocity received by the young neutron star at birth was also restricted and it was found that the kick vector should not strongly deviate from the orbital plane of the binary system.

3.4. *Ensemble pulsar time scale*

Alexander E. Rodin: The purpose of this work was to construct an algorithm of a new astronomical time scale based on the rotation of pulsars, which has comparable accuracy with the most precise terrestrial time scale TT.

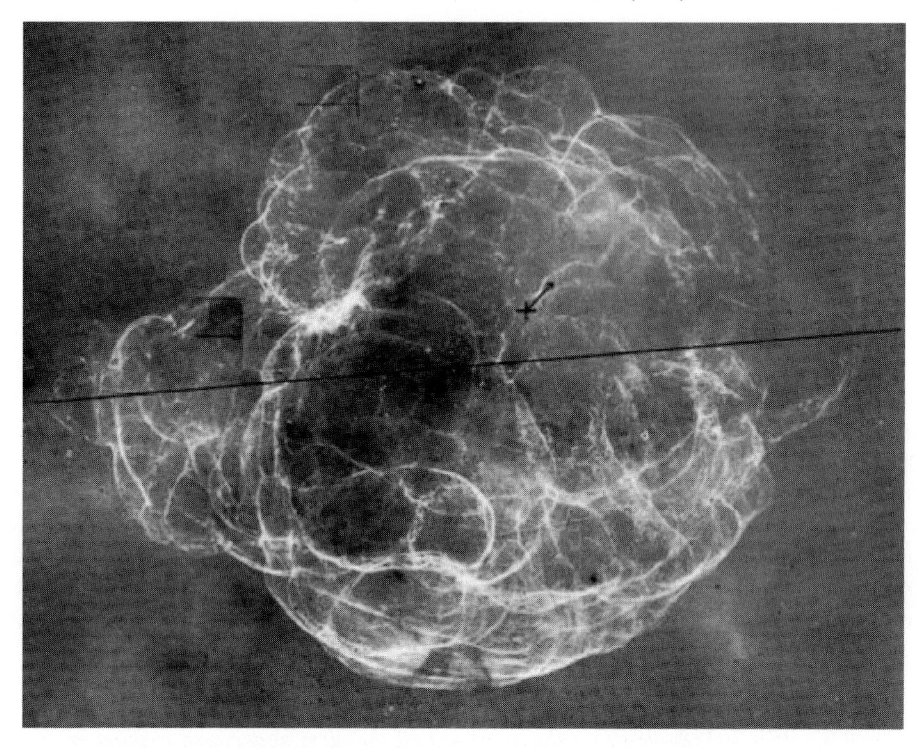

Figure 7. H_α image of the SNR S 147 (Drew *et al.* 2005; reproduced in the poster with the permission of the IPHAS collaboration). The position of PSR J0538+2817 is indicated by a cross. The arrow shows the direction of the pulsar proper motion vector (Kramer *et al.* 2003). The line drawn in the east-west direction shows the bilateral symmetry axis. North it up, east is west. Image from V.V. Gvaramadze (3.3).

This algorithm is based on the Wiener optimal filtering method and allows separating contributions to the post-fit pulsar timing residuals of the atomic clock used in pulsar timing and spin variations of the pulsar itself. The optimal filters were constructed with use of the cross and auto covariance functions (in time domain) or auto and cross power spectra (in frequency domain) of the post-fit timing residuals of pulsars participating in construction of the ensemble time scale.

The algorithm was applied to the timing data of millisecond pulsars PSR B1855+09 and B1937+21 (Kaspi *et al.* 1994) and allowed to obtain the corrections of UTC scale relative to ensemble pulsar time scale PTens. Comparison of the differences UTC PTens and UTC TT displays significant correlation between them at level 0.79. Subsequent analysis of TT and PTens shows that TT coincides with PTens within 0.40 ± 0.17 mcs and has a fractional stability 10^{-15} at the 7 years time interval.

Relatively close angular distance (15.5 degrees) on the sky between these pulsars gives grounds to expect that there is a correlated signal in the post-fit timing data caused by the stochastic gravitational wave background (GWB). A new limit of the fractional energy density of GWB based on the difference TT and PTens was established to be $\Omega_g h^2 \simeq 2 \times 10^{-10}$. This new value is by one order lower than the previously published one owing to application of the new algorithm that separates the proper pulsars spin and local atomic standard variations.

3.5. *Comparison of giant radio pulses in young pulsars and millisecond pulsars*

Agnieszka Slowikowska, Axel Jessner, Gottfried Kanbach, and Bernd Klein: Pulse-to-pulse intensity variations are a common property of pulsar radio emission. For some of the objects single pulses are often 10-times stronger than their average pulse. The most dramatic events are the so called giant radio pulses (GRPs). They can be 1000-times stronger than the regular single pulses from the pulsar. Giant pulses are a rare phenomenon, occurring in very few pulsars which split into two groups. The first group contains very young and energetic pulsars like the Crab pulsar, and its twin in the LMC (PSR B0540−69), while the second group is represented by old, recycled millisecond pulsars like PSR B1937+21, PSR B1821−24, PSR B1957+20, and PSR J0218+4232 – the only millisecond pulsar detected in γ-rays. The characteristics of GRP's for these two pulsar groups was discussed. In particular, the poster focused on the flux distributions of GRPs which were compared. Moreover, the latest findings of new features in the Crab GRPs were presented. Analysis of Effelsberg data taken at 8.35 GHz have shown that GRPs do occur in all phases of its ordinary radio emission, including the phases of the two high-frequency components (HFCs) visible only between 5 and 9 GHz. This suggests that a similar emission mechanism may be responsible for the main pulse, the inter pulse and the HFCs. Finally, the similarities and differences between both groups of pulsars in the context of timing, spectral and polarization properties of these pulsars were discussed. It was also attempted to answer the question why pulsars belonging to so different classes do show the same giant radio emission phenomena.

3.6. *Integral IBIS and JEM-X observations of PSR B0540−69*

Agnieszka Slowikowska, Gottfried Kanbach, Jurek Borkowski, and Werner Becker: The high-energy pulsar PSR B0540−69 in the Large Magellanic Cloud (d $\simeq$ 49.4 kpc), embedded in a synchrotron plerion in the center of SNR 0540−69.3 is often referred to as an extragalactic 'twin' of the Crab pulsar. Its pulsed emission has been detected up to about 48 keV so far. The results from the search for PSR B0540−69 up to 300 keV in the 1 Ms *INTEGRAL* data was presented. *INTEGRAL* was pointed to the LMC during 6 revolutions in January 2003, and 3 revolutions in January 2004. The events used for timing and spectral analysis of the source come from the *IBIS/ISGRI* and *JEM-X* detectors. Moreover, the details of data analysis technique used for this weak source were presented.

3.7. *Plasma modes along open field lines of neutron star endowed with gravitomagnetic NUT charge*

Bobomurat J. Ahmedov and Valeria G. Kagramanova: Electrostatic plasma modes along the open field lines of a rotating neutron star endowed with gravitomagnetic charge or NUT parameter have been considered. Goldreich-Julian charge density in general relativity was analyzed for the neutron star with non-zero NUT parameter. It was found that the charge density is maximal at the polar cap and remains almost the same in a certain extended region of the pole. For a steady state Goldreich-Julian charge density it was found that the usual plasma oscillation along the field lines; plasma frequency resembles the gravitational redshift close to the Schwarzschild radius. The results in studying the nonlinear plasma mode along the field lines were presented. The equation contained a term that described the growing plasma modes near Schwarzschild radius in a black hole environment. The term vanished with the distance far away from the gravitating object. For initially zero potential and field on the surface of a neutron star, Goldreich-Julian charge density was found to create the plasma mode which was en-

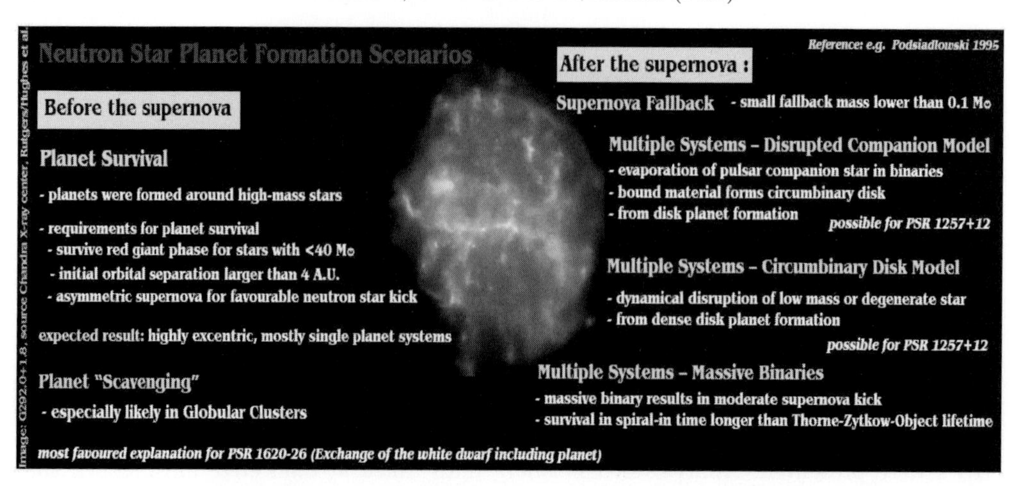

Figure 8. Image from Posselt, Neuhäuser and Haber (3.9).

hanced and propagated almost without damping along the open field lines of magnetized NUT star.

3.8. *The drift model of magnetars*

Igor F. Malov and George Z. Machabeli: It was shown that the drift waves near the light cylinder can cause the modulation of emission with periods of order several seconds. These periods explain the intervals between successive pulses observed in "magnetars" and radio pulsars with long periods. The model under consideration gave the possibility to calculate real rotation periods of host neutron stars. They are less than 1 sec for the investigated objects. The magnetic fields at the surface of the neutron star are of order 10^{11}–10^{13} G and equal to the fields usual for known radio pulsars.

3.9. *Sub-stellar companions around neutron stars*

Bettina Posselt, Ralph Neuhäuser, and Frank Haberl: Planets or sub-stellar companions around neutron stars can give valuable insights into a neutron star's formation history considering for example birth kicks or fallback disks. They may also help to derive neutron star masses which would be very welcome especially if the radius can be derived by other means as for the radio-quiet X-ray thermal neutron stars. Currently there are two planetary systems around millisecond pulsars known. They have been found by the pulse timing technique which is most sensitive to old millisecond pulsars. Some of the formation theories can already be ruled out for these systems. However, statistics are very poor and other search techniques are needed to cover also young, even radio-quiet neutron stars.

The first results of direct imaging search for sub-stellar companions around the closest and youngest neutron stars started three years ago with ESO's VLT were presented. Among the objects was the famous RX J1856.5−3754 for which a sub-stellar object could help to constrain the equation of state as the radius was already previously derived by its X-ray thermal emission.

3.10. *Detection of the individual pulses of the pulsars B0809+74, B0834+06, B0950+08, B0943+10, B1133+16 at decameter wave range*

O.M. Ulyanov, V.V. Zakharenko, Alexander A. Konovalenko, *et al.*: Radio emission of single pulses for five pulsars was found at frequencies 18 - 30 MHz. It was reported that the radio emission is caused by the strong subpulses that have peak intensity of more than 20 times larger than the peak intensity of average profiles. It was found that the intensity of single pulses has a strong variation in frequency and time. The probability of detection of the anomalous intense pulses thus does not exceed several percents at Decameter wave range. Usually such pulses are detected in short series (not more than 10 pulses). Typical band values of detection for the pulses with anomalous intensities lies in the range from 0.2 to 0.5 octaves.

3.11. *An analytical description of low-energy secondary plasma particle distribution in pulsar magnetospheres*

Victor M. Kontorovich and A.B. Flanchik: A simple analytical approximation of the form of low-energy cut-off of the secondary particle distribution was proposed. This approximation is acceptable for describing the known cascade numerical simulation. The distribution form and maximum position as function of pulsar parameters have been found theoretically by considering the curvature radiation process and the electron-positron pairs production. An influence of synchrotron radiation on the form and maximum of the low-energy distribution was considered.

3.12. *Timing irregularities and the neutron star stability*

Johnson O. Urama: Observations show that the different manifestations of neutron stars exhibit measurable departures from the predicted slow down. This has been largely attributed to rotational irregularities, timing noise, glitches, and precession. The analysis of the regular spin-down and jump parameters for a combined set of the pulsars (radio, optical, X-ray and γ-ray) and magnetars was presented. It was also attempted to quantify the stability of neutron stars in general.

3.13. *Force-free pulsar magnetosphere*

Andrey N. Timokhin: The properties of a force-free pulsar magnetosphere and address the role of electron-positron cascades in determining a particular configuration among other possible force-free magnetospheric configurations were discussed. Results of high resolution numerical simulations of the force-free magnetosphere of aligned rotator and analyze in details properties of an aligned pulsar were reported. It was argued that the closed field line zone should grow with time slower than the light cylinder; this yield the pulsar braking index less than 3. However, models of aligned rotator magnetosphere with widely accepted configuration of magnetic field, when the last closed field line lies in equatorial plane at large distances from pulsar, have serious difficulties. The solutions of this problem were discussed and it was argued that in any case, also for inclined pulsar, energy losses should evolve with time differently than it is predicted by the magneto-dipolar formula and the pulsar braking index should be different from the "canonical" value equal to 3.

3.14. *Crab pulsar optical photometry and spectroscopy with microsecond temporal resolution*

Gregory Beskin, Sergey Karpov, Vladimir Plokhotnichenko, *et al.*: The results of fast photometry and spectroscopy of the Crab pulsar with microsecond temporal resolution were presented. The observations have been performed on William Herschel 4 m and BTA 6 m telescopes using APD avalanche photon counter and PSD panoramic photon imager. The stability of the optical pulse was analyzed and the search for the variations of the pulse shape along with its arrival time stability was performed. Upper

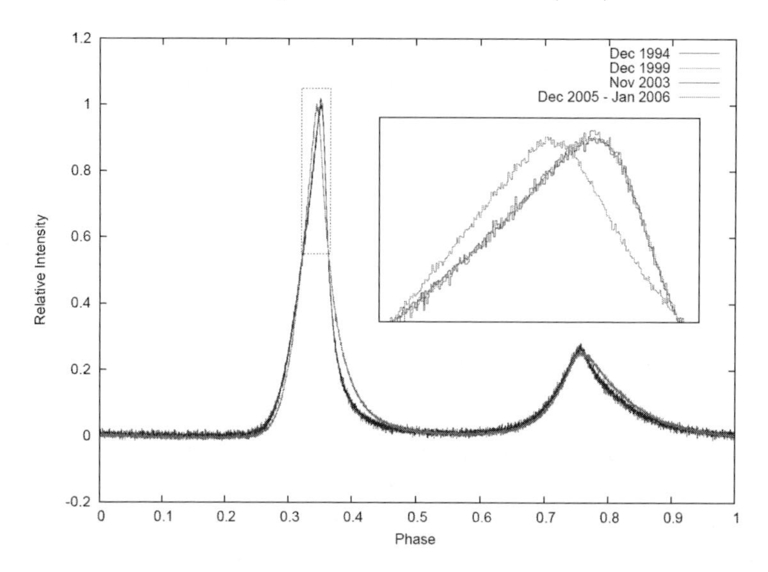

Figure 9. Phase-aligned light curves of the Crab pulsar from optical observations at different epochs all scaled to the same pulse height. In the observation at the 6 m SAO telescope in January 2007 the Crab pulsar light curve was similar to the one observed in 1994, 1999 and 2003. This means that the change of the light curve observed in December 2005/January 2006 was either a stochastic (rare) event or a not understood instrumental effect. Image from G. Beskin *et al.* (3.14).

limits on the possible short time scale free precession of the pulsar and the stochastic variable optical emission component were placed. The results of the low resolution (~ 300 Å) phase-resolved spectroscopy of the pulsar emission were discussed, first of all the distinction of the spectra of pulses and off-pulse phase intervals.

3.15. *Discovery of a large time scale cyclic evolution of radio pulsars rotational frequency*

Gregory Beskin, Anton Biryukov and Sergey Karpov: The recent massive measurements of pulsar frequency second derivatives have shown that they are 100–1000 times larger than expected for standard pulsar slowdown. Moreover, the second derivatives as well as braking indices were even negative for about half of the pulsars. It was explained that these paradoxical results from statistical analysis of the rotational parameters (frequency, its first and second derivatives) of the subset of 295 pulsars taken mostly from the ATNF database. The strong correlation of second and first frequency derivatives either for positive (correlation coefficient $r \simeq 0.9$) and negative ($r \simeq 0.85$) values of second derivative, and of the frequency and its first derivative ($r \simeq 0.7$) were found. These dependencies were interpreted as evolutionary ones due to the first frequency derivative being nearly proportional to the characteristic age. The derived statistical relations as well as "anomalous" values of the second frequency derivative were well explained in the framework of the simple model of cyclic evolution of the rotational frequency of the pulsars. It combined the secular change of the rotational parameters according to the power law with braking index $n \simeq 5$ and harmonic oscillations of 100–1000 years period with an amplitude from 10^{-3} Hz for young pulsars to 10^{-10} Hz for older ones. It was found that the physical nature of these cyclic variations of the rotational frequency may be similar to the well-known red timing noise, however, with much larger characteristic time scale.

3.16. *Abnormal phases in nuclear matter in supernova core collapse model*

D.J. Bora, Hira L. Duorah, and Kalpana Duorah: The role of abnormal phases in nuclear matter on the basis of the well-known Lee-Wick theory was studied for the determination of the shock strength at the core bouncing of type II supernova. Relativistic equation for collapse beyond nuclear density with a still stronger magnetic field was developed to study the effect of appearance of the abnormal phases. This lead to the softening of the equation of state giving rise to strong shock. The magnetar thus formed can be a store-house of many high-energy events including rapid γ-ray bursts. A strong magnetic field beyond some critical value sometimes can exponentially accelerate certain reactions producing rapid bursts of energy. Some seismic events on the solidified crusts of magnetars can also be alluded for the short γ-ray burst discovered in December 2004 at the site of SGR 1806−20.

3.17. *Instant radio spectra of giant pulses from the Crab pulsar over decimeter to decameter wave bands*

Michail V. Popov, Arkady D. Kuzmin, O.M. Ulyanov, *et al.*: The results of simultaneous multi-frequency observations of giant radio pulses (GPs) from the Crab pulsar PSR B0531+21 at frequencies of 23, 111 and 600 MHz were presented. For the first time GPs were detected at such low frequency as 23 MHz. Among 45 GPs detected in the overall observations time with 600 MHz, 12 GPs were identified as simultaneous ones at 600 and 23 MHz. At 111 MHz among 128 GPs detected in the overall observations time with 600 MHz, 21 GPs were identified as simultaneous ones at 600 and 111 MHz. Spectral indices for the power-law frequency dependence of GPs energy were enclosed between −3.1 and −1.6. The mean spectral index equals to -2.7 ± 0.1 and was the same for both frequency combinations 600 – 111 MHz and 600 – 23 MHz.

A big scatter in values of the individual spectral indexes and a large number of unidentified giant pulses indicated that a real form of spectra of individual giant pulses did not follow a simple power law. The shape of giant pulses at all three frequencies was governed by the scattering of radio waves on the inhomogeneities of the interstellar plasma. The pulse scatter broadening and their frequency dependence was measured as $\tau_{\rm SC} = 20(\nu/100)^{-3.5\pm0.1}$ ms, where frequency ν is in MHz.

3.18. *Pulsar nulling quantitative analysis*

John H. Seiradakis and Kosmas Lazaridis: Using long sequences of single pulses the nulling behavior of several pulsars was analyzed. Each pulsar has been characterized by a 'nulling parameter', which represents the average length of consecutive null pulses and a 'nulling max' parameter, which represents the maximum length of consecutive null pulses. These two parameters were compared to other pulsar parameters. Some interesting correlations were derived.

3.19. *Eclipse study of the double pulsar*

René P. Breton, Victoria M. Kaspi, Maura A. McLaughlin, *et al.*: The double pulsar system PSR J0737−3039 offers an unprecedented opportunity for studying General Relativity and neutron-star magnetospheres. This system has a favorable orbital inclination such that the millisecond pulsar A, is eclipsed when its slower companion B, passes in front. High time resolution light curves of the eclipses reveal periodic modulations of the radio flux corresponding to the fundamental and the first harmonic of pulsar B spin frequency. Eclipse modeling is highly sensitive to the geometrical configuration of the system and thus provides a unique probe for parameters like the inclination angle of pulsar B spin axis as well as their time evolution due to relativistic effects. Detailed fitting of the pulsar A eclipse light curves to a model that includes, for pulsar B, a simple dipolar magnetic field was presented. It was found that the eclipses can be reproduced very well, and one obtains precise measurements of pulsar B's orientation in space. Results on a search for secular changes caused by geodetic precession of pulsar B's spin axis were reported.

3.20. *Observations of southern pulsars at high radio frequencies*

Aris Karastergiou and Simon Johnston: A number of pulsars at 1.4, 3.1 and 8.4 GHz in full polarization at the Parkes radio-telescope was observed. The main objective was to study the frequency evolution of polarization by means of new high quality polarization data at high frequencies. Average polarization profiles with high-time resolution to update already existing previous observations were also obtained. Detailed description of a total of 97 polarization profiles at the three aforementioned frequencies was provided. This relatively large sample provided the opportunity to study effects related to the linear and circular polarization as well as the polarization position angle. An evidence was found that: (1) a simple model where two orthogonal polarization modes with competing spectral indices can account for many observational properties of the linear polarization and total power; (2) the position angle dependence on frequency depends on the different relative strength of profile components at different frequencies; (3) young, energetic pulsars remain highly polarized at high frequencies; and (4) highly polarized components may originate from higher up in the pulsar magnetosphere then unpolarized components. A summary of these results was presented.

3.21. *Pulsar braking indices*

Altan Baykal and M. Ali Alpar: Almost all pulsars with anomalous positive second derivative of angular acceleration measurements (corresponding to anomalous braking indices in the range $5 < n < 100$), including all the pulsars with observed large glitches as well as post glitch or inter-glitch second measurement obey the scaling between glitch parameters originally noted in the Vela pulsar. Negative second derivative values can be understood in terms of glitches that were missed or remained unresolved. The glitch rates and *a priori* probabilities of positive and negative braking indices according to the model developed for the Vela pulsar were discussed. This behavior supports the universal occurrence of a nonlinear dynamical coupling between the neutron star crust and an interior superfluid component.

3.22. *Electrodynamics of pulsar's electrosphere*

Jérôme A. Pétri: A self-consistent model of the magnetosphere of inactive, charged, aligned rotator pulsars with help of a semi-analytical and numerical algorithm, was presented. In this model the only free parameter was the total charge of the system. This "electro-sphere" is stable to vacuum breakdown by electron-positron pair production. However, it appears to be unstable to the so-called "diocotron" instability which is an electrostatic instability. Eigenspectra and eigenfunctions for different disc models, which differ by the total charge of the disc-star system were presented. The evolution of this instability on a long time-scale was studied in a fully non-linear description by means of numerical simulations. For multi-mode excitation, the average macroscopic response of the system could be described by a quasi-linear model. It was found that in the presence of an external source feeding the disk with positive charges, representing the effect of pair creation activity in the gaps, the diocotron instability may give rise to an efficient diffusion of charged particles across the magnetic field lines.

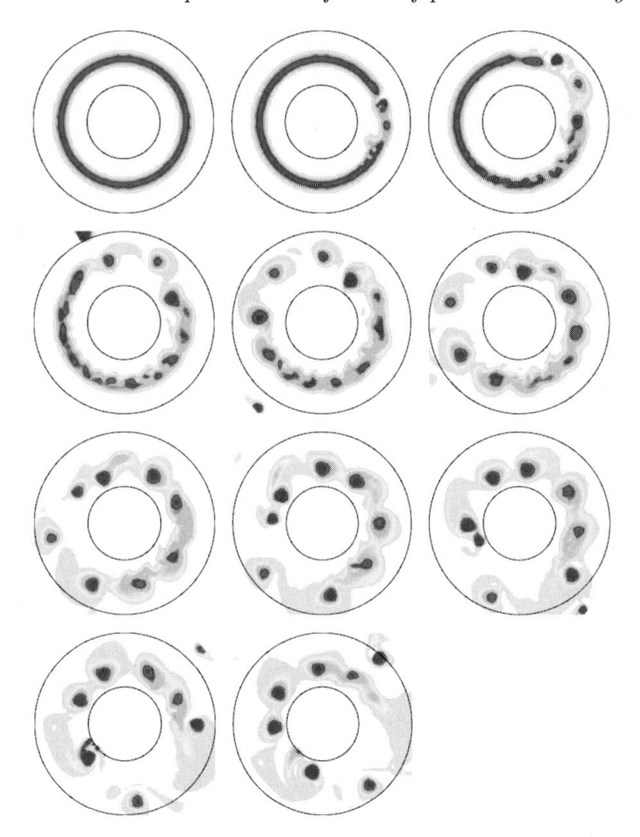

Figure 10. Non-linear evolution of the diocotron instability in an infinitely thin disk. Image from poster 3.22 of J.A. Petri.

3.23. *About one hypothesis on the origin of anomalous X-ray pulsars and soft γ-ray repeaters*

Fikret K.O. Kasumov, A. Allakhverdiev, and Abdul I. Asvarov: The possibility of realization of the scenario, according to which anomalous X-ray pulsars (AXPs) and soft γ-ray repeaters (SGRs) originate from the radio pulsars with the very close initial parameters (period, magnetic field, etc.), subjected to considerable and prolonged glitches, was analyzed. This scenario provides both an increase in the period of ordinary pulsars and the attainment of magnetic field strengths typical of these objects ($B \simeq 10^{13}$–10^{14} G), a new class of neutron stars, called magnetars, at an insignificant initial magnetic field value $B \simeq 3 \times 10^{10}$-$10^{11}$ G. With this aim, the criteria to be satisfied by a potential progenitor of AXPs and SGRs were determined and analyzed. So, taking into account the combined action of all factors (magnetic field, distance, birth place, satisfying to our criteria, etc.) we restricted our analysis to 100 pulsars with $B > 5 \times 10^{12}$ G and $P > 0.5$ s. The observed characteristics of such pulsars, their association with supernova remnants, and their evolution in the $P - \dot{P}$ diagram with allowance made for the actual age of the possible AXP and SGR progenitor were shown to be in conflict with the suggested scenario and can be better described in the framework of the standard magneto-dipole model of pulsar evolution.

3.24. *Combined models of evolution and real ages of pulsars*

A. Allakhverdiev, Fikret K.O. Kasumov, and Sevinc O. Tagieva: The test for checking the applicability of fall-back or propeller models for the pulsar evolution was proposed. This test was based on the comparison of the pulsars ages predicted by these models with the real kinematical ages of pulsars. With this goal two groups of pulsars, namely relatively young and old pulsars with the distances from the galactic plane $z < 100\,\mathrm{pc}$ and $z < 300\,\mathrm{pc}$, respectively, were selected. At the same time, the irregular character of deviation of the pulsars birthplaces from the geometrical plane of the Galaxy has been taken into account. The distribution of these groups of pulsars in $P - \dot{P}$ diagram was compared with the theoretical tracks of the evolution of the pulsars predicted by fall-back and combined dipole + propeller models at various values of the initial parameters of radio pulsars (magnetic field and accretion rate of matter). As it is well known, characteristic feature of combined model (unlike to the pure magneto-dipole) is the increase of period derivative up to some critical value with increase of the period of pulsars, i.e., the age of pulsars. It was shown that the distribution of selected pulsars in the mentioned diagram contradicts to this model and can be easily explained by standard model of pulsar evolution.

3.25. *X-ray emission from hot polar cap in pulsars with drifting subpulses*

Janusz Gil and George I. Melikidze: Within the framework of the partially screened inner acceleration region the relationship between the X-ray luminosity and the circulational periodicity of drifting subpulses was derived. This relationship was quite well satisfied in pulsars for which an appropriate radio and X-ray measurements exist. A special case of PSR B0943+10 was presented and discussed. The problem of formation of a partially screened inner acceleration region for all pulsars with drifting subpulses was also considered. It was argued that an efficient inner acceleration region just above the polar cap can be formed in a very strong and curved non-dipolar surface magnetic field.

3.26. *Pulsed radio emission from two XDINS*

Valerii M. Malofeev, O.I. Malov, and D.A. Teplykh: Investigations of two X-ray dim isolated neutron stars: J1308.6+212708 and J2143.03+065419 were reported. The observations were carried out on two sensitive transit radio telescopes in Pushchino at a few frequencies in the range 111–42 MHz. Mean pulse profiles, the flux density and the dispersion measures were presented. The measures of periods and their derivatives were reported, as well as the estimation of distances and integral radio luminosities. The comparison with X-ray observations was made.

3.27. *Magnetospheric eclipses in the double pulsar system J0737−3039*

Roman R. Rafikov and Peter Goldreich: The recently discovered double pulsar system, PSR J0737−3039, consisting of a millisecond and a normal pulsars in a 2.4 hr orbit, provides us with unprecedented tests of general relativity and magnetospheric effects. One of the most interesting phenomena observed in this system is the eclipse of the millisecond pulsar in the radio at its conjunction with the normal pulsar. A theory which explains this observation as a result of synchrotron absorption of the millisecond pulsar radio beam in the magnetosphere of the normal pulsar was presented. Absorption was induced in a sense that the intense radio beam of the millisecond pulsar itself strongly modifies the properties of the plasma in the closed part of the normal pulsar magnetosphere: absorption of high-brightness temperature radio emission heats up particles already present there and also allows additional pair plasma to be trapped in this region by magnetic bottling effect. This theory self-consistently predicts the size of the eclipsing region which agrees very well with the observed duration of eclipse. Recent observations of the variability of transmission during the eclipse modulated at the rotation period of the normal pulsar have been interpreted as resulting from the absorption by the rigidly rotating dipolar-shaped magnetosphere which is in perfect agreement with the presented theory.

3.28. *Electromagnetic fields of magnetized neutron stars in braneworld*

Bobomurat J. Ahmedov and F.J. Fattoyev: The dipolar magnetic field configuration in dependence on brane tension and present solutions of Maxwell equations in the internal and external background spacetime of a magnetized spherical neutron star in a Randall-Sundrum II type braneworld was studied. The star was modeled as a sphere consisting of perfect highly magnetized fluid with infinite conductivity and frozen-in dipolar magnetic field. With respect to solutions for magnetic fields found in the Schwarzschild spacetime brane tension introduces enhancing corrections both to the interior and the exterior magnetic field. These corrections could be relevant for the magnetic fields of magnetized compact objects as pulsars and magnetars and may provide the observational evidence for the brane tension through the modification of formula for magnetodipolar emission which gives amplification of electromagnetic energy loss up to few orders depending on the value of the brane tension.

3.29. *On dependence of some parameters of radio pulsars radiation on their age*

Viquen H. Malumian and Avetik N. Harutyunyan: The relationship between parameters of the radiation from pulsars and the dependence of the rates of the radiation periods of these objects on their characteristic ages was studied. The following results were obtained:

(*a*) The rate of change in the radiation periods (derivatives of periods, dP/dt) of pulsars depends on their characteristic age. These changes proceed more slowly with age. The rate of change of the radiation period of pulsars can in some way serve as an indicator of their age.

(*b*) The relationship between the rate dP/dt of change of the period and the period P has been demonstrated. For young pulsars this relationship is weak. In the course of evolution with age, the relationship between the derivative of the period and the period becomes closer. Whereas for young ($T < 10^6$ years) pulsars the correlation coefficient for the log dP/dt – log P plot is only 0.49 ($p < 0.0001$), for old ($T \geqslant 10^8$ yr) pulsars the correlation coefficient approaches unity ($p < 0.0001$). Since, as shown above, the rate of change of the radiation period decreases with age in pulsars, one can say that the lower the rate of change of the radiation period of a pulsar is, the closer is the relationship between the derivative of its period and its period.

The data in the catalog of Taylor *et al.*, which contains 706 objects, were examined separately. After eliminating the members of binary and multiple systems, as well as the members of the Magellanic Clouds, slightly more than 500 objects remain.

3.30. *The Nancy pulsar instrumentation: The BON coherent dedispersor*

Ismaël Cognard and Gilles Theureau: A summary of the Nancay pulsar instrumentation and the on going observational pulsar timing programs was presented. The BON coherent dedispersor is able to handle 128 MHz of bandwidth. It is made of a spectrometer, plus four data servers to spread data out to a 70-node cluster of PCs (with Linux Operating System). De-dispersion is done by applying a special filter in the complex Fourier domain. This backend has been designed in close collaboration with the UC Berkeley. It benefits from the many qualities of the large Nancy radio telescope (NRT, equivalent to a 94 m circular dish), which receivers were upgraded in 2000: a factor of

2.2 sensitivity improvement was obtained at 1.4 MHz, with an efficiency of 1.4 K/Jy for a system temperature of 35 K; a better frequency coverage was also achieved (from 1.1 to 3.5 GHz). The first two years of BON data acquisition demonstrates that the timing data quality is comparable with the Arecibo and Green Bank results. As an example, a Time of Arrival (ToA) measurement accuracy better than 200 ns (170 - 180 ns) is obtained in only 30 s of integration on the millisecond pulsar PSR B1937+21. With this up to date instrumentation, two main observational programs in pulsar timing with the Nancy antenna are operated: (1) the radio follow-up of X- and γ-ray pulsars for the building of a complete multi-wavelength sample and (2) the monitoring of both a millisecond pulsar timing array and a targeted list of binary or unstable pulsars for gravitational wave detection. Joining both list of targets, a total sample of 150 pulsars is then monitored regularly with a dense sampling in time.

3.31. *Relation of pulsars to the remnants of supernova bursts*

Viquen H. Malumian and Avetik N. Harutyunyan: Based on a large volume of statistical data it was shown that the spatial distributions of radio pulsars in the galaxy with characteristic ages $T < 10^6$ yr and $T > 10^6$ yr differ significantly. The overwhelming majority of the pulsars with $T < 10^6$ yr lie within a narrow band of width 400 pc around the Galactic plane. A large portion of the pulsars with $T > 10^6$ yr is concentrated outside this zone. In the case of younger pulsars, a larger fraction of them lies within the confines of the above mentioned zone. It is also shown that pulsars with $T < 10^6$ yr and the remnants of supernova explosions have essentially the same spatial distribution. These facts support the existence of a relationship between pulsars and supernova remnants, as well as the acquisition of high spatial velocities by pulsars during their birth.

3.32. *The multi-photon electron-positron pair production in the magnetosphere of pulsars*

Ara K. Avetissian: In general, the single-photon reaction $\gamma \rightarrow e^- + e^+$, as well as the inverse reaction of the electron-positron annihilation can proceed in a medium that must be a plasma-like. To provide a macroscopic refractive index $n(\omega) < 1$ necessary for pair production γ-frequencies one needs plasma densities $\rho > 10^{33}$ cm^{-3}. Such superdense matter exists in the core of the neutron stars/pulsars. At these densities the electron component of the superdense plasma is fully degenerated and taking also into account the Pauli principle the probabilities of these processes actually turn to zero. Hence, the possibility of multi-photon electron-positron pair production by strong electromagnetic radiation of soft frequencies in magnetosphere of pulsars is considered, which is possible at ordinary densities of plasma. Such multi-photon process occurs via nonlinear channels at high intensities of electromagnetic radiation in wide region of frequencies from radio to UV and soft X-ray in pulsars magnetosphere.

Numerical simulations for various pulsars ($\Omega \simeq 1$–$200\,\mathrm{s}^{-1}$) with the help of analytical distribution functions of magnetospheres plasma with densities $\rho \simeq 10^{20}$–10^{22} cm^{-3} and pair production probabilities have been made, and both energetic and angular distributions of produced electron-positron pairs were presented.

3.33. *Relativistic, electromagnetic waves in pulsar winds*

Olaf Skjaeraasen: Extremely nonlinear, coherent electromagnetic waves in the context of relativistic, expanding plasma flows, where a confining external medium triggers the formation of a shock, were considered. Using a combination of analytical methods and Particle-In-Cell simulations, the mechanisms of wave generation and dissipation, as well as how the waves affect the particle distribution were discussed. For a large-amplitude wave of general polarization, any given set of wave parameters uniquely fixes the particle and energy flux associated with the flow. In cases where the wave properties can be constrained, this can be used to estimate the flow parameters.

The prime application of this work is to pulsar winds and pulsar wind termination shocks, where our model provides a viable alternative to magneto-hydrodynamic models. Using canonical parameters for the Crab, the mode couplings and transitions between the inner and outer parts of the wind were discussed. The simulation data were explored to shed new light on the microphysics of the wave as it reaches the shock.

3.34. *Coupled spin, mass, magnetic field, and orbital evolution of accreting neutron stars*

M. Mirtorabi, A.J. Khasraghi, and S. Abdolrahimi: The presented study was mainly addressed to the coupled spin, mass, magnetic field, orbital separation, and orbital period evolution of a neutron star entering a close binary system with a low mass main sequence companion, which loses mass in form of homogenous stellar wind. Flux expulsion of the magnetic field from the superfluid superconductive core of a neutron star, based on different equation of states was applied, and its subsequent decay in the crust, which also depends on conductivity of the crust, and hence on the temperature, T, and the neutron star age. The initial core and surface magnetic field were of the same order of magnitude. To derive the rate of expulsion of the magnetic flux out of the core various forces which act on the fluxoids in the interior of a neutron star were considered, including a force due to their pinning interaction with the moving neutron vortices, buoyancy force, curvature force, and viscous drag force due to magnetic scattering of electrons. Various effects accompanying mass exchange in binaries can influence the evolution of spin and magnetic field of the neutron star. The orbital separation of the binary clearly affects the estimated value of, and it itself evolves due to mass exchange between the components, mass loss from the system, and two other sinks of the orbital angular momentum namely magnetic braking and gravitational waves. The neutron star passes through four evolutionary phases (isolated pulsar-propeller-accretion from the wind of a companion- accretion resulting from Roche-lobe overflow). Models for a range of parameters, and initial orbital period, magnetic field and spin period were constructed. The impurity parameter, Q, was assumed to be constant during the whole evolution of the star and range from 1 to 0.001. Final magnetic field, spin and orbital period were presented in this paper. The suggested mechanism could explain the lower magnetic field and faster spins of millisecond pulsars that have been recycled by accretion in close binaries.

3.35. *Investigating the magnetic field of the solar corona with pulsars*

Stephen M. Ord: A novel experiment to examine both the magnetic field and electron content of the solar corona was proposed. It was intended to measure the Faraday rotation and dispersion evident in observations of background pulsar sources as they are occulted by the Sun. With a number of simultaneous lines of sight that cut different paths through the corona as the Sun rotates, strong constrains on the global topology of both the plasma and the magnetic field should be obtained. Although similar experiments have been performed using other background radio sources and space probes, this experiment differs in that many lines of sight can be examined simultaneously, and the magnetic field and plasma density can be measured independently. The Parkes radio telescope is proposed to observe a number of pulsars as they are occulted by the Sun in December 2006. An outline of the experiment and a discussion of the expected results were presented.

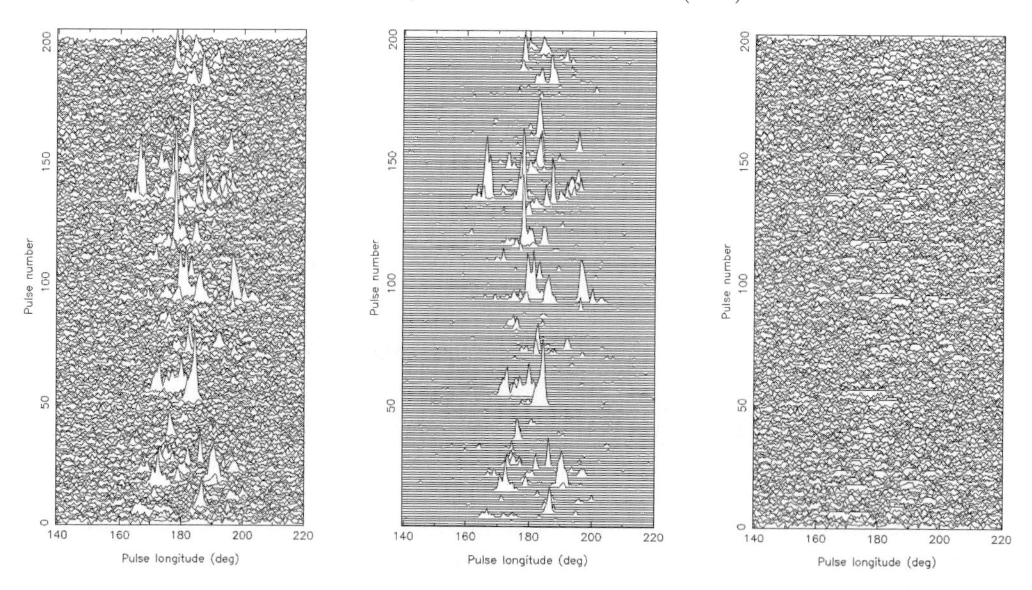

Figure 11. A typical sequence of successive pulses (left panel) from PSR B0656+14. The same pulses are shown in the middle and right panel, but there the emission is separated into the spiky and weak emission, respectively. See poster 3.37 of Weltevrede *et al.* and astro-ph/0701189 for further details.

3.36. *RRATs and PSR B1931+21*

Xiang Dong Li: The recent discovery of rotating radio transients and the quasi-periodicity of pulsar activity in the radio pulsar PSR B1931+24 has challenged the conventional theory of radio pulsar emission. It was suggested that these phenomena could be due to the interaction between the neutron star magnetosphere and the surrounding debris disk. The pattern of pulsar emission depends on whether the disk can penetrate the light cylinder and efficiently quench the processes of particle production and acceleration inside the magnetospheric gap. A precessing disk may naturally account for the switch-on/off behavior in PSR B1931+24.

3.37. *Is PSR B0656+14 a very nearby RRAT source?*

Patrick Weltevrede, Benjamin W. Stappers, Joanna M. Rankin, and Geoffrey A.E. Wright: The recently discovered RRAT sources are characterized by very bright radio bursts which, while being periodically related, occur infrequently. Bursts with the same characteristics for the known pulsar B0656+14 were found. These bursts represent pulses from the bright end of an extended smooth pulse-energy distribution and were shown to be unlike giant pulses, giant micro-pulses or the pulses of normal pulsars. The extreme peak-fluxes of the brightest of these pulses indicates that PSR B0656+14, were it not so near, could only have been discovered as an RRAT source. Longer observations of the RRATs may reveal that they, like PSR B0656+14, emit weaker emission in addition to the bursts.

The emission of PSR B0656+14 can be characterized by two separate populations of pulses: bright pulses have a narrow 'spiky' appearance consisting of short quasi-periodic bursts of emission with microstructure, in contrast to the underlying weaker broad pulses. The spiky pulses tend to appear in clusters which arise and dissipate over about 10 periods. It was demonstrated that the spiky emission builds a narrow and peaked profile,

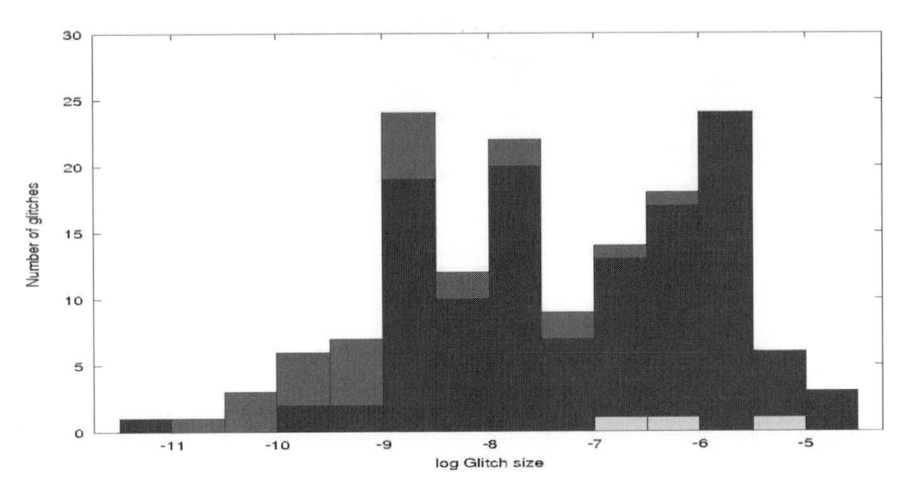

Figure 12. Glitch size histogram showing all known glitches. Glitches found in this study are shown in red, AXP glitches (overlapping) in yellow. See poster 3.38 of Janssen & Stappers and A&A 2006, 457, 611.

whereas the weak emission produces a broad hump, which is largely responsible for the shoulders in the total emission profiles at both high and low frequencies.

3.38. *Glitch observations in slow pulsars*

Gemma H. Janssen and Benjamin W. Stappers: An analysis of 5.5 years of timing observations of seven "slowly" rotating radio pulsars, made with the Westerbork Synthesis Radio Telescope was presented. The improved timing solutions were presented and 30, mostly small, new glitches were found. Particularly interesting were the results on PSR J1814−1744, which is one of the pulsars with similar rotation parameters and magnetic field strength to the anomalous X-ray pulsars (AXPs). Although the high-B radio pulsars don't show X-ray emission, and no radio emission is detected for AXPs, the roughly similar glitch parameters provide another tool to compare these classes of neutron stars. Furthermore, it was possible to detect glitches one to two orders of magnitude smaller than before, for example in our well-sampled observations of PSR B0355+54. The total number of known glitches in PSR B1737−30 was doubled, and improved statistics on glitch sizes for this pulsar individually and pulsars in general were obtained. No significant variations in dispersion measure for PSRs B1951+32 and B2224+65, two pulsars located in high-density surroundings, were detected. The effect of small glitches on timing noise was discussed. It was shown that it is possible to resolve timing-noise looking structures in the residuals of PSR B1951+32 by using a set of small glitches.

3.39. *Mode coupling in pulsar magnetospheres due to plasma gradients perpendicular to the magnetic field*

Alex C. Judge: Conventional ideas regarding plasma instabilities suggest that the polarization of pulsar radio emission should be dominated by that corresponding to the fastest growing mode. The presence of two distinct polarizations, indicating emission in two distinct modes, is, however, almost ubiquitous in observations of these objects. In order to reconcile the basic theory with the observations it has been proposed that energy is exchanged between the natural modes of the plasma as the radiation propagates through the magnetosphere of the pulsar.

The basic theory of mode-coupling in stratified media has already been developed in work relating to wave propagation in the ionosphere and the solar corona. This formalism was applied here to a relativistically streaming plasma and gradients in the plasma perpendicular to the direction of the local magnetic field were investigated as a possible mechanism for effective mode-coupling in a pulsar magnetosphere.

3.40. *Software aspects of WSRT-PUMA-II*

Ramesh Karuppusamy and Benjamin W. Stappers: The Pulsar Machine II (PuMa-II) is a state-of-the-art pulsar machine, installed at the Westerbork Synthesis Radio Telescope (WSRT), in December 2005. PuMa-II is a flexible instrument and is designed around an ensemble of 44 high-performance computers running the Linux operating system. Much of the flexibility of PuMa-II comes from the software that is being developed for this instrument. The radio signals reaching the telescope undergo several stages of electronic and software processing before a scientifically useful data product is generated. The electronic processing of signals includes the usual RF to IF conversion, analog to digital conversion and telescope dependent electronic digital delay compensation that happen in the signal chain of WSRT. Within PuMa-II, this data is acquired, stored and suitably processed. In this poster various aspects of PuMa-II software was presented and its pulsar signal processing capabilities were illustrated.

3.41. *High time resolution low-frequency pulsar studies*

Benjamin W. Stappers: Low frequency observations of radio pulsars have, to a certain extent, fallen out of favor in recent times. This is despite exciting and interesting work in Russia, Ukraine and India. The move to higher frequencies has mainly been due to the deleterious effects of the interstellar medium. However, with the increased availability of baseband recording and coherent de-dispersion techniques and new facilities such as the LFFEs at the Westerbork Synthesis Radio Telescope (WSRT) and in the future LOFAR/LWA, the interest in observations at frequencies below 300 MHz is growing again. Some exciting results on single pulse studies from observations at the WSRT (115–180 MHz) were presented here. These include the first full polarization observations of Crab giant pulses at these frequencies. The prospects for pulsar research with LOFAR. LOFAR will have unprecedented collecting area and bandwidth at frequencies below 220 MHz, allowing for a wide range of pulsar studies, in particular in emission physics, were also presented. The results from simulations which show that an all-sky survey with LOFAR could be expected to find up to 1500 new pulsars were shown. This survey would provide significant constraints on the low-end of the pulsar luminosity distribution which has important consequences for the total pulsar population. It was argued that LOFAR could detect pulsars in nearby galaxies.

3.42. *The 8gr8 Cygnus survey for New pulsars and RRATs*

Eduardo Rubio-Herrera, Robert Braun, Gemma H. Janssen, *et al.*: A survey to search for new pulsars and the recently found Rotating RAdio Transients (RRATs) in the Cygnus OB complex was currently undertaken. The survey uses the Westerbork Synthesis Radio Telescope in a unique mode which gives it the best sensitivity of any low-frequency wide-area survey. Few new pulsars were found so far. The program of using routines for the detection of RRATs. Some initial results on the new pulsars and possible transients were presented. It is expected to find a few tens of new pulsars and a similar number of RRATs. The latter discoveries should help to improve the knowledge about the population and properties of the poorly known objects as well as provide an

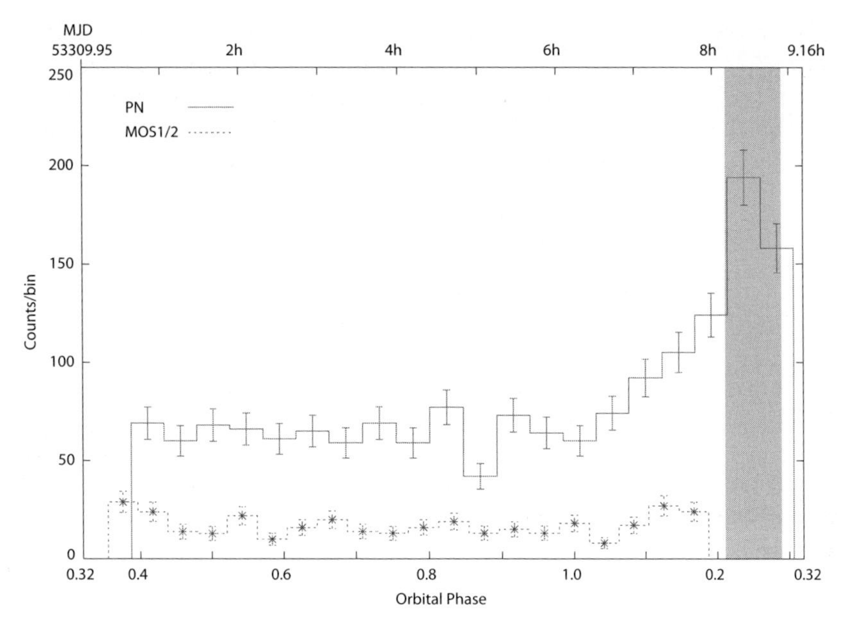

Figure 13. X-ray emission from PSR B1957+20 within 0.3–3.0 keV as function of the pulsar's orbital phase (ϕ). One complete orbital period of this system is mapped with the starting point MJD = 53309.95, i.e., $\phi = 0.32$. $\phi = 1.0$ corresponds to the ascending node of the pulsar orbit. The upper curve was obtained from the *XMM-Newton* EPIC-PN (background level at 65 cts/bin). The lower lightcurve is obtained from the MOS1/2 data. The gray strip between the orbital angle 0.21 - 0.29 indicates the eclipse of the pulsar. Phase bins with zero counts correspond to phase angles not covered in the observation. Image from poster 3.45 by Huang & Becker. (See also astro-ph/0701611.)

improved knowledge of the number of young pulsars associated with the OB complexes in the Cygnus region.

3.43. *Pulsar coherent de-dispersion observation at Urumqi Observatory*

Aili Yishamuding: Based on a Mark5A VLBI backend and a four node cluster, pulsar off-line coherent de-dispersion observations have been conducted by using Urumqi 25 m telescope. The observing system was described and the initial results were presented in this paper.

3.44. *X-ray monitoring of the pulsar PSR B1259–63*

Hsiu-Hui Huang and Werner Becker: PSR B1259−63, a rotation-powered radio pulsar with a $\sim$48 millisecond period, is in a highly eccentric ($e \simeq 0.87$) 3.4 yr orbit around the massive Be-type star SS 2883. The results of the *XMM-Newton* observations performed between 2001 and 2004 were summarized. Combining the *XMM-Newton* observations with the previous results from *ASCA*, it was found that the best-fit power-law models in 1.0 – 10.0 keV energy band show long term variations in the photon indices from $\sim$1.11 to $\sim$1.95. The X-ray flux was observed to increase by a factor of > 10 from apastron near to periastron. No X-ray pulsations at the pulsar's spin period were found in any observations so far. A model invoking the interaction between the pulsar and the stellar wind was likely to explain the observed orbital phase-dependent time variability in the X-ray flux and spectrum.

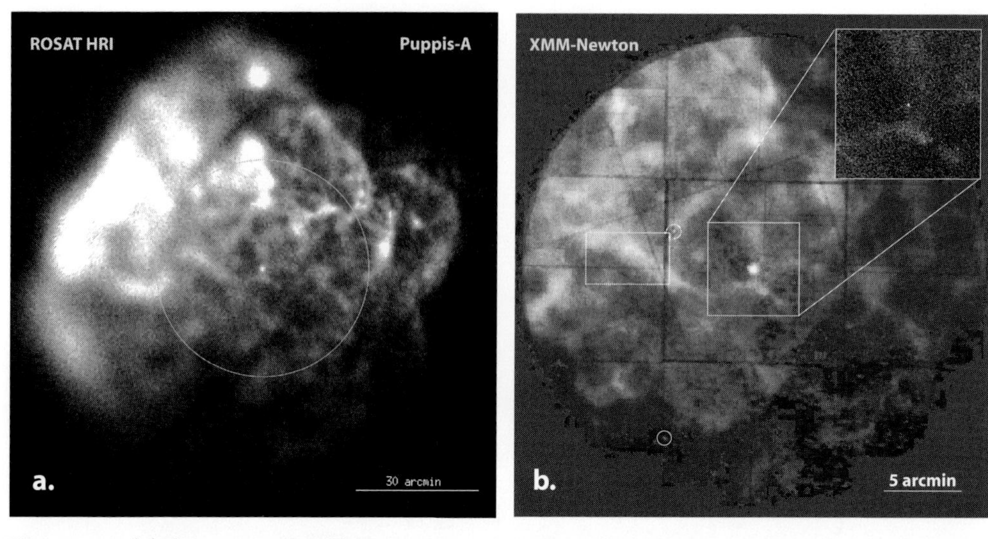

Figure 14. (*a*) Composite *ROSAT*-HRI image of the Puppis-A supernova remnant. The blue ring indicates the 30′ central region observed by *XMM-Newton*. (*b*) *XMM-Newton*-MOS1/2 false color image of the central region of Puppis-A (red: 0.3–0.75 keV, green: 0.75–2 keV, blue: 2–10 keV). The central source is the CCO RX 0822−4300. Inset: the squared region as observed by the *Chandra*-HRC-I. Further details in poster 3.47 by Hui & Becker and A&A 454, 543 (2006).

3.45. *XMM-Newton observation of PSR B1957+20*

Hsiu-Hui Huang and Werner Becker: The "Black Widow pulsar", PSR 1957+20, is a millisecond pulsar which is in a 9.16-hr binary system. Hα bow-shock nebula created by the interaction between the relativistic wind of the pulsar and the surrounding ISM and ablation of the low-mass companion star by the pulsar wind were observed. 30 ksec observation of PSR B1957+20, using the EPIC-MOS detector on-board the *XMM-Newton* observatory, were reported. The X-ray diffused emissions detected from this source was found to be consistent with the results derived from *Chandra* observations. The spectrum of the nebular emission was modeled with a single power-law spectrum of photon index $2.1^{+0.4}_{-0.3}$. This extended emission generated by accelerated particles in the post shock flow was considered to explain this result. For the first time, a significant X-ray flux modulation near to the pulsar's radio eclipse was detected.

3.46. *Optical observations of binary millisecond X-ray pulsars in quiescence*

Paul J. Callanan, Mark T. Reynolds, Alexei V. Filippenko, *et al.*: The discovery of accreting binary millisecond pulsars finally provided firm confirmation of the link between bright accreting Low Mass X-ray Binaries and millisecond pulsars. Little is known about their optical properties in quiescence, however. Here the optical observations of SAX J1808.4−3658 and IGR J00291+5934 in quiescence were presented, and comparison of them to other quiescent X-ray transients was made.

3.47. *X-Ray studies of the central compact objects in Puppis-A & RX J0852.0−4622*

Chung Yue Hui and Werner Becker: The supernova remnants (SNRs) Puppis-A and RX J0852.0−4622 (Vela-Junior) are located along the line of sight towards the outer rim of the Vela SNR. Central compact objects (CCOs) were discovered in each of them. Both CCOs are thought to be the compact stellar remnants formed in core-collapsed supernova explosions. Nevertheless, the emission properties observed from these sources

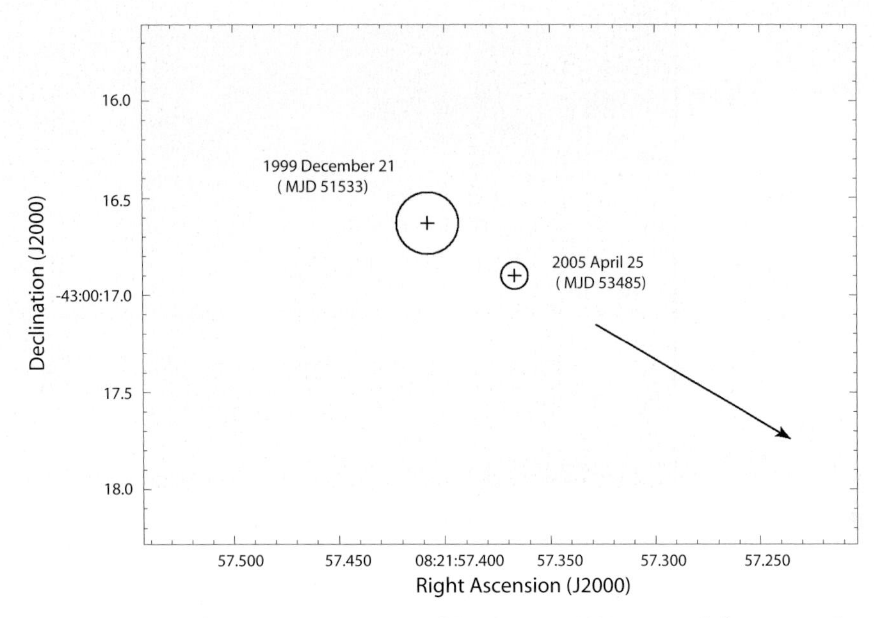

Figure 15. The best-fitted X-ray positions of RX J0822−4300 at two different epochs separated by 1952 d are marked by crosses. The circles indicate the 1 σ error. The arrow shows the direction of proper motion inferred from both positions. Image from poster 3.48 of Hui & Becker. See also Hui & Becker (2006, A&A,457, 33).

were found to be completely different from what is observed in other young canonical neutron stars. Based on observations with the X-ray observatories *Chandra* and *XMM-Newton*, the most recent results from a detailed spectro-imaging and timing analysis of these two enigmatic sources were presented.

3.48. *Probing the proper motion of the central compact object in Puppis-A*

Chung Yue Hui and Werner Becker: Using two observations taken with the High Resolution Camera (HRC-I) aboard the *Chandra* X-ray satellite, we have examined the central compact object RX J0822-4300 for a possible proper motion. The position of RX J0822−4300 was found to be different by 0.574 ± 0.184 arcsec, implying a proper motion of 107.49 ± 34.46 mas/yr with a position angle of 241 ± 24 deg. For a distance of 2.2 kpc, this proper motion is equivalent to a recoil velocity of 1121.79 ± 359.60 km/s. Both the magnitude and the direction of the proper motion are in agreement with the birth place of RX J0822−4300 being near to the optical expansion center of the supernova remnant. Although this is a promising indication of a fast moving compact object in a supernova remnant, the relative large error prevents any constraining conclusion.

3.49. *Exposing drifting subpulses from the slowest to the fastest pulsars*

Joeri van Leeuwen: Pulsar emission is surprisingly similar over a vast range of periods and magnetic fields: all the way from the 2-millisecond 10^8 G recycled pulsars to the 6-second 10^{14} G magnetar-like regular pulsars. It was investigated how the curious instabilities called 'drifting subpulses' can discern between different mechanisms for pulsar emission.

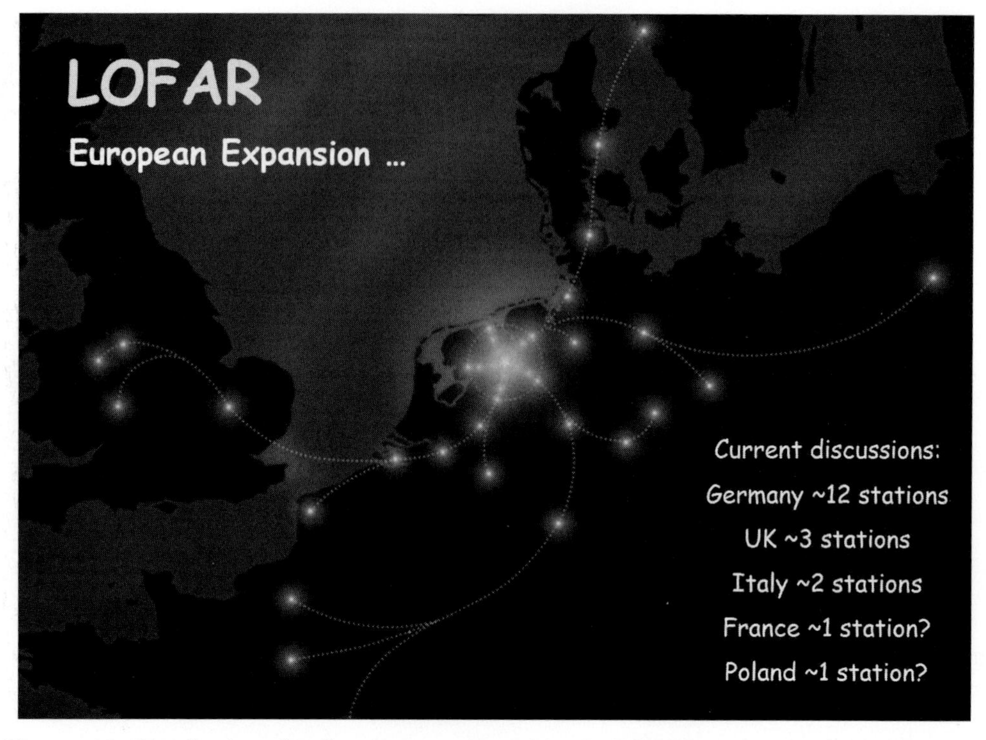

Figure 16. Distribution of radio antennas planned in the LOFAR project as presented in the talk by M. Kramer and in poster 3.50 by J. van Leeuwen and B.W. Stappers.

3.50. *Pulsar research with LOFAR, the first next-generation radio telescope*

Joeri van Leeuwen and Benjamin W. Stappers: LOFAR is a low-frequency radio telescope of revolutionary design that is currently being constructed and will become operational in 2007. In stark contrast to radio dishes, LOFAR is the first telescope that relies on a central supercomputer to combine the signals of ten thousand individual dipoles to form several extremely sensitive, independently steerable beams on the sky. It was discussed how LOFAR opens up a new frequency window with unprecedented sensitivity and why LOFAR will have considerable impact on radio pulsar research.

3.51. *Non-dipolar surface magnetic field of neutron stars: general approach and observational consequences*

George I. Melikidze, A. Szary, and Janusz Gil: It is widely accepted that the magnetic field structure near the surface of neutron stars may significantly differ from the star centered global dipole structure. Due to flux conservation of the open magnetic field lines, strong non-dipolar surface field results in significant shrinking of the canonical polar cap, in general. We have modeled different possible configurations and found out that for some configurations the pair creation is possible not only along the open field lines, but also in the region of closed field lines. Therefore, in this case, we can naturally explain some peculiarities of pulsar activities, such as unusual thermal x-ray emission, reversible radio emission and rotating radio transients.

The pairs created along the closed field lines can easily reach the stellar surface near the polar cap at the opposite side of the neutron star and heat the surface area that can even exceed that of the canonical polar cap. Both smaller (often) and larger (rarely) bolometric surface areas of the hot polar cap are observed.

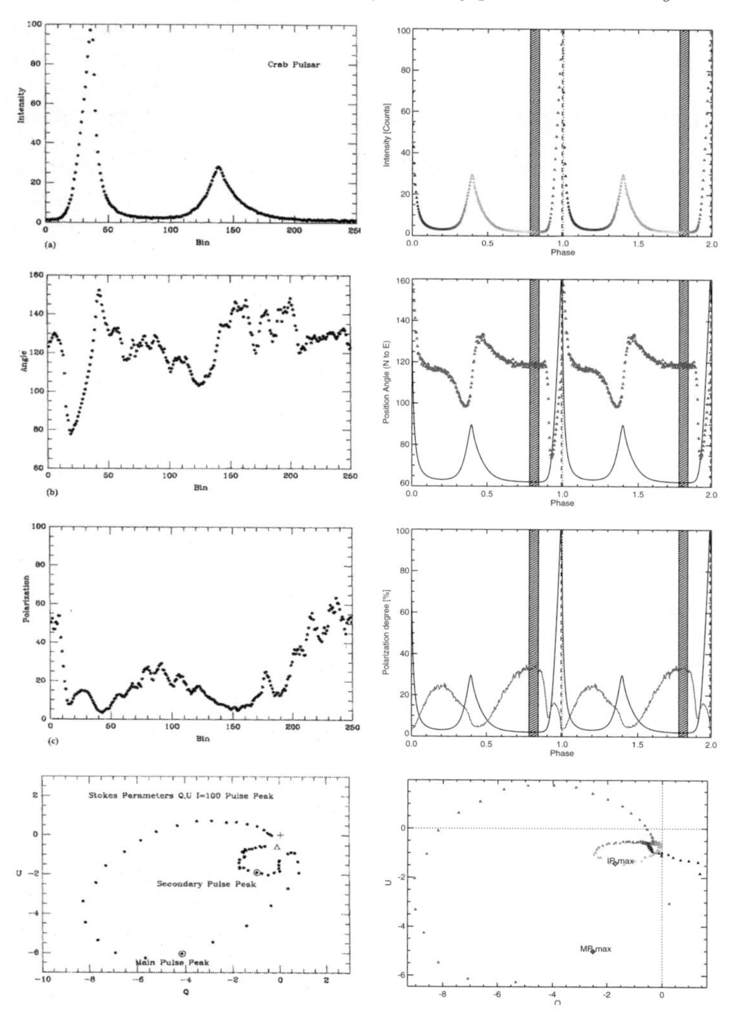

Figure 17. Comparison of the optical polarization of the Crab pulsar obtained by Smith *et al.* 1988 (*left*) and Slowikowska, Kanbach, and Stefanescu (*right*). From *top* to *bottom*: the intensity, position angle, and polarization degree are plotted as a function of the pulsar phase, while the Stokes parameters Q, U are plotted as a vector diagram. There are 250 bins per cycle in both cases, the only difference is that for clarity two periods are shown in the right chart. As a DC component Smith *et al.* took 50 out of 250 bins, whereas in the right chart only 7 % of the rotational period was taken. Figure and caption from Slowikowska *et al.* (3.53).

In the frame of this model, we can easily realize the configuration, which allows the pair creation near both polar caps (along the same field). In this case, two streams of the pair plasma penetrate each other creating a favorable condition for the two-stream instability to be developed. Such a process can lead to the radio emission generation, either in quasi-stationary or stochastic process. Consequently, either quasi-stationary reversible radiation, or stochastic emission of the transients can be observed.

3.52. *Glitches in the Vela pulsar*

Stephen J. Buchner, and Claire S. Flanagan: The Vela pulsar undergoes occasional sudden spin-ups in rotational frequency. The recovery from these glitches provides insight into the internal structure of the neutron star. The HartRAO was used to monitor Vela since 1984 and eight large glitches were observed. These data were presented in this paper.

3.53. *Optical polarization of the Crab pulsar with 10 μs time resolution*

Agnieszka Slowikowska, Gottfried Kanbach, and A. Stefanescu: The Crab nebula and pulsar were observed for about 25 hours with the high-speed photo-polarimeter OPTIMA in November 2003 at the Nordic Optical Telescope (NOT), La Palma. The instrument's sensitivity (white light) extends from about 450 nm to 950 nm and reaches about 60 %. Linear polarization is measured with a continuously rotating polaroid filter which modulates the incoming radiation. The astronomical target is viewed through the polaroid and imaged onto a hexagonal bundle of optical fibers which are coupled to single photon APD counters. The spacing and size of the fibers at the NOT correspond to about 2 arcsec. GPS based time tagging of single photons with 4 microsec resolution, together with the instantaneous determination of the angular position of the rotating polaroid filter, allowed to measure the phase dependent linear polarization state of the pulsar and the surrounding nebula simultaneously.

The Crab pulsar and its net optical polarization were determined at all phases of rotation with extremely high statistical accuracy. On time scales of a few tens of microseconds significant details of the polarization of the main emission peak became visible.

Scientific Organizing Committee

Jonathan Arons (USA), Werner Becker (Germany, co-chair), Kwong Sang Cheng (China HK), Ocker C. de Jager (South Africa), Janusz A. Gil (Poland, co-chair), Nobuyuki Kawai (Japan), Michael Kramer (UK), H. Jan M.E. Kuijpers (Netherlands), Roger W. Romani (USA), Bronislaw Rudak (Poland), David J. Thompson (USA), and Vladimir V. Usov (Israel/Russia).

Highlights of Astronomy, Volume 14
IAU XXVI General Assembly, 14-25 August 2006
Karel A. van der Hucht, ed.

doi:10.1017/S1743921307010071

Joint Discussion 3
Solar active regions and 3D magnetic structure

Debi Prasad Choudhary[1] and Michal Sobotka[2] (eds.)

[1]Department of Physics and Astronomy, California State University Northridge,
18111 Nordhoff Street, Northridge, CA, 19330, USA
email: debiprasad.choudhary@csun.edu

[2]Astronomical Institute, Academy of Sciences of the Czech Republic,
Fričova 298, 25165 Ondřejov, Czech Republic
email: msobotka@asu.cas.cz

Abstract. Keeping in view of the modern powerful observing tools, among others *Hinode* (formerly *SOLAR-B*), *STEREO* and Frequency-Agile Solar Radiotelescope, and sophisticated modelling techniques, Joint Discussion 3 during the IAU General Assembly 2006 focused on the properties of magnetic field of solar active regions starting in deep interior of the Sun, from where they buoyantly rise to the coronal heights where the site of most explosive events are located. Intimately related with the active regions, the origin and evolution of the magnetic field of quiet Sun, the large scale chromospheric structures were also the focal point of the Joint Discussion. The theoretical modelling of the generation and dynamics of magnetic field in solar convective zone show that the interaction of the magnetic field with the Coriolis force and helical turbulent convection results in the tilts and twists in the emerging flux. In the photosphere, some of these fluxes appear in sunspots with field strengths up to about 6100 G. Spectro-polarimetric measurements reveal that the line of sight velocities and magnetic field of these locations are found to be uncombed and depend on depth in the atmosphere and exhibit gradients or discontinuities. The inclined magnetic fields beyond penumbra appear as moving magnetic features that do not rise above upper photospheric heights. As the flux rises, the solar chromosphere is the most immediate and intermediary layer where competitive magnetic forces begin to dominate their thermodynamic counterparts. The magnetic field at these heights is now measured using several diagnostic lines such as Ca II 854.2 nm, H I 656.3 nm, and He I 1083.0 nm. The radio observations show that the coronal magnetic field of post flare loops are of the order of 30 G, which might represent the force-free magnetic state of active region in the corona. The temperatures at these coronal heights, derived from the line widths, are in the range from 2.4 to 3.7 million degree. The same line profile measurements indicate the existence of asymmetric flows in the corona. The theoretical extrapolation of photospheric field into coronal heights and their comparison with the observations show that there exists a complex topology with separatrices associated to coronal null points. The interaction of these structures often lead to flares and coronal mass ejections. The current MHD modelling of active region field shows that for coronal mass ejection both local active region magnetic field and global magnetic field due to the surrounding magnetic flux are important. Here, we present an extended summary of the papers presented in Joint Discussion 03 and open questions related to the solar magnetic field that are likely to be the prime issue with the modern observing facilities such as *Hinode* and *STEREO* missions.

Keywords. Solar activity, sunspots, magnetic field, corona, chromosphere, flare, radio observations, CME, magnetic field extrapolation

1. Introduction

The magnetic field of the Sun is responsible for most of its visible dynamic features, including the most energetic events that can affect the near earth space environment, producing space weather. The solar magnetic field is generated below the visible layer (the photosphere) and erupts into the solar atmosphere. The cross-section of the erupting field structure at the photosphere is observed as an active region, above which there exists a complex three-dimensional magnetic "dome". Many fundamental physical processes take place in and above active regions that govern the dynamics of the hot, magnetized plasma manifesting in the observed features. In order to understand these processes, a detailed knowledge of the origin and dynamics of magnetic field is essential. In the recent past there have been spectacular advances in various topics related to this field such as: magnetic helicity, temporal evolution of magnetic field creating large-scale structure; thermal and magnetic instabilities leading to fine-scale structure; wave dissipation and reconnection providing coronal heating; instability and non-equilibrium states leading to eruptions. The future observations in infrared wavelength and from space platforms combined with the sophisticated computer modeling are expected to make equally impressive advances.

The photospheric vector magnetic field of solar active regions has been measured on a synoptic basis for the last 30 years. From these, the 3D magnetic field is derived using numerical models. The spatial resolution of the measurements has improved steadily, and the models have been able to incorporate some departures from the force-free field approximation. Several of there models are now capable of reproducing the observed sheared coronal magnetic features.

Until now the field in the chromosphere and corona has been largely derived by extrapolating the photospheric measurements. Recently, the actual measurements of these fields have been carried out using both the Zeeman and Hanle effects. In the last few years, several exploratory measurements of magnetic fields in spectral lines originating at chromospheric and coronal heights have shown promising results. The technological advances in the field of detectors and polarizing optics make it possible to design and fabricate vector magnetographs that can be used to measure vector magnetic fields simultaneously in several layers above the photosphere. The longitudinal chromospheric field measurements using the Zeeman effect have been carried out on a regular basis at NSO/KP, Hawaii and Huairou in China and are being planned at NASA/MSFC and San Fernando Observatory. At the same time the Hanle effect has emerged as a new and powerful diagnostic tool. The Hanle effect is useful both for measuring weak horizontal magnetic fields in the solar chromosphere, and for exploring the sub-resolution tangled or turbulent magnetic fields. The turbulent magnetic field cannot be seen by the Zeeman effect, but has been found to carry a vast amount of "hidden" magnetic flux. The recently uncovered wealth of polarization phenomena throughout the "Second Solar Spectrum", formed by coherent scattering processes, has an exciting and entirely novel diagnostic potential.

The other aspect of solar magnetic field is inverting the observed polarization measurements in deriving the magnetic field. There is a considerable progress both in terms of atomic and molecular physics and radiative transfer analysis in this crucial topic. On the other hand, helioseismic analysis is beginning to provide interesting results in the sub-photospheric structure and flows in active regions. These results can be used to derive sub-photospheric magnetic fields in active regions and may in the near future also help us in refining models about how active regions emerge.

There are many new instruments that are now under development and poised to produce impressive results in near future. The solar optical space telescope with 50 cm aperture and its back-end instrumentation on *Hinode (Solar-B)*, successfully launched in September 2006, is poised to revolutionize our understanding the fine scale magnetic structure on the Sun. Other space initiatives such as *STEREO* will provide us the stereoscopic three dimensional vision of solar magnetic structures and their dynamics. Many ground-based instruments such as GREGOR, new 1-m SST at La Palma and NSO/SP adaptive optics are expected to yield new results in this field. There are other equally impressive instruments such as the chromospheric magnetograph at BBSO, Hawaii, IR spectropolarimeter at San Fernando Observatory of California State University Northridge and Norikura Solar Observatory, Japan, that are making important contribution in terms of magnetic field measurements. On the other hand radio emission is sensitive to magnetic fields throughout the solar atmosphere, and there is a long history of such measurements. A new radio facility, the Frequency-Agile Solar Radiotelescope (FASR) is now being designed with a main goal of routinely measuring coronal magnetic fields at high spatial resolution. In this symposium we shall focus to the measurement of solar magnetic field in different wavelengths and different techniques.

With this scientific rationale, the Joint Discussion 3 (JD03) was organized to summarize the progress in this field and explore the possible outstanding problems that can be addressed with the future instruments. The one and half day discussion consisted of 11 invited talks, 15 oral and 65 poster presentation demonstrating the overwhelming response. JD03 dealt with topics related to magnetic field in the solar convective zone, the observation and inversion techniques for photospheric, chromospheric and coronal magnetic field and small-scale structures. The role and properties of magnetic structure in eruptive processes leading to flares, filament eruptions and Coronal Mass Ejections were also the subject of JD03. This summary is divided into sections dealing with different parts of the Sun.

2. Magnetic field in the convection zone

The sunspots are the cross sections of the magnetic structures that protrude into the solar atmosphere through the photosphere. The kinematic dynamo model among others was reviewed by **Dibyendu Nandy**. He showed that the toroidal field is generated by strong differential rotation in the tachocline. The buoyant eruptions bring the flux to photospheric layers. The decay of tilted active regions produce the poloidal field. The entire evolution of the solar magnetic field is governed by α- and ω-effects, meridional circulation and magnetic buoyancy as shown in Fig. 1. Using observed large-scale flows it is now possible to reproduce the observed large-scale magnetic field evolution reasonably well. While the meridional circulation transports poloidal and toroidal flux, the diffusion transports and also destroys them. The three possible sources of α-effect that govern the surface flux dynamics are (1) Coriolis force during flux tube rise in convection zone, (2) diffusion of tilted dipole fields at surface, or (3) differential rotation in tachocline. The flux transport mediated time-delay dynamics introduces a memory mechanism in the Sun that may be used for predictive purposes. The question remains if the local and global dynamos are linked? The tilt and twist of the flux are generated by the Coriolis force and helical convection. The Coriolis force in the northern hemisphere would produce negative kinetic helicity that will distort the axis of rising flux tubes in a right handed sense (positive writhe), therefore negative tilt, and vice-versa in the south will generate a negative twist in the southern hemisphere. Analysis of twist distribution and its dependence on latitude suggests that the Solar Convection Zone (SCZ) turbulence is isotropic in latitude. The Active Region (AR) twist dependence on flux indicates that larger flux tubes are less affected by turbulent buffeting and hence pick up less twist

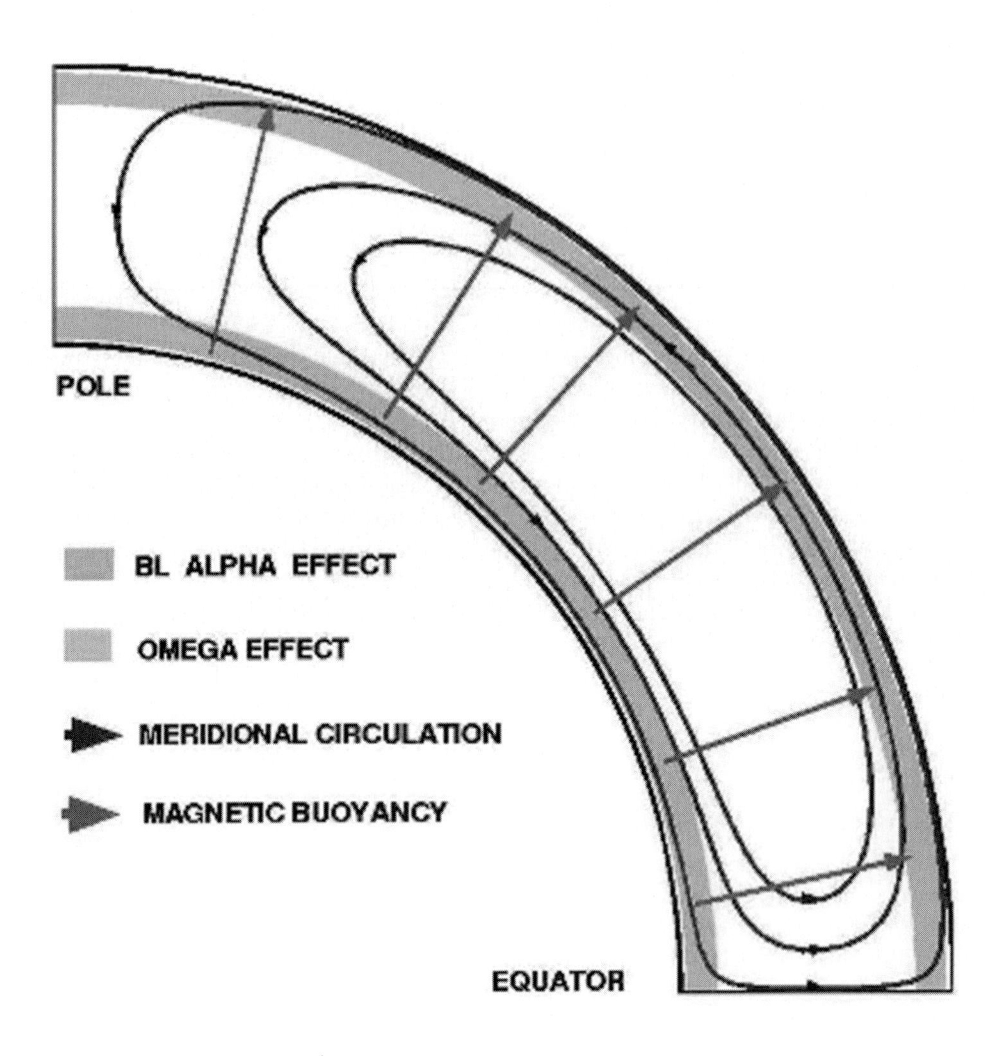

Figure 1. The sites of α and ω effects, buoyant force and meridional flow.

and scatter. Analysis of twist and tilt relationship supports a scenario where flux tubes at the base of the SCZ begin their outward journey with a wide range of initial twist, some with high enough twist to make them kink unstable. This may suggest that some physical process at the base of the SCZ, e.g., the dynamo or the process of flux tube formation, creates or imparts this initial twist.

The results of the numerical simulations flux ropes in the magnetic SCZ by **Soren B.F. Dorch** showed how twisted magnetic flux ropes interact with a magnetized model envelope. Buoyant magnetic flux ropes have been extensively studied through numerical simulations, cf. the review in Fan (2004). However, so far studies of the interaction of flux ropes with poloidal fields have solely dealt with flux ropes that emerge into the solar corona, e.g., Archontis *et al.* (2004), i.e., not with a magnetic SCZ. The question

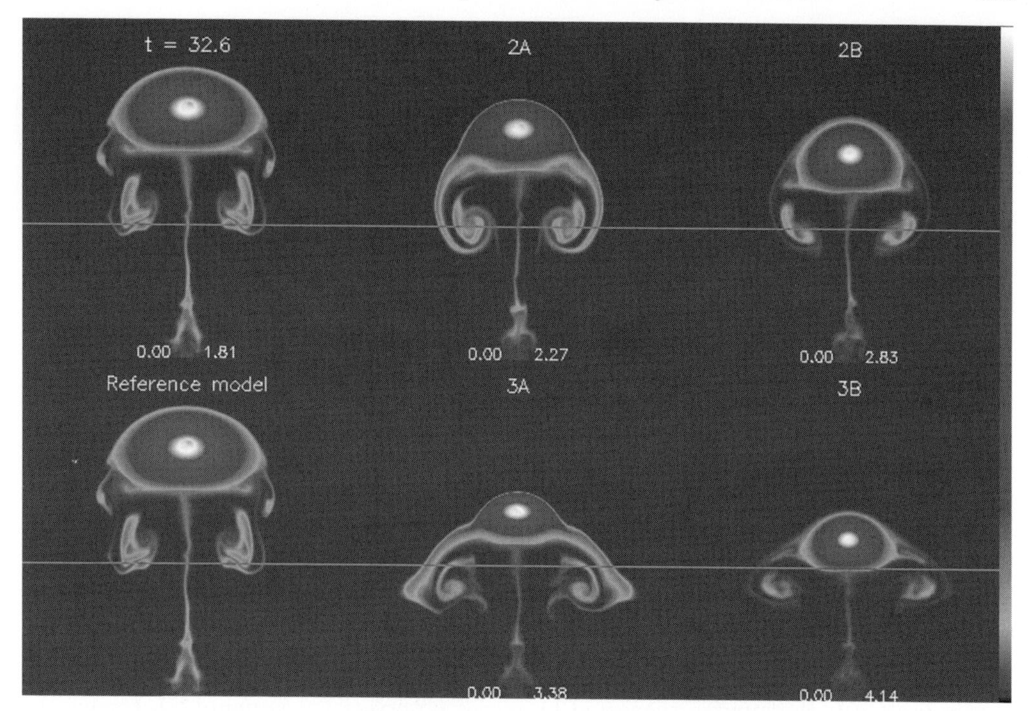

Figure 2. Grey-scale panel showing snapshots of parallel magnetic field B_z at $t \simeq 32.6$ time units for models with moderately magnetized envelopes. Minimum (zero) and maximum magnetic field strengths are stated for every snapshot (in units of 354 kG). Vertical dimension corresponds to x in the simulations. Shown are a reference model (shown twice), and in the upper row two models with initial $B_{\mathrm{SCZ}}/B_\phi = 18$ (but oppositely twisted) and in the lower row two models with initial $B_{\mathrm{SCZ}}/B_\phi = 9$ (and oppositely twisted). Also shown are horizontal lines indicating the initial position of the magnetized layer.

addressed here is how the polarity and strength of the field in a magnetized convection zone may affect the rise and evolution of twisted flux ropes. When the predominantly toroidal buoyant flux ropes rise through the SCZ, they encounter a poloidal magnetic field that has a component perpendicular to the ropes' axes. However, since the flux ropes are twisted, the transversal field components may be either parallel or anti-parallel to the average magnetic field in the SCZ. Using the MHD code by Galsgaard & Nordlund (1997) and others, Dorch performed several 2.5-dimensional numerical simulations of buoyant twisted flux ropes rising into a uniform magnetic field, cf. Dorch (2007). Some of the results are illustrated in Fig. 2 that shows 2-dimensional cross sections of the flux ropes (the parallel field strength B_z). There are both quantitative and qualitative differences depending on the ratio of the SCZ magnetic field to the rope's twisted field component, i.e., B_{SCZ}/B_ϕ. For a relatively strong poloidal field the rise is slower and therefore the flux ropes reach a lower height in the same amount of time. In certain cases the rise can be completely halted. It turns out that anti-parallel twisted ropes reach higher and faster. When it comes to the flux ropes' topology, the geometrical shape of ropes also depends on the strength and sign of the poloidal field: the apex of anti-parallel twisted ropes are flatter and have steeper magnetic gradients in their axial field.

Valeryj N. Kryvodubskyj presented the sign change of helicity parameter in the SCZ and observed meridional migration of the surface magnetic patterns. In the frames of the macroscopic magnetohydrodynamics in the mixing length approach, he investigated the helicity parameter of turbulent convection in the solar convection zone (SCZ) with taking into account the radial inhomogeneity of turbulent pulsations. Allowance for the sharp radial gradient in turbulent velocity results in a change of sign of the helicity parameter from positive to negative (in the northern hemisphere) and, as a consequence, in the formation of the negative α-effect layer near the bottom of the SCZ (which has thickness about 30 000 – 40 000 km). The mean negative helicity parameter amplitude (averaged over the negative values region) amounts to $2\,\mathrm{m\ s^{-1}}$. When taking into consideration a negative sign of α-effect in deep layers, then the physical conditions at mid- to low-latitudes domain (where a radial gradient of angular velocity, derived from recent helioseismological measurements, has positive sign) are favourable for α - ω dynamo waves to propagate toward equator producing the sunspots migration as the solar cycle progresses. Since at high latitudes the radial gradient of angular velocity has opposite sign, then here the excited dynamo waves must move toward the poles.

The effects of the shape and width of the tachocline on the butterfly diagram of the solar cycle was investigated by **Gianantonio Guerrero, and Elisabete M. de Gouveia Dal Pino**. Flux-dominated kinematic solar dynamo models, which have demonstrated to be quite successful in reproducing most of the observed features of the solar magnetic cycle, generally produce an inappropriate latitudinal distribution of the toroidal magnetic fields, showing fields of large magnitude in polar regions where the radial shear has a maximum amplitude. Employing a kinematic solar dynamo model, they explore the contribution of both the radial and the latitudinal shear in the generation of the toroidal magnetic fields by varying the shape and the width of the solar tachocline. They find that a prolate tachocline (as suggested by recent observations) is unable to resolve the aforementioned problem. On the other hand, they find that the latitudinal component of the shear term of the dynamo equation is always dominant over the radial component for producing toroidal field amplification, and that this field is suppressed at high latitudes if the tachocline has a width $d \leqslant 0.03\,\mathrm{R_\odot}$ or $d \geqslant 0.1\,\mathrm{R_\odot}$. The best fit to the observed butterfly diagram is found for a tachocline with width smaller than 3 % of the solar radius. The authors note, however, that this result is somewhat sensitive to the adopted magnetic diffusivity. In the present work, they have adopted a diffusion profile for which the tachocline is approximately half sub-adiabatic and half turbulent. If taking, for instance, a tachocline with a smaller turbulent portion, then its width could be a slightly larger than the value above.

3. Photospheric magnetic field

The fine structure of the sunspot penumbra is an ideal target to study the effects of radiative magnetoconvection in inclined magnetic field and to compare theoretical models with observed properties. **Rolf Schlichenmaier** presented the spectroscopic and spectro-polarimetric measurements, which are interpreted to reconstruct the thermal stratification and the complex topology of the magnetic field and the flow field. Such measurements reveal that the line-of-sight velocity and the magnetic field depend on depth in the atmosphere and exhibit gradients or discontinuities. The magnetic field is found to be uncombed, with an essentially horizontal component that carries a radial outward flow (Evershed flow), and a less inclined magnetic field component. He referred to the discovery of dark-cored bright filaments and presented spectroscopic measurements, which demonstrate that the dark cores are associated with the Evershed flow.

Debi P. Choudhary, and K.S. Balasubramaniam showed that the inclined magnetic fields beyond the penumbra appear as Moving Magnetic Features (MMF) that do not rise above upper photospheric heights. The dynamical properties of the moving magnetic features of the active region NOAA 9662 observed on 16 and 17 October 2001 using the Advanced Stokes Polarimeter (ASP) at Richard B. Dunn Solar Telescope of National Solar Observatory showed that spatial and temporal properties of the MMF filed is convoluted.

William C. Livingston, John W. Harvey, Olena Malanushenko, and Larry Webster showed that the sunspot magnetic fields can be as high as 6000 G. Their estimate is based on the fact that the photoelectric magnetographs underestimate field strength in sunspot umbrae because of scattered light. Two techniques that do work are photographic spectra or visual observations where the Zeeman σ-components are matched by the observer with the Hale-Nicholson tipping plate and the plate angle is converted to the field strength. They studied field measures of 32 000 spots, years 1917 to 2004, from the archives of Mt. Wilson, Potsdam, Rome, and Crimea and found 58 spot groups with fields of 4000 G or more. A careful look at the time history of equipment used at Mt. Wilson reveals that the non-linearity of the glass plate at large tilt angles was not taken properly into account, perhaps because of their rarity. When they correct the Mt. Wilson values for strong fields they find several cases of 5000 G and one of 6100 G. Often such strong fields are associated with the presence of light bridges, but not always.

The simultaneous full-Stokes imaging polarimetry observations in Fe I 6303 and Hα was presented by **Yoichiro Hanaoka**. The Solar Flare Telescope of NAOJ, Mitaka, has been observing the photospheric magnetic field in the Fe I 6303 line with the imaging polarimetry technique for more than 10 years. Recently, they replaced the polarimeter for the Fe I 6303 observation with a high-sensitive one, using ferroelectric liquid crystals (FLC). Another FLC polarimeter was also installed into the Hα imager of the Solar Flare Telescope. Their FLC polarimeters have the polarization sensitivity of the order of 10^{-4}, and with this sensitivity not only substantial improvement of the quality of photospheric vector magnetograms, but also detection of the full-Stokes polarization signals in the Hα line were realized. Now, they are operating simultaneous full-Stokes imaging polarimetry observations in the photosphere and in the chromosphere, and the three-dimensional information of the structure of the magnetic field can be obtained.

The other important aspect of determining the magnetic field from polarimetry observations are the inversion techniques. **Julio Cesar Ramírez Vélez, Arturo López Ariste, and Meir Semel** presented the spectropolarimetric observational data and the correspondent magnetic field inversion of the solar photosphere in quiet regions. The presence of atomic coupling between the nuclei momentum I and the total electron momentum $J_{\rm LS}$, known as hyperfine structure, atomic regime where the total momentum $F = I + J$, was found in the observed data for the atom Mn I at 553.7 nm. The signature of these hyperfine structure is clearly detected in circular polarisation in form of protuberance in the central part of the profile. They have developed an inversion code of the magnetic field using the Principal Component Analysis (PCA). A statistical test is applied to the code obtaining satisfactory inversion results. The inversion of the observed profiles, using the circular polarisation V and the intensity I, shows mixed regions with values of B higher than 800 G and lower than 600 G associated with the network and internetwork respectively.

Lofti Yelles Chaouche, Mark Cheung, Andreas Lagg, and Sami K. Solanki presented simulations of flux emergence in the solar photosphere diagnostics based on 3D radiation-MHD. They investigate flux tube emergence in the solar photosphere using a diagnostic procedure based on analyzing Stokes signals from different spectral lines calculated in 3D radiation-MHD simulations. The simulations include the effects of radiative transport and partial ionization and cover layers both above and below the solar surface.

The simulations consider the emergence of a twisted magnetic flux tube through the solar surface. They consider different stages in the emergence process, starting from the early appearance of the flux tube at the solar surface, and following the emergence process until the emerged flux looks similar to a normal bipolar region. At every stage they compute line profiles by numerically solving the Unno-Rachkovsky equations at every horizontal grid point. Then, following observational practice, they apply Milne-Eddington-type inversions to the synthetic spectra in order to retrieve different atmospheric parameters. They include the influence of spatial smearing on the deduced atmospheric parameters to identify signatures of different stages of flux emergence in the solar photosphere.

The quiet Sun magnetic field is an important ingredient in the solar magnetism, which was reviewed by **Sami K. Solanki**. The amount of magnetic flux in the quiet Sun, its distribution on the solar surface, its origin and evolution are topics that have been strongly debated in recent years. In JD03 he presented a critical overview of some of the results obtained from high-sensitivity polarimetric data and from numerical 3D radiation-MHD simulations. He posed several important questions that are given in the concluding remarks of this article.

Continuing the topic of formation of small-scale plasma structures in the photosphere, **Yu. V. Kyzyurov** considered the possibility of formation of such structures in the turbulent flows of photospheric gas and to analyse the dependence of their spectrum and intensity on height and on the strength of magnetic field. Results of observations with high-spatial resolution clearly show that the solar photospheric flows include both organized and stochastic motions. The spectra associated with the stochastic velocity fields obey power laws, which are close to the law of Kolmogorov turbulence. Since the degree of photospheric gas ionization is small, electrically charged particles can be regarded as passive contaminants embedded in motions of the gas. Taking these facts into account, the present consideration was based on the macroscopic description of the charged particle behaviour in the ambipolar diffusion approximation, and under the given velocity field of the neutral gas. Length-scales of the plasma structures were restricted to an inertial range of the turbulence, where the random velocity field of the gas is homogeneous and isotropic one with the known statistical properties. An analytic expression for the spatial spectrum of the structures was derived as well as the formula for estimation of the RMS level of their intensity. The plasma structure formation was analysed in the height range from 150 to 350 km. It was shown that in spite of changes in parameters of the photosphere and the turbulence with altitude (e.g., an increase in the kinematic viscosity of the gas, and decrease in the rate of turbulent energy dissipation), the shape of the spectrum and the intensity of the structures remain almost unchanged in the calculations. At the same time, the increase in the magnetic field from 5 to 250 G results in the increase in the intensity of deviation from the mean plasma density from 2.5 to 6 % for the length-scales of structures smaller than 300 km, and in the slope of the spectrum, if it is approximated by a power law, then power index takes values from -1.2 to -1.6. The obtained result seems to be important for better understanding of basic solar phenomena, such as generation of the random component of magnetic field or chaotic excitation of solar oscillations.

Transport of solar magnetic elements in the intergranular lanes was presented by **A.A. Stanislavsky, and Dmytro V. Mukha**. The Sun is an excellent natural laboratory for studying the nonlinear plasma phenomena. Photospheric convective motions of magnetic flux elements cause much of the solar surface phenomena. Using the observed solar data to estimate diffusion properties of photospheric bright points for times periods less than 20 minutes, they construct a transport model describing a sub-diffusive motion of solar magnetic elements in the intergranular lanes. The anomalous diffusion arises from the probabilistic distribution of waiting times attracted to a stable law. This suggests an analytical explanation of this observations and a revision of the Leightons model.

Observations of magnetic fields in the quiet Sun indicate that kilogauss-strength fields can be found in the intergranular lanes. Since the magnetic energy of these localised features greatly exceeds estimates of the kinetic energy of the surrounding granular convection, it is difficult to see how these features could be formed simply by convective flux concentration. **Paul J. Bushby** presented super-equipartition fields in simulations of photospheric magnetoconvection. Idealised, high-resolution simulations of three-dimensional compressible magnetoconvection are used to investigate the formation of these features numerically. Initially he takes a fully developed non-magnetic convective state into which he inserts a weak, uniform, vertical magnetic field. Magnetic flux is rapidly swept into the convective downflows, where it is concentrated into localised regions. As the field strength within these regions becomes dynamically significant, the high magnetic pressure leads to partial evacuation (via the convective downflows). This process continues until the field is strong enough to locally suppress convection. Provided that the magnetic Reynolds number is large enough, the strength of the resulting magnetic fields significantly exceeds the (so called) "equipartition" value, with the dynamical effects of the surrounding convection playing an important role in confining these magnetic features to localised regions. These results can be related to the well-known convective collapse instability, although there are some important differences between the two models. Although this is only an idealised representation of photospheric magnetoconvection, super-equipartition magnetic fields are a robust feature of these simulations. Therefore, the existence of kilogauss-strength magnetic fields in the quiet Sun is certainly qualitatively consistent with the findings of this model.

Katarzyna Mikurda, and Christian Beck presented the observational evidence for the "hot wall" effect in small magnetic flux concentrations. When lacking polarimetric observations, the bright points (BPs) visible in the G-band at 430 nm are commonly used as tracers for magnetic fields. Observations were taken on October 11, 2005 at the German Vacuum Tower Telescope and involved the Tenerife Infrared Polarimeter (TIP) at 1.5 μm, the Telecentric Etalon Solar Spectrometer (TESOS) in the Fe I spectral line at 557.6 nm, and a speckle setup in G-band. The area scanned by TIP was $75'' \times 33''$ and covered a pore surrounded by network. The TIP spectra were inverted with the SIR (Stokes Inversion based on Response functions) code to retrieve the magnetic field vector. They find that G-band BPs are not co-spatial with the central part of the flux concentrations. Even at the small heliocentric angle of 12 degree, BPs appear projected on the limb side walls of the granules, whereas the fields are concentrated in the intergranular lanes. Their findings indicate that the G-band bright points are a result of the hot wall effect. The downward shift of the optical depth scale in the presence of magnetic fields allows to see deeper and hotter layers, where the CH molecule dissociates, in the granules next to the field concentrations. Thus, information drawn from the observations of BPs cannot be used to conclude on the actual variation of the magnetic field structure, as only the outer parts of the flux concentrations are seen in the BPs.

3.1. *Photospheric magnetic field and flares*

The photospheric magnetic field has been extensively used to study the solar flares. The magnetic field configuration and evolution of a highly flare-productive region NOAA 10808 was presented by **Takako T. Ishii, Kaori Nagashima, Hiroki Kurokawa, Reizaburo Kitai, Satoru Ueno, Shoichi Nagata, and Kazunari Shibata**. Active regions on the Sun have different flare productivities. Some regions produce many large flares, while others produce no flares. The key factor of a high flare productivity is the complexity of magnetic field configuration of the region. They found that the twisted structure of emerging magnetic flux bundles is the essential feature of flare-productive active regions. The active region NOAA 10808 showed the highest flare activity during the current solar cycle (cycle 23) in September 2005. They studied the formation process of delta-type magnetic configuration using SOHO/MDI magnetograms and flares using *TRACE* data. They also studied the evolution of magnetic shear and Hα filaments using Hα full disk images and full disk vector magnetograms obtained with *SMART*. They summarize the characteristics of magnetic field configuration of this region and discuss the relation between the configuration and the high flare activity.

The flare-related magnetic field evolution was studied by **S.N. Chornogo, and N.N. Kondrashova**, who modelled the photospheric layers to study their physical state in two sites of an active region about 30 minutes before and in the onset phase of the solar two-ribbon 2N/M2 flare. The soft X-ray enhancement and microwave bursts occurred prior to and during the flare. The models derived from the inversion reproduce the spectral observations in several Fraunhofer lines. All the lines are weakened to a various degree before the flare relative to those in the quiet photosphere spectrum. In the initial phase of the flare they became still more weak and wide. The models consist of two atmospheres: the magnetic flux tube and the ambient medium. The inferred models show the changes in all photospheric parameters both before and at the onset of the flare. A comparative analysis reveals the difference between photospheric physical conditions in two kernels. Before the flare the inner kernel located near the filament was cooler than the outer one. The flux-tube models show strong downflows in the outer kernel and upflows in the inner kernel in the deep photospheric layers. Upward motions of the surrounding photospheric material were found. In the impulsive phase of the flare the variations of the photospheric parameters manifested itself in the inner kernel more than in the outer one.

Magnetic fields and thermodynamic conditions at photospheric level of two solar flares, on July 19, 2000 and August 4, 2005 of importance M6.4/3N and C8.4/1N, respectively, were studied by **V.G. Lozitsky, E.V. Kurochka, and O.B. Osyka** using the echelle Zeeman spectrograms obtained with horizontal solar telescope at the Astronomical Observatory of the Kyiv Shevchenko National University. The Stokes $I \pm V$ profiles were compared for about ten lines of Fe I, Fe II, Cr II, Sr II, and Ti II. The direct magnetic field measurements using the "center of gravity" method has shown the non-monotonous magnetic field gradient in the area of the flare, in contrary to usual weak negative gradient outside flare emission. Semi-empirical models of flares, calculated close to the maxima using Baranovsky's program (Baranovsky 1993), are in good agreement with these direct measurements. In particular, for the flare of July, 2000, a very narrow ($\leqslant 100$ km) and sharp magnetic field peak at the photospheric level ($\log \tau_{500} \approx -2$) was discovered. This peculiarity indicates likely the local magnetic field amplification due to some specific process during the flare. Another interesting detail are two discrete hot flare layers related to the middle photosphere and temperature minimum zone. The turbulent velocities had one or two peaks placed close to the mentioned peaks of temperature. In the weaker flare on August 4, 2005 the magnetic field increased with the depth in the photosphere without peculiarities.

Graham Barnes, Dana W. Longcope, Colin Beveridge, B. Ravindra, and Kimberly D. Leka employed the Minimum Current Corona (MCC) model to estimate the amount of magnetic free energy and helicity injected into the coronal magnetic field of an active region. In the MCC model, each concentration of photospheric magnetic flux is represented by a point source, greatly simplifying the magnetic topology. Advecting an initial partitioning of the flux through a long time series of magnetograms results in a persistent set of sources. They show that the centroid velocity of a partition compares well with the flux-weighted average over the partition of the local correlation tracking velocities. Flux domains, bundles of field lines interconnecting pairs of sources, are surrounded by separatrix surfaces. The intersection of two separatrices is a separator field line, which is the site of reconnection in this model. The evolution of the photospheric field causes the sources to also evolve, which would lead to changes in the domain fluxes to maintain a potential field configuration if reconnection could proceed rapidly. However, in the absence of reconnection, currents begin to flow to maintain the initial distribution of domain fluxes. The minimum energy state occurs when currents flow along the separators. The magnitude of the separator currents can be estimated and combined with geometrical properties of the separators to give a lower bound to the magnetic free energy of the system. The motion of sources about one another adds braiding helicity to the system, while the internal rotation of a partition adds spin helicity. Starting from an initial potential field configuration, changes in the free energy are presented for a time series of data for NOAA AR 8210 on May 1, 1998.

Build-up of a CME and its interaction with large-scale magnetic structures was presented by **Lidia van Driel-Gesztelyi, Christopher P. Goff, Pascal Démoulin, J. Len Culhane, Karl-Ludwig Klein, Cristina H. Mandrini, Sarah A. Matthews, Louise K. Harra, and Hiroki Kurokawa**. A series of flares (*GOES* class M, M and C) and a CME were observed on January 20, 2004, occurring in close succession in NOAA 10540. Types II, III and a N radio bursts were associated. They investigate the link between the flares (two impulsive flares followed by an LDE) and the CME as well as the origin of the rare decametric N-burst by using the combined observations from *TRACE*, *SOHO*-EIT, Hα images from Kwasan Observatory, *SOHO*-MDI magnetograms, *GOES* and radio observations from Culgoora and *Wind*-WAVES as well as magnetic modelling to understand the complex development of this event. They find that there is a link between the first two impulsive flares to tether-cutting reconnections and the launch of the CME, while the last of the flares, an LDE, in the relaxation phase forced reconnections between the erupting flux rope and neighbouring magnetic field lines. They show that reconnection with the magnetic structure of a previous CME, launched about 8 hours earlier, injects electrons into open field lines having a local dip and apex of about 6 solar radii height. The dipped shape of these field lines was due to large-scale magnetic reconnection between expanding magnetic loops and open field lines of a neighbouring streamer. This particular situation explains the observed decametric N burst. This complex observation shows that impulsive quadrupolar flares can be eruptive, while an LDE may remain a confined event. They find that the reconnection forced by the expanding CME structure is followed by a relaxation phase, when reconnection reverses and restores some of the pre-eruption magnetic connectivities. The observed decametric N-burst was caused by the interaction of two CMEs and reconnection of their expanding magnetic field with neighbouring streamer field lines – a very particular interplay, which explains why N-bursts are so rare.

Yu Liu presented a study of halo coronal mass ejections and configuration of the ambient magnetic fields. He examined 104 halo CMEs in the time interval 2000–2004 with respect to the configuration of their ambient magnetic fields. He defines three types of halo CMEs in terms of characteristics of the ambient magnetic field where the CMEs occur: the CMEs occur under the heliospheric current sheet (type 1), in the open flux areas (type 2), and under the plasma sheet (type 3) and find that type 2 and 3 CMEs appear to be faster than type 1, which implies that configuration of the ambient magnetic field play a role in determining halo CMEs' speed. He also find a weak correlation between speed of type 3 halo CMEs and the peak X-ray flux of the associated flares, but no such correlations are found for types 1 and 2. This suggests that kinetic characteristics of type 3 CMEs are more directly related with properties of the associated flares. Considering the configuration of the ambient magnetic field of type 3 CMEs that possess much less closed flux for confining propagation of CMEs, it is suggested that the overall closed field works as an additional factor to shape propagation of a CME, while the associated flare might define the speed of this CME in the first place, as demonstrated in several current studies. It is not clear so far what roles the ambient field may play in initiating a CME. He also presented configuration of magnetic field related to fast halo CMEs. He found that the fast halo CMEs in October-November 2003 initiated from areas with special configuration of magnetic field such as open flux or plasma sheet. He concludes that the configuration of the background magnetic field plays an additional role in shaping the propagation of the fast halo CMEs, and thus ensure a high speed when special configurations apply. In his work he extends the study to a large sample: including all halo CMEs detected by *SOHO*-LASCO and identified by Gopalswamy group, to explore more general relationship between halo CMEs' propagation and configuration of background magnetic field.

Ayumi Asai, Takako T. Ishii, Kazunari Shibata, and Nat Gopalswamy showed the anemone structure of active region NOAA 10798 and related geo-effective flares/CMEs. They reported the evolution and the coronal features of an active region NOAA 10798, and the related magnetic storms by examining in detail the photospheric and coronal features of the active region by using observational data in soft X-rays, in extreme ultraviolet images, and in magnetogram obtained with *GOES*, *SOHO* satellites. They also examined the interplanetary disturbances from the ACE data. This active region was located in the middle of a small coronal hole, and generated 3 M-class flares. The flares are associated with high speed CMEs up to 2000 $\mathrm{km\,s^{-1}}$. The interplanetary disturbances also show a structure with southward strong magnetic field. These produced a magnetic storm on August 24, 2005. They conclude that the anemone structure may play a role for producing the high-speed and geo-effective CMEs even at the near-limb locations.

Hong-Qi Zhang presented an observational study of solar magnetic field and eruption phenomena and the development of new optical instrumentation at Huairou Solar Observing Station, National Astronomical Observatories of China. They find that the synthetical analysis between photospheric vector magnetic field and the morphological configuration in solar atmosphere provides the essential information on the developments of magnetic energy in source regions of flare-CME eruptions. For example the observational study on the evolution of photospheric vector magnetic field provides the basic information for the emergence, storage, relaxation of non-potential magnetic energy in solar active regions (such as, NOAA 9077, 10488 and 10720), and the relationship with the trigger of solar flare-CMEs. The photospheric vector magnetic field, inferred from the observations, also provides the possible formation mechanism of helical magnetic configuration and transferring form in the solar atmosphere into the interplanetary space.

4. Chromospheric magnetic field

As the flux rises into the chromosphere it expands to maintain the pressure balance in a decreasing gas pressure environment. The expanded chromospheric magnetic field creates a canopy. **John W. Harvey** reviewed the properties of chromospheric magnetic field by

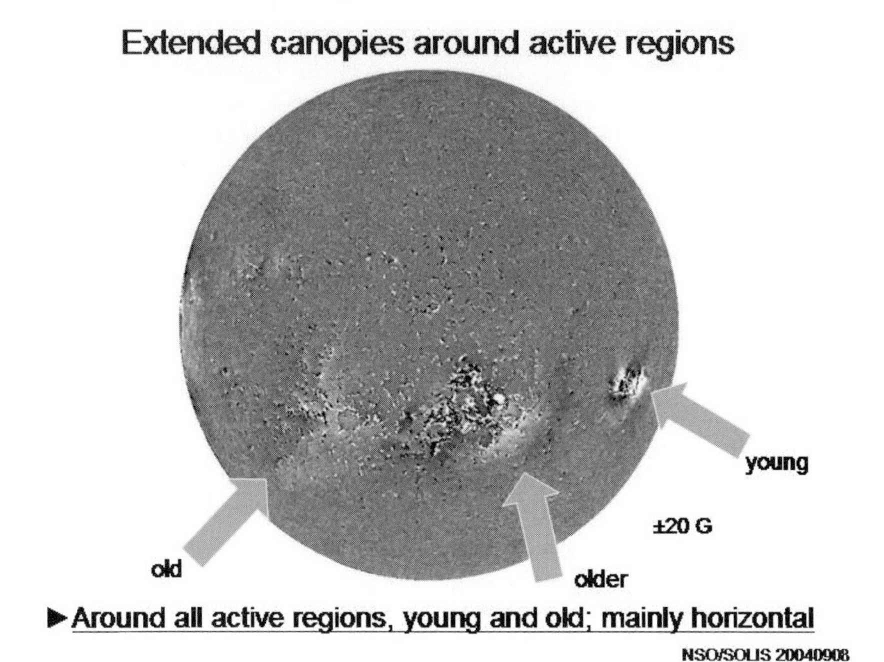

Figure 3. The extended canopy in chromospheric magnetogram.

using a decade of regular full-disk observations of the longitudinal Zeeman splitting of the chromospheric 854.2 nm Ca II line that reveals many interesting magnetic phenomena. The most prominent is a large-scale, mainly horizontal diffuse field surrounding active regions. This extended canopy field is not present in the photosphere and is closely related to chromospheric fibrils and the circumfacule phenomenon. Chromospheric canopy fields are also found around small magnetic concentrations. The larger patterns endure as long as active regions can be recognized as distinct entities. The typical field strength is a few Gauss. The direction is mainly horizontal with azimuth usually, but not always, directed away from the strongest nearby fields. The atmosphere in chromospheric canopy regions has different radiative, dynamic and structural properties than other regions of the chromosphere. Also prominent are magnetic channels which arc closely related to filament channels. Their properties are similar to the chromospheric canopies but the azimuth of the field frequently has a strong component along the axis of the channel. Filaments are frequently but not always seen in magnetic channels. On the largest scale, there are weak areas (~ 1 G) of radially oriented fields that are observed when many active regions are simultaneously present. The 854.2 nm line shows strong spatial variations including core emission. Calculating the longitudinal field with the traditional center-of-gravity scheme gives incorrect results. The field is estimated by comparing the observed local V profile with the wavelength derivative of the local I profile at each pixel. As shown in Fig. 3 the chromospheric magnetogram shows extended canopy where the magnetic field is mostly horizontal irrespective of whether the active region is old or young. How does the quiet region flux rise in solar atmosphere? Do quiet Sun flux tubes poke upward through the horizontal canopy fields that extend far away from active regions (more inter-mingling of different field components)? The future vector magnetic field observations with high spatial resolution might hold the answer.

The chromospheric small-scale solar magnetic features were studied with high-resolution CN spectroscopy by **Vladimir V. Zakharov, Achim Gandorfer, and Sami K. Solanki.** The high-resolution spectroscopic observations of the Sun have been carried out with the TRIPPEL spectrograph installed at the new 1-m Swedish Solar Telescope (SST) using real-time adaptive optics correction. A detailed spectroscopic analysis of individual photospheric bright points (BP) and faculae-like structures simultaneously in two spectral domains, i.e., $387.588 < \lambda < 388.473$ nm (violet CN band) and in a blue spectral band at $436.1 < \lambda < 436.9$ nm, containing absorption lines of CH, obtained at the disc center and near the limb is presented. The estimated spatial resolution of the obtained spectra is around $0.25''$ while the spectral resolving power is around 130.000 in the first domain, and 76.000 in the second spectral region, respectively. The first spectral band covers absorption lines of both, CH and CN molecules, as well as many atomic lines. This made it possible to perform a quantitative comparison of their absorption and Doppler shifts in the different photospheric features. The absorption lines of the CN molecule and many atoms are depressed in a BPs interior with respect to those in the quiet Sun. A quantitative comparison of the relative line depression of CH lines with respect to CN lines showed that the latter have weaker absorption by a factor of 1.28 at the disc centre and 1.32 near the limb. The CN line-core intensity, at the disc centre, has higher BP contrast than the contrast in the CH line-core by a factor of 1.9, and the ratio of these contrasts is decreasing with increasing continuum intensity of the BPs. This trend is similar to that obtained from previous simultaneous G-band and violet CN-band imaging observations. Measurements of contrasts and rms contrasts of line-core, integrated and local continuum intensities are provided. Analysis of Doppler shifts and line broadening of an Fe I line at 387.777 nm revealed an increase of the FWHM in the BPs interior and in dark intergranular lanes and a decrease with increasing intensity of the granules. The first results of a direct comparison of observed CN spectra with those simulated in MHD models in different photospheric features are presented.

The challenges for chromospheric spectropolarimetry were reviewed by **K.S. Balasubramaniam**. The solar chromosphere is the most immediate and intermediary layer where competitive magnetic forces begin to dominate their thermodynamic counterparts. The chromosphere quickly responds to magnetic energy release processes, and is one of the early indicators of this energy release. Hence, understanding the nature of the magnetic field at the chromospheric layer is vital. Measurements of the chromospheric magnetic field using spectropolarimetry are necessary to constrain 3D models of the magnetic field geometry via extrapolation methods. In the review he described the challenges for measuring and inferring the chromospheric magnetic field in solar-disk measurements.

One of the important chromospheric lines used for the magnetic diagnostic is He I 1083.0 nm. **Yulia S. Zagaynova, Pavel G. Papushev, and Sergey A. Chuprakov** investigated the He I 1083.0 triplet profile in sunspot umbrae. This triplet gives an unique view of the chromosphere in active regions. Several articles at early time had indicated the enhancement of the absorption at this line in sunspot umbra among other features. The line depths of the He I 1083.0 nm triplet are compared with plages and other AR features reliably registered in sunspot umbrae. It is a good opportunity for examination of the chromospheric heating processes connected with quite different magnetic structures. Up to date, systematic investigation of helium spectrum in sunspot umbrae has not been conducted. Statistical results of the He I 1083.0 nm triplet line observations in sunspot umbrae are presented. The observations were made at the large Nicolsky coronagraph at Sayan mountain observatory. The setup for the multi-spectral observation was used. This setup consists of the CCD spectrograph and the two bandwidth (He/Hα) birefringent filter. The lens system of the coronagraph provides the possibility for synchronous registration of the spectral line and two monochromatic images in quite different spectral regions. A method to avoid blurring effects on line profiles based on the point spread function (PSF) has been developed. PSFs have been derived from a pore continuum contrast profile, registered at the same time and place, and from a set of empirical data. These PSFs were used to restore monochromatic images. More than 35 umbral spectra of sunspots with areas from 1 to 10^6 millionths have been recorded. The currently available data is not rich enough but some implications are available. It can be stated that He I 1083.0 triplet really exists in sunspot umbrae and its line depth, equivalent line width, and half-width are increasing with sunspot area. For leading and individual unipolar sunspots, the triplet dependences between line parameters and spot areas have been found which are quite different from those obtained for following spots.

Inferring the magnetic field from He I 1083.0 nm is complex. **Clementina Sasso, Andreas Lagg, and Sami K. Solanki** successfully used the multi-component analysis of a flaring region in the chromospheric He I 1083.0 nm triplet. They present infrared spectropolarimetric observations of the solar active region NOAA 10763 obtained with the new Tenerife Infrared Polarimeter (TIP 2) at the German Vacuum Tower Telescope (VTT) during May 2005. The region was in the initial phase of a C2.0 flare. They observe up to 4 atmospheric components within the spatial resolution element. The components are clearly separated in wavelength, the largest separation corresponding to downflow velocities of up to 64 km s^{-1}. They give preliminary results on the magnetic vector and the line-of sight velocity obtained by inverting the Stokes I, Q, U, and V profiles of the He I 1083.0 nm triplet.

Even quiescent solar prominences may become active and sometimes erupt. These events are occasionally linked to coronal mass ejections. However we know very little about the plasma properties during the activation and eruption processes. The spectral line of helium in chromosphere and transition heights is an important part of magnetic diagnostic of these phenomena. The first computations of the helium line profiles emitted by an eruptive prominence were presented by **Nicolas Labrosse, Jean-Claude Vial, and Pierre Gouttebroze**. The prominence is modelled as a plane-parallel slab standing vertically above the solar surface and moving upward as a solid body. The helium spectrum is computed with a non local thermodynamic equilibrium radiative transfer code. The effect of Doppler dimming or brightening is investigated in the resonance lines of He I and He II formed in the EUV, as well as in the He I 1083.0 nm and He I 587.6 nm lines. They focus on the line profile properties and the resulting integrated intensities. They also study the effect of frequency redistribution in the formation mechanisms of the resonance lines. It is shown that the helium lines are very sensitive to Doppler dimming effects. Together with the hydrogen lines they offer the possibility of a powerful diagnostic of the active and eruptive prominence plasma.

Pavel Papushev, and Rafik Salakhutdinov presented the evolution of chromospheric spicules. This well-known phenomenon observed in the solar chromosphere has been subject of a long-standing investigation both observationally and theoretically. At last few years, another theoretical approach has been proposed. It is the leakage of p-mode into the chromosphere. This approach is based on the observation of spicules on the solar disk in the enhanced network. They demonstrated wave-like fluctuations with duration much less than spicules' lifetime (with a period less than 100 s), anomalous half-width of spectral lines, and limb spicules' transversal motion variations, which time spectrum is very similar to that of a line-of-sight velocity. These phenomena need to be explained.

Merging of filaments in a dual-filament system was presented by **Katarzyna Mikurda, and Sara F. Martin**. The dual filament system merges to form one extended filament. The filaments were observed at Helio Research in multiple wavelengths around Hα using a tunable filter and a narrow band Fabry-Perot etalon as part of the Joint Observing Campaign. These observations are used to create two-dimensional Dopplergrams. The Hα images are compared with data taken onboard the *SOHO (Solar and Heliospheric Observatory)* spacecraft (EIT at 304 Å and LASCO C2). The *GONG* magnetograms provide the information on photospheric magnetic fields. The filaments were observed on the solar disk between October 10 and 16, 2004. The morphology of the filament system in Hα and He II line at 304 Å was compared and it was found that there is no clear evidence of an eruption associated with the merging of the filaments from either EIT or LASCO C2 in contrary to some previous findings.

As the magnetic structure explodes, the energy is released in chromosphere. **Jing P. Li, and Ming-De Ding** showed the detection of explosive chromospheric evaporation during a two-ribbon flare. They presented two-dimensional spectral observations of an X1.2 two-ribbon flare in which blue asymmetry of Hα profiles are found at one kernel for more than 1 minute. It is interesting that this kernel seems to move along a flare loop seen on the *Transition Region and Coronal Explorer (TRACE)* images since its appearance. While on the loop footpoints, the Hα profiles are found to be red shifted. The statistical distribution and the spatial distribution of the blue asymmetry at the kernel are also studied. There seems to be a counterpart of the kernel on TRACE 1600 Å images. They calculate the moving speed of the kernel by measuring its moving front on *TRACE* images and find that it has a good temporal correlation with the line-of-sight velocity calculated from the Hα profiles using a wing bisect method. The maximum speed of the kernel measured from *TRACE* images is $\sim$200 km s^{-1}, which is similar to the chromospheric evaporation speed deduced from soft X-ray lines. Suppose that this kernel may be the plasma evaporated from the chromosphere, it will be the first time to get a direct evidence that supports the scenario of chromospheric evaporation and condensation of solar flares in Hα line.

A study relating the solar differential rotation and helicity of magnetic clouds was presented by **Katya Georgieva**. Magnetic clouds – a subclass of coronal mass ejections distinguished by the smooth rotation of the magnetic field inside the structure – are the most geo-effective solar drivers. The rotation of the magnetic field in magnetic clouds reflects the rotation in their source region on the Sun whose quantitative measure is the magnetic helicity. The helicity in the corona is determined by two factors: the helicity transferred from the solar interior and the helicity generated by the surface differential rotation. The helicity transferred from the interior into the corona is proportional to the squared magnetic flux (maximum in sunspot minimum and minimum at sunspot maximum), and is equal in magnitude and opposite in sign in the two solar hemispheres: positive in the south and negative in the north, independent of the magnetic polarity cycle. The surface differential rotation, fastest at the solar equator and decreasing toward the poles, generates helicity with the same sign: negative in the north and positive in the south. Therefore, magnetic clouds originating from the northern solar hemisphere should contain negative helicity, while the ones originating from the southern hemisphere should have positive helicity, with the total negative helicity over the solar cycle equal to the total positive helicity. However, this "hemispheric helicity rule" is true in only 70 – 80 % of the cases. Moreover, in the last solar cycle the net helicity carried away by magnetic clouds was found to be negative, and the number of the left-handed magnetic clouds was higher than of the right-handed ones. She studies the solar differential rotation and finds that in cycle 23 the rotation rate was higher and the rotation was more differential in the northern solar hemisphere, consistent with the excess negative helicity carried away by magnetic clouds. She investigates 39 magnetic clouds associated with major geomagnetic storms in the last solar cycle (1997–2001) whose solar sources have been identified. In 27 of them (73%) the magnetic cloud observed at the Earths orbit with the expected helicity. In 10 out of the 12 cases of violation of the hemispheric helicity rule, the differential rotation in the source region of the magnetic cloud is found to be of the "anti-solar type".

Masaoki Hagino, Reizaburo Kitai, and Kazunari Shibata studied the helicity sign of 239 intermediate filaments (IFs) on the solar disk from 2005 July 1 to 2006 May 15. The intermediate filament usually locates between two active regions. Hα images used were obtained with the Solar Magnetic Activity Research Telescope (SMART), located at the Hida Observatory of Kyoto University. The distribution of the chromospheric helicity shows a hemispheric tendency; namely, the filaments in the northern (southern) hemisphere tend to show a dextral (sinistral) barbs, respectively, in agreement with previous studies. On the other hand, the EUV filament helicity was determined from the *SOHO*-EIT data. Using both data, normalized helicity of intermediate filaments was derived. The latitude distribution of normalized helicity shows negative slope. Moreover, 12 filament eruptions were found in the data. The 7 of 12 filaments show the clear opposite sign to the hemispheric tendency of the magnetic helicity. Figs. 4*a* and *b* show a latitude distribution of dextral and sinistral IFs in Hα. The helicity of 239 IFs (110 in the north hemisphere, 129 in the southern hemisphere) was studied and it was found that 71 % (78 of 110) of IFs in the northern hemisphere have dextral barbs and 67 % (87 of 129) of filaments in the southern hemisphere have sinistral, which agreed with the well-known hemispheric tendency of the filament barbs. Figs. 4*c* and *d* show a latitude distribution of positive and negative chirality in EUV IFs. The helicity pattern of EUV filaments obtained with *SOHO*-EIT 171 Å also shows a similar hemispheric tendency. Namely, 65 % (71 of 110) of EUV filaments in the northern hemisphere exhibit negative helicity and the 65 % (84 of 129) of filaments in the southern hemisphere show positive helicity. Using both data above, normalized helicity of IFs was derived. The normalized helicity depends on the ratio L/ϖ and the pitch angle γ. $\langle L/\varpi \rangle \approx 30$ in the IFs data. Due to the long sinuous structure, these IFs aspect ratio may be smaller (Rust & Kumar (1994)). $\langle |\gamma| \rangle \approx 35°$. The normalized helicity was derived for 181 IFs where it was possible to determine the helicity sign. The slope of linear fit is $dH/d\theta = -0.11 \pm 0.06\,\mathrm{deg}^{-1}$, which is negative, in agreement withprevious studies measured in active regions (Hagino & Sakurai 2004, etc.).

5. Coronal magnetic field and eruptive phenomena

For a given photospheric magnetic field distribution, the field topology can be derived using the models. The magnetic field is thought to be the source of the energy released in diverse observed coronal phenomena, from the less energetic coronal heating to the most violent flares and prominence eruptions. These phenomena involve not only very different scales from the energetic, but also from the temporal point of view. Magnetic field reconnection, which is efficient only at very small spatial scales, has been the energy release mechanism so far proposed. **Cristina H. Mandrini** reviewed the progress in understanding the magnetic field at higher solar atmospheric layers with photospheric

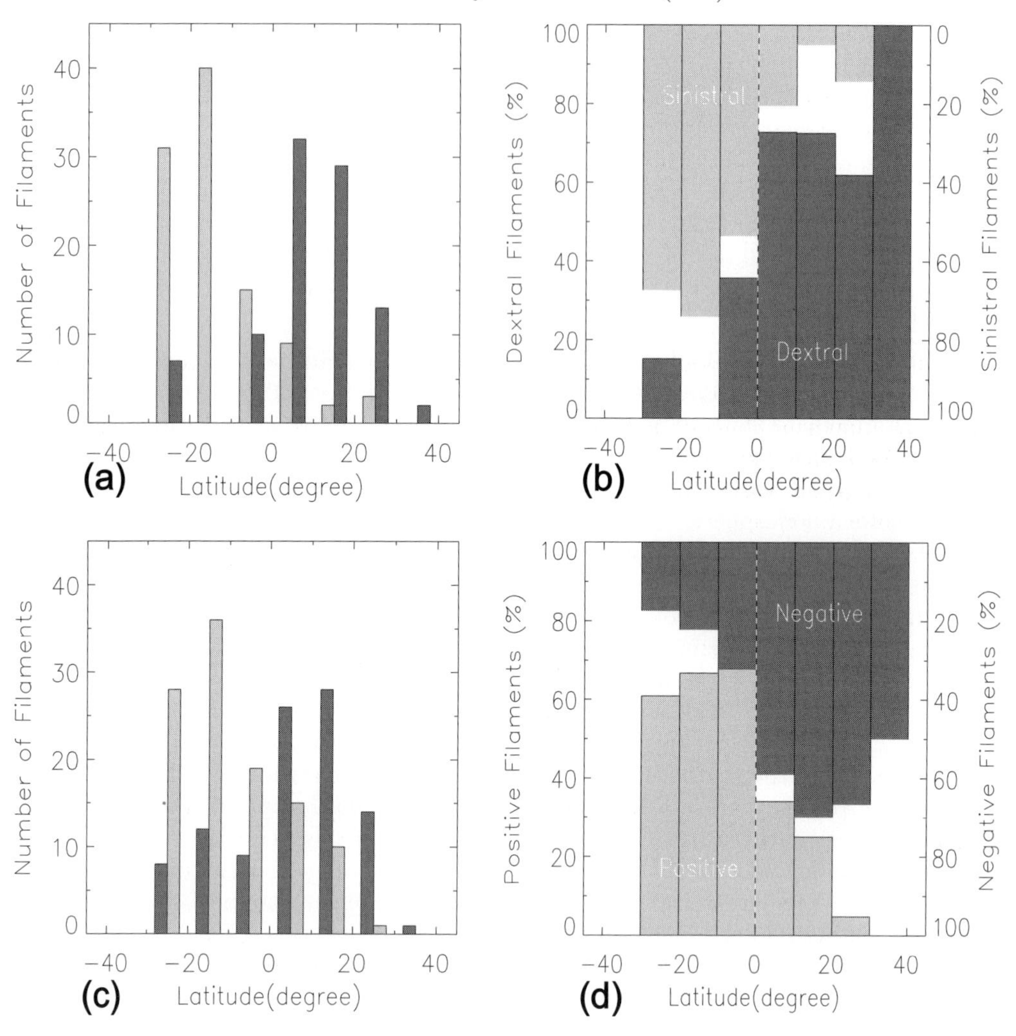

Figure 4. Latitude distribution of helicity sign of intermediate filaments. (*a*) Number of dextral (dark-shade bar) and sinistral (light-shade bar) Hα filaments observed by SMART Hα data. (*b*) Relative frequency of helicity for every latitudinal zone. (*c*) Number of negative (dark-shade bar) and positive (light-shade bar) EUV filaments observed by *SOHO*-EIT. (*d*) Same as in (*b*), but from EUV filaments.

boundary condition. From a theoretical point of view, magnetic configurations with a complex topology, i.e., having separatrices, are the ones where current sheets can form in 2D. When going to 3D, and if the photospheric magnetic field is described by a series of isolated polarities (surrounded by field free regions), a complete topological description is given by the skeleton formed by null points, spines, fans and separators, and associated separatrices (see the review by Longcope 2006 and the examples therein discussed). However, if the photosphere is fully magnetized, most of the above topological structures disappear: only separatrices associated to coronal magnetic nulls remain. Separatrices of a different origin are linked to the field lines curved up at the photosphere, defining the bald-patch locations (Titov *et al.* 1993). For some observed magnetic configurations, those topological structures are enough to understand where flare brightenings appear as a result of magnetic field reconnection. Few examples have been found where coronal null points, computed using either subphotospheric sources or magnetic field extrapolations to represent the coronal field, were associated with coronal activity (see e.g. Mandrini *et al.* 1991; Gaizauskas *et al.* 1998; Aulanier *et al.* 2000; Fletcher *et al.* 2001a; Mandrini *et al.* 2006). Concerning bald patches, they have been only found in connection with low energy release events: small flares (Aulanier *et al.* 1998), EUV brightenings (Fletcher *et al.* 2001b), and chromospheric events (Mandrini *et al.* 2002; Pariat *et al.* 2004).

The analysis of the topological structure of numerous active regions has shown that flares occur in a larger variety of configurations than those just discussed (see, e.g., Démoulin *et al.* 1994). Quasi-separatrix layers (QSLs) (Priest & Démoulin 1995; Démoulin *et al.* 1996; Titov *et al.* 2002), which are regions where there is a drastic change in field-line linkage, generalize the concept of separatrices to magnetic configurations without magnetic null points and/or bald patches. Using coronal magnetic field models, QSLs have been computed in the largest variety of observed magnetic configurations (Démoulin *et al.* 1997; Mandrini *et al.* 1997; Fletcher *et al.* 2001b; Bagalá *et al.* 2000, and references therein). QSLs have been found located in coincidence with chromospheric and coronal loop brightenings of varied intensity, these brightenings could be also connected by field lines in the way expected by magnetic reconnection theory. Moreover, in cases where vector magnetograms were available, photospheric current concentrations were also located at the photospheric trace of QSLs. Recently, to relate QSLs with the formation of strong current concentrations and study the characteristic of the reconnection process occurring at QSLs, 3D MHD simulations of characteristic observed field distributions and photospheric motions have been developed by Aulanier *et al.* (2005) and Aulanier *et al.* (2006). The results from these simulations imply that electric currents at QSLs are amplified in time only if the QSLs are broader than the dissipative scale length, what was suggested in the previous observational analysis. Magnetic reconnection at QSLs occurs when the self-pinching of the current layers is strong enough to efficiently enhance the dissipation term in the induction equation. A property of this reconnection process is the continuous slippage of field lines along each other as they pass through the current layers. Their reviewed examples of observed regions showing varied levels of activity and their topologies, teach us that magnetic reconnection can occur in magnetic configurations with much wider topological characteristics than traditionally thought.

The 3D topology structure of coronal magnetic field was studied by **Hui Zhao, Jing Xiu Wang, Jun Zhang, and Chi Jie Xiao**. They used the Poincaré index of isolated null-points in vector field to strictly identify the 3D null-points in coronal magnetic fields reconstructed from observed vector magnetograms of several important active regions. Based on the null-points they identified, they revealed the essential topology skeletons of 3D coronal magnetic fields. Comparing these skeletons with images of Hα, *SOHO*-EIT, *TRACE*, and X-ray, they found that the 3D topology structures of coronal magnetic fields are closely associated with solar flares and CMEs. They also found an indication of 3D magnetic reconnection by studying temporal series of the 3D topology structures. The thorough investigation of the 3D topology structures of coronal magnetic fields is a key to understand the physical mechanism of solar activity.

The effect of strong MHD discontinuities on the solar magnetic field was discussed by **S.A. Grib** who studied different cases of the collisions between the solar otational MHD discontinuities, going through the flux tubes, and the contact discontinuity, describing the solar transition region. It is shown in the frame of non-linear MHD problem of Riemann that both a fast and a slow shock waves may appear in the solar corona. By the way the first falling rotational discontinuity being 3-dimensional helps the 2-dimensional (described by the theorem of complanarity) shock wave to appear as a refracted one. Thus the behaviour of the magnetic field is changed in the result of the splitting of an arbitrary discontinuity. The catastrophic case of the interactions of the solar MHD discontinuities due to the nonlinear character is indicated.

L.T. Song, Cheng Fang, Yu-Hua Tang, Shi Tsan Wu, and Yang A. Zhang described a fast extrapolation method for reconstructing the coronal nonlinear force-free field which is an improved method of upward integration to reconstruct the nonlinear force-free field (NFFF) in the solar corona. The method of upward integration can be modified to a well-posed one in the following way: Instead of using finite difference to express partial derivatives, smooth continuous functions are used to approximate the magnetic field. Three field components are expressed by amplitude functions multiplying morphology functions, and the four basic NFFF equations can be reduced to ordinary differential ones. They are then solved in an asymptotic manner. Considering the physical meaning of the force-free parameter α, they found a self-consistent compatibility condition for the boundary values. Furthermore, they propose a computation algorithm, similar to the usual time-dependent 2D MHD simulation scheme. This algorithm is steady and not sensitive to the noise in the magnetic field (in particular the transverse field) measurement, and is able to deal with concentrated photospheric currents. The code runs very fast in an usual PC-computer and lasts only 6 minutes for the mesh of 80×60 ($\mathrm{x:y}$) up to a height of 80 (216 000 km). This new method provides a powerful tool to analyze the magnetic field property of solar active regions, and is useful for the prediction of solar activity.

The quasi-static sequences of 2D magnetic potential fields in barotropic ideal MHD flows was discussed by **Dieter H. Nickeler, Miroslav Bárta, and Marian Karlický** taking into account the equation of motion. The equation of motion is often neglected as the plasma β is assumed to be very small in the regions above the photosphere. In contrast to that they analyse how magnetic potential fields do evolve in the frame of barotropic ideal MHD flows. Here neither the pressure gradient nor the equation of motion can be neglected. They show special solutions by solving the set of ideal MHD equations in the case of a quasi-static approach. This implies that the non-linear term in the equation of motion is neglected.

Jun Zhang, and Guiping Zhou reported the magnetic field evolution in a coronal hole using Big Bear Solar Observatory (BBSO) and *SOHO* observations exploring the magnetic evolution and temperature variation in a coronal hole during its appearing and disappearing phases. The results are: (*a*) In the early phase of the coronal hole formation, more than 65 % of the magnetic flux is negative, and the coronal temperature, estimated from the ratio of *SOHO*-EIT 195 and 171 Å images, is 1.05×10^6 K in the hole; (*b*) In the end phase of the hole, about 55 % of the magnetic flux is negative, and the temperature increases to 1.10×10^6 K. The results display that in a coronal hole there is a connection between magnetic evolution and coronal temperature and that the coronal temperature increases while the magnetic imbalance of the both polarities decreases.

5.1. *Coronal heating*

Coronal heating was another point of intense discussion. The adequacy of different models of the corona heating is discussed by **Olga G. Badalyan, and Vladimir N. Obridko**. The correlation between the brightness of the coronal green line Fe XIV 530.5 nm and the calculated strength of the magnetic field in the corona is shown to be a very useful instrument to solve the problem. This correlation manifests a very strong dependence both on the solar cycle phase and on the heliographic altitude. It seems reasonable that the efficiency of the mechanisms depends strongly on the relative area of low and high (including open field) loops. The models based on slow field dissipation (DC) are more efficient in the low-latitude zone and the wave models based on dissipation of Alfvén waves (AC), in the high-latitude zone.

A.D. Voitsekhovska, and A.K. Yukhimuk discussed the transformation of MHD Alfvén waves and solar corona heating. One of the most interesting phenomena appearing on the Sun is the high temperature of the solar corona, which reaches 10^6 K (compared with 5×10^3 K at the level of the photosphere). To maintain such a high temperature and compensate for radiative cooling, the constant inflow of energy is needed. The necessary energy is considered to be carried by Alfvén waves. Alfvén waves also play an important role in the heating of the solar wind and in transporting energy from the Sun to the Earth. However, MHD Alfvén waves are weakly damping and there is a problem of the transformation of energy from waves to plasma particles. Therefore, solar corona heating by Alfvén waves can be connected with transformation of long-waved Alfvén waves to short-waved kinetic Alfvén waves. Kinetic Alfvén waves are rapid damped waves and can effectively transfer energy to plasma particles. The non-linear mechanism of Alfvén wave transformation to kinetic Alfvén waves and magneto-acoustic waves in the solar coronal plasma with small plasma parameter $\beta < 1$ is investigated. The parametric instability, where the Alfvén wave is a pump wave, is considered as the generation mechanism. Using a two-fluid model and Maxwell's equation, they derive a non-linear dispersion relation governing these three-wave interaction processes, the instability growth rate, and the time of instability development. Their theoretical investigations show that the decay of Alfvén waves into kinetic Alfvén and magnetoacoustic waves can cause the solar corona heating.

Paul G. Watson, and I.J.D. Craig explored the role of viscous and resistive damping of shear Alfvén waves in magnetized plasmas. Perpendicularly polarized shear waves that ride on spatially varying horizontal background fields are remarkably efficient at dissipating wave energy via phase mixing. A combination of analytical and numerical treatments confirm that solutions in simple 1D Cartesian and cylindrical geometries allow energy release rates that are only weakly dependent on the visco-resistive damping coefficients. Numerical simulations also confirm that more general 2D solutions exhibit similar behaviour. This process could provide a fast and efficient method for heating the solar corona.

The statistics of dissipation events from reduced MHD simulations was discussed by **Daniel O. Gomez, and Pablo Dmitruk**. Within the reduced MHD approximation, they numerically simulate the dynamics of a coronal loop driven by a stationary velocity field at the photospheric boundaries. After several photospheric turnover times, a turbulent stationary regime is reached, characterized by a broad-band power spectrum and heating rate levels compatible with the heating requirements of active region loops. The energy dissipation rate as a function of time displays a complex superposition of impulsive events, which they associate to the so-called nanoflares. A statistical analysis of these events yields a power law distribution as a function of their energies, which is consistent with those obtained for flare energy distributions reported from X-ray observations. They also study the distributions of peak dissipation rates, durations, and waiting times between events.

In order to understand the plasma processes in the solar corona, **Evgeny V. Derishev, Vitaly V. Kocharovsky, Vladimir V. Kocharovsky, and V. Yu. Martyanov** studied a new class of self-consistent current sheets and filaments in collisionless plasma. They consider a continuous set of stationary current sheets and filaments in collisionless multi-component plasma found analytically using integrals of two-dimensional motion of particles in the self-consistent magnetic field. In their solutions, which are relativistic in general, the magnetic energy density can be comparable to that of particles, and the spatial scale can be arbitrary compared to a typical gyroradius of the particles. They also consider the properties of newly found stationary solutions and their possible applications to the analysis of magnetic field configurations emerging in various astrophysical plasmas, including coronal structures, shocks and jets. The results are used for interpretation of recent observations and numerical simulations. By choosing a particular dependence of particle distribution function on the integrals of motion they are able to obtain various profiles of magnetic field and self-consistent current, including non-monotone. The obtained solutions describe a much more general class of equilibrium configurations as compared to known generalizations of Harris current sheets. On this basis, they suggest a way to describe slow dynamics and filamentation of collisionless current configurations in coronal plasma and in active galactic nuclei, γ-ray bursts, and microquasars.

Dynamics and structure of the active region NOAA 10139 on the basis of microwave observations at RT-22, SSRT, and RATAN-600 have been investigated by **Yury T. Tsap, L.I. Tsvetkov, Yu.F. Yurovsky, Natalia G. Peterova, T.B. Borisevich, and B.V. Agalakov**. As appears from the wavelet analysis, the observed behavior of the S-component of microwave emission is determined by elementary flare events. It has been shown that accelerated electrons can play an important role in heating the plasma of the transition region and upper chromosphere.

The Fabry-Perot interferometric study of the green coronal line during the total eclipse of 2001 from Zambia was presented by **Thyagarajan Chandrasekhar, Nagarhalli M. Ashok, B.G. Ananda Rao, Jay M. Pasachoff, and Terry-Ann Suer**. They reported an interesting ground-based Fabry-Perot interferometric experiment on the green coronal line at 530.3 nm carried out successfully during the total eclipse of 21 June 2001 from Lusaka. Unlike as in earlier experiments, a cooled CCD was used to record as many as 17 interferograms during 194 seconds of totality. The instrumental profile is well determined by a green He-Ne laser and has a FWHM of 20 pm. The Fabry-Perot was off centred with respect to the solar disk to permit wider fringe coverage of the corona. Radial scans from fringe centre of only one interferogram number over 500 and each scan has several fringes. The data base spread over 17 interferograms is huge and has been only partially analysed. The example of the interferogram is given in Fig. 5. Line-width temperatures derived from fringes analysed so far range from 2.4×10^6 to 3.7×10^6 K and many profiles are asymmetric. The data base permits a search of line width oscillations at many positions in the corona with a temporal resolution of a few seconds which has implications for wave heating of the corona.

6. Other phenomena

6.1. *Coronal loops*

Robert F. Wilson presented the VLA, *SOHO*, and *RHESSI* observations of magnetic interactions and particle propagation across large-scale coronal loops. Among the different types of soft X-ray and EUV coronal structures, detected by *Yohkoh*, *SMM*, and *SOHO*-EIT, are large-scale loops that appear to interconnect active regions across the solar equator, which are known as trans-equatorial loops. These trans-equatorial interconnecting loops are of particular interest, because they may play a role in heating and re-structuring the quiet Sun magnetic field while also being potential source regions of coronal mass ejections (CMEs). The physical mechanism that gives rise to these structures is not well understood; it is thought that the anti-parallel magnetic fields of the two active regions on opposite sites of the solar equator reconnect and form large-scale

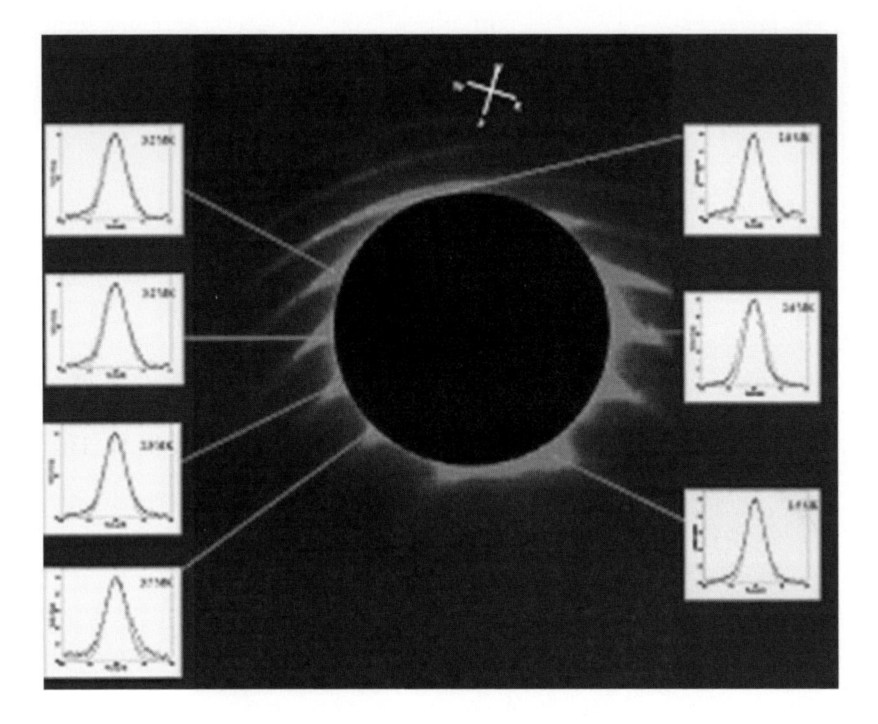

Figure 5. Coronal interferogram in $\lambda\,530.3\,\mathrm{nm}$ with line profiles at selected positions.

trans-equatorial loops (TELs) in the corona, rather than coming directly from below the surface in the convective zone. In this way, the formation of these loops involves a global reconfiguration of the solar magnetic field that may play a role in heating the quiet Sun coronal plasma. In Fig. 6 is shown a series of VLA 20.7 cm and 91.6 cm wavelength snapshot maps (2 minute duration) made on July 1, 2005 that have been overlaid on *SOHO*-EIT images of the Fe XII 195 Å lines taken at about the same time. The VLA 20.7 cm images delineate slowly-varying coronal loops that extend above and between active regions 10783 (S02E46) and 10784 (N16E57) on the solar disk. These sources have peak brightness temperatures of $T_b = 1.5 \times 10^6$ K and are most likely attributed to thermal *bremsstrahlung* or thermal gyroresonance emission. Beginning at 17:15 UT, another 20.7 cm source is seen to extend and separate from the western end of AR 10783 and then move northward across the solar equator with an apparent velocity of $v \simeq 55\,\mathrm{km\,s^{-1}}$, finally reaching AR 10784 at 18:50 UT. The path of this moving source lies along a transequatorial EUV loop connecting the two active regions. This new source appears to emerge following a gradual increase in the brightness temperature of the 20.7 cm source associated with AR 10783 (Fig. 6 – *top*) beginning 16:30 UT; it also coincides with a multi-peaked hard X-ray burst detected by *RHESSI* (arrow) from the eastern part of this region. The 91 cm emission is contained within three time-variable sources, A, B and C, located between the two active regions and along the TELs detected by *SOHO*-EIT. The long-lasting and eastern-most of the sources, source A, decreases in intensity and is replaced by a second source, B, located close to the eastern edge of AR 10783 following a second *RHESSI* hard X-ray burst at 17:44 UT. Finally, source B fades and is replaced by an intense burst source C, located $\sim 1'$ west of source A. The appearance of the 91 cm

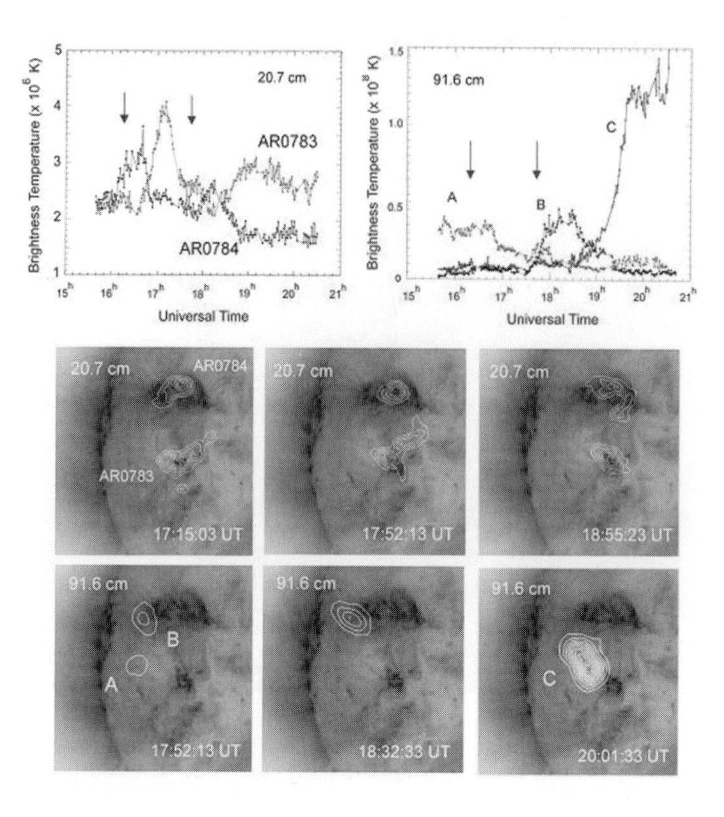

Figure 6. A plot of the peak brightness temperature (Stokes parameter I) from the sources associated with active regions AR0783 and AR0784 at 20.7 cm wavelength and the sources A, B and C labeled at 91.6 cm on 1 July 2005. A series of VLA snapshot maps (1 minute intervals) at 20.7 cm (*middle row*) and 91.6 cm (*bottom row*) are overlaid on *SOHO*-EIT Fe XII 195 Å images taken at the same time. Here, the contours of the VLA maps denote levels of equal brightness temperature with an outermost contour and contour interval of $T_b = 5.0 \times 10^5$ K. The arrows denote the onset times of hard X-ray bursts detected by *RHESSI*.

source C is also accompanied by a 50 % increase in the brightness temperature of the 20.7 cm emission from AR 10783. Both of the 91 cm sources B and C are nearly 100 % right-circularly polarized (source A is 15 % left-circularly polarized).

These changes in the structure and brightness of the 20.7 and 91.6 cm sources that may be related to morphological changes in the large-scale EUV TELs that join the two active regions and to hard X-ray bursts in one of them. The hour-long increase in the brightness of the 20.7 cm emission associated with AR 10783 and the subsequent northward movement of a source across the solar equator along interconnecting loops seems to have followed a weak hard X-ray burst at 16:14 UT from AR 10783. Although there was no obvious 20.7 or 91.6 cm impulsive counterpart to this burst, it is possible that the magnetic reconnection event that produced the hard X-ray burst may have resulted in gradual heating of the overlying coronal loop; the TEL then acted as a conduit for hot thermal plasma which moved along the interconnecting loop to its northern footpoint. Changes in the brightness of three long-lasting, highly circularly-polarized 91 cm sources, possibly attributed to Type I noise storm emission, or to decimetric Type IV-like lare

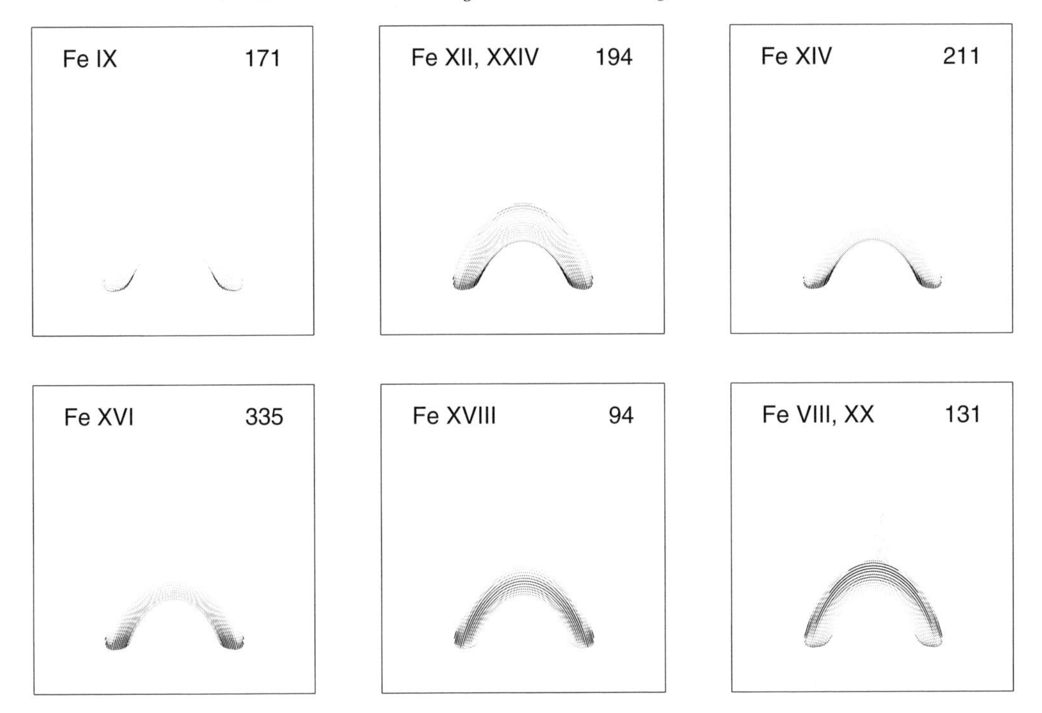

Figure 7. Simulated flare emission for the six *SDO*-AIA EUV wavelengths.

continuum radiation, were accompanied by continuous restructuring of the TELs along which these 91 cm sources appear to lie. One of these sources, B, appears at about the same time as a second hard X-ray burst from AR 10783 and may have been produced by nonthermal particles injected into the transequatorial loops during this impulsive hard X-ray event at 17:40 UT. The most intense 91 cm source, C, was associated with a *GOES* soft X-ray burst and a subsequent northward extension of the hot coronal plasma at 18:51 UT (*RHESSI* was in the night-time part of its orbit), and may have been triggered by continuous restructuring of the coronal magnetic environment. If the 91 cm sources are due to plasma emission, then the disappearance of the 91 cm sources A and B may be due to the fact that the emitting particles may have been displaced upward or downward in the corona by the evolving loops and therefore out of the plasma frequency level corresponding to 327 MHz , i.e., a region where the electron density is equal to $N_{\rm e} = 1.3 \times 10^9$ cm^{-3}.

Katharine K. Reeves, Harry P. Warren, and Terry G. Forbes presented a study, using a loss-of-equilibrium model for solar eruptions to calculate the thermal energy input into a system of flare loops. In this model, the flare consists of a system of reconnecting loops below a current sheet that connects the flare to an erupting flux rope. The thermal energy is calculated by assuming that all of the Poynting flux into the current sheet is thermalized. The density, temperature and velocity of the plasma in each reconnected loop are then calculated using a 1D hydrodynamic code. These parameters are coupled with the instrument response functions of various solar instruments to calculate flare emissions. They simulate spectra from the Bragg Crystal Spectrometer (BCS) on Yohkoh and find that the strong blueshifts that should be present due to the chromospheric evaporation during flare initiation are difficult to observe with BCS, but may be better observed with a more sensitive instrument. They also find that a density enhancement occurs at the top of a loop when evaporating plasma fronts in each loop leg collide there. This enhancement gives rise to bright loop-top intensities in simulated *Transition Region and Coronal Explorer (TRACE)* and *Yohkoh* Soft X-ray Telescope (SXT) images. These loop-top features have been observed in *TRACE* and *Yohkoh*-SXT images, and are not explained by single-loop flare models. The Atmospheric Imaging Assembly (AIA) on the *Solar Dynamics Observatory* should be able to observe these features in detail and this model will be used to help to develop flare observing programs for *SDO*-AIA. A snapshot of a simulated flare viewed through the filters on the *SDO*-AIA telescope is shown in Fig. 7.

Hossein Safari, Sadollah Nasiri, and Yousef Sobouti discussed the effect of density stratification on a modal structure of solar coronal loops. Since the earliest identifications of the kink oscillations in coronal loops, a considerable amount of data has been analysed and possible factors causing the shift in frequency and affecting the oscillation properties of the loops were investigated by different authors. Here, analytical and numerical methods are used to understand the effects of longitudinal density variation on the coronal loops oscillations. The authors assume that the longitudinal density stratification have mathematically the same functional form at the inside and outside of the loops. They also assume that the mean density of the stratified loop is the same as that of the unstratified one, thus, for two identical loops the mass of unstratified loop is the same as of the stratified one. The density distribution along the radius of the loop is assumed to be a step function with different values for inside and outside of the loop. Equations of motion are expressed by second-order differential equations that are separable into radial and transverse components. The radial equation is solved in thin tube regime. The transverse equation is solved both by perturbation method, for small density scale heights, and numerically, otherwise. The summary of results are: (*a*) the ratio of periods of fundamental and first overtone modes decreases markedly with increasing density scale parameter. This is in accord with the observational data of *TRACE*; (*b*) the behavior of the eigenfunctions for different longitudinal wave numbers is completely different for unstratified and stratified loops.

6.2. *Radio and microwave observations*

The radio and microwave observations are powerful tools to measure the magnetic field of solar corona. **Kiyoto Shibasaki** reviewed the microwave measurements of coronal magnetic field. The solar corona is filled with highly ionised plasma and magnetic field. Moving charged particles interact with magnetic field due to Lorentz force. This results in gyration motion perpendicular to the magnetic field and free motion along the magnetic field. Circularly polarized electromagnetic waves interact with gyrating electrons efficiently and the interaction depends on the sense of circular polarization (right-handed or left-handed). This is the reason why we can measure magnetic field strength through microwave observations. This process does not require complicated quantum physics but the classical treatment is enough. Hence the inversion of measured values to magnetic field strength is simpler than in the case of optical and infrared measurements. He presented the examples of distribution of magnetic field strength in the solar corona measured by the Nobeyama Radioheliograph.

Yeon-Han Kim, Yong-Jae Moon, Kyung-Seok Cho, Su-Chan Bong, and Young-Deuk Park reported on the first near-simultaneous observation of an X-ray plasma ejection (XPE), type II solar radio burst, and a coronal mass ejection (CME) on October 26, 1999. First, an XPE was observed from 21:12 UT to 21:23 UT in the *Yohkoh*-SXT field of view (1.1 to $1.4\,R_{\odot}$). The XPE was initially accelerated and then constantly propagated with a speed of about 350 km $^{-1}$. Second, a type II solar radio burst was observed at 21:30 UT by the Culgoora solar radio spectrograph. The burst started at the height of about 1.5 $R_\odot$, which is estimated from its starting frequency assuming the one-fold Newkirk coronal density model. From the frequency drift rate of the burst, the propagation speed is estimated to be about 400 km s^{-1}. Third, the associated CME was observed by the Mauna Loa Mk 4 coronameter (1.2 to 2.8 $R_\odot$). The CME front was identified at 21:20 UT and its speed was 500 km s^{-1} (around 2 $R_\odot$) at the type II starting time. By comparing these three phenomena, they found that (*a*) there is a remarkable difference (0.4 $R_\odot$) between the CME front and the XPE front at 21:23 UT, (*b*) the type II formation height is not consistent with the CME front but with the trajectory extrapolated from the XPE front, and (*c*) the three speeds are comparable with one another. Regarding the type II origin, their results suggest two possibilities: a coronal shock generation by the XPE front or by the CME flank.

Timothy S. Bastian reviewed radio diagnostics of magnetic fields in the solar corona. A measurement of the coronal magnetic field has been an elusive goal for many years. In recent years, however, progress has been made in developing techniques at radio and IR wavelengths to measure or constrain the magnetic field in the quiet and active corona. He discussed the radio diagnostic techniques in a variety of contexts, including the quiet corona, active regions, flares, and coronal mass ejections. Techniques that exploit emission intrinsic to the phenomena themselves - e.g., gyrosynchrotron radiation - and those which exploit external probes such as spacecraft beacons or extragalactic background sources are both discussed. Prospects for exploiting these techniques with future generations of radio instrumentation were also considered.

Stephen M. Ord presented the investigations of the solar corona with radio observations of pulsars. He proposed to examine both the magnetic field and electron content of the solar corona via measurement of the Faraday rotation and a dispersion evident in observations of background pulsar sources as they are occulted by the Sun. Utilizing a number of simultaneous lines of sight that cut different paths through the corona as the Sun rotates, it will be possible to strongly constrain the global topology of both the plasma and the magnetic field. Pulsars are periodic, broadband, polarised point sources of radio waves. In December 2006 the Sun will occult, or pass close to, a number of pulsars in the Galactic plane. He has proposed to use the Parkes radio telescope in New South Wales, Australia, to observe these pulsars throughout a ten day period at two frequencies simultaneously. The purpose is to detect any excess Faraday rotation and pulse dispersion associated with the solar corona. Although background sources have been observed through the corona successfully before, the proposed experiment differs due to the accessibility of different simultaneous lines of sight, and the ability to measure the integrated electron content and integrated magnetic field independently. He presented an outline of the intended experiments and some predictions of expected results.

Search for high-energy particle acceleration signatures in the submillimetre-visible solar flare emission spectrum was presented by **Pierre Kaufmann, Antonio M. Melo, R. Marcon, A.S. Kudaka, Adolfo Marun, Pablo Pereyra, Jean-Pierre Raulin, and Hugo Levato**. Recent results obtained at sub-millimeter waves indicated that key questions regarding physical mechanisms at the origin of solar flares are expected to become better understood with measurements in the far to mid-IR range. A new spectral component discovered with fluxes increasing for shorter sub-millimeter wavelengths indicates emissions by particles accelerated to very high energies. The nature of emission is not known. The observed parameters rule out a thermal interpretation, and the first emission models recently suggested assume three different mechanisms which may become comparable in importance: (*a*) synchrotron radiation by beams of ultrarelativistic electrons; (*b*) synchrotron radiation by positrons produced by nuclear reactions arising from energetic beams interactions at dense regions close to the photosphere; and (*c*) Langmuir waves emission from deep photosphere excited by high energy electron beams. The spectral band where observed features would be critically defined is in the far-infrared to visible range, where the terrestrial atmosphere is highly opaque. New experiments are being considered to observe solar flares from the ground in the remaining high frequency atmospheric "windows" at 670, 850 GHz, and in mid- and near-infrared. Space experiments are planned for discrete frequencies between 1 – 20 THz. They show the first results obtained with a new setup developed to measure solar activity at 10 μm (or 30 THz), using a camera with a focal plane array of uncooled microbolometers coupled to a celostat by an adequate optical arrangement.

7. Concluding remarks and open questions

7.1. *Solar databases*

In next few years, several space and ground-based observations of the solar magnetic field and associated activity will help us to understand the role of magnetic field in solar activity. Also, as Helen E. Coffey presented, the NOAA data rescue of key solar databases and digitization of historical solar images will play an important role in understanding the long term behaviour of the Sun. Over a number of years, the staff at NOAA National Geophysical Data Center (NGDC) has worked to rescue key solar databases by converting them to digital format and making them available via the World Wide Web. NOAA has had several data rescue programs where staff competed for funds to rescue important and critical historical data that are languishing in archives and at risk of being lost due to deteriorating condition, loss of any metadata or descriptive text that describe the databases, lack of interest or funding in maintaining databases, etc. The Solar-Terrestrial Physics Division at NGDC was able to obtain funds to key in some critical historical tabular databases. Recently the NOAA Climate Database Modernization Program (CDMP) funded a project to digitize historical solar images, producing a large on-line database of historical daily full disk solar images. The images include the wavelengths Ca II K, Hα, and white light photographs, as well as sunspot drawings and the comprehensive drawings of a multitude of solar phenomena on one daily map (Fraunhofer maps and Wendelstein drawings). Included in the digitization are high resolution solar Hα images taken at the Boulder Solar Observatory in 1967 – 1984. The scanned daily images document many phases of solar activity, from decadal variation to rotational variation and daily changes. Smaller versions are available online. Larger versions are available by request. See <`http://www.ngdc.noaa.gov/stp/SOLAR/ftpsolarimages.html`>.

7.2. *The questions*

The eruption or emergence of solar active regions will be the focus of observational and theoretical research of solar magnetism in the next decade. The key questions that must be addressed are:

(1) What is the height of triggering sites of explosive events? Is the magnetic field reconnection at coronal height always responsible for the destabilization of magnetic structures? It will be useful to make simultaneous observations of magnetic field at multiple heights and to compare with models to investigate this issue.

(2) How do different magnetic structures interact?

(3) How does the field propagate beyond the chromosphere and how does it respond to photospheric flux movements?

The magnetic field in the quiet Sun has been the subject of intense study in recent years, with numerous questions waiting to be answered. Here a couple of such questions are listed as examples.

(1) What is the origin of the quiet Sun magnetic flux and how does it evolve? This question is related to the currently hotly debated issue of whether the Sun has, besides the main dynamo located near the base of the convection zone (which produces the active regions and the 11-year cycle of activity), also a turbulent dynamo within the convection zone. Of particular interest is to what extent a local dynamo acting near the surface plays a role in producing the flux present in the quiet Sun.

(2) What is the amount of quiet Sun flux and in what state is it found? This question can be subdivided into numerous sub-questions. For example: Which fraction of the magnetic flux is in the network and which fraction is in the internetwork? What is the field strength distribution in these regions? Is it equal or different? What is the exact location of the features (are all found in the intergranular lanes, or does a significant amount of flux exist also inside the upflowing bodies of granules)? Which fraction of the flux is in the form of structures that can be described by strong magnetic flux tubes, which fraction is in the form of turbulent fields?

None of these (and of many other) questions has so far found a final answer and there is a considerable controversy surrounding almost each of them. This has partly to do with the fact that outside the magnetic network, the magnetic field in the quiet Sun produces a signal in polarized light that is small and difficult to measure. New instrumentation that will start to operate in the coming months and years will play a big role in advancing our knowledge of the quiet Sun magnetic field.

Acknowledgements

We would like to acknowledge the scientific organizing committee members Drs. Takashi Sakurai, Hongqui Zhang, John M. Davis, Axel Hofmann, Sami Solanki, Tom Berger, Rob Rutten, Brigitte Schmieder, S. Ananthakrishnan, Oddbjorn Engvold, Eric Priest, Jan O. Stenflo, Donald B. Melrose and Dale Gary. Our special thanks to Dr. David F. Webb for his valuable guidance during the initial stage of preparation of this meeting. We thank the the invited speakers and the participants of the JD03 for their contributions. We also thank Dr. James A. Klimchuk for giving a summary talk at the end of this JD03.

References

Archontis, V., Moreno Insertis, F., Galsgaard, K., Hood, A., & O'Shea, E. 2004, *A&A*, 426, 1047
Aulanier, G., Démoulin, P., Schmieder, B., Fang, C., & Tang, Y. H. 1998, *A&A*, 183, 369
Aulanier, G., DeLuca, E. E., Antiochos, S. K., McMullen, R. A., & Golub, L. 2000, *ApJ*, 540, 1126
Aulanier, G., Pariat, E., & Démoulin, P. 2005, *A&A*, 444, 961
Aulanier, G., Pariat, E., Démoulin, P., & DeVore C. R. 2006, *Solar Phys.*, 238, 347
Bagalá, L. G., Mandrini, C. H., Rovira, M. G., & Démoulin, P. 2000, *A&A*, 363, 779
Baranovsky E.A. 1993, *Contrib. Astron. Obs. Skalnaté Pleso*, 23, 107
Démoulin, P., Hénoux, J. C., & Mandrini, C. H. 1994, *A&A*, 285, 1023
Démoulin, P., Hénoux, J. C., Priest, E. R., & Mandrini, C. H. 1996, *A&A*, 308, 643
Démoulin, P., Bagalá, L. G., Mandrini, C. H., Hénoux, J. C., & Rovira, M. G. 1997, *A&A*, 325, 305
Dorch, S. B. F. 2007, *A&A*, 461, 325
Fan, Y. 2004, *Living Reviews in Solar Physics* 1, 1

Fletcher, L., López Fuentes, M. C., Mandrini, C. H., Schmieder, B., Démoulin, P., Mason, H. E., Young, P. R., & Nitta, N. 2001, *Solar Phys.*, 203, 255

Fletcher, L., Metcalf, T. R., Alexander, D., Brown, D. S., & Ryder, L. A. 2001b, *ApJ*, 554, 451

Gaizauskas, V., Mandrini, C. H., Démoulin, P., Luoni, M. L., & Rovira, M. G. 1998, *A&A*, 332, 353

Galsgaard, K., Nordlund, Å. 1997, *Journ. Geoph. Res.*, 102, 219

Hagino, M., & Sakurai, T. 2004, *Publ. Astron. Soc. Japan*, 56, 831

Longcope, D. W. 2006, *Living Reviews in Solar Physics* 2, ... (`http://www.livingreviews.org/lrsp-2005-7`)

Mandrini, C. H., Démoulin, P., Hénoux, J.-C., & Machado, M. E. 1991, *A&A*, 250, 541

Mandrini, C. H., Démoulin, P., Bagalá, L.G., van Driel-Gesztelyi, L., Hénoux, J. C., Schmieder, B., & Rovira, M. G. 1997, *Solar. Phys.* 174, 229

Mandrini, C. H., Démoulin, P., Schmieder, B., Deng, Y. Y., & Rudawy, P. 2002, *A&A* 391, 317

Mandrini, C. H., Démoulin, P., Schmieder, B., DeLuca, E. E., Pariat, E., & Uddin, W. 2006, *Solar Phys.*, 238, 293

Pariat, E., Aulanier, G., Schmieder, B., Georgoulis, M. K., Rust, D. M., & Bernasconi, P. N. 2004, *A&A*, 614, 1099

Priest, E. R., & Démoulin, P. 1995, *Journ. Geoph. Res.*, 100, 23443

Rust, D. M., & Kumar, A. 1994, *Solar Phys.*, 155, 69

Titov, V. S., Priest, E. R., & Démoulin, P. 1993, *A&A* 276, 564

Titov, V. S., Hornig, G., & Démoulin, P. 2002, *Journ. Geoph. Res.*, 107, SSH 3, 1–13.

Highlights of Astronomy, Volume 14
IAU XXVI General Assembly, 14-25 August 2006
Karel A. van der Hucht, ed.

doi:10.1017/S1743921307010083

Joint Discussion 4
UV astronomy: stars from birth to death

Ana I. Gómez de Castro[1] and Martin A. Barstow[2] (eds.)

[1]Instituto de Astronomía y Geodesia (CSIC-UCM), Facultad de Matemáticas,
Universidad Complutense de Madrid, Madrid, Spain
email: aig@mat.ucm.es
[2]Department of Physics and Astronomy, University of Leicester,
University Road, Leicester LE1 7RH, UK
email: mab@star.le.ac.uk

Abstract. The scientific program is presented as well a the abstracts of the contributions. An extended account is published in *"The Ultraviolet Universe: stars from birth to death"* (Ed. Gómez de Castro) published by the Editorial Complutense de Madrid (UCM), that can be accessed by electronic format through the website of the Network for UV Astronomy (www.ucm.es/info/nuva).

There are five telescopes currently in orbit that have a UV capability of some description. At the moment, only *FUSE* provides any medium- to high-resolution spectroscopic capability. *GALEX*, the *XMM* UV-Optical Telescope (UVOT) and the *Swift*. UVOT mainly delivers broad-band imaging, but with some low-resolution spectroscopy using grisms. The primary UV spectroscopic capability of *HST* was lost when the Space Telescope Imaging Spectrograph failed in 2004, but UV imaging is still available with the *HST*-WFPC2 and *HST*-ACS instruments.

With the expected limited lifetime of sl FUSE, UV spectroscopy will be effectively unavailable in the short-term future. Even if a servicing mission of *HST* does go ahead, to install COS and repair STIS, the availability of high-resolution spectroscopy well into the next decade will not have been addressed. Therefore, it is important to develop new missions to complement and follow on from the legacy of *FUSE* and *HST*, as well as the smaller imaging/low resolution spectroscopy facilities. This contribution presents an outline of the UV projects, some of which are already approved for flight, while others are still at the proposal/study stage of their development.

This contribution outlines the main results from Joint Discussion 04 held during the IAU General Assembly in Prague, August 2006, concerning the rationale behind the needs of the astronomical community, in particular the stellar astrophysics community, for new UV instrumentation. Recent results from UV observations were presented and future science goals were laid out. These goals will lay the framework for future mission planning.

Keywords. ultraviolet-general, ultraviolet-solar system, ultraviolet-stars, ultraviolet-ISM, space vehicles-instruments

1. Preface

This joint discussion was organized to provide a forum during the IAU General Assembly where the accomplishments of UV astrophysics could be highlighted and a new road map for the future discussed.

The UV range is of prime interest for astrophysics since the resonance lines of the most abundant atoms and ions at temperatures between 3 000 K and 300 000 K, together with the electronic transitions of the most abundant molecules (H_2, CO, OH, CS, S_2, CO_2^+, C_2, O_2, O_3, ...) are at UV wavelengths. After enjoying more than 30 years of continuous access to this range, the astronomical community has been facing uncertain times and provision during the decade 2010–2020 remains so. Coordination is required to define the science goals for the future and the resulting requirements for future UV instrumentation.

This meeting focused in particular on stellar astrophysics. The understanding of stellar physics is at the very base of our understanding of the Universe. The chemical evolution of the Universe is controlled by stars. Supernovae are prime distance indicators that have allowed to measure the evolution of the curvature of the Universe and to detect the existence of *dark energy*. The development of life sustaining systems depends strongly on the evolution of stars like our Sun. Some of the most extreme forms of matter in the Universe, the densest and more strongly magnetized, are the magnetars, debris of stellar evolution.

UV instrumentation is required to make progress in these fields. The excellent contributions presented in this *Joint Discussion* dealt with the many aspects of stellar astrophysics from the analysis of dissipative processes in the atmosphere of cool stars and their impact on the evolution of the planetary systems to the study of the atmospheres and winds of the hot massive stars or the determination of the abundances in white dwarfs. The physics of disks, its role in the evolution of binary systems, and the formation of supernovae were among the main topics treated in the meeting. We should also not forget the role of starbursts and, in general, high-mass stars in the chemical evolution of galaxies. The metallicity gradient in the Galaxy is traced in the UV spectrum of planetary nebulae. The evolution of young planetary disks and the role of the central stars in the photo-evaporation of the giant gaseous planets that have been detected recently. The numerous and high quality contributions to the JD could not fit within the space allowed for these proceedings. However, the full proceedings will be published in extended form in *"The Ultraviolet Universe: stars from birth to death"* (Ed. Gómez de Castro & Barstow) published by the Editorial Universidad Complutense de Madrid (UCM) that can be accessed in electronic format through the website of the Network for UV Astronomy (www.ucm.es/info/nuva).

These pages concentrate on reporting on the status of current UV facilities and on the approved and proposed projects for new UV instrumentation. Also, there was an important and interesting discussion about the general needs of the astronomical community that is summarized in the proceedings.

This conference was originally proposed and organized by Willem Wamsteker, who wished to dedicate it to Professor (Emeritus) Cornelius de Jager. Over the years, Professor de Jager contributed significantly to the understanding of stellar atmospheres. Professor de Jager's work on the Sun itself has had a major impact on our understanding of the stellar atmospheres in general. He also played an important role in the IAU through a period when it had to adjust to the new possibilities generated by opening up the space windows.

Sadly, Willem Wamsteker passed away on November 24th 2005, a great loss for all of us. Willem was the director of ESA's *IUE* Observatory until the mission terminated in 1996. Through his leadership, the *IUE* data archive became the first fully internet driven astronomical archive. Willem was also a promotor of the NUVA and a key initiator of the *World Space Observatory UV (WSO/UV)* project; a 1.7 m UV telescope equipped with state of the art instrumentation providing a factor of 10 improvement on the high resolution spectroscopic capabilities of *HST*-STIS. The project is driven by a broad international collaboration led by Russia (ROSCOSMOS). His wonderful enthusiasm for science remains with us.

In December 2005, we took over the duty of shaping the scientific program and preparing these proceedings. This would not have been possible without the support of the SOC: T.R. Ayres, W.P. Blair, D. De Martino, M.A. Dopita, J. Grygar, J.B. Hutchings, C. Jordan, Y. Kondo, E.M. Leibowitz, A. Niedzielski, B.M. Shustov, J. Sahade, and Z.R. Wang.

Members of the NUVA helped with the logistics of the conference organization and the proceedings preparation.

The scientific program is presented as well a the abstracts of the contributions. An extended account of JD04 is published as *The Ultraviolet Universe: Stars from Birth to Death*, Eds. Ana Ines Gómez de Castro & Martin A. Barstow (Madrid: Editorial Complutense, ISBN 978-84-7491-852-6) and can be downloaded from the website of the Network for UV Astronomy <www.ucm.es/info/nuva>.

2. A brief historical introduction to UV astronomy

by **Cornelis de Jager**

In 1957 *Sputnik 1* was launched, followed by a long and ever more sophisticated series of spacecraft. This impressive development in the USSR and USA triggered other nations to follow that example and also to embark in space research. In Europe a preparatory European space research organisation was founded, named COPERS (Commision Prparatoire Europenne pour la Recherche Spatiale). After the preparatory phase the name was changed in ESRO (European Space Research Organisation) and presently the name is ESA.

COPERS established an impressive scientific program that included, among other things, the launching of a 60 cm UV telescope. Critical questions such as, how Europe with a budget that was less than one tenth of that of NASA, could embark in such an ambitious project, were countered with a reference to the 'Atlantic Factor'. That was kind of a mystical way of reasoning saying that salaries in Europe were considerably less than those in the US and that therefore, Europeans could do much more than NASA with the same amount of money. After a number of years we learned, muddling through, that such is not the case: high-quality labour costs the same. But around 1964 we were not yet at that point. Three competing consortia came into being; each of them made a proposal. These were a French-Belgian-Italian consortium, a German-Dutch one while the British decided to do it all by themselves.

I was involved in the German-Dutch project. It was our first experience in a large international undertaking and I value highly the fine co-operation with our German colleagues. Until a few years ago I kept on a drawer the three voluminous books that resulted from our study.

Evidently the Large European Space-telescope, LAS as it was called, did not come to realisation. The Atlantic Factor appeared not to exist in space research. But the enterprise had its fruits. For Bob Wilson and his British colleagues the study yielded sufficient experience and knowledge to propose to NASA a project that would later become the *International Ultraviolet Explorer, IUE*. We, in The Netherlands, decided to propose to ESRO the development of a 35 cm Cassegrain-type UV spectrograph for inclusion in a spacecraft that would later become the *TD1A* satellite. That spacecraft was a polar orbiter moving in a plane at right angles to the solar direction. An instrument on board of that spacecraft and pointing away from the direction to Earth would thus scan nearly the whole sky in half a year. The telescope could deflect over an angle of some 15 degrees in a direction at right angles to the direction of motion of the spacecraft and over a larger angle in the direction of motion. When a star of sufficient brightness would appear in its field of view the telescope would then automatically point at the star and during some five minutes it would stay so. During that time the UV spectrum should be scanned in three wavelength regions. The data were to be stored on a tape recorder and twice a day, when the satellite would pass over the European ground station the data would be collected.

Satellite launching (12 March 1972) went excellent and the first spectra came streaming in; they were fine.

A week or so later, at a conference in Madrid, I met Ernst Trendelenburg, who later would become scientific director of ESRO. "How are you feeling", he called at me. "Fine", I said, "enjoying our wonderful spectra". "Fine???" he wondered, "Don't you know that the tape recorder broke down? No more data." What a misery!

ESRO could partially cure the situation by establishing a number of improvised ground stations over the globe. One of then was on a location as remote as Easter Island. That way, in the course of two yeas we still succeeded in getting spectra of some 200 stars.

In the mean time things had developed. Morton and his colleagues had obtained the first UV rocket spectra of some stars in Orion. These were marvellous. For the first time we got direct evidence of the large rate of mass loss of early type stars. Lyman Spitzer developed his idea for a very large astronomical telescope that would later become the *Hubble Space Telescope*. Not everyone believed him. In 1969 we organised the first international symposium on UV astronomy, IAU Symposium No. 36. It was in Lunteren, the Netherlands, and among the 150 participants were famous astrophysicists as well as many youngsters like young Rashid Sunyaev and young Roger Bonnet. At the closing dinner there were a number of speakers telling funny stories. Laughter all over the place. Then Spitzer's talk came. People were again expecting a funny story and when he told of his idea for a 2.5 m telescope in space the attendance became hilarious. "Expected lifetime 10 years". Bursts of laughter again. A youngster asked me "Who is that fellow. He is the most funny of all". I had to explain to him that he was listening to one of the great astronomers of the century and that this 'funny fellow' was deadly serious. Gradually the audience became to realize that they were actually listening to a visionary scientist with far reaching ideas.

About half a year after *TD1A* (21 August 1972) the *Orbiting Astronomical Observatory Copernicus* was launched. It was designed to obtain UV spectra in several bands between 100 and > 300 nm.

After the termination of our *TD1A* experiment we realized that in some three, four, years IUE would be launched and we decided to fill that gap with a balloon experiment, together with US colleagues. The Balloon Ultra-violet Stellar Spectrometer BUSS had a spectral resolution that was slightly better than *IUE*. It was launched a dozen of times during several launching campaigns in the US.

I am pleased to see in this room several of the participants in the BUSS campaigns. Thank you again for this fine co-operation. The period that I described here was for me really a great time.

There is one person in particular that I am greatly missing. Willem Wamsteker, who has served the community so immensely and thoughtfully, told me last year that this meeting would be dedicated to me. While I felt honoured I told him that there are others who more deserve this honour. I wish that this conference could be dedicated to Willem.

Willem, we will not forget you.

3. Star formation and young stellar objects

3.1. *Young Stellar Objects: accretion and outflow*

Ana I. Gómez de Castro: Young stellar objects are an excellent laboratory to study the physics of accretion and outflow. The relevance of ultraviolet astronomy to study the physics of these objects is reviewed in this contribution. Special emphasis is made in the following aspects:

(*a*) The magnetic interaction between the star and the disk produces high energy radiation and particles that heat the inner disk. The spectral output form this region is described and compared with the theoretical predictions.

(*b*) H_2 and CO electronic transitions are observed in the spectra of the T-Tauri stars allowing an analysis of the circumstellar/disk warm material and its interaction with radiation.

3.2. *High-mass stars: starbursts*

Rosa M. González Delgado: Starbursts are the preferred place where massive stars form; the main source of thermal and mechanical heating in the interstellar medium, and the factory where the heavy elements form. Thus, starbursts play an important role in the origin and evolution of galaxies. Starbursts are bright at ultraviolet (UV) wavelengths, and after the pioneering *IUE* program, high spatial and spectral resolution UV observations of local starburst galaxies, mainly taken with *HST* and *FUSE*, have made relevant contributions to the following issues: (*a*) the determination of the initial mass function in violent star forming systems in low- and high-metallicity environments, and in dense (e.g., in stellar clusters) and diffuse environments; (*b*) the modes of star formation: Starburst clusters are an important mode of star formation; (*c*) the role of starbursts in AGN; (*d*) the interaction between massive stars and the interstellar and intergalactic media; and (*e*) the contribution of starbursts to the re-ionization of the universe. Despite the very significant progress obtained over the past two decades of UV observations of starbursts, there are important problems that still need to be solved. High-spatial resolution UV observations of nearby starbursts are crucial to further progress in understanding the violent star formation processes in galaxies, the interaction between the stellar clusters and the interstellar medium, and the variation of the IMF. Thus, a new UV mission furnished with an intermediate spectral resolution long-slit spectrograph with high spatial resolution and high UV sensitivity is required to further progress in the study of starburst galaxies and their impact on the evolution of galaxies.

4. Life in main sequence

4.1. *Atmospheres and winds in cool stars*

Isabella Pagano: Cool star outer atmospheres represent an important laboratory in which solar-like magnetic activity phenomena can be studied under a wide variety of conditions, allowing us to gain insight into the fundamental processes involved.

The UV range is especially useful for such studies because it contains powerful diagnostics extending from warm (10 000 K) chromospheres out to hot (1–10 MK) coronae. Also very weak coronal winds from cool stars have been identified and characterized thanks to high resolution UV spectra.

Here I give a review on UV spectroscopy main achievements for cool star chromospheres, transition regions, coronae and coronal winds. I also outline the requirements for future experiments able to allow progresses in this field.

4.2. *Evolution over time of magnetic dynamo driven UV emissions of dG - dM stars and effects on hosted planets*

Eduwrd F. Guinan, Ignasi Ribas, and S.G. Engle The evolution over time of the magnetic activity and the resulting X-ray and UV coronal and chromospheric emissions of main-sequence dG, dK, and dM stars with widely different ages will be discussed.

Young cool stars spin rapidly and have correspondingly very robust magnetic dynamos and strong coronal and chromospheric X-ray-UV (XUV) emissions. However, these stars spin-down with time as they lose angular momentum via magnetized winds and their magnetic generated activity and emission decrease. For example, the study of solar proxies shows that the young Sun was rotating more than ten times faster than today and had correspondingly very high levels of magnetic activity and very intense X-ray-UV (XUV) emissions. Studies of dK-dM stars over a wide range of ages and rotations show similar (but not identical) behavior. Particular emphasis will be given to discussing the effects that UV emissions have on the atmospheres and evolution of solar system planets as well as the increasing number of exoplanets found hosted by dG - dM stars. The results from modeling the early atmospheres of Venus, Earth and Mars using recently determined XUV irradiances and winds of the young Sun are also briefly discussed. For example, the loss of water from juvenile Venus and Mars can be explained by action of the strong XUV emissions and robust winds of the young Sun. We also examine the effects of strong X-ray and UV coronal and chromospheric emissions (and frequent flares) that dM stars may have on possible planets orbiting within their habitable zones (HZ) – located close to the low luminosity host stars ($0.05 < \mathrm{HZ} < 0.4$ AU). Dwarf M stars make interesting targets for further study because of their deep convective zones, efficient dynamos and strong XUV emissions. Furthermore, a large fraction of dM stars are very old (> 5 Gyr), which present possibilities for the development of highly advanced modes of intelligent life on planets that may orbit them. This research is supported by grants from NASA and utilizes data from the *IUE, FUSE, HST, EUVE, ROSAT, XMM*, and the *Chandra* missions. We are very grateful for this support.

4.3. *Systematics of OB spectra*

Nolan R. Walborn: The systematics of OB spectra are reviewed in the optical domain, dominated by photospheric lines, and in the far ultraviolet (both *IUE* and *FUSE* ranges), in which the stellar-wind profiles dominate. First, the two-dimensional (temperature, luminosity) trends in normal spectra are surveyed. Then, the normal reference frame having been established, various categories of peculiar objects can be distinguished relative to it, which reveal several phenomena of structural and/or evolutionary significance. Included are CNO anomalies at both early and late O-type stars, three varieties of rapid rotators, hot and cool Of/WN transition objects, and the recently discovered second known magnetic O-type star. The importance of both optical and UV observations to understand these phenomena is emphasized; for instance, progress in understanding the structure of the new O-type magnetic oblique rotator is hampered by the current lack of a UV spectrograph. While progress in the physical interpretation of these trends and anomalies has been and is being made, increased attention to modeling the systematics would accelerate future progress in this author's opinion. Finally, preliminary results from a *Chandra* high-resolution survey of OB-type star X-ray spectra (PI: W. Waldron) are presented. They provide evidence that, just as emerged earlier in the UV, systematic morphological trends exist in the X-ray domain that are correlated with the optical spectral types, and hence the fundamental stellar parameters, contrary to prevailing opinion.

4.4. *High mass stars: stellar populations*

Luciana Bianchi: Massive stars dominate the chemical and dynamical evolution of the ISM, and ultimately of their parent galaxy and the universe, because of their fast evolution and intense supersonic winds. Decades ago, the advent of space-born UV spectroscopy marked the discovery of mass loss in massive stars and began to change our understanding of their evolution. Recently, significant advances in stellar modeling, and the observation of crucial ions in the far-UV spectral range disclosed by *FUSE*, led to resolve long-standing issues in modeling massive star winds and atmospheres, and to assess of the roles of shocks and clumping. A revised (downwards) calibration of Teff for early spectral types emerged as a result. Meanwhile, *HST* imaging had opened the possibilty of detailed resolved studies of stellar populations in Local Group galaxies, sampling a variety of metallicity and environment conditions. More recently, *GALEX* is providing the first global, deep view of the massive star populations for hundreds of nearby galaxies. The unprecedented coverage and sensitivity of the *GALEX* UV imaging, easily detecting extremely low levels of SF, is once again revolutionizing some of our views of massive star formation in galaxies.

5. Star death

5.1. *Planetary Nebula*

Letizia Stanghellini: The asymptotic giant branch (AGB) phase of stellar evolution is common to most stars of low and intermediate mass. Most of the carbon and nitrogen in the Universe is produced by AGB stars. The final fate of the AGB envelopes are represented by planetary nebulae (PN). By studying PN abundances and compare them with the yields of stellar evolution is possible to quantify carbon and nitrogen production, and to study cosmic recycling in galactic and Magellanic Cloud populations. In this paper we present the latest results in PN chemical abundance analysis and their implication to the chemical evolution of the galaxy and the Magellanic Clouds, with particular attention to carbon abundance, available only thanks to ultraviolet spectroscopy.

5.2. *Planetary Nebulae abundances and stellar evolution*

Stuart R. Pottasch: A summary is given of planetary nebulae abundances from *ISO* measurements. It is shown that these nebulae show abundance gradients (with galactocentric distance), which in the case of neon, argon, sulfur and oxygen (with four exceptions) are the same as H II regions and early type star abundance gradients. The abundance of these elements predicted from these gradients at the distance of the Sun from the center are exactly the solar abundance. Sulfur is the exception to this; the reason for this is discussed. The higher solar neon abundance is confirmed; this is discussed in terms of the results of helioseismology. Evidence is presented for oxygen destruction via ON cycling having occurred in the progenitors of four planetary nebulae with bilobal structure. These progenitor stars had a high mass, probably greater than $5\,M_{\odot}$. This is deduced from the high values of He/H and N/H found in these nebulae. Formation of nitrogen, helium and carbon are discussed. The high-mass progenitors which showed oxygen destruction are shown to have probably destroyed carbon as well. This is probably the result of hot bottom burning.

5.3. *Supernovae and GRB progenitors*

Stephen J. Smartt: Over the past five years we have been attempting to directly detect the massive progenitor stars of nearby supernovae (<20 Mpc) in archive *HST* images, combined with an *HST* T-o-O programme in Cycles 10-14. We have detected the progenitors of three type II-P SNe in M 74 and M 51, showing them to be red supergiants with initial masses close to the theoretical limit for core-collapse ($8\,M_{\odot}$). We have set robust upper luminosity and mass limits on another 12 progenitor stars (from type II-P, and Ib/c) supernovae. There are three striking results so far from this project. One is that we have observationally confirmed for the first time that the common type II-P SNe do indeed come from red supergiants. The second is that II-P SNe only appear to arise in stars with initial masses less than $15\,M_{\odot}$. Thirdly, we find that faint SNe which have been thought to come from black-hole forming SNe are more likely to have low-mass progenitors. We present the results of this survey and review the work from other competing groups in the field.

6. Compact Objects

6.1. *The importance of UV observations for the study of white dwarfs and the local interstellar medium*

Martin A. Barstow: The development of far-UV astronomy has been particularly important for the study of hot white dwarf stars. A significant fraction of their emergent flux appears in the far-UV and traces of elements heavier than hydrogen or helium are, in general, only detected in this waveband or at shorter wavelengths that are also only accessible from space. Although white dwarfs have been studied in the far-UV throughout the past ~ 25 yr, since the launch of *IUE*, only a few tens of objects have been studied in great detail and a much larger sample is required to gain a detailed understanding of the evolution of hot white dwarfs and the physical processes that determine their appearance. This talk reviews the current knowledge regarding hot white dwarfs and outlines what work needs to be carried out by future far-UV observatories.

6.2. *UV observations of interacting binaries*

Knox S. Long: Virtually all of the exotic objects in the Galaxy are the products of binary evolution, including binary pulsars, black-hole candidates, low mass X-ray binaries, cataclysmic variables and symbiotic stars. Type Ia supernovae, the standard candle of modern cosmology, are produced by them. UV spectroscopy of interacting binaries obtained with *HST* and *FUSE* have dramatically improved our understanding of interacting binaries and of the wide range of physical processes that characterize their emission. UV imaging has made it possible to isolate binaries and the products of binary evolution in old stellar populations and thereby test directly models of binary evolution in dense stellar systems. In this review, I will summarize some of the highlights of what we have learned about binaries and their evolution from UV observations, and suggest how, if suitable instrumentation were to exist, furthur observations could resolve some of the important questions that remain.

6.3. *RS Ophiuchi in outburst, seen with XMM-Newton and Chandra*

M. Orio, R. Gonzalez-Riestra, Elia Leibowitz, T. Nelson, and E. Tepedenlelioglu: We present X-ray observations of the recurrent nova RS Ophiuchi in outburst, obtained with the *Chandra* and *XMM-Newton* observatories on six different dates post discovery. An important goal of the observations was to study the central UEV-supersoft X-ray source. X-ray grating spectra are the best way to place constraints on the physical parameters of the white dwarfs in these systems. In the first month after the outburst, we detected emission lines of H and He -like ions of metals including Si, Mg and Ne. The presence of these lines indicates a range of plasma temperatures, and most likely we were detecting emission from red giant wind material shocked by the nova ejecta. Later spectra show a large increase in emission at longer wavelengths, peaking two months after the outburst. We attribute this to the emergence of the underlying white dwarf. We used NLTE white dwarf atmosphere models to fit the spectra, finding $\log g = 9$, and $T \simeq 800\,000$ K. This indicates a white dwarf mass of $> 1.2\,M_{\odot}$. We also present a preliminary timing analysis of the *XMM-Newton* light curves. We find a period of ~ 35 s both in March and April. Although the origin of this period is still a question for debate, it may be the spin period of the white dwarf. If this is the case, the white dwarf must again be very massive, making RS Ophiuchi a likely supernova Type Ia progenitor.

7. The impact of stellar astrophysics on cosmological evolution

7.1. *Missing baryons: the Local Group and beyond*

F. Nicastro, S. Mathur, M. Elvis, Roy Williams, and F. Fiore: More than half ($\sim 54\%$) of the expected number of baryons in the local Universe have eluded detection until very recently, due to their high temperature ($T \simeq 10^5$ - 10^7 K) and low density ($n_b \simeq 10^{-6} - 10^{-5}$ cm^{-3}). At redshift $z > 2$ most of the baryons lie in the Intergalactic Medium, in a mildly photo-ionized (by the metagalactic radiation field) phase, and they are easily observed and counted in the branches and trees of the so called Ly-alpha forest. However, at $z < 2$, intervening H I Ly-alpha absorption becomes progressively rarer, while already virialized structures, or structures in the way of forming, account for only about 20 - 30 % of the baryons. Theory predicts that the majority of baryons get shock-heated during the continuous process of structure formation and so becomes invisible to optical or UV photons at $z \leqslant 2$. Metals in such high-temperature medium, however, would still produce some opacity to soft X-ray and Far-Ultraviolet (FUV) photons, mainly through $K\alpha$ resonant transitions from Li-like (FUV/X-rays), He-like and H-like (X-rays) ions of C, N, O and Ne.

In this contribution, I review recent observational efforts towards the detections of the so-called missing baryons in the local Universe, both in our own Local Group and at cosmological distances. I will show how both X-ray (*Chandra*) and FUV (*FUSE*) data clearly point toward the presence of a large amount of hot, shock-heated, gas ($T \simeq 10^6$ K) either forming an extended Galactic halo ($\geqslant 200$ kpc) or an even more extended (2 – 4 Mpc) Warm-Hot Intergalact Medium filament of metal-enriched primordial material, connecting our own Local Group possibly with the Virgo Cluster, and in which our Galaxy is embedded. I present the first detections of WHIM filaments at cosmological distance, and show how these first measurements are consistent with both theoretical predictions and the number of missing baryons, within the still large uncertainties due to the low number statistics.

8. The impact of stellar astrophysics in understanding the formation of life sustainable systems

8.1. *Young planetary disks*

Alain Lecavelier des Étangs: Young planetary disks have been widely observed in the ultraviolet during the last two decades. The sensitivity to electronic transition of atoms and ions in the near ultraviolet, and of molecules like CO and H_2 in the far ultraviolet, lead to detailed observations of the gaseous content of disks surrounding pre-main sequence and main sequence stars.

The main results are the measurements of the physical characteristics of the molecular disk in proto-planetary systems, and the description of the gas phase in few debris disks.

In the very particular case of the β Pictoris disk, UV and optical spectroscopy lead to the extraordinary discovery of extra-solar comets falling onto the star. We are witnessing a very strong cometary activity in a young planetary system, likely similar to our own Solar system, four billion years ago.

8.2. *Extrasolar planets atmospheres*

Alfred Vidal-Madjar: Almost 200 extrasolar planets have been discovered since 1995, when the first one was detected orbiting the solar type star 51 Peg. The extrasolar planets known so far show a variety of characteristics, in some cases not at all represented in our solar system. The "hot Jupiters", in particular, were not predicted by models before being discovered: they are gas giant planets orbiting at only few stellar radii. This peculiarity gives us the unique opportunity of exploring their upper atmosphere while they transit in front of their parent star. Among the atmospheric species detected so far with this method, we can list Na I, in the visible, and H I, O I and C II in the far UV. These UV detections, in particular, unveiled a even more striking feature: the upper atmosphere of these exoplanets is evaporating at a very fast pace, in a "blow off" process.

9. Poster contributions

9.1. *AGB Stars: testing carbon loss via the ultraviolet lines*

Yulia V. Milanova, and Alexander F. Kholtygin: A method is proposed to determine the realistic abundances of carbon in planetary nebulae based on the actual distribution functions of errors in measuring line intensities. Fluctuations both in temperature and in mass density in a nebula are taken into account. The C abundances and the amplitudes of temperature and density fluctuations for the large sample of PNe are given. The intensity of the ultraviolet lines of C ions are used for determining the more exact abundances. These abundances are probably the most reliable at the present time and give estimations of the primordial CNO abundances at the epoch when the progenitors of PNs are formed. Based on the newly carbon abundances, the total mass losses of carbon during the AGB stage of evolution are estimated.

9.2. *On the origin of two-shell supernova remnants*

Vasilii V. Gvaramadze: It is known that proper motion of massive stars causes them to explode far from the geometric centers of their wind-driven bubbles and thereby affects the symmetry of the resulting diffuse supernova remnants (SNRs). We use this fact to explain the origin of SNRs consisting of two partially overlapping shells (e.g., 3C 400.2, Cygnus Loop, Kes 32, etc.), whose unusual morphology is usually treated in terms of the collision (or superposition) of two separate SNRs or breakout phenomena in a region with a density discontinuity. We propose that a SNR of this type is a natural consequence of an off-centered cavity supernova (SN) explosion of a moving massive star, which ended its evolution near the edge of the main-sequence (MS) wind-driven bubble. Our proposal implies that one of the shells is the former MS bubble reenergized by the SN blast wave. The second shell, however, could originate in two somewhat different ways, depending on the initial mass of the SN progenitor star. It could be a shell swept-up by the SN blast wave expanding through the unperturbed ambient interstellar medium if the massive star ends its evolution as a red supergiant (RSG). Or it could be the remainder of a pre-existing shell (adjacent to the MS bubble) swept-up by the fast progenitor's wind during the late evolutionary phases if after the RGS phase the star evolves through the Wolf-Rayet (WR) phase. In both cases the resulting (two-shell) SNR should be associated only with one (young) neutron star (thus one can somewhat improve the statistics of neutron star/SNR associations since the two-shell SNRs are quite numerous). We discuss several criteria to discern the SNRs formed by SN explosion after the RGS or WR phase.

9.3. *Analysis of the high-temperature region in Be-type stars*

Andrea F. Torres, Adela E. Ringuelet: The High Temperature Region (HTR) that surrounds the photospheres of Be-type stars is studied in order to derive observational constraints for modelling Be-type stars, in particular for the region where superionization takes place. 50 Be-type stars, representative of a considerable range of temperature, were chosen. From archival, high-dispersion *IUE* spectra, different lines that originate in the HTR region were considered, namely the resonance lines of Si IV, C IV and Al III, and He II 1640. Equivalent widths (corrected for photospheric contribution), optical depths, atom columns and expansion velocities were measured. From this observational data several correlations between different observables were obtained. These correlations permit us to discuss the geometry, density distribution and heat input of the lines formation regions (LFRs). The major results can be summarised as follows:

(*a*) The circumstellar material contributes to the resonance lines of Si IV, C IV, Al III and to the He II 1640 at all inclination angles.

(*b*) In Si IV, C IV and Al III the equivalent widths have a tendency to increase in objects with high rotational velocities.

(*c*) Si IV and C IV equivalent widths are also correlated to the kinetic energy of the expansion velocity. This means that dissipation of mechanical energy is one of the heating mechanisms.

(*d*) On the basis of the expansion velocities and the line profiles, we establish a sequence for the LFRs: The LFR of He II is at the base of the wind and the closest to the central star. The LFRs of Si IV and C IV are inmersed in the stellar wind. The LFR of Al III is an interface between the HTR and the cool envelope.

The analysis followed in this work is completely model-independent. Consequently, these results could be useful to decide which are the facts that are to be considered when modelling Be-type stars.

9.4. *High resolution spectroscopy of halo stars in groundbased UV*

Valentina Klochkova, Gang Zhao, S. Ermakov, and Vladimir Panchuk: For the first time an atlas of high-spectral resolution ($R = 60\,000$) CCD-spectra in the low studied wavelength range 3 500 - 5 000 Å is presented for four stars with values of metallicity $-3.0 < [\mathrm{Fe/H}] < -0.6$, temperatures $4750 < T_{\mathrm{e}} < 5900$ K, and surface gravity $1.6 < \log g < 5.0$. Based on these spectral data we determined model atmosphere parameters and calculated abundances of 29 chemical elements or their ions.

9.5. *Hyper ionization phenomena in the C IV, N IV, and N V regions of 20 Oe-type stars, including HD 93521*

Antonios Antoniou, E. Danezis, Evaggelia Lyratzi, D. Nikolaidis, L.C. Popovic, and M.S. Dimitrijevic: As it is already known, the spectra of many Oe- and Be-type stars present Discrete Absorption Components (DACs) which, due to their profiles' width as well as the values of the radial velocities, create a complicated profile of the main spectral lines. In this poster paper we detect the presence of this phenomenon (DACs or SACs) in the C IV resonance lines, the N IV spectral line, and the N V resonance lines of 20 Oe-type stars of different spectral subtypes. In particular we discuss these lines in the spectrum of the star HD 93521 which is a relatively bright, very rapidly rotating O9.5V star.

Method: In our study we apply the method proposed by Danezis *et al.* on the *IUE* spectra of 20 Oe-type stars, including the star HD 93521 observed with *IUE* from 1979 until 1995, and we examine the time variations of the physical parameters, stated below, as a function of the spectral subtype.

Results: As a first result we detect that the C IV resonance lines, the N IV spectral line, and the N V resonance lines each consist of one to five Satellite Absorption Components (SACs or DACs). With the above method we calculate the values of the apparent rotational and radial velocities, the Gaussian standard deviation of the random motions of the ions, the random velocities of these motions, as well as the optical depth, the column

density, the Full Width at Half Maximum (FWHM), the absorbed and the emitted energy of the independent regions of matter which produce the main and the satellites components of the studied spectral lines.
Discussion: We point out that the new and important aspect of our study is the values' calculation of the above parameters, their time scale variations and their variations as a function of spectral subtype,using the DACs or SACs theory. Our results are a successful test of this theory and of Danezis *et al.* proposed method. This study is a part of a Ph.D. Thesis.

9.6. *Study of Hα regions in 120 Be-type stars, and the complex structure of the Si* IV *1393.755, 1402.77 Å regions of 68 Be-type stars*

Evaggelia Lyratzi, E. Danezis, Antonios Antoniou, D. Nikolaidis, L.C. Popovic, and M.S. Dimitrijevic: As it is already known, the spectra of many Oe- and Be-type stars present Discrete Absorption Components (DACs) which, due to their profiles' width as well as the values of the radial velocities, create a complicated profile of the main spectral lines. In this poster paper we detect the presence of this phenomenon (DACs or SACs) in the shape of Hα line in the spectra of 120 Be-type stars, and in the Si IV resonance lines in the spectra of 68 Be-type stars of all the spectral subtypes and luminosity classes.
Method: In our study we apply the method proposed by Danezis *et al.* on the stellar spectrographs of 120 Be-type stars which were taken by Fehrenbach and Andrillat (resolution 5,5 and 27 Å with the telescope of 152 cm in the Observatory of Haute Provence), and on the spectra of 68 Be stars observed with *IUE*, and we examine the variations of the physical parameters, stated below, as a function of spectral subtype and luminosity class.
Results: We find that in the Be-type stellar atmospheres, there are two regions that can produce the Hα Satellite Absorption Components (SACs or DACs). The first one lies in the chromosphere and the second one in the cool extended envelope. With the above method we calculate: (*a*) For the chromospheric absorption components we calculated the optical depth as well as the rotational and radial velocities of the independent regions of matter which produce the main and the satellites components. *b*) For the emission and absorption components which are created in the cool extended envelope we calculated the FWHM, the optical depth and the radial velocities of the independent regions of matter which produce the main and the satellites components.
We find that the absorption atmospherical regions where the Si IV resonance lines originated may be formed of one to five independent density layers of matter which rotate with different velocities, producing one to five Satellite Absorption Components (SACs or DACs). With the above method we calculate the values of the apparent rotational and radial velocities, as well as the optical depth of the independent regions of matter which produce the main and the satellites components of the studied spectral lines.
Discussion: We point out that the new and important aspect of our study is the values' calculation of the above parameters and their variations as a function of spectral subtype and luminosity class, using the DACs or SACs theory. Our results are a successful test of this theory and of Danezis *et al.* (2003, 2005) proposed method. This study is a part of a Ph.D. Thesis.

9.7. *A new approach for DACs and SACs phenomena in the atmospheres of hot emission-line stars*

D. Nikolaidis, E. Danezis, Evaggelia Lyratzi, L.C. Popovic, M.S. Dimitrijevic, Antonios Antoniou, and E. Theodossiou: As it is already known, the spectra of

many Oe- and Be-type stars present Discrete Absorption Components (DACs) which, due to their profiles' width as well as the values of the radial velocities, create a complicated profile of the main spectral lines. This fact is interpreted by the existence of two or more independent layers of matter, in the region where the spectral lines are formed. Such a structure is responsible for the formation of a series of satellite components (DACs or SACs) for each spectral line (Bates & Halliwell, 1986, Danezis *et al.* 2003, 2005).
Method: In this paper we present a mathematical model reproducing the complex profile of the spectral lines of Oe-type and Be-type stars that present DACs or SACs. This model presupposes that the regions, where these spectral lines are formed, are not continuous but consist of a number of independent absorbing or emitting density layers of matter and an external general absorption region. In this model we assume that the line broadening is due to the random motion of the ions and the rotation of the density regions that produce the spectral line and its satellite components. With this method we can calculate the values of the apparent rotational and radial velocities, the Gaussian standard deviation of the random motions of the ions, the random velocities of these motions, as well as the optical depth, the Full Width at Half Maximum (FWHM), the absorbed and the emitted energy and finally the column density of the independent regions of matter which produce the main and the satellites components of the studied spectral lines.
Results: In order to check the above spectral line function, we calculated the rotational velocity of He I 4387.928 Å absorption line in the spectra of five Be-type stars, using two methods, the classical Fourier analysis and our model. The values of the rotational velocities, calculated with Fourier analysis, are the same with the values calculated with our method.
Discussion: We point out that the new and important aspect of this method is the values' calculation of the above parameters using the DACs or SACs theory.

9.8. *Eta Carinae: what we have learned from HST-STIS in the UV*

Theodore R. Gull: The Luminous Blue Variable η Carinae is revealing many answers to its mysteries by high spatial resolution in the visible and the ultraviolet. Studies with the *HST*-STIS from 1998.0 to 2004.3 show major changes in the stellar and nebular spectra that track with the 2024-day period first noted by A. Damineli in the visible and followed by M. Corcoran via *RXTE* x-ray monitoring. We will show examples of the stellar and nebular spectra indicating changes in the central source, likely a massive binary system and indicating the response of the nebular ejecta, which is the $> 12\,M_\odot$ Homunculus, the $0.5\,M_\odot$ Little Homunculus, both bipolar structures, with intervening skirts. Within the interior skirt are located the Weigelt blobs, B, C and D, plus the Strontium Filament, all of which respond to the strong UV emission originating from the hot, less massive companion. Narrow-line absorption systems correlate with the Homunculus and Little Homunculus and are seen in hundreds of metal lines. For the Homunculus, the metal energy level populations correspond to 760 K, but the OH, CH, NH and CH+ to 60 K, while nearly a thousand H_2 lines are visible during the broad maximum. The Little Homunculus has a kinetic temperature of $\sim$6400 K during the broad maximum, but drops to 5000 K during the short minimum. Much is being learned about the N-rich, C,O-poor chemistry of this ejecta from a massive star in the late stages of CNO-processing. Recent GRB spectra show similar hot metal absorption gases likely being the ejecta from progenitor stars. Were they Wolf-Rayet stars?

9.9. *High resolution echelle spectrograph NES for visible and groundbased UV regions*

Vladimir E. Panchuk, Valentina G. Klochkova, I.D. Najdenov, and Maxim V. Yushkin: We present the high-resolution echelle spectrograph NES of the 6 m telescope.

The NES is located at the Nasmyth focus (1:30) platform of the telescope. The NES in combination with the image slicer provides a spectral resolution of $R = 75\,000$ within the spectral region 3 200 - 10 000 Å.

9.10. *Absorption profiles of Lyman-γ of atomic hydrogen perturbed by collisions with protons*

Nicole F. Allard, and I. Noselidze: A strong feature near 995Å in the Lyman-γ wing of hot white dwarfs has been demonstrated to be due to quasi-molecular absorption of H_2 molecule. We present new theoretical calculations of the line profile of Lyman gamma that include the variation of the radiative dipole moment during the collision.

9.11. *Activity of supersoft X-ray sources: the case of V Sge*

Vojtech Simon, and Janet A. Mattei: Super-soft X-ray sources (SSXSs) are unique binary systems with a very high mass transfer rate onto the white dwarf (WD) (dm/dt $\simeq 10^{-}7\,M_{\odot}$/yr). Such a high dm/dt gives rise to a steady nuclear burning of accreted hydrogen on the surface of the WD (e.g., the model by van den Heuvel *et al.* 1992), and consequently to an intense soft X-ray emission. Its reprocessing on the disk and donor makes SSXS bright source in the ultraviolet and optical. Most SSXS are observed in the Magellanic Clouds. V Sge is one of a few SSXSs located in our Galaxy and is called the representative of the V Sge-type class. This system is very active in various spectral passbands. Here we concentrate on the long-term optical activity of this system in the context of SSXSs. We show that this activity in V Sge underwent dramatic evolution during the recent decades. We also find that in so-called active segments (usually displaying high/low states) cycles are often apparent but their length undergoes large, often gradual variations. We give the implications for the accretion wind evolution (model by Hachisu and Kato 2003): an increase of the mass outflow rate from the donor is in plausible agreement with the observations. We also analyse the color indices and absolute magnitudes of V Sge and compare them with those of the 'classical' (SSXBs), the V Sge-type class, the symbiotic SSXBs, and the classical novae in the soft X-ray phase. This approach helps comparing the properties and configuration of the reprocessing medium in the individual systems. We also show that there is no systematic difference between the classical SSXBs and the V Sge-type class, as regards their V- and U-band absolute magnitudes. Since soft X-rays are easily absorbed, only a small fraction of SSXSs is identified as such. The activity and properties in the ultraviolet and optical can thus serve as a useful method for identification of SSXSs.

9.12. *Distance to Wolf-Rayet Star WR 134*

Krzysztof Czart, and Andrzej Strobel: Distances to Wolf-Rayet stars are still the subject of research. The only WR star with reliable *Hipparcos* parallax is WR 11 (van der Hucht *et al.* 1997). Other Wolf-Rayet stars have photometric distances estimations (van der Hucht 2001). Recently, a correlation between equivalent width of interstellar Ca II H (3968 A) and K (3933 A) lines and *Hipparcos* parallaxes has been found for OB stars (Megier *et al.* 2005). These relations are given with formulas: $\pi = 1/[2.78EW(K) + 95]$ and $\pi = 1/[4.58EW(H) + 102]$, where π is a parallax in arcseconds and EW is the equivalent width in milliangstroms. We have used these correlations to estimate distance to Wolf-Rayet star WR 134. Equivalent widths of Ca II lines have been measured in low resolution spectra with $R = 2000$. We have got: 1.17 kpc from the Ca II H, and 1.18 kpc the Ca II K line. The distance to this star as cited in the VIIth Catalogue of Galactic Wolf-Rayet Stars (van der Hucht 2001) is 1.74 kpc. If WR 134 belongs to the complex of associations around Cyg OB1, then its distance agrees very well with distances of other members of these associations – also obtained from Ca II line intensities.

9.13. *SpectroWeb: an interactive graphical database of digital stellar spectral atlases*

Alex Lobel: SpectroWeb is an online maintained interactive graphical database of digital spectral atlases of bright stars at <`http://spectra.freeshell.org`>. It is an efficient and user-friendly research tool for accurate analyses of stellar spectra observed with high-spectral resolution, including the solar spectrum. The web-interface displays observed and theoretical stellar spectra, and comprehensively provides detailed atomic and molecular line information via user interaction. It fully integrates interactive spectrum visualization tools for the analysis, management, and maintenance of large volumes of spectral line-identification, -transition, and -property data. SpectroWeb 1.0 currently offers optical (3300 - 6800 Å) flux normalized high-resolution spectra of Betelgeuse (M2), Arcturus (K1), the Sun (G2), Beta Aqr (G0), Procyon (F5), and Canopus (F0). The provided line identifications are based on state of the art spectrum synthesis calculations. The graphical database is under permanent development as an online repository of identified (absorption) lines in bright star reference spectra, covering a broad range of stellar spectral types. Its object-oriented (Java) implementation offers future expansion capabilities to link and read stellar spectral atlases from various public internet sites.

9.14. *Three dimensional radiative transfer in winds of massive stars: Wind3D*

Alex Lobel, and Ronny Blomme: We discuss the development of the new radiative transfer code Wind3D. It solves the non-LTE radiative transport problem in moving stellar atmosphere models in three geometric dimensions. The code accepts arbitrary 3D velocity fields in Cartesian geometry without assumptions of axial symmetry. Wind3D is currently implemented as a fully parallelized (exact) accelerated lambda iteration scheme with a two level atom formulation. The numerical transfer scheme is efficient and sufficiently accurate to trace small variations of local velocity gradients on line opacity in strongly scattering dominated extended stellar winds. We investigate the detailed formation of P-Cygni line profiles observed in ultraviolet spectra of massive stars. We compute the detailed shape of these resonance lines to model local enhancements of line opacity that can for instance be caused by clumping in supersonically expanding winds. Wind3D is applied to preliminary hydrodynamic disk models to investigate physical properties of discrete absorption line components.

9.15. *White Dwarfs in the GALEX survey*

Adela Kawka, and Stephane Vennes: We have cross-correlated the 2dF QSO Redshift Survey (2QZ) white dwarf catalog with the GALEX 2nd data release and the Sloan Digital Sky Survey (SDSS) data release 4 to obtain ultraviolet photometry (FUV, NUV) for approximately 700 objects and optical photometry (*ugriz*) for approximately 800 objects. We have compared the optical-ultraviolet colours to synthetic white dwarf colours to obtain temperature estimates for approximately 250 of these objects. These white dwarfs have effective temperatures ranging from 10 000 K (cooling age of about 1 Gyr) up to about 40 000 K (cooling age of about 3 Myr), with a few that have even higher temperatures. We have found that to distinguish white dwarfs from other stellar luminosity classes both optical and ultraviolet colours are necessary, in particular for the hotter objects where there is contamination from B- and O-type main-sequence stars. Using this sample of white dwarfs we will determine the fraction of helium-rich white dwarfs (DO, DB) relative to hydrogen-rich white dwarfs (DA), and make luminosity functions for both hot DA and DO white dwarfs. The temperatures of the 2QZ white dwarfs obtained from the UV colours will be used to obtain luminosities of the objects and hence their distance. From this we will be able to determine the scale height of these stars.

9.16. *Relative strengths of Raman scattered He* II *6545 and Hα wings in the symbiotic star V1016 Cyg and the young planetary nebulae IC 5117*

Hee-Won Lee, and Suna Kang: Many symbiotic stars and some young planetary nebulae are known to exhibit very prominent and broad wings around Hα emission. Lee (2000) proposed that these Hα wings are formed from Raman scattering by atomic hydrogen of far UV radiation around Lyβ. However, broad wings may also arise from other mechanisms including fast gas flows around the hot star and Thomson scattering. Furthermore, it is unclear whether continuum around Lyβ or Lyβ line photons themselves are responsible for the broad Hα wings in the proposal of Lee (2000). In order to confirm the Raman scattering origin of Hα wings, we present our high resolution spectrum around Hα of the symbiotic star V1016 Cyg and the young planetary nebula IC 5117 obtained with the Bohyunsan Optical Echelle Spectrograph installed on the 1.8 m telescope. The two objects show relatively strong Raman scattered He II 6545 features but Hα wings are significantly weak in IC 5117 compared with those in V1016 Cyg. In the spectra, we note that the Hα and He II emission lines in IC 5117 are much narrower than the counterparts in V1016 Cyg. Applying Gaussian fits to these emission lines and inferring the far UV radiation compatible with the recombination theory, we perform Monte Carlo simulations of Raman scattered He II 6545 and Hα wings in an assumed neutral region with H I column density of $1022\,\mathrm{cm}^{-2}$. Our numerical results show relatively weak Hα wings compared with He II 6545 in IC 5117 and quite strong Hα wings in V1016 Cyg, which is consistent with our observational spectra. This result strongly indicates that the broad Hα wings are originated from the Lyβ line photons, not continuum photons around Lyβ.

9.17. *A flare-induced mass transfer/accretion event in AM Her?*

Steven H. Saar, Vinay L. Kashyap, and Frederick A. Ringwald: We report observations of a mass transfer/accretion event in AM Her which appears to have been induced by a strong flare on the secondary. UV observations of the magnetic CV star AM Her were made with *HST*-STIS towards the end of a deep photometric minimum in late 2003. Our goal was to search for evidence of magnetic activity on the secondary (as seen in the Doppler-shifted hot emission lines of C IV and Si IV), one of the fastest rotating cool stars ever studied in the UV. There was little evidence for quiescent C IV emission at the velocity of the M star secondary, placing useful limits on its steady transition region emission. One strong flare was seen in C IV (and initially, in Si IV and He II as well), with a velocity consistent with the M-type dwarf. This was accompanied by a near-simultaneous increase in continuum emission from the white dwarf; the excess had a temperature of about 100 000 K. We argue that this strong flare may have ejected mass in the form of CME, or disturbed a large prominence system, exciting an accretion event on the white dwarf. We discuss implications of the results for low-state CV activity, mass transfer, and cool star activity at extreme rotation rates.

9.18. *White Dwarfs with strange matter cores: an analysis of candidates*

Grant J. Mathews, Nguyen Quynh Lan, In-Saeng Suh, Wolfgang Zech, Kaori Otsuki, and Fridolin Weber: We summarize masses and radii for a number of white dwarfs as deduced from a combination of proper motion studies, *Hipparcos* parallax distances, effective temperatures, and binary or spectroscopic masses. We construct a projection of white-dwarf radii for fixed effective mass and conclude that there is at least marginal evidence for bimodality in the radius distribution for white dwarfs. Some stars appear to have radii which are significantly smaller than that expected for a standard electron-degenerate white-dwarf equations of state. We argue that if such compact white dwarfs exist it is unlikely that they contain an iron core. We show that the data exhibit several features consistent with the expected mass-radius relation of strange dwarfs. We identify eight nearby white dwarfs which are possible candidates for strange matter cores and suggest observational tests of this hypothesis.

9.19. *New powerful diagnostics for hot evolved stars: constraining the hottest temperatures, the faintest winds and the neon abundance*

Luciana Bianchi, and James Herald: We have recently identified a strong P Cygni feature ($\lambda\,973$ Å) in the far-UV spectra of some very hot ($T_{\rm eff} > 85\,000$ K) central stars of planetary nebulae (CSPN) as originating from Ne VII. By including highly ionized neon calculations in stellar atmosphere models, we could reproduce this feature as it is observed in hydrogen-deficient CSPN, as well as in PG 1159 objects. The discovery of highly ionized neon diagnostic features in CSPN spectra is important because an overabundance of this element is indicative of processed material that has been dredged up to the surface as predicted in the 'born-again' scenario, a possible explanation for hydrogen-deficient CSPN. We show the potential of this strong feature as well as weaker neon features as diagnostics in stellar atmosphere analyses for extremely hot post-ABG objects (where other diagnostics are scarce). This diagnostic is potentially useful also in the analysis of massive stars with winds such as WO stars.

9.20. *A revised temperature scale for massive stars*

Luciana Bianchi, James Herald, and M. Antonieta Garcia: A consistent analysis, using non-LTE codes with sphericity and mass loss, of different ionization species in O-type star spectra observed from the far-UV to UV allowed us to constrain the long-time uncertain ionization equilibrium in the stellar winds, to quantify the role of soft X-rays from shocks and to reconcile otherwise discrepant UV line diagnostics. The modelling initially performed with the WM-basic code led to a revised (downwards) $T_{\rm eff}$ scale for Galactic O-type stars, a result with important implications for the energy-balance calculations in H II regions and for massive star evolution. A further analysis using the CMFGEN code, including clumping effects, allowed us to resolve the remaining UV line discrepancies, in particular the P V line in the far-UV, and to finally reconcile mass loss and photospheric parameters derived from optical and UV lines.

9.21. *Classification and properties of Milky Way UV sources from the GALEX surveys*

Luciana Bianchi, *et al.*: We used *Galaxy Evolution Explorer (GALEX)* imaging survey data matched to the Sloan Digital Sky Survey (SDSS) in the overlapping areas to explore the nature of the UV sources. We present statistical properties of the *GALEX*/SDSS matched catalogs and we classify the stellar sources by comparing their seven bands photometry (*GALEX* far-UV, near-UV, SDSS u, g, r, i, z) to model colours computed for different astrophysical objects, taking nto account the effects of extinction. The present sample covers mostly high galactic latitudes. The *GALEX* photometric surveys detect hot White Dwarfs all the way throughout the MW halo, providing an unprecedentedly unbiased and possibly complete census of this elusive component of the MW. The results are compared to MW model predictions.

10. Updates on current UV facilities

10.1. *Hubble Space Telescope*

Martin A. Barstow (University of Leicester): The *Hubble Space Telescope* has had a remarkable existence. Through the ability to service it in space, it has survived what would otherwise have been a fatal flaw in it telescope figure and its instrumentation has been successively enhanced with new technologies to keep it at the forefront of astronomical research in much the same way as a ground-based facility. For UV astronomy, it has become the work-horse facility, particularly since the termination of *IUE*, with the *HST*-FOS and *HST*-GHRS spectrographs initially followed by *HST*-STIS. Direct UV imaging has been provided by the series of *HST*-Wide Field Cameras and, latterly, *HST*-ACS.

Nevertheless, space is a harsh environment. Spacecraft support systems such as gyros fail from time to time and the observatory is currently operating in a '2-gyro' mode (compared to the normal 3-gyro operations) to preserve spacecraft life time by shutting down one working gyro and saving it 'for a rainy day'. In addition to instrument changes, important subsystems have been replaced during servicing missions. At the moment, *Hubble* is probably instrumentally at its lowest ebb since launch. Anything other than very low resolution UV/optical spectroscopy came to an end when STIS failed in August 2004, leaving only imaging or grism spectroscopy available in the UV. Just two years later in summer 2006, the prime *HST* instrument ACS side-1 electronics failed. Fortunately, recovery was swift and the camera is now operating on the back-up side-2 electronics. However, as second failure, as in the case of STIS, would render the ACS unusable.

Following the tragic loss of *Columbia* in 2003, a further servicing mission to *Hubble* seemed very unlikely. Indeed, it was formally cancelled in 2004. However, with a successful return-to-flight of the *Space Shuttle*, the SM-4 mission is now back on the agenda and the go ahead was formally announced on 31 October 2006, following the successful pre-requisite of two problem free shuttle missions. The SM-4 mission is expected no earlier than January 2008, but planning for this had to begin in 2005 to ensure readiness by then. The exact plans and priorities are still to be decided, but full range of activities includes:

- Replacement of failed gyros
- Replacement of batteries
- Removal of WFC2 and installation of WFC3
- Installation of COS
- Replacement of a Fine Guidance Sensor
- Repair of STIS electronics

Completion of all these tasks will be very challenging and, in particular, the STIS repair will be very difficult. Furthermore, the servicing mission depends on a continuing problem-free record of *Shuttle* flights, comprising $\sim 6-7$ before SM-4. In addition, it will be necessary to have a second *Space Shuttle* ready for launch in case of problems with the servicing vehicle, since the latter cannot dock with the *International Space Station* from *HST*s orbit. Nevertheless, a successful SM-4 will give *Hubble* a completely new lease of life, taking it well into its 3rd decade in space.

10.2. *Far Ultraviolet Spectroscopic Explorer*

Luciana Bianchi (Johns Hopkins University): Launched in 1999, the *Far Ultraviolet Spectroscopic Explorer (FUSE)* has extended the available ultraviolet spectroscopic range from *HST*'s cutoff at 1150 Å down to the Lyman limit at 912 Å. The satellite (Moos *et al.* 2000; Sahnow *et al.* 2000) consists of four co-aligned telescopes/spectrographs mounted on a three axis stabilized spacecraft with arcsecond pointing capability. It provides spectroscopic data in the wavelength range 905 – 1187 Å at a spectral resolution of $R \approx 20\,000$. *FUSE* is operated from the Department of Physics and Astronomy of the Johns Hopkins University, and the mission has been extended through 2008.

Science highlights from *FUSE* include the discovery of an extended, tenuous halo of very hot gas surrounding the Milky Way and evidence of similar hot gas haloes around other galaxies; detection of molecular hydrogen in the atmosphere of the planet Mars; detection of molecular nitrogen in dense interstellar gas and dust clouds, at levels well below what astronomers had expected, prompting revision for theories of interstellar chemistry; and the discovery of highly ionized neon in extremely hot central stars of planetary nebulae (see also Bianchi & Herald, this JD04, section 9.19) which has great implications for the yield of chemical elements and late evolutionary phases. Lines in the *FUSE* range also enabled scientists to solve long-standing uncertainties and inconsistencies in modeling atmospheres and winds from hot massive stars, which lead to a revision of effective temperatures and luminosities for O-type stars.

10.3. *GALEX*

Luciana Bianchi (Johns Hopkins University): The *Galaxy Evolution Explorer (GALEX)* is a NASA Small Explorer mission launched in April 2003, performing imaging and spectroscopic surveys of the sky in two UV bands: near-UV and far-UV, at a depth comparable to the existing or on-going deepest surveys at other wavelengths. With a large field of view and great sensitivity, *GALEX* is filling a crucial gap in the spectrum, so that our knowledge of the sky from 100 μm to 10 Å is complete.

GALEX is surveying the sky simultaneously in two bands, far ultraviolet (FUV, 1344 – 1786 Å, $\lambda_{\rm eff} \simeq 1516$ Å), and near ultraviolet (NUV, 1771 – 2831 Å, $\lambda_{\rm eff} \simeq 2267$ Å). *GALEX* consists of a Ritchey-Chretien telescope with a 50-cm aperture. It uses a dichroic beam splitter, large-format detectors, and a novel optical design to provide a circular field of view of 1.2 degrees. Observations are performed during 25 minute orbital nights (one eclipse per 98 minute orbit, the basic *GALEX* observational unit) outside the South Atlantic Anomaly. In imaging mode, *GALEX* obtains a resolution of ~ 5 arcsec FWHM, with 5 s photon-limited sensitivity for point sources of $m_{\rm AB} \simeq 20.5$ in 100 s. In grism mode, spectral resolution $R \simeq 100 - 250$ is obtained for most objects in the field of view. The grism is rotated at different position angles to mitigate effects of spectrum overlap. Time-tagged single photon data are stored on board and telemetered to the ground during 3 or 4 daily ground station contacts. All pointings are stationary except for a 1 arcminute spiral dither pattern to uniformly distribute detector fatigue. Night sky backgrounds are extremely low: the FUV[NUV] is dominated by diffuse galactic light [zodiacal light], with typical levels 27.5[26.5] mag arcsec^{-2}, corresponding to 3[30] photon/PSF in one orbital night. Astrometric accuracy is < 1 arcsec (rms).

GALEX is providing an unprecedented, statistically powerful database of UV images and spectra of nearby and distant galaxies, and Milky Way objects (several million objects), linked to a multi-wavelength archive. *GALEX* has recently delivered its second major data release (GR2), which constitutes 36% of the final baseline mission data set. The *GALEX* data are hosted by MAST (http://galex.stsci.edu/). GR2 includes 7200 deg^2 of sky from nearly 4 million seconds of observing time.

GALEX is performing a series of imaging and spectroscopic surveys in the UV band (1300 – 3000 Å) that map the history, and probe the causes, of star formation over the $0 < z < 2$ range, 80% of the life of the universe, a period of dramatic evolution when most stars, elements, and galaxies were formed. The baseline mission focused mainly on two questions: (1) how does the UV trace global star formation?; and (2) what is the star formation rate (SFR) in galaxies, and how does it evolve over $0 < z < 2$?

With projected completion of the baseline mission by the end of FY07, *GALEX* is well on its way to fulfilling Goals 1 and 2. The proposed Extended Mission, starting in FY08, will address two other, more complex questions: (3) what are the physical drivers of star formation and its evolutionary history in galaxies?; and (4) what is the nature of the static and time-variable UV universe?

Several additional science questions are being addressed by the science team, and by the scientific community through GI programs. The extensive *GALEX* science data archive will allow many more investigations by the scientific community, and will provide a road map to other space-based and ground-based observing programs.

(*a*) The goal of the all-sky Imaging Survey (AIS) is to survey the sky to a sensitivity of $m_{\rm AB} = 20.5$ (~ 100 s exposure time), comparable to the SDSS spectroscopic ($m_{\rm AB} = 17.6$) and POSS II ($m_{\rm AB} = 21$) limits. *GALEX* has surveyed $> 13\,000\,\rm deg^2$ to date, and completion of $28\,000\,\rm deg^2$ is planned by the end of 2007.

(*b*) The Medium Imaging Survey (MIS) has covered $1000\,\rm deg^2$ with single-orbit ($\sim$ 1500 s) exposures ($m_{\rm AB} < 23.5$), overlapping SDSS, 2dF, *ROSAT*, and *FIRST* surveys. The MIS is $\sim 85\%$ complete.

(*c*) The Deep Imaging Survey (DIS) consists of 20-orbit (30 ks, $m_{\rm AB} \simeq 25$) exposures, over $80\,\rm deg^2$, of regions where major multi-wavelength efforts are already underway. 25 observations of the DIS are $\sim$80% complete.

(*d*) The Nearby Galaxy Survey (NGS) targets well-resolved nearby galaxies for 1-2 orbits. Surface brightness limits are $m_{\rm AB} \simeq 27.5\,\rm deg^{-2}$, or a star formation rate of $\sim 10^{-3}\,\rm M_{\odot} yr^{-1} kpc^{-2}$. Targets included most galaxies from the *Spitzer* IR Nearby Galaxy Survey, and many other surveys. The NGS is nearly complete (few hundreds galaxies)

(*e*) Spectroscopic surveys. Include a suite of three nested surveys

Some of the proposed surveys for the extended mission are:

(*f*) Galactic Cap Survey (GCS) Survey [8000 orbits]: to increase the survey at MIS depths by an order of magnitude in area over the SDSS N and S galactic caps. (see Bianchi, section 9.21 in this JD04)

(*g*) SDSS Deep Survey (SDS) [2000 orbits]: to obtain deep images (15 000 s)over $100\,\rm deg^2$ of the SDSS survey. This will permit a definitive study of the UV upturn and residual star formation in elliptical galaxies, star formation and chemical evolution of very low mass galaxies, and the properties of extremely red (extincted) star forming galaxies.

(*h*) Legacy Spectroscopy Project (LSP) [2000 orbits]: in fields that already have publicly available optical spectroscopy (SDSS), extending the coverage to $25\,\rm deg^2$ (factor of 6 increase in SDSS overlap).

(*i*) Time Domain Survey (TDS) designed to exploit the excellent *GALEX* UV sensitivity, dark UV sky, wide field-of-view, photon-counting, repeated visits, and the contrast sensitivity of the UV emission from most variable phenomena. It will also provide a Ultra-Deep Imaging Survey (UDIS), [2000 dedicated orbits (not in other surveys); 6000 orbits total].

11. Future UV instrumentation

11.1. *World Space Observatory/UV*

Ana I. Gómez de Castro (UCM), Boris Shustov (INASAN), and the WSO Team: The *World Space Observatory/UV (WSO/UV)* is a 170 cm telescope equipped with three UV spectrometers covering the spectral band from Lyman-α (Lyα) to the atmospheric cut-off with $R \simeq 55\,000$ and offering longslit capability over the same band with $R \simeq 3\,000$. In addition, a number of UV and optical imagers view adjacent fields to that sampled by the spectrometers. The imaging performance compares well with that of *HST*-ACS while the spectral capabilities are comparable to the *HST*-STIS echelle modes. However, with a smaller number of instruments in the focal plane, compared to *HST*, the required number of optical elements in each subsystem is reduced. Hence, the *WSO* delivers considerably enhanced effective area. The planned instrument sensitivity will exceed that of *HST*-STIS by a factor of 10 - 20. From a scientific point of view, this more than an order of magnitude improvement in UV capability will allow significant opportunities in three general directions:

(*a*) observe objects 4 – 5 magnitudes fainter than possible with *HST*, providing completely new opportunities in extragalactic astronomy and cosmology;

(*b*) carry out large scale, high resolution spectroscopic surveys of galactic sources;

(*c*) map the evolution of dissipative phenomena.

The *WSO/UV* project is led by ROSCOSMOS and it is being carried out by a broad international colaboration with participating countries: Russia, Ukraina, Germany, Italy, Spain, China, United Kingdom, Israel, France, the Netherlands, México, South Africa, Baltic Nordic Countries and Argentina (<www.inasan.rssi.ru/rus/WSO/index.html>).

11.2. *Tel Aviv University UltraViolet EXplorer*

Noah Brosch (Wise Observatory): *TAUVEX* is the acronym for the *Tel Aviv University UltraViolet EXplorer*, a space telescope array designed and constructed in Israel by El-Op, Electro-Optical Industries, Ltd (a division of Elbit systems), for the exploration of the ultraviolet (UV) sky. *TAUVEX* was selected in 1988 by the Israel Space Agency (ISA) as its first priority scientific payload. Although originally slated to fly on a national Israeli satellite of the Ofeq series, *TAUVEX* was shifted to fly as part of the *Spectrum Roentgen-Gamma (SRG)* international observatory, a collaboration of a large number of countries with the Soviet Union (Space Research Institute IKI) leading.

Due to repeated delays of the *SRG* project, ISA decided in 2001 to shift *TAUVEX* to a different satellite. In early-2004 ISA signed an agreement with the Indian Space Research Organization (ISRO) to launch *TAUVEX* on board the Indian technology demonstrator satellite GSAT-4. *TAUVEX* is a scientific collaboration between Tel Aviv University and the Indian Institute of Astrophysics in Bangalore. Its Principal Investigators are Noah Brosch at Tel Aviv University and Jayant Murthy at the Indian Institute of Astrophysics. *TAUVEX* is slated to be launched in the second half of 2007.

TAUVEX consists of three bore-sighted 20 cm diameter telescopes on a single bezel. Each telescope images the same sky area of 0.9 degree, with an angular resolution of 7 – 11 arcseconds. The imaging is onto position-sensitive detectors (CsTe cathodes on calcium fluoride windows) equipped with multi-channel plate electron intensifiers. The detectors oversample the point-spread-function by a factor of approximately 3. The output is detected by position-sensitive anodes (wedge-and-strip) and is digitized to 12 bits. The full image of each telescope has about 300 resolution elements across its diameter.

The CsTe cathode assures sensitivity from longward of Lyman-α to the atmospheric limit with a peak quantum efficiency of approximately 10%. The operating spectral range is separated in a number of segments selectable with filters. Each telescope is equipped with a four-position filter wheel. Each wheel contains one blocked position (shutter) and three band-selection filters.

TAUVEX is mounted to the spacecraft on a plate that can rotate around its axis (the MDP), enabling to point the telescopes' line-of-sight to any desired declination. Being on a geostationary satellite, the observation is therefore of a scanning type. A 'ribbon' of a constant declination, 0.9 degree wide, is scanned as time advances, completing an entire 360 degree circuit during one sidereal day. The dwell time of a source within the detector field of view is a function of the pointing declination and of the exact location in the FOV relative to the detector diameter. The closer a source is to one of the celestial poles, the longer it will reside in the *TAUVEX* field of view during a single scan. The longest theoretically-possible exposure is for sources at $\delta > 89^\circ\, 30'$; these can be observed for 86.4 ksec during one day.

Each photon event hitting the detectors is transmitted to the ground in real time and processed in a near-real-time pipeline. In-between the photon events a time tag is added every 128 ms. The time between the adjacent time tags is sufficiently short so that the orbital motion of the nadir-pointing platform is much smaller than the *TAUVEX* virtual pixel. Given that *TAUVEX* operates from a geo-synchronous platform, *TAUVEX* enjoys a dedicated 1 Mbit/s downlink to the ISRO control station near Bangalore. Command sequences will be uplinked after being generated by IIA and ISRO and the downlink will be analyzed on-line to monitor the payload state of health.

In most situations, *TAUVEX* will be able to download all the detected photon events. However, in case of strong straylight or of many bright sources in the field of view, the collected event rate may overload the capacity of the telemetry link. In this case, *TAUVEX* stores the photon events in a solid state memory module (4 GByte), from which the events are transmitted at the nominal 1 Mbit/s rate.

The science of *TAUVEX* is based on its unique characteristics: three bore-sighted and independent telescopes able to operate independently, with different filters but measuring the same sources, and reasonably fine time resolution since every detected photon is time-tagged.

11.3. *Ultraviolet Imaging Telescopes*

Jayant Murthy (IIA): The *Ultraviolet Imaging Telescopes* are a set of two telescopes which will simultaneously image the sky in three bands from the FUV to the visible. Both telescopes are essentially identical Ritchey-Chretian designs with 38 cm primary mirrors. A dichroic splits the beam in one of the telescopes onto a NUV detector (RbTe photocathode) and a VIS detector (S20 photocathode) with the other telescope dedicated to the FUV (CsI photocathode). With a FOV of 0.5 deg and a spatial resolution of better than $1.''5$, *UVIT* is intended to observe relatively large areas of the sky with a resolution comparable to ground-based observatories.

UVIT is an international collaboration led by the Indian Institute of Astrophysics (IIA). The detectors (MCP + CMOS readout) are being fabricated by Photek (UK) and integrated with photon-counting electronics by Routes (Canada) under the direction of John Hutchings and the Canadian Space Agency; the optics are fabricated by LEOS, a division of ISRO; and IIA is responsible for the structure and the instrument integration. A full calibration of the detectors (only) will be done in Canada with the focusing, characterization and integration done at new facilities built at IIA.

UVIT will launch as part of the *ASTROSAT* satellite which includes several X-ray telescopes (see `<http://www.rri.res.in/astrosat>` for more information). *ASTROSAT* is India's first satellite dedicated to astronomy and is scheduled for launch in late 2008. The mission is driven by proposals with time shared between the different collaborators with some part reserved for international proposers.The data will be archived at and disseminated from the Indian Space Science Data Centre located in Bangalore. After an initial proprietary period, the data will be released to the astronomical community.

UVIT is intended to be the beginning of an active thrust into UV astronomy at IIA. A significant investment is being made into an astronomical instrumentation program with further space programs under planning. An integration and calibration facility is almost complete at which future space payloads can be developed and made ready for flight.

11.4. *Stellar Imager*

Kenneth G. Carpenter (NASA GSFC), Carolus J. Schrijver (LMATC), Margarita Karovska (SAO), and the SI Vision Mission Study Team: The *Stellar Imager (SI)* is a UV-optical, space-based interferometer designed to enable 0.1 milliarcsec spectral imaging of stellar surfaces, and of their interiors via asteroseismology, and of the Universe in general. SI is identified as a "Flagship and Landmark Discovery Mission" in the 2005 NASA Sun Solar System Connection (SSSC) Roadmap and as a candidate for a "Pathways to Life Observatory" in the NASA Exploration of the Universe Division (EUD) Roadmap (May, 2005). The ultra-sharp images of the *Stellar Imager* (<`http://hires.gsfc.nasa.gov/si/`>) will revolutionize our view of many dynamic astrophysical processes. The 0.1 mas resolution of this deep-space telescope will transform point sources into extended sources, and snapshots into evolving views. SI's science focuses on the role of magnetism in the Universe, particularly on magnetic activity in stars like the Sun. *SI*'s prime goal is to enable long-term forecasting of solar activity and the space weather that it drives in support of the Living With a Star program in the Exploration Era. *SI* will also revolutionize our understanding of the formation of planetary systems, of the habitability and climatology of distant planets, and of many magneto-hydrodynamically controlled processes in the Universe. Concept development for the full-up mission and for a possible smaller "pathfinder" mission continues and needed technology development, driven by *SI* and numerous other missions (*SPECS, BHI/MAXIM, LF, PI*), proceeds in the areas of precision formation flying (e.g., the *Synthetic Imaging Formation Flying Testbed*), a GSFC/MIT/MSFC collaboration) and closed-loop optical control of sparse arrays (e.g., the Fizeau Interferometer Testbed (FIT) at GSFC). The Pathfinder mission could be ready for launch in the next decade and the full mission by 2025, if they were selected for flight in future NASA and/or ESA opportunities.

11.5. *The Super-Earths Explorer*

Jean Schneider (LUTH - Paris Observatory), Pierre Riaud (LESIA, Paris Observatory), Giovanna Tinetti (IAP, Paris) and the SEE-COAST Team: The *Super-Earths Explorer Coronagraphic Off Axis Space Telescope (SEE-COAST)* is a space mission project to be submitted to ESA. It is mainly devoted to the diected imaging, in the visible and near UV (down to Lyα), of extrasolar planetary companions reflecting the light of their parent star. It will also measure the polarization of planets. As a complementary programme, it will detect circumstellar disks, AGNs and investigate stellar activity.

12. Summary of JD 4

12.1. *Introduction*

The scientific contributions to this JD have shown the broad range of problems within stellar astrophysics that require development of new UV instrumention. There are two ways to approach this need: observatory class facilities and space missions that have a much more narrowly defined, and often single, science goal.

The observatory approach assumes that continued access to the UV-range has to be granted to the world-wide community in routine manner (similar to the access to radio-infrared or optical ranges). In this sense, multiple purpose instrumentation needs to be defined that suits most of the needs of the astronomical community. The *World Space Observatory/UV* represents one example of this approach, as described in Section 11.1. A handicap is the high cost of space facilities. There are, however, some relevant processes whose understanding requires the development of specific instrumentation, often at the edge of the technological capabilities of modern engineering. At the end of this JD04 there was a lively discussion on future UV instrumentation, which we report in Section 11.

12.2. *Observatory-class access to the UV-range*

At the end of 2002, a group of European astronomers coming from a broad range of disciplines: from fundamental astrophysical research to observational expertise in the optical, UV and X-ray ranges, as well as space instrument development teams, joined efforts to evaluate the need to develop new UV instrumentation for the coming decade. This group set the seed for the Network for UltraViolet Astrophysics (NUVA) to establish some of the major scientific objectives of the astronomical community and to fully exploit the large astronomical facilities planned for other spectral ranges.

At that time, *HST*-STIS was still working in and the Cosmic Origins Spectrograph (COS) was being built for replacement. *HST* was expected to last till 2010/12, *FUSE* was working nominally, and *GALEX* was close to launch. However, the major space agencies had no plans for new UV spectroscopic missions, apart from the large optical/UV telescope included in the NASA Origins plan, expected to enter development phase in 2015 - 2020 ready for a launch in the second quarter of this century. On Friday 16 January 2004 NASA decided that the planned *Shuttle* mission to service and upgrade *HST* (SM-4), including the repair of STIS and installation of COS, would be cancelled. Just a few months later STIS failed. No access to high resolution UV spectroscopy was available during most of 2005, until *FUSE* resumed observations in November 2005 after the failure of the third of the four onboard reaction wheels in December 2004. Therefore, the need for new missions to fly during the next decade was becoming urgent. Fortunately, the earlier decision on the SM-4 (the final) *Hubble* servicing mission was reversed on October 31st, 2006. The complete plan covers the installation of two new instruments *HST*-COS and *HST*-Wide Field Camera 3, both with UV capability. In addition, gyros and batteries will be replaced and *HST*-STIS will be repaired. This will lead to a new lease of life for *HST* as a UV observatory and should allow the mission to continue at least to its current projected end ~2012.

The NUVA (<www.ucm.es/info/nuva>) was officially established in January 2004. A detailed study of the needs for UV instrumentation in the many fields of astrophysical research can be found in a special issue of Astrophysics and Space Science, entitled *Fundamental problems in astrophysics: requirements for future UV observatories* (Gómez de Castro *et al.* 2006) which summarizes the thoughts and work of the NUVA and the UV community at-large. A list of the fields in stellar astrophysics requiring access to this range was elaborated; these are:

- Solar System: planets atmospheres, magnetospheres and interaction with the Solar wind.
- Cool Stars: Mapping dissipative processes (heating of the chromosphere and the corona), winds, evolution of solar-like dynamos, extending the knowledge to brown-dwarfs structure.
- Massive stars: wind clumpiness and DACs evolution, relation with magnetic fields, massive stars in the Local Group and further out.

• ISM: disk halo interaction and HVCs, energy input into the ISM (mechanical – bubble, supernovae; radiative), chimneys, halo heating and galactic winds. Extending this work to the Local Group.

• Star Formation-I: outflows-powering by accretion, accretion geometry, mapping dissipative processes, evolution to the main sequence, mapping dissipative processes, influence of dissipative processes in the evolution of the MRI and the inner disk.

• Planets: Evolution of young planetary disks, comets in extrasolar systems, evolution of comets-hive and its impact on the evolution of young planetary systems. Detection and characterization of the atmospheres of extrasolar planets.

• Star formation II: violent star formation processes in galaxies, the interaction between the stellar clusters and the ISM and the variation of the IMF, contribution of starbursts to the reionization of the Universe, chemical enrichment of the IGM, role of starburst on AGNs

• White dwarfs: the origin and relationships of the H and He-rich groups, the initial-final mass relation for white dwarfs and their progenitors and the 3D structure of the ISM; to increase the sample of WD and extend it to the globular clusters and nearby galaxies.

• Interacting binaries: nature of SN Ia progenitors exploring both single and double-degenerate channels, the physics of accretion discs, in particular the role of viscosity and its time-dependence, and the development of winds, the fundamental properties of white dwarfs in CVs, the nature of the IB population in globular clusters.

• Globular clusters and stellar evolution.

• Planetary nebulae.

• Supernovae: SN explosions, characterize SNIa across the Universe.

Most of science described there could be carried out with two basic instruments:

(*a*) A high-resolution ($R = 50\,000$–$100\,000$) spectrograph covering the whole 90- 320 nm spectral range.

(*b*) A low-spectral resolution ($R = 1000$ - 5000) high-sensitivity spectrograph allowing integral field spectroscopy (long-slit in its simplest version) with spatial resolution (50 mas) and wavelength coverage from 110 – 450 nm.

These instruments should provide an improvement by a factor of ~ 20 in effective area over the *HST*-STIS capabilities. Additional features demanded by the community are high-resolution imaging (better than 80 mas) and an orbit that allows mapping the temporal evolution of dissipative phenomena and a rapid response to 'Targets of Opportunity'. High-time resolution is also required to map the evolution of rapid phenomena. Groups working on the atmospheres of extrasolar planets also expressed interest in some overlap with the optical range (from 3200 Å to 3500 Å).

This projected improvement is somewhat conservative from the technological point-of-view, since it is mostly achieved by improving optical designs and coatings and making use of detectors with enhanced sensitivity. Larger detector formats are also available and fast read-out electronics gives improved dynamic range. It is amazing this progress that can be achieved with a relatively modest investment coupled to a modest sized telescope. A good example of this is the proposed *World Space Observatory* project.

However, looking into the far future, it is clear that the frontier of UV astronomy will require construction of much larger aperture facilities that increase the effective collecting surface by 2 orders of magnitude or more. A properly instrumented 4 – 6 m telescope in space will be essential for future UV observations.

12.3. *Single science objective facilities*

This approach requires the identification of the fields of astrophysics which are the most relevant for the progress of physics. There is a general agreement on those being:

- The characterization of dark energy
- Physics in extreme environments: black holes, SNe or the polarized vacuum.

Another important field is to understand life sustainability in the Universe and among it,

- The magnetic 'stability' of the Sun and solar-like stars and its role in the evolution of life in the associated planetary systems.
- The formation of planetary systems
- The characterization of planetary systems.

All these projects require extension of the effective area of the telescopes by at least two orders of magnitude over *HST*-STIS (or one order of magnitude over *HST*-COS, *WSO/UV*), since most of these science goals require absorption spectroscopy of weak/distant far UV sources.

12.4. *Discussion*

The situation of the various on-going projects was described during the discussion, especially, the possible (now approved) mission to service and upgrade *HST* and the *WSO/UV* project.

Once detector and instrument systems reach their ultimate efficiency, as is now occurring, the factor of 100 increase in effective area can only be realised by increasing the collecting surface. This implies that these future projects, if feasible, are going to be very expensive and will need to be able to answer at least one of the key questions of science to justify their cost. The most realistic approach seems to be that followed by the *Stellar Imager* in proposing a cluster of small coordinated telescopes. Another possible option a Moon-based observatory, where a large telescope could be constructed using technologies currently applied to large ground-based telescopes.

An important issue raised during the meeting is the difficulty in gaining the support of the community at large for very focussed projects as well as the need to have teams of instrumentalists and observers working together. Concern was expressed about the diminishing number of these teams.

Acknowledgements

It is a great pleasure to acknowledge the support of the members of the IAU XXVI General Assembly LOC who dealt brilliantly with the difficult task of organizing a plethora of small and large meetings, joint discussions and discourses involving around 2 500 astronomers.

References

Gómez de Castro, A. I., Wamsteker, W., Barstow, M. A., *et al.* 2000, *Ap&SS*, 303, 133
van der Hucht, K. A., Schrijver, H., Stenholm, B., *et al.* 1997, *New Astronomy*, 2, 245
van der Hucht, K. A. 2001, *New Astronomy Reviews*, 45, 135
Moos, H. W., Cash, W. C., Cowie, L. L., *et al.* 2000, *ApJ* (Letters), 538, L1
Sahnow, D. J., Moos, H. W., Ake, T. B., *et al.* 2000, *ApJ* (Letters), 538, L7

Highlights of Astronomy, Volume 14
IAU XXVI General Assembly, 14-25 August 2006
Karel A. van der Hucht, ed.

doi:10.1017/S1743921307010095

Joint Discussion 5 Calibrating the top of the stellar M - L relation

Claus Leitherer[1], Anthony F. J. Moffat[2] and Joachim Puls[3] (eds.)

[1]Space Telescope Science Institute, Baltimore, MD, USA, email: leitherer@stsci.edu
[2]Université de Montréal, Canada, email: moffat@astro.umontreal.ca
[3]Universitäts-Sternwarte München, Germany, email: uh101aw@usm.uni-muenchen.de

Rationale and overview

The goal of this Joint Discussion is to bring together theorists and observers from the stellar and extragalactic communities to discuss the properties of the most massive stars and the implications for cosmological studies. We will focus on a set of themes that follow from fundamental stellar astronomy, such as mass determinations in binary stars, to recent modeling of atmospheres and evolution, to the significance of massive stars for the ecology of the host galaxy, and finally to a critical assessment of the properties of the first generation of stars in the universe.

Until now, few efforts have been devoted to aggressively searching for the most massive stars. Massey *et al.* (2007) have looked at the most luminous O-type stars on the main sequence, with mass estimates for several O3-type stars in the surprisingly broad, low range from some 35 – 60 $M_{\odot}$. Rauw *et al.* (2004) and Rauw *et al.* (2005) have concentrated on luminous WR stars in the 7th Galactic WR Catalogue (van der Hucht 2001), and recent light curve measurements for the WN6ha + WN6ha system WR20a lead to component masses close to 83 and 82 $M_{\odot}$, the highest binary-inferred masses to date. Moffat *et al.* (2007) are concentrating on the most luminous, H-rich WNL stars mainly in the LMC and NGC 3603 in the Galaxy. It thus remains an open issue how much above the binary-inferred maximum 80 $M_{\odot}$ the most massive stars may actually be.

Optical interferometry is entering a new age with several ground-based long-baseline observatories now making observations of unprecedented resolution. Interferometers bring a new level of resolution to bear on massive spectroscopic binaries, enabling the full extraction of the physical parameters for the component stars with high accuracy. This will be an opportunity to determine hundreds of fully three-dimensional orbits in absolute units, when interferometric results are combined with spectroscopic orbits, thus accurately determining masses, diameters, and distances (hence luminosities) of tens of massive binary stars through their orbital parallaxes.

In contrast to such direct observations of masses, based on Keplerian orbits in binaries, which have never yielded values above 85 $M_{\odot}$, less direct techniques based on spectroscopic analyses tend to give a fairly large spread of masses. For extremely luminous stars like η Carinae and the Pistol Star, keeping within the Eddington limit requires masses near 100 $M_{\odot}$. Up-to-date atmospheric models accounting for mass-loss and line-blanketing seem to indicate that the long-standing mass discrepancy (i.e., the problem that atmospheric masses are systematically lower than the evolutionary ones) has still not been fully resolved. Since mass is the most fundamental stellar parameter, such a discrepancy is intolerable, and its theoretical origin needs to be explored.

While determining the mass of the most massive stars and calibrating the top of the mass-luminosity relation are meritorious in their own right, stellar masses are a central theme in current extragalactic astronomy as well. The top end of the stellar initial mass function is imprinted in the observed spectra of distant populations, and our knowledge of the stellar content relies on local calibrations. A comparison of the properties of the most massive stars in the Galaxy and the Magellanic Clouds might indicate a trend of higher masses with lower metal abundance. The subject has received particular interest from research aimed at understanding and predicting the properties of distant stellar populations, which are thought to be the powering sources of Lyman-break or SCUBA galaxies.

A common theme of contemporary cosmology is the quest for the first generation of stars formed in the universe. Such stars are thought to be the supermassive cousins of the massive stars studied locally. Local calibrations of the fundamental stellar parameters will provide much sought guidance for cosmological models. Modeling massive metal-free stars poses major challenges due to yet unexplored phenomena such as rotation and mass loss in such extreme environments. A particularly important issue regards the importance of episodes of super-Eddington, continuuum-driven mass loss (such as occurs in η Carinae and other Luminous Blue Variable stars), which can reduce the evolutionary mass even in the absence of a substantial metal-line-driven stellar wind. With rapid rotation a bipolar shaping of the mass ejection (such as seen in the Homunculus nebula of η Carinae) can reduce the associated angular momentum loss, with potentially important consequences for leaving the rapidly rotating core central to the collapsar model for γ-ray bursts.

Scientific Organizing Committee

Norbert Langer (Netherlands), Claus Leitherer (USA, chair), Anthony F.J. Moffat (Canada), Stanley P. Owocki (USA), and Joachim Puls (Germany).

References

van der Hucht, K. A. 2001, *New Astronomy Reviews*, 45, 135

Massey, P., Morrell, N., Eastwood, K. D., Gies, D. R., & Penny, L. R. 2007, in preparation

Moffat, A. F. J., Schnurr, O., Chené, A.-N., St-Louis, N., & Casoli, J. 2004, this JD05, p. 197

Rauw, G., De Becker, M., Nazé, Y., Crowther, P. A., Gosset, E., Sana, H., van der Hucht, K. A., Vreux, J.-M., & Williams, P. M. 2005, *A&A*, 420, L9

Rauw, G., Crowther, P. A., De Becker, M., Gosset, E., Nazé, Y., Sana, H., van der Hucht, K. A., Vreux, J.-M., & Williams, P. M. 2005, *A&A*, 422, 177

Highlights of Astronomy, Volume 14
IAU XXVI General Assembly, 14-25 August 2006
Karel A. van der Hucht, ed.

doi:10.1017/S1743921307010101

Keplerian masses for the most massive stars: the very luminous, hydrogen-rich Wolf-Rayet stars

Anthony F. J. Moffat, Olivier Schnurr, André-Nicolas Chené, Jules Casoli, Nicole St-Louis and Alfredo Villar-Sbaffi

Département de physique, Université de Montréal,
C.P. 6128, Succ. C-V, Montréal, QC, H3C 3J7, Canada
email: moffat, schnurr, chene, casoli, stlouis, alfredo@astro.umontreal.ca

Abstract. We propose that the most massive stars in the local Universe start out on the main sequence being extremely luminous and already exhibiting wind-like emission lines. Such stars can probably be identified as hydrogen-rich WN5-7 stars. Masses are now being determined for these stars using the least model-dependent technique of binary orbits.

Keywords. Binaries: spectroscopic, stars: early-type, fundamental parameters, stars: Wolf-Rayet

---

Locating the most massive stars and measuring their masses is key to understanding the formation and nature of massive stars in general. This is best done by using the least model-dependent technique, i.e. Kepler's laws (combined with techniques such as photospheric or wind eclipses, polarimetry, or wind-wind collision effects, to extract the orbital inclination) for binary systems that contain very luminous stars on or near the main sequence, where interaction effects are not (yet) important. In the local Universe, it appears that the potentially most massive, main-sequence stars are the most luminous hydrogen-rich WR stars of type WN5-7ha, which are even more luminous than, and probably at least as hot as, the recently recognized hottest main-sequence stars of type O2. WN5-7ha stars exhibit strong, broad WR-like emission lines because of the relatively strong winds they drive as a result of their high luminosity, not because of their compactness and high ratio of L/M as classical He-burning, WR stars. It is probably no coincidence that the currently most massive stars measured this way are in WR 20a, an eclipsing binary system containing two identical WN6ha stars of 83 and 82 $M_{\odot}$ (Rauw *et al.* 2004; Bonanos *et al.* 2004; Rauw *et al.* 2005). Other WN5-7ha binary stars (e.g., NGC 3603/A1, 30 Dor/R145), some even more luminous than WR 20a and with masses over 100 $M_{\odot}$, are being measured currently by different groups, including ourselves.

Acknowledgements

We are grateful to NSERC (Canada) and FQRNT (Québec) for financial support.

References

Bonanos, A. Z., Stanek, K. Z., Udalski, A., Wyrzykowski, L., Żebruń, K., Kubiak, M., Szymański, M.K., Szewczyk, O., Pietrzyński, G., & Soszyński, I. 2004, *ApJ* (Letters), 611, L33

Rauw, G., De Becker, M., Nazé, Y., Crowther, P. A., Gosset, E., Sana, H., van der Hucht, K. A., Vreux, J.-M., & Williams, P. M. 2004, *A&A* (Letters), 420, L9

Rauw, G., Crowther, P. A., De Becker, M., Gosset, E.; Nazé, Y., Sana, H., van der Hucht, K. A., Vreux, J.-M., & Williams, P. M. 2005, *A&A*, 432, 985

Highlights of Astronomy, Volume 14
IAU XXVI General Assembly, 14-25 August 2006
Karel A. van der Hucht, ed.

doi:10.1017/S1743921307010113

Massive extragalactic eclipsing binaries

Alceste Z. Bonanos
Carnegie Institution of Washington, 5241 Broad Branch Road, Washington, DC 20015, USA
email: bonanos@dtm.ciw.edu

Keywords. stars: binaries: eclipsing, stars: Wolf-Rayet, stars: individual (WR 20a)

Masses, radii and luminosities of distant stars can only be measured accurately in eclipsing binaries. The most massive eclipsing binary currently known is WR 20a, which consists of two $\sim 80\,M_{\odot}$ stars in a 3.7 d orbit. Analogs of WR 20a are bound to exist both in massive stellar clusters in our Galaxy and in nearby galaxies. The nearest ones are located in the clusters near the Galactic Center: the Center, Arches, and Quintuplet clusters. The severe amount of reddening in the galactic disk makes the study of galactic clusters challenging. However, with current 8-m class telescopes, the study of massive stars in nearby galaxies is also feasible. The nearest Local Group galaxies (LMC, SMC, M 31, M 33) provide the perfect laboratory for studying massive stars and determining their properties as a function of metallicity. Such studies will constrain models, confirm the dependence of evolution on metallicity and help understand the rate and nature of supernovae and gamma-ray bursts.

Finding massive binaries requires a variability survey from which the brightest and thus most massive eclipsing binaries are selected for spectroscopic follow-up. Given both light and radial velocity curves, one can determine the size of the orbit, the component radii and masses. Furthermore, by fitting synthetic spectra to the observed ones, one can infer the effective temperature, therefore the luminosity, and additionally solve for the distance. Several surveys have and are continuing to produce candidate massive binaries in nearby galaxies: the All Sky Automated Survey (Pojmanski 2002), the DIRECT Project surveying M31 and M33 for detached eclipsing binaries (see Bonanos *et al.* 2003, and references therein), the Araucaria project surveying nearby galaxies (e.g. NGC 6822 and NGC 300, Pietrzynski *et al.* 2004; Gieren *et al.* 2004) and the microlensing surveys towards the Magellanic Clouds and M31 (e.g. MACHO, OGLE, MEGA, POINT-AGAPE).

In this paper, I present preliminary results of a project to follow up the brightest eclipsing binaries found by OGLE in the LMC and SMC. Spectra were obtained over 13 nights using the Echelle spectrograph on the 2.5 m DuPont telescope and 2 nights on MIKE on the Clay Telescope at Las Campanas Observatory (LCO), Chile. Each binary was visited 4 - 10 times, depending on the number of nights during the observations that it was not in eclipse. Cross correlation with a template spectrum of a galactic early type star, excluding the blended Balmer lines was used to derive preliminary radial velocities for one of the LMC systems. The semi-amplitudes yielded minimum masses of $\sim 40\,M_{\odot}$ for the primary and $\sim 20\,M_{\odot}$ for the secondary, thus confirming that the brightest eclipsing binaries in the LMC do contain very massive stars. Future analysis with 2D cross correlation (Zucker *et al.* 1994) and spectral disentangling will provide more accurate velocities. This survey will yield accurate to $\sim 5\%$ fundamental parameters for ~ 20 massive stars of low metallicity.

Highlights of Astronomy, Volume 14
IAU XXVI General Assembly, 14-25 August 2006
Karel A. van der Hucht, ed.

doi:10.1017/S1743921307010125

The masses of late-type WN stars

Götz Gräfener and Wolf-Rainer Hamann

Department of Physics, University of Potsdam, D-14469 Postdam, Germany
email: goetz,wrh@astro.physik.uni-potsdam.de

Abstract. We present recent results for galactic WNL stars, obtained with the new Potsdam Wolf-Rayet (PoWR) hydrodynamic model atmospheres. Based on a combination of stellar wind modeling and spectral analysis we identify the galactic WNL subtypes as a group of extremely luminous stars close to the Eddington limit. Their luminosities imply progenitor masses around $120\,M_\odot$ or even above, making them the direct descendants of the most massive stars in the galaxy. Because of the proximity to the Eddington limit our models are very sensitive to the L/M ratio, thus allowing for a direct estimate of the present masses of these objects.

Keywords. stars: mass loss, stars: winds, outflows, stars: Wolf-Rayet.

In the recent re-analysis of the galactic WN sample with line-blanketed atmosphere models (Hamann *et al.* 2006) the WN stars turned out to form two distinct groups in the HR diagram, which are divided by their luminosities. Among these, the H-rich WNL stars, with luminosities above $10^6\,L_\odot$, are found to the right of the ZAMS, whereas early to intermediate subtypes show lower luminosities and hotter temperatures. The relatively large number of extremely luminous WNL stars already implies that many of these objects might be very massive stars in the phase of central H-burning.

Our hydrodynamic atmosphere models, on the other hand, imply that the formation of WR-type stellar winds is caused by the proximity to the Eddington limit (Gräfener & Hamann 2006; Gräfener, & Hamann 2005). In fact, our models reveal a rather strong dependence of the WR mass loss rates on the Eddington factor Γ_e or, equivalently, on the M/L ratio. Weak-lined WNL stars, with their relatively low mass loss rates, thus should have considerably higher M/L ratios than their strong-lined counterparts.

Detailed spectral modeling of weak-lined WNL stars in Carina OB 1 indeed indicates very high masses for these objects. Note, however, that the results depend on the adopted distance. For WR 22 (WN7h) we find a luminosity of $10^{6.3}\,L_\odot$ (for $m-M=12.1$) and a mass of $78\,M_\odot$, in agreement with the mass estimate by Rauw *et al.* 1996. For WR 25 (WN6ha) we determine values between $110\,M_\odot/10^{6.4}\,L_\odot$ (for $m-M=11.8$), and $210\,M_\odot/10^{6.7}\,L_\odot$ (for $m-M=12.55$). These masses are in agreement with H-burning stars in a late pase of their main-sequence evolution. Our models thus suggest an evolutionary sequence of the form O → WNL → LBV → WN → WC for very massive stars. Interestingly, WR 25 is the only evolved object in the young OB cluster Tr 16, apart from the LBV prototype η Car. Its location at the top of the main-sequence of this cluster, with a slightly lower luminosity than η Car (see Hillier *et al.* 2001), strongly supports its evolutionary stage as an LBV progenitor.

References

Gräfener, G., & Hamann, W.-R. 2005, *A&A*, 432, 633
Gräfener, G., & Hamann, W.-R. 2006, *A&A*, submitted
Hamann, W.-R., Gräfener, G., & Liermann, A. 2006, *A&A*, 457, 1015
Rauw, G., Vreux, J.-M., Gosset, E., *et al.* 1996, *A&A*, 306, 771
Hillier, D. J., Davidson, K., Ishibashi, K., & Gull, T. 2001, *ApJ*, 553, 837

Highlights of Astronomy, Volume 14
IAU XXVI General Assembly, 14-25 August 2006
Karel A. van der Hucht, ed.

doi:10.1017/S1743921307010137

Inflation of luminous metal-rich WR stars – a clue to mass loss?

Onno R. Pols

Sterrenkundig Instituut Utrecht, Postbus 80000, NL-3508TA Utrecht, the Netherlands
email: o.r.pols@astro.uu.nl

Abstract. We investigate the influence of metallicity and wind mass loss on the radii of luminous Wolf-Rayet stars.

Keywords. stars: mass loss

We investigate the influence of metallicity and wind mass loss on the radii of luminous Wolf-Rayet stars. We have calculated chemically homogeneous models of Wolf-Rayet stars of 10 to 200 $M_{\odot}$ without mass loss for two metallicities ($Z = 0.02$ and $Z = 0.001$), using OPAL opacities. We also constructed theoretical helium main sequences of 15 to 30 $M_{\odot}$ with stellar wind mass loss rates between 10^{-4} and 10^{-5} $M_{\odot}\,yr^{-1}$.

Our models confirm the radius extension of luminous, metal-rich Wolf-Rayet stars reported previously by Ishii *et al.* (1999), i.e., the inflation of the hydrostatic stellar radius. We also show that for small values of the stellar wind mass loss rate, an extended envelope structure is still present. However, for mass loss rates above a critical value Wolf-Rayet radii decrease and the stellar structure becomes compact. We discuss possible evolutionary and observational consequences of an inflated envelope for a mass-losing Wolf-Rayet star.

Reference

Ishii, M., Ueno, M., & Kato, M. 1999, *PASJ*, 51, 417

Highlights of Astronomy, Volume 14
IAU XXVI General Assembly, 14-25 August 2006
Karel A. van der Hucht, ed.

doi:10.1017/S1743921307010149

Spectroscopic *versus* evolutionary masses

Artemio Herrero

Instituto de Astrofisica de Canarias,
C/ Via Lactea s/n, E-38200, La Laguna, Tenerife, Spain
email: ahd@iac.es

Abstract. We describe the present status of the mass discrepancy in the Upper HR Diagram.

Keywords. stars: fundamental parameters (masses)

The mass determination of isolated stars is particularly difficult in the Upper HR Diagram, and yet it is a fundamental parameter for our understanding of the stellar structure and evolution and the interpretation of integrated spectra, stellar populations and the IMF. The two main methods are both indirect: analysis of the stellar spectra by means of model atmospheres (spectroscopic masses, SM) and comparison with evolutionary tracks, after placing the star on a HRD (evolutionary masses, EM). Herrero *et al.* (1992) noted that for O stars these two methods gave different results, and called this situation *the mass discrepancy*, with the EM being systematically larger than the SM.

Model improvements during the next decade, including line-blanketing and mass-loss led to new, higher SM (due to the influence of mass-loss) and lower EM (due to lower effective temperatures and luminosities). These changes solved most of the mass discrepancy (see Repolust *et al.* 2004). Other recent analyses confirmed that both masses now agree within the limits of the error bars. However, second order effects might still be present: (*a*) for SM between 20 - 60 $M_{\odot}$ there might be a milder form of the mass discrepancy, with EM being still about 30 % larger; (*b*) for SM below 20 $M_{\odot}$, some discrepancy may still be present; and (*c*) for SM above 60 $M_{\odot}$ an inverted form of the mass discrepancy (with SM *larger* than EM) may appear. These results are consistent with analyses of O and B stars in the Magellanic Clouds and the Milky Way by different authors.

Mokiem (2006) has presented a plot of the mass discrepancy versus the stellar helium abundance. Their findings (He-enriched O-supergiants, show no mass discrepancy; He-enriched O-dwarfs show mass discrepancy) are consistent with the analyses of Herrero & Lennon (2004), who proposed that the more massive stars have an evolution dominated by mass-loss and the intermediate massive stars have an evolution dominated by rotation.

In conclusion, there is no clear supporting evidence for a general mass-discrepancy problem. We find good agreement between EM and SM in most cases, (and with dynamical masses when comparing with binary systems). Second order effects may still be present. More sophisticated model atmospheres (clumping, 2-D effects) and more homogeneous evolutionary tracks at different Z are needed for the comparisons.

References

Herrero, A., Kudritzki, R.-P., Vilchez, J. M., *et al.* 1992, *A&A*, 261, 209
Herrero, A., & Lennon, D. J. 2004, in: A. Maeder & P. Eenens (eds.), *Steller Rotation*, Proc. IAU Symp. No. 215 (San Francisco: ASP), p. 209
Mokiem, R. M. 2006, PhD thesis, University of Amsterdam
Repolust, T., Puls, J., & Herrero, A. 2004, *A&A*, 415, 349

Highlights of Astronomy, Volume 14
IAU XXVI General Assembly, 14-25 August 2006
Karel A. van der Hucht, ed.

doi:10.1017/S1743921307010150

Continuum-driven *versus* line-driven mass loss and the Eddington limit

Stanley P. Owocki

Bartol Research Institute, University of Delaware, Newark, DE 19716, USA
email: owocki@bartol.udel.edu

Abstract. Basic stellar structure dictates that stars of $\sim 100\,\mathrm{M}_{\odot}$ or more will be close to the Eddington limit, with luminosities in excess of $10^6\,\mathrm{L}_{\odot}$, and radiation pressure contributing prominently to the support against gravity. Although it is formally possible to generate static structure models of even more massive stars, recent studies of dense clusters show there is a sharp cutoff at masses above $\sim 150\,\mathrm{M}_{\odot}$. This talk examines the role of extreme mass loss is limiting the masses of stars, emphasizing in particular that continuum driving, possibly associated with structural instabilities of radiation dominated envelope, can lead to much stronger mass loss than is possible by the usual line-scattering mechanism of steady stellar winds.

However, population studies of very young, dense stellar clusters now suggest quite strongly that there is a sharp cutoff at masses above ca. $150 M_{\odot}$ (see, e.g., the talk by Sally Oey, in this JD 05, p. 206). This is sometimes attributed to a mass limit on star formation by accretion processes, though there are competing formation scenarios by binary or cluster merging that would seem likely to lead to formation of even higher mass stars (see talks in JD14 and S237).

So given the above rough coincidence of the observational upper mass limit with the Eddington-limit domain of radiation-pressure dominance, it seems associated instabilities in stellar structure might actually be a more important factor in this upper mass limit, leading to extreme mass loss in LBV and/or giant eruption events, much as inferred from circumstellar nebulae observed around high mass stars like eta Carinae and the Pistol star.

Keywords. stars: early-type, stars: winds, outflows, stars: mass loss, stars: activity

---

In this Joint Discussion on *Calibrating the Top of the Stellar M - L Relation*, it is perhaps worth recalling that the steep overall scaling of stellar luminosity with mass was already worked out nearly a century ago by Eddington, Schwarzschild, and others, well before there were extensive atomic opacity tables, or even a full understanding of the core nuclear burning needed to sustain a star's luminosity. In fact, from just the two basic equations of stellar structure, namely hydrostatic equilibrium and radiative diffusion, one can readily find from homology relations (or from an even simpler dimensional analysis) the approximate basic scaling $L \propto M^3$ for gas-pressure-supported stellar envelopes at low or moderate mass, $M < 20\,\mathrm{M}_{\odot}$. If one follows this relation to higher masses, then for stars above about $100\,\mathrm{M}_{\odot}$ the outward radiation force from just electron opacity ($\propto L$) exceeds the inward force of gravity ($\propto M$), which defines the classical Eddington limit. Inclusion of radiation-pressure terms in the stellar structure equations still allows formal derivation of a bound, static envelope at an even much higher masses, with now a linear scaling of $L \propto M$ that keeps the stars just below the Eddington limit.

However, population studies of very young, dense stellar clusters now suggest quite strongly that there is a sharp cutoff at masses above ca. $150\,\mathrm{M}_{\odot}$ (see, e.g., the talk in this JD 05 by Sally Oey, p. 206). This is sometimes attributed to a mass limit on star formation by accretion processes. But given its rough coincidence with the Eddington-limit domain of radiation-pressure dominance, it seems that associated instabilities in stellar structure might also be an important factor in explaining this upper mass limit. Such instabilities could induce extreme mass loss in LBV and/or giant eruption events, much as inferred from circumstellar nebulae observed around high mass stars like eta Carinae and the Pistol star. This could reduce the stellar mass in a fraction of a nuclear burning time, effectively enforcing an upper mass limit.

Indeed, even in stars above only 0.1% this Eddington limit, the bound-bound opacity of metal ions can result in a steady outflow or 'wind' from the stellar surface layers, where the Doppler shift associated with flow acceleration implies an effective line desaturation that allows the line-force to sustain this accelerating outflow. But the saturation of lines in denser, deeper layers means such *line*-driving is unlikely to be the mechanism for more extreme mass loss events. By contrast for opacity from electron scattering or other continuum processes, the associated radiative flux and radiative remains nearly constant (without saturation) even at large optical depths. This implies that *continuum*-driving could indeed propel the much stronger mass loss of LBV or eruptive events, which are inferred to sometimes approach the energy or 'photon-tiring' limit of the available total luminosity.

Recent work has focused on how the 'porosity' associated with spatial clumping could reduce the net continuum force in the dense inner layers and so keep them gravitationally bound, but then lead to a net outflow from the surface layers where the clumps become optically thin and are thus subject to the full radiative force. Current efforts are focused on understanding the nature of both local and global instabilities in a radiation-pressure-dominated envelope near the Eddington limit, and particularly on how this could lead to envelope structure and/or super-Eddington luminosities for continuum-driven eruptions. A key issue is whether, unlike line driving, such continuum-driven mass loss might be relatively insensitive to metallicity, and thus might also be a central process in limiting the masses of Population III stars formed in the early Universe.

Highlights of Astronomy, Volume 14
IAU XXVI General Assembly, 14-25 August 2006
Karel A. van der Hucht, ed.

doi:10.1017/S1743921307010162

The ejecta of η Carinae

Theodore R. Gull

Exploration of the Universe Division, Laboratory for Extrasolar Planets and Stellar Astrophysics, Code 667, Goddard Space Flight Center, Greenbelt, MD 20771, USA
email: theodore.r.gull@nasa.gov

Abstract. High-dispersion spectroscopic observations of the neutral Homunculus and the ionized Little Homunculus, ejecta of η Car, are being analyzed to determine the relative abundances of metals. Thousands of lines of neutral and singly-ionized metals and molecules seen in the Homunculus suggest that this oxygen-, carbon-poor, nitrogen-, helium-rich gas contains very different dust grains likely devoid of metal oxides. The gas to dust ratio is likely much larger than the canonical 100:1 implying that the 12 $M_\odot$ estimate of the ejecta is a lower limit.

Keywords. stars: individual (η Car), winds, mass loss, abundances, binaries

---

High-dispersion spectra of η Car recorded with the *HST*-STIS and the VLT-UVES are filled with thousands of narrow absorption lines with several different velocity systems (Gull *et al.* 2006). Two well-isolated, dominant velocity systems correspond to the Homunculus ($-513\,\mathrm{km\,s^{-1}}$, 760 K, $10^{6-7}\,\mathrm{cm^{-3}}$), ejected in the 1840s and the Little Homunculus ($-146\,\mathrm{km\,s^{-1}}$, 6400 K, $10^7\,\mathrm{cm^{-3}}$) associated with the 1890s lesser event (Gull *et al.* 2005). In the $-513\,\mathrm{km\,s^{-1}}$ gas, lines of Fe I, Fe II, Ni II II, Cr II, Ti II, V II, Sr II, Sc II, Mn II, Mg II, Na I, etc., (with ionization potentials less than 8 eV), are found along with nearly a thousand lines of H_2, plus several lines of CH, OH, NH, and CH+ ($-513\,\mathrm{km\,s^{-1}}$, but 60 K, $10^7\,\mathrm{cm^{-3}}$), but no CO. The H_2 absorption lines originate from very high rotational levels and are characteristic of UV photo-excitation/dissociation, are present during η Car's periodic broad (5 year) maximum but disappear during the few month long low-excitation minimum. This suggests a layered structure with transition from partially ionized gas to a neutral gas with molecular and dust formation. IR flux measures (Smith *et al.* 2003) imply a total ejection mass of $\sim 12\,M_\odot$.

Initial measures of ionic and neutral column densities of this gas and the Strontium Filament, a partially ionized emission nebula in the skirt of the Homunculus, demonstrate that gaseous Ti/Ni, Cr/NI, V/Fe are much more than solar, indicating that most metals remain suspended in be of silicates and alumina (Chesneau *et al.* 2005). Additional abundance ratios, but are constrained by the need for improved laboratory measurements and/or theoretical modeling of gf-values for these species.

Acknowledgements

Observations for this project were accomplished with the NASA/ESA *Hubble Space Telescope*-STIS and the ESO Very Large Telescope-UVES.

References

Chesneau, O., Verhoelst, T., Lopez, B., *et al.* 2005, *A&A*, 435, 563
Gull, T. R., Vieira Kober, G., Bruhweiler, F., *et al.* 2005, *ApJ*, 620, 442
Gull, T. R., Vieira Kober, G., & Nielsen, K. E. 2006, *ApJS*, 163, 173
Smith, N., Gehrz, R. D., Hinz, P. M., *et al.* 2003, *AJ*, 125, 1458

Highlights of Astronomy, Volume 14
IAU XXVI General Assembly, 14-25 August 2006
Karel A. van der Hucht, ed.

doi:10.1017/S1743921307010174

Supernova impostors: LBV outbursts from the most massive stars

Schuyler D. Van Dyk

Spitzer Science Center/Caltech,
Mail Code 220-6, 1200 E. California Blvd., Pasadena, CA 91125 , USA
email: vandyk@ipac.caltech.edu

Abstract. Several recent luminous events that have been identified initially as supernovae (SNe) are probably not genuine SNe at all. Instead we argue that these events are more likely the outbursts, or super-outbursts, of very massive stars in the luminous blue variable (LBV) phase. At least two of these events are analogous to the Great 1843 Eruption of η Carinae.

Keywords. supernovae: general, supernovae: individual (SN 1954J/Variable 12, SN 1961V, SN 1997bs, SN 2000ch, SN 2002kg/Variable 37)

The fate of the most massive stars in galaxies is not well known. Such stars have been linked to black hole formation, to some GRBs, and to early metal enrichment. Today, very massive stars are extremely rare, so studying known examples, and finding new ones, in our Galaxy and other nearby galaxies affords us with essential insight into these processes. The Great Eruption of 1843 for η Carinae demonstrated that some very massive stars go through spectacular eruptive phases of mass ejection prior to exploding as a SN. For some extragalactic stars observed to date, the energetics of these pre-SN outbursts are comparable to those of SNe themselves, and, hence, these objects have been dubbed SN "impostors". The classical examples are SN 1961V in NGC 1058 and SN 1954J/Variable 12 in NGC 2403. (Zwicky actually considered η Car an underluminous SN!) More recent examples include SN 1997bs in NGC 3627 and SN 2000ch in NGC 3432. Although underluminous and photometrically erratic relative to normal, core-collapse SNe, spectroscopically the impostors generally resemble the Type II-narrow SNe, with H emission exhibiting narrow line profiles atop broader bases.

In the case of SN 1954J/V12, using Keck and *Hubble Space Telescope* data, we have identified the survivor of the super-outburst within a dusty ($A_V \simeq 4$ mag) nebula, analogous to the Homunculus around η Car, and have estimated that the star had an initial mass $M_{\rm ini} > 20\,{\rm M}_\odot$ (Van Dyk, Li, & Filippenko 2005, PASP, 117, 553). For SN 1961V it is still unclear whether or not a survivor has been located; a star with a spectrum resembling that of η Car has been detected at the SN location, but the presence of nonthermal radio emission, possibly declining in flux, may indicate we are simply seeing an old SN (see Van Dyk 2005, ASP Conf. v. 332, 49, and references therein). For SN 1997bs we actually identified the outburst precursor – however without color information, we can only infer that the highly luminous star ($M_V = -8.1$ mag) also had $M_{\rm ini} > 20\,{\rm M}_\odot$ (Van Dyk *et al.* 2000, PASP, 112, 1532). Far less is known about SN 2000ch and any putative outburst survivor. Not all of these outburst events are as extreme as those examples above: SN 2002kg/Variable 37, also in NGC 2403, has properties closer to the more normal, luminous blue variable S Doradus than η Car. We have been able to limit the outburst precursor's mass to $M_{\rm ini} > 40\,{\rm M}_\odot$ (Van Dyk *et al.* 2007, PASP, submitted).

The few known SN impostors so far exhibit a broad range of properties and clearly represent an interesting pre-SN evolutionary phase for very massive stars.

Highlights of Astronomy, Volume 14
IAU XXVI General Assembly, 14-25 August 2006
Karel A. van der Hucht, ed.

doi:10.1017/S1743921307010186

Empirical evidence suggesting a stellar upper-mass limit

M. Sally Oey

University of Michigan, Department of Astronomy,
830 Dennison Building, Ann Arbor, MI 48109-1042, USA
email: msoey@umich.edu

Abstract. While theoretical understanding remains to be clarified regarding the mechanisms that may or may not limit stellar masses, it is possible to empirically evaluate the existence of an upper-mass limit. ZAMS masses of the most massive stars have been estimated for a range of environments in our local Milky Way neighborhood and the Magellanic Clouds. Various statistical techniques demonstrate the existence of an upper-mass limit in this stellar sample.

Keywords. stars: fundamental parameters, stars: mass function, stars: statistics

---

The stellar IMF may be treated as either a *deterministic* or *probabilistic* relation (Koen 2006). Extremely rich clusters like R136a (Oey & Clarke 2005; Weidner & Kroupa 2004) and the Arches cluster (Figer 2005) have enough stars that a deterministic approach can evaluate whether an upper-mass cutoff $m_{\rm up}$ exists, provided that the cluster is young enough to preclude any supernovae. Studies of these two objects find $m_{\rm up} \simeq 150\,{\rm M}_\odot$. Koen (2006) uses a probabilistic approach, which assumes that the IMF is a probability density function, to evaluate the existence of a cutoff in R136a. Using distribution fitting and maximum likelihood techniques, he finds $m_{\rm up} \simeq 140-160\,{\rm M}_\odot$ if the IMF slope is -2.35; and a steep slope of -4 to -5 if $m_{\rm up} = \infty$. Weidner & Kroupa (2004) also emphasize that the data are consistent with no upper limit if the IMF has a much steeper slope. Ensembles of lower-mass OB associations can also be evaluated for the existence of a cutoff using a probabilistic approach. Oey & Clarke (2005) demonstrate $m_{\rm up} = 120\text{ - }150\,{\rm M}_\odot$ for 9 young OB associations in the Galaxy and LMC, by compiling the distribution of probabilities of each cluster having the observed maximum stellar mass $m_{\rm max}$. The expected uniform distribution of these probabilities occurs only for $m_{\rm up} \simeq m_{\rm max}$. This illustrates results by Aban *et al.* (2006) who derive an estimator for the maximum value of a truncated Pareto distribution that is equal to the maximum data value. It is remarkable that these extremely different environments sampled by the Galactic center, the LMC super star cluster R136a, and ordinary OB associations, all show the same $m_{\rm up} \simeq 150\,{\rm M}_\odot$ to date. Finally, a reminder that the H II region luminosity function offers a probe of $m_{\rm up}$ in more distant galaxies: the existence of a flattening in its power-law slope is the manifestation of stochastic effects that are induced by an IMF having a maximum mass (Oey & Clarke 1998).

References

Aban, I. B., Meerschaert, M. M., & Panorska, A. K. 2006, *J. Am. Stat. Assoc.*, 101, 270
Figer, D. 2005, *Nature*, 434, 192
Koen, C. 2006, *MNRAS*, 365, 590
Oey, M. S., & Clarke, C. J. 1998, *AJ*, 115, 1543
Oey, M. S., & Clarke, C. J. 2005, *ApJ* (Letters), 620, L43
Weidner, C., & Kroupa, P. 2004, *MNRAS*, 348, 187

Highlights of Astronomy, Volume 14
IAU XXVI General Assembly, 14-25 August 2006
Karel A. van der Hucht, ed.

doi:10.1017/S1743921307010198

Stellar and wind properties of massive stars in the central parsec of the Galaxy

Fabrice Martins, Reinhard Genzel, Frank Eisenhauer, Thibaut Paumard, Thomas Ott, Sascha Trippe and Stefan Gillessen

MPE, Postfach 1312, D-85741 Garching bei München, Germany
email: martins@mpe.mpg.de

Abstract. The stellar and wind properties of the new population of massive stars in the central parsec of the Galaxy are derived through quantitative analysis with atmosphere models.

Keywords. stars: atmospheres - stars: early-type - stars: Wolf-Rayet - Galaxy: center

1. The central cluster

The central cluster of the Galaxy, which harbors the supermassive black hole SgrA*, is a unique environment to study massive stars. In this particularly hostile environment where tidal forces are large, the presence of massive stars tracing recent star formation is puzzling: this is the "paradox of youth" of the Galactic Center.

The first stars observed in the central cluster were evolved massive stars: the so-called He I stars showing strong emission lines in the K-band (Najarro *et al.* 1997). But they were only the tip of the iceberg: in the last years, the development of 3D spectroscopy has lead to the discovery ~ 80 massive stars in the central parsec, including various types of Wolf-Rayet stars (see also van der Hucht 2006) as well as O- and B-type supergiants and dwarfs. This population results from a burst of star formation ~ 6 Myr ago and is spatially distributed in two counter-rotating disks (Paumard *et al.* 2006).

2. Stellar and wind properties

Quantitative spectroscopic analysis of the Wolf-Rayet population with atmosphere models reveal that these stars are similar to other Galactic WR stars, possibly indicating a common formation process in spite of the peculiar environment of the Galactic Center. Their metallicity appears to be about solar. The new population of early type stars accounts for the H and He I ionizing fluxes required to ionize the local ISM. This means that contrary to the claim of Lutz (1999), standard stellar evolution does not face any serious problem in the Galactic Center. The He I stars are less chemically evolved than late type WN stars (especially WN8) and we suggest that they may be their precursors. Their mass range is 20–70 $M_{\odot}$, one of them being in a binary system and having a mass of $\sim 50\, M_{\odot}$ (Martins *et al.* 2006).

References

van der Hucht, K. A. 2006, *A&A*, 458, 453
Lutz, D. 1999, in: P. Cox & M. F. Kessler (eds.), *The Universe as Seen by ISO*, ESA SP-427, 623
Martins, F., Trippe, S., Paumard, T., *et al.* 2006, *ApJ* (Letters), 649, L103
Najarro, F., Krabbe, A., Genzel, R., *et al.* 1997, *A&A*, 325, 700
Paumard, T., Genzel, R., Martins, F., *et al.* 2006, *ApJ*, 643, 1011

Highlights of Astronomy, Volume 14
IAU XXVI General Assembly, 14-25 August 2006
Karel A. van der Hucht, ed.

doi:10.1017/S1743921307010204

The high-mass stellar IMF in different environments

Pavel Kroupa

Argelander Institute for Astronomy, University of Bonn,
Auf dem Hügel 71, D-53121 Bonn, Germany
email: pavel@astro.uni-bonn.de

Abstract. The massive-star IMF is found to be invariable. However, integrated IMFs probably depend on galactic mass.

Keywords. stars: luminosity function, mass function; stars: pre–main-sequence

---

To set the stage it is useful to emphasis that the Salpeter (1955) mass function (MF), $\xi(m) = k\,m^{-\alpha}, \alpha = 2.35$, is strictly valid only for stellar masses with $0.4 \lesssim m/\mathrm{M}_\odot \lesssim 10$. The number of stars in the mass interval $m, m + dm$ is $dN = \xi(m)\,dm$. Given statistical noise in the IMF of otherwise equal systems the question we need to answer is whether there is any significant empirical evidence for systematic variation of the IMF with the physical conditions of star formation. This question is of fundamental importance for star-formation theory because an observed systematic variation poses constraints on the theory. The question is also of fundamental importance for cosmology because the physical conditions of star formation have changed dramatically over a cosmological epoch implying possible systematic changes of the young stellar populations with cosmic time and therefore systematic changes in the properties of galaxies. Very different theoretical approaches lead to one common result, namely that the average stellar mass ought to shift towards larger values with decreasing metallicity. Thus, Adams & Fatuzzo (1996) develop a model of the origin of the IMF based on the notion that stars regulate their own masses through feedback, while Larson (1998) investigates the systematic changes of the IMF as a result of the temperature-dependence of the Jeans mass. Recent developments based on a change of the equation of state as a result of dust processes may be an avenue of explaining the general absence of a variation of the stellar mass at which the IMF peaks (Bonnell *et al.* 2006). Observationally, the IMF has been constrained above a few $\mathrm{M}_\odot$ by Massey (2003) for populations in the Milky Way, the Large and Small Magellanic Clouds and for different densities. The massive-star IMF has been found to have an invariable "Salpeter/Massey" index $\alpha = 2.3 \pm 0.2$. This insensitivity to the physical conditions needs to be understood, and perhaps suggests that massive star formation is dominated by scale-free processes such as coagulation and/or competitive accretion in the very dense environments where massive stars form (Bonnell *et al.* 2006). For entire galaxies the composite IMF probably depends on galaxy type such that low-mass galaxies have very steep indices (Weidner & Kroupa 2006), despite the invariability of the IMF.

References

Adams, F. C., & Fatuzzo, M. 1996, *ApJ*, 464, 256
Bonnell, I. A., Larson, R. B., & Zinnecker, H. 2006, in: B. Reipurth, D. Jewitt & K. Keil (eds.), *Protostars and Planets V* (Tucson: Univ. Arizona Press), p. 951
Weidner, C., & Kroupa, P. 2006, *MNRAS*, 365, 1333
Larson, R. B. 1998, *MNRAS*, 301, 569
Massey, P. 2003, *ARAA*, 41, 15
Salpeter, E. E. 1955, *ApJ*, 121, 161

Highlights of Astronomy, Volume 14
IAU XXVI General Assembly, 14-25 August 2006
Karel A. van der Hucht, ed.

doi:10.1017/S1743921307010216

Rotation, mass loss and nucleosynthesis

Georges Meynet
Geneva Observatory, University of Geneva, CH-1290 Sauverny
email: georges.meynet@obs.unige.ch

Abstract. We list below a few problems whose resolution might bring new lights on the impacts of the first stellar generations on the early phases of the evolution of the galaxies.

Keywords. Stars: evolution, rotation

For metallicities between those of the Small Magellanic Cloud and of the solar neighborhood, rotating models better reproduce the observed characteristics of stars than non-rotating models (see, e.g., Meynet *et al.* 2006). When the same physical processes as those necessary to obtain good fits at high metallicity are accounted for in metal poor stars, one notes that stars are strongly mixed and may lose large amounts of material. These features might be helpful for explaining puzzling observational facts, as the synthesis of important amounts of primary nitrogen by massive and intermediate mass stars (Ciappini *et al.* 2006), the existence of C-rich extremely metal poor stars (Meynet *et al.* 2006), the chemical inhomogeneities observed in globular clusters (Decressin *et al.* 2007), the presence of very He-rich stars in ω Cen (Maeder & Meynet 2006).

To go ahead in these areas of research, we need to answer the following questions:

(*a*) How the initial distribution of the rotational velocities vary with the metallicity?

(*b*) What is the main physical mechanism responsible for determining the rotation of massive stars on the ZAMS?

(*c*) For a given initial mass star, initial rotation velocity, at a given age, the surface enrichments are higher at lower Z. Is this prediction confirmed by the observations?

(*d*) How the surface velocity vary at the surface of stars of different initial masses and metallicities along the Main-Sequence phase?

(*e*) What are the conditions for the surface velocity of a star to reach the critical velocity? (N.B. The critical velocity is the equatorial rotation velocity such that the centrifugal acceleration exactly compensates for the gravity.) When does it happen? What happens at the critical limit?

(*f*) What happens for what concerns mass loss, when important amounts of CNO elements appear at the surface as a consequence of rotational mixing?

(*g*) What is the role of the magnetic field and of the gravity waves in stellar interiors?

The list is not exhaustive, but these few questions already represent serious challenges for the years to come.

References

Decressin, T., Meynet, G., Charbonnel, C., Prantzos, N., & Ekström, S. 2007, *A&A*, 464, 1029
Chiappini, C. , Hirschi, R., Meynet, G., Ekström, S., Maeder, A., & Matteucci, F. 2006, *A&A* (Letters), 449, L27
Maeder, A., & Meynet, G. 2006, *A&A* (Letters), 448, L37
Meynet, G., Ekström, S., & Maeder, A. 2006, *A&A*, 447, 623

Highlights of Astronomy, Volume 14
IAU XXVI General Assembly, 14-25 August 2006
Karel A. van der Hucht, ed.

doi:10.1017/S1743921307010228

Evolution of ultra-massive stars

Lev R. Yungelson

Institute of Astronomy, Pyatnitskaya 48, 119017 Moscow, Russia
email: lry@inasan.ru

Abstract. We discuss the evolution of ultra-massive stars.

Keywords. stars: evolution, stars: mass-loss

We have calculated evolution of $M_0 = (60 - 1001)\,\mathrm{M}_\odot$ non-rotating solar-metallicity stars from main-sequence through core helium-burning stage. The value of $1001\,\mathrm{M}_\odot$ is the upper mass limit for ZAMS of solar-metallicity stars. Assuming that Γ-instability is responsible for mass loss we applied an *ad hoc* mass-loss law $\dot{M} = \frac{L}{v_\infty c}\frac{1}{(1-L/L_{\mathrm{Edd}})^\alpha}$. Here L_{Edd} is the Eddington luminosity in the outermost mesh point of the model. The value of $\alpha = 0.25$ was chosen such that stars with mass below $120\,\mathrm{M}_\odot$ spend less that 1 % of their lifetime to the right of Humphreys-Davidson limit in the HR-diagram. Results of computations are relevant to the evolution of the most massive stars in the Galactic Center clusters or for (still hypothetical) products of runaway stellar collisions in young dense clusters that are thought to be progenitors of black hole accretors in some ultraluminous X-ray sources.

Mass-loss rates both for hydrogen- and helium-rich stars that follow from above given expression for $\dot{M}$ are consistent with the limits implied for line-driven winds and with $\dot{M}$ found for the most massive stars, e.g., in Arches, Quintuplet, He I-clusters in the Galactic Center or in R 136.

The end-products of the evolution of ultra massive stars have oxygen-neon cores with minor admixture of carbon and thin helium-carbon envelopes ($Y_s \approx 0.2 - 0.4$). Initial-final mass relation obtained in calculations and the fate of stellar remnants (as inferred from model calculations for helium stars) are shown in the Figure. These results suggest that runaway stellar collisions hardly can produce black holes with masses $\gtrsim 100\,\mathrm{M}_\odot$.

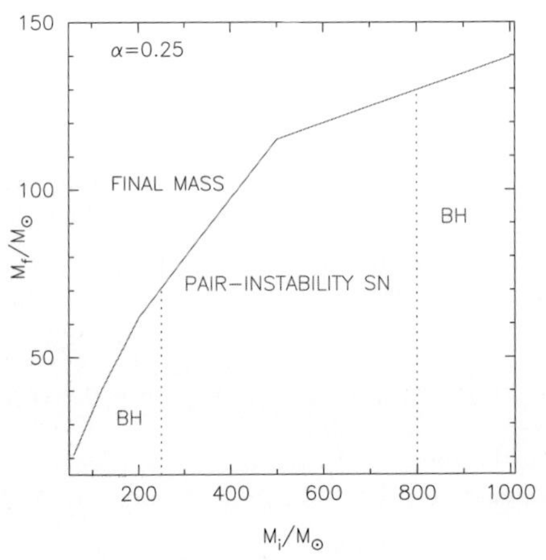

Figure 1. Relation between initial and final masses of stars and the nature of stellar remnants.

Highlights of Astronomy, Volume 14
IAU XXVI General Assembly, 14-25 August 2006
Karel A. van der Hucht, ed.

doi:10.1017/S174392130701023X

Properties of massive Population III and metal-poor stars

Daniel Schaerer[1,2]

[1]Geneva Observatory, 51 Ch. des Maillettes, CH–1290 Sauverny, Switzerland
email: daniel.schaerer@obs.unige.ch

[2]Laboratoire d'Astrophysique, OMP, 14 Avenue E. Belin, F-31400 Toulouse, France

Abstract. We review the properties of massive Population III and very metal-poor stars, including briefly their formation, IMF, their main sequence evolution, possible mass loss mechanisms, atmosphere modeling etc. For detailed predictions concerning the properties of these stars we refer to Schaerer (2002) and Schaerer (2002) and references therein. Extending these calculations, Schaerer & Fall (2007) present new calculations concerning the ionizing power, Ly-α strength and related properties for different metallicities as well as for a range of power-law and log-normal IMFs. For illustrations from these studies see the Figures below. New detailed calibrations for solar metallicity O-type stars have recently been presented by Martins *et al.* (2005).

Keywords. galaxies:high-redshift, stars: Population III, stars: general

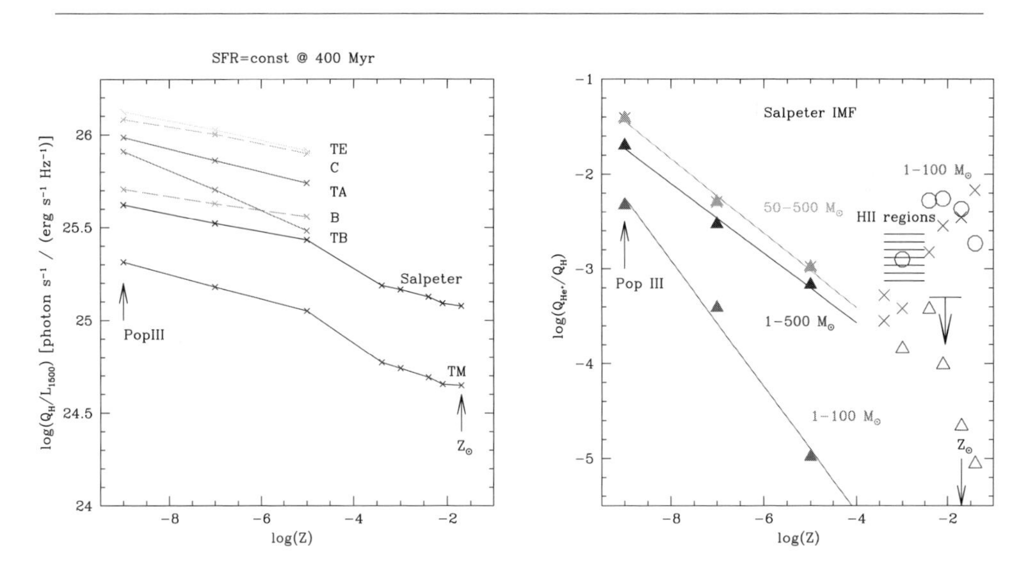

Figure 1. *Left:* Predicted ionizing to UV power of starbursts as a function of metallicity, from $Z \simeq 0$ (Pop.III) to solar. Different curves illustrate the predictions for various IMFs, including the Salpeter IMF and several log-normal IMFs (for details see Schaerer & Fall 2007) *Right:* Predicted hardness (as measured by the ratio of the He^+ to H ionizing photons) of the ionizing spectrum of starbursts from Pop.III to normal metallicities (from Schaerer 2003).

References

Martins, F., Schaerer, D., & Hillier, D. J. 2005, *A&A*, 436, 1049
Schaerer, D. 2002, *A&A*, 382, 28
Schaerer, D. 2003, *A&A*, 397, 527
Schaerer, D., & Fall, M. 2007, in preparation

Highlights of Astronomy, Volume 14
IAU XXVI General Assembly, 14-25 August 2006
Karel A. van der Hucht, ed.

doi:10.1017/S1743921307010241

Influence of X-ray radiation on the non-LTE wind models of O-type stars

Jiří Krtička

Institute for Theoretical Physics and Astrophysics, Faculty of Sciences, Masaryk University, Kotlarska 2, CZ – 611 37 Brno, Czech Republic
email: krticka@physics.muni.cz

Abstract. We study the influence of shock X-ray emission on the structure of non-LTE wind models of O-type stars.

Keywords. stars: early-type, stars: atmospheres, stars: winds, outflows

---

We study the influence of shock X-ray emission on the structure of non-LTE wind models of O-type stars. For this purpose we use our own non-LTE wind code with an artificial X-ray source to understand the influence of X-rays on the wind ionization structure and, consequently, on the basic wind parameters of O-type stars (mass-loss rate and the terminal velocity). The inclusion of X-ray sources into wind models enables to obtain wind ionization structure which is more consistent with the observed one. However, the presence of strong X-ray radiation does not significantly influence the mass-loss rate, but it may influence the terminal velocity for cooler O-type stars. Wind models with constant X-ray filling factor are able to roughly explain the observed relation between the X-ray luminosity and the total luminosity. Star exhibiting so-called 'weak wind problem' produce X-rays apparently more efficiently than the 'normal' stars. We discuss the implications of wind models with X-ray emission on the current problems of hot star wind research, namely the discrepancy between wind properties derived from observation and theoretical wind models. (*Poster Presentation.*)

Highlights of Astronomy, Volume 14
IAU XXVI General Assembly, 14-25 August 2006
Karel A. van der Hucht, ed.

doi:10.1017/S1743921307010253

M - L relations for rapid and slow rotators

Oleg Yu. Malkov

Center for Astronomical Data, Institute of Astronomy of the Russian Academy of Sciences, Pyatnitskaya ul. 48, 119017 Moscow, Russian Federation
email: malkov@inasan.ru

Abstract. Comparing the radii of eclipsing binaries components and single stars we have found a noticeable difference between observational parameters of B0V - G0V components of eclipsing binaries and those of single stars of the corresponding spectral type. This difference was confirmed by re-analyzing the results of independent investigations published in the literature.

Keywords. binaries: eclipsing

Comparing the radii of eclipsing binaries components and single stars we have found a noticeable difference between observational parameters of B0V - G0V components of eclipsing binaries and those of single stars of the corresponding spectral type. This difference was confirmed by re-analyzing the results of independent investigations published in the literature.

Larger radii and higher temperatures of A - F eclipsing binaries can be explained by synchronization of such stars in close systems that prevents them to rotate rapidly.

So, we have found that the M - L relation based on eclipsing binary data cannot be used to derive the initial mass function of single stars. While our current knowledge of the empirical M - L relation for intermediate-mass (1.5 to $10\,M_\odot$) stars is based exclusively on data from eclipsing binaries, knowledge of the M - L relation should come from dynamical mass determinations of visual binaries, combined with spatially resolved precise photometry. Then the initial mass function should be revised for $M > 1.5\,M_\odot$.

Data were collected on fundamental parameters of stars with masses $M > 1.5\,M_\odot$. They are components of binaries with $P > 15$ d and consequently are not synchronized with the orbital periods and presumably are rapid rotators. These stars are believed to evolve similarly with single stars, so these data allow us to construct M - L and other relations, that can more confidently be used for statistical and astrophysical investigations of single stars, than so called standard relations, based on data on detached main-sequence double-lined short-period eclipsing binaries. Mass-luminosity, mass-temperature and mass-radius relations of single stars are presented, as well as their HR diagram. (*Poster Presentation.*)

Highlights of Astronomy, Volume 14
IAU XXVI General Assembly, 14-25 August 2006
Karel A. van der Hucht, ed.

doi:10.1017/S1743921307010265

Joint Discussion 6
Neutron stars and black holes in star clusters

Frederic A. Rasio[1] (ed.), Holger Baumgardt[2], Alessandro Corongiu[3], Francesca D'Antona[4], Giuseppina Fabbiano[5], John M. Fregeau[1], Karl Gebhardt[6], Craig O. Heinke[1], Piet Hut[7], Nataliya Ivanova[8], Thomas J. Maccarone[9], Scott M. Ransom[10] and Natalie A. Webb[11]

[1]Northwestern University, Dept of Physics and Astronomy, Evanston, Illinois, USA
email: rasio@northwestern.edu

[2]Argelander Institut für Astronomie, University of Bonn, Germany

[3]INAF, Osservatorio di Cagliari e Università di Cagliari, Italy

[4]Osservatorio Astronomico di Roma, Monteporzio, Italy

[5]Harvard-Smithsonian Center for Astrophysics, Cambridge, Massachusetts, USA

[6]Astronomy Department, University of Texas at Austin, USA

[7]Institute for Advanced Study, Princeton, New Jersey, USA

[8]Canadian Institute for Theoretical Astrophysics, Toronto, Ontario, Canada

[9]School of Physics and Astronomy, University of Southampton, UK

[10]NRAO, Charlottesville, Virginia, USA

[11]Centre d'Etude Spatiale des Rayonnements, Toulouse, France

1. Introduction

This article was co-authored by all invited speakers at Joint Discussion 6 on *Neutron Stars and Black Holes in Star Clusters*, which took place during the IAU General Assembly in Prague, Czech Republic, on August 17 and 18, 2006. Each section presents a short summary of recent developments in a key area of research, incorporating the main ideas expressed during the corresponding panel discussion at the meeting.

Our meeting, which had close to 300 registered participants, was broadly aimed at the large community of astronomers around the world working on the formation and evolution of compact objects and interacting binary systems in dense star clusters, such as globular clusters and galactic nuclei. The main scientific topics cut across all traditional boundaries, including Galactic and extragalactic astronomy, environments from young starbursts to old globular clusters, phenomena from radio pulsars to gamma-ray bursts, and observations using ground-based and space-based telescopes, with a significant component of gravitational-wave astronomy and relativistic astrophysics.

Great advances have occurred in this field during the past few years, including the introduction of fundamentally new theoretical paradigms for the formation and evolution of compact objects in binaries as well as countless new discoveries by astronomers that have challenged many accepted models. Some of the highlights include: a nearly complete census of all the millisecond pulsars in 47 Tucanae; first detections of many new radio pulsars in other clusters, particularly Terzan 5; detailed studies of X-ray binary populations and their luminosity functions in many galaxies and extragalactic globular clusters; increasing evidence for intermediate-mass black holes in clusters and greatly improved theoretical understanding of their possible formation processes.

The next few years will prove at least as exciting, with many more data sets coming from recently or soon-to-be launched satellites, many new objects found in extensive deep radio and X-ray surveys, and follow-up spectroscopy and photometry with optical telescopes. On the theoretical side, advances in computer codes and special-purpose hardware will allow for more and more realistic modeling of whole large clusters including fairly complete treatments of all the relevant physics.

2. Direct N-body simulations

Direct N-body simulations follow stars individually. This is important when modeling star clusters, where specific interactions between single stars and binaries, as well as more complex multiple systems, play a central role (Aarseth 2003; Heggie & Hut 2003). In contrast, for larger-scale simulations of encounters between galaxies, and cosmological simulations in general, the stars are modeled as a fluid in phase space, and the individual properties of the stars are no longer important.

Traditionally, the term 'collisionless stellar dynamics' has been used for the latter case, and 'collisional stellar dynamics' for the former case. When these terms were coined in the nineteen sixties, they were perhaps appropriate, but now that we have started to simulate physical collisions between stars in a serious way, the use of the word 'collision' in the old sense has become rather confusing, since it was meant to denote relatively distant encounters that contribute to the two-body relaxation of a system.

It may be useful to introduce a new expression for the study of star clusters, shorter than 'collisional stellar dynamics,' and broader in the sense of including stellar evolution and hydrodynamics as well. One option would be to use *smenology*, or in Greek $\sigma\mu\eta\nu o\lambda o\gamma\iota\alpha$, after $\sigma\mu\eta\nu o\sigma$ (*smenos*), swarm, which is the word in use in modern Greek for a cluster; a star cluster is called $\sigma\mu\eta\nu o\sigma$ $\alpha\sigma\tau\epsilon\rho\omega\nu$ (*smenos asteroon*), literally a swarm of stars (Dimitrios Psaltis, personal communication).

Simulations of dense stellar systems, such as globular clusters (hereafter GCs) and galactic nuclei, have never yet been very realistic. Simplifying assumptions, such as those used in gas models or Fokker-Planck and Monte Carlo codes, have allowed us to model large particle numbers at the expense of a loss of detail in local many-body interactions and the imposition of global symmetry constraints. Conversely, direct n-body integration, while far more accurate, has labored under a lack of computer speed needed to model a million stars.

The good news is that we will soon be approaching effective computer speeds in the Petaflops range (Makino 2006), which will allow us to model the gravitational million-body problem with full realism, at least on the level of point particles. Adding equally realistic stellar evolution and hydrodynamics will be no problem as far as the necessary computer speed is concerned.

When the hardware bottleneck will thus be removed, the software bottleneck for realistic cluster simulations will become painfully obvious (Hut 2007). This is the bad news. While some serious uncertainties remain in the science needed to improve the software, currently the main bottleneck is neither science nor computer speed, but rather a sufficiently robust implementation of already available knowledge.

The main two codes currently being used for direct N-body simulations, NBODY4 and Kira, are both publicly available: NBODY4 at <www.ast.cam.ac.uk/~sverre/web/pages/nbody.htm>, and Kira at <www.ids.ias.edu/~starlab/>.

NBODY4 and other related codes form the results of a more than forty-year effort by Sverre Aarseth, as documented in Aarseth (2003). These codes are written in Fortran and they can be run stand-alone. A parallel version has been developed, named NBODY6++ (Spurzem 1999; Spurzem & Baumgardt 2003), publicly available at <ftp://ftp.ari.uni-heidelberg.de/pub/staff/spurzem/nb6mpi/>. In addition to stellar dynamics, the version of NBODY4 developed by Jarrod Hurley and collaborators (see Hurley *et al.* 2005 and references therein) includes a treatment of stellar evolution for both single stars (named SSE) and binary stars (named BSE), using fitting formulae and recipes (Hurley *et al.* 2002).

The Kira code forms an integral part of the Starlab environment (Portegies Zwart *et al.* 2001). Kira and Starlab are written in C++. The basic data structure of Kira consists of a flat tree containing leaves representing single stars as well as nodes that hold center of mass information for small clumps of interacting stars. Each clump is represented by a binary tree, where each node determines a local coordinate system. The Kira code has built-in links to Seba, a stellar evolution module using fitting formulae developed by Tout *et al.* (1996) and recipes developed by Portegies Zwart & Verbunt (1996). In addition to Kira and Seba, the Starlab environment contains tools for setting up initial conditions for star clusters, using various models, and for analyzing the results of N-body simulations. Starlab also contains packages for binary–single-star and for binary–binary scattering.

Within the next ten years, multi-Petaflops computers will enable us to follow the evolution of star clusters with up to a million stars (Makino 2006). To make efficient use of this opportunity, while including increasingly realistic treatments of stellar evolution and stellar hydrodynamics, a number of new developments are required.

On the purely stellar dynamics level, some form of tree code may be useful for speeding up the long-range force calculations, as pioneered by McMillan & Aarseth (1993). In addition, guaranteeing accurate treatments of local interactions will become more challenging, especially for extreme mass ratios; designing good algorithms for following the motions of stars in the neighborhood of a massive black hole is currently an area of active research.

The largest challenge, however, will be to develop robust stellar evolution and stellar hydrodynamics codes, that can interface reliably with stellar dynamics codes, without crashing. The MODEST initiative (for MOdeling DEnse STellar systems) was started in 2002 with the intention to provide a forum for discussions concerning this challenge (Hut *et al.* 2003). A pilot project, MUSE (for MUlti-scale MUlti-physics Scientific Environment), was initiated recently to develop a modular software environment for modeling dense stellar systems, allowing packages written in different languages to interoperate within an integrated software framework (see the MODEST web site at <www.manybody.org/modest.html> and click on "projects").

Finally, for any large software project that involves a team of code developers, good documentation is essential. For most astrophysical simulation codes, documentation has come mainly as an afterthought. An attempt to develop a new code for modeling dense stellar systems, using an almost excessive amount of documentation can be found at <www.ArtCompSci.org>, the web site for ACS (the Art of Computational Science).

3. Monte Carlo methods

Hénon's Monte Carlo method has given rise to an industry in the business of simulating the evolution of dense stellar systems, providing fast and accurate simulations of large-N systems. Its computational speed, coupled with the physical assumptions it requires (notably spherical symmetry and dynamical equilibrium) make it a very natural complement to 'direct' N-body simulations (Sec. 2), which are computationally much more expensive (or, equivalently, allow for smaller N) and generally require the use of special-purpose (GRAPE) hardware. Here we briefly discuss the method and the primary Monte Carlo

Table 1. Comparison of the capabilities of different methods for simulating the evolution of dense stellar systems. The first column lists the different physical processes at work in stellar systems, column 'NB' lists the capabilities of the N-body method, column 'MC' lists what the Monte Carlo method is in principle capable of, columns 'NU', 'F', 'G', and 'GS', list the current capabilities of the Northwestern, Freitag, and Giersz Monte Carlo codes, as well as the Giersz & Spurzem hybrid gas/Monte Carlo code. A filled circle means the code is fully capable of treating the physical process, while an open circle means it is capable subject to some limitations.

Physics	NB	MC	NU	F	G	GS
two-body relaxation	•	•	•	•	•	•
stellar evolution	•	•	○	○	○	
stellar collisions	•	•	•	•		
binary interactions	•	•	•		○	•
external effects	•	○	○	○	○	○
central BH	•	•		•		
rotation	•					
violent relaxation	•					
large-angle scattering	•	•				
three-body binaries	•	•			•	•
large N, f_b		•	•	•	•	•

codes before focusing on recent progress on the technique and recent contributions to the study of dense stellar systems made by those codes.

At the core of the modern Monte Carlo method is the Hénon technique, which amounts to the following. The evolution of each particle's orbit in a dense stellar system is influenced by all other particles in the system, although one cannot realistically sum all two-body scattering interactions and achieve reasonable computational speed on standard hardware. Instead, for each particle one performs a 'super'-encounter with a nearby particle, with the deflection angle chosen so as to represent the effects of relaxation due to the whole cluster (see Freitag & Benz 2001 for a pedagogical discussion and original references). This 'trick' makes the Monte Carlo method scale with particle number as $N \log N$ per time step, instead of N^2 with direct N-body methods.

There are currently three primary Monte Carlo evolution codes in use, along with a fourth hybrid gas model/Monte Carlo code. The Northwestern code uses the Hénon technique with a timestep shared among all particles (Fregeau & Rasio 2006, and references therein). Notably, it incorporates dynamical integration of all binary interactions in a cluster, allowing for the study of the long-lived binary-burning phase. The Freitag code uses the Hénon technique with a radius-dependent timestep (Freitag *et al.* 2006b, and references therein). It is notable for its inclusion of physical stellar collisions drawn from a library of SPH simulations, as well as a treatment of loss cone physics. The Giersz code uses the Hénon technique with radial timestep zones (Giersz 2006, and references therein), and includes stellar evolution of single stars. The Giersz & Spurzem hybrid code couples a gas dynamical model for the single star population with a Monte Carlo treatment of binary interactions (Giersz & Spurzem 2003, and references therein). Table 1 lists the capabilities of the Monte Carlo codes just discussed, as well as those of the direct N-body method.

In the past year or so, six papers relying on the Monte Carlo codes mentioned above have been published. Gürkan *et al.* (2006) studied the process of runaway collisional growth in young dense star clusters, and found the exciting new result that, when the runaway process operates in young clusters with primordial binaries, generally two very massive stars (VMSs) are formed. The VMSs formed may quickly undergo collapse after formation to become intermediate-mass black holes (IMBHs), yielding the exotic possibility of IMBH–IMBH binaries forming in young clusters. Giersz (2006) performed simulations of clusters with $N = 10^6$ stars subject to the tidal field of their parent galaxy. The large particle number allowed a detailed study of the evolution of the cluster mass function. Freitag *et al.* (2006c) performed a comprehensive study of the process of mass segregation in galactic nuclei containing supermassive black holes, with implications for the distribution of X-ray binaries (XRBs) at the Galactic Center. Freitag *et al.* (2006a) and Freitag *et al.* (2006b) studied in great detail the process of runaway collisional growth in young dense star clusters. Their study yielded several key results. First, a comparison of approximate physical stellar collision prescriptions with the detailed results of SPH simulations showed that the simple 'sticky-star' approximation – in which stars are assumed to merge without mass loss when their radii touch – is sufficiently accurate for clusters with velocity dispersions less than the typical stellar surface escape velocity to faithfully model the physics of runaway collisions. Second, runaway collisional growth of a VMS to $\sim 10^3\,M_\odot$ is generic for clusters with central relaxation times sufficiently short ($\lesssim 25$ Myr) and for clusters which are initially collisional. Fregeau & Rasio (2006) presented the first Monte Carlo simulations of clusters with primordial binaries to incorporate full dynamical integration of binary scattering interactions (the work of Giersz & Spurzem (2003) performed integration of binary interactions, but used a gas dynamical model for the single star population). They performed detailed comparisons with direct N-body calculations, as well as with semi-analytical theory for cluster core properties as a function of the binary population, and found good agreement with both. They then simulated an ensemble of systems and compared the resulting cluster structural parameters (r_c/r_h and the concentration parameter, for example) during the binary burning phase with the observed Galactic GC population. The interesting result is that the values of r_c/r_h predicted by the simulations are roughly a factor of 10 smaller than what is observed. The most likely explanation is that physical processes ignored in the simulations (such as stellar evolution and collisions) are at work in the Galactic sample, generating energy in the cores and causing them to expand. However, more detailed simulations should be performed to test this hypothesis.

4. Stellar and binary evolution in globular clusters

Much of the recent work in the area of stellar evolution in GCs has concentrated on the evolution of low-mass X-ray binaries (LMXBs) and their likely remnants, the millisecond pulsars (MSPs). In particular, several new studies have considered the possible effects of a 'radio-ejection phase' initiated when the mass transfer temporarily stops during the secular evolution of the systems.

The much larger fraction of binary MSPs and LMXBs in GCs, with respect to their fraction in the Galactic field, is regarded as a clear indication that binaries containing neutron stars (NSs) in GCs are generally not primordial, but are a result of stellar encounters due to the high stellar densities in the GC cores. On the other hand, it is still not clear how the LMXBs are formed in the Galactic field, as the result of a supernova explosion in a binary in which the companion is a low-mass star will generally destroy the binary. Many possible processes have been invoked to explain LMXBs: (*i*) accretion-induced collapse of a white dwarf primary into a neutron star; (*ii*) supernova kicks due to asymmetric neutrino energy deposition during the supernova event; (*iii*) formation of LMXBs as remnants of the evolution of binaries with intermediate-mass donors; (*iv*) LMXBs formed by capture in the dense environment of GCs and later released when the GC is destroyed. This last hypothesis, originally due to Grindlay *et al.* (1984), was recently re-evaluated in the literature (Podsiadlowski *et al.* 2002; see also Sec. 11).

The 'standard' secular evolution of LMXBs as progenitors of binary MSPs is reasonable in the context of the evolution of binaries above the so-called 'bifurcation period' $P_{\rm bif}$ (Tutukov *et al.* 1985; Pylyser & Savonije 1988), in which the donor star begins the mass transfer phase after it has finished the phase of core hydrogen burning, and the system ends up as a low-mass white dwarf (the remnant helium core of the donor) in a relatively long or very long period orbit with a radio MSP (e.g., Rappaport *et al.* 1995. Some of these systems may also be the remnants of the evolution of intermediate-mass donors, with similar resulting orbital periods (Rasio *et al.* 2000; Podsiadlowski *et al.* 2002. Recently, D'Antona *et al.* (2006) showed that the secular evolution at $P > P_{\rm bif}$ may need to take into account the detailed stellar evolution of the giant donor, in order to explain the orbital period gap of binary MSPs between ~ 20 and ~ 60 days. During the evolution along the RGB, the hydrogen burning shell encounters the hydrogen chemical discontinuity left by the maximum deepening of convection: the thermal readjustment of the shell causes a luminosity and radius drop, which produces a well known 'bump' in the luminosity function of the RGB in GCs. In semi-detached binaries, at the bump, the mass transfer is temporarily stopped, following the sudden decrease in radius. We consider it possible that, when mass transfer starts again, a phase of 'radio-ejection' begins (Burderi *et al.* 2001; Burderi *et al.* 2002), in which mass accretion onto the NS is no longer allowed because of the pressure from the radio pulsar wind. In this case, the matter is lost from the system at the inner lagrangian point, carrying away angular momentum and altering the period evolution. This will occur for magnetic moments of the NS in a range ~ 2–$4 \times 10^{26}\,{\rm G\,cm^3}$, which is the most populated range for binary MSPs.

Turning now to the evolution below $P_{\rm bif}$, it is well known that, if the secular evolution of LMXBs is similar to that of cataclysmic variables (CVs), one should expect many systems at $P \lesssim 2\,{\rm hr}$, and a minimum orbital period similar to that of CVs, namely $\sim 80\,{\rm min}$. On the contrary, there are very few of these systems, and instead several 'ultrashort' period binaries. In particular, three LMXBs in the field (which are also X-ray MSPs) and one in a GC (X1832–330 in NGC 6652) are concentrated near $P_{\rm orb} \simeq 40\,{\rm min}$. In addition there are two other ultrashort period systems in GCs. While for GCs we may think that these systems were formed by capture of a white dwarf by the NS, the field systems should have arrived at this period by secular evolution. Models have been constructed by Nelson & Rappaport (2003) and Podsiadlowski *et al.* (2002), and all imply that the donors began Roche lobe overflow at periods just slightly below $P_{\rm bif}$, so that the donor evolved to become a degenerate dwarf predominantly composed of helium, but having a residual hydrogen abundance $< 10\%$. Until hydrogen is present in the core of the donor star, in fact, the evolution proceeds towards short $P_{\rm orb}$ (convergent systems). If the hydrogen content left is very small, the mass radius relation when these objects become degenerate is intermediate between that of hydrogen-dominated brown dwarfs and that of helium white dwarfs, so that smaller radii and shorter $P_{\rm orb}$ will be reached before radius and period increase again.

One problem of this scenario is that *there is a very small interval of initial* $P_{\rm orb}$ which allows this very peculiar evolution: in most cases, either a helium core is already formed before the mass transfer starts, and the system evolves towards long $P_{\rm orb}$, or there is enough hydrogen that the system is convergent, but the minimum period is similar to that of CVs, and cannot reach the ultrashort domain. Podsiadlowski *et al.* (2002) notice that, for a $1\,{\rm M_\odot}$ secondary, the initial period range that leads to the formation of ultracompact systems is 13 – 18 hr. Since systems that start mass transfer in this period range might be naturally produced as a result of tidal capture, this could perhaps explain the large fraction of ultracompact LMXBs observed in GCs. However, quantitatively, this appears highly unlikely (van der Sluys *et al.* 2005).

In any case, this does not apply to LMXBs in the field. In her PhD thesis, Anamaria Teodorescu (2005) derived the period distribution expected for LMXBs from convergent systems under several hypotheses, and compared it with the available observed period distribution. An expected result was that the range $P_{\rm orb} < 2\,{\rm hr}$ is very populated and the distribution is inconsistent with observations, unless we can suppress the secular evolution of all the systems below the 'period gap', which should occur at about the same location as in CVs. One can consider several different possibilities to do this:
(*a*) The lack of systems at $P_{\rm orb} < 2\,{\rm hr}$ is again a consequence of radio-ejection: after the period gap is traversed by the detached system, when the mass transfer resumes, it is prevented by the pulsar wind pressure, the matter escapes from the system with high specific angular momentum, and the evolution is accelerated. Indeed, this is probably occurring in the system containing pulsar W in 47 Tucanae, which has $P_{\rm orb} = 3.2\,{\rm hr}$. This system exhibits X-ray variability which can be explained by the presence of a relativistic shock within the binary that is regularly eclipsed by the secondary star (Bogdanov *et al.* 2005). The shock can then be produced by the interaction of the pulsar wind with a stream of gas from the companion passing through the inner Lagrange point (L1), a typical case of what is expected in radio-ejection (Burderi *et al.* 2001). This mechanism could affect all the systems which enter a period gap. Notice that only systems which end up at ultrashort periods *do not* detach during the secular evolution, and they only might have a 'normal' secular evolution. Thus both the lack of systems at $P_{\rm orb} < 2\,{\rm hr}$ and the presence of ultrashort periods could be due to this effect.
(*b*) 'Evaporation' of the donor, due to the the pulsar wind impinging on, and ablating material from, the surface of the companion (Ruderman *et al.* 1989) is another possible mechanism, with results not so different from the previous case.
(*c*) It is possible that the secular evolution almost never begins when the donor is not significantly evolved. This can be true only if binaries are mostly formed by tidal capture, in which the NS captures a main-sequence star only at separations $\lesssim 3\,R_*$ (Fabian *et al.* 1975). This might happen in GCs, but we need to explain the $P_{\rm orb}$ distribution of *all* the LMXBs in the Galaxy. We could then reconsider the possibility that *most* of the field LMXBs were in fact formed in GCs, which were later destroyed (e.g., by tidal interactions with the Galactic bulge; but see Sec. 11).

There are other specific cases that we must take into account when discussing evolutions starting close to $P_{\rm bif}$. The famous interacting MSP binary in NGC 6397, PSR J1740−5340 is such a case. At $P_{\rm orb} = 35.5\,{\rm hr}$, it is in a radio-ejection phase and the companion has certainly not been captured recently in a stellar encounter: it is an evolved subgiant, as predicted by the secular evolution models (Burderi *et al.* 2002), and as confirmed by the CN cycled chemistry of the donor envelope, observed by Sabbi *et al.* (2003b) and predicted by Ergma & Sarna (2003). We suspect that PSR J1748−2446ad in Terzan 5 is also in a radio-ejection phase (Burderi *et al.* 2006), but the lack of information on the donor precludes a very secure interpretation. At $P_{\rm orb} = 26.3\,{\rm hr}$, again, the donor should be in an early subgiant stage, and have evolved very close to $P_{\rm bif}$.

Finally, the whole period distribution of binary MSPs in GCs is consistent with a very high probability of the onset of mass transfer being close to $P_{\rm bif}$. In fact, there is a large group having $P_{\rm orb}$ from 0.1 to 1 day, a range not covered at all by the 'standard' evolutions in Podsiadlowski *et al.* (2002), but which results easily from the range of initial periods between those leading to ultrashort period binaries and those above the bifurcation (Teodorescu 2005). In addition, there are several binary MSPs in GCs for which the white dwarf mass is very low ($0.18 - 0.20\,{\rm M}_{\odot}$), close to the minimum mass which can be formed by binary evolution (Burderi *et al.* 2002), indicating again evolution starting at a period slightly larger than $P_{\rm bif}$.

5. Population synthesis with dynamics

Ivanova and collaborators have developed a new simulation code to study the formation and retention of NSs in clusters, as well as the formation and evolution of all compact binaries in GCs. This code is described in Ivanova *et al.* (2005) and Ivanova *et al.* (2006). The method combines the binary population synthesis code `StarTrack` (Belczynski *et al.* 2002; Belczynski *et al.* 2007) and the `Fewbody` integrator for dynamical encounters (Fregeau *et al.* 2004). Compared to other numerical methods employed to study dense stellar systems, this method can deal with very large systems, up to several million stars, and with large fractions of primordial binaries, up to 100%, although the dynamical evolution of the cluster is not treated in a fully self-consistent manner.

In addition to the formation of NSs via core collapse, these simulations take into account NSs formed via electron-capture supernovae (ECS). When a degenerate ONeMg core reaches a mass $M_{\rm ecs} = 1.38\,{\rm M}_\odot$, its collapse is triggered by electron capture on ^{24}Mg and ^{20}Ne before neon and subsequent burnings start and, therefore, before the formation of an iron core (see, e.g., Nomoto 1984). The explosion energy of such an event is significantly lower than that inferred for core-collapse supernovae (Dessart *et al.* 2006), and therefore the associated natal kick velocities may be much lower. There are several possible situations when a degenerate ONeMg core can reach $M_{\rm ecs}$:

- During the evolution of single stars: if the initial core mass is less than that required for neon ignition, $1.37\,{\rm M}_\odot$, the core becomes strongly degenerate. Through the continuing He shell burning, this core grows to $M_{\rm ecs}$. The maximum initial mass of a single star of solar metallicity that leads to the formation of such a core is $8.26\,{\rm M}_\odot$, and the minimum mass is $7.66\,{\rm M}_\odot$. This mass range becomes 6.3 to 6.9 ${\rm M}_\odot$ for single stars with a lower GC metallicity $Z = 0.001$. The range of progenitor masses for which an ECS can occur depends also on the mass transfer history of the star and therefore can be different in binary stars, making possible for more massive progenitors to collapse via ECS (Podsiadlowski *et al.* 2004).
- As a result of accretion onto a degenerate ONeMg white dwarf (WD) in a binary: accretion-induced collapse (AIC). In this case, a massive ONeMg WD steadily accumulates mass until it reaches the critical mass $M_{\rm ecs}$.
- When the total mass of coalescing WDs exceeds $M_{\rm ecs}$: merger-induced collapse (MIC). The product of the merger, a fast rotating WD, can significantly exceed the Chandrasekhar limit before the central density becomes high enough for electron captures on ^{24}Mg and ^{20}Ne to occur, and therefore more massive NSs can be formed through this channel (Dessart *et al.* 2006).

Both metal-poor ($Z = 0.001$) and metal-rich ($Z = 0.02$) stellar populations have been studied by Ivanova *et al.* (2005) and Ivanova *et al.* (2006), who find that the production of NSs via core-collapse SNe (CC NSs) is 20% lower in the metal-rich population than in the metal-poor population. In a typical cluster (with total mass $2 \times 10^5\,{\rm M}_\odot$, age $\sim 10\,$Gyr, 1-dimensional velocity dispersion $\sigma = 10\,$km/s and central escape velocity 40 km/s), about 3000 CC NSs can be produced, but less than 10 will typically be retained in the cluster.

ECS in single stars in a metal-rich population are produced from stars of higher masses, but the mass range is the same as in metal-poor populations. As a result, the number of ECS from the population of single stars in the metal-rich case is 30% smaller than in the metal-poor population, in complete agreement with the adopted initial mass function (IMF). The total number of NSs produced via this channel is several hundreds (and depends on the initial binary fraction), but the number of retained NSs is higher than in the core-collapse case: about 150 NSs in a typical cluster. The binarity smoothes the mass range where ECS could occur, and there are fewer differences between the production of

NSs via ECS in binary populations of different metallicity. The number of retained NSs produced via AIC and MIC is comparable to the number of ECS, about 100 in a typical metal-poor cluster. Overall, one finds that, if a metal-rich GC has the same IMF and initial binary properties as a metal-poor GC, it will contain 30 – 40% fewer NSs.

These simulations can also be used to examine the spatial distribution of pulsars and NSs in clusters, although the present method only distinguishes between a central 'core' (where all interactions are assumed to take place) and an outer 'halo'. For a typical half-mass relaxation time $t_{\rm rh} = 10^9$ yr, about 50% of all NSs and 75% of pulsars should be located in the core, and for a longer $t_{\rm rh} = 3 \times 10^9$ yr, these fractions decrease to about 25% and 50% respectively. Such predicted spatial distributions are in good agreement with observations of pulsars in many GCs (Camilo & Rasio 2005).

Ivanova *et al.* analyzed three main mechanisms for the formation of close binaries with NSs: tidal captures, physical collisions with giants, and binary exchanges. Very few primordial binaries with a NS can survive, except for those that were formed via AIC. Typically $\sim 3\%$ of all NSs in a metal-poor GC can form a binary via physical collision and $\sim 2\%$ via tidal captures, while 40% of dynamically formed binary systems will start mass transfer (MT) in a Hubble time. These number are slightly higher in the case of a metal-rich cluster, and can be as much as two times higher in a cluster of the same metallicity but with a lower velocity dispersion, down to $\sigma = 5\,{\rm km/s}$. The binary exchange channel is more important for binary formation, as up to 50% of all NSs will be at some point members of binary systems, but only about 8% of these systems will start MT.

Overall, taking into account the formation rates of MT binaries with a NS and a MS star, and the duration of the MT phase, the probability that a cluster contains a NS-MS LMXB is almost unity, although most of them will be in quiescence. For NS-WD binaries, the probability is $\sim 50\%$, but only a few percent of these will be in the bright phase, when $L_{\rm x} > 10^{36}$ erg/s. More LMXBs per NS are formed in metal-rich clusters, but since fewer NSs are produced and retained, no significant difference in the resulting LMXB formation rate is found.

Finally, we note that if all ECS channels indeed work, too many NSs and pulsars (more than observed) are produced in these models. Therefore, either one or more of the ECS channels (standard ECS, AIC or MIC) does not work, or they have smaller allowed physical ranges where they can occur, or the kick associated with ECS could be larger. Our current understanding of stellar evolution and NS formation and retention in GCs of different metallicities, coupled with the dynamical formation of mass-transferring binaries with NSs, cannot explain the statistically significant overabundance of LMXBs in more metal-rich clusters. Instead, different physics for the MT with different metallicities or different IMFs are required (Ivanova 2006).

6. Green Bank observations of millisecond pulsars in clusters

Since its first scientific observations five years ago, the Green Bank Telescope (GBT), has uncovered at least 60 GC pulsars, almost doubling the total number known†. Almost all of these systems are MSPs, and the majority are members of binaries. Incredibly, 30 of these new MSPs, including many strange systems, are in the dense and massive bulge GC Terzan 5 (with a total of 33), while another 10 are in the bulge cluster M 28 (for a total of 11). Other clusters with new pulsars (and the numbers new/total) are M 30

† There are at least 133 known GC pulsars, of which 129 are currently listed in Paulo C.C. Freire's catalog at `<http://www2.naic.edu/~pfreire/GCpsr.html>`. For a recent review of GC pulsars, see Camilo & Rasio (2005).

(2/2), M 62 (3/6), NGC 6440 (5/6), NGC 6441 (3/4), NGC 6522 (2/3), NGC 6544 (1/2), and NGC 6624 (4/5).

Most of the GCs with new pulsars are in the Galactic bulge, with large columns of ionized gas along the lines of sight. Almost all of the new pulsars have been found using wide bandwidth (600 MHz) observations centered near 2 GHz, a relatively high radio frequency for pulsar searches. Such observations are much less affected by interstellar dispersion and scattering than traditional searches (at 1.4 GHz or 430 MHz), resulting in greatly improved search sensitivities, particularly for the fastest MSPs.

Pulsar timing solutions using the GBT now exist for almost 50 of the new MSPs, as well as for an interesting binary MSP found with the GMRT (NGC 1851A; Freire *et al.* 2004). These timing solutions provide precise spin and orbital parameters, which are useful for probing many aspects of NS physics, binary evolution, and cluster dynamics. In addition, the highly precise astrometric positions (with typical errors of $\lesssim 0.1''$) allow additional probes of cluster dynamics and possible identification of pulsar companions at optical or X-ray wavelengths.

Using the ensemble of 32 Terzan 5 (Ter 5) MSP timing solutions, the positions of the pulsars with respect to the cluster center allow a statistical measurement of the average NS mass ($\sim 1.35-1.4\,M_\odot$). Similarly, the pulsar positions and dispersion measures (DMs; the integrated electron column density along the line of sight to the pulsar) provide a unique probe of interstellar medium electron density variations over $0.2-2$ pc scales and show that they are not inconsistent with Kolmogorov turbulence. Several of the brighter Ter 5 pulsars with timing solutions encompassing older Parkes observations are beginning to show evidence for proper motions. Average proper motion values from GC MSPs may provide the best proper motion measurements of highly reddened clusters like Ter 5. Finally, comparisons of the spin-period and luminosity distributions of the 33 pulsars in Ter 5 and 22 in 47 Tuc show that they are significantly different, and hence may be related to the properties and dynamics of the GCs.

Among the interesting new pulsars are Ter 5E, a 2.2 ms pulsar in a 60 d orbital period (the 2$^{\rm nd}$ longest of any cluster MSP, the longest is in the low-density cluster M53); Ter 5N, an 8 ms pulsar with a likely CO white dwarf companion, the first known in a GC; five 'black-widow'-like systems (M 62E, Ter 5O, Ter 5ae, M 28G, and M 28J) with few-hour circular orbits and $\sim 10-40\,M_{\rm Jupiter}$ companions; and at least seven eclipsing binaries (M 30A, Ter 5O, Ter 5P, Ter 5ad, NGC 6440D, NGC 6624F, and M 28H).

Several of the above systems hint at production mechanisms involving stellar interactions. But there are two other classes of very interesting pulsars that are almost certainly produced via exchange interactions: pulsar-'main-sequence' binaries, and highly eccentric ($e > 0.25$) binaries. Recent 2-GHz GBT searches have uncovered at least two of the former, and (amazingly) nine of the latter.

Ter 5ad is the fastest MSP known (P=1.396 ms; Hessels *et al.* 2006) and finally beats the 23-yr-old 'speed' record established by the first MSP discovered Backer *et al.* (1982). Ter 5P is the 5$^{\rm th}$ fastest MSP known. Both systems are in circular binaries ($P_{\rm orb} = 26$ hr for Ter 5ad and 8.7 hr for Ter 5P) with companions of mass $\gtrsim 0.14\,M_\odot$ for Ter 5ad and $\gtrsim 0.36\,M_\odot$ for Ter 5P. Both systems are eclipsed for $\sim 40\%$ of their orbit, yet on some occasions the eclipses appear to be irregular (of different duration or possibly of variable depth). These systems appear to be very similar to the fascinating MSP J1740−5340 in NGC 6397 (D'Amico *et al.* 2001b).

Timing solutions for Ter 5ad and P associate both pulsars with hard X-ray point sources detected in a *Chandra* observation of the cluster. In addition, both systems exhibit extremely large orbital period derivatives ($\dot{P}_{\rm orb} \simeq 7 \times 10^{-9}$) and numerous (4 or more) higher-order period derivatives, likely due to tidal interactions with the companion stars. Upcoming *HST*-ACS and near-IR VLT adaptive-optics observations may identify 'bloated' companions, as for PSR J1740−5340 (Ferraro *et al.* (2001)).

These two systems raise many questions: Why has it taken so long to find a new 'fastest MSP'? Do faster systems exist? Why does Ter 5 have 5 of the 10 fastest MSPs known in the Galaxy and *the* 5 fastest-spinning pulsars known in the GC system? Can the large orbital period variations constrain tidal circularization theory? Is the X-ray emission from magnetospheric pulsations, an intra-binary shock, or some combination of both?

The second class of exchange products are the highly eccentric binaries. M 15C, a double NS system, was the first highly-eccentric binary discovered in a GC (Anderson *et al.* 1990), but it took ten years to find the next, NGC 6441A (Possenti *et al.* 2001). Soon afterwards, M 30B (found with the GBT; Ransom *et al.* 2004) and then NGC 1851A (currently being timed with the GBT; Freire *et al.* 2004) were detected. The recent GBT 2-GHz surveys, though, have uncovered *nine* additional highly-eccentric binaries: six in Terzan 5 (I, J, Q, U, X, and Z), two in M28 (C and D), and one in NGC 6440 (B).

Eccentric MSP binaries systems can be important probes of NS physics, as they provide a way to constrain (or even directly measure) the masses of fully-recycled pulsars. Given the angular reference that an ellipse provides, pulsar timing can easily measure the orbital advance of periastron. If the companion star is compact, the advance is dominated by general relativistic effects and determines the total system mass ($M_{\rm tot}$). The amount of mass required to spin-up a MSP is currently unknown: the double NS systems with precisely determined masses are only partially recycled, and there are only a handful of mass measurements for fully recycled pulsars ((Stairs *et al.* 2004; Lattimer & Prakash 2004). Since the recycling scenario in general creates binary MSPs in circular orbits (due to tidal circularization during the accretion phase), these systems are *only* produced during interactions in dense stellar systems (Rasio & Heggie 1995).

Timing solutions from the GBT are now available for all of the known highly eccentric binaries except for M 15C (although, see Jacoby *et al.* 2006), M 30B (which has only been detected once, likely due to strong scintillation effects), and Ter 5U (a very strange system with $P_{\rm orb} = 1.8$ d, $e \simeq 0.27$, and a minimum companion mass of only 25 $M_{\rm Jupiter}$). From these 10 timing solutions, the advance of periastron is highly significant in 9, indicating total system masses between 1.6 - 2.5 $M_\odot$. Such values are expected for recycled pulsars (with the NS mass being 1.4 - 2 $M_\odot$) with white-dwarf-like companions, indicating that the periastron advance is likely dominated by general relativity and not by classical effects.

Two of these systems (Ter 5I and J) appear to have 'massive' NSs ($\sim 1.7\,M_\odot$), which constrain the equation of state (EOS) of matter at nuclear densities, possibly ruling out very soft EOSs or those with strange-matter components (Ransom *et al.* 2005). Over the next couple of years, measurements of the relativistic γ parameter for Ter 5I and possibly of the Shapiro delay for M 28C are likely. These measurements, if the companions are white dwarfs, will provide accurate masses for the NSs. Several of the other eccentric binaries are interesting as well: Does Ter 5Q (with $P = 2.8$ ms and $P_{\rm orb} = 30$ d) have a NS companion? How do you create a highly eccentric binary like Ter 5U with a 25 – 30 $M_{\rm Jupiter}$ companion? Why was Ter 5Z not ejected from the core when the interaction that made it eccentric occurred? Why does M28D (with $P = 79.8$ ms and $P_{\rm orb} = 30$ d) appear to be a 'young' pulsar (characteristic age $t_c \simeq 1 \times 10^7$ yr)? Was it really formed (and possibly partially-recycled) only recently?

Timing observations are ongoing for most of the new pulsars mentioned here, and will continue to refine known parameters, to allow searches for planetary companions, to measure secular effects possibly due to unseen companions or stellar encounters, and likely to determine the proper motions of the clusters. In addition, searches of all timing data and GBT observations of other clusters are underway. We fully expect the GBT to uncover many more GC MSPs, including some new surprises, in the coming years.

7. Parkes observations of radio pulsars in clusters

The Parkes Globular Cluster Pulsar Search is a project started in the mid 1990s as a side search project of the Parkes Multibeam Pulsar Survey. Observations with the Parkes radio telescope have been extensively performed with the central beam of the multibeam receiver and the collected data have been processed with dedicated algorithms. This project has so far led to the discovery of twelve new MSPs in six GCs for which associated pulsars were previously unknown. This section summarizes some recent results obtained from timing these sources.

7.1. *NGC 6266*

The cluster NGC 6266 hosts six MSPs. The first three of them, PSR J1701−3006A, B and C, have been discovered in the framework of the Parkes Globular Cluster Pulsar Search by D'Amico *et al.* (2001a), while the other three have been discovered with further GBT observations by Jacoby *et al.* (2002).

All six pulsars in this cluster are members of binary systems (D'Amico *et al.* 2001a; Possenti *et al.* 2003; Jacoby *et al.* 2002). This unusual occurrence is very unlikely to be due to chance. It is possible that the absence of isolated pulsars in this cluster is related to the peculiar dynamical state of the cluster that lowers the rate at which binaries are disrupted via dynamical encounters.

A peculiar object in this cluster is the MSP PSR J1701−3006B (Possenti *et al.* 2003). It is in the emerging family of eclipsing MSPs with relatively massive companions. Several observations of this pulsars as it transits at the superior conjunction show distortions in the signal that can be ascribed to the pulsar motion through the companion wind matter. The ablation timescale for the companion is compatible with the pulsar's age only if less than 10% of the released mass is ionized, which appears unlikely. The alternate possibility is that mass loss is due to nuclear evolution of the companion, analogous to the case of the eclipsing system in NGC 6397 (Possenti *et al.* 2003).

7.2. *NGC 6397*

PSR J1740−5340 is a peculiar MSP in NGC 6397 (D'Amico *et al.* 2001a; D'Amico *et al.* 2001b). It is a binary source, whose radio signal suffers eclipses as the pulsar approaches superior conjunction (D'Amico *et al.* 2001b). The extent of the eclipses strongly depends on the observing frequency. At 1.4 GHz eclipses last up to 40% of the orbit, and signal distortion is observed at all orbital phases. Such distortions become less prominent at an observing frequency of 2.3 GHz and become nearly absent at 3.0 GHz.

Optical observations of the companion have revealed several phenomena that occur as the pulsar emission interacts with the companion surface:

(*i*) The study of Hα lines indicates that matter is swept away in a cometary tail by a radio ejection mechanism (Sabbi *et al.* 2003a);

(*ii*) The presence of He lines in absorption may indicate that hot barbecue-like strips on the companion surface are heated by a highly anisotropic pulsar flux (Sabbi *et al.* 2003b);

(*iii*) The signature of an enhanced lithium abundance on the companion surface may be perhaps ascribed to lithium production from nuclear reactions triggered by accelerated particles flowing from the pulsar (Sabbi *et al.* 2003b).

The strong distortion suffered by the radio signal from the pulsar makes it very difficult to obtain a fully coherent timing solution for the pulses. This is illustrated by a recent optical determination of the companion position (Bassa & Stappers 2004), which is inconsistent with the previous determination obtained from the pulsar timing (D'Amico *et al.* 2001b). The measurement of the companion position by Bassa & Stappers (2004) allowed the determination of a new timing solution (Possenti *et al.* 2005), which will be updated with timing observations at 3.0 GHz.

7.3. *NGC 6441*

The cluster NGC 6441 hosts PSR J1750−3703 (D'Amico *et al.* 2001a), a binary pulsar in a highly eccentric orbit ($e = 0.712$) with a relatively massive companion. Since a data span of about five years is now available, the periastron advance for this system is actually measured with a precision of about 20 σ (Possenti *et al.* 2006), which gives a total mass for this binary $M_{\rm tot} = 2.20 \pm 0.17\,{\rm M}_\odot$. This result can be combined with the mass function for this system and the minimum measured mass for a NS to obtain a range for the mass of the companion $0.6\,{\rm M}_\odot \leqslant M_{\rm C} \leqslant 1.17\,{\rm M}_\odot$ (Possenti *et al.* 2006). This makes it unlikely that the companion is another NS.

7.4. *NGC 6752*

NGC 6752 contains five known MSPs. The discovery of PSR J1910−5959A as a binary pulsar (D'Amico *et al.* 2001a) allowed the subsequent discovery of four more isolated pulsars (D'Amico *et al.* 2002).

All five pulsars in this cluster show peculiar features. PSR J1910−5959B and E, located within a few arcseconds from the cluster center, show a large negative value for the spin period derivative (D'Amico *et al.* 2002). These negative values are ascribed to the motion of these objects inside the cluster potential well. PSR J1910−5959D is also located close to the cluster core (D'Amico *et al.* 2002). Its spin period derivative is positive and of the same order of magnitude as the values for PSR J1910−5959B and PSR J1910−5959E, implying that, for this pulsar as well, the spin period derivative is affected by the cluster potential (D'Amico *et al.* 2002). These measurements allow us to investigate the mass-to-light ratio in the central region of the cluster. A lower limit $M/L_V \geqslant 5.5\,{\rm M}_\odot/{\rm L}_\odot$ has been obtained by Ferraro *et al.* (2003). Such a high value indicates the presence of a large number of low-luminosity objects in the cluster core.

PSR J1910−5959A and C are located in the outskirts of the cluster, namely $\theta_{\perp,{\rm PSR\,A}} = 6.3'$ and $\theta_{\perp,{\rm PSR\,C}} = 2.7'$ (D'Amico *et al.* 2001a; D'Amico *et al.* 2002; Corongiu *et al.* 2006), values that put these two objects in first and second place, respectively, among GC pulsars that show a large offset from the cluster center. These unusual positions have been investigated in detail by Colpi *et al.* (2002) and Colpi *et al.* (2003). The most probable explanation invokes the ejection of these objects from the cluster core by dynamical interactions with a central massive object that may be either a single massive black hole or a binary black hole of intermediate mass (Colpi *et al.* 2003).

The recent measurement of proper motions for the pulsars PSR J1910−5959A and PSR J1910−5959C (Corongiu *et al.* 2006) shows that they are compatible with each other, but they are not in agreement with the proper motion of the cluster as determined from optical observations. Further observations of this cluster will soon allow us to determine the proper motion of the pulsars in the cluster core (certainly belonging to the cluster) and the comparison between these proper motions and those of pulsars A and C will then establish whether the more distant pulsars are truly associated with the cluster.

8. *Chandra* observations of X-ray sources in clusters

The *Chandra X-ray Observatory* has provided fundamental new insights into the nature of faint ($L_X \simeq 10^{30-34}$ ergs/s) GC X-ray sources, through its superb spatial resolution and moderate spectral resolution. See Verbunt & Lewin (2006) for a fuller (but dated) review, and the next section for some *XMM-Newton* results. The best-studied cluster (and the main focus in this section) is 47 Tucanae (47 Tuc), which has been well observed with *Chandra* (detecting 300 X-ray sources), *HST* in the optical and UV, and Parkes for pulsar timing observations (detecting 22 pulsars; see Sec. 7).

8.1. *Low-mass X-ray binaries*

The bright sources in GCs have long been known to be accreting neutron stars; X-ray bursts have now been detected from all Galactic GCs hosting luminous X-ray sources (in't Zand *et al.* 2003). *Chandra*'s resolution resolved a longstanding puzzle about the M 15 LMXB: the optically identified LMXB in M 15 is seen edge-on, not allowing direct view of the accreting (likely) neutron star; and yet X-ray bursts from a neutron star surface have been seen from M 15. This puzzle is resolved by the identification of a second LMXB in M 15 (White & Angelini 2001). This second LMXB has a period of 22.6 min (Dieball *et al.* 2005); this is the third neutron star accreting from a white dwarf (an ultracompact XRB; see Sec. 4) known in a GC.

8.2. *Transient LMXBs*

Chandra has allowed the identification of the quiescent counterparts to three transient LMXBs. Quiescent LMXBs, at $L_X \simeq 10^{32-34}$ ergs/s, tend to show soft spectra, dominated by a $\sim$0.3 keV blackbody-like spectrum. This component can be fit with a neutron star hydrogen atmosphere model, with implied radius of 10 - 15 km (Rutledge *et al.* 2002a). This radiation is commonly thought to be produced by heating of the core during accretion, which will slowly leak out over 10^4 yr (Brown *et al.* 1998). A second harder component (of unknown origin) is often required above 2 keV, typically fit by a power-law with photon index 1 - 2. Two of the transient LMXBs observed in quiescence fit this model; those in NGC 6440 (in't Zand *et al.* 2001) and Terzan 1 (Cackett *et al.* 2006). In contrast, the spectrum of the transient in Terzan 5 requires only a power-law component, with a photon index of $1.8^{+0.5}_{-0.4}$, indicating that quiescent LMXBs may also have relatively hard spectra (Wijnands *et al.* 2005).

8.3. *Quiescent LMXBs*

In addition to the known quiescent counterparts of transient LMXBs, additional X-ray sources are seen in clusters with spectra and luminosities characteristic of quiescent LMXBs. Spectral fitting of the brightest of these with neutron star atmosphere models gives inferred radii consistent with 10 - 12 km (Rutledge *et al.* 2002b). Two such systems in the cluster 47 Tuc show regular eclipses at periods of 8.7 and 3.1 hr (Heinke *et al.* 2005a), and two systems have faint optical counterparts (Haggard *et al.* 2004). Pooley *et al.* (2003) and Heinke *et al.* (2003) showed that the numbers of quiescent LMXBs in different clusters scaled with the stellar interaction rate in those clusters, implying they are formed dynamically.

If the distance to the GC is reasonably well-known, then it is possible to constrain the radius (or a combination of radius and mass) of the glowing neutron star through fits to hydrogen atmosphere models. This has the potential to improve our understanding of the composition of neutron star interiors, and thus the behavior of matter at high density. In units of $R_\infty (= R * (1+z))$, constraints have been placed on the neutron star in ω Cen ($R_\infty = 14.3 \pm 2.1$ km, Rutledge *et al.* 2002b), and on X7 in 47 Tuc ($R_\infty = 18.3^{+3.8}_{-1.2}$ km,

Heinke *et al.* 2006b); see below for *XMM-Newton* results. Perhaps the largest remaining source of uncertainty in these calculations is the distance to the GCs; recent authoritative determinations of the distance to 47 Tuc by the subdwarf main-sequence fitting method and direct geometry give results which differ by 20% (Gratton *et al.* 2003; McLaughlin *et al.* 2006).

8.4. *Cataclysmic Variables*

Optical counterpart searches using *HST* have identified 22 cataclysmic variables (CVs) in 47 Tuc through blue optical/UV colors and variability, eight of which have secure orbital periods (Edmonds *et al.* 2003a). Ten CVs have also been identified in NGC 6752 (Pooley *et al.* 2002) and nine in NGC 6397 (Grindlay *et al.* 2001b). These CVs have blue $U-V$ colors, but $V-I$ colors that are on or near the main sequence. This indicates that the secondaries dominate the optical light, which is in agreement with the identification of ellipsoidal variations in several of these systems. Comparison of these and other cluster CVs with Galactic CVs shows that cluster CVs have fainter accretion disks than Galactic CVs with similar periods (Edmonds *et al.* 2003b). This suggests that cluster CVs have relatively low mass transfer rates. However, the lack of dwarf nova outbursts from cluster CVs (Shara *et al.* 1996) may be an indication that cluster CVs tend to be strongly magnetic (e.g., Dobrotka *et al.* 2006).

CVs may be formed in GCs either dynamically or from primordial binaries. Several *Chandra* observational studies (Pooley *et al.* 2003; Heinke *et al.* 2003; Pooley & Hut 2006; Heinke *et al.* 2006b; Kong *et al.* 2006) as well as population synthesis studies (Ivanova *et al.* 2006) point to contributions by both mechanisms to the existing CV population in clusters.

8.5. *Active binaries*

Numerous chromospherically active binaries (mostly close main-sequence binaries) have been identified in several GCs. Sixty have been identified with *Chandra* sources in 47 Tuc alone (Heinke *et al.* 2005b). Bassa *et al.* (2004) and Kong *et al.* (2006) have recently shown that the population of active binaries in clusters, unlike CVs and LMXBs, is produced from primordial binaries; in the densest clusters (such as NGC 6397), these binaries have been largely destroyed (Cool & Bolton 2002).

8.6. *Radio millisecond pulsars*

MSPs have been detected in X-ray in several GCs (e.g., Bassa *et al.* 2004). The deep observations of 47 Tuc have detected all 19 MSPs with known positions (Bogdanov *et al.* 2006), showing that in most cases their X-ray spectra are dominated by thermal emission from their hot polar caps. Comparison of the X-ray spectra of unidentified sources in 47 Tuc with known MSPs and active binaries reveals that the majority of the unknown sources are active binaries, and constrains the total number of MSPs in 47 Tuc to < 60, most likely $\leqslant 30$ (Heinke *et al.* 2005b). This helps to resolve a suggested discrepancy between the birthrates of LMXBs and MSPs in Galactic GCs.

A few of the MSPs in 47 Tuc (and elsewhere) show harder X-ray spectra, suggestive of nonthermal synchrotron or shock emission. One of these, 47 Tuc-W, shows long X-ray eclipses, indicating the X-rays are produced in a shock near the companion from matter that continues to overflow the companion's Roche lobe (Bogdanov *et al.* 2005) – making this a 'missing link' between LMXBs and MSPs. These discoveries have greatly improved our understanding of the evolution of LMXBs into MSPs in clusters.

9. *XMM-Newton* observations of X-ray sources in clusters

Observations of Galactic GC faint X-ray sources made with the two X-ray satellites *XMM-Newton* (e.g., Webb *et al.* 2006; Webb *et al.* 2004; Gendre *et al.* 2003a; Gendre *et al.* 2003b; Webb *et al.* 2002) and *Chandra* (Heinke *et al.* 2006a; Heinke *et al.* 2006b; Pooley *et al.* 2003; see Sec. 8) have revealed that 25 are neutron star XRBs. *XMM-Newton* spectra of these systems are of sufficiently high quality, even with only 30 ks observations, to well constrain the mass and radius of the neutron star, using neutron star atmosphere models (e.g., Zavlin *et al.* 1996; Heinke *et al.* 2006a; Heinke *et al.* 2006b) and taking advantage of the fact that their distances and interstellar absorptions are well constrained due to their situation in a GC (Servillat *et al.* in preparation; Gendre *et al.* 2003a; Gendre *et al.* 2003b). The masses and radii are essential for constraining the (poorly known) equation of state of the nuclear matter in these very compact stars.

Gendre *et al.* (2003b), Pooley *et al.* (2003) and Heinke *et al.* (2003) have also used the GC observations of faint X-ray sources, coupled with the result that the bright X-ray sources ($L_x > 10^{36}\,\mathrm{erg\,s^{-1}}$; Hertz & Grindlay 1983) are also neutron star XRBs (see Verbunt & Hut 1987 and references therein), to confirm, through observations, the theory that these objects are formed mainly through dynamical encounters. This implies a total population of approximately 100 neutron star XRBs distributed throughout the 151 Galactic GCs (Pooley *et al.* (2003)). This population is wholly insufficient to slow down the inevitable core collapse of these self-gravitating stellar clusters if the energy liberated by binaries interacting with other cluster stars is indeed the internal energy source necessary to counter the tendency of clusters to collapse (see Hut *et al.* 1992 for a review).

Cataclysmic variables (CVs) exist in much greater numbers in GCs. Indeed Di Stefano & Rappaport (1994) predict of the order one hundred CVs in a single Galactic GC. This prediction is born out by observations, for example more than 30 CVs have been detected in 47 Tuc using X-ray observations (Heinke *et al.* 2005a; Heinke *et al.* 2005b) and approximately 60 candidate CVs have been identified in NGC 2808 using UV observations (Dieball *et al.* 2005). Thus, although we do not yet know the whole population size of CVs in the GCs observed (unlike for the brighter, soft neutron stars) with which to determine their formation mechanisms and numbers, it is apparent that they exist in large numbers and thus it is possible that they are important to the cluster's fate.

With more and more cataclysmic variables identified in Galactic GCs, one striking and unexplained difference has become clear between cluster CVs and field CVs. Cluster CVs show a distinct lack of outbursts (characterized by a steep rise in the flux by several orders of magnitude) compared to field CVs. Due to the proximity of the white dwarf and its companion in a CV, material is accreted from the companion star and stored in the accretion disk around the white dwarf whilst it loses sufficient angular momentum to fall onto the compact object. Outbursts are believed to occur when too much material builds up in the disc, increasing both the density and the temperature, until the hydrogen ionizes and the viscosity increases sufficiently for the material to fall onto the white dwarf (Osaki 1974; Meyer & Meyer-Hofmeister 1981; Bath & Pringle 1981. Many types of field CVs show such outbursts every few weeks to months. However, only very few GC outbursts have been observed (e.g., Paresce & de Marchi 1994; Shara *et al.* 1996; Shara *et al.* 1987) and it is unclear why this should be.

It was originally suggested that cluster CVs may be mainly magnetic (see the five CVs in Grindlay 1999). Magnetic CVs have accretion discs that are either partially or totally disrupted by the strong white dwarf magnetic fields and these two types are known as intermediate polars and polars, respectively. Material is channelled along the field lines onto the white dwarf, although in the case of intermediate polars, a truncated disk can exist and these systems can undergo a limited number of outbursts (e.g., Norton & Watson 1989). Recently it has been proposed that it may not simply be the magnetic field that is responsible for the lack of outbursts. Dobrotka *et al.* (2006) suggest that it may be due to a combination of low mass transfer rates ($\lesssim 10^{14-15}\,\mathrm{g\,s^{-1}}$) and moderately strong white dwarf magnetic moments ($\gtrsim 10^{30}\,\mathrm{G\,cm^3}$) which could stabilize the CV discs in globular clusters and thus prevent most of them from experiencing *frequent outbursts.* This result suggests that the brightest globular cluster CVs in Ter 5s should be intermediate polars. Ivanova *et al.* (2006) have also proposed that the lack of outbursts is due to higher white dwarf masses (higher mean masses are observed amongst strongly magnetic isolated white dwarfs; Wickramasinghe & Ferrario 2000). This is likely to be due to the difference in the formation mechanisms of GC and field CVs, since a substantial fraction of cluster CVs are likely to be formed through encounters, rather than from their primordial binaries (Ivanova *et al.* 2006).

Intermediate polars show modulation on the spin period (typically $\sim 10^{2-3}$ s) of the accreting white dwarf which can be detected through Fourier analysis. For example, Parker *et al.* (2005) showed that 70% of the intermediate polars that were observed with *ASCA* and *RXTE* showed this modulation. Thanks to the sensitivity of the *XMM-Newton* satellite, observations made with this observatory of the cluster NGC 2808 revealed that the brightest CV in this cluster shows evidence for a modulation with a 430 s period (Servillat *et al.*, in preparation). This is likely to be the modulation on the spin period, supporting an intermediate polar identification. Low-resolution spectra of the brightest CV (candidate) in the cluster M 22 (Webb *et al.* 2004; Webb *et al.* in preparation) also show some evidence for the He 4686 Å line in emission, indicative of a magnetic white dwarf (e.g., Szkody *et al.* 2005). As the CV has already been observed to outburst (Anderson *et al.* 2003; Bond *et al.* 2005; Pietrukowicz *et al.* 2005), it would indicate that this source is also an intermediate polar, again supporting the idea that cluster CVs have moderate magnetic field strengths, in part responsible for their lack of outbursts.

We now turn briefly to possible formation mechanisms for cluster CVs. It is now believed that there are two populations of CVs in GCs, those formed dynamically (as the neutron star LMXBs), thought to be located in the dense cluster cores, and those that have evolved from a primordial binary without undergoing any significant encounter. This latter population may reside outside the cluster core (Davies 1997), where the stellar density is much lower than near the center. Naturally we expect that the more concentrated GCs, which have higher core densities, have higher encounter rates, thus increasing the number of CVs formed through encounters. In addition, the timescales of encounters between primordial binaries and single stars are shortened, thus decreasing the number of primordial CVs. The GCs that have been observed with *XMM-Newton* are particularly well adapted to searching for a primordial binary population, as they are low-density clusters, chosen to ensure that we can resolve all the X-ray sources (the angular resolution of the *XMM-Newton*-EPIC cameras is approximately 6″ FWHM of the PSF). In addition, *XMM-Newton*'s large collecting area ensures that there are enough photons for a full spectral study of about 20% of the sources detected, advantageous for identifying CVs using X-ray data alone. Several CVs have already been detected in the cores of GCs, like AKO 9 in 47 Tuc (Auriére *et al.* 1989), which Knigge *et al.* (2003) state was almost certainly formed dynamically, either via tidal capture or in a three-body encounter. Such dynamically formed CVs exist in other GCs, like ω Cen (e.g., Carson *et al.* 2000; Gendre *et al.* 2003a) and M 22 (Webb *et al.* 2004).

Several X-ray sources in GCs studied with *XMM-Newton* lie outside the half-mass radius and have X-ray luminosities, spectra, colors, and lightcurves that indicate they may be CVs. Recently, Pietrukowicz *et al.* (2005) confirmed, using optical photometry, that one of these X-ray sources (Webb *et al.* 2004) lying at 3.9 core radii from the centre of M 22 is indeed a CV. It is possible that this CV was formed from a primordial binary. Ivanova *et al.* (2006) predict that as many as 37% of the CVs in a cluster like 47 Tuc should be formed from the primordial binaries, thus one would expect an even greater percentage for a lower-concentration cluster such as M 22, supporting the primordial formation mechanism.

10. X-ray luminosity functions

Populations of X-ray sources have been discovered with *Chandra* in all kinds of galaxies. These populations provide a novel approach to study the evolution of XRBs. This section summarizes recent results from the study of the X-ray Luminosity Functions (XLFs) in these extragalactic populations.

XLFs (in either differential or cumulative form) provide a useful tool for characterizing and comparing XRB populations. Cumulative XLFs are typically described by functional slope(s), breaks, and normalization. Each of these parameters is potentially related to the formation and evolution of XRBs in a given stellar population: the distribution of luminosities (slope) has been found to be related to the age of the population (see below); breaks in the XLF are a possible indication of multiple or evolving XRB populations in the same galaxy; the normalization is a measure of the total number of XRBs. Grimm *et al.*(2002) first reported differences in the XLFs of different types of binaries, by deriving the 'young-short-lived' high-mass X-ray binary (HMXB) and 'old LMXB' XLFs for the Milky Way. They found that the HMXB XLF is well fitted by a single power-law, while the LMXB XLF may show both high- and low-luminosity breaks. More recent studies, based on *XMM-Newton* and *Chandra* observations, are in general agreement with these early results, but also show a more complex reality (see review by Fabbiano 2006).

Early studies of the integrated X-ray luminosity of star-forming galaxies pointed to a tight connection between the number of XRBs and star formation activity (e.g., Fabbiano *et al.* 1988; Fabbiano & Shapley 2002)). More recently, comparisons of XLFs have suggested a dependence of the normalization on the star formation rate (SRF) of the galaxy (Kilgard *et al.* 2002; Zezas & Fabbiano 2002; Grimm *et al.* 2003). Grimm *et al.* (2003), in particular, propose that all HMXB XLFs follow a similar cumulative slope of -0.6, and have normalization strictly proportional to the SFR. The XLFs of individual spiral galaxies (see Fabbiano 2006 and references therein) not always agree with this conclusion. However, deviations can be understood if the effect of XRB populations of different ages is considered. A particularly illuminating case is that of M83, a grand-design spiral with a nuclear starburst (Soria & Wu 2003). In M83, the XLF of the nuclear region is a power-law with a cumulative slope -0.7, reasonably consistent with Grimm *et al.* (2003); instead, the XLF of the outer disk is complex, suggesting several XRB populations. This XLF has a break (and becomes steeper) at luminosities above 8×10^{37} erg/s, suggesting an older XRB population than that of the nuclear region; below the break it follows a -0.6 power-law, but this power-law is interrupted by a dip. Soria & Wu suggest that this complex disk XLF may result from the mixing of an older disk XRB population mixing with a younger (but aging) population of spiral-arm sources.

The best-studied example of an extreme young HMXB population so far is given by the deep *Chandra* study of Antennae galaxies (Fabbiano *et al.* 2003), which led to the discovery of 120 X-ray sources down to a limiting luminosity near 2×10^{37} erg/s (Zezas *et al.* 2006), and including about 14 ultra-luminous X-ray sources (ULXs). The cumulative XLF is well fitted with a single power-law of slope of ~ -0.5. There is a small deviation from this power-law near 2×10^{38} erg/s, suggesting a NS Eddington limit effect (however this result is not statistically significant; Zezas *et al.* 2007).

Two papers, reporting the analysis of samples of early-type galaxies, give a good picture of the LMXB XLF at luminosities greater than a few 10^{37} erg/s. Kim & Fabbiano (2004) analyzed 14 E and S0 XLFs, corrected each individual XLF for incompleteness by means of simulations, and found that the corrected XLFs could be fitted (above 6×10^{37} erg/s) with similar steep power-laws. Given this similarity, the data were co-added resulting in a significantly higher signal-to-noise XLF, which shows a break, formally at 4.5×10^{38} erg/s, and marginally consistent with the NS Eddington limit. Gilfanov (2004) using a sample of early-type galaxies and spiral bulges reached a similar conclusion.

Gilfanov (2004) suggested that the global stellar mass of the galaxy or bulge is the driving factor for the normalization of the XLF (the total number of slowly-evolving LMXBs in a given stellar population). A somewhat different conclusion was reached by Kim & Fabbiano (2004; see also Kim *et al.* 2006). They reported a correlation between the total LMXB luminosity of a galaxy and the K-band (stellar) luminosity, in agreement with the stellar-mass-normalization link, but the scatter of this correlation is large, and while independent of total K-band luminosity, is correlated with the specific frequency of GCs in the galaxies (a link previously suggested by White *et al.* 2002). The conclusion, similar to that suggested for Galactic LMXBs (Clark 1975), is that GCs have a special effect on the formation and evolution of LMXBs. These results, however, do not exclude the possibility that evolution of native field binaries is also important, a point stressed by Irwin (2005).

LMXBs as short-lived ultra-compact binaries formed in GCs, have been discussed by Bildsten & Deloye (2004) (; see also Ivanova *et al.* 2005), who point out that their model can reproduce the observed XLF. Disruption of (or expulsion from) GCs could then give rise to LMXBs in the stellar field. Alternatively, LMXBs may form and evolve in the field (Verbunt & van den Heuvel 1995; Kalogera 1998, and references therein). Field source evolution models (e.g., Piro & Bildsten 2002) have not set constraints on the XLF, but predict that most high-luminosity LMXBs should be detached binaries with large unstable disk, and therefore recurrent transients. While more time monitoring observations are needed, transients are indeed detected in some cases (e.g., in NGC 5128; Kraft *et al.* 2001).

A large body of work has addressed the association of individual X-ray sources with GCs in E and S0 galaxies, and the properties of observed field and GC LMXBs (Sec. 11). A particularly relevant result from the recent paper by Kim *et al.* (2006), based on the analysis of the LMXB populations of 6 galaxies observed with *Chandra*, is worth mentioning here: no significant difference can be seen in the XLFs of LMXBs with and without a GC counterpart. Moreover, both XLFs extend to luminosities above 5×10^{38} erg/s, a regime where the accreting object is likely to be a black hole. Ivanova & Kalogera (2006) have pointed out that high-luminosity LMXBs are likely to be field transients populating the XLF in outburst, and that the XLF can be considered as the footprint of the black-hole mass function, with a differential slope of -2.5 and upper mass cut-off at $20\,M_\odot$. However, the field and cluster LMXB XLFs are similar and both extend to high luminosities. Are black-hole X-ray binaries therefore present in GCs, despite their very low expected formation probability (Kalogera *et al.* 2004)?

Very recent work (Kim *et al.* 2006), based on deep *Chandra* observations of two nearby elliptical galaxies with well studied old stellar populations, NGC 3379 and NGC 4278, is addressing the low-luminosity LMXB XLF, at luminosities below a few times 10^{37} erg/s, which are typical of the majority of LMXBs in the bulge of the Milky Way and M 31. The LMXB XLF of the Milky Way (Grimm *et al.* 2002) becomes flatter at these lower luminosities. Gilfanov (2004) suggested a significant 'universal' flattening below 5×10^{37} erg/s in the LMXB XLF. This flattening is suggested by a number of models (see, e.g., Bildsten & Deloye 2004; Pfahl *et al.* 2003). Kim *et al.* (2006) demonstrate that there is no universal low-luminosity flattening of the LMXB XLF. The XLF of NGC 4278 is very well fitted with a continuous power-law with cumulative slope -1, down to 1×10^{37} erg/s. The XLF of NGC 3379 (extending down to near 10^{36} erg/s) is also well represented by a similar power-law, but it presents a statistically marginal localized excess near 4×10^{37} erg/s.

11. Extragalactic globular cluster X-ray sources

Other galaxies, particularly ellipticals, have proved to be excellent grounds for learning about what kind of GCs produce bright X-ray sources. These galaxies have GC systems up to two orders of magnitude larger than the Milky Way's. Furthermore, they often show much more diversity than the Milky Way in terms of metallicities and ages of the clusters. On top of this all, there are dozens of galaxies within 20 Mpc, so even if one galaxy fails to provide a sufficiently large or diverse population of clusters for studying a particular effect, one can co-add many galaxies. Studies of elliptical galaxies have been especially fruitful, since the specific frequencies of GCs are larger in more massive, later type galaxies. Additionally, the GC samples are better understood in elliptical galaxies than in spiral galaxies because of the smoother field star backgrounds in elliptical galaxies are easier to subtract off than the knotty emission in spiral galaxies.

It was determined early in the *Chandra* era that about half of all X-ray sources in elliptical galaxies are in GCs (see, e.g., Sarazin *et al.* 2001 and Angelini *et al.* 2001 for the first few studies, and Kim *et al.* 2006 for an analysis of an ensemble of galaxies). This compares with matching fractions of 10% in the Milky Way (van Paradijs & van den Heuvel 1995) or $\sim 20\%$ in M 31 (Supper *et al.* 1997). The fraction was found to increase continuously through the Hubble tuning fork diagram from spirals to lenticulars to ellipticals to cD galaxies (Maccarone *et al.* 2004). It has been suggested that a substantial fraction of non-GC X-ray sources were originally formed in clusters (e.g., White *et al.* 2002), but recent work has shown both that the fraction of X-ray sources in clusters increases with specific frequency of GCs (Juett 2005), and that the ratio of X-ray to optical luminosity increases more slowly than linearly with the specific frequency (Irwin 2005).

One of the key areas of interest for extragalactic GC studies is the determination of which cluster properties most affect the likelihood that a cluster will contain an XRB. The most significant parameter is the cluster mass (Kundu *et al.* 2002), with several studies finding that the probability a cluster will be an X-ray source scales in a manner consistent with $M^{1.1 \pm 0.1}$ (Kundu *et al.* 2003; Jordan *et al.* 2004; Smits *et al.* 2006).

The normalization in the number of XRBs per unit stellar mass is still considerably larger than in galaxies' field populations, where the number of LMXBs seems to be linearly proportional to the stellar mass as well (Gilfanov 2004; Kim & Fabbiano 2004), so clearly the number of XRBs must be related to the stellar interaction rates in the clusters. Determining the stellar interaction rate requires an understanding of the radial profile of the cluster as well as its total mass, and this is rather difficult to determine for most GCs at the 10 – 16 Mpc distances of the best-studied elliptical galaxies. Kundu *et al.* (2002) found for NGC 4472 that the half-light radius of a cluster was a marginally significant parameter for predicting whether a GC would have an X-ray source. Jordan *et al.* (2004) attempted to fit King models to the GCs in M 87 and found no statistically significant difference between the predictive power of cluster mass and of inferred cluster collision rate for whether a cluster would contain an X-ray binary. There is only a weak correlation between cluster core radius and cluster half-light radius, except in the least concentrated clusters, and since the core radii of nearly all Milky Way GCs are too small to be resolved, even with the *Hubble Space Telescope*, at distances exceeding a few Mpc. It is thus not surprising that we cannot obtain any quantitative information about the relation between cluster collision rate and probability of containing an XRB by looking at Virgo Cluster elliptical galaxies.

It was found by Smits *et al.* (2006) that the observed $P(\text{LMXB}) \propto M^{1.1}$ is consistent with bimodal pulsar kick-velocity distributions (e.g. Arzoumanian *et al.* 2002; Brisken *et al.* 2003), but not with a single Maxwellian kick-velocity distribution around 200 km/s (e.g., Hobbs *et al.* 2005). Because binary evolution effects can produce low kick-velocity pulsars (e.g., Pfahl *et al.* 2002; Dewi *et al.* 2005) it is not clear whether the cluster XRB results have direct implications for the controversies concerning isolated pulsar velocity distributions.

Aside from mass, the other key parameter which helps determine whether a cluster will contain an X-ray source is its metallicity. There were some indications from the pre-*Chandra* era that this was the case, based on the Milky Way and M 31 (Grindlay 1999), but the strong correlation in spiral galaxies between metallicity and galactocentric radius, along with the relatively strong tidal forces in the centers of spiral galaxies, left some doubt about which was the underlying physical cause of the enhancement of X-ray sources in metal-rich bulge GCs. It has since been proven clearly, in numerous elliptical galaxies, that metallicity really is a strong predictor of whether a cluster will have an X-ray source (Kundu *et al.* 2002; Di Stefano *et al.* 2003 ; Jordan *et al.* 2004; Minniti *et al.* 2004; Xu *et al.* 2005; Posson-Brown *et al.* 2006; Chies-Santos *et al.* 2006).

Attempts have also been made to determine whether cluster ages affect X-ray binary production, especially in light of theoretical suggestions that there should be a peak in the X-ray source production rate at ages of about 5 Gyr (Davies & Hansen 1998). In NGC 4365, which has a substantial sub-population of intermediate-age clusters (Puzia *et al.* 2002; Larsen *et al.* 2003; Kundu *et al.* 2005), it is clear that there is an effect of metallicity on the probability a GC will be an X-ray source (Kundu *et al.* 2003); roughly the same effect of metallicity is seen in NGC 3115 (Kundu *et al.* 2003), which has only old GCs (Puzia *et al.* 2002). This argues in favor of the idea that the metallicity effect is causual.

Two viable possibilites have been suggested for this effect. One is irradiation-induced stellar winds (Maccarone *et al.* 2004), which should be stronger in low-metallicity environments (Iben *et al.* 1997) since the energy deposited by irradiation in a low metallicity star cannot easily be dissipated by line cooling. As a result, the metal-poor stars will lose much of their mass to the interstellar medium, rather than to the accreting star, yielding effectively lower duty cycles as bright sources. The other model depends on the smaller convection zones of metal-poor stars compared with metal-rich stars (Ivanova 2006). This leads to reduced cross sections for formation of X-ray binaries through tidal capture, and to less efficient magnetic braking, and hence lower accretion rates. A distinguishing characteristic of the models is that the irradiation wind model should leave behind absorbing material which will leave an absorption signature in X-ray spectra, while the convection zone model should not. Spectra of M 31 clusters in the 0.1 – 2.4 keV *ROSAT* band are harder in the more metal-poor clusters (Irwin & Bregman 1999), while *Chandra* spectra showed no correlation between X-ray spectrum and cluster metallicity (Kim *et al.* 2006 Kim *et al.* 2006). It is thus not clear whether the M 31 results are a statistical fluke, or the *Chandra* data, with very little sensitivity to X-rays below 0.7 keV, are simply not sensitive to this effect.

12. Massive black holes in globular clusters

There has been considerable debate as to whether evidence supports the existence of massive central black holes in GCs. M15 has been the focus for decades, and the latest observational results by van den Bosch *et al.* (2006) show no significance for a black hole ($1000 \pm 1000\,\mathrm{M}_{\odot}$). The same is true of 47 Tuc where McLaughlin *et al.* (2006) report $700 \pm 700\,\mathrm{M}_{\odot}$. In both of these cases, the value reported, while not significant, is consistent with that mass expected from an extrapolation of the correlation between black hole mass and host dispersion as reported in Gebhardt *et al.* (2002) and Tremaine *et al.* (2002).

The case is different in G1, the largest cluster in M 31. In G1, Gebhardt *et al.* (2002) reported a mass of $2 \times 10^4\,\mathrm{M}_{\odot}$, which was subseqently challenged by Baumgardt *et al.* (2003), who argued against a central black hole. However, the latest observations by Gebhardt *et al.* (2005) continue to argue for a massive black hole using newer data. Whether GCs contain black holes has significant effects on both the evolution of the cluster and on how supermassive black holes grow. Thus it is very important to understand possible number densities for these black holes, and the current observational situation is not satisfying. Theoretically, there are reasons to expect massive black holes in clusters, although observations are required.

We now turn to very recent observations of ω Cen. This cluster is an ideal candidate to look for a central black hole. It has one of the largest velocity dispersions among GCs, implying it may have a large black hole mass. It is nearby both allowing for any black hole influence to be well resolved and allowing access to many stars used to trace the gravitational potential. With an integrated velocity dispersion of $18\,\mathrm{km\,s^{-1}}$, the expected black hole is $10^4\,\mathrm{M}_{\odot}$ which has a sphere of influence of $6''$ at the 4.8 kpc distance of the cluster. The issue with ω Cen is that it may not be a globular cluster, but has been suspected to be the nucleus of an accreted dwarf galaxy (Freeman 1993; Meza *et al.* 2005). Thus, while a massive black hole in ω Cen would not necessarily answer the question as to whether GCs contain black holes, it would help establish the existence and frequency of intermediate-mass black holes in general.

Kinematic data on the cluster come from two sources. Noyola *et al.* (2007) used Gemini/GMOS-IFU data to measure the integrated light in the central $3''$ and at $14''$ radius. Gebhardt & Kissler-Patig (2007) used individual velocities in the central $8''$ to measure the dispersion. Both observations are consistent, and, since one uses integrated light and the other individual velocities, argue for a robust central dispersion estimate. The central velocity dispersion for the cluster is $24 \pm 2\,\mathrm{km\,s^{-1}}$. The dispersion at $14''$ is $20\,\mathrm{km\,s^{-1}}$. Beyond $25''$, data have been compiled by van de Ven *et al.* (2006), coming primarily from radial velocities of Xie *et al.* (2007). Noyola *et al.* and Gebhardt & Kissler-Patig use orbit-based dynamical models and require a central black hole mass of $4(\pm 0.8) \times 10^4\,\mathrm{M}_{\odot}$.

The main arguments against having a black hole are allowing radial orbital anisotropy and having a significant population of heavy remnants. For the orbital anisotropy, there are two considerations. Van de Ven *et al.* (2006) model ω Cen using both radial velocities and proper motions, at radii beyond $25''$. They find a distribution function very close to isotropic. The amount of radial anisotropy required to increase the central dispersion is extreme (see Noyola *et al.*) and very inconsistent with the van de Ven *et al.* model. Furthermore, orbit-based models have been constructed which allow for any orbital distribution consistent with the Jeans equations. For these models, the no-black-hole model is radially biased in the central region, but not enough to make a significant improvement to the fit to the data. In other words, the black hole model is still a better fit even given the maximum radial bias that the radial velocities can tolerate. Presumably, including the proper motion will lead to an even poorer fit for the no-black-hole model

To have the increase be caused by heavy remnants, there are two main problems: having the required number of remnants in the first place, and having all remnants well within the observed core with a very steep density profile. The core radius of ω Cen is 50″. The total mass inside 50″ is 8×10^4 $M_\odot$. If the dispersion increase seen inside 10″ is due to remnants, then for a cluster with a r^{-2} profile all remnants would have to be inside 10″; having 4×10^4 $M_\odot$ of material clustered inside 10″ within a core of 50″ would cause the cluster to evaporate on very short timescales (Maoz 1998). Furthermore, the total mass in heavy remnants (neutron stars and white dwarfs over 1 $M_\odot$) would require a very top heavy initial mass function, inconsistent with what is generally observed. The main problem with alternatives to a black hole is that the velocity dispersion rises inside the core of ω Cen.

The black hole model fits the ω Cen data the best and is consistent with an extrapolation of the black hole $M_{\rm bh}$ - σ correlation. The same situation is true in G1. However, for M 15 and 47 Tuc, the black hole model is preferred but not statistically significant. A main observational point is that there is no black hole mass estimate for a GC that is below the expected value from the $M_{\rm bh}$ - σ correlation.

13. *N*-body simulations of massive black hole formation

The first N-body simulations of the formation of IMBHs in young, dense clusters and their later interactions with passing stars have been performed recently. In some cases, it is found that a massive (>1000 $M_\odot$) object can form as a result of collisions between young stars and that, if turning into an IMBH, this star will capture passing stars through tidal energy dissipation. Gas accretion from circularized stars onto the IMBH may be sufficient to create ultra-luminous X-ray sources (ULXs) in the cluster. These simulations therefore strengthen the connection between ULXs and IMBHs.

ULXs are point-like X-ray sources with isotropic X-ray luminosities in excess of $L = 10^{40}$ erg s^{-1}. Since the Eddington luminosity of a star of mass M is given by $L_{\rm Edd} = 1.3 \times 10^{38}$ erg s$^{-1} (M/M_\odot)$, where M is the mass of the accreting object, most low-luminosity ULXs are probably stellar-mass black holes. However, there is mounting evidence that the brightest ULXs with luminosities exceeding 10^{40} erg s^{-1} could be IMBHs.

The starburst galaxy M 82 for example hosts a ULX with brightness in the range $L = (0.5 - 1.6) \times 10^{41}$ erg s^{-1} (Matsumoto *et al.* 2001; Kaaret *et al.* 2001), corresponding to a black hole with mass 350 - 1200 $M_\odot$ if emitting photons at the Eddington luminosity. The case for an IMBH in M 82 is supported by a 54 mHz quasi-periodic oscillation found in the X-ray flux (Strohmayer & Mushotzky 2003) and also by the soft X-ray spectrum of this source (Fiorito & Titarchuk 2004). Additional observational support for an IMBH comes from the observation of a 62-d period in the X-ray luminosity (Kaaret, Simet & Lang 2006; Patruno *et al.* 2006). The position of the ULX in M 82 coincides with that of the young star cluster MGG-11. Recent N-body simulations, summarized below, have showed how runaway merging of young stars could have led to the formation of an IMBH in MGG-11 and how this IMBH later could have captured passing stars to became a ULX.

The evolution of MGG-11 was simulated through N-body simulations of star clusters containing $N = 131{,}072$ (128K) stars using Aarseth's collisional N-body code NBODY4 on the GRAPE computers in Bonn and Tokyo (see Sec. 2). The initial set-up was given by King models with various central concentrations in the range $3 \leqslant W_0 \leqslant 12$ and half-mass radius $r_h = 1.3$ pc. The initial mass function of cluster stars was given by a Salpeter power-law between $1.0\,M_\odot \leqslant m \leqslant 100\,M_\odot$. These clusters have a projected half-mass radius, mass-to-light ratio and total cluster mass after 12 Myr (the age of MGG-11) that are consistent with the properties of MGG-11 as observed by McCrady *et al.* (2003). In the simulations, stars were merged if their separation became smaller than the sum of their radii. Orbital energy loss by tidal interactions between a star and the IMBH was implemented in the N-body simulations using the prescription of Portegies Zwart & Meinen (1993). More details on these simulations are presented in Portegies Zwart *et al.* (2004) and Baumgardt *et al.* (2006).

For low-concentration models ($W_0 < 8$), the core radii hardly change with time in the first few Myrs and expand at later times due to stellar evolution mass loss from massive stars. Few stellar collisions are observed in these models and no IMBH is formed. For clusters with higher concentration, the central relaxation time is short enough that massive stars spiral into the cluster core while still being in the hydrogen burning stage. Once in the cluster core, they can collide with each other due to the high central density and the large stellar radii of massive stars. Repeated collisions lead to the formation of a VMS with $m > 300\,\mathrm{M}_\odot$. Once such a VMS is produced, all further collisions are predominantly with this star and its mass grows up to $m = 500\,\mathrm{M}_\odot$ to a few $1000\,\mathrm{M}_\odot$ (see also Sec. 3). If this VMS collapses to an IMBH at the end of its lifetime, the presence of a ULX in MGG-11 could be explained by this IMBH. Furthermore, simulations of other young star clusters in M 82 show that runaway merging of stars can happen only in MGG-11 and in none of the other clusters, explaining why only MGG-11 hosts a ULX.

In another set of runs the further dynamical evolution of the IMBH in M 82 was studied. These simulations show that a cusp develops in the stellar density distribution around the IMBH. Inside this cusp, high-mass stars are enriched due to dynamical-friction-driven inspiral. Encounters of stars with the IMBH could lead to tidal capture of the star if its pericenter distance is only slightly larger than the tidal radius, $r_p/r_t \lesssim 3$. If the orbit is unperturbed by other stars, repeated pericenter passages will further decrease its orbital semimajor axis until it circularizes near the black hole. Angular momentum conservation requires this circularization to end at an orbital radius equal to twice the initial pericenter distance. Although perturbations by other stars can either scatter the inspiraling star away from the IMBH or onto an orbit with $r_p < r_t$ where it is destroyed, the simulations show that on average ~ 3 successful inspirals leading to circularization happen within the lifetime of MGG-11.

Once circularized, stars will sooner or later fill their Roche lobe due to stellar evolution and start to transfer mass onto the IMBH. The combined star-IMBH system will then become visible as a ULX. In total, in 10 out of the 12 performed runs a ULX source was produced at least once between 3 and 12 Myrs. Furthermore, 4 runs created an X-ray source brighter than 2×10^{39} erg/s within the age range of MGG-11. Hence, the performed N-body simulations provide a good explanation for the ULX source seen in MGG-11 (cf. Blecha *et al.* 2006).

A further test of this scenario might come from the stellar and orbital evolution of the IMBH binaries. Since the runs show that mostly massive stars are captured and circularize near the IMBH, stellar-mass black holes or NSs will be formed out of the donor stars after they have undergone a supernova. The further evolution of the IMBH binaries will then be driven by encounters with cusp stars, which harden the binaries. In the later stages, emission of gravitational waves will also be important and will lead to the merger of the stars with the IMBH. Hopman & Portegies Zwart (2005) have shown that the event rate for this is likely to be high enough to be detectable by *LISA*, in particular if the IMBH mass is larger than $\sim 3 \times 10^3\,\mathrm{M}_\odot$. Observations of gravitational waves from such binaries would therefore give further support to the scenario discussed here.

Scientific Organizing Committee

Tomasz Bulik (Polska), V. Jorge Casares (Spain), Philip A. Charles (South Africa), Monica Colpi (Italy), Robert P. Fender (UK), Pranab Ghosh (India), Eric Gourgoulhon (France), Jonathan E. Grindlay (USA), Victoria M. Kaspi (Canada), Andrew R. King (UK), Richard N. Manchester (Australia), I. Felix Mirabel (Chile), Frederic A. Rasio (USA, co-chair), Hans Ritter (Germany), Masura Shibata (Japan), and Ingrid H. Stairs (Canada, co-chair).

Acknowledgements

G.F. acknowledges support from NASA contract NAS8-39075 (CXC) and *Chandra* GO grant G06-7079A. J.M.F. acknowledges support from NASA Grant NNG06GI62G and a *Chandra* Theory grant. C.O.H. acknowledges support from the Lindheimer Fellowship at Northwestern University and thanks his many collaborators, especially J. Grindlay, P. Edmonds, and G. Rybicki. S.R. thanks the whole GBT cluster pulsar search and timing team, especially I. Stairs, J. Hessels, P. Freire, and S. Begin. F.A.R. thanks Ingrid Stairs for help with the organization of IAU JD06, particularly in Prague.

References

Aarseth, S. J. 2003, *Gravitational N-Body Simulations* (Cambridge: CUP)
Anderson, S. B., Gorham, P. W., Kulkarni, S. R., *et al.* 1990, *Nature*, 346, 42
Anderson, J., Cool, A. M. & King, I. R. 2003, *ApJ*, 597, 137
Angelini, L., Loewenstein M., & Mushotzky R. F. 2001, *ApJ* (Letters), 557, L35
Arzoumanian, Z., Chernoff D. F., & Cordes J. M. 2002, *ApJ*, 568, 289
Aurière, M., Koch-Miramond, L., & Ortolani, S. 1989, *A&A*, 214, 113
Backer, D. C., Kulkarni, S. R., Heiles, C., Davis, M. M., & Goss, W. M. 1982, *Nature*, 300, 615
Bassa, C. G., & Stappers, B. W. 2004, *A&A*, 425, 1143
Bassa, C. G., Pooley, D. A, Homer, L., *et al.* 2004, *ApJ*, 609, 755
Bath, J. T., & Pringle, J. E. 1981, *MNRAS*, 194, 976
Baumgardt, H., Makino, J., Hut, P., *et al.* 2003, *ApJ* (Letters), 589, L25
Baumgardt, H., Hopman, C., Portegies Zwart, S., & Makino, J. 2006, *MNRAS*, 372, 467
Belczynski, K., Kalogera, V., & Bulik, T. 2002, *ApJ*, 572, 407
Belczynski, K., Kalogera, V., Rasio, F. A., *et al.* 2007, *ApJ*, 662, 504
Bildsten, L., & Deloye, C. J. 2004, *ApJ* (Letters), 607, L119
Blecha, L., Ivanova, N., Kalogera, V., *et al.* 2006, *ApJ*, 642, 427
Bogdanov, S., Grindlay, J. E., & van den Berg, M. 2005, *ApJ*, 630, 1029
Bogdanov, S., Grindlay, J. E., Heinke, C. O., *et al.* 2006, *ApJ*, 646, 1104
Bond, I. A., Abe, F., Eguchi, S., *et al.* 2005, *ApJ* (Letters), 620, L103
Brisken, W. F., Fruchter A. S., Goss W. M., Hernstein R. M., & Thorsett S. E. 2003, *AJ*, 126, 3090
Brown, E. F., Bildsten, L., & Rutledge, R. E. 1998, *ApJ* (Letters), 504, L95
Burderi, L., Possenti, A., D'Antona, F., *et al.* 2001, *ApJ* (Letters) 560, L71
Burderi, L., D'Antona, F. & Burgay, M. 2002, *ApJ* 574, 325
Burderi, L., D'Antona, F. & Di Salvo, T., in preparation
Cackett, E. M., Wijnands, R., Heinke, C. O., *et al.* 2006, *MNRAS*, 369, 407
Camilo, F., & Rasio, F. A. 2005, in: F. A. Rasio & I. H. Stairs (eds.), *Binary Radio Pulsars*, *ASP-CS*, 328, 147
Carson, J. E., Cool, A. M., & Grindlay, J. E. 2000, *ApJ*, 532, 461
Chies-Santos, A. L., Pastorizia M. G., Santiago B. X., & Forbes D. A. 2006, *A&A*, 455, 453
Clark, G. W. 1975, *ApJ* (Letters), 199, L143
Colpi, M., Possenti, A., & Gualandris, A. 2002, *ApJ*, 570, L85
Colpi, M., Mapelli, M., & Possenti, A. 2003, *ApJ*, 599, 1260

Cool, A. M., & Bolton, A. S. 2002, in: M. M. Shara (ed.) *Stellar Collisions, Mergers and their Consequences*, *ASP-CS*, 263, 163
Corongiu, A., *et al.* 2006, *ApJ*, 653, 1417
D'Amico, N., Lyne, A. G., Manchester, R. N., *et al.* 2001, *ApJ* (Letters), 548, L171
D'Amico, N., Possenti, A., Manchester, R. N., *et al.* 2001, *ApJ* (Letters), 561, L89
D'Amico, N., Possenti, A., Fici, L., *et al.* 2002, *ApJ* (Letters), 570, L89
D'Antona, F., Ventura, P., Burderi, L., *et al.* 2006, *ApJ*, 640, 950
Davies, M. B. 1997, *MNRAS*, 288, 117
Davies, M. B., & Hansen B. M. S. 1998, *MNRAS*, 301, 15
Dessart, L., Burrows, A., Ott, C. D., *et al.* 2006, *ApJ*, 644, 1063
Dewi, J. D. M., Podsiadlowski P., & Pols O. 2005, *MNRAS* (Letters), 363, L71
Dieball, A., Knigge, C., Zurek, D. R., Shara, M. M., & Long, K. S. 2005, *ApJ*, 625, 156
Di Stefano, R., & Rappaport, S. 1994, *ApJ* 423, 274
DiStefano, R., Kong, A. K. H., VanDalfsen, M. L., *et al.* 2003, *ApJ*, 599, 1067
Dobrotka, A., Lasota, J.-P., & Menou, K. 2006, *ApJ*, 640, 288
Edmonds, P. D., Gilliland, R. L., Heinke, C. O., & Grindlay, J. E. 2003, *ApJ*, 596, 1177
Edmonds, P. D., Gilliland, R. L., Heinke, C. O., & Grindlay, J. E. 2003, *ApJ*, 596, 1197
Ergma, E., & Sarna, M. J. 2003, *A&A*, 399, 237
Fabbiano, G., Gioia, I. M., & Trinchieri, G. 1988, *ApJ*, 324, 749
Fabbiano, G., & Shapley, A. 2002, *ApJ*, 565, 908
Fabbiano, G., Zezas, A., King, A. R., *et al.* 2003, *ApJ* (Letters), 584, L5
Fabbiano, G. 2006, *ARAA*, 44, 323
Fabian, A.C., Pringle, J. E., & Rees, M. J. 1975, *MNRAS*, 172, 15P
Ferraro, F. R., Possenti, A., D'Amico, N., & Sabbi, E. 2001, *ApJ* (Letters), 561, L93
Ferraro, F. R., Possenti, A., Sabbi, E., *et al.* 2003, *ApJ*, 595, 179
Fiorito, R., & Titarchuk, L. 2004, *ApJ* (Letters), 614, L113
Freeman, K. 1993, in: Graeme H. Smith & Jean P. Brodie (eds.), *The Globular Cluster-Galaxy Connection*, *ASP-CS*, 48, 608
Fregeau, J. M., Cheung, P., Portegies Zwart, S. F., & Rasio, F. A. 2004, *MNRAS*, 352, 1
Fregeau, J., & Rasio, F. 2006, *ApJ*, 658, 1047
Freire, P. C., Gupta, Y., Ransom, S. M., & Ishwara-Chandra, C. H. 2004, *ApJ* (Letters), 606, L53
Freitag, M., & Benz, W. 2001, *A&A* 375, 711
Freitag, M., Rasio, F. A., & Baumgardt, H. 2006, *MNRAS*, 368, 121
Freitag, M., Gürkan, M. A., & Rasio, F. A. 2006, *MNRAS*, 368, 141
Freitag, M., Amaro-Seoane, P., & Kalogera, V. 2006, *ApJ*, 649, 91
Gebhardt, K., Rich, R. M., & Ho, L. C. 2002, *ApJ* (Letters), 578, L41
Gebhardt, K., Rich, R. M., & Ho, L. C. 2005, *ApJ*, 634, 1093
Gebhardt, K. *et al.* 2000, *ApJ* (Letters), 539, L13
Gebhardt, K., & Kissler-Patig, M. 2007, in preparation
Gendre, B., Barret, D., & Webb, N. A. 2003a, *A&A*, 400, 521
Gendre, B., Barret, D., & Webb, N. A. 2003b, *A&A* (Letters), 403, L11
Giersz, M., & Spurzem, R. 2003, *MNRAS*, 343, 781
Giersz, M. 2006, *MNRAS*, 371, 484
Gilfanov, M., 2004, *MNRAS*, 349, 146
Gratton, R. G., Bragaglia, A., Carretta, E., *et al.* 2003, *A&A*, 408, 529
Grimm, H. J., Gilfanov, M., & Sunyaev, R. 2002, *A&A*, 391, 923
Grimm, H. J., Gilfanov, M., & Sunyaev, R. 2003, *MNRAS*, 339, 793
Grindlay, J. E., Hertz, P., Steiner, J. E., *et al.* 1984, *ApJ* (Letters), 282, L13
Grindlay, J. E. 1984, in: *High Energy Transients in Astrophysics*, *AIP-PC*, 115, 306
Grindlay, J. E. 1999, in: C. Hellier & K. Mukai (eds.), *Annapolis Workshop on MCVs*, *ASP-CS*, 157, 377
Grindlay, J. E., Heinke, C., Edmonds, P. D., & Murray, S. S. 2001, *Science*, 292, 2290
Grindlay, J. E., Heinke, C.O., Edmonds, P. D., *et al.* 2001, *ApJ* (Letters), 563, L53

Gürkan, M. A., Fregeau, J. M., & Rasio, F. A. 2006, *ApJ*, 640, L39
Haggard, D., Cool, A. M., Anderson, J., *et al.* 2004, *ApJ*, 613, 512
Heggie D. C., & Hut, P. 2003, *The Gravitational Million Body Problem* (Cambridge: CUP)
Heinke, C. O., Grindlay, J. E., Lugger, P. M., *et al.* 2003, *ApJ*, 598, 501
Heinke, C. O., Grindlay, J. E., & Edmonds, P. D. 2005, *ApJ*, 622, 556
Heinke, C. O., Grindlay, J. E., Edmonds, P. D., *et al.* 2005, *ApJ*, 625, 796
Heinke, C. O., Rybicki, G. B., Narayan, R., & Grindlay, J. E. 2006, *ApJ*, 644, 1090
Heinke, C. O., Wijnands, R., Cohn, H. N., *et al.* 2006, *ApJ*, 651, 1098
Hertz, P., & Grindlay, J. 1983, *ApJ*, 275, 105
Hessels, J. W. T., Ransom, S. M., Stairs, I. H., *et al.* 2006, *Science*, 311, 1901
Hobbs, G., Lorimer D. R., Lyne A. G., & Kramer M. 2005, *MNRAS*, 360, 974
Hopman, C., & Portegies Zwart, S. 2005, *MNRAS* (Letters), 363, L56
Hurley, J., Tout, C., & Pols, O., 2002, *MNRAS*, 329, 897
Hurley, J., Pols, O, Aarseth, S. & Tout, C. 2005, *MNRAS*, 363, 293
Hut, P. 2007, *Prog. Theor. Phys.*, in press [astro-ph/0610222]
Hut, P., McMillan, S., Goodman, J., *et al.* 1992, *PASP*, 104, 981
Hut, P., Shara, M. M., Aarseth, S. J., *et al.* 2003, *NewA*, 8, 337
Hut, P., & Makino, J. 2003, `http://www.ArtCompSci.org`
Iben, I., Tutukov A. V., & Fedorova, A. V. 1997, *ApJ*, 486, 955
in't Zand, J. J. M., van Kerkwijk, M. H., Pooley, D., *et al.* 2001, *ApJ* (Letters), 563, L41
in't Zand, J. J. M., Hulleman, F., Markwardt, C. B., *et al.* 2003, *A&A*, 406, 233
Irwin, J. A., & Bregman, J. N. 1999, *ApJ* (letters), 510, L21
Irwin, J. A. 2005, *ApJ*, 631, 511
Ivanova, N., Belczynski, K., Fregeau, J. M., & Rasio, F. A. 2005, *MNRAS*, 358, 572
Ivanova, N. 2006, *ApJ*, 636, 979
Ivanova, N., & Kalogera, V. 2006, *ApJ*, 636, 985
Ivanova, N., Heinke, C. O., Rasio, F. A., Taam, R. E., *et al.* 2006, *MNRAS*, 372, 1043
Ivanova, N., Rasio, F. A., Lombardi, J. C., *et al.* 2005, *ApJ* (Letters), 621, L109
Jacoby, B. A., Chandler, A. M., Backer, D. C., *et al.* 2002, *IAU Circ* 7783, 1
Jacoby, B. A., Cameron, P. B., Jenet, F. A., *et al.* 2006, *ApJ* (Letters), 644, L113
Jordan, A., Côté, P., Ferrarese, L., *et al.* 2004, *ApJ*, 613, 279
Juett, A. M., 2005, *ApJ* (Letters) 621, L25
Kaaret, P., Prestwich, A. H., Zezas, A., *et al.* 2001, *MNRAS* (Letters), 321, L29
Kaaret, P., Simet, M. G. & Lang, C. C. 2006, *ApJ*, 646, 174
Kalogera, V. 1998, *ApJ*, 493, 368
Kalogera, V., King, A. R., & Rasio, F. A. 2004, *ApJ* (letters), 601, L171
Kilgard, R. E., Kaaret, P., Krauss, M. I., *et al.* 2002, *ApJ*, 573, 138
Kim D. W., & Fabbiano, G. 2004, *ApJ*, 611, 846
Kim, D.-W., Kim, E, Fabbiano, G., Trinchieri, G. 2007, *ApJ*, in press (2007arXiv0706.4254)
Kim, E., Kim, D.-W., Fabbiano, G., *et al.* 2006, *ApJ*, 647, 276
Knigge, C., Zurek, D. R., Shara, M.M., Long, K.S., & Gilliland, R.L. 2003, *ApJ*, 599, 1320
Kong, A. K. H., Di Stefano, R., Garcia, M. R., & Greiner, J. 2003, *ApJ*, 585, 298
Kong, A. K. H., Bassa, C., Pooley, D., *et al.* 2006, *ApJ*, 647, 1065
Kraft, R. P., Kregenow, J. M., Forman, W. R., Jones, C., & Murray, S. S. 2001, *ApJ*, 560, 675
Kundu, A., Maccarone, T. J., & Zepf, S. E. 2002, *ApJ* (Letters), 574, L5
Kundu, A., Maccarone, T. J., Zepf, S. E., & Puzia, T. H. 2003, *ApJ* (Letters), 598, L81
Kundu, A., Zepf, S. E., Hempel, M., *et al.* 2005, *ApJ* (Letters), 634, L41
Larsen, S., Brodie, J.P., Beasley, M. A., *et al.* 2003, *ApJ*, 585, 767
Lattimer, J. M., & Prakash, M. 2004, *Science*, 304, 536
Lugger, P. M., Cohn, H. N., Heinke, C. O., *et al.* 2007, *ApJ*, 657, 286
Maccarone, T. J., Kundu, A., & Zepf, S. E. 2004, *ApJ*, 606, 430
Makino, J. 2006, IAU JD 14, in this volume, p. 424
Maoz, E. 1998, *ApJ*, 494, 181
Matsumoto, H., Tsuru, T.G., Koyama, K., *et al.* 2001, *ApJ* (Letters), 547, L25

McCrady, N., Gilbert, A. M., & Graham, J. R. 2003, *ApJ*, 596, 240
McLaughlin, D. E., Anderson, J., Meylan, G., *et al.* 2006, *ApJS*, 166, 249
McMillan, S. L. W., & Aarseth, S. J. 1993, *ApJ*, 414, 200
Meyer, F., & Meyer-Hofmeister, E. 1981, *A&A* (Letters), 104, L10
Meza, A., Navarro, J., Abadi, M., & Steinmetz, M. 2005, *MNRAS*, 359, 93
Minniti, D., Rejkuba M., Funes J. G., & Akiyama, S. 2004, *ApJ*, 600, 716
Nelson, L. A., & Rappaport, S. 2003, *ApJ*, 598, 431
Nomoto, K. 1984, *ApJ*, 277, 791
Norton, A. J., & Watson, M. J. 1989, *MNRAS*, 237, 853
Noyola, E., Gebhardt, K., & Bergmann, M. 2007, *ApJ*, submitted
Osaki, Y. 1974, *PASJ*, 26, 429
Paresce, F., & de Marchi, G. 1994, *ApJ*, 427, 33
Parker, T. L., Norton, A. J., & Mukai, K. 2005, *A&A*, 439, 213
Patruno, A., Portegies Zwart, S., Dewi, J., & Hopman, C. 2006, *MNRAS* (Letters), 370, L6
Pfahl, E., Rappaport, S., Podsiadlowski, P., & Spruit H. 2002, *ApJ*, 574, 364
Pfahl, E., Rappaport, S., & Podsiadlowski, P. 2003, *ApJ*, 597, 1036
Pietrukowicz, P., Kaluzny, J., Thompson, I. B., *et al.* 2005, *AcA*, 55, 261
Piro, A. L., & Bildsten, L. 2002, *ApJ* (Letters), 571, L103
Podsiadlowski, P., Rappaport, S., & Pfahl, E. D. 2002, *ApJ*, 565, 1107
Podsiadlowski, P., Langer, N., Poelarends, A. J. T., *et al.* 2004, *ApJ* 612, 1044
Pooley, D., Lewin, W. H. G., Homer, L., *et al.* 2002, *ApJ*, 569, 405
Pooley, D., Lewin, W. H. G.. Anderson, S. F., *et al.* 2003, *ApJ* (Letters), 591, L131
Pooley, D., & Hut, P. 2006, *ApJ* (Letters), 646, L143
Portegies Zwart, S. F., & Meinen, A. T. 1993, *A&A*, 280, 174
Portegies Zwart, S. F., & Verbunt, F. 1996, *A&A*, 309, 179
Portegies Zwart, S. F., McMillan, S. L. W., Hut, P., & Makino, J. 2001, *MNRAS*, 321, 199
Portegies Zwart, S. F., Baumgardt, H., Hut, P., *et al.* 2004, *Nature*, 428, 724
Possenti, A., *et al.* 2001, presented at ACP Summer 2001 Workshop on *Compact Objects in Dense Star Clusters* [astro-ph/0108343]
Possenti, A., D'Amico, N., Manchester, R. N., *et al.* 2003, *ApJ*, 599, 475
Possenti, A., D'Amico, N., Corongiu, A., *et al.* 2005, in F. A. Rasio & I. H. Stairs (eds.), *Binary Radio Pulsars*, *ASP-CS*, 328, 189
Possenti, A., *et al.* 2006, in preparation
Posson-Brown, J., Raychaudhury, S., Forman, W., Donnelly, R. H., & Jones, C. 2006, *ApJ*, submitted [astro-ph/0605308]
Puzia, T. H., Zepf, S. E., Kissler-Patig, M., *et al.* 2002, *A&A*, 391, 453
Pylyser, E., & Savonije, G. J. 1988, *A&A*, 191, 57
Ransom, S. M., Stairs, I. H., Backer, D. C., *et al.* 2004, *ApJ*, 604, 328
Ransom, S. M., Hessels, J, W. T., Stairs, I. H. *et al.* 2005, *Science*, 307, 892
Rappaport, S., Podsiadlowski, P., Joss, P. C., Di Stefano, R., & Han, Z. 1995, *MNRAS*, 273, 731
Rasio, F. A., & Heggie, D. C. 1995, *ApJ* (Letters), 445, L133
Rasio, F. A., Pfahl, E. D., & Rappaport, S. 2000, *ApJ* (Letters), 532, L47
Ruderman, M., Shaham, J., & Tavani, M. 1989, *ApJ*, 336, 507
Rutledge, R. E., Bildsten, L., Brown, E. F., Pavlov, G. G., & Zavlin, V. E. 2002, *ApJ*, 577, 346
Rutledge, R. E., Bildsten, L., Brown, E. F., Pavlov, G. G., & Zavlin, V. E. 2002, *ApJ*, 578, 405
Sabbi, E., Gratton, R., Ferraro, F. R., *et al.* 2003a, *ApJ* (Letters), 589, L41
Sabbi, E., Gratton, R. G., Bragaglia, A., *et al.* 2003b, *A&A*, 412, 829
Sabbi, E., Ferraro, F. R., Sills, A., & Rood, R. T. 2004, *ApJ*, 617, 1296
Sarazin, C. L., Irwin, J. A., & Bregman, J. N. 2001, *ApJ*, 556, 553
Shara, M. M., Potter, M., & Moffat, A. F. J. 1987, *AJ*, 94, 357
Shara, M. M., Bergeron, L. E., Gilliland, R. L., Saha, A., & Petro, L. 1996, *ApJ*, 471, 804
Smits M., Maccarone T. J., Kundu A., & Zepf S. E. 2006, *A&A*, 458, 477
Soria, R., & Wu, K. 2003, *A&A*, 410, 53

Spurzem, R. 1999, *JCAM*, 109, 407

Spurzem, R. & Baumgardt, H. 2003, preprint, submitted to *MNRAS*, also available via `<ftp://ftp.ari.uni-heidelberg.de/pub/staff/spurzem/edinpaper.ps.gz>`

Stairs, I. H. 2004, *Science*, 304, 547

Strohmayer, T. E., & Mushotzky, R. F. 2003, *ApJ* (Letters), 586, L61

Supper, R., Hasinger, G., Pietsch, W., *et al.* 1997, *A&A*, 317, 328

Szkody, P., Henden, A., Fraser, O. J., *et al.* 2005, *AJ*, 129, 2386

Teodorescu, A. 2005, *The Evolution of Short and Ultrashort LMXBs*, PhD thesis, University of Roma - Tor Vergata (XVIII cycle)

Tout, C. A., Pols, O. R., Eggleton, P. P., & Han, Z. 1996, *MNRAS*, 281, 257

Tremaine, S., Gebhardt, K., Bender, R., *et al.* 2002, *ApJ*, 574, 740

Tutukov, A. V., Fedorova, A. V., Ergma, E. V., & Yungelson, L. R. 1985, *Sov. Astron Letters*, 11, 52

van de Ven, G., van den Bosch, R. C. E., Verolme, E. K., & de Zeeuw, P. T. 2006, *A&A*, 445, 513

van den Bosch, R. C. E., de Zeeuw, P. T., Gebhardt, K., *et al.* 2006, *ApJ*, 641, 852

van der Sluys, M. V., Verbunt, F., & Pols, O. R. 2005, *A&A*, 440, 973

van Paradijs, J., & van den Heuvel, E. P. J. 1995, in: W. H. G. Lewin, J. van Paradijs, & E. P. J. van den Heuvel (eds.), *X-ray Binaries* (Cambridge: CUP)

Verbunt, F., & Hut, P. 1987, in: D. J. Helfand & J.-H. Huang (eds.), *The Origin and Evolution of Neutron Stars*, Proc. IAU Symposium No. 125 (Dordrecht: Reidel), p. 187

Verbunt, F., & van den Heuvel, E. P. J. 1995, in: W.H.G. Lewin, J. van Paradijs, & E.P.J. van den Heuvel (eds.), *X-ray Binaries* (Cambridge: CUP)

Verbunt, F., & Lewin, W. H. G. 2006, in: W. H. G. Lewin & M. van der Klis, *Compact Stellar X-ray Sources* (Cambridge: CUP), p. 341

Webb, N. A., Gendre, B., & Barret, D. 2002, *A&A*, 381, 481

Webb, N. A., Serre, D., Gendre, B., Barret, D., Lasota, J.-P., & Rizzi, L. 2004, *A&A*, 424, 133

Webb, N. A., Wheatley, P. J., & Barret, D. 2006, *A&A*, 445, 155

White, N. E., & Angelini, L., 2001, *ApJ* (Letters), 561, L101

White, R. E., Sarazin, C. L., & Kulkarni, S. R. 2002, *ApJ* (Letters), 571, L23

Wickramasinghe, D. T., & Ferrario, L. 2000, *PASP*, 112, 873

Wijnands, R., Heinke, C. O., Pooley, D., *et al.* 2005, *ApJ*, 618, 883

Xie, B., Gebhardt, K., Pryor, C., & Williams, T. 2007, in preparation

Xu, Y., Xu, H., Zhang, Z., al. 2005, *ApJ*, 631, 809

Zavlin, V. E., Pavlov, G. G., & Shibanov, Y. A. 1996, *A&A*, 315, 141

Zezas, A., Fabbiano, G., Rots, A. H., & Murray, S. S. 2002, *ApJ*, 577, 710

Zezas, A., & Fabbiano, G. 2002, *ApJ*, 577, 726

Zezas, A., Fabbiano, G., Baldi, A., *et al.* 2006, *ApJS*, 166, 211

Zezas, A., Fabbiano, G., Baldi, A., *et al.* 2007, *ApJ*, 661, 135

Highlights of Astronomy, Volume 14
IAU XXVI General Assembly, 14-25 August 2006
Karel A. van der Hucht, ed.

doi:10.1017/S1743921307010277

Joint Discussion 7
The Universe at $z > 6$

Daniel Schaerer[1] and Andrea Ferrara[2] (eds.)

[1]Observatoire de Geneve, 51, Ch. des Maillettes, CH-1290 Sauverny, Switzerland
email: daniel.schaerer@obs.unige.ch
[2]SISSA/ISAS, via Beirut 2-4, I-34014 Trieste, Italy
email: ferrara@sissa.it

Preface

The exploration of the earliest phase of star and galaxy formation after the Big Bang remains an important challenge of contemporary astrophysics and represents a key science driver for numerous future facilities. During this phase the first stars and galaxies appear and start to light up and ionize the then neutral Universe ending thereby the so called cosmic dark ages and leading progressively to the complete reionization we observe now at redshift $z \simeq 6$ or earlier.

Important theoretical and numerical advances have been made in the modeling of the early Universe. Also, it has recently become possible to obtain direct observations of galaxies, quasars, and the intergalactic medium (IGM) out to redshift $z \simeq 7$. Furthermore new optical to near-IR observations using ground-based and space-borne telescopes are now opening up the view to even higher redshift, directly probing for the first time galaxies during the first billion years after the Big Bang.

Motivated by these facts Joint Discussion 07 on "The Universe at $z > 6$" was organised during the IAU General Assembly as a forum to present and discuss the latest results both from numerical modeling and from observations in this rapidly advancing field.

Judging at least from the attendance – a very full room with up to ~ 150 persons! – this Joint Discussion was a success. We hope that all participants, speakers, poster presenters, and the 'audience' felt the same. In any case we thank all the persons who have contributed to this Joint Discussion 07, the SOC members, all the local organisers, and the participants.

Scientific Organizing Committee

Andrea Ferrara (Italy, co-chair), Esther M. Hu (USA), Matthew D. Lehnert (Germany), Roser D. Pelló (France), Daniel Schaerer (Switzerland), and Yoshiaki Taniguchi (Japan).

Daniel Schaerer and Andrea Ferrara, co-chairs SOC and editors
Geneva and Trieste, 22 November 2006

Highlights of Astronomy, Volume 14
IAU XXVI General Assembly, 14-25 August 2006
Karel A. van der Hucht, ed.

doi:10.1017/S1743921307010289

Lyman break galaxies at $z > 6$

Rychard J. Bouwens and Garth D. Illingworth
UCO/Lick Observatory and Department of Astronomy,
University of California Santa Cruz, Santa Cruz, California 95064, USA
email: bouwens,gdi@ucolick.org

Abstract. Extending the study of star-forming galaxies to $z > 6$ is extremely difficult due to the faintness of the sources and the challenging nature of deep near-infrared observations. Nevertheless, current observations are now just good enough that we can begin drawing some conclusions about the nature of galaxies at $z \gtrsim 7$. At present, deep near-infrared observations with NICMOS (reaching $\gtrsim 27$ AB mag at $5\,\sigma$) cover more than 20 arcmin^2 of area with deep optical coverage and allow us to identify four strong $z \simeq 7-8$ candidates. Comparing this sample with dropout samples at later times ($z \simeq 4-6$), we are able to study evolution in the rest-frame *UV* LF over the range $z \simeq 8$ to $z \simeq 4$. We find strong evidence for significant evolution in the characteristic luminosity with time (brightening by ~ 2 mag, from $z \simeq 8$ to $z \simeq 4$). The observed evolution appears to be the direct result of hierarchical growth in the galaxy population.

Keywords. galaxies: evolution, galaxies: formation, galaxies: high-redshift

---

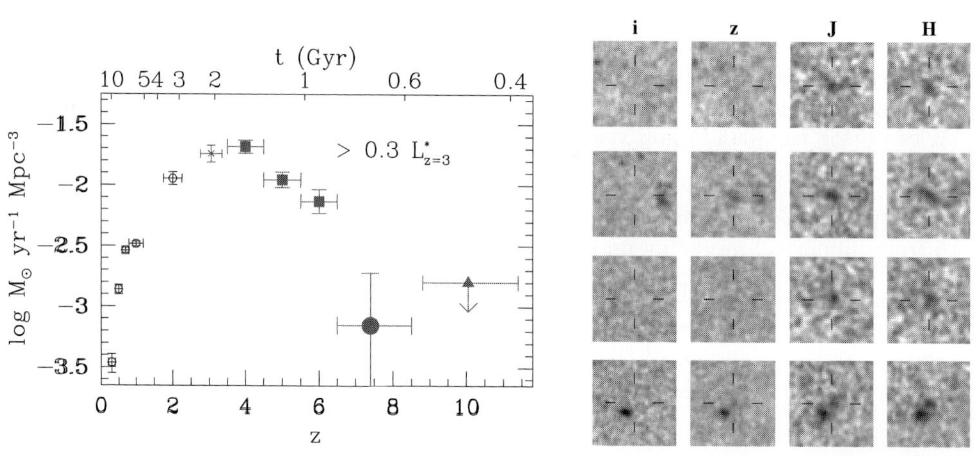

Figure 1. *Left:* Star formation history of the universe (uncorrected for extinction and integrated down to $0.3\,L^*_{z=3}$). Shown are our determinations at $z \simeq 4-6$ (large red squares: Bouwens *et al.* 2006 and Bouwens *et al.* 2006, in preparation), our recent determination at $z \simeq 7.4$ (large red circle: Bouwens & Illingworth 2006), and upper limits at $z \simeq 10$ (red triangle: Bouwens *et al.* 2005). Included are also determinations at $z \simeq 0-2$ (Schiminovich *et al.* 2005) and $z \simeq 3$ (Steidel *et al.* 1999). The star formation rate density is observed to increase rather dramatically from $z \simeq 8$ to $z \simeq 4$. *Right:* Optical and near-infrared images of four candidate star-forming galaxies at $z \simeq 7-8$. These galaxies were found in deep NICMOS imaging available over the Ultra Deep Field and GOODS fields (Bouwens & Illingworth 2006).

References

Bouwens, R. J., & Illingworth, G. D. 2006, *Nature*, 443, 189
Bouwens, R. J., Illingworth, G. D., Blakeslee, J. P., & Franx, M. 2006, *ApJ*, 653, 53
Bouwens, R. J., Illingworth, G. D., Thompson, R. I., & Franx, M. 2005, *ApJ* (Letters), 624, L5
Steidel, C. C., Adelberger, K. L., Giavalisco, M., *et al.* 1999, *ApJ*, 519, 1
Schiminovich, D., Ilbert, O., Arnouts, S., *et al.* 2005, *ApJ* (Letters), 619, L47

Highlights of Astronomy, Volume 14
IAU XXVI General Assembly, 14-25 August 2006
Karel A.van der Hucht, ed.

doi:10.1017/S1743921307010290

Subaru surveys for the high-redshift universe

Yoshiaki Taniguchi

Department of Physics and Engineering, Graduate School of Science, Ehime University,
Bunkyo-cho 2-5, Matsuyama 790-8577, Japan
email: tani@sgr.phys.sci.ehime-u.ac.jp

Abstract. We present a brief summary of Subaru observations of high-z galaxies. We discuss our future plan to probe galaxies beyond $z = 10$.

Keywords. galaxies: evolution, galaxies: formation, galaxies: high-redshift

---

Among the most pressing issues in modern observational cosmology is the question of finding and studying the first generation of galaxies – their formation epoch, their early evolution, and their contribution to cosmic re-ionization. This provides clues towards early star formation activity in the Universe and mass assembly in galaxies as well as the long standing problem of understanding the physical processes responsible for cosmic reionization of intergalactic medium. Furthermore, this provides an independent test for WMAP results which identifies the reionization epoch to be at $z \simeq 11$. This is the only direct observation currently available constraining the epoch of re-ionization.

The Subaru telescope has been used to search for very high-z galaxies for these several years. The very wide field of view of Subaru-SUPRIME-CAM allows us to identify more than 100 galaxies with $z > 5$; $z \simeq 5.7$ (e.g., Shimasaku *et al.* 2006), $z \simeq 6.6$ (e.g., Taniguchi *et al.* 2005; Kashikawa *et al.* 2006), and $z \simeq 7$ (Iye *et al.* 2006; Hu *et al.* 2007). In order to extend the search for galaxies at $z > 10$, we are promoting to use our custom made near-infrared narrow-band filter (NB2296) on Subaru-MOIRCS to carry out an ultra-deep survey to search for Lyα emitting galaxies at $z \simeq 18$. The central wavelength of this filter is 2296 nm, with a bandwidth corresponding to 23 nm. Therefore, this filter can detect Lyα emission from galaxies at $z \approx 17.9$; $z_{\rm min} = 17.8$ and $z_{\rm max} = 18.0$. The relatively large field of view of Subaru-MOIRCS ($4' \times 7'$) allows a wide-area and ultra-deep survey, rather unprecedented at near-IR wavelengths. Although a re-ionization redshift of $z \simeq 9$ is found by WMAP, it is expected that there are Population III sources out to $z \simeq 30$.

References

Hu, E. M., Cowie, L. L., & Kakazu, Y. 2007, these JD07 proceedings, p. 252
Iye, M., Ota, K., Kashikawa, N., *et al.* 2006, *Nature*, 443, 186
Kashikawa, N., Shimasaku, K., Malkan, M. A., *et al.* 2006, *ApJ*, 648, 7
Shimasaku, K., Kashikawa, N., Doi, M., *et al.* 2006, *PASJ*, 58, 313
Taniguchi, Y., Ajiki, M., Nagao, T., *et al.* 2005, *PASJ*, 57, 165

Highlights of Astronomy, Volume 14
IAU XXVI General Assembly, 14-25 August 2006
Karel A. van der Hucht, ed.

doi:10.1017/S1743921307010307

Galaxies in the first billion years: implications for re-ionization and the star formation history at $z > 6$

Andrew J. Bunker[1], Elizabeth R. Stanway[2], Laurence P. Eyles[1], Richard S. Ellis[3], Richard G. McMahon[4], Mark Lacy[5] and Daniel P. Stark[3]

[1]University of Exeter, School of Physics, Stocker Road, Exeter, EX4 4QL, UK
email: bunker@astro.ex.ac.uk

[2]Astronomy Dept., University of Wisconsin, 475 N. Charter Street, Madison, WI 53706, USA

[3]California Institute of Technology, Mail Stop 169-327, Pasadena, CA 91109, USA

[4]Institute of Astronomy, Madingley Road, Cambridge, CB3 0HA, UK

[5]Spitzer Science Center, 1200 E. California Blvd., Pasadena, CA 91125, USA

Abstract. We discuss the selection of star-forming galaxies at $z \simeq 6$ through the Lyman-break technique. *Spitzer* imaging implies many of these contain older stellar populations (> 200 Myr) which produce detectable Balmer breaks. The ages and stellar masses ($\sim 10^{10}$ $M_\odot$) imply that the star formation rate density at earlier epochs may have been significantly higher than at $z \simeq 6$, and might have played a key role in re-ionizing the universe.

Keywords. galaxies: high-redshift; evolution; formation; stellar content

---

We have discovered a population of star-forming galaxies at $z \simeq 6$ and beyond (within the first billion years) through the i'-drop technique. The first application of this to *HST*-ACS imaging was presented in Stanway *et al.* (2003), using the public GOODS survey. We were able to prove this technique through Keck-DEIMOS spectroscopy (Bunker *et al.* 2003). Using the same i'-drop selection, our first analysis of the Hubble Deep Field revealed 50 star forming galaxies at redshifts around six with magnitudes $z_{AB} > 28.5$ (Bunker *et al.* 2004). *Spitzer* observations with IRAC enable us to estimate the stellar masses and luminosity-weighted ages for this population; we find in some cases that there are Balmer breaks, indicating ages of > 200 Myr and formation redshifts of $z \simeq 10$ (Eyles *et al.* 2005). From the whole sample of v-drops and i'-drops we estimate the stellar mass density at $z \approx 5$ (Stark *et al.* 2007) and at $z \approx 6$ (Eyles *et al.* 2007). The implications of this work are that the previous star formation history was higher prior to $z \simeq 6$, and might have played a key role in generating the UV photons necessary to re-ionize the universe at $z \simeq 8-10$. Our work is the strongest constraint to date on the star formation history at $z > 6$.

References

Bunker, A. J., Stanway, E. R., Ellis, R. S., *et al.* 2003, *MNRAS* (Letters), 342, L47
Bunker, A. J., Stanway, E. R., Ellis, R. S., & McMahon, R. G. 2004, *MNRAS*, 355, 374
Eyles, L. P, Bunker, A. J., Stanway, E. R., *et al.* 2005, *MNRAS*, 364, 443
Eyles, L. P, Bunker, A. J., Ellis, R. S., *et al.* 2006, *MNRAS*, 374, 910
Stanway, E. R., Bunker, A. J., & McMahon, R. G. 2003, *MNRAS*, 342, 439
Stark, D. P., Bunker, A. J., Ellis, R. S., Eyles, L. P., & Lacy, M. 2007, *ApJ*, 659, 84

Highlights of Astronomy, Volume 14
IAU XXVI General Assembly, 14-25 August 2006
Karel A. van der Hucht, ed.

doi:10.1017/S1743921307010319

A *Spitzer*-IRS search for the galaxies that re-ionized the Universe

Mark Lacy[1], Andrew J. Bunker[2], Jean-Paul Kneib[3] and Harry I. Teplitz[1]

[1]Spitzer Science Center, California Institute of Technology, Pasadena, CA 91125, USA
email: mlacy,hit@ipac.caltech.edu

[2]School of Physics, University of Exeter, Stocker Road, Exeter, EX4 4QL, UK
email: bunker@astro.ex.ac.uk

[3]Observatoire Midi-Pyrénées, UMR 5572, 14 Avenue Edouard Belin, F-31000 Toulouse, France
email: jean-paul.kneib@oamp.fr

Abstract. We describe an observation designed to find Hα emission from galaxies at $z \simeq 7-12$ made using the InfraRed Spectrograph (IRS) on the *Spitzer Space Telescope*.

Keywords. galaxies: formation,galaxies: clusters: individual (Abell 2218)

Spectra of the most distant quasars at $z \simeq 6.5$ show the Gunn-Peterson trough caused by absorption by neutral hydrogen in the ISM (White *et al.* 2003). The most recent analysis of the WMAP dataset is consistent with re-ionization at $z \simeq 11$ (Page *et al.* 2007). These observations suggest that starburst galaxies or quasars with large amounts of escaping UV emission were present at $z \simeq 7-12$, but such a population is inconsistent with a straightforward extrapolation of the evolution of known galaxy or quasar populations at $z \simeq 6$ (e.g., Bunker *et al.* 2004), unless the dominant stellar populations at $z \simeq 6$ is very metal poor (Stiavelli *et al.* 2004). Prior to re-ionization, Lyα emission is expected to be extinguished by a large (though uncertain) factor $\sim 10-100$ by the damping wing of the Gunn-Peterson trough (e.g., Santos 2006). This makes blank field Hα searches with *Spitzer* competative with ground-based near-infrared searches for Lyα.

We used the on *Spitzer*-IRS spectrograph to observe a single slit position aligned along the critical line for $z \gtrsim 7$ galaxies in the cluster Abell 2218 to demonstrate the feasibility of such observations. We used the second order short-low spectrum to search for Hα emission lines redshifted into the $5-9\,\mu$m range. We achieved a $3\,\sigma$ detection limit $\sim 2 \times 10^{-19}\,\mathrm{WHz^{-1}m^{-2}}$, corresponding to star formation rates $\sim 1000/\mu \mathrm{M_{\odot}yr^{-1}}$ where $\mu \simeq 10-100$ is the magnification factor. We have one candidate emission line detection at this level, which, if it is indeed Hα, is at a wavelength corresponding to $z = 9.8$.

Acknowledgements

This work is based on observations made with the *Spitzer Space Telescope*, which is operated by the Jet Propulsion Laboratory, California Institute of Technology under a contract with NASA.

References

Bunker, A. J., Stanway, E. R., Ellis, R. S., & McMahon, R. G. 2004, *MNRAS*, 355, 374
Page, L., Hinshaw, G., Komatsu, E., *et al.* 2007, *ApJS*, 170, 335
Santos, M. R. 2006, *MNRAS*, 349, 1137
Stiavelli, M., Fall, S. M., & Panagia, N. 2004, *ApJ* (Letters), 610, L1
White, R. L., Becker, R. H., Fan, X., & Strauss, M. A. 2003, *AJ*, 126, 1

Highlights of Astronomy, Volume 14
IAU XXVI General Assembly, 14-25 August 2006
Karel A. van der Hucht, ed.

doi:10.1017/S1743921307010320

High-redshift lensed galaxies

Roser Pelló[1], Daniel Schaerer[2,1], Johan Richard[3], Jean-Franc Le Borgne[1], Jean-Paul Kneib[4], Angela Hempel[2], Eiichi Egami[5], Frédéric Boone[6], Franqise Combes[6], Jean-Gabriel Cuby[4], Andrea Ferrara[7] and Michael W. Wise[8]

[1]Laboratoire d'Astrophysique, OMP, 14 Avenue E. Belin, F-31400 Toulouse, France
email: roser@ast.obs-mip.fr

[2]Geneva Observatory, 51 Ch. des Maillettes, CH-1290 Sauverny, Switzerland

[3]Caltech Astronomy, MC105-24, Pasadena, CA 91125, USA

[4]OAMP, Laboratoire d'Astrophysique de Marseille, F-13012 Marseille, France

[5]Steward Observatory, U. of Arizona, 933 North Cherry Avenue, Tucson, AZ 85721, USA

[6]Observatoire de Paris, LERMA, 61 Av. de l'Observatoire, F-75014 Paris, France

[7]SISSA, International School for Advanced Studies, Via Beirut 4, I-34100 Trieste, Italy

[8]M.I.T., Center for Space Research, Cambridge, MA 02139-4307, USA

Abstract. We present the results obtained from our deep survey of lensing clusters aimed at constraining the abundance of star-forming galaxies at $z \sim 6-11$.

Keywords. galaxies:high-redshift, infrared: galaxies, early universe

---

A first attempt was made to address the properties of star-forming galaxies at $z \geqslant 6$ using lensing clusters. High-z candidates were selected among optical dropouts using near-IR photometry. The luminosity funtion (LF) derived at $z \simeq 6$ - 10 is consistent with the LF for LBGs at $z \simeq 3$–4, and also compatible in the low-luminosity regime with the $z \simeq 6$ sample in the UDF and GOODS fields (e.g., Bouwens & Illingworth 2006), but we don't see the turnover observed by other authors towards the bright end. Taken at face value, our results are consistent with a constant SFR density up to $z \simeq 10$. Spectroscopic follow-ups are underway to determine the efficiency of our selection. Additional multi-wavelength photometry is being collected (*HST*, *Spitzer*-IRAC) to improve the characterization of high-z candidates. Lensing clusters seem more efficient than present blank fields to explore the $z \simeq 6$ - 12 domain (within the same photometric depth and FOV). Positive magnification bias is expected from simulations, and seems to be confirmed by our results in lensing clusters. The follow up with the new generation of near-IR spectrographs is also optimized in lensing fields, because of their typical FOV and multiplexing capabilities (e.g., EMIR at GTC). Wide and deep optical+ near-IR surveys in blank fields are also needed to set robust constraints on the bright end of the LF.

More details are given in Pelló *et al.* (2005), Richard *et al.* (2006), Schaerer *et al.* (2006), and references therein.

References

Bouwens, R. J., & Illingworth, G. D. 2006, *New Astronomy Reviews*, 50, 152

Pelló, R., Schaerer, D., Richard, J., *et al.* 2005, in: Y. Mellier, & G. Meylan (eds.), *Gravitational Lensing Impact on Cosmology*, Proc. IAU Symp. No. 225 (Cambridge: CUP), p. 373

Richard, J., Pelló, R., Schaerer, D., *et al.* 2006, *A&A*, 456, 861

Schaerer, D., Pelló, R., Richard, J., *et al.* 2006, *ESO Messenger*, vol. 125, p. 20

Highlights of Astronomy, Volume 14
IAU XXVI General Assembly, 14-25 August 2006
Karel A. van der Hucht, ed.

doi:10.1017/S1743921307010332

New constraints on the co-moving star formation rate in the redshift interval $6 < z < 10$

Richard S. Ellis[1], Daniel P. Stark[1], Johan Richard[1], Andrew J. Bunker[2], Eiichi E. Egami[3] and Jean-Paul Kneib[4]

[1]California Institute of Technology, Mail Stop 169-327, Pasadena, CA 91109, USA
email: rse@astro.caltech.edu
[2]University of Exeter, School of Physics, Stocker Road, Exeter, EX4 4QL, UK
[3]University of Arizona, Tucson, USA
[4]Laboratoire D'Astrophysique, Marseille, France

Abstract. Recent progress in measuring the optical depth of neutral hydrogen in distant quasars and that of electron scattering of microwave background photons suggests that most of the sources responsible for cosmic re-ionisation probably lie in the redshift interval 6 to 10. We present two new observational results which, together, provide valuable constraints on the contribution from star-forming sources in this redshift interval. First, using a large sample of v-band dropouts with unconfused *Spitzer*-IRAC detections, we determine the integrated stellar mass density at $z = 5$. This provides a valuable 'integral constraint' on past star formation. It seems difficult to reconcile the observed stellar mass at $z = 5$ with the low abundance of luminous i-, z-, and J-band dropouts in deep *Hubble Space Telescope* data. Accordingly, we explore whether less luminous star-forming sources in the redshift interval 6 to 10 might be the dominant cause of cosmic re-ionization. In the second component of our research, we report on the results of two surveys for weak Lymanα emitters and z- and J-band dropouts highly-magnified by foreground lensing clusters. Although some promising $z = 8-9$ candidates are found, it seems unlikely that low luminosity sources in this redshift interval can dominate cosmic reionization. If our work is substantiated by more extensive and precise surveys, the bulk of the re-ionizing photons may come from yet earlier sources lying at redshifts $z > 10$.

Keywords. galaxies:high-redshift, infrared: galaxies, early universe

---

Highlights of Astronomy, Volume 14
IAU XXVI General Assembly, 14-25 August 2006
Karel A. van der Hucht, ed.

doi:10.1017/S1743921307010344

Observations of galaxies at $z > 6$. The properties of large, spectroscopic samples

Esther M. Hu, Lennox L. Cowie, and Yuko Kakazu

Institute for Astronomy, Honolulu, USA
email: hu@ifa.hawaii.edu

Abstract. Observed properties of spectroscopically confirmed galaxies at $z \gg 5$ and $z \gg 6$ based on selection from deep, multi-wavelength wide-field samples provide a picture of the current status of the properties of high-redshift galaxies and their evolution to yet higher redshifts.

In the current presentation, we use results of deep, wide-field spectroscopy with the multi-object DEIMOS spectrograph on Keck in combination with deep, wide-field multi-color imaging studies using the SUPRIMECAM CCD camera of Subaru for a number of fields, to evaluate the luminosity function of high-redshift galaxies and its evolution at $z > 6$. High-redshift candidates are selected using both narrow-band Lyman alpha emission and broad-band colors with a high success-rate from a number of SUPRIMECAM (0.5 degree FOV) fields.

Luminosity functions and Lymanα emission line profiles and equivalent widths appear similar between samples at $z \simeq 5.7$ and $z \simeq 6.5$, and the galaxy distribution is structured both spatially and in redshift. A large amount of cosmic variance is seen in the distribution of $z \gg 6$ galaxies from field to field.

The observed properties are discussed in relationship to their impact on strategies for complementary optical surveys of high-redshift galaxies, and in relationship to surveys at very different wavelengths (X-ray, far-infrared, and submillimeter) that cover the same regions.

Keywords. galaxies:high-redshift, infrared: galaxies, early Universe

Highlights of Astronomy, Volume 14
IAU XXVI General Assembly, 14-25 August 2006
Karel A. van der Hucht, ed.

doi:10.1017/S1743921307010356

Properties of Lyα emitters at $z \simeq 6$

Crystal L. Martin[1], Marcin Sawicki[1], Alan Dressler[2] and Patrick J. McCarthy[2]

[1]Department of Physics, UCSB, Santa Barbara, CA 93106, USA
email: cmartin@physics.ucsb.edu

[2]OCIW, 813 Santa Barbara Street, Pasadena, CA 91101-1292, USA

Abstract. We confirm the redshift of several $z \simeq 6$ objects discovered by our IMACS multislit emission-line survey. Their Lyα luminosities are lower than those of galaxies previously discovered using narrow-band imaging, as expected due to the excellent sky-supression inherent to this technique. Based on the line profiles of these objects, we argue that they are extremely young starbursts and find strong evidence for prominent galactic winds. This population of young galaxies is largely beyond the reach of current large surveys that use continuum selection.

Keywords. galaxies:high-redshift, line: identification, techniques: miscellaneous

---

Using the Inamori Magellan Areal Camera and Spectrograph (IMACS) on the Baade Telescope, we carried out a spectroscopic, emission-line survey through the OH-free atmospheric window at 8200 Å. A custom blocking filter and multislit mask allowed us to search 50 square arcminutes of *blank sky* per pointing. The total survey area to date is 200 square arcminutes, a significantly larger area than that covered by previous multislit surveys (Martin & Sawicki 2004; Tran *et al.* 2004).

We discovered nearly 300 emission-line galaxies and identified $\sim 90\,\%$ of them as foreground objects in our follow-up observations. From the spectra of the confirmed Lyα emitters, we derive physical properties of galaxies at redshift 5.7. For example, we detect N V $\lambda\lambda 1239, 43$ in the spectrum of MSDM 29.5 + 5.1 in addition to a 280 km s^{-1} wide Lyα line. We argue that this object is either a very young starburst (light dominated by O and WR stars) or a Type II AGN. Higher resolution spectra of this object and that of MSDM 80.0+3 reveal structure in the Lyα line profiles characteristic of radiative transfer effects in galactic winds (Hansen & Oh 2006; Verhamme *et al.* 2006).

These Lyα emitters are the first discovered using the multislit search technique. They are drawn from a much larger volume than lensed searches. Typical line fluxes, $F \approx 6 \times 10^{-18}$ ergs s^{-1} cm^{-2} or $\log L \approx 42.32$ erg s^{-1}, are fainter than those from narrow-band imaging surveys. Addition of our results to those from these techniques will better determine the luminosity of the knee in the Lyα luminosity distribution near redshift 6. Measurement of the Lyα luminosity function will impact the outstanding question of whether the objects that ionize the intergalactic medium have been identified and make it possible to trace the progression of re-ionization via evolution in the luminosity function.

References

Hansen, M., & Peng Oh, S. 2006, *MNRAS*, 367, 979
Martin, C. L., & Sawicki, M. 2004, *ApJ*, 603, 414
Tran, K. H., Lilly, S. J., Crampton, D., & Brodwin, M. 2004, *ApJ*, 612, 89
Verhamme, A., Schaerer, D., & Maselli, A. 2006, *A&A*, 460, 397

Highlights of Astronomy, Volume 14
IAU XXVI General Assembly, 14-25 August 2006
Karel A. van der Hucht, ed.

doi:10.1017/S1743921307010368

High-redshift Lyman-α galaxies

Sangeeta Malhotra[1] and James E. Rhoads[2]

[1]Department of Physics and Astronomy, Arizona State University,
Tempe, AZ 85287-1504, USA
email: san@stsci.edu
[2]School of Earth and Space Exploration, Arizona State University,
Tempe, AZ 85287-1504, USA
email: james.rhoads@asu.edu

A strong Lyman-α line enables relatively easy detection of high redshift galaxies. Lyman-α galaxies are now known from $z=3$ to 6.6. No evolution is discerned in the Lyman-α line luminosity function in this redshift range. This implies that the intergalactic medium at $z=6$ is at least 50 % ionized over more than 50 % of the volume. Recent continuum detections of these galaxies from *HST*, MMT and *Spitzer* are now allowing us to address questions about the nature of these Lyman-α emitters, their stellar populations and ages. We find that by and large the Lyman-α galaxies are young galaxies dominated by stellar populations that are less than 25 Myr old.

doi:10.1017/S1743921307012392

Cosmic microwave background: probing the universe from $z=6$ to 1100

David N. Spergel

Princeton University, Peyton Hall, Ivy Lane, Princeton, NJ 08544-1001, USA
email: dns@astro.princeton.edu

Observations of cosmic microwave background temperature and polarization fluctuations are sensitive to both physical conditions at recombination ($z=1100$) and physical process along the line of sight. I will discuss recent results from the Wilkinson Microwave Anisotropy Probe and planned ground and space-based observations. The talk will emphasize the role of CMB observations in determining the initial conditions for the growth of structure and as a probe of the physics of re-ionization.

Highlights of Astronomy, Volume 14
IAU XXVI General Assembly, 14-25 August 2006
Karel A. van der Hucht, ed.

doi:10.1017/S1743921307010381

An observational pursuit for Pop III stars in a Lyα emitter at $z > 6$ through He II emission

Tohru Nagao

National Astronomical Observatory of Japan, Mitaka, Tokyo 181-8588, Japan
email: tohru@optik.mtk.nao.ac.jp

Abstract. We report our on-going observational project to search for population III (Pop III) stars in high-z galaxies. We searched Lyα emitters (LAEs) with a large equivalent width (EW), by our new selection technique 'NB921-depressed i'-dropout selection'. We found eight photometric candidates and spectroscopically identified five LAEs with $EW_0(\mathrm{Ly}\alpha) > 100$ ÅWe then carried out a very deep near-infrared spectroscopy for a LAE among the above five, to search for the redshifted He II λ1640 emission from Pop III stars in the galaxy, but obtained only an upper limit.

Keywords. early Universe, galaxies: evolution, galaxies: starburst, stars: early

---

The detection and investigation of the first-generation stars, Population III (PopIII) stars, will be one of the main goals of astronomy in the next decade. Since galaxies with massive Pop III stars are expected to show a very strong Lyα emission and a detectable He II emission in their spectra (e.g., Schaerer 2002; Schaerer 2003), we are promoting a project to search for such spectroscopic signatures of Pop III stars.

We developed a new method to select Lyα emitters (LAEs) with a large equivalent width (EW) at a wide redshift range, $6.0 < z < 6.5$, by focusing 'NB921-depressed i'-dropout' objects (Nagao *et al.* 2004). Through the follow-up spectroscopic observations with Subaru and Keck telescopes, we identified five strong LAEs with $EW_0(\mathrm{Ly}\alpha) > 100$ Å among eight photometric candidates (Nagao *et al.* 2005; Nagao *et al.* 2007).

Among the identified NB921-depressed i'-dropout galaxies, we focused on a LAE at $z = 6.33$ and with $EW_0(\mathrm{Ly}\alpha) = 130$ Å and carried out a very deep J-band spectroscopic observation to search for the redshifted He II λ1640 emission from Pop III stars in this LAE. Even after 42 ksec of integration with the Subaru-OHS spectrograph, no emission-line features are detected in the J-band. We obtained a 2σ upper limit of 9.06×10^{-18} erg s^{-1}cm^{-2} on the He II λ1640 flux, which corresponds to a luminosity of 4.11×10^{42} erg s^{-1}. This upper limit implies that the upper limit on the Pop III star formation rate is in the range 4.9 - 41.2 $M_\odot$yr^{-1} if Pop III stars suffer no mass loss, and in the range 1.8 - 13.2 $M_\odot$yr^{-1} if strong mass loss is present. The non-detection of He II in the target LAE may thus disfavor weak feedback models for Pop III stars.

Acknowledgements

This study was done under a large collaboration and the names of collaborators cannot be given in the author list due to the limited space. TN is financially supported by JSPS.

References

Nagao, T., Taniguchi, Y., Kashikawa, N., *et al.* 2004, *ApJ* (Letters), 613, L9
Nagao, T., Kashikawa, N., Malkan, M. A., *et al.* 2005, *ApJ*, 634, 142
Nagao, T., Murayama, T., Maiolino, R., *et al.* 2007, *A&A*, 468, 877
Schaerer, D. 2002, *A&A*, 382, 28
Schaerer, D. 2003, *A&A*, 397, 527

Highlights of Astronomy, Volume 14
IAU XXVI General Assembly, 14-25 August 2006
Karel A. van der Hucht, ed.

doi:10.1017/S1743921307010393

The end of the re-ionization epoch probed by Lyα emitters at $z = 6.5$

Nobunari Kashikawa, *et al.*

National Astronomical Observatory of Japan, Mitaka, Tokto, 181-8588, Japan
email: kashik@zone.mtk.nao.ac.jp

Abstract. We report an extensive search for Lyα emitters (LAEs) at $z = 6.5$ in the Subaru Deep Field (SDF). We carried out spectroscopic observations with Subaru/Keck to identify LAEs at $z = 6.5$ that were selected by narrow-band excess at 920 nm. We have identified eight new LAEs based on their significantly asymmetric Lyα emission profiles. This increases the sample of spectroscopically confirmed $z = 6.5$ LAEs in the SDF to 17. Based on this spectroscopic sample of 17, complemented by a photometric sample of 58 LAEs, we have derived a more accurate Lyα luminosity function (LF) of LAEs at $z = 6.5$, which reveals an apparent deficit at the bright end, of ~ 0.75 mag fainter $L*$, compared with that observed at $z = 5.7$. The difference has $3\,\sigma$ significance, which is reduced to $2\,\sigma$ when cosmic variance is taken into account. Several LAEs with high Lyα luminosity have been actually identified by spectroscopy at $z < 5.7$, while our LAE sample at $z = 6.5$ has no confirmed object having such a high Lyα luminosity. The LF of the rest UV continuum, which is not sensitive to neutral IGM, of our LAE sample has almost the same as those of LAEs at $z = 5.7$ and i-dropouts at $z \simeq 6$, even at their bright ends. This result may imply that the reionization of the universe has not been completed at $z = 6.5$. The decline of the Lyα LF implies the cosmic neutral fraction $x_{\rm HI} = 0.45$ based on a theoretical IGM model, although this predicted value is strongly model dependent. The spatial distribution of our LAE sample was found to be homogeneous over the field, based on three independent methods to quantify the clustering strength. The composite spectrum of our LAE sample clearly reveals an asymmetric Lyα profile with an extended red wing, which can be explained by either a galactic wind model composed of double Gaussian profiles, or by a reionization model expected from the damping wing profile. Although our result has uncertainties in LAE evolution and large cosmic variance, it can be interpreted that LAEs at $z = 6.5$ are at the end of the reionization epoch.

Keywords. cosmology: observations, early universe

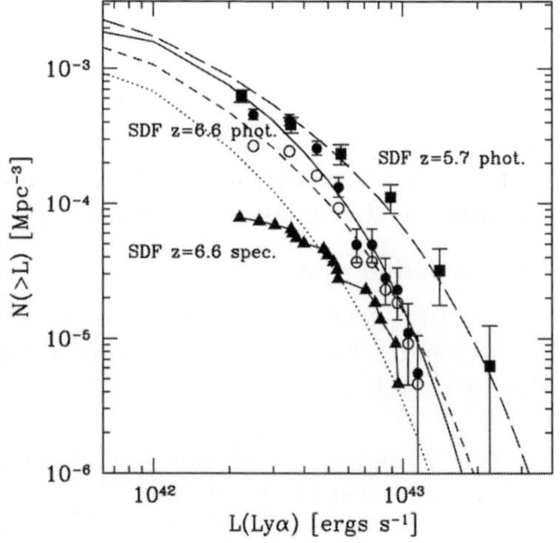

Figure 1: Cumulative Lyα LF of our LAE sample at $z = 6.5$. The open circles denote the raw counts of our spectroscopic sample + additional photometric sample, and the filled circles are corrected for detection completeness. The triangles denote the raw counts of the pure spectroscopic sample. The squares indicate the LF of LAEs at $z = 5.7$ evaluated from the SDF. The short-dashed and dotted lines show the Schechter LFs, in which the Lyα luminosities are reduced by a factor of 0.6 ($L^* \times 0.6$) and 0.4 from the $z = 5.7$ LF, respectively.

Reference

Kashikawa, N., Shimasaku, K., Malkan, M. A., *et al.* 2006, *ApJ*, 648, 7

Highlights of Astronomy, Volume 14
IAU XXVI General Assembly, 14-25 August 2006
Karel A. van der Hucht, ed.

doi:10.1017/S174392130701040X

Metallicity and black hole masses of $z \simeq 6$ quasars

Jaron D. Kurk[1], Fabian Walter[1], Dominik Riechers[1], Hans-Walter Rix[1], Stefan Wagner[2], Laura Pentericci[3] and Xiaohui Fan[4]

[1]Max-Planck-Institut für Astronomie, Königstuhl 17, D-69117, Heidelberg, Germany

[2]Landessternwarte Heidelberg, Königstuhl 12, D-69117, Heidelberg, Germany

[3]Osservatorio Astronomico di Roma, Via di Frascati 33, I-00040, Monte Porzio Catone, Italy

[4]Steward Observatory, The University of Arizona, 933 N. Cherry Av., Tucson, AZ 85721, USA

Abstract. We present NIR spectroscopy of emission lines of a sample of five $z \simeq 6$ quasars, including fainter objects than observed before. The measured Fe II / Mg II ratios are around solar and consistent with a lack of evolution of the metallicity of quasar BLRs up to $z \simeq 6$, suggesting that stars in their hosts formed at $z \gg 6$. The BH masses, measured from both Mg II and C IV line widths are within the range 2 - 16 $\times 10^8$ $M_{\odot}$, the smallest found in such distant objects.

Keywords. galaxies: high-redshift, galaxies: fundamental parameters, galaxies: abundances

---

Quasars at $z \simeq 6$ provide direct probes of galaxy hosts and the IGM close to the epoch of reionization. In the last few years, more than twenty such objects at $z > 5.7$ have been discovered. We have selected five of these for follow-up NIR spectroscopy. The sample (J0836+0054, J1030+0524, J1306+0356, J0005−0006, and J1411+1217, see Fan *et al.* 2001, 2004) contains all published quasars observable by the VLT with $z > 5.8$ and includes quasars with magnitudes in the range $18.7 < i^* < 20.5$ at $5.8 < z < 6.3$, extending this kind of research by about a factor two down the luminosity function (e.g., Maiolino *et al.* 2004). Observations were carried out with VLT-ISAAC in low resolution mode in the *SZ*-, *J*- and *K*-bands with exposure times of one to three hours per spectrum. The emission lines were fitted by a Gaussian curve (Mg II), Lorentzian curve (C IV) or iron template (Fe II, from Vestergaard & Wilkes 2001), while the underlying continuum was fitted by a power law. In the region around 3000 Å, the Balmer pseudo-continuum was modeled. The measured Fe II / Mg II ratios are in the range 2.2 - 4.7, consistent with those found in lower redshift quasars. Applying the relations found by McLure & Jarvis (2002) and Vestergaard (2002) to the width of the Mg II and C IV lines and the continuum fluxes at $\lambda_0 = 3000$ Å and 1350 Å, respectively, we find black hole (BH) masses of 2, 3, 12, 13 and 16×10^8 $M_{\odot}$. These results imply that the metallicity of BLRs in quasars does not evolve up to $z \simeq 6$, suggesting a period of intense star formation at $z > 10$ in the host galaxy. We also find that quasars at $z \simeq 6$ do no only contain BHs more massive than 10^9 $M_{\odot}$.

Acknowledgements

JK is supported by the DFG under grant SFB/439. Based on observations carried out at ESO, Paranal, Chile.

References

Fan, X., Narayanan, V. K., Lupton, R. H., *et al.* 2001, *AJ*, 122, 2833
Fan, X., Hennawi, J. F., Richards, G. T., *et al.* 2004, *AJ*, 128, 515
Maiolino, R., Oliva, E., Ghinassi, F., *et al.* 2004, *A&A*, 420, 889
McLure, R. J., & Jarvis, M. J. 2002, *MNRAS*, 337, 109
Vestergaard, M., & Wilkes, B. J. 2001, *ApJS*, 134, 1
Vestergaard, M. 2002, *ApJ*, 571, 733

Highlights of Astronomy, Volume 14
IAU XXVI General Assembly, 14-25 August 2006
Karel A. van der Hucht, ed.

doi:10.1017/S1743921307010411

Populations of candidate black holes at redshift 7 or above

Anton M. Koekemoer

Space Telescope Science Institute, Baltimore, MD 21218, USA
email: koekemoe@stsci.edu

Abstract. I will describe recent results on constructing samples of candidate active galactic nuclei (AGN) at or beyond redshift 7, probing several orders of magnitude fainter than the top end of the quasar luminosity function at redshift 6. These advances have been made possible by the advent of deep, wide multi-waveband surveys that enable the selection of samples of sources that are detected at radio or X-ray wavelengths but completely undetected at optical wavelengths to very deep limits. A variety of multi-band selection criteria are used to identify the high-redshift candidates and eliminate lower-redshift interlopers by means of extensive spectral energy distribution modelling. The resulting constraints on the numbers of high-redshift AGN at or above redshift 7 are used to examine the evolution of the AGN luminosity function at high redshift, and help understand the properties of the first supermassive black holes in the universe.

Keywords. galaxies: active, quasars: general

Highlights of Astronomy, Volume 14
IAU XXVI General Assembly, 14-25 August 2006
Karel A. van der Hucht, ed.

doi:10.1017/S1743921307010423

Re-ionization imprints in high-z QSO spectra

Simona Gallerani, T. Roy Choudhury and Andrea Ferrara

SISSA/ISAS, via Beirut 2-4, 34014 Trieste, Italy
email: galleran,chou,ferrara@sissa.it

Abstract. We use a semi-analytical approach to simulate absorption spectra of QSOs at high redshifts with the aim of constraining the cosmic reionization history. More details are given in Gallerani *et al.* (2006) and references therein.

Keywords. intergalactic medium quasars: absorption lines, cosmology: theory large-scale structure of Universe.

We consider two physically motivated and detailed re-ionization histories: (i) an Early Re-ionization Model (ERM) in which the intergalactic medium is re-ionized by Pop III stars at $z > 10$ (Choudhury & Ferrara 2005), and (ii) a more standard Late Re-ionization Model (LRM) in which overlapping, induced by QSOs and normal galaxies, occurs at $z \simeq 6$. From the analysis of current Lyα forest data at $z < 6$, we conclude that it is impossible to disentangle the two scenarios, which fit equally well the observed Gunn-Peterson optical depth, flux probability distribution function and dark gap width distribution. At $z > 6$, however, clear differences start to emerge which are best quantified by the dark gap width distribution, as can be seen from Figure 1.

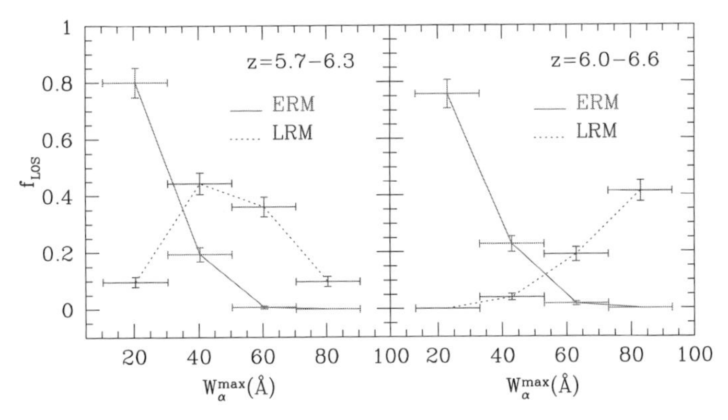

Figure 1. Distribution of the largest dark gap widths $W_\alpha^{\max}$ for 300 lines of sight in the redshift range 5.7 - 6.3 (*left panel*) and 6.0 - 6.6 (*right panel*) for ERM (*solid line*) and LRM (*dotted line*). The vertical error bars denote the cosmic variance; the horizontal error bars show the bin size.

We find that 35 (zero) per cent of the lines of sight within $5.7 < z < 6.3$ show dark gaps widths > 50 Å in the rest frame of the QSO if reionization is not (is) complete at $z \gtrsim 6$. We conclude that the dark gap width statistics represent a superb probe of cosmic re-ionization if about ten QSOs can be found at $z > 6$.

References

Gallerani, S., Choudhury, T. R., & Ferrara, A. 2006, *MNRAS*, 370, 3
Choudhury, T. R., & Ferrara, A. 2005, *MNRAS*, 361, 577

Highlights of Astronomy, Volume 14
IAU XXVI General Assembly, 14-25 August 2006
Karel A. van der Hucht, ed.

doi:10.1017/S1743921307010435

On the size of H II regions around high-redshift quasars

Antonella Maselli[1], Simona Gallerani[2], Andrea Ferrara[2] and T. Roy Choudhury[3]

[1]Max-Planck-Institut für Astrophysik,
Karl-Schwarzschild-Straße 1, D-85748 Garching, Germany
[2]SISSA/International School for Advanced Studies, via Beirut 2-4, I-34014 Trieste, Italy
[3]Centre for Theoretical Studies, Indian Institute of Technology, Kharagpur 721302, India

Abstract. We investigate the possibility of constraining the ionization state of the Intergalactic Medium (IGM) close to the end of re-ionization ($z \approx 6$) by measuring the size of the H II regions in high-z quasars spectra, via a combination of SPH and 3D radiative transfer (RT) simulations and a statistical analysis of mock quasar spectra through the simulated cosmological volume.

Keywords. cosmology: theory, radiative transfer, methods: numerical, intergalactic medium, cosmology: large scale structure of Universe

---

The size of H II regions around high-z luminous quasars prior to complete re-ionization is strongly dependent on the mean neutral hydrogen fraction of the IGM, $x_{\rm HI}$. Previous studies (e.g., Wyite & Loeb 2004) have tried to constraints $x_{\rm HI}$ using the size of high-z QSOs H II regions, measurable in their own spectra as the extent of the transmitting region between the quasar emission redshift and the onset of the Gunn-Peterson trough.

The aim of our work is to assess the robustness of the above method. We have performed a combination of state-of-art multiphase SPH and 3D radiative transfer (RT) simulations to accurately predict the properties of a typical high-z quasars H II region (eg. extent, geometrical shape, inner opacity), assuming an initial $x_{\rm HI} = 0.1$.

The simulation results show that RT effects do not induce strong deviations from spherical symmetry; we find a mean dispersion in the H II region size along different LOS of the order of roughly 6 % of the mean radius.

By deriving and analyzing mock spectra through the simulated quasar environment we have found that the H II region size deduced from quasar spectra typically underestimates the physical one by 30 %. This effect, to which we refer as *apparent shrinking*, results to be almost completely due to resonant absorption of residual H I inside the ionized bubble.

Additional maximum likelihood analysis shows that this offset induces an overestimate of the neutral hydrogen fraction, $x_{\rm HI}$, by a factor ~ 3. By applying the same statistical method to a sample of six observed QSOs spectra analyzed by Fan *et al.* (2006), our study favors a mostly ionized ($x_{\rm HI} < 0.06$) Universe at $z = 6.1$.

All together the results of our work suggest that measurements of the H II size in quasar spectra can only provide rough constraints on $x_{\rm HI}$, as far as the knowledge of intrinsic properties of observed QSOs remains incomplete.

More details are given in Maselli *et al.* (2007) and references therein.

References

Fan, X., Carilli, C. L., & Keating, B. 2006, *ARA&A*, 44, 415
Maselli, A., Gallerani, S., Ferrara, A., & Choudhury, T. R. 2007, *MNRAS* (Letters), 376, 34
Wyithe, J. S. B., & Loeb, A. 2004, *Nature*, 427, 815

Highlights of Astronomy, Volume 14
IAU XXVI General Assembly, 14-25 August 2006
Karel A. van der Hucht, ed.

doi:10.1017/S1743921307010447

Constraints on very faint high-z quasars

Mark Dijkstra
The School of Physics, The University of Melbourne, Parkville, VIC 3066, Australia
email: dijkstra@physics.unimelb.edu.au

Abstract. I discuss the constraints that can be derived on the abundance of high redshift ($z > 6$) (mini)quasars from the unresolved soft X-Ray background. Furthermore, I will show how existing Lyα surveys can be used to probe the very faint $M_B \gtrsim -21$ mag (i.e., $\gtrsim 7$–8 mag fainter than the SDSS quasars) end of the $z \geqslant 4.5$ quasar luminosity function.

Keywords. galaxies: high-redshift, X-rays: diffuse background, cosmology: theory, quasars: emission lines.

1. Constraints from the soft X-ray background

A population of black holes (BHs) at high redshifts ($z > 6$) that contributes significantly to the ionization of the intergalactic medium would be accompanied by copious production of hard (> 10 keV) X-ray photons. The resulting hard X-ray background would redshift and be observed as a present-day soft X-ray background (SXB). In Dijkstra *et al.* (2004) we show how existing models, in which BHs are the main producers of re-ionizing photons in the high-redshift universe, contribute more to the present-day SXB, than its unresolved component. This suggests that accreting BHs (be it luminous quasars or their lower-mass ‘miniquasar’ counterparts) did not dominate re-ionization. These results depend most sensitively on the exact spectrum emitted by these accreting BHs.

2. Constraints from Lyα surveys

There is good evidence that low numbers of Active Galactic Nuclei (AGN, or quasars) are among observed faint Lyα emitters at $z = 4.5$–6.5. Combining this observations with an empirical relation between the intrinsic Lyα and B-band luminosities of AGN, we obtain an upper limit on the number density of AGN with absolute magnitudes $M_B \in [-16, -19]$ at $z = 4.5$–6.5 (Dijkstra & Wyithe 2007). These AGN are up to two orders of magnitude fainter than those discovered in the Chandra Deep Field, resulting in the faintest observational constraints on the quasar luminosity function at these redshifts to date. We believe that existing and future Lyα surveys could make a significant contribution to our understanding of the formation and evolution of high redshift BHs and AGN.

Acknowledgements

MD is supported by the Australian Research Counsil. I thank my collaborators Zoltan Haiman, Abraham Loeb and Stuart Wyithe for many useful discussions and the organisers of Joint Discussion 7, particularly Daniel Schaerer, for the opportunity to present this work.

References

Dijkstra, M., Haiman, Z., & Loeb, A. 2004, *ApJ*, 613, 646
Dijkstra, M., & Wyithe, J. S. B. 2007, *MNRAS*, 379, 1589

Highlights of Astronomy, Volume 14
IAU XXVI General Assembly, 14-25 August 2006
Karel A. van der Hucht, ed.

doi:10.1017/S1743921307010459

Dust at $z > 6$. Observations and theory

Roberto Maiolino

INAF - Osservatorio Astronomico di Roma, Monte Porzio Catone, 00142, Italy
email: maiolino@oa-roma.inaf.it

Abstract. I shortly review the investigation of dust in the early universe. I discuss the possible evolution of the dust properties, as well as the possible dust production mechanisms at $z > 6$.

Keywords. Dust, galaxies: ISM, galaxies: active, early Universe

---

Dust plays a crucial role in the early universe. It affects the formation of high mass stars, it allows the formation of low mass stars in low metallicity environments (Schenider *et al.* 2006), and it also greatly enhances the formation of molecules. However, according to the standard scenarios we expect little or no dust at $z > 6$. Indeed, in the local universe dust is mostly produced in the envelopes of evolved, low mass stars (mostly AGB stars) and, as a consequence, most of the dust is produced with a delay of ~ 1 Gyr with respect to the onset of star formation. At $z > 6$ the age of the universe is less than 1 Gyr, thus falling short of time to produce dust through AGBs.

Therefore, although other production mechanisms have been proposed, a strong decrease of the average dust mass in galaxies is expected at $z > 6$, just because the main dust contributor is missing. Currently there are only hints of this effect, observed in the IR data of high-z QSOs (e.g., Jiang *et al.* 2007) and in the optical-UV SEDs of star forming galaxies (e.g., Eyles *et al.* 2007).

The most sensitive tracer of the dust mass is its far-IR emission, which at $z > 6$ is observed in the submm-mm range. However, currently there are only a few submm-mm detections, of hyperluminous QSOs, at $z > 6$, preventing any meaningful investigation on the evolution of the dust mass at high-z. However, these observations reveal that in these few objects huge masses of dust ($\sim 10^8 M_\odot$) have already formed by $z = 6$ (Beelen *et al.* 2006). Such large masses cannot be accounted for by AGB stars at such early times in the Universe.

An alternative, rapid channel of dust production are the ejecta of type II SNe (Todini & Ferrara 2001). The dust properties inferred from the extinction curve in QSOs at $z > 6$ are consistent with the presence of SN dust (Maiolino *et al.* 2004). Whether dust production in SNe can be efficient enough to account for the whole dust mass observed in bright QSOs at $z > 6$ is still under debate. The constraints on the SN dust yield inferred from Galactic SNRs are not conclusive (*Spitzer* mid-IR observation only probe the warm component of dust, while sub-mm observations provide only loose upper limits). However, the recent *Spitzer* detection of dust forming in SN 2003gd (Sugerman *et al.* 2006) indicate a dust production efficiency which could account for the dust masses observed at $z > 6$.

References

Beelen, A., Cox, P., Benford, D. J., *et al.* 2006, *ApJ*, 642, 694
Eyles, L. P, Bunker, A. J., Ellis, R. S., *et al.* 2007, *MNRAS*, 374, 910
Jiang, R., Fan, X., Hines, D. C., *et al.* 2007, *AJ*, in press [astro-ph/0608006]
Maiolino, R., Schneider, R., Oliva, E., *et al.* 2004, *Nature*, 431, 533
Schneider, R., Omukai, K., Inoue, A. K., & Ferrara, A. 2006, *MNRAS*, 369, 1437
Sugerman, B. E. K., Ercolano, B., Barlow, M. J., *et al.* 2006, *Science*, 313, 196
Todini, P. & Ferrara, A. 2001, *MNRAS*, 325, 726

Highlights of Astronomy, Volume 14
IAU XXVI General Assembly, 14-25 August 2006
Karel A. van der Hucht, ed.

doi:10.1017/S1743921307010460

The first galaxies at cm and mm wavelengths

Fabian Walter[1], Chris L. Carilli[2], Frank Bertoldi[3], Pierre Cox[4] and Karl M. Menten[5]

[1]Max Planck Insitut für Astronomie, Heidelberg, Germany
[2]National Radio Astronomy Observatory, USA. [3]Bonn University, Germany
[4]IRAM, France. [5]Max Planck Insitut für Radioastronomie, Bonn, Germany

Abstract. Observations of the most distant ($z \simeq 6$) QSOs in the centimetre and millimetre regime currently serve as the only direct probe of the host galaxies of these extreme systems in the Epoch of Re-ionization. Such observations reveal that about 1/3 of the hosts contain massive reservoirs of dust ($> 10^8$ M$_\odot$) and molecular gas ($> 10^10$ M$_\odot$) – the fuel for galaxy formation, and also indicate coeval starbursts at a rate $> 10^3$ M$_\odot$ yr^{-1}, adequate to form a large elliptical galaxy in a dynamical timescale. These data imply that a highly metal enriched, molecular ISM, can be generated in galaxies within 870 Myr of the Big Bang. High-resolution imaging of the gas also provide an estimate of the host galaxy dynamical mass. However, current observations are restricted to rare, hyper-luminous IR galaxies. I will close by considering the prospects of observing the gas, dust, and star formation in the first 'normal' galaxies (e.g., the Ly-α galaxies) into cosmic reionization ($z > 6$), using ALMA and the EVLA.

Keywords. radio lines: ISM,radio lines: galaxies, quasars: general

---

We present new results of our ongoing efforts to study the properties of the molecular gas and dust in high-redshift quasars. Our ongoing survey using MAMBO at the IRAM 30 m to detect dust emission in $z \simeq 6$ QSOs has resulted in a number of new detections (Wang *et al.* 2007). As the detected objects are dust-rich, they represent good targets to perform follow-up observations of the molecular gas phase. We also report on observations of the (rest-frame) FIR continuum of the GRB 050904 at $z = 6.29$ with MAMBO, which have resulted in a non-detection of the GRB host galaxy (Walter *et al.* 2006).

Sensitive radio continuum observations of the quasars using the VLA have resulted in a number of new detections which reveal that at least some of the objects appear to be radio loud (Wang *et al.* 2007). These observations provide a glimpse of what will be possible with the expanded (E)VLA, which will improve the continuum sensitivity of similar observations by an order of magnitude. Observations of the molecular gas phase in $z \simeq 6$ objects are still limited by (*i*) redshift uncertainties, and (*ii*) the sensitivity of current instruments. This is the main reason why to date there is only one $z > 6$ CO detection (the $z = 6.42$ QSO J1148+5251, Walter *et al.* 2003, 2004).

This situation will improve dramtically with the advent of new telescopes with larger bandwidths and collecting area (most notably ALMA), which will also enable us to map the CO emission in the host galaxies at these extreme redshifts. Of additional interest are detection and mapping experinents of the emission from ionized carbon which appears to be bright in the $z = 6.42$ QSO (Maiolino *et al.* 2005).

References

Maiolino, R., Cox, P., Caselli, P., *et al.* 2005, *A&A* (Letters) , 440, L51
Walter, F., Bertoldi, F., Carilli, C. L., *et al.* 2003, *Nature*, 424, 406
Walter, F., Carilli, C. L., Bertoldi, F., *et al.* 2004, *ApJ* (Letters), 615, L17
Walter, F., *et al.* 2006, *GCN Circular* 5300
Wang, R., Carilli, C. L., Beelen, A., *et al.* 2007, *AJ*, 134, 617

Highlights of Astronomy, Volume 14
IAU XXVI General Assembly, 14-25 August 2006
Karel A. van der Hucht, ed.

doi:10.1017/S1743921307010472

Spectroscopy of the near-infrared afterglow of GRB 050904 at $z = 6.3$

Nobuyuki Kawai

Department of Physics, Tokyo Institute of Technology,
2-12-1 Ookayama, Meguro-ku, Tokyo 152-8551, Japan
email: nkawai@hp.phys.titech.ac.jp

Abstract. We present the optical/NIR spectrum of the afterglow of GRB 050904 obtained with the Faint Object Camera And Spectrograph on the Subaru 8.2 m telescope taken 3.4 days after the burst. It is, as of June 2006, the only GRB with a known redshift larger than 6. The spectrum shows a clear continuum at the long wavelength end of the spectrum with a sharp cutoff at around 900 nm due to Lyα absorption at a redshift of 6.3 with a damping wing. Little flux is present in the waveband shortward of the Lyα break. A system of absorption lines of heavy elements at redshift $z = 6.295 \pm 0.002$ were also detected, yielding a precise measurement of the largest known redshift of a GRB. Analysis of the silicon and sulphur absorption lines suggests a dense environment around the GRB with the metallicity larger than 0.1 solar, providing unique information on the galaxy and star forming region at $z > 6$. This observation has shown that GRB is a powerful probe of the early universe.

Keywords. galaxies:high-redshift, gamma rays: bursts

Highlights of Astronomy, Volume 14
IAU XXVI General Assembly, 14-25 August 2006
Karel A. van der Hucht, ed.

doi:10.1017/S1743921307010484

Implications for the cosmic re-ionization from the optical afterglow spectrum of GRB 050904 at $z = 6.3$

Tomonori Totani[1], Nobuyuki Kawai[2], George Kosugi[3], Kentaro Aoki[4], Toru Yamada[3], Masanori Iye[3], Kouji Ohta[1] and Takashi Hattori[4]

[1]Department of Astronomy, Kyoto University, Sakyo-ku, Kyoto 606-8502, Japan
email: totani@th.nao.ac.jp

[2]Department of Physics, Tokyo Institute of Technology, 2-12-1 Ookayama, Meguro-ku, Tokyo 152-8551, Japan

[3]National Astronomical Observatory of Japan, 2-21-1 Osawa, Mitaka, Tokyo 181-8588, Japan

[4]Subaru Telescope, National Astronomical Observatory of Japan, Hilo, HI 96720, USA

Abstract. We discuss the implications for the cosmic re-ionization from the optical afterglow spectrum of GRB 050904 at $z = 6.3$

Keywords. cosmology: observations, cosmology: theory, early Universe

---

The gamma-ray burst GRB 050904 at $z = 6.3$ provides the first opportunity of probing the intergalactic medium (IGM) by GRBs at the epoch of the reionization. Here we present a spectral modeling analysis of the optical afterglow spectrum taken by the Subaru Telescope, aiming to constrain the reionization history. The spectrum shows a clear damping wing at wavelengths redward of the Lyman break, and the wing shape can be fit either by a damped Lyα system with a column density of $\log(N_{\rm HI}/{\rm cm}^{-2}) \simeq 21.6$ at a redshift close to the detected metal absorption lines ($z_{\rm metal} = 6.295$), or by almost neutral IGM extending to a slightly higher redshift of $z_{\rm IGM,u} \simeq 6.36$. In the latter case, the difference from $z_{\rm metal}$ may be explained by acceleration of metal absorbing shells by the activities of the GRB or its progenitor.

However, we exclude this possibility by using the light transmission feature around the Lyβ resonance, leading to a firm upper limit of $z_{\rm IGM,u} \leqslant 6.314$. We then show an evidence that the IGM was largely ionized already at $z = 6.3$, with the best-fit neutral fraction of IGM, $x_{\rm HI} = 0.00$, and upper limits of $x_{\rm HI} < 0.17$ and 0.60 at 68 and 95 % C.L., respectively. This is the first direct and quantitative upper limit on $x_{\rm HI}$ at $z > 6$. Various systematic uncertainties are examined, but none of them appears large enough to change this conclusion.

To get further information on the re-ionization, it is important to increase the sample size of $z \gtrsim 6$ GRBs, in order to find GRBs with low column densities ($\log N_{\rm HI} \lesssim 20$) within their host galaxies, and for statistical studies of Lyα line emission from host galaxies.

See Kawai *et al.* (2006) and Totani *et al.* (2006) for details of our work.

References

Kawai, N., Kawai, N., Kosugi, G., Aoki, K., Yamada, T., Totani, T., Ohta, K., Iye, M., Hattori, T., Aoki, W., Furusawa, H., *et al.* 2006, *Nature*, 440, 184

Totani, T., Kawai, N., Kosugi, G., Aoki, K., Yamada, T., Iye, M., Ohta, K., & Hattori, T. 2006, *PASJ*, 58, 485

Highlights of Astronomy, Volume 14
IAU XXVI General Assembly, 14-25 August 2006
Karel A. van der Hucht, ed.

doi:10.1017/S1743921307010496

Anisotropies of the infrared background and primordial galaxies

Asantha R. Cooray

Center for Cosmology, Department of Physics and Astronomy, University of California, Irvine, CA 92697, USA
email: asante@caltech.edu

Abstract. We discuss anisotropies in the near-IR background between 1 to a few microns. This background is expected to contain a signature of primordial galaxies. We have measured fluctuations of resolved galaxies with Spitzer imaging data and we are developing a rocket-borne instrument (the *Cosmic Infrared Background ExpeRiment*, or *CIBER*) to search for signatures of primordial galaxy formation in the cosmic near-infrared extra-galactic background.

Keywords. large scale structure of Universe, diffuse radiation, infrared: galaxies

---

The intensity of the cosmic near-infrared background (IRB) is a measure of the total light emitted by stars and galaxies in the universe. While the absolute background has been estimated by space-based experiments, such as the *Diffuse Infrared Background Experiment (DIRBE)*, the total IRB intensity measured still remains fully unaccounted for by sources. Primordial galaxies at redshifts 8 and higher, especially those involving Population III stars, are generally invoked to explain the missing IR flux between 1 μm and 2 μm, with most of the intensity associated with red-shifted Lyman-α emission during re-ionization, though there are difficulties with such an assumption.

As pointed out in Cooray *et al.* (2004), if a high-redshift population contributes significantly to the IRB, then these sources are expected to leave a distinct signal in the anisotropy fluctuations of the near-IR intensity, when compared to the anisotropy spectrum associated with low-redshift sources. In Sullivan *et al.* (2007), we presented clustering measurements at 3.6 μm in several fields of *Spitzer*-IRAC data and we refer the reader to this work for more details and implications.

We are also developing a rocket-borne instrument (the *Cosmic Infrared Background ExpeRiment*, or *CIBER*) to search for signatures of primordial galaxy formation in the cosmic near-infrared extra-galactic background. *CIBER* consists of a wide-field two-color camera, a low-resolution absolute spectrometer, and a high-resolution narrow-band imaging spectrometer. The cameras will search for spatial fluctuations in the background on angular scales from 7 arcseconds to 2 degrees. In a short rocket flight, *CIBER* has sensitivity to probe fluctuations 100 times fainter than *DIRBE*. By jointly observing regions of the sky studied by *Spitzer* and *Akari*, *CIBER* will build a multi-color view of the near-infrared background, allowing a deep and comprehensive survey for first-light galaxy background fluctuations. The low-resolution spectrometer will search for a redshifted Lyman cutoff feature between 0.8 - 2.0 μm. The high-resolution spectrometer will trace zodiacal light using the intensity of scattered Fraunhofer lines, providing an independent measurement of the zodiacal emission.

References

Cooray, A. R., Bock, J. J., Keatin, B., *et al.* 2004, *ApJ*, 606, 611
Sullivan, I., Cooray, A. R., Chary, R.-R., *et al.* 2007, *ApJ*, 657, 37

Highlights of Astronomy, Volume 14
IAU XXVI General Assembly, 14-25 August 2006
Karel A. van der Hucht, ed.

doi:10.1017/S1743921307010502

Additional sources of ionization in the early universe and the 21 cm line

Evgenii O. Vasiliev[1,2] and Yuri A. Shchekinov[3]

[1]Tartu Observatory, 61602 Tõravere, Estonia
[2]Institute of Physics, University of Rostov, Stachki Ave. 194, Rostov-on-Don, 344090 Russia
email: eugstar@mail.ru

[3]Department of Physics, University of Rostov, Sorge St. 5, Rostov-on-Don, 344090 Russia
email: yus@phys.rsu.ru

Abstract. We consider the influence of decaying dark matter particles and ultra-high energy cosmic rays (UHECRs) on the ability of neutral gas at redshifts $z = 10 - 50$ to emit and absorb in the 21 cm line . We show that the signal in 21 cm is sensitive to properties of decaying particles and UHECRs, and conclude that future radio telescopes (LOFAR, LWA and SKA) are able not only to detect 21 cm signal originated from decaying particles and UHECRs, but discriminate between them as well.

Keywords. early Universe, dark matter, diffuse radiation

Decaying dark matter particles can strongly affect the reionization of the universe (e.g., Chen & Kamionkowski 2004). Additional ionization and heating from such particles before the re-ionization affects also the ability of neutral gas to absorb or emit in 21 cm line. Thus the decaying dark matter can influence the cosmological 21 cm background. We consider three sorts of decaying dark matter particles: long- and short-living particles, as well as ultra-high energy cosmic rays (UHECRs), if they form from decaying superheavy dark matter particles (e.g., Berezinsky *et al.* 1997).

The long-living particles provide permanent heating, so the gas kinetic temperature grows towards lower redshifts. Contrary, in case of the short-living particles the injection rate of heat decreases fastly, when the lifetime of particles becomes comparable with the comoving age of the universe, which manifests in a relatively fast decrease of the kinetic temperature at low redshfts.

The UHECRs produce only Ly-c and Ly-α photons, which give negligible heating, and the major influence on the 21 cm brightness temperature history is through the Wouthuysen-Field effect. We show that long-living and short-living unstable dark matter particles and UHECRs produce fairly distinct dependences of brightness temperature on redshift $T_b(z)$ – the first and the third give negative and positive second derivatives of the curves $T_b(z)$, while the second has $T_b(z)$ with an inflection point. This circumstance may have a principal significance for choosing a strategy for observational discrimination between these sources of ionization.

In the presence of UHECRs 21 cm can be seen in absorption with the brightness temperature $T_b = -(5–10)$ mK in the range $z = 10–30$. Decaying particles can stimulate a 21 cm signal in emission with $T_b \simeq 50–60$ mK at $z = 50$, and $T_b \simeq 10$ mK at $z \simeq 20$. Future radio telescopes (such as LOFAR, LWA and SKA) seem to have sufficient flux sensitivity for detection the signal in 21 cm influenced by decaying particles and UHECRs.

References

Berezinsky, V. S., Kachelrieß, M., & Vilenkin, A. 1997, *Phys. Rev. Lett.*, 79, 4302
Chen, X., & Kamionkowski, M. 2004, *Phys. Rev. D* 70, 043502

Highlights of Astronomy, Volume 14
IAU XXVI General Assembly, 14-25 August 2006
Karel A. van der Hucht, ed.

doi:10.1017/S1743921307010514

Assessing the influence of metallicity on fragmentation of proto-galactic gas

Anne-Katharina Jappsen[1,2], Simon C. O. Glover[2], Ralf S. Klessen[3] and Mordecai-Mark Mac Low[4]

[1]Canadian Institute for Theoretical Astrophysics, University of Toronto, Toronto, ON M5S 3H8, Canada
email: jappsen@cita.utoronto.ca

[2]Astrophysikalisches Institut Potsdam, An der Sternwarte 16, D-14482 Potsdam, Germany
email: sglover@aip.de

[3]Insitut für Theoretische Astrophysik/Zentrum für Astronomie der Universität Heidelberg, Albert-Überle-Str. 2, D-69120, Germany
email: rklessen@ita.uni-heidelberg.de

[4]Department of Astrophysics, American Museum of Natural History, 79th Street at Central Park West, New York, NY 10024-5192, USA
email: mordecai@amnh.org

Abstract. In cold dark matter cosmological models, the first stars to form are believed to do so within small protogalaxies. We study the influence of low levels of metal enrichment on the cooling and collapse of ionized gas in these protogalactic halos using three-dimensional, smoothed particle hydrodynamics simulations.

Keywords. stars: formation, methods: numerical, hydrodynamics, early Universe

---

We wish to understand how the evolution of early protogalaxies changes once the gas forming them has been enriched with small quantities of heavy elements, which are produced and dispersed into the intergalactic medium by the first supernovae. Adding heavy elements to the gas increases its ability to radiate heat and to control its temperature. It has been argued that enrichment beyond a certain 'critical metallicity' allows the first solar-mass stars to form, while protogalaxies with fewer metals form only massive stars, with masses greater than a hundred times solar. This idea has been accepted as a working hypothesis by many cosmologists, but it has yet to be rigorously tested. Although observational tests will not be feasible until the next generation of telescopes become available, we can begin to test this idea numerically, using high-resolution hydrodynamic simulations that incorporate the effects of the appropriate chemical and thermal processes. Our initial conditions represent protogalaxies forming within a fossil H II region – a previously ionized region that has not yet had time to cool and recombine. We vary the initial redshift between $z = 15$ and $z = 30$ and the dark matter halo masses between 5×10^4 and $10^7\,\mathrm{M}_\odot$. The gas mass resolution lies between 20 and $400\,\mathrm{M}_\odot$ (Jappsen *et al.* 2007). Our simulations demonstrate that for metallicities $Z \leqslant 10^{-3}\,\mathrm{Z}_\odot$, metal line cooling alters the density and temperature evolution of the gas by less than 1% compared to the metal-free case at low densities ($n < 1\,\mathrm{cm}^{-3}$) and high temperatures ($T > 2000\,\mathrm{K}$). We also present the results of high-resolution simulations using particle splitting (Kitsionas & Whitworth 2002) to improve resolution in regions of interest.

References

Kitsionas, S., & Whitworth, A. P. 2002, *MNRAS*, 330, 129
Jappsen, A.-K., Glover, S. C. O., Klessen, R. S., & Mac Low, M.-M. 2007, *ApJ*, 660, 1332

Highlights of Astronomy, Volume 14
IAU XXVI General Assembly, 14-25 August 2006
Karel A. van der Hucht, ed.

doi:10.1017/S1743921307010526

UV radiative feedback on high-redshift proto-galaxies

Andrei Mesinger, Greg L. Haiman, and Zoltán Haiman

Department of Astronomy, Columbia University,
550 West 120th Street, New York, NY 10027, USA

Abstract. We discuss the UV radiative feedback on high-redshift proto-galaxies.

Keywords. cosmology: theory, early Universe , galaxies: high-redshift, evolution

---

We use three-dimensional hydrodynamic simulations to investigate the effects of a transient photo-ionizing ultraviolet (UV) flux on the collapse and cooling of pregalactic clouds. These clouds have masses in the range 10^{5-7} $M_{\odot}$, form at high redshifts ($z \gtrsim 18$), and are assumed to lie within the short-lived cosmological H II regions around the first generation of stars. In addition, we study the combined effects of this transient UV flux and a persistent Lyman–Werner (LW) background (at photon energies below 13.6 eV) from distant sources. In the absence of a LW background, we find that a critical specific intensity of $J_{\rm UV} \simeq 0.1 \times 10^{-21}$ erg s^{-1}cm^{-2}Hz^{-1}sr^{-1} demarcates a transition from net negative to positive feedback for the halo population. A weaker UV flux stimulates subsequent star formation inside the fossil H II regions, by enhancing the H_2 molecule abundance. A stronger UV flux significantly delays star-formation by reducing the gas density, and increasing the cooling time, at the centers of collapsing halos.

At a fixed $J_{\rm UV}$, the sign of the feedback also depends strongly on the density of the gas at the time of UV illumination. Regardless of whether the feedback is positive or negative, we find that once the UV flux is turned off, its impact starts to diminish after $\sim 30\%$ of the Hubble time. In the more realistic case when a LW background is present, with $J_{\rm LW} \gtrsim 0.01 \times 10^{-21}$ erg s^{-1}cm^{-2}Hz^{-1}sr^{-1}, strong suppression persists down to the lowest redshift ($z = 18$) in our simulations.

Finally, we find evidence that heating and photoevaporation by the transient UV flux renders the $\sim 10^6 M_{\odot}$ halos inside fossil H II regions more vulnerable to subsequent H_2 photo-dissociation by a LW background.

See Mesinger *et al.* (2006) for details of our work.

Reference

Mesinger, A., Bryan, G. L., & Haiman, Z. 2006, *ApJ*, 648, 835

Highlights of Astronomy, Volume 14
IAU XXVI General Assembly, 14-25 August 2006
Karel A. van der Hucht, ed.

doi:10.1017/S1743921307010538

Posters presented at Joint Discussion 7

A high-accuracy method for the removal of point sources from maps of the cosmic microwave background
A.T. Bajkova

A universe with both decelaration and acceleration
L.N. de Silva

CMB quadrupole induced polarisation from large scale structure.
G. Liu, A. da Silva, and N. Aghanim

Modeling $z \simeq 3$ Lyman-alpha spectra to gain insight into high-redshift Lyα emitters
A. Verhamme and D. Schaerer

News from $z \simeq 6$ - 10 galaxy candidates found behind gravitational lensing clusters
A. Hempel, D. Schaerer, R. Pelló, E. Egami, J. Richard, J.-P. Kneib, and M. Wise

Nucleosynthesis without a beginning
G.R. Burbidge

Observational test on a generalized theory of gravity
S. Rahvar, M.M. Sheikh Jabbari, S. Baghram, and F. Habibi

Probing weakest extragalactic magnetic fields with γ-ray bursts
K. Ichiki, K. Takahashi, and S. Inoue

SDSS J0836+0054: a radio quasar at $z \simeq 6$ with a resolution of 40 pc
S. Frey, Z. Paragi, L. Mosoni, and L.I. Gurvits

Space/time, supernovae, and the faint young sun
L.M. Riofrio

Stability and formation of cluster of galaxies, galaxies, clusters of stars and stars in some cosmological models
M.I. Wanas, A.B. Morcos, and M.A. Bakry

Subaru deep imaging of the field of QSO 1508+5714 at $z = 4.28$
Y.P. Wang, Y.P. Wang, T. Yamada, I. Tanaka, and M. Iye

The 21-cm signature of the early radiation sources
L. Chuzhoy, M.A., Alvarez, and P.R. Shapiro

The mass loss from hot Pop III stars
J. Krticka and J. Kubat

The spectrum of primordial gravitational waves
M. Soares-Santos and E.M. de Gouveia Dal Pino

Highlights of Astronomy, Volume 14
IAU XXVI General Assembly, 14-25 August 2006
Karel A. van der Hucht, ed.

doi:10.1017/S174392130701054X

Joint Discussion 8
Solar and stellar activity cycles

Alexander G. Kosovichev[1] and Klaus G. Strassmeier[2] (eds.)

[1]W.W. Hansen Experimental Physics Laboratory, Stanford University, Stanford, CA 94305-4085, USA
email: sasha@quake.stanford.edu

[2]Astrophysical Institute Potsdam, D-14482 Potsdam, Germany
email: kstrassmeier@aip.de

Preface

The solar magnetic field and its associated atmospheric activity exhibits periodic variations on a number of time scales. The 11-year sunspot cycle and its underlying 22-year magnetic cycle are, besides the 5-minute oscillation, the most widely known. Amplitudes and periods range from a few parts per million (ppm) and 2 - 3 minutes for p-modes in sunspots, a few 10 ppm and 10 minutes for the granulation turn around, a few 100 ppm and weeks for the lifetime of plages and faculae, 1000 ppm and 27 days for the rotational signal from spots, to the long-term cycles of 90 yr (Gleissberg cycle), 200 - 300 yr (Wolf, Spörer, Maunder minima), 2,400 yr from ^{14}C tree-ring data, and possibly in excess of 100,000 yr.

The enormous complexity of atmospheric structure as observed on the Sun and some other cool stars makes it hard, if not impossible, to compare the Sun with 'solar-type

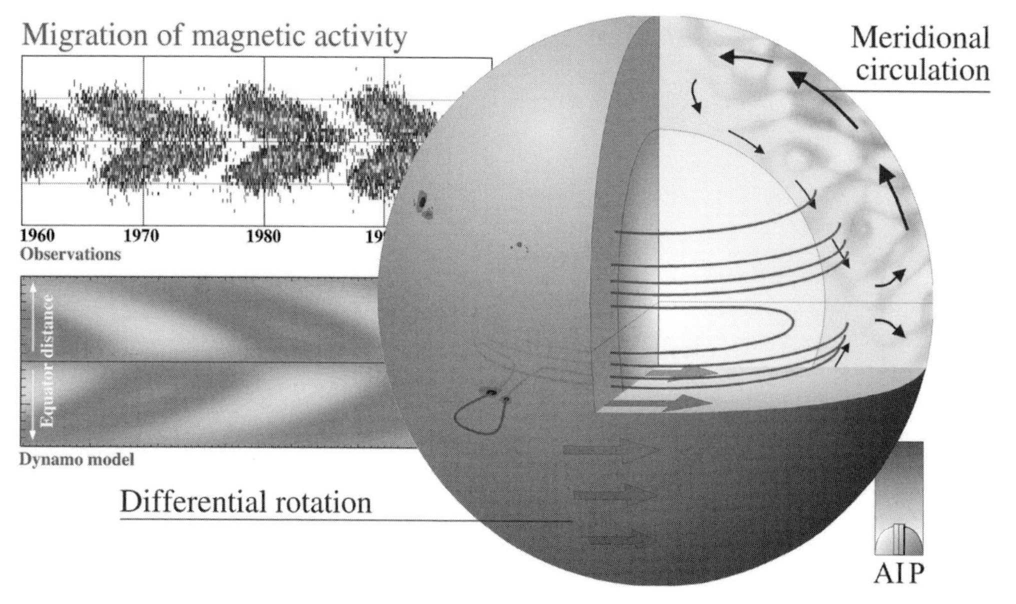

Figure 1. A graphical summary of some of the processes of magnetic activity and its observables. Courtesy of R. Arlt, AIP.

stars' in great detail. Clearly, we need to identify parameters that can be observed and interpreted unambiguously. The most accepted dynamo signature is the presence of an activity cycle, well documented for the Sun as well as for solar-type stars by the Mount Wilson H&K project. Just about recently, we are also detecting differential rotation and meridional flows on other stars. Again, the picture is not unambiguous yet, despite that there is just a single main ingredient that acts as the driving mechanism for activity in all atmospheric layers and the convective envelope: the dynamo-driven magnetic field.

Therefore, a Joint Discussion (JD) on the topic of *"Solar and Stellar Activity Cycles"* was held during the IAU General Assembly in Prague on August 17-18, 2006. Organized by Divisions II and IV with participating commissions 10, 12, 29, 35, 36, and 49, the meeting was made up by 18 invited talks in three subtopics. A total of approximately 120 astronomers participated in this one-and-a-half day gathering with many and lively discussions.

Scientific Organizing Committee

Alexander Kosovichev (USA) (co-chair), Klaus G. Strassmeier (Germany) (co-chair), Pavel Ambroz (Czech Republic), Martin Asplund (Australia), Svetlana V. Berduygina (Switzerland), Andrew Collier Cameron (UK), Dainis Dravins (Sweden), Lidia van Driel-Gesztelyi (France), Cheng Fang (China), John D. Landstreet (Canada), Valentin I. Makarov (Russia), Gautier Mathys (Chile), Roberto Pallavicini (Italy), Fernando Moreno-Insertis (Spain), Takashi Sakurai (Japan), and Parameswaran Venkatakrishnan (India). We thank all of them for their help and many suggestions.

We thank all of the SOC members for their help and many suggestions. Special thanks go to our chairmen Nicolai Piskunov and John Landstreet who kept a strict eye on the watch. Rainer Arlt from the AIP produced and maintained the meeting web site and Katrin Götz managed the LOC and related issues.

Sasha Kosovichev and Klaus Strassmeier, co-chairs SOC,
Potsdam, October 30, 2006

Highlights of Astronomy, Volume 14
IAU XXVI General Assembly, 14-25 August 2006
Karel A. van der Hucht, ed.

doi:10.1017/S1743921307010551

Polar magnetic field reversals on the Sun

Elena E. Benevolenskaya[1,2]

[1]W.W. Hansen Experimental Physics Laboratory, Stanford University, CA 94305, USA
email: elena@quake.stanford.edu

[2]Pulkovo Astronomical Observatory, 196140, St. Petersburg, Russia

Abstract. The polar magnetic fields on the Sun have been an attractive subject for solar researches since Babcock measured them in solar cycle 19. One of the remarkable features of the polar magnetic fields is their reversal during the maxima of 11-year sunspot cycles. I have present results of the investigations of the polar magnetic field using *SOHO*-MDI data. It is found, that the polar magnetic field reversal is detected with MDI data for polar region within 78°–88°. The North Pole has changed polarity in CR1975 (April 2001). The South reversed later in CR1980 (September 2001). The total unsigned magnetic flux does not show the dramatic decreasing during the polar reversals due to omnipresent bi-polar small-scale magnetic elements. The observational and theoretical aspects of the polar magnetic field reversals are discussed.

Keywords. Sun: magnetic fields, Sun: photosphere, Sun: activity

---

1. Introduction

The investigation and further discussions about Polar Magnetic Field Reversals on the Sun began since a small report in Science was published by Harold D. Babcock and William C. Livingston in 1958. They announced that observations of solar magnetic fields at South pole had reversed their polarity by June 1957. To understand the origin of the polar magnetic field reversals many investigators employed the mean-field dynamo theory (e.g., Dikpati *et al.* 2004). The surface-diffusion or transport models explain the polar magnetic field reversals as a result of turbulent diffusion, differential rotation and meridional circulation (Leighton 1964; Leighton 1969; Wang *et al.* 1989). Fox *et al.* (1998) described the evolution of the large-scale fields and their association with polar coronal holes. Their question was whether the polar fields resulted from the local polar dynamo or not. There is no a certain answer to this question. However, Durrant *et al.* (2002) have observed that high-latitude flux emergence can effect the evolution of individual high-latitude plumes, but this flux does not seriously affect the whole reversal times of the polar magnetic field. Nowadays, we understand that a process of polar magnetic field reversals is a complex phenomena including different physical processes in the convection zone, photosphere and corona (Schrijver *et al.* 2002; Fisk & Schwadron 2001; Benevolenskaya *et al.* 2001 Benevolenskaya *et al.* 2002; Gopalswamy *et al.* 2003).

2. Polar reversals in cycle 23 and small-scale magnetic elements

The total magnetic flux $F_r = F_+ + F_-$for polar caps ($\pm$ (78° - 88°) is present in Figure 1a (left panels). The positive and negative fractions of the magnetic flux are $\frac{F_+}{F_r}$ and $\frac{F_-}{F_r}$. The time of reversals can be determined at $\frac{F_+}{F_r} = 0.5$ 1975 $\pm$ 2 or $\frac{F_-}{F_r} = 0.5$ or using value of total signed magnetic flux (Figure 1c, left panels)). This was in CR 1980 $\pm$ 2 for the southern magnetic field, it is about 1975 $\pm$ 2 in the North (Figure 1a, left panel). This is close to the periods obtained by Durrant & Wilson (2003): CR 1975 $\pm$ 2 in North and CR 1981 $\pm$ 1 in South.

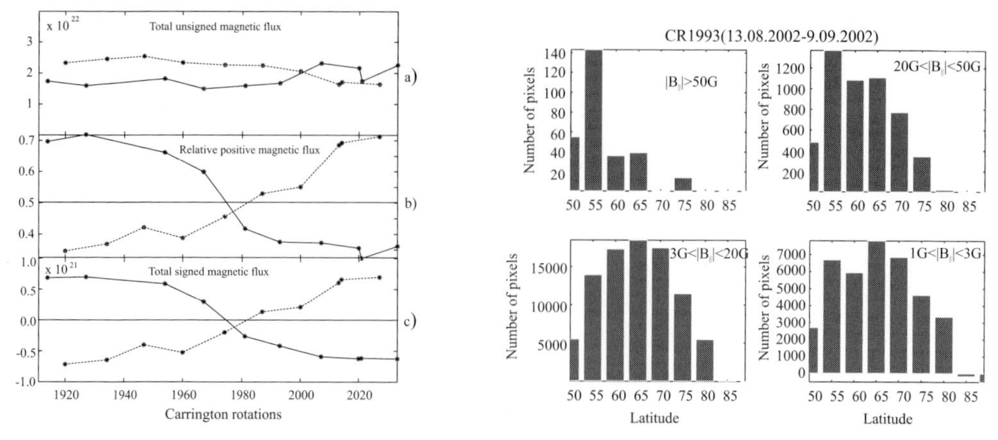

Figure 1. *Left panels:* (a) The total unsigned magnetic flux of the radial field component in the latitude zones from 78° to 88° in Northern (*solid line*) and Southern Hemispheres (*dash lines*); (b) The relative positive polarity parts of magnetic flux in Northern (*solid line*) and Southern (*dash line*) hemispheres; (c) The total signed magnetic flux. The polar magnetic field reversal was in CR 1975 ± 2 (March 2001) in the North and in CR 1980 ± 2 (September 2001) in South. *Right panels:* Differences between the number of pixels with negative and positive polarity after the polar magnetic field reversals for carrington rotation 1993 (CR 1993)in North.

The solar magnetic field and, in particular, the polar magnetic field, is dominated by small magnetic elements of mixed polarities as has been shown by Severny (1965), Lin *et al.* (1994), and Benevolenskaya (2004). Magnetic flux of small scale bi-polar elements contribute to the total unsigned magnetic flux, but how they affect the polar magnetic field reversals is not yet clearly understand. The difference between the number of magnetic elements of negative and positive polarity as a function of latitude shows the dynamic of the magnetic elements. The evolution of the elements of strong and the weak magnetic field after the polar magnetic field reversals in North is considered in Figure 1 (right panels). The number of elements of weak magnetic field ($1\,G < |B_{||}| < 3$ G) displays a delay in the changing of their polarity to compare with elements of strong magnetic field (Figure 1, right panels) in the line-of-sight component.

References

Babcock, H. W., & Babcock, H. D. 1955, *ApJ*, 121, 349
Babcock, H. W., & Livingston W. C. 1958, *Science*, 127, 1058
Babcock, H. D. 1959, *ApJ*, 130, 364
Benevolenskaya, E. E., Kosovichev, A. G., & Scherrer, P. H. 2001, *ApJ* (Letters), 554, L107
Benevolenskaya, E. E., Kosovichev, A. G., Lemen, J. R., *et al.* 2002, *ApJ* (Letters), 571, L181
Benevolenskaya, E. E. 2004, *A&A* (Letters), 428, L5
Dikpati, M., de Toma, G., Gilman, P. A., Arge, C. N., & White, O. R. 2004, *ApJ*, 601, 1136
Durrant, C. J., Turner, J., & Wilson, P. R. 2002, *Solar Phys.*, 211, 103
Durrant, C. J., & Wilson, P. R. 2003, *Solar Phys.*, 214, 23
Fisk, L. A., & Schwadron, N. A. 2001, *ApJ*, 560, 425
Fox, P. A., McIntosh, P. S., & Wilson, P. R. 1998, *ApJ*, 117, 375
Gopalswamy, N., Lara, A., Yashiro, S., & Howard, R. A. 2003, *ApJ* (Letters), 598, L63
Leighton, R. B. 1964, *ApJ*, 140, 1547
Leighton, R. B. 1969, *ApJ*, 156, 1
Lin, H., Varsik, J., & Zirin, H. 1994, *Solar Phys.*, 155, 243
Severny, A. B. *Soviet Astron. Let*, 9, 171
Schrijver, C. J., DeRosa, M. L., & Title, A. M. 2002, *ApJ*, 577, 1006
Wang, Y.-M., Nash A. G., & Sheeley N. R. 1989, *ApJ*, 347, 529

Highlights of Astronomy, Volume 14
IAU XXVI General Assembly, 14-25 August 2006
Karel A. van der Hucht, ed.

doi:10.1017/S1743921307010563

Flip-flop cycles in solar and stellar activity

Svetlana V. Berdyugina[1,2]

[1]Institute of Astronomy, ETH Zurich, CH-8092 Zurich, Switzerland
email: sveta@astro.phys.ethz.ch

[2]Tuorla Observatory, University of Turku, Väisäläntie 20, FI-21500 Piikkiö, Finland

Abstract. We discuss flip-flop cycles in solar and stellar activity.

Keywords. Sun: activity, stars: activity, stars: late-type, stars: spots

---

Decades of continuous photometric monitoring of RS CVn-type stars (binaries with cool active giants or subgiants) revealed that large spots maintained their identities for years which was interpreted as a signature of one or two active longitudes. Berdyugina & Tuominen (1998) showed that active longitudes on these stars are persistent structures which can however continuously migrate in the orbital reference frame. The active longitudes are separated by 180° on average and differ in their activity level. Periodic switching of the dominant activity from one active longitude to the other results in a so-called flip-flop cycle. This was detected in both light curve variations and Doppler images (Fig. 1; Berdyugina & Tuominen 1998; Berdyugina *et al.* 1998). Two active longitudes and flip-flops seem to be typical patterns of stellar activity. In addition to RS CVn stars they have been found on FK Com-type stars (Jetsu *et al.* 1993; Korhonen *et al.* 2002) and very active young solar analogues (Berdyugina & Järvinen 2005).

Persistent active longitudes and a flip-flop cycle have also been found in the distribution of sunspots (Berdyugina & Usoskin 2003). Large sunspot groups in both northern and southern hemispheres are preferably formed around two active longitudes which are separated by 180° and persistent for at least 120 years. Similar to young solar-type dwarfs, the two active longitudes on the Sun are long-lived quasi-rigid structures which are not fixed in any reference frame due to differential rotation. They continuously migrate, with a variable rate, with respect to the Carrington system, especially in the beginning of the solar cycle (Fig. 2). The migration of active longitudes is caused by changes in the mean latitude of the sunspot formation and by differential rotation.

More than a dozen of active stars exhibiting flip-flop cycles enable a statistical analysis of their properties. There is a noticeable trend for stars with longer rotational periods to have longer flip-flop cycles (Fig. 3, left panel). The trend is prominent for RS CVn binaries as they have a wide range of rotation periods which are synchronized with their orbital motion. In connection to the spot activity cycle (analogous to the 11-yr sunspot cycle), there are clearly two groups of stars with cycle ratios 2:1 and 1:3 (Fig. 3, right panel). This implies different dominant dynamo mechanisms operating in these stars. The presence of binaries in both groups excludes possible effects of binarity on this ratio. However, it appears that the two groups greatly differ by the differential rotation rate, which might provide a clue to the nature of their dynamos.

References

Berdyugina, S. V., & Järvinen, S. P. 2005, *AN*, 326, 283
Berdyugina, S. V., & Tuominen, I. 1998, *A&A* (Letters), 336, L25
Berdyugina, S. V., & Usoskin, I. G. 2003, *A&A*, 405, 1121
Berdyugina, S.V., Berdyugin, A. V., Ilyin, I., & Tuominen, I. 1998, *A&A*, 340, 437
Jetsu, L., Pelt, J., & Tuominen, I. 1993, *A&A*, 278, 449
Korhonen, H., Berdyugina, S. V., & Tuominen, I. 2002, *A&A*, 390, 179

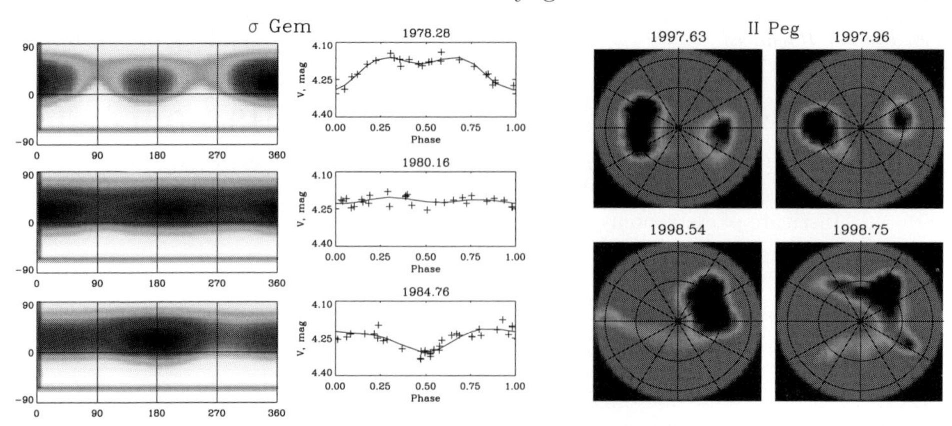

Figure 1. The flip-flop phenomenon on cool active binary components: on the left, as observed in light curves of σ Gem, and on the right, in Doppler images of II Peg. The images on the left show the distribution of the spot filling factor on the stellar surface obtained via inversions of the light curves (plots in the middle). The II Peg images obtained from inversions of spectral line profiles show the temperature distribution on the stellar surface as seen from the pole. Flip-flops appear as a switch of the dominant activity to the opposite longitude.

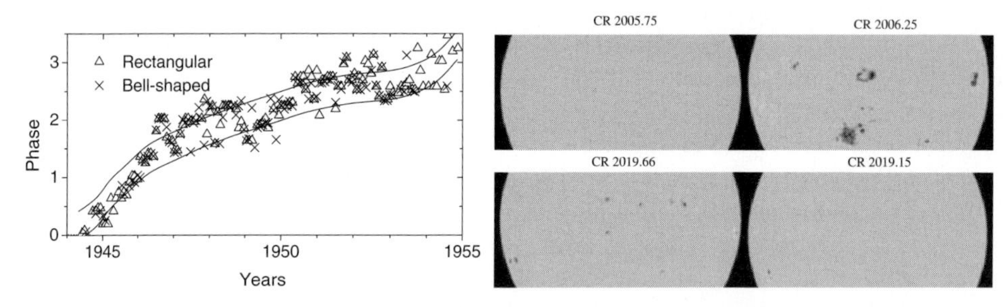

Figure 2. The flip-flop phenomenon on the Sun. On the left, rotational phases (in the Carrington system) of the biggest sunspot-cluster in the North are shown for cycle 18. The phases obtained with two different methods are denoted by different symbols (see Berdyugina & Usoskin 2003). On the right, illustration of a flip-flop in cycle 23: two opposite sides of the Sun (*SOHO*-MDI, only equatorial region) are shown for two Carrington rotations (CR) one year apart. The maximum activity switched from the phase ~0.25 to the phase ~0.66.

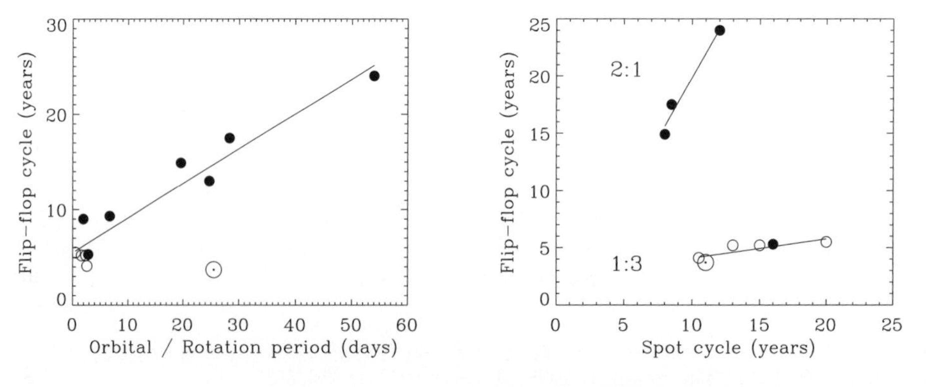

Figure 3. Flip-flop cycles on cool active stars: binaries (filled circles), single young stars (open circles), and the Sun (big circle with dot). On the left, correlation of the flip-flop cycles versus the orbital (binaries) or rotational (single stars) periods. On the right, the relation with the sunspot-like cycles are shown. Two groups with the cycle ratios of 1:3 and 2:1 are clearly seen.

Highlights of Astronomy, Volume 14
IAU XXVI General Assembly, 14-25 August 2006
Karel A. van der Hucht, ed.

doi:10.1017/S1743921307010575

Large-scale patterns, complexes of solar activity and 'active longitudes'

Pavel Ambrož

Astronomical Institute of the Academy of Sciences of the Czech Republic,
CZ-25165 Ondřejov, Czech Republic
email: pambroz@asu.cas.cz

Abstract. Typical latitude zones and longitude sectors with a dominant occurrence of newly emerging magnetic flux were systematically detected during three last activity cycles. Two long time persistent longitude sectors with the preferred occurrence of a new strong magnetic flux are characteristic for magnetic flux distribution and their rotation rate is latitude dependent in the relationship with differential rotation rate. Recent new information about the large-scale flows in convection zone relates to a concept of the expected giant cells and jets and show a new relationship with different scales of the large-scale circulation. Non-axially symmetric horizontal flow in upper part of convection zone gives a good motivation for an extension of the existing axially symmetric 2D models into new 3D concept.

Keywords. Sun, Sun: magnetic field, solar activity

The motivation for the present contribution is: which kind of symmetry in large-scale solar activity occurrence is observed on the Sun? Well known latitude dependence of solar activity occurrence is considered as a substantial criterion for the solar cycle temporal evolution with distinct symmetry according to equatorial plane. However, during the last half century some studies showed the non random distribution of the active phenomena were also present in the heliographic longitude. It was shown, that the distribution of the activity on the Sun is generally non-axially symmetric.

Large-scale processes on the Sun are generally connected with characteristic dimension, which is greater than 100 Mm. In this case the physical quantities are generally parameterized and lose their original physical meaning. Usual objects studied in this connection are the active regions and their complexes (Gauzaiskas *et al.* 1983), large-scale magnetic fields (Bumba & Howard 1969) and corresponding magnetic flux, large-scale coronal structures, their evolution and relationship with interplanetary magnetic field.

In the second half of the last century many studies describing '*Sonnefleckenherd*' (Becker 1955), active nests (Castenmiller *et al.* 1986), complexes of activity (Gaizauskas *et al.* 1983; Bumba & Howard 1965), active longitudes (Berdyugina & Usoskin 2003) and 'hot spots' (Bai 1988) were prepared. Many authors used different methods and specific terminology, however, a very similar phenomenon is considered. A recent study of the strong magnetic field confirmed the existence of preferred longitudes for the emergence of big sunspot groups. These longitudes occur in pairs, rotate at the same rates and are separated by 180° in longitude. The northern and southern hemispheres behave differently regarding the occurrence of activity in longitude. Hot spots with different rotation periods coexist in the same hemisphere during the same solar cycle. Active longitudes are affected by the differential rotation and migrate in Carrington longitude according to the mean latitude of sunspot manifestation. Usually one or two active longitudes are dominant and dominance switches in time, known as the 'flip-flop' effect (Berdyugina & Usoskin 2003). The mean time between two consecutive switches is of the order of

20 Carrington rotation periods. In fact, the rotation period is mixed with the period of switches, and creates two period side lobes, below and above of the period of rotation. Formation of preferred longitudes is closely connected with another phenomenon in which a position of newly emerging strong magnetic flux is nearly identical with previous in the place of just decaying magnetic flux. The scatter in position is usually only a few heliographic degrees. Terms of activity nests or hot spot were introduced. This mechanism supports creation of the persistent in time regions with dominant occurrence of the solar active phenomena. Lifetime of such processes vary from one to several activity cycles, therefore this process is not always continuously active throughout its lifetime. The differential rotation responsible for the migration of active longitudes is different from that obtained from individual spots.

The newly emerging magnetic flux is expected to be transported in the upper part of the solar convection zone due to turbulent diffusion and large-scale motions. In both directions, in addition to global differential rotation and meridional circulation also the large-scale departure from axially symmetric motions were observed. Large-scale horizontal motions with a characteristic dimension of about 40° in both zonal and meridional directions and with characteristic velocity about $10\,\mathrm{m\,s^{-1}}$ were detected in the photosphere by Ambrož (2001) and also below it up to depth 12 Mm by Zhao & Kosovichev (2004).

Departure of zonal velocity from axial symmetry near the bottom of the convection zone can, according to theoretical considerations, initiate specific instabilities in tachocline leading to the tipping of the basic toroidal fields in both solar hemispheres. Study of this process with determination of a tip angle and a position angle of the tilted toroid still confirm Norton & Gilman (2005) the possibility of this mechanism . A tipped toroid is characterized by its position angle in longitude and can be a way to a 3D non-axially symmetric model of the solar dynamo. Excellent representation of stable evolving active longitudes was obtained by analysis of the total dipole component of the photospheric magnetic field. Position of the north pole, evolving in latitude and longitude during activity cycle confirm that all previous conclusions about active longitudes, the 'flip-flop' effect also in the upper corona and interplanetary magnetic field. All arguments show that the above mentioned large-scale processes are typical symptoms of the function of the solar dynamo and solar cycle mechanism and can be typical of the activity processes also other Sun-like type stars.

Acknowledgements

This study was supported by Academy of Sciences of the Czech Republic under grant GA/AVCR A300030506.

References

Ambrož, P. 2001, *Solar Phys.*, 198, 253
Bai, T. 1988, *ApJ*, 328, 860
Becker, U. 1955, *Z. Astrophys*, 37, 47
Berdygina, S. V., & Usoskin, I. G. 2003, *A&A*, 405, 1121
Bumba, V., & Howard, R. 1965, *ApJ*, 141, 1502
Bumba, V., & Howard, R. 1969, *Solar Phys.*, 7, 28
Castenmiller, M. J. M, Zwaan, C., & van der Zalm, E. B. J. 1986, *Solar Phys.*, 105, 237
Gaizauskas, V., Harvey, K. L., & Zwaan, C. 1983, *ApJ*, 265, 1056
Norton, A. A., & Gilman, P. A. 2005, *ApJ*, 630, 1194
Zhao, J., & Kosovichev, A. G. 2004, *ApJ*, 603, 776

Highlights of Astronomy, Volume 14
IAU XXVI General Assembly, 14-25 August 2006
Karel A. van der Hucht, ed.

doi:10.1017/S1743921307010587

Solar irradiance variability

Sami K. Solanki
Max-Planck-Institut für Sonnensystemforschung, D-37191 Katlenburg-Lindau, Germany
email: solanki@mps.mpg.de

Abstract. We study solar irradiance variability. The current generation of models show that the irradiance since then has increased by between 0.9 and 1.5 $\mathrm{W\,m^{-2}}$.

Keywords. Sun: activity, solar-terrestrial relations, stars: activity

---

The total irradiance of the Sun (the wavelength and solar disk integrated radiative output as seen above the Earth's atmosphere) is observed to vary by approximately 1 $\mathrm{W\,m^{-2}}$ over the solar cycle. Unfortunately, the instruments that measure this variation have not survived for the full length of time that observations have been carried out, so that cross-calibration issues become important. Different composites of irradiance have been put together from the observational data which show different trends.

There are different reasons for carrying out a detailed modelling of the Sun's irradiance. Firstly, such models are needed to uncover the physical causes of the irradiance variability. This may also be important for stellar astrophysics, since similar (and stronger) variability is also seen in many Sun-like stars, in particular in more active ones. Secondly, the time span over which solar irradiance has been directly observed is too short for a meaningful comparison with the Earth's climate. Models are needed in order to extend the irradiance further back in time.

Such models have now been constructed by various groups, based on a variety of assumptions and techniques. The most successful have been those that reproduce variations of both the total as well as the most sensitive measurements of spectral irradiance of the Sun with just a single free parameter. These models show that over 90 more of the irradiance variations of the Sun on time scales of days to decades are due to the magnetic field at the solar surface. The models have also been used to extend the irradiance back in time until the Maunder minimum. The current generation of models show that the irradiance since then has increased by between 0.9 and 1.5 $\mathrm{W\,m^{-2}}$.

Highlights of Astronomy, Volume 14
IAU XXVI General Assembly, 14-25 August 2006
Karel A. van der Hucht, ed.

doi:10.1017/S1743921307010599

Cyclic fluctuations in the differential rotation of active stars

Pascal Petit

Observatoire Midi-Pyrénées, 14 avenue Edouard Belin, F-31400 Toulouse, France
email: petit@ast.obs-mip.fr

Abstract. Differential rotation is described in stellar dynamo models as one of the fundamental phenomena governing the amplification of magnetic fields in active stars. Using indirect imaging methods, the measurement of photospheric differential rotation is now achieved on a growing number of very active stars, a fraction of which exhibit temporal fluctuations of potentially large amplitude in their latitudinal shear, on a time-scale of a few years. I first describe the modeling tools on which such analysis is based, then discuss the implications of this observational work on our understanding of stellar dynamos and of the impact stellar magnetic fields may have on the dynamics of convective envelopes.

Keywords. Stars: magnetic fields, stars: late-type, stars: rotation

---

Differential rotation is one of the basic ingredients invoked to explain the generation of the solar magnetic field, through its ability to transform a large-scale poloidal field into a stronger toroidal component. However, much details of this general principle are still poorly understood, and a major aim for stellar differential rotation measurements is to evaluate, in the stellar parameter space, how various properties of stellar activity (magnetic field intensity, existence of activity cycles) can be connected to differential rotation.

The spatial distribution of photospheric magnetic fields can be reconstructed for fast rotating stars (with a significant rotational broadening of spectral lines) by means of tomographic inversion techniques very similar to Doppler imaging (Vogt & Penrod 1983), but based on the inversion of polarized light and therefore called Zeeman-Doppler Imaging (Semel 1989; Donati & Brown 1997). For cool stars, circularly polarized signal is generally used alone. The location of magnetic regions is obtained very similarly to that of star-spots in classical Doppler mapping. Some information about the orientation of field lines can also be determined by following the distortion of Zeeman signatures during the transit of magnetic regions over the stellar disc.

The short-term temporal evolution of surface structures can be analyzed with Zeeman-Doppler Imaging. Measurements of differential rotation can in particular be performed, using cool spots or magnetic regions as tracers of the large-scale surface flows (Petit *et al.* 2002). A solar-like surface shear (the equator rotating faster than the pole) has been detected on several stars (see, e.g., Donati *et al.* 2003; Petit *et al.* 2004a; Petit *et al.* 2004b).

An exciting result is also the recent detection, on the young dwarf AB Dor, of secular fluctuations of differential rotation (Collier Cameron & Donati 2002; Donati *et al.* 2003; Marsden *et al.* 2005), about 40 times stronger in amplitude than fluctuations reported for the Sun (Howe *et al.* 2000; Vorontsov *et al.* 2002). This observation may unveil, for the first time on a star other than the Sun, the feedback effect of magnetic fields on the dynamics of convective zones (through Lorentz forces) during stellar activity cycles (Applegate 1992).

References

Applegate, J. H. 1992, *ApJ*, 385, 621
Collier Cameron, A., & Donati, J.-F. 2002, *MNRAS*, 329, 23
Donati, J.-F., & Brown, S. F. 1997, *A&A*, 326, 1135
Donati, J.-F., Collier Cameron, A., & Petit, P. 2003, *MNRAS*, 345, 1187
Howe, R., Christensen-Dalsgaard, J., Hill, F., *et al.* 2000, *Science*, 287, 2456
Marsden, S. C., Waite, I.A., Carter, B. D., & Donati, J.-F. 2005, *MNRAS*, 359, 711
Petit, P., Donati, J.-F., & Collier Cameron, A. 2002, *MNRAS*, 334, 374
Petit, P., Donati, J.-F., Wade, G. A., *et al.* 2004a, *MNRAS*, 348, 1175
Petit, P., Donati, J.-F., Oliveira, J., *et al.* 2004b, *MNRAS*, 351, 826
Semel, M. 1989, *A&A*, 225, 456
Vogt, S. S., & Penrod, G. D. 1983, *PASP*, 95, 565
Vorontsov, S. V., Christensen-Dalsgaard, J., Schou, J., *et al.* 2002, *Science*, 296, 101

Highlights of Astronomy, Volume 14
IAU XXVI General Assembly, 14-25 August 2006
Karel A. van der Hucht, ed.

doi:10.1017/S1743921307010605

Maunder Minimum stars revisited: recalibrating Ca II H & K measures

Jason T. Wright

Department of Astronomy, University of California, Berkeley, CA 94720, USA
email: jtwright@astro.berkeley.edu

Abstract. We discuss a recalibration of Ca II H & K measures in sun-like stars.

Keywords. stars: activity, emission lines, Sun: activity

Baliunas & Jastrow (1990) analyzed the distribution of activity among apparent solar analogs (selected by $B - V$ color) in the Mount Wilson H & K survey (Baliunas *et al.* 1995) and found a population of stars with activity levels significantly below that of the Sun at solar minimum. They suggested that these stars reside in a state analogous to the Maunder Minimum (Eddy 1976) and that their fraction of the sample was representative of the time Sun-like stars spend in such a state.

Wright (2004) (W04 hereafter) used *Hipparcos* parallaxes to show that this population of apparently very inactive stars is composed almost entirely of subgiants or slightly evolved dwarfs, and found an activity floor as a decreasing function of ΔM_V, the height a star above the main sequence. Wright (2004) argued that this effect is due to the inadequate calibration of the Mt. Wilson S-index of Noyes *et al.* (1984), which neglects the effects of gravity and metallicity on the photospheric flux in the H and K line cores.

Errors in estimates of the photospheric component of the H and K flux are only important for old, inactive stars such as the Sun, where the photospheric contribution can comparable to the chromospheric contribution. Its improper subtraction causes photospheric contamination in chromospheric flux measurements, making chromospheric ages beyond ~ 2 Gyr unreliable.

Analysis of the abundances and gravities in the SPOCS catalog Valenti & Fischer (2005) confirms the suspicions of Wright (2004) that not only evolved stars but metal-rich stars also show a significantly lower activity 'floor' than do solar analogs, consistent with their height above the main sequence. In fact, the activity floor is a surprisingly clean function of $T_{\rm eff}$ and ΔM_V independent *reason* a star sits above the main sequence – metallicity or evolution. Recalibrating the S-index (and the related $R'_{\rm HK}$ index) for the effects of metallicity and gravity will first require measurement of the photospheric contribution to S as a function of those quantities using high-resolution spectra. Further correction for the effects of metallicity on line blanketing in the continuum in the color-correction factor will require flux-calibrated spectra of the continuum near the H and K lines for stars spanning a range of [Fe/H], $\log g$, and $T_{\rm eff}$ values. These efforts will be the subject of a forthcoming work.

References

Baliunas, S. L., Donahue, R. A., Soon, W. H., *et al.* 1995, *ApJ*, 438, 269
Baliunas, S., & Jastrow, R. 1990, *Nature*, 348, 520
Eddy, J. A. 1976, *Science*, 192, 1189
Noyes, R. W., Hartmann, L. W., Baliunas, S. L., *et al.* 1984, *ApJ*, 279, 763
Valenti, J. A., & Fischer, D. A., *ApJS*, 159, 141
Wright, J. T. 2004, *AJ*, 128, 1273

Highlights of Astronomy, Volume 14
IAU XXVI General Assembly, 14-25 August 2006
Karel A. van der Hucht, ed.

doi:10.1017/S1743921307010617

Photospheric stellar activity cycles

Katalin Oláh
Konkoly Observatory, P.O. Box 67, HU-1525 Budapest, Hungary
email: olah@konkoly.hu

Abstract. Cycle lengths of active stars are derived from long-term photometric monitoring of their secular light variability. With the help of photographic data archives the lengths of the datasets are extended for HK Lac and V833 Tau. Using time-frequency analysis it is shown that the cycles are continuously changing in time. Thus, the reported cycle lengths derived by simple Fourier analysis are mean values that are valid only for a given time interval.

Keywords. stars: spots , stars: activity, stars: atmospheres, stars: late-type, stars: magnetic fields

---

The photospheric solar activity cycles, like the ~ 11 yr long Schwabe and the ~ 80 yr long Gleissberg cycles, as the sunspot data reveal, are continuously changing in time, e.g., the Gleissberg cycle is getting steadily longer during the last few hundred years (Oláh & Strassmeier 2002). Activity cycles, sometimes multiple cycles, are derived for several active stars, from long-term photometric data, see Oláh *et al.* (2000) and Oláh & Strassmeier (2002) for more. The resulting cycle lengths are usually just 'quasiperiods', with high scatter around the mean period values, or are just marginal detections. The reason of these results well could be that the cycle periods are not stable but changing in time like on the Sun. To find such changes, however, very long uninterrupted datasets are necessary. We show results of analyzing long-term datasets which are compiled from archival photographic and continuous photoelectric observations using the time-frequency analysis program package TiFrAn created by Kolláth (2006). This method uses evenly sampled data, therefore, the observations first should be averaged and interpolated to get a suitable dataset for the analysis.

On the left side of Fig. 1 the time behavior of the cycle periods of HK Lac during 48 yr, is given. Recently, Fröhlich *et al.* (2006) found 6.8, 9.65 and 13.0 yr cycle periods from the same dataset as was presently used. Earlier, from a shorter dataset, and without the archival photographic data Oláh & Strassmeier (2002) derived 6.4 and 13.2 yr long cycles. The difference between the two results is the markedly present 9.65 yr long cycle, which is dominant in the beginning of the dataset, and which later gradually changed to about 13 yr as is seen in the bottom panel of Fig. 1.

The 108 yr long dataset of V833 Tau shows continuous presence of about 5.8 and 8.6 yr long cycles found earlier by Oláh & Strassmeier (2002) who also derived the shortest cycle as of 2.3 yr. However, this shortest cycle was around 3.4 yr in the first half of the 20th century, and later it showed continuous change between 2 - 2.7 yr, lasting till the present. Thus it is clear why the photometric data alone, which span only for the last 20 years, result in a 2.3 yr long cycle from the simple Fourier analysis.

All cycle periods, even marginal detections, that were derived earlier for the studied active stars by Oláh & Strassmeier (2002) are confirmed. However, it is found that the cycle lengths are varying in time on HK Lac and on V833 Tau. This result shows that the detection of cycles could be problematic with conventional methods, like the simple periodogram analysis. On the other hand, the time-frequency analysis needs more continuous and longer datasets than those existing at present for most active stars.

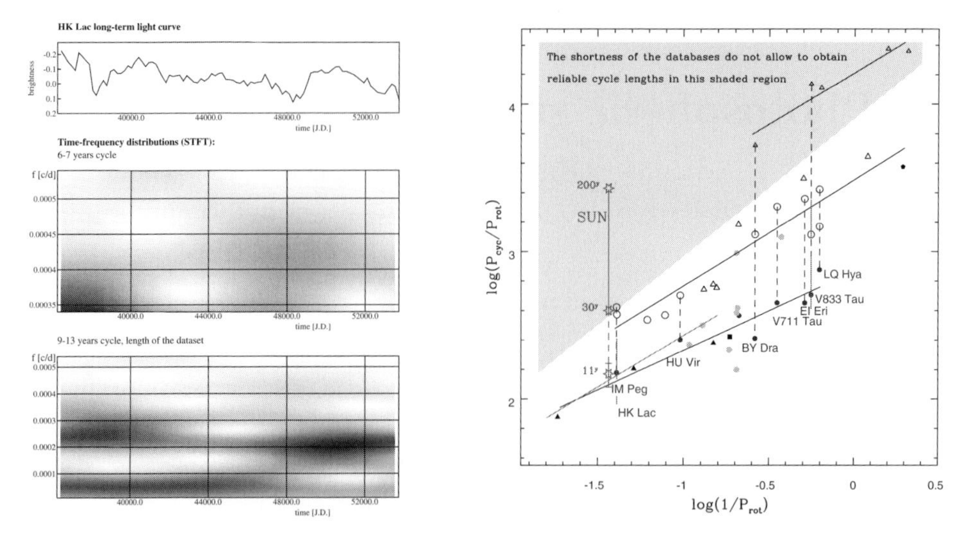

Figure 1. *Left:* time-frequency result for HK Lac. *Upper panel:* data; *middle panel:* cycle around 6 - 7 yr (0.00045 - 0.00030 cycle/d); *lower panel:* continuously growing cycle between 9 - 13 years (0.0003 - 0.0002 cycle/d). *Right:* rotation-activity cycle diagram based on Oláh & Strassmeier (2002). Also given are the cycles from Messina & Guinan (2002) for single solar type stars. The relation from Baliunas *et al.* (1996) is represented by a thick line at the long period part of the diagram. The continuous vertical lines show the ranges of the cycle length variations for the Sun, HK Lac and V833 Tau.

The right side of Fig. 1 shows the rotation - activity cycle diagram based on the work of Oláh & Strassmeier (2002). Additional cycles for solar type single stars are added from the work of Messina & Guinan (2002), which agree with the general trend between the rotational period and activity cycle. A thick line in the longer period regime represent the relation from Baliunas *et al.* (1996), based on Ca II H & K measurements for a large sample of single lower main sequence stars. Vertical continuous lines show the range of the derived cycle lengths for HK Lac, V833 Tau and for the Sun. The extended ranges of the cycle lengths may explain the large scatter towards the longer cycles. Oláh & Strassmeier (2002) suggested that the cycle lengths derived by simple Fourier analysis could be just snapshots valid for a given time interval. The present results seem to confirm this idea. Stars can have a wide range of possible cycle lengths, thus the determination of the shortest value is very important, since that would show clearly the relation between the rotational and cycle periods.

Acknowledgements

Supports from the Hungarian Research Grants OTKA T-043504 and T-048961 is acknowledged.

References

Baliunas, S. L., Nesme-Ribes, E., Sokoloff, D., & Soon, W. H. 1996, *ApJ*, 460, 848
Fröhlich, H.-E., Kroll, P., & Strassmeier, K. G. 2006, *A&A*, 454, 295
Kolláth, Z. 2006, *The program package TiFrAn*, `http://www.konkoly.hu/tifran/index.html`
Messina, S., & Guinan, E. F. 2002, *A&A*, 393, 225
Oláh, K., Kolláth, Z., & Strassmeier, K. G. 2000, *A&A*, 356, 643
Oláh, K., & Strassmeier, K. G. 2002, in: K. G. Strassmeier (ed.), Proc. 1st Potsdam Thinkshop on Sunspots and Starspots, *AN*, 323, 3/4, 361

Highlights of Astronomy, Volume 14
IAU XXVI General Assembly, 14-25 August 2006
Karel A. van der Hucht, ed.

doi:10.1017/S1743921307010629

Global stellar magnetic fields: the crossover from fossils to dynamos

Gautier Mathys

European Southern Observatory, Casilla 19001, Santiago 19, Chile
email: gmathys@eso.org

Abstract. Ap stars comprise the majority of early-type stars in which magnetic fields have been definitely detected. Their observational properties are reviewed, with emphasis on those relevant to the understanding of the origin and evolution of their fields.

Keywords. Stars: magnetic fields, stars: early-type, stars: chemically peculiar

---

While the presence of magnetic fields is ubiquitous in lower main sequence stars, only about 10% of early-type stars are observably magnetic. Most of them are main-sequence A- and B-type stars in the atmospheres of which a number of chemical elements are strongly over- or under-abundant. These Ap and Bp stars (often conveniently referred to collectively as Ap stars) generally show periodic variations of one or more of their observed properties: photometry, spectral line intensities, and magnetic fields. They are explained by the Oblique Rotator Model, in which the aspect of the visible stellar hemisphere changes as the star rotates because the distribution of the elemental abundances, brightness and magnetic field over the stellar surface is not symmetric about its rotation axis. Hence the variation period is the rotation period of the star. Rotation periods of Ap stars range from half a day to several decades: Ap stars rotate in average slower than superficially normal A- and B-type dwarfs in the same temperature range. For more details, see Mathys (2004a).

Magnetic fields of Ap stars cover their whole surface and show a considerable degree of large-scale organization; their strength varies from place to place on the star within a limited range. In first approximation, their structure often resembles a single dipole. The properties of Ap star magnetic fields have been recently reviewed by Mathys (2004b); here I shall only mention some of them, with emphasis on their relevance for the understanding of the origin and evolution of the fields. For a more complete recent presentation of the theoretical studies of the latter, see Moss (2004).

The favoured (albeit not undisputed) view is that the large-scale magnetic fields of the Ap stars have a fossil origin: they are either the relic of a field present in the interstellar medium at the time of star formation, and somehow locked into the star when it contracted, or the relic of a pre-main sequence dynamo. One merit of the fossil field hypothesis is that it provides a natural explanation of the fact that fields are observed in only a fraction of the early-type stars, and of the co-existence of observably magnetic and non-magnetic stars in the same environment (e.g., binaries, clusters). It has long been questioned, though, whether field configurations with sufficient long-term stability are possible: recent theoretical work by Braithwaite & Spruit (2004) and by Braithwaite & Nordlund (2006) now appears to have solved this difficulty.

The evolutionary status of magnetic Ap stars, and in particular the stage at which they become observably magnetic, has been hotly debated in recent years. There now seems to be some consensus emerging, according to which magnetic Ap stars with $M > 3\,\mathrm{M}_\odot$ appear homogeneously distributed across the whole width of the main sequence band, while lower mass magnetic Ap stars tend to be concentrated around the centre of this band (Hubrig *et al.* 2000; Kochukhov & Bagnulo 2006). This concentration may possibly be explained by the secular evolution of stable field configurations (Braithwaite & Nordlund 2006).

For a long time, observers have tried to identify the progenitors of magnetic Ap stars. The latter may finally have been found with the recent detections of magnetic fields in Herbig Ae/Be and related stars by Hubrig *et al.* (2004) and by Wade *et al.* (2005).

More than 10% of the Ap stars whose periods (or their lower limits) have been determined have periods longer than 100 days (Mathys 2004a). This leaves little doubt that these long periods are genuinely part of the Ap phenomenon, and that any theory describing how Ap stars form and acquire their properties must be able to account for them. Convincing arguments have been presented supporting the view that the variations of the Ap stars with extremely long periods do result from rotation, like for their shorter period counterparts. Furthermore, differences in the magnetic properties between extremely slowly rotating and faster rotating Ap stars have been identified. One of them is based on consideration of the mean magnetic field modulus, that is, the average over the visible stellar hemisphere of the modulus of the magnetic vector, weighted by the local emergent line intensity. This field moment can be measured only in those stars that rotate slowly enough and that have a sufficiently strong field so that their spectral lines can be observationally resolved into their Zeeman split components. At present, about 50 such stars are known, of which approximately 50% have a rotation period longer than 150 d. None of these long-period stars have a mean field modulus exceeding 7.5 kG, while in more than half of the stars with periods shorter than 150 days, the mean field modulus is greater than 7.5 kG (Mathys *et al.* 1997). On the other hand, Landstreet & Mathys (2000) and Bagnulo *et al.* (2002), using models with a predominant dipole component to represent the structure of Ap star magnetic fields, showed that in stars with periods shorter than 30 d, the angle between the dipole axis and the stellar rotation axis is generally large, while the two axes are nearly aligned in longer-period stars. This result may possibly be explained by the accretion braking mechanism proposed by Mestel & Moss (2005).

In summary, while in recent years significant progress has been achieved in the understanding of the origin and evolution of the magnetic fields of early-type stars, many questions remain open, which require further observational and theoretical studies.

References

Bagnulo, S., Landi Degl'Innocenti, M., Landolfi, M., & Mathys, G. 2002, *A&A*, 394, 1023

Braithwaite, J., & Nordlund, Å. 2006, *A&A*, 450, 1077

Braithwaite, J., & Spruit, H. C. 2004, *Nature*, 431, 819

Hubrig, S., North, P., & Mathys, G. 2000, *ApJ*, 539, 352

Hubrig, S., Schöller, M., & Yudin, R. V. 2004, *A&A* (Letters), 428, L1

Kochukhov, O., & Bagnulo, S. 2006, *A&A*, 450, 763

Landstreet, J. D., & Mathys, G. 2000, *A&A*, 359, 213

Mathys, G. 2004a, in: A. Maeder & P. Eeenens (eds.), *Stellar Rotation*, Proc. IAU Symp. No. 215 (San Francisco: ASP), p. 270

Mathys, G. 2004b, in: J. Zverko, J. Žižňovský, S. J. Adelman & W. W. Weiss (eds.), *The A-Star Puzzle*, Proc. IAU Symp. No. 224 (Cambridge: CUP), p. 225

Mathys, G., Hubrig, S., Landstreet, J. D., Lanz, T., & Manfroid, J. 1997, *A&AS*, 123, 353

Mestel, L., & Moss, D. 2005, *MNRAS*, 361, 595

Moss, D. 2004, in: J. Zverko, J. Žižňovský, S. J. Adelman & W. W. Weiss (eds.), *The A-Star Puzzle*, Proc. IAU Symp. No. 224 (Cambridge: CUP), p. 245

Wade, G. A., Drouin, D., Bagnulo, S., *et al.* 2005, *A&A* (Letters), 442, L31

Highlights of Astronomy, Volume 14
IAU XXVI General Assembly, 14-25 August 2006
Karel A. van der Hucht, ed.

doi:10.1017/S1743921307010630

Recent X-ray studies of stellar cycles and long-term variability

Giovanni Peres[1], Giuseppina Micela[2], and Fabio Favata[3]

[1]Dipartimento di Scienze Fisiche e Astronomiche, Sezione di Astronomia, Università di Palermo,, Piazza Parlamento 1, Palermo, I-90134, Italy
email: peres@astropa.unipa.it

[2]INAF – Osservatorio di Palermo, Piazza Parlamento 1, Palermo, I-90134, Italy
email: giusi@astropa.unipa.it

[3]Astrophysics Mission Division, RSSD of ESA/ESTEC, Postbus 299, NL-2200AG Noordwijk, the Netherlands
email: ffavata@rssd.esa.int

Abstract. We discuss recent X-ray studies of stellar cycles and long-term variability.

Keywords. stars: coronae, stars: activity, stars: atmospheres, stars: late-type, stars: magnetic fields

---

X-ray observations of the Sun, most recently with *Yohkoh*-SXT, have clearly shown that the X-ray luminosity of the corona changes by more than an order of magnitude along the solar cycle, albeit the exact value depends on the specific X-ray band of interest. One may thus expect that equally or more luminous X-ray stars show similar X-ray cycles. Detecting similar X-ray cycles on other stars, however, is not easy.

First of all, the idea that X-ray luminosity is due to magnetic activity, in turn tied to dynamo effect which is known to cause cyclic behavior on the Sun, cannot simply be extended to other stars, especially those much more luminous than the Sun, given our incomplete understanding of the highly non-linear dynamo action; some stars, furthermore, are believed to host a turbulent dynamo, different from the solar one, and hardly conducive to cyclic behavior.

Second, detecting any X-ray cycle demands several observational requirements, e.g., having a high enough photon count rate, making relatively short observations (a few thousand seconds) frequently enough (roughly every six months) and over 10 - 20 yr, gaining enough time at highly-on-demand X-ray telescopes (*Chandra*, *XMM-Newton*), being able to disentangle short-term variability from long-term/cyclic one.

Despite all these caveats, Marino *et al.* (2000, 2002, 2003) gave statistical evidence of long term variability in dF7 - dK2 stars in field and in Pleiades. More important, Hempelmann *et al.* (2003) in 61 Cyg A and B, Favata *et al.* (2004) in HD 81809, and Robrade *et al.* (2005) in Alpha Cen A, are accumulating clear evidence for long term X-ray variations, possibly stellar X-ray cycles. Indeed the variability in some cases resemble the solar one (e.g., large luminosity variations and small temperature changes) and long time with additional data points are still needed to prove whether these are cycle.

However, if we want a scenario of stellar X-ray cycles instead of a few case-studies, we need a dedicated project or satellite.

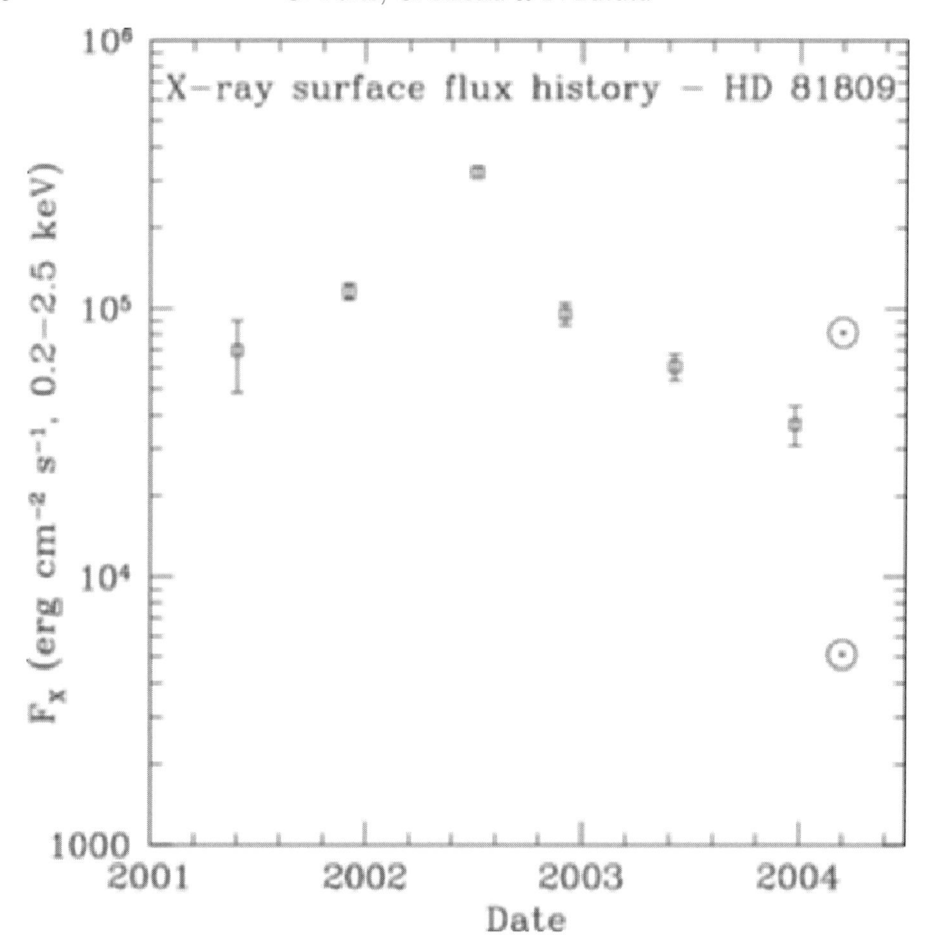

Figure 1. Evolution of the X-ray surface flux (in the 0.2 - 2.5 keV band) of HD 81809 from April 2001 to November 2003. At the right of the plot also the typical X-ray surface flux of the Sun at minimum and maximum of the cycle are plotted (from Favata *et al.* 2004).

References

Favata, F., Micela, G., Baliunas, S. L., Schmitt, J. H. M. M., Guedel, M., Harnden, F. R., Sciortino, S., & Stern, R. A. 2004, *A&A* (Letters), 418, L13

Hempelmann, A., Schmitt, J. H. M. M., Baliunas, S. L., & Donahue, R. A. 2003, *A&A* (Letters), 406, L39

Marino, A., Micela, G., & Peres, G. 2000, *A&A*, 353, 177

Marino, A., Micela, G., Peres, G., & Sciortino, S. 2002, *A&A*, 383, 210

Marino, A., Micela, G., Peres, G., & Sciortino, S. 2003, *A&A*, 406, 629

Robrade, J., Schmitt, J. H. M. M., & Favata, F. 2005, *A&A*, 442, 315

Highlights of Astronomy, Volume 14
IAU XXVI General Assembly, 14-25 August 2006
Karel A. van der Hucht, ed.

doi:10.1017/S1743921307010642

Topology of stellar coronae

Moira M. Jardine

SUPA, School of Physics and Astronomy, North Haugh, St Andrews, Fife, KY16 9UA, UK
email: mmj@st-andrews.ac.uk

Abstract. While the existence of cycles in stellar chromospheric flux has been known for some time, the nature of the corresponding coronal response has been more elusive. We describe recent results on the relationship between cyclic variations in surface magnetic flux and coronal structure and re-assess the role of prominence observations in understanding the topology of stellar coronas. We present a new paradigm for prominence support which allows for extended prominences to co-exist with a compact corona. We discuss briefly the recent results on coronal structure in high- and low-mass stars and their implications for dynamo theory.

Keywords. stars: activity, stars: imaging,stars: magnetic fields.

---

1. Introduction

In the case of the Sun, the relationship between the surface and coronal structure of the magnetic field is well known, but in the case of other stars the nature of this relationship is less clear. In addition to the surface field, other factors such as the rotation rate may be important. Thus for example, Jardine (2004) showed that the (super) saturation of coronal X-ray luminosity at high rotation rates can be explained by the effect of centrifugal forces stripping away closed field regions. This predicts that supersaturated stars should have a high rotational modulation of their X-ray emission, as is indeed observed in VXR 45 (Marino *et al.* 2003).

Recent advances have however been made in relating the surface magnetic structure (obtained using Zeeman-Doppler imaging, see e.g., Brown *et al.* 1991) and the coronal structure (as implied by X-ray observations, see e.g., Hussain *et al.* 2005). The surface field can be extrapolated into the corona using a *Potential Field Source Surface* model (Altschuler & Newkirk 1969) originally developed to model the solar corona. By assuming a hydrostatic, isothermal corona it is then possible to model the coronal density structure and hence the X-ray emission.

The best-observed example of the application of this technique is to AB Dor, a young, marginally pre-main sequence K-type dwarf of rotation period 0.514 d Donati *et al.* 1997, 1999, 2003). The maps of its magnetic field over the last 10 yr show a complex, multipolar structure at all latitudes, including the pole. Extrapolations of the coronal field (Jardine *et al.* 2002a,b) show a similarly complex structure with many closed-loop regions at high latitudes, some even spanning the pole. Since much of the magnetic flux is at high latitudes, the corresponding X-ray emission modelled using this field distribution also shows many bright regions at the pole, leading to a low rotational modulation in X-rays, which is consistent with the analysis of contemporaneous *Chandra* spectra (Hussain *et al.* 2005). This picture of a compact corona with loops extending at most 1 stellar radius above the surface is at odds, however, with the observation of prominences trapped in co-rotation out to some 5 or 6 stellar radii, far beyond the corotation radius (Jardine *et al.* 1997). These must be confined against centrifugal ejection by closed magnetic field lines. One possible way to resolve this contradiction is to confine the prominences in closed loops that extend *beyond* the X-ray corona into the wind region. Jardine &

van Ballegooijen (2005) have developed a model for the support of such loops on rapid rotators and predict a maximum height $y_m = 0.5(-1 + \sqrt{1 + 8R_k^3})$ where heights are in units of a stellar radius and R_k is the co-rotation radius. This agrees well with observed prominence distributions (Dunstone *et al.* 2006).

The relationship between cycles in surface magnetic field and the corresponding coronal X-ray emission is also of great interest for stars. Mackay *et al.* (2004) and later McIvor *et al.* (2006) showed that different types of surface cycles can lead to very different X-ray cycles (or apparently no cycles at all). If the presence (or absence) of cycles on solar-mass stars is as yet an open question, this is even more true in the case of high and low mass stars. Recent results on the coronal structure of the $15\,M_\odot$ star τ Sco and the fully-convective $0.3\,M_\odot$ star V374 Peg (Donati et al 2006a,b) show surprising results. The complex field topology of τ Sco, and the almost dipolar field of V374 Peg combined with a small or vanishing differential rotation for both presents a challenge for theories of magnetic field generation in stellar interiors and for studies of rotational evolution in low mass stars particularly. The variations in coronal topology along the main sequence, and particularly in the pre-main sequence stage are questions that should be addressed in the next few years.

References

Altschuler, M.D., & Newkirk, G. 1969, *Solar Phys.*, 9, 131
Brown, S. F., Donati, J.-F., Rees D., & Semel, M. 1991, *A&A*, 250, 463
Donati, J.-F., & Collier Cameron, A. 1997, *MNRAS*, 291, 1
Donati, J.-F., Collier Cameron, A., Hussain, G., & Semel, M. 1999, *MNRAS*, 302, 437
Donati, J.-F., Collier Cameron, A., Semel, M., *et al.*, 2003, *MNRAS*, 345, 1145
Donati, J.-F., Forveille, T., Collier Cameron, A., *et al.* 2006, *Science*, 311, 633
Donati, J.-F., Howarth, I. D., Jardine, M. M., *et al.*, 2006, *MNRAS*, 370, 629
Dunstone, N. J., Barnes, J. R., Collier Cameron, A., Jardine, M. M. 2006, *MNRAS*, 365, 530
Hussain, G., Brickhouse, N., Dupree, A., *et al.* 2005, *ApJ*, 621, 999
Jardine, M. M., Barnes, J., Unruh, Y. C., & Collier Cameron, A. 1997, in: D. Webb, D. Rust & B. Schmieder (eds.), *New Perspectives on Solar Prominemces*, Proc. IAU Coll. No. 167, *ASP-CS*, 150, 235
Jardine, M. M., Collier Cameron, A., & Donati, J.-F. 2002, *MNRAS*, 333, 339
Jardine, M. M., Wood, K., Collier Cameron, A., *et al.* 2002, *MNRAS*, 336, 1364
Jardine, M. M. 2004, *A&A* (Letters), 414, L5
Jardine, M. M., & van Ballegooijen, A. A. 2005, *MNRAS*, 361, 1173
Mackay, D. H., Jardine, M. M., Collier Cameron, A., *et al.* 2004, *MNRAS*, 354, 737
Marino, A., Micela, G., Peres, G., & Sciortino, S. 2003, *A&A* (Letters), 407, L63
McIvor, T., Jardine, M. M., Mackay, D., & Holzwarth, V. 2006, *MNRAS*, 367, 592

Highlights of Astronomy, Volume 14
IAU XXVI General Assembly, 14-25 August 2006
Karel A. van der Hucht, ed.

doi:10.1017/S1743921307010654

Why coronal mass ejections are necessary for the dynamo

Axel Brandenburg[1,2]

[1]Nordita, Blegdamsvej 17, DK-2100 Copenhagen Ø, Denmark
email: brandenb@nordita.dk

[2]AlbaNova University Center, SE-106 91 Stockholm, Sweden

Abstract. Large scale dynamo-generated fields are a combination of interlocked poloidal and toroidal fields. Such fields possess magnetic helicity that needs to be regenerated and destroyed during each cycle. A number of numerical experiments now suggests that stars may do this by shedding magnetic helicity. In addition to plain bulk motions, a favorite mechanism involves magnetic helicity flux along lines of constant rotation. We also know that the sun does shed the required amount of magnetic helicity mostly in the form of coronal mass ejections. Solar-like stars without cycles do not face such strong constraints imposed by magnetic helicity evolution and may not display coronal activity to that same extent. I discuss the evidence leading to this line of argument. In particular, I discuss simulations showing the generation of strong mean toroidal fields provided the outer boundary condition is left open so as to allow magnetic helicity to escape. Control experiments with closed boundaries do not produce strong mean fields.

Keywords. hydrodynamics (magnetohydrodynamics:) MHD, turbulence, Sun: coronal mass ejections (CMEs), Sun: magnetic fields, stars: magnetic fields, stars: mass loss

---

All known large scale dynamos ($\alpha\Omega$, shear-current, and α^2 dynamos) produce magnetic

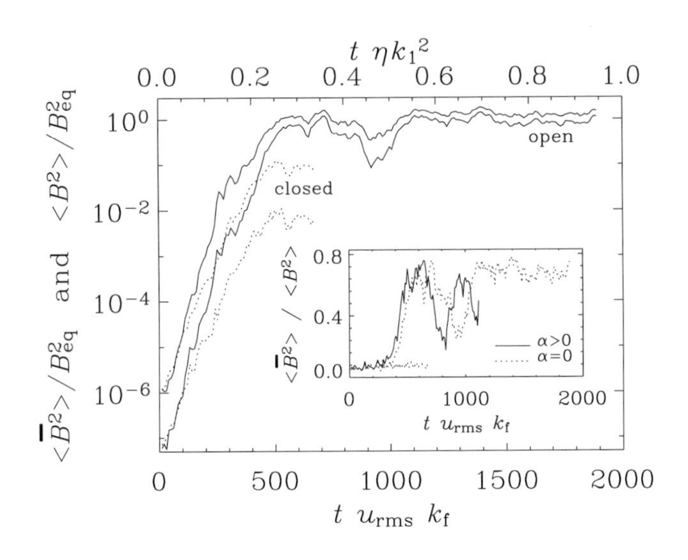

Figure 1. Evolution of the energies of the total field $\langle \mathbf{B}^2 \rangle$ and of the mean field $\langle \overline{\mathbf{B}}^2 \rangle$, in units of B_{eq}^2, for runs with non-helical forcing and open or closed boundaries; see the solid and dotted lines, respectively. The inset shows a comparison of the ratio $\langle \overline{\mathbf{B}}^2 \rangle / \langle \mathbf{B}^2 \rangle$ for nonhelical ($\alpha = 0$) and helical ($\alpha > 0$) runs. For the nonhelical case the run with closed boundaries is also shown (dotted line near $\langle \overline{\mathbf{B}}^2 \rangle / \langle \mathbf{B}^2 \rangle \approx 0.07$). Adapted from Brandenburg (2005).

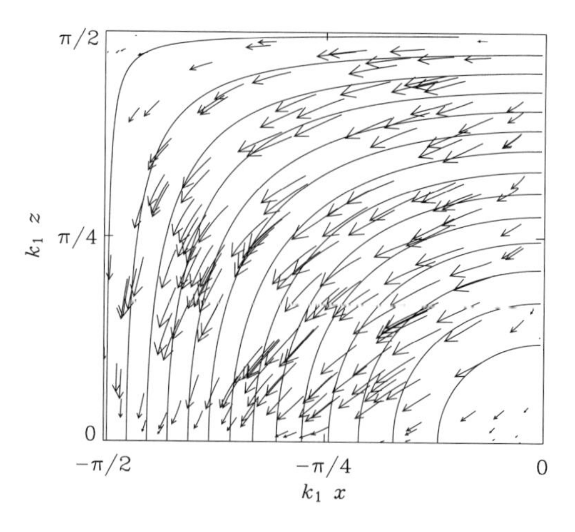

Figure 2. Vectors of the Vishniac & Cho (2001) flux together with contours of the mean flow which also coincide with the streamlines of the mean vorticity field. The orientation of the vectors indicates that negative current helicity leaves the system at the outer surface at $x = 0$. The equator corresponds to $z = 0$. Note that the vectors indicate positive flux, so vectors pointing away from the outer surface correspond to negative helicity leaving the sun in the northern hemisphere. Adapted from Brandenburg *et al.* (2005).

helicity, which reacts back on the dynamo. As a consequence, the mean field saturates at a low value, $\overline{\mathbf{B}}^2 \ll B_{\rm eq}^2 \equiv \langle \mu_0 \rho \mathbf{u}^2 \rangle$. By allowing for magnetic helicity fluxes out of the domain, the large scale field is able to saturate at equipartition field strength (Fig. 1). The results of simulations are qualitatively, and in some cases also quantitatively, well reproduced by mean field models where the effect of magnetic helicity fluxes enters into the dynamical feedback formula for the magnetic alpha effect (even when there is no kinetic alpha effect!). For closed boundary conditions, the field saturates at much lower strength and no large scale field is being produced.

Contributions to the magnetic helicity flux include the shear-driven Vishniac & Cho (2001) flux, which can be written in the form $\overline{\mathbf{F}} \propto (\overline{\mathsf{S}}\,\overline{\mathbf{B}}) \times \overline{\mathbf{B}}$, where $\overline{\mathsf{S}}$ is the strain rate of the mean flow (Subramanian & Brandenburg 2006), and an advectively driven flux (Shukurov *et al.* 2006) of the form $\overline{\mathbf{F}} \propto \alpha_{\rm M} \overline{\mathbf{U}}$, where $\alpha_{\rm M}$ is the magnetic α effect. The former is the one operating predominantly in the simulations (Fig. 2).

A connection between dynamo-generated magnetic helicity fluxes and coronal activity was first predicted by Blackman & Field (2000). There are at present no direct simulations of turbulent dynamos that also include the relevant physics behind coronal mass ejections. On the other hand, the observed magnetic helicity fluxes from coronal mass ejections (Berger & Ruzmaikin 2000) of around $10^{46}\,{\rm Mx}^2$ per cycle agrees with what is predicted from simulations (Brandenburg & Sandin 2004).

References

Berger, M. A., & Ruzmaikin, A. 2000, *JGR*, 105, 10481
Blackman, E. G., & Field, G. B. 2000, *MNRAS*, 318, 724
Brandenburg, A. 2005, *ApJ*, 625, 539
Brandenburg, A., & Sandin, C. 2004, *A&A*, 427, 13
Brandenburg, A., Haugen, N. E. L., Käpylä, P. J., & Sandin, C. 2005, *AN*, 326, 174
Shukurov, A., Sokoloff, D., Subramanian, K., & Brandenburg, A. 2006, *ApJ* (Letters), 448, L33
Subramanian, K., & Brandenburg, A. 2006, *ApJ* (Letters), 648, L71
Vishniac, E. T., & Cho, J. 2001, *ApJ*, 550, 752

Highlights of Astronomy, Volume 14
IAU XXVI General Assembly, 14-25 August 2006
Karel A. van der Hucht, ed.

doi:10.1017/S1743921307010666

Tachocline, dynamo and the meridional flow connection

Günther Rüdiger

Astrophysikalisches Institut Potsdam, An der Sternwarte 16, D-14482 Potsdam
email: gruediger@aip.de

Abstract. Our model for global circulations in the solar convective zone leads to an equatorial drift at its base with an amplitude of about 7 m/s. Its penetration into the solar tachocline is too weak to play any role in the solar dynamo. It confines, however, the internal magnetic decay modes to the radiative core so that the tachocline can be explained as a magnetic Hartmann layer. In the tachocline for the toroidal fields the magnetic Tayler instability exists. We found stability limits for toroidal fields of only 100 G.

Keywords. MHD, turbulence, Sun: rotation, Sun: magnetic fields

---

The transition region between the differential rotation in the solar convection zone and the rigid rotation of the solar interior is called the 'tachocline' which is known from helioseismology as thinner than 5% of the solar radius. Sofar such a thin tachocline could not be explained without magnetic fields but the inclusion of a weak poloidal magnetic field of order 10^{-4} G, which is basically confined to the radiative solar interior easily produces a thin enough transition zone between outer differential rotation and inner rigid rotation (Fig. 1, see Kitchatinov & Rüdiger 2005).

The meridional circulation associated with such a rotation law flows at the base of the convection zone towards the equator with an amplitude of about 7 m/s. This flow is fast enough to realize an advection-dominated α-Ω dynamo within the convection zone if the turbulent magnetic diffusivity there is low enough ($\sim 10^{11}\text{cm}^2\text{s}^{-1}$, see Küker *et al.* 2001; Bonanno *et al.* 2005).

The penetration of the meridional flow into the solar tachocline strongly depends on the viscosity values beneath the convection zone . For viscosity values of (say) $\sim 10^{11}\text{cm}^2\text{s}^{-1}$ the penetration depth is about 4,000 km. This is much too weak to play any role in the solar dynamo mechanism.

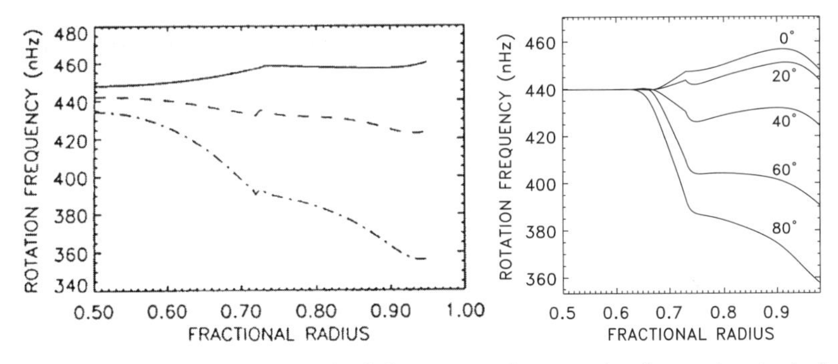

Figure 1. *Left:* Hydrodynamical model of the inner solar rotation law using turbulent values within the convection zone and microscopic viscosity values beneath. *Right:* In the radiative zone a fossil magnetic field is confined inducing a toroidal magnetic field which effectively destroys the latitudinal rotational shear.

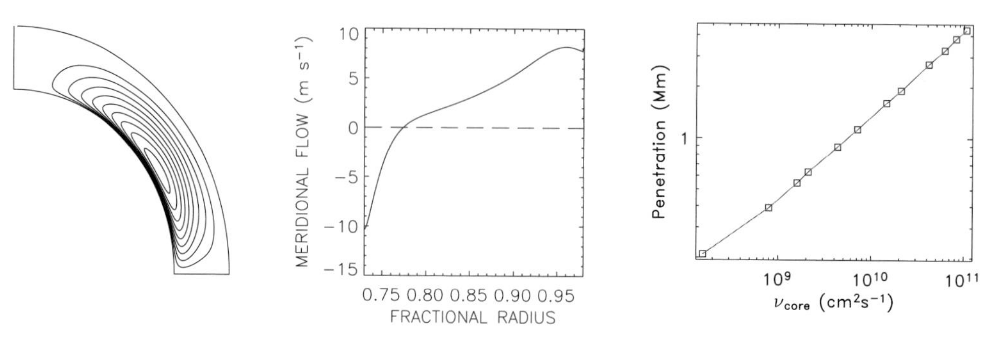

Figure 2. *Left:* The flow pattern of the meridional circulation which belongs to rotation law given in Fig. 1 (*right*). *Middle:* The flow amplitude over radius; at the base of the convection zone the flow drifts equatorwards. *Right:* The penetration of the meridional flow into the radiative solar core for the viscosity, $\nu_{\rm core}$ (Rüdiger *et al.* 2005).

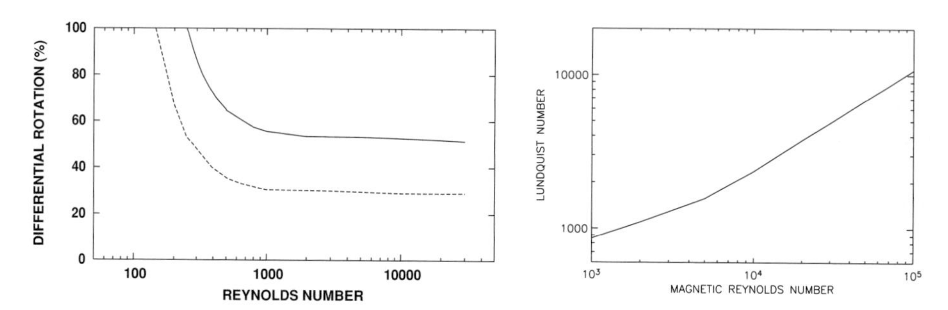

Figure 3. *Left:* the hydrodynamical stability of the solar tachocline (*dashed:* no radial shear; *solid:* real solar rotation law. Note the stabilizing action of the radial rotation shear. *Right:* the critical toroidal magnetic field measured by its Lundquist number for given magnetic Reynolds number (for the observed rotation law).

The flow is sufficiently fast, however, to induce strong latitudinal magnetic fields so that finally the internal magnetic field has a strong latitudinal field component. Only by this effect of (slight) penetration of the meridional flow into the solar core the tachocline can be explained as a magnetic Hartmann layer (Kitchatinov & Rüdiger 2006).

Also the *stability* of toroidal magnetic fields in the differentially rotating solar tachocline is important for our tachocline concept. The tachocline is the location where both the hydrodynamic Rayleigh instability and the magnetic Tayler instability can interact. If the magnetic field is strong enough then the hydrodynamically stable tachocline becomes unstable against nonaxisymmetric perturbations. We found stability limits for toroidal fields of only 100 G for a stably stratified sphere (Fig. 3, see Arlt *et al.* 2007). The tachocline is thus *not* a suitable site for the storage of strong magnetic fields.

References

Arlt, R., Sule, A., & Rüdiger, G. 2007, *A&A*, 461, 295
Bonanno, A., Elstner, D., Belvedere, G., & Rüdiger, G. 2005, *AN*, 326, 170
Kitchatinov, L. L., & Rüdiger, G. 2005, *AN*, 326, 379
Kitchatinov, L. L., & Rüdiger, G. 2006, *A&A*, 453, 329
Küker, M., Rüdiger, G., & Schultz, M. 2001, *A&A*, 374, 301
Rüdiger, G., Kitchatinov, L. L., & Arlt, R. 2005, *A&A* (Letters), 444, L53

Highlights of Astronomy, Volume 14
IAU XXVI General Assembly, 14-25 August 2006
Karel A. van der Hucht, ed.

doi:10.1017/S1743921307010678

The impact of stellar activity on planets

Ignasi Ribas
Institut de Ciències de l'Espai (CSIC/IEEC), Campus UAB, Facultat de Ciències,
Torre C5-parell, 2a planta, E-08193 Bellaterra, Spain
email: iribas@ieec.uab.es

Abstract. The results of the *Sun in Time* program indicate that the X ray, far ultraviolet and ultraviolet fluxes of the young Sun were significantly higher than today. Similarly, the solar wind may have been much stronger in the past. Such environment of intense energy and particle emissions could have influenced the paleo-atmospheres of Solar System planets and, by extension, the habitability and stability of exoplanets.

Keywords. Sun: evolution, Sun: activity, Sun: rotation, Sun: UV radiation, Sun: solar wind, planets and satellites: general, stars: activity, stars: rotation

---

It has long been established that young late-type stars rotate faster than their older counterparts (Skumanich 1972; Soderblom 1982). As a consequence, the interplay between the high rotation rate and the convective motions in the stellar envelope gives rise to high levels of dynamo-generated magnetic activity. From our privileged view of the active Sun and the observation of stars it is found that such activity is manifested through strong high-energy fluxes and particle emissions. The decrease in rotation rate between the 'young' and 'old' stellar populations does not appear to be abrupt but rather a gradual one that follows a power-law relationship of exponent ~ -0.6 as a function of age. Following the correlation between rotation rate and activity, Zahnle & Walker (1982) concluded using *IUE* observations that the UV emissions of late-type stars strongly decrease with increasing stellar age.

Inspired by these findings, the *Sun in Time* program was established to investigate the magnetic evolution of the Sun and associated high-energy emissions across its main sequence lifetime. To do so, a sample of stars within a narrow spectral range (G0 - G5) and well-determined properties (including age) were selected as stellar proxies for the Sun at different stages of its evolution. The five studied stars have ages covering from 130 Myr to 6.7 Gyr. Data obtained with the *ASCA*, *ROSAT*, *EUVE*, *FUSE* and *IUE* satellites, covering the wavelength range between 1 and 170 nm, indicate that the Sun had emissions in X rays, far ultraviolet and ultraviolet that were stronger than today's by factors of 1000 - 100, 60 - 20, and 20 - 0, respectively. Ribas *et al.* (2005) also propose a time-evolution relationship for emissions in the interval 1 - 120 nm. Such relationship indicates that the Sun had high-energy emissions some 2.5 times stronger 2.5 Gyr ago and 6 times stronger 3.5 Gyr ago. Near the ZAMS (age 0.1 Gyr), the Sun could have had fluxes up to 100 times larger than today (albeit during a short period of time).

Besides high-energy radiation, particle emissions are also an integral component of the active Sun. From simple arguments, it is reasonable to assume that the strong X-ray emissions and hot coronal temperatures of young solar analogs are also correlated with more massive stellar winds. Direct detection of such winds has been very elusive but Wood *et al.* (2002) succeeded in estimating indirectly the mass loss rate (i.e., stellar wind) of a number of main sequence stars. The results in that study indicated that the Solar wind may have decreased with time following a power law relationship with an exponent of roughly -2, which would seem to predict a mass loss rate some 1000 times higher than today for the young ZAMS Sun. More recent results (Wood *et al.* 2005), however, question the suitability of an extrapolation back in time to very young ages and suggest a change in regime for an age of about 0.7 Gyr. In any case, it is now clear that the wind of the young Sun could have been some two or three orders of magnitude stronger than today.

These high-energy and particle emissions of the young Sun could have played an important role in the evolution of the atmospheres (and even the surfaces) of the Solar System planets. The evolution of the Martian water inventory is one example. Because of its low gravity and the lack of a protecting magnetic field, Mars is (and has been) subject to strong losses of light atmospheric constituents from thermal and non-thermal processes. Simulations indicate that Mars could have lost a global ocean with a depth of about 10 m from photolysis of the water molecule and subsequent escape of hydrogen to space. The remaining oxygen would have been incorporated into the ground giving its characteristic rusty appearance. A similar process may have occurred in Venus, which could have lost an amount of water comparable to a terrestrial ocean, although results are still inconclusive because of uncertainties with the history of the magnetic field of the planet. In the case of the Earth, its protecting magnetic field and strong gravity have prevented massive evaporation processes from taking place. But the strong UV emissions of the young Sun could have played a role in the development of life, for example, by influencing the abundance of greenhouse gases such as CH_4 and NH_3, both prone to photo-dissociation processes, and by triggering photochemical reactions.

The results of the *Sun in Time* program can also be applied to exoplanets detected around solar-type stars. The most obvious case is that of the so-called Hot Jupiters. In this case, the high-energy emissions of the parent star heat the exosphere of the planet and produce strong thermal loss processes (of hydrogen and other constituents) that could evaporate a significant fraction of the planet's mass (Lammer *et al.* 2003). Such loss processes have indeed been detected observationally (Vidal-Madjar *et al.* 2003).

A logical extension of the study is the investigation of the evolution of high-energy and particle emissions in stars of other spectral types. This is done by using the X-ray flux as an overall proxy for stellar activity. With such scaling and a normalization using the bolometric luminosity, preliminary results suggest that the integrated 1 - 120 nm emissions of early K-type stars are some 3 - 4 times stronger than the emissions of solar-type stars of the same age and in the case or early M-type stars the emissions are larger by factors of 10 - 100. Accounting for these high emissions is critical to understand both the evolution and even the stability of exoplanets orbiting such stars. Because of its immediate application to the design of missions such as *Darwin* and *TPF*, the question of habitability of planets around M-type stars is especially relevant (Segura *et al.* 2005).

References

Lammer, H., Selsis, F., Ribas, I., Guinan, E. F., Bauer, S. J., & Weiss, W. W. 2003, *ApJ* (Letters), 598, L121

Ribas, I., Guinan, E. F., Güdel, M., & Audard, M. 2005, *ApJ*, 622, 680

Segura, A., Kasting, J. F., Meadows, V., Cohen, M., Scalo, J., Crisp, D., Butler, R. A. H., & Tinetti, G. 2005, *Astrobiology*, 5, 706

Skumanich, A. 1972, *ApJ*, 171, 565

Soderblom, D. R. 1982, *ApJ*, 263, 239

Vidal-Madjar, A., Lecavelier des Etangs, A., Désert, J.-M., Ballester, G. E., Ferlet, R., Hébrard, G., & Mayor, M. 2003, *Nature*, 422, 143

Wood, B. E., Müller, H.-R., Zank, G., & Linsky, J. L. 2002, *ApJ*, 574, 412

Wood, B. E., Müller, H.-R., Zank, G. P., Linsky, J. L., & Redfield, S. 2005, *ApJ* (Letters), 628, L143

Zahnle, K. J. & Walker, J. C. G. 1982, *Rev. Geophys. Space Phys.*, 20, 280

Highlights of Astronomy, Volume 14
IAU XXVI General Assembly, 14-25 August 2006
Karel A. van der Hucht, ed.

doi:10.1017/S174392130701068X

Future facilities for solar and stellar activity research

Klaus G. Strassmeier

AIP, Astrophysical Institute Potsdam, D-14482 Potsdam, Germany
email: kstrassmeier@aip.de

Abstract. I try to list the currently ongoing instrumental activities for solar and stellar activity research. Only projects that lead to operational ground-based facilities no later than 2013 and to operational space-based observatories no later than 2016 are considered. Any facility already in operation or any instrument under construction but with a very wide range of usage is excluded from this listing (like e.g. ALMA, *Herschel* or SOFIA). No details on science programs are given. The text is organized according to space, radio, and optical/IR projects.

Keywords. telescopes, space vehicles, instrumentation: miscellaneous

---

SOLAR-B (now named *Hinode*) has just been launched (Sept. 22, 2006).

CoRoT is up for launch late November 2006. Its science mission is twofold. Firstly, it will focus on ultra-high precision photometry of very bright stars for asteroseismology of all kinds of non-radially pulsating stars. Secondly, it will perform photometry of some 30,000 stars down to 15th magnitude to search for transits of extrasolar planets.

SDO (*Solar Dynamic Observator*) is getting ready for launch in 2008. It is the anticipated *SOHO* successor and will employ, besides other instruments, the Helioseismic Magnetic Imager which will allow a full inversion of the internal solar magnetic field.

Kepler is planned for launch in early 2009. Its aim is to detect earth-sized planets from a massive transit search in a dedicated 100-sqr degree field in Cygnus with $>100{,}000$ stars. It employs high precision integral-light photometry over the period of 4 yr.

Gaia is one of the ESA cornerstone missions planned for launch in 2012. Its aim is to measure positions, parallaxes and proper motions of up to 1 billion stars with precisions of up to μ-arcseconds. 7-channel photometry and km/s-level radial velocities of a smaller subsample are also among its output. Final data are expected no earlier than 2016.

WSO-HIRDES (*World Space Observatory* - HIgh Resolution Double Echelle Spectrograph) is a mission concept for possible launch in 2016 or later. It is based on a 1.7 m UV-optimized telescope with a double spectrograph for the wavelength range 103 - 310 nm at a resolution of 50,000.

ATA (Allen Telescope Array) is a 0.5 - 11.2 GHz array of 350 (final configuration) 6 m dishes at Hat Creek, California. Its FOV is 2.45° at 21 cm. Operated by the SETI Institute it will monitor 1 million stars in the 1 - 10 GHz range and some 40 billion stars at 1.450 GHz. It could provide the most comprehensive view of the galactic magnetic field.

LOFAR (Low Frequency Array) is a digital new-technology ground-based radio spectro-interferometer for the 10 - 240 MHz range. In its final configuration in 2009 it consists of dipole antenna fields across northern Europe, mainly in the Netherlands. The array observes the visible sky all the time, pointings will be done posteriori.

STELLA (Stellar Activity) is a fully robotic facility at the Teide Observatory, Tenerife. It consists of two 1.2 m telescopes that feed a high-resolution echelle spectrograph (390 - 870 nm, $R=55{,}000$) and a wide-field imaging photometer ($22' \times 22'$, $0.3''$/px, 17 filters), respectively. It will be fully operational by 2007.

APF (Automated Planet Finder). A 2.4 m automated telescope currently under construction at Lick Observatory. Its sole instrument is a 'planetometer' designed for 1m/s-precision radial velocity measurement based on an iodine cell.

Pan-STARRS (Panoramic Survey Telescope and Rapid Response System) will consist of four 1.8 m telescopes on Haleakala, Maui, Hawaii. The first telescope is already being commissioned. All four telescopes are wide-field imaging systems with a 1.4-Gigapixel CCD camera each (based on 64 600×600 orthogonal-transfer CCD arrays). Its main scientific goal is to search for Near Earth Objects.

DCT (Discovery Channel Telescope). Installed and operated by Lowell Observatory in Arizona by 2012, the 4.2 m DCT will provide wide-field prime-focus imaging with a scientific focus towards faint Kuiper-belt objects.

LAMOST (Large-sky Area Multi-Object fibre Spectroscopy Telescope) is a 4 m segmented mirror telescope for the Xinglong station in China for 2009. 4000 fibers feed stellar light from a 5-degree FOV to 16 spectrographs covering a wavelength range of 370 900 nm at resolutions of $R=1$ - 5,000 and 5 - 10,000, respectively.

LBT (Large Binocular Telescope) consists of two 8.4 m mirrors on a joint mount with altogether 6 foci per telescope, two of which are interferometric. One of its instruments is PEPSI, the Potsdam Echelle Polarimetric and Spectroscopic Instrument. It will provide full Stokes-vector spectropolarimetry at spectral resolution 120,000 for the wavelength range 450 - 1050 nm. Its integral light modus enables resolving powers of 40,000 and 300,000 for the 390 - 1050 nm range. Located on Mt Graham in Arizona, the LBT and PEPSI will go into routine operation in 2009.

LSST (Large Synoptic Survey Telescope). Built upon the LBT mirror experience, the LSST is a 8.4 m ultra-wide field telescope with a FOV of 3.5°. Its three-reflection design provides a 64 cm diameter focal plane that will be paved with 200 4k $\times$ 4k CCDs with altogether 3.2 Gpx. It is expected to provide a full image of the visible sky every night. The telescope is expected to be located at Cerro Pachon in Chile by 2013.

GREGOR is the new German solar telescope on Tenerife with first light in 2007. With an aperture of 1.5 m and integrated multi-conjugate adaptive optics (at second light) it is designed for high spatial resolution spectropolarimetry for the optical and IR range (up to 10 μm). Its night time is dedicated to the solar-stellar connection.

ATST (Advanced Technology Solar Telescope). The next-generation solar telescope to be build on Maui, Hawaii around 2012. A 4 m off-axis parabolic mirror will enable unprecedented image quality and shall provide enough light for high time resolution IR spectropolarimetry.

CONCORDIA (the French-Italian Antarctic station at Dome C). Located at 3280 m on the east Antarctic plateau with an average night temperature of $-63°$ and practically zero (absolute) humidity, Dome C is 'between Earth and Space'. A number of astrophysics projects are foreseen for the years 2008 - 2013. Among them are IRAIT (a 80 cm *JHKLM* imaging telescope for 2008), A-STEP (a 40 cm optical telescope for planet transit photometry for 2008), ICE-T (a dual 60 cm ultra-wide-field ultra-high-precision optical photometer for 2011), and PILOT (a 2.4 m IR and optical high-resolution imager for 2013).

Acknowledgements

Contributions to this talk came from my AIP colleagues M.I. Andersen, G. Mann, M. Weber and M. Woche, and Xiangqun Cui (NAO, China), A. Kosovichev (Stanford), S. Vogt (Lick) and R. Kudritzki (Hawaii) to whom I am thankful. I also acknowledge the many www sites that provide the pictorial information beyond the mouth-to-ear approach.

Highlights of Astronomy, Volume 14
IAU XXVI General Assembly, 14-25 August 2006
Karel A. van der Hucht, ed.

doi:10.1017/S1743921307010691

Joint Discussion 9
Supernovae: one millennium after SN 1006

P. Frank Winkler[1] and Wolfgang Hillebrandt[2] (eds.)

[1]Department of Physics, Middlebury College, Middlebury, VT 05753, USA
email: winkler@middlebury.edu

[2]Max-Planck-Institut für Astrophysik, Karl-Schwarzschild-Strasse 1, D-85748 Garching, Germany
email: wfh@mpa-garching.mpg.de

Preface

The year 2006 marks the 1000th anniversary of the supernova of 1006 C.E., the brightest supernova in all of recorded human history. This is also a time of great excitement in the supernova community: Observations from space observatories including *Hubble*, *Chandra*, *XMM-Newton*, and *Spitzer*, together with ones from powerful new ground-based telescopes and instruments, are revealing supernova remnants in the Galaxy and beyond in unprecedented detail. Fully three-dimensional computational codes and simulations running on powerful new machines are providing insight into the physics of supernovae freed from the simplifying assumptions that have restricted past understanding. Automated supernova searches are discovering hundreds of new supernovae every year, some at redshifts of 1 or beyond. And supernovae have revolutionized cosmology through the discovery of an accelerating universe, and they hold promise for deepening our understanding of the 'dark energy' that drives the acceleration.

Joint Discussion 09 on *Supernovae: one millennium after SN 1006*, brought together observers, theorists, and some historians – to celebrate the SN 1006 millennium by reviewing recent progress in understanding supernovae, their remnants, and their application to cosmology. In a stimulating day and a half there were 25 (mostly invited) oral papers, as well as some two dozen posters touching on many observational and theoretical aspects of supernova research. The oral papers focused primarily on Type Ia supernovae and their remnants – including observations of SN 1006 itself and its more recent cousins, models for SN Ia progenitors, explosion mechanisms, and how they interact with the interstellar and circumstellar medium on the way to becoming remnants, through the application of SN Ia's to cosmology.

In addition to stimulating discussions among participants both inside and outside the conference hall, there was culinary stimulation in the form of a delicious 1000th birthday cake for SN 1006 (Figure 1).

The organizers thank all who presented oral or poster papers at JD09 for their contributions. We hope that they as well as the many other astronomers who took part in the IAU XXVI General Assembly, and readers of this volume, will glimpse the excitement that attends supernova research today. We also wish to thank all the members of the local organizing committee for all that they did to make IAU XXVI General Assembly and JD09 a success.

Figure 1. Participants of JD09 eagerly awaiting the cutting of a Birthday Cake to celebrate the Millennium of SN 1006.

Scientific Organizing Committee

Gloria Dubner (Argentina), Claes Fransson (Sverge), Wolfgang Hillebrandt (Germany, co-chair), Katsuji Koyama (Japan), Ken'ichi Nomoto (Japan), Robert Petre (USA), Pilar Ruiz-Lapuente (Spain), Brian P. Schmidt (Australia, co-chair), Virginia L. Trimble (USA), J. Graig Wheeler (USA), and P. Frank Winkler (USA, co-chair).

P. Frank Winkler, Wolfgang Hillebrandt, co-chairs SOC,
Middlebury, Vermont, USA, and Garching, Germany, December 2006

Highlights of Astronomy, Volume 14
IAU XXVI General Assembly, 14-25 August 2006
Karel A. van der Hucht, ed.

doi:10.1017/S1743921307010708

SN 1006: a thousand-year perspective

P. Frank Winkler

Department of Physics, Middlebury College, Middlebury, VT 05753, USA
email: winkler@middlebury.edu

Abstract. We review some of the extensive historical observations of SN 1006, emphasizing estimates of its brightness at maximum. An estimate of $V_{max} \approx -7.5$ is consistent with what may be the most reasonable interpretation of these records and with an *a posteriori* calculation based on typical peak magnitudes for Type Ia supernovae together with the distance and extinction to SN 1006. We also give a brief overview of the discovery of the SN 1006 remnant in 1965, and contrast the earliest radio, optical, and X-ray observations of the remnant with recent ones, as reported in more detail by other papers in this JD09 review.

Keywords. stars: supernovae, individual (SN 1006), ISM: supernova remnants, history and philosophy of astronomy

---

The supernova of 1006 CE, whose 1000th anniversary we mark through JD09, is generally regarded as the brightest supernova (and quite likely the brightest stellar event of any kind) ever recorded in human history. This star was so spectacular that contemporary observers throughout the northern hemisphere recorded the event, despite its southern location at Dec (1006) = 38°. (See papers by Stephenson and Sun for more details.) The *most* northerly sighting was from the Benedictine Abbey of St. Gall, in Switzerland at latitude 47.5° N. The chronicles of the abbey record a star *"dazzling to the eyes ... in the extreme limits of the south"*. These chronicles record the single most significant event of each year for more than three centuries. It is noteworthy that the writer gives such an extensive entry to this star – far more than he devotes to a 'famine more severe than any of our age' of the year before, or a deadly plague the year after (Fig. 1).

Records from around the world agree that the star first appeared on 1 May 2006, ± 1 day. But since these same records compare its brightness with Mars, Venus, the Moon, and even the Sun, there has been widespread disagreement about its peak brightness. The most explicit record is that of the Egyptian physician and astrologer Ali ibn Ridwan, who described it as *"2.5 or 3 times the magnitude of Venus, ... and a little more than a quarter of the brightness of the Moon"* (Goldstein 1965).

Two weeks after the star first appeared, when it should have been near peak brightness, Venus, the full Moon, and SN 1006 would all have appeared $\sim 15°$ above the horizon in different directions. Let us assume that ibn Ridwan was indeed referring to the *full* Moon ($V \approx -12.5$), and was using it and Venus ($V \approx -4$) as benchmarks for a logarithmic scale, similar to the modern magnitude scale (and to the response of the human eye). The result gives a magnitude $-7.3 > V_{max} > -7.8$ for SN 1006 (Winkler *et al.* 2003 = WGL03).

According to Chinese records, SN 1006 remained visible for three years (Stephenson paper following). Records of its appearance continued to appear through the following centuries, but at some point European records conflated the 1006 C.E. event with one from 837 C.E., and as a result it became described as a comet, rather than a star. All of this was sorted out through the painstaking tracing of Arabic and other records by Bernhard Goldstein (1965), a young historian of astronomy at Yale, whose attention was drawn to the event by his friend Alistair Cameron. In a rocket flight the year before, Herbert Friedman's NRL group had identified the Crab Nebula (= SN 1054) as a strong X-ray

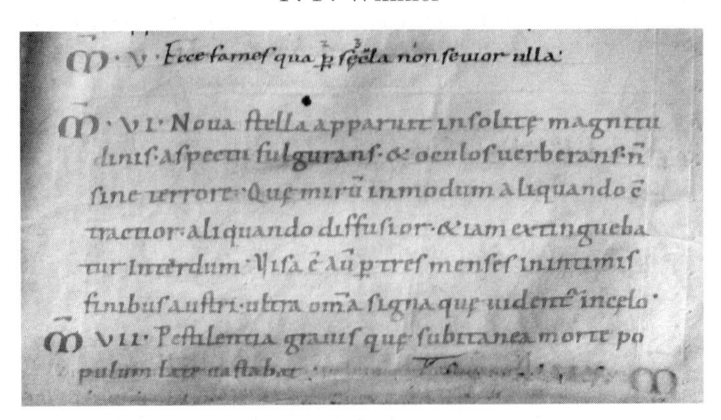
M · V · Ecce fames qua p secla non seuior ulla:
M · VI · Noua stella apparuit insolitę magnitu
dinis, aspectu fulgurans, & oculos uerberans n̄
sine terrore; Quę mirū inmodum aliquando c̄
tractior, aliquando diffusior, & iam extinguebatur interdum. Visa ē aūt p tres menses in intimis
finibus austri, ultra om̄a signa quę uidentur in celo.
M VII · Pestilentia grauis quę subitanea morte po
pulum late uastabat.

Figure 1. Entries from the years 1005'-1007 from the chronicles of the Abbey of St. Gall, including an extensive reference to SN 1006. (Stiftsbibliothek St. Gallen)

source – the first cosmic X-ray source to be positively identified (Bowyer *et al.* 1964), and this had attracted speculation by Cameron and others that other historical supernovae might also be X-ray sources. Only a few months after Goldstein's paper appeared, Gardner & Milne (1965) found that the radio source PKS 1459−41 = G327.6+14.6 was a polarized, non-thermal, shell-type radio source, and concluded that it was the probably remnant of SN 1006. (For a far fuller account of the recovery of SN 1006 as a remnant, see Devorkin 1985.)

A decade later X-ray (Winkler & Laird 1976) and optical (van den Bergh 1976) emission from SN 1006 was detected. Since then it has been much studied at all wavelengths, as several papers in this JD09 review. Among recent observations is a precise geometric measurement of the distance to SN 1006: 2.18 ± 0.08 kpc, based on measurement of the proper motions for the Balmer-dominated optical filaments delineating the NW region of the shell (WGL03), together with a measurement of the shock velocity based on spectra of the same filament by Ghavamian *et al.* (2002). WGL03 used this distance, the extinction of only $A_V = 0.31$ to the background Schweizer & Middleditch (1980) star, and the peak absolute magnitude for Type Ia supernovae of $M_V(\mathrm{max}) = -19.4 \pm 0.4$ to estimate that at its peak SN 1006 would have had $V(\mathrm{max}) \approx -7.5 \pm 0.5$, in remarkable agreement with the estimate based on the interpretation of the historical record outlined above.

In May of 1006, the star we now know as SN 1006 would have indeed been dazzling to the eyes, "*oculos verberans*," and as several other papers in this Joint Discussion will show, observations of its unique remnant continue to dazzle (and sometimes confound) modern astronomers.

Acknowledgements

I would like to acknowledge support from the National Science Foundation through grant AST-0307613, and from Middlebury College's Gamaliel Painter enrichment fund.

References

Bowyer, S., Byram, E. T., Chubb, T. A., & Friedman, H. 1964, *Science*, 146, 912
Devorkin, D. H. 1985, *Astronomy Quarterly*, 5, 71
Gardner, F. F., & Milne, D. K. 1965, *AJ*, 70, 754
Ghavamian, P., Winkler, P. F., Raymond, J. C., & Long, K. S. 2002, *ApJ*, 572, 888
Goldstein, B. R. 1965, *AJ*, 70, 105
Schweizer, F., & Middleditch, J. 1980, *ApJ*, 241, 1039
van den Bergh, S. 1976, *ApJ* (Letters), 208, L17
Winkler, P. F., Gupta, G., & Long, K. S. 2003 (WGL03), *ApJ*, 585, 324
Winkler, P. F., & Laird, F. N. 1976, *ApJ* (Letters), 204, L111

Highlights of Astronomy, Volume 14
IAU XXVI General Assembly, 14-25 August 2006
Karel A. van der Hucht, ed.

doi:10.1017/S174392130701071X

SN 1006 and other historical supernovae

F. Richard Stephenson

Department of East Asian Studies, Durham University, Elvet Hill, Durham DH 1 3TH, UK
email: f.r.stephenson@durham.ac.uk

Abstract. The supernova which appeared in AD 1006 is unique in history for its brilliance, duration of visibility, and the interest it aroused. Almost thirty separate records of the star are preserved from various parts of the world. This paper briefly summarizes historical records of SN 1006 and discusses the prospects of uncovering further historical records of supernovae.

Among the many temporary stars observed over the past 1500 years, only five Galactic supernova can be confidently identified. These events occurred in the years AD 1006, 1054, 1181, 1572 and 1604. All five stars remained visible for many months. SN 1006 was by far the most brilliant of these objects. Several observers compared its brightness with that of the Moon (!), and the star continued to seen for more than three years after it was first sighted on 30 April in AD 1006. No other Galactic supernova attracted so much attention worldwide; almost thirty separate records are preserved from East Asia, the Arab dominions and Europe.

Chinese astronomers measured the RA of the supernova as 3 degrees east of the reference star α Lib, while in Egypt the longitude of the star was estimated as the 15th degree of Scorpio. As seen from St Gallen in Switzerland, the supernova appears to have skimmed the mountainous horizon, thus setting a southerly limit to its declination. Combining these various observations and adjusting for precession, the location of the supernova at the epoch J2000 may be estimated as within 1 or 2 degrees of a site with RA $15^h 15^m$ and declination $-43°$. The galactic latitude is thus very high (approximately $+13°$) and only two known supernova remnants lie in this vicinity. These are G330.0+15.0 and G327.6+14.6. The former remnant (the Lupus Loop) is very large and faint and is much too old to be associated with SN 1006. However, in the case of the latter remnant (= PKS 1459−41), an age of a thousand years is quite acceptable.

The locations and changing brightness of the supernovae of AD 1572 and 1604 were carefully measured by European astronomers and some valuable Korean observations of the latter event are also preserved. There can be no doubt about the identity of their remnants (respectively G120.1+1.4 and G4.5+6.8). Both of the the supernovae occurring in AD 1054 and 1181 were almost exclusivly observed in China and Japan, but in each case a substantial number of records is preserved leading to confident identification of the remnants (G184.6−5.8 and G130.7+3.1). Studies of historical records of temporary stars of long duration which appeared in earlier centuries have revealed a few further supernova candidates, notably those of AD 185 and 393. Regrettably the reported positional information is poor and so far it has not proved possible to locate a unique remnant.

More than two hundred Galactic supernova remnants have been catalogued, based on their radio and X-ray properties. In recent years several attempts have been made to select a specific supernova remnant and search historical records for possible observations of the original outburst. However, this technique is of limited utility. Most early records of "new stars" give no indication of the duration of visibility, while positional descriptions are usually fairly vague. As a result, the probability of chance coincidences in position – leading to misidentification – is high.

Highlights of Astronomy, Volume 14
IAU XXVI General Assembly, 14-25 August 2006
Karel A. van der Hucht, ed.

doi:10.1017/S1743921307010721

SN 1006 in context: from Type Ia supernova to remnant

Roger A. Chevalier
Department of Astronomy, University of Virginia, Charlottesvile, VA 22904, USA
email: rac5x@virginia.edu

Abstract. The evolutionary phases in going from a Type Ia supernova explosion to a 1000-yr-old remnant are reviewed. The explosion sets up the density and composition structure of the ejecta and can have radiative effects on the surroundings. The early shock interactions may be observable at radio and X-ray wavelengths if the circumstellar density is sufficiently high. The later interaction with the interstellar medium is affected by the collisionless shock physics and hydrodynamic instabilities.

Keywords. supernovae: general, supernova remnant

In the explosion of a white dwarf, as expected in a SN Ia (Type Ia supernova), the shock wave accelerates through the outer layers with rapidly declining density. The resulting density profile is a steep power law in radius that can extend out to relativistic velocities. The lack of detectable radio emission from SNe Ia suggests a low surrounding density and the shock interaction region is expected at a high velocity ($\gtrsim 40\,000\,\mathrm{km\,s^{-1}}$) at an age of ~10 days. Based on experience with SNe Ib/c, which are observed as radio and X-ray sources, it is likely that synchrotron self-absorption is the dominant absorption mechanism for radio emission if their emission is close to current upper limits. X-ray emission is likely to be non-thermal, but is very uncertain. In this interpretation of the strongest radio limits, the circumstellar density is lower than a mass loss rate of $10^{-8}\,\mathrm{M_{\odot} yr^{-1}}$ for a wind velocity of $10\,\mathrm{km\,s^{-1}}$.

The recent apparent detection of SN 2005ke at X-ray wavelengths suggests a higher density for that case, but the detection cannot be regarded as completely secure. SN 2002ic appears to be running into very dense gas at some distance from the supernova and is the prototype of a number of similar objects; however, recent studies indicate it may not have been a SN Ia.

The expansion rate and deduced surroundings of SN 1006 are consistent with interstellar interaction. The same is true for other young SN Ia remnants, except for SN 1604 (Kepler). In this case, there is interaction with relatively dense surroundings, indicated by radiative shock waves, but there are some H I clouds at high Galactic altitude. The type of SN 1604 has been uncertain, but it is becoming regarded as a likely SN Ia. The general lack of evidence in SNe Ia for wind interaction is consistent with progenitor systems in which the companion star to the exploding white dwarf is a main sequence star or another white dwarf. This result is consistent with the possible stellar companion found in the central region of the SN 1572 (Tycho) remnant.

The SN 1006 remnant shows X-ray emission from ejecta close to the outer shock front, as does the SN 1572 remnant, where this property has been interpreted as indicating a narrow shocked region because of efficient cosmic ray acceleration. However, an inhomogeneous ejecta structure could give rise to a broad region of ejecta in the shocked region. The combination of non-radiative shock optical emission and synchrotron X-ray emission in SN 1006 make it an excellent laboratory for the study of fast shock physics.

Highlights of Astronomy, Volume 14
IAU XXVI General Assembly, 14-25 August 2006
Karel A. van der Hucht, ed.

doi:10.1017/S1743921307010733

Radio observations of SN 1006

Estela M. Reynoso†

Instituto de Astronomía y Física del Espacio, C.C. 67, Suc. 28, 1428 Buenos Aires, Argentina
email: ereynoso@iafe.uba.ar

Abstract. In this article I review our present knowledge of SN 1006 based on radio observations since the identification of the radio remnant four decades ago. I also report preliminary results of a new radio expansion study which combines VLA 1991/92 data with new VLA and ATCA observations performed in 2003.

Keywords. (ISM:) supernova remnants, individual: SN 1006, radio continuum: ISM, radio lines: ISM, polarization.

Based on a polarization of about 10% and a non-thermal spectral index of −0.6, Gardner & Milne (1965) concluded that the catalogued radio source PKS 1459−41 was the remnant of the SN recorded in 1006. SN 1006, the faintest of the historical SNRs in radio wavelengths, depicts a $\sim 30'$ diameter shell with a remarkable bilateral symmetry, where the symmetry axis is perpendicular to the Galactic plane. Estimates of the distance to SN 1006 based on different arguments yield values varying approximately from 1.4 to 2.2 kpc. Some correspondence between the SNR and H I features suggest a systemic velocity of ~ -20 km s^{-1}, which corresponds to a distance of 1.7 kpc (Dubner *et al.* 2002). High-resolution polarization maps of SN 1006 (Reynolds & Gilmore 1993) show a radial distribution of the magnetic field, typical of young SNRs, with a mean polarization of 13% and peaks of 30%.

The expansion of the radio remnant of SN 1006 was measured by Moffett, Goss & Reynolds (1993), comparing two VLA (NRAO) images obtained within an interval of ~ 8 yr. They obtained a mean expansion parameter $m = 0.48 \pm 0.13$, where m is defined as $R \propto t^m$. A new radio expansion study is currently being carried out based on archival 1991/92 VLA data and new observations performed simultaneously in 2003 with the ATCA (Australia) and the VLA. We obtain an average $m = 0.53 \pm 0.40$, with pronounced azimuthal variations. The brightest lobes are expanding much faster than the faintest regions, with m typically between 0.6 and 0.8. Such values are compatible with a remnant in which the ejecta, with density described by an exponential or a r^{-7} power law, expand into a circumstellar density profile of r^{-2} (Dwarkadas & Chevalier 1998). Alternatively, the ambient density could be constant if the SNR is still in the ejecta-dominated phase. The velocities inferred for the shock front are in excellent agreement with recent numerical simulations (Velázquez, private communication).

References

Dubner, G. M., Giacani, E. B., Goss, W. M., Green, A. J., & Nyman, L.-Å. 2002, *A&A*, 387, 1047
Dwarkadas, V. V., & Chevalier, R. A. 1998, *ApJ*, 497, 807
Gardner, F. F., & Milne, D. K. 1965, *AJ*, 70, 754
Moffett, D. A., Goss, W. M., & Reynolds, S. P. 1993, *AJ*, 106, 1566
Reynolds, S. P., & Gilmore, D. M. 1993, *AJ*, 106, 272

† Member of the Carrera del Investigador Científico of CONICET, Argentina.

Highlights of Astronomy, Volume 14
IAU XXVI General Assembly, 14-25 August 2006
Karel A. van der Hucht, ed.

doi:10.1017/S1743921307010745

SN 1006 at optical and UV wavelengths

Knox S. Long

Space Telescope Science Institute, 3700 San Martin Drive, Baltimore, MD 21218, USA
email: long@stsci.edu

Abstract. Optical and UV emission from SN 1006 is observed on the periphery of the SNR, particularly in the NW, and arises as material from interstellar gas is ionized behind the shock front. The shapes of the emission lines have been used to infer a shock velocity of about 2900 km s^{-1} and to show that the electron and ion temperatures are close to that expected from standard shock theory. The combination of the shock velocity and the proper motion of the filaments in the NE accurately locates SN1006 at a distance of 2.2 kpc, some 550 pc above the Galactic plane. At UV wavelengths, ejecta from SN 1006 have also been observed as broad absorption lines from Si, Ca and Fe in spectra of the Schweitzer-Middleditch star and two quasars. The observations appear to limit the mass of Fe in SN 1006 to less than 0.16 M$_{\odot}$, much less than expected from models of Type Ia supernovae. In this brief review of the UV and optical properties of SN 1006, I will summarize how these observations yield a fairly consistent description of SN 1006, and suggest what further observations might be undertaken to extend our understanding of SN 1006 in its second millennium.

Sidney van den Bergh (1976) was the first to observe the remnant of SN 1006 optically. With modern CCDs, Winkler *et al.* (2003) imaged Hα emission from the entire periphery of the SNR and measured the proper motion of the NW filament as 0.280 ± 0.008 arcsec yr^{-1}. As suggested by Chevalier *et al.* (1980), the optical emission is from the primary shock interacting with the ISM, and is due to H atoms being ionized. From the line shapes, Ghavamian *et al.* (2002) estimated a shock velocity of 2890 ± 80 km s^{-1}. As a result, the distance to the SN is geometrically determined to be 2.18 ± 0.18 kpc. Raymond *et al.* (1995) observed emission lines of H, He, C, N, and O from the NW limb in the far UV; the line widths are consistent with velocity randomization in the post-shock gas, implying that ion-ion equilibration is governed mainly by Coulomb processes.

Schweizer & Middleditch (1980) discovered a sub-dwarf OB-type star that is behind SN 1006. Using *IUE*, Fesen *et al.* (1988) and other subsequent observers obtained UV spectra from this star and found broad Fe, Si, and Ca absorption lines from freely expanding ejecta with speeds up to 5000 km s^{-1} in SN 1006. Using *HST*, Winkler *et al.*(2005) observed additional UV lightbulbs behind SN 1006. The line shapes indicate that Si is located mainly outside of the Fe and that the SNR is expanding asymmetrically. This is the only SN whose ejecta have been probed in this fashion. The amount of Fe is less that predicted, but the general characteristics support a Type Ia origin for SN 1006.

References

Chevalier, R. A., Kirshner, R. P., & Raymond, J. C. 1980, *ApJ*, 235, 186
Fesen, R. A., Wu, C. C., Leventhal, M., & Hamilton, A. J. S. 1988, *ApJ*, 327, 164
Ghavamian, P., Winkler, P. F., Raymond, J. C., & Long, K. S. 2002, *ApJ*, 572, 888
Raymond, J. C., Blair, W. P., & Long, K. S. 1995, *ApJ* (Letters), 454, L31
Schweizer, F., & Middleditch, J. 1980, *ApJ*, 241, 1039
van den Bergh, S. 1976, *ApJ* (Letters), 208, L17
Winkler, P. F., Gupta, G., & Long, K. S. 2003, *ApJ*, 585, 324
Winkler, P. F., Long, K. S., Hamilton, A. J. S., & Fesen, R. A. 2005, *ApJ*, 624, 189

Highlights of Astronomy, Volume 14
IAU XXVI General Assembly, 14-25 August 2006
Karel A. van der Hucht, ed.

doi:10.1017/S1743921307010757

Supernova remnants and their progenitors

John R. Dickel

Department of Physics and Astronomy, University of New Mexico,
800 Yale Blvd. NE, Albuquerque, NM 87131, USA
email: johnd@phys.unm.edu

Abstract. Young supernova remnants (SNRs) show obvious differences that can be related to characteristics of the progenitors and supernova types as tabulated in Fig. 1. Questions remain.

Keywords. ISM: supernova remnants, supernovae: general, acceleration of particles

---

Questions include: why do some Type Ia supernova remnants show a definite gap between the leading-shock rim and the main shell (e.g., Tycho), while others show a single continuous shell out to the sharp rim (SN 1006)?

Why do some SNe leave only a cooling neutron star with no pulsar wind nebula (PWN) (Cas A), or erratic point X-ray sources (RCW 103), while others have pulsars? Is it the magnetic field, spin rate or what?

What physical conditions can discriminate between mixed morpology remnants with internal thermal X-rays (W 44) and those with more standard shells (Vela)? Is it just the surrounding medium?

The composite SNR G11.2−0.3 has been identified with SN 386. It contains a central pulsar with a period of 65 ms and a calculated spin-down age about three times longer than the age since its explosion. In contrast, the Crab Nebula, an extended pulsar wind nebula with no apparent shell, has a close match between the two ages. Is the presence or absence of a shell related to differences in the moment of inertia of the neutron star?

Why do some composite SNRs (MSH 15−56) show a radio pulsar wind nebula somewhat offset from the center, and then a point X-ray source with an apparent X-ray PWN out near the shell and not necessarily aligned with the radio PWN?

	Low Mass Type Ia	**High Mass Type Ib or II**		
Examples (approximate explosion date if known)	**Kepler (1604)? Tycho (1572) AD1006 (1006) 0519-690 (350-1500) SN1885 in M31 (1885)**	**Cas A (~1680) RCW103 (~0) E0102-723 Puppis A**	**3C58(1181) Crab(1054)**	**G11.2-0.3(386) W44 Vela 0540-693 (~0) N157B IC443**
Morphology	**Reasonably round shells**	**Broken up shells, shrapnel**	**Only pulsar wind nebula**	**Composite PWN plus shell**
Compact core	**--**	**Cooling neutron star, no pulsar wind nebula**	**Pulsar plus pulsar wind nebula (PWN)**	
Spectral results optical X-ray	**Balmer dominated, N,S Thermal Fe, Si, S**	**O (FMKs), N (QSFs) Fe, metals**	**Ni, N, S Power law (synchrotron) in PWN**	
Pre point-blast expansion	**Near point blast $R \propto t^{0.5}$**	**Near free expansion $R \propto t^{0.9}$**	**--**	
Emvironment	**Isolated object but often complex circumstellar medium**	**Pre-explosion mass loss**	**Pre-explosion mass loss**	

Figure 1. Properties of young SNRs

Highlights of Astronomy, Volume 14
IAU XXVI General Assembly, 14-25 August 2006
Karel A. van der Hucht, ed.

doi:10.1017/S1743921307010769

On the road to consistent Type Ia supernova explosion models

Friedrich K. Röpke[1,2]

[1]Department of Astronomy and Astrophysics, University of California Santa Cruz, 1156 High Street, Santa Cruz, CA 95064, USA

[2]Max-Planck-Institut für Astrophysik, Karl-Schwarzschild-Str. 1, D-85741 Garching, Germany
email: fritz@mpa-garching.mpg.de

Abstract. Keeping up with ever more detailed observations, Type Ia supernova (SN Ia) explosion models have seen a brisk development over the past years. The aim is to construct a self-consistent picture of the physical processes in order to gain the predictive power necessary to answer questions arising from the application of SNe Ia as cosmological distance indicators. We review recent developments in modeling these objects focusing on three-dimensional simulations.

Keywords. supernovae: general, hydrodynamics, methods: numerical

---

Despite the importance of SNe Ia for astrophysics and cosmology, a fully consistent description of the explosion mechanism is still lacking. Yet several ideas exist (Hillebrandt & Niemeyer 2000), and the interplay of modeling and observation has helped to shape a picture of the thermonuclear explosions of the progenitor C+O white dwarfs. Recent work has focused on modeling the single-degenerate Chandrasekhar-mass scenario. Here, the white dwarf commences nuclear burning as it approaches the limiting Chandrasekhar mass due to accretion from a non-degenerate binary companion. After about a century of convective carbon burning, a thermonuclear runaway leads to the formation of a flame near the star's center. It propagates outward giving rise to the explosion.

Hydrodynamically, two modes of flame propagation are admissible – a sub-sonic deflagration and a supersonic detonation. These two burning modes provide different options when building an explosion model. The signature of intermediate-mass elements in the spectra rules out a prompt detonation as a valid SN Ia model (Arnett 1969). Thus, the thermonuclear flame has to start out as a deflagration (Nomoto *et al.* 1976). Since laminar flame propagation is far too slow to explode the star, flame acceleration is the key issue of all models. The subsonic deflagration flame is subject to buoyancy and shear instabilities and strong turbulence is expected to be generated. The interaction of the flame propagation with turbulence provides an efficient way of accelerating its propagation.

This phenomenon has been studied in detail over the past years in elaborate three-dimensional numerical simulations (Reinecke *et al.* 2002; Gamezo *et al.* 2003; Röpke *et al.* 2005, 2006; Schmidt & Niemeyer 2006). In a Large Eddy Simulation approach, it is possible to implement a self-consistent model of the turbulent flame propagation (Reinecke *et al.* 1999). In such simulations, explosions were found that seem capable of reproducing gross features of observed SNe Ia (e.g., Blinnikov *et al.* 2006).

However, it seems unlikely that the turbulent deflagration model can cover the full sample of SNe Ia. The ultimate way of accelerating the thermonuclear burning would be a transition of the deflagration flame propagation mode to a supersonic detonation. The problems of this 'delayed detonation' model (Khokhlov 1991) arise from the unknown mechanism of the deflagration-to-detonation transition (DDT), and potentially from the fact that a detonation wave cannot cross ash regions (Maier & Niemeyer 2006).

Thus, it has to propagate around the complex deflagration structure and may not be capable of burning out pockets of fuel. Moreover, if ignited off-center, it has to compete with the expansion of the star, diluting the fuel, and may not reach the far side of the deflagration structure. These issues can be tested with multi-dimensional simulations parametrizing the DDT (Gamezo *et al.* 2005; Golombek & Niemeyer 2005).

Apart from the DDT, a major uncertainty of the explosion models is the configuration of the igniting flame. The ignition process is hard to address both analytically and numerically. Most of the simulations assumed a central ignition, but recent studies indicated that an asymmetric, off-center ignition may be possible (Kuhlen *et al.* 2006).

As a consequence, less material would be consumed and the nuclear energy release may not be sufficient to gravitationally unbind the white dwarf. A failure to explode the star results in gravitationally bound ashes erupting from the surface, sweeping around a core consisting of fuel and colliding on the far side of the star. Plewa *et al.* (2004) suggested that the compression of fuel in the collision region may be sufficient to trigger a detonation which would then burn the remaining fuel and give rise to a supernova explosion ('Gravitationally Confined Detonation' scenario). However, in a recent parameter study Röpke *et al.* (2007) showed that the conditions in the compressed region may not allow for a spontaneous detonation in realistic models. The bound configuration will then start to pulsate – a second chance to trigger a detonation that needs further investigation (Arnett & Livne 1994; Bravo & García-Senz 2006).

We conclude that the two major uncertainties in modeling SN Ia explosions – the DDT and the flame ignition – admit different scenarios that can be explored in large-scale three dimensional simulations. An evaluation of the scenarios needs to be carried out on the basis of comparison with observations (e.g., Kozma *et al.* 2005). At the same time, the physical processes underlying the DDT and the flame ignition need to be explored in separate numerical approaches.

References

Arnett, W. D. 1969, *Ap&SS*, 5, 180

Arnett, D., & Livne, E. 1994, *ApJ*, 427, 330

Blinnikov, S. I., Röpke, F. K., Sorokina, E. I., Gieseler, M., Reinecke, M., Travaglio, C., Hillebrandt, W. & Stritzinger, M. 2006, *A&A*, 453, 229

Bravo, E., & García-Senz, D. 2006, *ApJ* (Letters), 642, L157

Gamezo, V. N., Khokhlov, A. M., Oran, E. S., Chtchelkanova, A. Y. & Rosenberg, R. O. 2003, *Science*, 299, 77

Gamezo, V. N., Khokhlov, A. M. & Oran, E. S. 2005, *ApJ*, 623, 337

Golombek, I., & Niemeyer, J. C. 2005, *A&A*, 438, 611

Hillebrandt, W., & Niemeyer, J. C. 2000, *ARAA*, 38, 191

Khokhlov, A. M. 1991, *A&A*, 245, 114

Kozma, C., Fransson, C., Hillebrandt, W., Travaglio, C., Sollerman, J., Reinecke, M., Röpke, F. K., & Spyromilio, J. 2005, *A&A*, 437, 983

Kuhlen, M., Woosley, S. E., & Glatzmaier, G. A. 2006, *ApJ*, 640, 407

Maier, A., & Niemeyer, J. C. 2006, *A&A*, 451, 207

Nomoto, K., Sugimoto, D., & Neo, S. 1976, *Ap&SS* (Letters), 39, L37

Plewa, T., Calder, A. C. & Lamb, D. Q 2004, *ApJ*(Letters) , 612, L37

Reinecke, M., Hillebrandt, W., & Niemeyer, J. C. 2002, *A&A*, 391, 1167

Reinecke, M., Hillebrandt, W., Niemeyer, J. C., Klein, R., & Gröbl, A. 1999, *A&A*, 347, 724

Röpke, F. K., & Hillebrandt, W. 2005, *A&A*, 431, 635

Röpke, F. K., Hillebrandt, W., Niemeyer, J. C., & Woosley, S. E. 2006, *A&A*, 448, 1

Röpke, F. K., Woosley, S. E., & Hillebrandt, W. 2007, *ApJ*, 660, 1344

Schmidt, W., & Niemeyer, J. C. 2006, *A&A*, 446, 627

Highlights of Astronomy, Volume 14
IAU XXVI General Assembly, 14-25 August 2006
Karel van der Hucht, ed.

doi:10.1017/S1743921307010770

Global parameters of Type Ia supernovae

Bruno Leibundgut[1] and Maximilian Stritzinger[2]

[1]ESO, Karl-Schwarzschild-Strasse 3, D-85748 Garching, Germany
email: bleibundgut@eso.org

[2]Dark Cosmology Centre, Niels Bohr Institute, University of Copenhagen,
Juliane Maries Vej 30, DK-2100 Copenhagen Ø, Denmark
email: max@dark-cosmology.dk

Abstract. Using simple physical assumptions we derived fundamental parameters including: the mass of synthesised ^{56}Ni, the ejecta mass, and the explosion energy of Type Ia supernovae. The methods have been described in several recent publications and comparison with other methods shows general consistency.

Keywords. supernovae: general, stars: evolution, stars: fundamental parameters

---

Our understanding of thermonuclear supernovae is mostly shaped by the forward comparison of model calculations with observations. While the observational record is normally described in a comparative manner between individual events, the physical parameters governing the explosions have not been directly derived. Explosion models start from very basic and essentially untested assumptions about the progenitor mass, the explosion energy, and the amount of ^{56}Ni ($M_{\rm Ni}$) produced in the explosion. Considering the importance of Type Ia supernovae (SNe Ia) for cosmology and for the metal enrichment of the universe, it is important to justify these parameters. As exemplified in the contribution by Howell *et al.* (2006), surprises are not excluded.

We investigated the explosion characteristics of SNe Ia through the detailed study of their bolometric light curves. The governing factors for their light curve shapes are $M_{\rm Ni}$, the $\gamma-$ray opacity, the total ejecta mass, and the explosion energy. For a small sample of well-observed SNe Ia we have determined $\rm M_{Ni}$ (Stritzinger & Leibundgut 2005). Our method has been tested by closing the circle from explosion simulations through the radiative transport calculations to synthetic light curves from which $M_{\rm Ni}$ was derived to within 15 % of the $M_{\rm Ni}$ in the input model (Blinnikov *et al.* 2006). Assuming all calculations in this chain were correct, a small correction to the determination of the peak bolometric flux has to be applied to use Arnett's rule (Arnett 1982) to find $M_{\rm Ni}$ for individual events. Comparison of $M_{\rm Ni}$ derived from late-phase observations of the iron emission lines (and a density structure taken from models) with results obtained via Arnett's rule yields a consistent picture (Stritzinger *et al.* 2006b). Current models typically predict smaller amounts of $M_{\rm Ni}$ than what is derived from observations for most SNe Ia. There is also a wide range of $M_{\rm Ni}$ produced in these explosions. Simple arguments also yield a lower limit for the Hubble constant – found to be 50 km s^{-1}Mpc^{-1} (Stritzinger & Leibundgut 2005).

The late-time evolution of the bolometric light curves were used to estimate the total ejecta mass for a number of SNe Ia. The ejecta mass is a function of both the opacity and explosion energy, as expressed through their expansion velocity (Stritzinger *et al.* 2006a; see Fig. 1). Again, there is a rather large range of ejecta masses. None of our objects show the canonical Chandrasekhar mass typically assumed for these thermonuclear explosions (Stritzinger *et al.* 2006a).

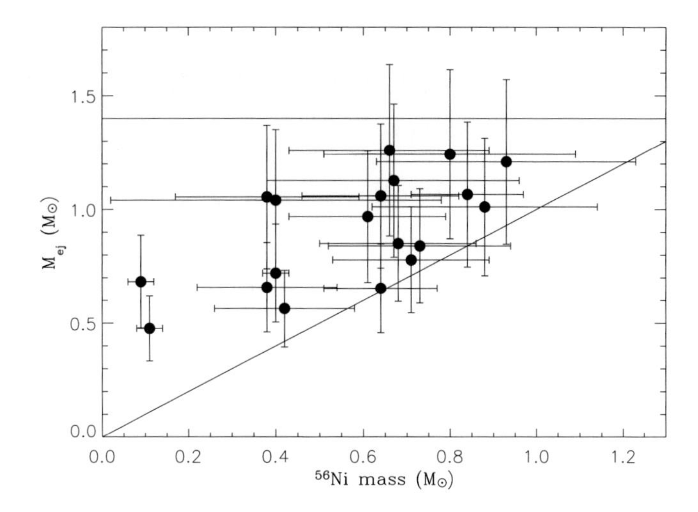

Figure 1. The derived nickel and ejecta masses for a sample of well-observed nearby supernovae. There is a significant scatter amongst the measured masses. While the spread in nickel masses is most likely due to different conditions in the explosions, the scatter in ejecta masses is not understood. Taken from Stritzinger *et al.* (2006a).

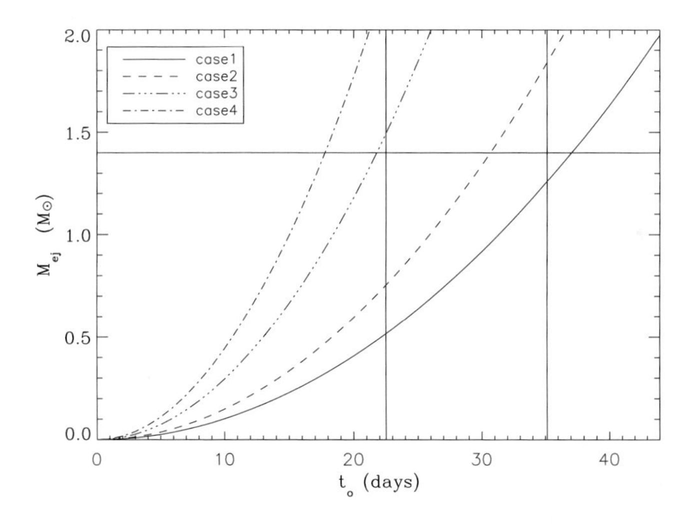

Figure 2. The dependencies of the derived ejecta masses on the assumed γ−ray opacity and the explosion energy inferred from the expansion velocity. Details can be found in Stritzinger *et al.* (2006a).

References

Arnett, W. D. 1982, *ApJ*, 253, 785

Blinnikov, S., Röpke, F. K., Sorokina, E., I., Gieseler, M., Reinecke, M., Travaglio, C., Hillebrandt, W., & Stritzinger, M. 2006, *A&A*, 453, 229

Howell, A., *et al.* 2006, these proceedings

Stritzinger, M., & Leibundgut, B. 2005, *A&A*, 431, 423

Stritzinger, M., Leibundgut, B., Walch, S., & Contardo, G. 2006a, *A&A*, 450, 241

Stritzinger, M., Mazzali, P., Sollerman, J., & Benetti, S. 2006b, *A&A*, 460, 793

Highlights of Astronomy, Volume 14
IAU XXVI General Assembly, 14-25 August 2006
Karel A. van der Hucht, ed.

doi:10.1017/S1743921307010782

Exploring the global properties of Type Ia supernovae

Paolo Mazzali

Max-Planck Institute for Astrophysics,
Karl-Schwarzschild-Strasse 1, D-85741 Garching bei München, Germany
email: mazzali@mpa-garching.mpg.de

Abstract. While Type Ia supernovae are widely used as distance indicators, the reasons for the correlation between luminosity and light curve width that allows SNe Ia to be calibratable standard candles is not yet fully understood, and in particular the details of the explosion mechanism are still the subject of heated debate. We present the results of a systematic approach that uses the high-quality data collected by the European Research Training Network "The Physics of Type Ia Supernova Explosions" to map the supernova ejecta and to infer the properties of the explosion.

Keywords. Supernovae, photometry, spectroscopy, abundance tomography

---

The distribution of the elements in the outer layers of the ejecta of a (Type Ia) supernova can be derived by modelling a closely spaced time series of spectra, while the nebular spectra give information about the inner ejecta (Stehle *et al.* 2005). Results for SN 2002bo show that the composition stratification that is imprinted at the time of the explosion is largely preserved. Using the abundance distribution obtained as above, and assuming that all NSE elements contribute equally to the opacity while only radioactive nickel contributes to the radiative output, the light curve of SN 2002bo can be reproduced very accurately.

Examining the spectra of a large sample of SNe Ia, we find that the outer extent of the burning, as indicated by the velocity of the strong Si II 6355Å line, is approximately the same in all SNe Ia, regardless of their luminosity. We confirm that the amount and distribution of ^{56}Ni is proportional to the luminosity of the SN, and find that SNe Ia produce a roughly constant amount of stable NSE isotopes. We show that the sum of all NSE elements gives a better correlation with the width of the light curve. These results suggest that the kinetic energy of all SNe Ia is similar, and that the mass of the progenitor white dwarf is constant, which supports the single degenerate scenario.

A variation in the ratio of stable *versus* unstable NSE species can generate a dispersion about the mean brightness-decline rate relation. Such a variation may be the consequence of different progenitor metallicities. A systematic evolution of the metallicity of SN Ia progenitors may cause a redshift-dependent shift of SNe in the brightness-decline rate plane which may to a certain extent mimic the accelerated expansion of the Universe.

Reference

Stehle, M., Mazzali, P. A., Benetti, S., & Hillebrandt, W. 2005, *MNRAS*, 360, 1231

Highlights of Astronomy, Volume 14
IAU XXVI General Assembly, 14-25 August 2006
Karel A. van der Hucht, ed.

doi:10.1017/S1743921307010794

Light echoes of SNe in the LMC

Armin Rest[1], Nicholas B. Suntzeff[2], R. Chris Smith[1], Knut Olsen[1], Alfredo Zenteno[1], Guillermo J. Damke[1], Jose L. Prieto[1], Christopher Stubbs[3], Arti Garg[3], Peter Challis[2], Andrew C. Becker[4], Gajus A. Miknaitis[4], Antonino Miceli[4], Douglss L. Welch[5], Alejandro Clocchiatti[6], Dante Minniti[6], Lorenzo Morelli[6], Kem H. Cook[7], Sergei Nikolaev[7], Mark E. Huber[7], and Andrew Newman[8]

[1]National Optical Astronomy Observatory (NOAO)/Cerro Tololo Inter-American Observatory (CTIO), La Serena, Chile
email: arest@ctio.noao.edu

[2]Department of Physics & Astronomy, Texas A & M University, 4242 TAMU, College Station, TX 77843 USA

[3]Physics Department, Harvard University, 17 Oxford Street, Cambridge, MA 02138, USA

[4]Department of Astronomy, University of Washington, Box 351580, Seattle, WA 98195, USA

[5]Dept. of Physics and Astronomy, McMaster University, Hamilton, Ontario, L8S 4M1, Canada

[6]Dept. of Astronomy, Pontificia Universidad Católica de Chile, Casilla 306, Santiago 22, Chile

[7]Lawrence Livermore National Laboratory, 7000 East Ave., Livermore, CA 94550, USA

[8]Department of Physics, Washington University, Saint Louis, MO 63130, USA

Abstract. The SuperMACHO project has discovered light echoes from 3 ancient SNe in the LMC. These SNRs are three of the six youngest in the LMC, and are classified as likely SN Ia based on X-ray data.

Keywords. stars: supernovae: general, ISM: supernova remnants, ISM: reflection nebulae

---

In recent years, light echoes have been discovered around some nearby extragalactic supernovae well after the explosion, most notably the light echoes from SN 1987A (Crotts 1988). However, to date no light echoes of historical SNe of Galactic or extragalactic origin have been discovered. In the SuperMACHO project, we have imaged the bar of the LMC repeatedly and used an automated pipeline to subtract point-spread-function matched template images from the recent epoch images. The resulting difference images are remarkably clean of the constant (in time) stellar background and are ideal for searching for variable objects. Using these difference images, we have mapped the extensive light echo complex around SN 1987A further out, and deeper, than has been previously possible. Besides the SN 1987A light echoes, we found three other groups of light echoes associated with known supernova remnants (SNRs). These SNRs are three of the six youngest in the LMC, and are classified as likely SN Ia based on X-ray data (Hughes *et al.* 1995). By combining the position and apparent proper motions of the light echoes, we derive ages for the SNRs (Rest *et al.* 2005). Spectra of the light echoes taken with the GMOS spectrograph on the Gemini South telescope suggests that the explosion causing one of the SNRs was an overluminous Type Ia supernova explosion.

References

Crotts, A. 1988, *IAUC* 4561
Hughes, J. P., Hayashi, I., Helfand, D., *et al.* 1995, *ApJ* (Letters), 444, L81
Rest, A., Suntzeff, N. B., Olsen, K., & SuperMACHO collaboration 2005, *Nature*, 438, 1132

Highlights of Astronomy, Volume 14
IAU XXVI General Assembly, 14-25 August 2006
Karel A. van der Hucht, ed.

doi:10.1017/S1743921307010800

Observing SNe Ia at $z > 1$ with the *HST* and at $z \simeq 0.2$ with the SDSS-II

Hubert Lampeitl[1], Adam G. Riess[1], and the SDSS-II Supernova Collaboration and Higher-z Team

[1]Space Telescope Science Institute, 3700 San Martin Drive, Baltimore, MD 21218, USA

Abstract. SNe Ia are currently providing the most direct measurements of the accelerated expansion of the Universe and also put constraints on the nature and evolution of the so-called 'dark energy'. Despite major efforts to increase the number of known high-redshift SNe Ia with reliable distance estimates, two regions in the Hubble diagram remain only sparsely observed. At redshifts $z > 1$ the limitations of ground-based instruments require the *Hubble Space Telescope* and its superior angular resolution to get meaningful distance estimates, while at intermediate redshifts ($z \simeq 0.2$) the large solid angle necessary presents an obstacle to most surveys that can be overcome with the Sloan Digital Sky Survey, SDSS-II.

Keywords. cosmological parameters, cosmology: observations, supernovae: general

---

The *HST* has been used in conjunction with the GOODS survey (Giavalisco *et al.* 2004) to search for supernovae at a redshifts above $z > 1$ (Riess *et al.* 2004). In total ~ 900 orbits have been used in 2002 – 2005 to discover and follow up very high redshift SNe. Photometric observations are carried out using the *HST*-ACS instrument (Advanced Camera for Surveys) and spectra are obtained by the ACS/grism filter. A preselection of possible SN Ia candidates for spectroscopic follow-up is done based on the UV deficit caused by the metal-rich atmospheres of SNe Ia. In total the program revealed 135 SN events of which 50 turned out to be of Type Ia, including 19 SNe Ia at a redshift larger than 1. These SNe increase the known sample by more than a factor of two compared to Riess *et al.* (2004). This dataset further supports the previously reported deceleration at high redshifts and puts more stringent constraints on the equation of state of the dark energy'', attributed as $w(z)$, at $z > 1$. Nevertheless a cosmological constant is still compatible with the data.

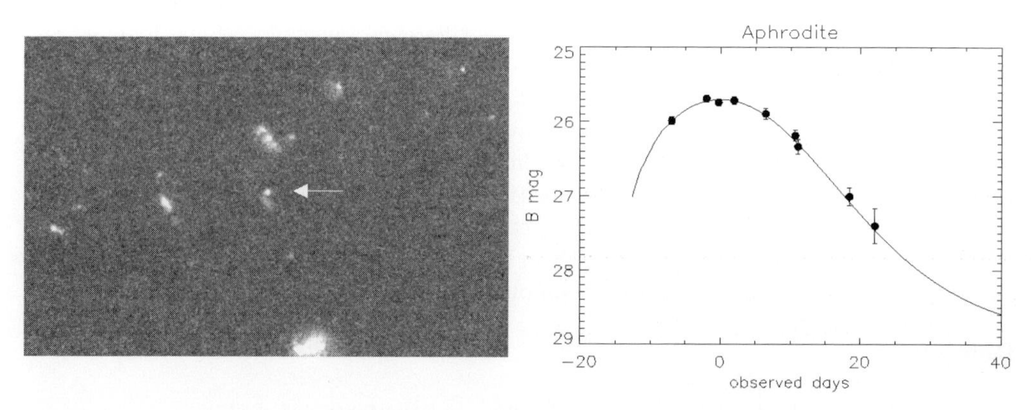

Figure 1. The discovery image of one of the very high-redshift SN, at a redshift of $z = 1.3$. The right panel shows the obtained lightcurve in the B-band.

In the intermediate redshift range between $z = 0.1$ and $z = 0.4$ the SDSS supernova survey (SDSS-SN webpage; Sako *et al.* 2005; Lampeitl *et al.* 2006) is currently finding a significant number of SN Ia by observing in repeated scans a 120 degree wide portion of the galactic equator. In total an area of 280 sq. degrees is surveyed. The SDSS camera measures simultaneously light curves in the SDSS filters (u,g,r,i and z). Photometrically identified SNe candidates are then spectroscopically observed on a variety of instruments including the ARC (3.5 m), HET (9.2 m), Subaru (8.2 m), WHT (4.2 m), NTT (3.6 m) and MDM (2.4 m). All new discovered SN are reported to the IAU (SDSS telegrams) and also made public available through the SDSS-SN webpage. The program will last over a three year period – each year using the time between September and November when there is the least conflict with other on-going projects. The 2005 campaign is already completed and two additional years of observations are coming up in 2006 and 2007. During the first year of data collection 129 SNe of Type Ia have been discovered and spectroscopically confirmed by the SDSS SN survey (SDSS telegrams). These data also provide a unique opportunity to search for peculiar SNe events as demonstrated, for example, by the discovery of SN2005gj (Prieto *et al.* 2005).

With respect to cosmology, approximately two-thirds of the identified SNe Ia will provide light curves suitable for cosmology. This dataset is of particular value because it will connect the low-redshift SNe Ia sample to the high-redshift sample and test for a smooth transition between these two regions. Ultimately this dataset will constrain the equation of state of 'dark energy' in the nearby universe and will, in combination with other datasets like SNLS (Astier *et al.* 2006), ESSENCE (Matheson *et al.* 2005) and the *HST data* improve the overall knowledge of the equation of state w for 'dark energy'.

Acknowledgements

Based on observations made with the NASA/ESA *Hubble Space Telescope*, obtained from the data archive at the Space Telescope Institute. STScI is operated by the association of Universities for Research in Astronomy, Inc. under the NASA contract NAS 5-26555. Funding for the SDSS and SDSS-II has been provided by the Alfred P. Sloan Foundation, the Participating Institutions, the National Science Foundation, the U.S. Department of Energy, the National Aeronautics and Space Administration, the Japanese Monbukagakusho, the Max Planck Society, and the Higher Education Funding Council for England. The SDSS Web Site is `http://www.sdss.org/`.

References

Astier, P., Guy, J., Regnault, N., *et al.* 2006, *A&A*, 447, 31

Giavalisco, M., Ferguson, H. C., Koekemoer, A. M., *et al.* 2004, *ApJ* (Letters), 600, L93

Lampeitl, H., SDSS-SN Collaboration 2006, *AAS Meeting Abstracts*, 208, #58.01

Matheson, Th., Blondin, S., Foley, R. J., *et al.* 2005, *AJ*, 129, 2352

Prieto, J., Garnavich, P., Depoy, D., *et al.* 2005, *IAU Circular* 8633, 1

Riess, A. G., Strolger, L.-G., Tonry, J., *et al.* 2004, *ApJ*, 607, 665

Sako, M., Romani, R., Frieman, J. (SDSS collaboration), *et al.* 2005, in: Proc. 22nd Texas Symp. on Relativistic Astrophysics [astro-ph/0404455]

SDSS-SN webpage: `http://sdssdp47.fnal.gov/sdsssn/sdsssn.html`

SDSS telegrams: CBET 304, CBET 611, CBET 621, CBET 623, CBET 624, CBET 627, CBET 629, IAU Circular 8359, IAU Circular 8481

Highlights of Astronomy, Volume 14
IAU XXVI General Assembly, 14-25 August 2006
Karel A. van der Hucht, ed.

doi:10.1017/S1743921307010812

Sternberg Astronomical Institute Supernova Catalogue, and radial distribution of supernovae in host galaxies

O. S. Bartunov, D. Yu. Tsvetkov, and N. N. Pavlyuk
Sternberg Astronomical Institute, Universitetskii pr. 13, 119992 Moscow, Russia

Abstract. We present a new version of the Sternberg Astronomical Institite Supernova Catalogue and the results of our investigation of the supernova radial distribution in their host galaxies based on the new data.

Keywords. supernovae: general

We present the results of our ongoing work on statistics of SNe and a new version of SAI SN Catalogue. The previous results and descriptions of earlier versions of the catalogue can be found in Tsvetkov & Bartunov (1993) and Tsvetkov *et al.* (2004).

The new catalogue presents an important step towards creation of a modern database on SNe, conforming with the principles of the Virtual Observatory and allowing for new methods for the organization of SN research.

The new investigation of SN radial distributions reveals the following conclusions:

- There is a decrease of SNe Ia density in the central parts of spiral galaxies, which is not observed in ellipticals.
- SNe Ib/c show a high concentration towards the centers of spiral galaxies. Their distribution is different from that for SNe II, in spite of the generally accepted similarity of masses and ages of their progenitors.
- In the outer parts of the spiral galaxies the distributions of SNe ,Ia and SNe II are very similar.

These results are in general agreement with the previous works by van den Bergh (1997), Wang *et al.* (1997), and Tsvetkov *et al.* (2004) We should note that they still lack a consistent interpretation and present some difficulties for current theories on the nature of progenitors for different types of SNe.

The complete version of this paper is available at
< `http://www.sai.msu.su/sn/pubs/sncatdistr.pdf` >.

Acknowledgements

This work was partly supported by the grants 05-02-17480 and 05-07-90225 of the Russian Foundation for Basic Research.

References

Tsvetkov, D. Yu., & Bartunov, O. S. 1993, *Bull. Inform. CDS* 42, 17
Tsvetkov, D. Yu, Pavlyuk, N. N., & Bartunov, O. S. 2004, *Astron. Lett.* 30, 729
van den Bergh, S. 1997, *AJ*, 113, 197
Wang, L., Hoflich, P., & Wheeler, J. C. 1997, *ApJ* (Letters), 483, L29

Highlights of Astronomy, Volume 14
IAU XXVI General Assembly, 14-25 August 2006
Karel A. van der Hucht, ed.

doi:10.1017/S1743921307010824

Summary of JD 9
Supernovae: past, present, and future

Virginia L. Trimble

Physics Department & Las Cumbres Observatory, University of California Irvine,
4129 Frederick Reines Hall, Irvine, CA 92697-4575, USA
email: vtrimble@uci.edu

SN 1006 was the first (and for a very long time the only) event to be caught before peak light. A passage which, according to Stephenson, does not actually pertain to the SN, nevertheless makes clear that, even then, a hypothesis was more likely to be accepted if it made a prediction later verified, though the prediction was that something bad would happen to Emperor Sanjo In. The hypothesis was that the star was not new, but related to behavior of existing stars in Qichen Jianjun. According to the poster by P.J. Boner, Kepler made the opposite choice for 'his' SN, calling it a genuinely new star formed out of the ether, rather than mere change in appearance. Indeed the distinction between true novae and variable stars was not drawn correctly until Hevelius's 1662 study of Mira, after Tycho had shown that his event (and the comet of 1577) were more distant than our Moon, a point disputed by many of his contemporaries, but accepted by Galileo, who applied a very early statistical method to many different observations of SN 1572. Tycho's main advantages were better equipment and hard work, again not so different from present conditions.

The next 'golden moment' was perhaps the recording of the Crab nebula (Bevis 1731), followed by its cataloguing as M 1 (1758) and its naming by Rosse (1844). Handing over to the theorists, we find Milne attributing novae to stars collapsing to white dwarfs, and Pickering and Seelinger blaming collisions with a planet or another star (for the brightest events) in the 1920s. That decade also saw Lundmark's identification of the Crab with the 1054 event, along with his and Curtis's conclusion that there were likely to be two sorts of novae, the second, brighter, class including S And (SN 1885) with peak M_V near -15 (on an old distance scale). Baade and Zwicky, with core collapse, neutron stars, and cosmic rays, belong to the 1930s, along with Humason & Baade's recognition of Balmer lines in SNe 1926a and 1936a (both indeed probably Type IIs), and the suggestion from Olin C. Wilson and Zwicky (separately!) that supernovae could serve as better beacons for cosmological studies than whole galaxies. To the 1940s belong Minkowski's distinction of Types I and II (only the latter showing H features) and the beginning of radio astronomical searches and discoveries of SNRs. X-ray,γ-ray, and pulsar identifications began in the 1960s.

Turning again to the theorists, we find B^2FH on nucleosynthesis, also pointing out in 1957 the need for continuing energy input to SN light curves. Truran picked out the right nucleus for this (Ni-56) in 1969, after Hoyle and Fowler had put forward nuclear explosions as an alternative energy source to core collapse in 1960. The idea that binaries with accretion onto white dwarfs were a very good bet was 'in the air' in the early 1970s, the first published paper I've found coming from Wheeler & Hansen in 1971. That it appeared in a non-prestigious journal (*Ap&SS*) probably means something.

February 23, 1987 was a *very* good day for core-collapse events, and whoever figured out that Types' Ib and Ic belong to that group, but with progenitors stripped of their hydrogen was, I think, exceedingly clever. He also broke the old anti-degeneracy of Type I's happening in Population II and Type II's happening in Population I.

Going into Joint Discussion 09, I had a list of eight wishes and questions:
(*a*) good statistics on rates and hosts and distributions in hosts;
(*b*) identification of the correct progenitors (private prejudice in favor of WDX2);
(*c*) possible ignition triggers apart from reaching $M_{\rm Ch}$;
(*d*) identification of the explosion mechanism(s);
(*e*) a good balance sheet for nucleosynthesis, in which the amount of Ni-56 calculated would equal the amount of Fe-56 seen later (not to mention the various other elements fed into chemical evolution by SNe Ia);
(*f*) ways of describing the interactions of ejecta with interstellar and circumstellar matter, including that shed by a donor companion (if there is one);
(*g*) contribution of nuclear events to acceleration of cosmic rays, heating and stirring of the ISM, and various other kinds of feedback in early and late star formation; and
(*h*) certainty that the candles could be at least standardized for cosmological application.

And I came out with more or less the same list of wishes and questions, plus the cold comfort that they are the same ones worrying every one else.

A few personal thoughts:
(*i*) very possibly there is more than one progenitor type and more than one explosion mechanism represented in the events we see;
(*ii*) in light of Alan Title's remarks in his Invited Discourse about solar X-ray spectroscopy, the task of measuring Fe or anything else in SNRs seems more daunting than ever;
(*iii*) if M_2 survives, find it!
(*iv*) what produces the magnetic fields in Type Ia remnants (Gerardy talk); and
(*v*) theorists are cheap, while telescopes are expensive.

Highlights of Astronomy, Volume 14
IAU XXVI General Assembly, 14-25 August 2006
Karel A. van der Hucht, ed.

doi:10.1017/S1743921307010836

Joint Discussion 10
Progress in planetary exploration missions

Guy Consolmagno (ed.)

Specola Vaticana, VA-00120 Citta del Vaticano, Vatican City State
email: gjc@as.arizona.edu , gjc@specola.va

Preface

The astronomical study of planets is as old as Galileo's telescope, but in a profound way it was reborn with the advent of the Space Age. By constructing probes capable of leaving the surface of the Earth and traveling to other places in our solar system, sending back data collected from the very places that the astronomers wished to study, for the first time we were freed from the restrictions of observing astronomical objects from afar. These *in-situ* measurements, in their turn, have inspired countless new research projects back on Earth, from laboratory studies of materials to telescopic observations, of objects and in wavelengths now known to be of astronomical interest, thanks to those probes.

The year 2006 marked the end of the first fifty years of the Space Age, begun with the launch of Sputnik in October, 1957. It also concluded a remarkable triennium which saw the first sample returns of materials from space since the end of the Soviet and American lunar programs more than thirty years ago; the first major planetary missions from Western Europe; and the first successful interplanetary space mission from Japan.

At the Joint Discussion 10 of the triennial General Assembly of the IAU in Prague, August 2006, the results of planetary spacecraft missions that returned their first results since 2003 were summarized by the project scientists and other principal investigators of the missions. In organizing this Joint Discussion, we recognized that planetary astronomy extends far beyond the results of spacecraft missions alone. Thus, we encouraged the speakers to relate their new results to highlight the body of work in Planetary Sciences that has proceeded in parallel with the spacecraft missions.

By concentrating on the results from recent spacecraft missions in these proceedings, we hope we have accomplished three goals of interest to the larger IAU community:
(*i*) First and foremost, we present results that are intrinsically interesting, even for those of us not directly involved in planetary sciences.
(*ii*) In addition, these papers provide a context in which the wider accomplishments of the planetary astronomy community can be described, and results that can have an impact on other fields of astronomy ranging from nucleosynthesis, to star-formation, to the evolution of solar-type stars.
(*iii*) And finally, by concentrating specifically on the high-profile space missions which have already been widely publicized in the general media, the proceedings of this Joint Discussion provide in-depth information which our fellow astronomers can take back to their home institutions and help them answer the kinds of questions that the general public might ask about these missions.

The brief papers presented here could not possibly summarize all of the interesting results of those missions. Instead, it has been our intent to provide, in one place, an overview of the outlines of each mission; and a guide to the literature that can serve as a starting point for anyone wishing to look more deeply into the results of these programs.

In the Joint Discussion itself, results from the *Cassini-Huygens* mission to Saturn and Titan were also presented. Unfortunately, the principle investigators of that mission were not able to provide a written summary of their mission in time to be published here.

Missions reviewed

Mission	Target	Principal Scientist	Agency	Launch	Arrival	Return
Spirit/Opportunity	Mars	W. Bruce Banerdt	NASA	2003	2004	
Mars Express	Mars	Agustin F. Chicarro	ESA	2003	2004	
Cassini	Saturn	Dennis L. Matson	NASA	1997	2004	
Stardust	Comet	Donald E. Brownlee	NASA	1999	2004	2006
Genesis	Solar Wind	Donald S. Burnett	NASA	2001	2004	2004
Huygens	Titan	Jean-Pierre Lebreton	ESA	2004	2004	
SMART-1	Moon	Bernard H. Foing	ESA	2003	2005	
Hayabusa	Asteroid	Akira Fujiwara	JAXA	2003	2005	2010?
Deep Impact	Comet	Michael F. A'Hearn	NASA	2005	2005	
Venus Express	Venus	L. Hakan Svedhem	ESA	2005	2006	

Scientific Organizing Committee

Guy J. Consolmagno (Vatican City State, chair), Michael F. A'Hearn (USA), Carlo Blanco (Italy), Regis Courtin (France), Dale P. Cruikshank (USA), Walter F. Huebner (USA), Peter M.M. Jenniskens (the Netherlands), H. Uwe Keller (Germany), Leonid V. Ksanfomality (Russia), Mikhail Ya. Marov (Russia), Melissa A. McGrath (USA), Keith S. Noll (USA), Maarten C. Roos-Serote (Portugal), Edward F. Tedesco (USA), and Iwan P. Williams (UK).

Guy J. Consolmagno, chair SOC
Vatican City State, November 30, 2006

Highlights of Astronomy, Volume 14
IAU XXVI General Assembly, 14-25 August 2006
Karel A. van der Hucht, ed.

doi:10.1017/S1743921307010848

Genesis discovery mission science results

Donald S. Burnett

Division of Geological and Planetary Sciences, California Institute of Technology,
Pasadena, CA 91125, USA
email: burnett@gps.caltech.edu

Abstract. Results of the Genesis mission to sample the solar wind are summarized.

Keywords. Solar wind, space vehicles

1. Results

Genesis returned samples of high purity materials exposed to the solar wind for 27 months, including separate samples of 3 solar wind regimes. The science objectives of the mission are to measure solar isotopic and elemental abundances to a precision sufficient to address planetary science problems.

A crash upon Earth return resulted in massive breakage of collector materials, serious losses of large amounts of materials, and particulate surface contamination. However, atoms are not destroyed by impact. We have over 15,000 pieces of collector materials greater than 3 mm in size. We can show that the implanted solar wind resides safely beneath the surface of the collector materials, while the contamination is on the surface. We have a margin of only 100 Å, but, because we are a sample return mission, we can use all of the twenty-first century's technology to exploit this margin. We were fortunate in that three-fifths of the materials for dedicated experiments survived in relatively good shape.

A major issue is the extent to which the acceleration of the solar wind has modified elemental and isotopic abundances from the photospheric values. Spacecraft studies show that elements with first ionization potential (FIP) greater than 9 eV are depleted in the solar wind relative to lower FIP elements, but the lower FIP elements, which make up most of terrestrial planet material, appear unfractionated. Our preliminary *Genesis* Fe/Mg is 0.78, which is in good agreement with both spacecraft data and the photospheric absorption line ratio.

Little is known about isotopic fractionation in the solar wind, but our regime samples represent materials formed by different solar processes. The isotopic compositions of Ne and Ar in the different regime samples agree with permil level precision. There is no evidence for isotope fractionation in the solar wind relative to the photosphere.

A sample of bulk metallic glass (BMG) which etches uniformly with nitric acid vapor was recovered intact. This sample provides a check on the presence of higher energy (SEP) solar ions with different isotopic compositons. Extensive lunar sample analyses indicated an SEP ^{20}Ne/^{22}Ne ratio of 11.4 compared to 13.7 for the solar wind. BMG etching produced a continuously dropping 20/22 ratio down to at least 10.8. The trend matches exactly what is seen in lunar samples; however, the trend can be quantitatively reproduced by allowing for the differential implantation of ^{22}Ne and ^{20}Ne, and differential implantation appears to explain the lunar results as well.

2. Mission details

Launch date: 8 August 2001

Collector opened: 30 November 2001

Collector closed: 1 April 2004

Return to Earth: 8 September 2004

Payload mass: 494 kg

Primary science instrument:

Three distinct collector arrays, each a grid of ultra-pure wafers of silicon, gold, sapphire, diamond and other materials, designed to collect solar wind particles during different periods of solar wind activity.

An electrostatic mirror (concentrator): focuses incoming O and N ions from a diameter of about 20 cm onto a 6 cm diameter set of target materials, to increase the signal for these particular elements.

3. For further information

A detailed description of the *Genesis* mission can be found in Burnett *et al.* (2003). Results from the mission have yet to be published in one place, but can be found in abstract form in the 2006 meetings of the American Geophysical Union and the Meteoritical Society. One result recently published in Grimberg *et al.* (2003) compares the solar wind neon measured by Genesis with that found in the lunar noble gas record, concluding that the solar neon isotope composition seen in lunar samples may be due to the solar wind being fractionated during implantation.

References

Burnett, D. S., Barraclough, B. L., Bennett, R., Neugebauer, M., Oldham, L. P., Sasaki, C. N., Sevilla, D., Smith, N., Stansbery, E., Sweetnam D., & Wiens R. C. 2003, *Space Sci. Revs*, 105, 509

Grimberg, A., Baur, H., Bochsler, P., Bühler, F., Burnett, D. S., Hays, C. C., Heber, V. S., Jurewicz, A. J. G., & Wieler, R. 2006, *Science*, 314, 1133

Highlights of Astronomy, Volume 14
IAU XXVI General Assembly, 14-25 August 2006
K.A. van der Hucht, ed.

doi:10.1017/S174392130701085X

Hayabusa and its adventure around the tiny asteroid Itokawa

Makoto Yoshikawa, Akira Fujiwara, and Jun'ichiro Kawaguchi

ISAS/JAXA, 3-1-1 Yoshinodai, Sagamihara, Kanagawa 229-8510, Japan
email: makoto@isas.jaxa.jp

Abstract. Results of the *Hayabusa* mission to image and sample the asteroid Itokawa are summarized.

Keywords. minor planets, asteroids, space vehicles

1. Results

The *Hayabusa* spacecraft, which was launched in May 2003, arrived at asteroid Itokawa in September, 2005. The size of Itokawa is about 540 m in length, so it is the smallest celestial object that manmade spacecraft has ever visited.

The view of Itokawa was totally unexpected. The surface of Itokawa is covered with numerous boulders and we could see only a few craters on it.

Hayabusa carried out scientific observations of Itokawa in detail for about two months by using four instruments; AMICA(Asteroid Multiband Imaging Camera), NIRS (Near-Infrared Spectrometer), XRS (X-ray fluorescence Spectrometer), and LIDAR (Light Detection and Ranging instrument).

The surface of Itokawa is basically divided into two parts, smooth terrain and rough terrain, and we found many geographical and geological features form the images of AMICA with a resolution of less than 1 cm at the most. Also we found from the data of NIRS and XRS that the surface material of Itokawa is similar to LL-chondrite, ordinary chondrite meteorites.

The mass of Itokawa was estimated by the orbit analysis of *Hayabusa* using the optical navigation cameras, LIDAR, and radiometric data. Then the bulk density was calculated as $1.9\,g/cm^3$ using the obtained shape model of Itokawa. This low density indicates a macroporosity of about 40 %. From these results, we have concluded that Itokawa is a rubble pile asteroid.

After these scientific observations, *Hayabusa* tried to approach Itokawa closely several times in November 2005, and finally *Hayabusa* executed a touchdown on the surface of Itokawa twice. The second touchdown was executed as scheduled but after that we had troubles in the operation of *Hayabusa*. We are not sure if some surface materials were taken or not. However, expecting to have some samples, we are now attempting to return *Hayabusa* to the Earth in June 2010.

2. Mission details

Launch date: 9 May 2003

Arrival at Itokawa: 12 September 2005

Dry mass: 374 kg, wet mass: 502 kg

Primary science instruments:

Hayabusa-AMICA: collects scientific images with seven filters in the ECAS system bands: *ul*, *b*, *v*, *w*, *x*, *p*, and *zs*.

Hayabusa-LIDAR: uses time-of-flight measurement of a laser pulse reflected from the surface to determine the surface topography, with a footprint at 7 km altitude of about 5×12 m.

Hayabusa-NIRS: studies the mineralogy and physical properties of the surface with a 64 channel InGaAs photodiode array detector and a diffraction grating combined with a prism, covering wavelengths from 0.76 to 2.1 μm with a dispersion per pixel of 23.6 nm.

Hayabusa-XRS: studies the elemental composition of the surface using a CCD X-ray detector with an energy resolution of 160 eV at 5.9 keV.

3. In the literature

A detailed description of the *Hayabusa* mission and the first results from the mission were published in the 2 June 2006 issue of Science. Fujiwara *et al.* (2006) give an overview of the rubble-pile asteroid Itokawa as observed by *Hayabusa*. M. Abe *et al.* (2006) discuss the near-infrared spectral results from the *Hayabusa* spacecraft. Okada *et al.* (2006) report on the X-ray fluorescence spectrometry by *Hayabusa*. Saito *et al.* (2006) present detailed images of asteroid 25143 Itokawa from *Hayabusa*. S. Abe *et al.* (2006) discuss mass and local topography measurements of Itokawa. Demura *et al.* (2006) report the pole and global shape of asteroide 25143 Itokawa. And Yano *et al.* (2006) describe the touchdown of the *Hayabusa* spacecraft at the Muses Sea on Itokawa.

References

Abe, M., Takagi, Y., Kitazato, K., Abe, S., Hiroi, T., Vilas, F., Clark, B. E., Abell, P. A., Lederer, S. M., Jarvis, K. S., Nimura, T., Ueda, Y., & Fujiwara, A. 2006, *Science*, 312, 1334

Abe, S., Mukai, T., Hirata, N., Barnouin-Jha, O. S., Cheng, A. F., Demura, H., Gaskell, R. W., Hashimoto, T., Hiraoka, K., Honda, T., Kubota, T., Matsuoka, M., Mizuno, T., Nakamura, R., Scheeres, D. J., & Yoshikawa, M. 2006, *Science*, 312, 1344

Demura, H., Kobayashi, S., Nemoto, E., Matsumoto, N., Furuya, M., Yukishita, A., Muranaka, N., Morita, H., Shirakawa, K., Maruya, M., Ohyama, H., Uo, M., Kubota, T., Hashimoto, T., Kawaguchi, J., Fujiwara, A., Saito, J., Sasaki, S., Miyamoto, H., & Hirata, N. 2006, *Science*, 312, 1347

Fujiwara, A., Kawaguchi, J., Yeomans, D. K., Abe, M., Mukai, T., Okada, T., Saito, J., Yano, H., Yoshikawa, M., Scheeres, D. J., Barnouin-Jha, O. S., Cheng, A. F., Demura, H., Gaskell, R. W., Hirata, N., Ikeda, H., Kominato, T., Miyamoto, H., Nakamura, A. M., Nakamura, R., Sasaki, S., & Uesugi, K. 2006, *Science*, 312, 1330

Okada, T., Shirai, K., Yamamoto, Y., Arai, T., Ogawa, K., Hosono, K., & Kato, M. 2006, *Science*, 312, 1338

Saito, J., Miyamoto, H., Nakamura, R., Ishiguro, M., Michikami, T., Nakamura, A. M., Demura, H., Sasaki, S., Hirata, N., Honda, C.,. Yamamoto, A, Yokota, Y., Fuse, T., Yoshida, F., Tholen, D.J., Gaskell, R. W., Hashimoto, T., Kubota, T., Higuchi, Y., Nakamura, T., Smith, P., Hiraoka, K., Honda, T., Kobayashi, S., Furuya, M., Matsumoto, N., Nemoto, E., Yukishita, A., Kitazato, K., Dermawan, B., Sogame, A., Terazono, J., Shinohara, C., & Akiyama H. 2006, *Science*, 312, 1341

Yano, H., Kubota, T., Miyamoto, H., Okada, T., Scheeres, D.J., Takagi, Y., Yoshida, K., Abe, M., Abe, S., Barnouin-Jha, O. S., Fujiwara, A., Hasegawa, S., Hashimoto, T., Ishiguro, M., Kato, M., Kawaguchi, J., Mukai, T., Saito, J., Sasaki, S., & Yoshikawa, M. 2006, *Science*, 312, 1350

Highlights of Astronomy, Volume 14
IAU XXVI General Assembly, 14-25 August 2006
Karel A. van der Hucht, ed.

doi:10.1017/S1743921307010861

Deep Impact: excavating comet Tempel 1

Michael F. A'Hearn

Department of Astronomy, University of Maryland, College Park, MD 20742-2421, USA
email: ma@astro.umd.edu

Abstract. Results of the *Deep Impact* mission to impact Comet Tempel 1 are summarized.

Keywords. comets, Comet Tempel 1, space vehicles

---

1. Results

Deep Impact delivered 19 GJoules of kinetic energy to comet 9P/Tempel 1 on 4 July 2005. We present results both from the impact event and from the observations made prior to the impact.

The impact itself was oblique. An initial, hot plume carried away most of the delivered energy as the kinetic energy of roughly a ton of material. Most ejecta followed that and were cold, slow-moving, few-micron sized particles. These carried most of the momentum of the ejecta and totalled roughly 10^4 tons. After the first two seconds, the ejecta include small crystals of ordinary ice, indicating excavation without heating and thus without chemical alteration. The ejected gases included a large amount of CO_2 and a very large amount of organics in addition to water and species yet unidentified. The refractory to volatile ratio in the ejecta is of order unity but is sensitive to assumptions about the cutoff at the large end of the size distribution of the ejecta.

The ejecta enable us to show that the strength of the surface layers, at scales from microscopic to a few hundred meters is remarkably weak and also to show that the bulk density of the nucleus is so low that the entire nucleus must be extremely porous.

On approach, we learned that outbursts by comets are far more common than previously realized and that they can be associated with specific regions on the surface, and in one case with sunrise on that surface. We can confidently rule out exogenic sources (other than sunrise) for these outbursts. Although there are similarities, the geology of the surface is clearly different from that of the few other cometary nuclei visited and very puzzling. There are clearly distinct layers, which are likely not concentric shells but rather discrete blocks. Surface photometric properties are reasonably uniform except in a few small areas. We also showed that, although there are some small spots of ice on the surface, the bulk of the water is released near the subsolar point from sub-surface ice.

2. Mission details

Launch date: 12 January 2005

Arrival at Tempel 1: 4 July 2005

Payload mass: Impactor, 370 kg; Flyby Spacecraft, 90 kg

Primary science instruments:

Flyby spacecraft: High Resolution Instrument (HRI) and Medium Resolution Instrument (MRI): imaging, infrared spectroscopy, and optical navigation.

Impactor: Impactor Targeting Sensor (ITS): nearly identical to the MRI, as it uses the same type of telescope and CCD camera but without the filter wheel.

3. In the literature

A detailed description of the *Deep Impact* mission was presented in a special issue of *Space Science Reviews*; see Russell (2005). The first results from the mission were published in the 14 October 2005 and 10 March 2006 issues of *Science*; see A'Hearn *et al.* (2005) and Sunshine *et al.* (2006). A special issue of *Icarus* (Volume 187, Issue 1, 2007), is dedicated to the *Deep Impact* mission.

References

A'Hearn, M. F., Belton, M. J. S., Delamere, W. A., Kissel, J., Klaasen, K. P., McFadden, L. A., Meech, K. J., Melosh, H. J., Schultz, P. H., Sunshine, J. M., Thomas, P. C., Veverka, J., Yeomans, D. K., Baca, M. W., Busko, I., Crockett, C. J., Collins, S. M., Desnoyer, M., Eberhardy, C. A., Ernst, C. M., Farnham, T. L., Feaga, L., Groussin, O., Hampton, D., Ipatov, S. I., Li, J.-Y., Lindler, D., Lisse, C. M., Mastrodemos, N., Owen, Jr., W. M., Richardson, J. E., Wellnitz, D. D., & White, R. L. 2005, *Science*, 310, 258

Russell, C. T. (ed.) 2005, *Space Sci. Rev.*, 117, 1

Sunshine, J. M., A'Hearn, M. F., Groussin, O., Li, J-Y., Belton, M. J. S., Delamere, W. A., Kissel, J., Klaasen, K. P., McFadden, L. A., Meech, K. J., Melosh, H. J., Schultz, P. H., Thomas, P. C., Veverka, J., Yeomans, D. K., Busko, I. C., Desnoyer, M., Farnham, T. L., Feaga, L. M., Hampton, D. L., Lindler, D. J., Lisse, C. M., & Wellnitz, D. D. 2006, *Science*, 311, 1453

Highlights of Astronomy, Volume 14
IAU XXVI General Assembly, 14-25 August 2006
Karel A. van der Hucht, ed.

doi:10.1017/S1743921307010873

Preliminary examination of the comet Wild 2 samples returned by the *Stardust* spacecraft

Michael E. Zolensky[1], Donald E. Brownlee[2], Peter Tsou[3], Friedrich P. Hörz[1], Scott A. Sandford[4], George J. Flynn[5], Kevin D. McKeegan[6], and Lindsay P. Keller[1]

[1]NASA Johnson Space Center, Houston, TX 77058, USA
email: michael.e.zolensky@nasa.gov

[2]Department of Astronomy, University of Washington, Seattle, WA 98195, USA

[3]Jet Propulsion Laboratory, CalTech, Pasadena, CA 91109-8099, USA

[4]NASA Ames Research Center, Moffett Field, CA 94035, USA

[5]Dept. of Physics, State University of New York at Plattsburgh, Plattsburgh, NY 12901, USA

[6]Dept. of Earth and Space Sciences, University of California, Los Angeles, CA 90095, USA

Abstract. Results of the *Stardust* mission to sample dust from comet Wild 2 are summarized.

Keywords. comets, comet Wild 2, space vehicles

---

1. Results

The sample return capsule of the *Stardust* spacecraft was successfully recovered in northern Utah, USA, on January 15, 2006, and its cargo of coma grains from comet Wild 2 has now been the subject of intense investigation.

The period since spacecraft recovery been sufficient to permit numerous analyses by over 200 researchers to have been performed and to permit some understanding of the following fundamental sample issues:

(1) Comet nucleus composition, mineralogy, petrology, isotopic composition and grain physical properties.

(2) Sample variability.

(3) Type and degree of sample alteration by the collection process, and subsequent sample handling.

(4) Sample documentation and handling procedures.

(5) Comparisons to the *Deep Impact* mission to comet Tempel I.

Following the close of sample preliminary examination, *Stardust* samples will be made available to the larger community, as are lunar samples, IDPs, and Antarctic meteorites. A sample catalog will be available at the NASA Johnson Space Center Curation website. A dedicated peer review committee will consider all sample requests. The *Stardust* interstellar tray is being scanned in the Cosmic Dust Lab; when this operation is complete (approximately by the end of 2006) the Cosmic Dust Lab will be re-opened for business.

2. Mission details

Launch date: 7 February 1999

Arrival at comet Wild 2: 2 January 2004

Return to Earth: 15 January 2006

Payload mass: 270 kg.

Primary science instruments:

Stardust-SRC (Sample Return Capsule) and Aerogel Collector: a compact system, consisting primarily of a sample canister with an aeroshield/basecover, plus navigation recovery aids, an event sequencer and a small parachute system.

Stardust-CIDA (Cometary and Interstellar Dust Analyzer): intercepts dust and performs real-time compositional analysis for transmission back to Earth.

Stardust-NC (Navigation Camera): used to navigate the *Stardust* spacecraft upon approach to the comet, the camera also served as an imaging camera to capture high-resolution color images of the comet nucleus on approach and on departure, and broadband images at various phase angles while nearby.

Stardust-DFMI (Dust Flux Monitor Instrument): records the impact of small particles; in addition, two separate acoustic impact sensors monitor strikes by larger particles.

3. In the literature

A detailed description of the *Stardust* mission encounter with comet Wild 2 was published in the 18 June 2004 issue of *Science.* Brownlee *et al.* (2004) describe the surface of comet 81P/Wild 2 as seen from the *Stardust* spacecraft. Sekanina *et al.* (2004) modeled the nucleus and jets of the comet based on data from the *Stardust* encounter. Kissel *et al.* (2004) discuss results from the cometary and interstellar dust analyzer during the encounter. And Tuzzolino *et al.* (2004) report on the dust Measurements in the comet's coma as measured by the dust flux monitor instrument.

The first reports of analysis from the returned samples has to date been presented in special sessions of the Meteoritical Society and the American Geophysical Union annual meetings. A full refereed report is in preparation.

References

Brownlee, D. E., Horz, F., Newburn, R. L., Zolensky, M., Duxbury, T. C., Sandford, S., Sekanina, Z., Tsou, P., Hanner, M. S., Clark, B. C., Green, S. F., & Kissel, J. 2004, *Science*, 304, 1764

Kissel, J., Krueger, F. R., Silén, J., & Clark, B. C. 2004, *Science*, 304, 1774

Sekanina, Z., Brownlee, D. E., Economou, T. E., Tuzzolino, A. J., & Green, S. F. 2004, *Science*, 304, 1769

Tuzzolino, A. J., Economou, T. E., Clark, B. C., Tsou, P., Brownlee, D. E., Green, S. F., McDonnell, J. A. M., McBride, N., & Colwell, M. T. S. H. 2004, *Science*, 304, 1776

Highlights of Astronomy, Volume 14
IAU XXVI General Assembly, 14-25 August 2006
Karel A. van der Hucht, ed.

doi:10.1017/S1743921307010885

Results from the *SMART-1* lunar mission

Bernhard H. Foing

Space Science Department, EsTec, ESA, Noordwijk, the Netherlands
email: bernard.foing@esa.int

Abstract. Results of the *SMART-1* mission to Earth's Moon are summarized.

Keywords. Moon, space vehicles

1. Results

SMART-1 is the first ESA mission that reached the Moon. It demonstrated Solar Electric Primary Propulsion (SEP) and tested new technologies for spacecraft and instruments.

Launched on 27 Sept. 2003, as Ariane-5 auxiliary passenger, *SMART-1* spiraled out towards lunar capture on 15 November 2004, and then towards lunar science orbit, which was reached on 1 March 2005. The mission has been extended, to end with an impact on 2-3 September 2006.

This mission's purpose is not only to perform science but also to prepare future international lunar exploration, in collaboration with upcoming missions.

The 19 kg payload includes a miniaturized high-resolution camera AMIE, a near-infrared point-spectrometer SIR for mineralogy investigation, and a very compact X-ray spectrometer D-CIXS for surface elemental composition. There is also an experiment (KaTE) aimed at demonstrating deep-space telemetry and telecommand communications in the X and Ka-bands, a radio-science experiment RSIS, a deep space optical link (Laser-Link Experiment), using the ESA Optical Ground station in Tenerife, and the validation of a system of autonomous navigation (OBAN) based on image processing.

SMART-1 lunar science investigations include studies of the chemical composition of the Moon, of geophysical processes (volcanism, tectonics, cratering, erosion, deposition of ices and volatiles) for comparative planetology, and high resolution studies in preparation for future steps of lunar exploration. The mission addresses several topics such as the accretional processes that led to the formation of rocky planets, and the origin and evolution of the Earth-Moon system.

2. Mission details

Launch date: 27 September 2003

Arrival date: 15 November 2004

Payload mass: 19 kg.

Primary science instruments:

SMART-1-AMIE camera: surface images in visible and near-infrared light.

SMART-1-SIR infrared spectrometer: maps lunar minerals.

SMART-1-D-CIXS X-ray spectrometer: identifies key chemical elements in the lunar surface.

SMART-1-XSM X-ray detector: monitors variations in solar X-ray emissions, studies solar variability.

SMART-1-SPEDE electric field experiment: monitors the Moon's wake in the solar wind.

SMART-1-RSIS radio science experiment: measures lunar librations.

3. In the literature

A detailed description of the *SMART-1* mission plan was published in the January 1999 issue of *Earth, Moon, and Planets.* Racca *et al.* (1999) outline the mission itself, while Foing *et al.* (1999) discusses the science goals of the mission.

The first results from the mission were published in *Advances in Space Research* in January 2006. Foing *et al.* (2006) describe the status, first results and goals of the *SMART-1* mission, while Josset *et al.* (2006) describe the science objectives and first results from the *SMART-1*/AMIE multicolour micro-camera.

References

Foing, B. H., Heather, D. J., Almeida, M., & the *SMART-1* Science Technology Working Team 1999, *Earth, Moon, and Planets*, 85-86, 379

Foing, B. H., Racca, G. D., Marini, A., Evrard, E., Stagnaro, L., Almeida, M., Koschny, D., Frew, D., Zender, J., Heather, J., Grande, M., Huovelin, J., Keller, H. U., Nathues, A., Josset, J. L., Malkki, A., Schmidt, W., Noci, G., Birkl, R., Iess, L., Sodnik, Z., & McManamon, P. 2006, *Adv. Sp. Res.*, 37, 6

Josset, J.-L., Beauvivre, S., Cerroni, P., de Sanctis, M. C., Pinet, P., Chevrel, S., Langevin, Y., Barucci, M. A., Plancke, P., Koschny, D., Almeida, M., Sodnik, Z., Mancuso, S., Hofmann, B. A., Muinonen, K., Shevchenko, V., Shkuratov, Yu., Ehrenfreund, P., & Foing, B. H. 2006, *Adv. Sp. Res.*, 37, 14

Racca, G. D., Foing, B. H., & Coradini M. 1999, *Earth, Moon, and Planets*, 85-86, 379

Highlights of Astronomy, Volume 14
IAU XXVI General Assembly, 14-25 August 2006
Karel A. van der Hucht, ed.

doi:10.1017/S1743921307010897

Venus Express

L. Hakan Svedhem

Space Science Department, EsTec, ESA, Noordwijk, the Netherlands
email: hakan.svedhem@esa.int

Abstract. Results of the *Venus Express* mission to Venus are summarized.

Keywords. planets, space vehicles

---

1. Results

After having been the 'forgotten planet' for more than a decade, Venus has again become a target for intense studies to better understand the many problems not answered by the more than twenty US and Soviet probes launched in the previous decades.

After a launch from the Baikonur cosmodrome, Kazakhstan, 9 November 2005, and a five-month cruise, the *Venus Express* spacecraft reached Venus on 11 April 2006. The spacecraft has now reached its final operational, 24 hr polar orbit, with apocentre altitude of 66 000 km and pericentre altitude of 250 km. The pericentre is located at about 80 ° northern latitude and drifts only slowly towards the pole. During the first weeks of routine operation the spacecraft has already sent back a wealth of exiting information.

The objective of the *Venus Express* mission is to carry out a comprehensive study of the atmosphere of Venus, the plasma environment and its interaction with the solar wind, and to study certain aspects of the surface of the planet. A well optimised payload, composed of two multi channel spectrometers, an IR-Vis-UV imaging spectrometer, a wide angle camera, a multi-sensor energetic particle instrument, a magnetometer, and a radio science experiment, allows all aspects of the objectives to be addressed at a sufficient depth.

Venus Express has been developed in record time, less than four years, using an efficient concept of re-using elements of recently developed spacecraft, mainly *Mars Express* and *Rosetta*. Significant savings for both the space and ground segments have been possible by using existing teams in industry, in ESA and in several of the science institutes involved.

The first data has shown a highly dynamic atmosphere, including close-ups of the double vortex at the South Pole and evidence of the superrotation, indeed topics of high interest and among the top priority objectives. The high-resolution spectrometers are finding several minor species at various depths of the atmosphere. *Venus Express* is the first mission fully exploiting the spectral 'windows' at infrared wavelengths, in order to map the atmosphere in three dimensions. The dynamic behaviour is studied by making measurements at a high-repetition rate. The surface will be studied through the 1 μm 'window' in order to search for active volcanism and hot lava fields. *Venus Express* will return more data to Earth than all previous Venus missions together, not counting the *Magellan* radar mapper. The first data returned from the mission promises an exiting time to come with analysis of very high-quality data from state-of-the-art instrumentation.

2. Mission details

Launch date: 9 November 2005

Arrival at Venus: 11 April 2006

Payload mass: 93 kg

Primary science instruments:

Venus Express-ASPERA (Analyser of Space Plasma and Energetic Atoms): investigates the interaction between the solar wind and the atmosphere of Venus by measuring outflowing particles from the planet's atmosphere and the particles making up the solar wind.

Venus Express-MAG (Magnetometer): studies the magnetic field that exists around the planet due to the interaction between the solar wind and the atmosphere, and the effect this has on Venus' atmosphere. This is an important measurement for improved understanding of the atmospheric escape processes.

Venus Express-PFS (Planetary Fourier Spectrometer): measures the temperature of the atmosphere between altitudes of 55 – 100 km at a high resolution, the surface temperature (to search for volcanic activity), and the abundance of the minor species of the atmosphere. (This instrument is presently out of operation due to a blocked mechanism)

Venus Express-SPICAV/SOIR (Ultraviolet and Infrared Atmospheric Spectrometer): assists in the analysis of Venus' atmosphere, to search for the small quantities of water expected to exist in the atmosphere, sulphur compounds and molecular oxygen in the atmosphere, and determine the density and temperature of the atmosphere at 80 - 180 km altitude. The new additional SOIR channel has a spectral resolution $R > 20\,000$.

Venus Express-VeRa (Venus Radio Science Experiment): uses the powerful radio link between the spacecraft and Earth to investigate the conditions prevalent in the ionosphere and atmosphere of Venus; to study the electron density between 100 km and the ionopause, the density, temperature, and pressure of the atmosphere from 35 - 40 km up to 100 km from the surface. It will also determine roughness and electrical properties of the surface and investigate the conditions of the solar wind in the inner part of the Solar System.

Venus Express-VIRTIS (Ultraviolet/Visible/Near-Infrared mapping Spectrometer): studies the composition of the lower atmosphere between 40 km altitude and the surface, tracks the clouds in both ultraviolet and infrared wavelengths and studies atmospheric dynamics at different altitudes. Virtis will be able to recover parts of the objectives of the presently non-operating PFS instrument

Venus Express-VMC (Venus Monitoring Camera): a wide-angle multi-channel camera for imaging of the planet in the near-infrared, ultraviolet and visible wavelengths for global and local investigations, with emphasis on cloud dynamics and surface imaging.

3. In the literature

A series of overview papers of the *Venus Express* mission was published in the November 2005 issue of the *ESA Bulletin*. An insight into the project development is given by McCoy *et al.* (2005) and a description of the spacecraft and subsystems design is summarised by Winton *et al.* (2005). The scientific rationale for the mission and a description of the payload is given by Svedhem *et al.* (2005). A detailed description of specific scientific topics is given in the November 2006 issue of *Planetary and Space Science*. *Venus Express* science planning is outlined by Titov *et al.* (2006). Baines *et al.* (2006) describe how *Venus Express* will explore the deep atmosphere and surface of Venus. Formisano *et al.* (2006) describe the Planetary Fourier Spectrometer (PFS) onboard *Venus Express*. Häusler *et al.* (2006) report on the radio science investigations by VeRa onboard the *Venus Express* spacecraft. Russell *et al.* (2006) describe lightning detection on the *Venus Express* mission. Zhang *et al.* (2006) report on the magnetic field investigation of the Venus plasma environment by *Venus Express*. Müller-Wodarg *et al.* (2006) discuss the thermosphere of Venus and its exploration by the *Venus Express* Accelerometer Experiment. Luhmann *et al.* (2006) discuss the expectations for *Venus Express* for analyzing Venus O+ pickup ions.

References

Baines, K. H., Atreya, S., Carlson, R. W., Crisp, D., Drossart, P., Formisano, V., Limaye, S. S., Markiewicz, W. J., & Piccioni, G. 2006, *Planetary and Space Science*, 54, 1263

Formisano, V., Angrilli, F., Arnold, G., Atreya, S., Baines, K. H., Bellucci, G., Bezard, B., Billebaud, F., Biondi, D., Blecka, M. I., Colangeli, L., Comolli, L., Crisp, D., D'Amore, M., Encrenaz, T., Ekonomov, A., Esposito, F., Fiorenza, C., Fonti, S., Giuranna, M., Grassi, D., Grieger, B., Grigoriev, A., Helbert, J., Hirsch, H., Ignatiev, N., Jurewicz, A., Khatuntsev, I., Lebonnois, S., Lellouch, E., Mattana, A., Maturilli, A., Mencarelli, E., Michalska, M., Lopez Moreno, J., Moshkin, B., Nespoli, F., Nikolsky, Yu., Nuccilli, F., Orleanski, P., Palomba, E., Piccioni, G., Rataj, M., Rinaldi, G., Rossi, M., Saggin, B., Stam, D., Titov, D., Visconti, G., & Zasova, L. 2006, *Planetary and Space Science*, 54, 1298

Häusler, B., Pätzold, M., Tyler, G. L., Simpson, R. A., Bird, M. K., Dehant, V., Barriot, J.-P., Eidel, W., Mattei, R., Remus, S., Selle, J., Tellmann, S., & Imamura, T. 2006, *Planetary and Space Science*, 54, 1315

Luhmann, J. G., Ledvina, S. A., Lyon, J. G., & Russell, C. T. 2006, *Planetary and Space Science*, 54, 1457

McCoy, D., Siwitza, T., & Gouka, R. 2005, *ESA Bulletin*, 124, 10

Müller-Wodarg, I. C. F., Forbes, J. M., & Keating, G. M. 2006, *Planetary and Space Science*, 54, 1415

Russell, C. T., Strangeway, R. J., & Zhang, T. L. 2006, *Planetary and Space Science*, 54, 1344

Svedhem, H., Witasse, O., & Titov, D. 2005, *ESA Bulletin*, 124, 24

Titov, D. V., Svedhem, H., Koschny, D., Hoofs, R., Barabash, S., Bertaux, J.-L., Drossart, P., Formisano, V., Häusler, B., Korablev, O., Markiewicz, W. J., Nevejans, D., Pätzold, M., Piccioni, G., Zhang, T. L., Merritt, D., Witasse, O., Zender, J., Accomazzo, A., Sweeney, M., Trillard, D., Janvier, M., & Clochet, A. 2006, *Planetary and Space Science*, 54, 1279

Winton, A. J., Schnorhk, A., McCarthy, C., Witting, M., Sivac, P., Eggel, H., Pereira, J., Verna, M., & Geerling, F. 2005, *ESA Bulletin*, 124, 17

Zhang, T. L., Baumjohann, W., Delva, M., Auster, H.-U., Balogh, A., Russell, C. T., Barabash, S., Balikhin, M., Berghofer, G., Biernat, H. K., Lammer, H., Lichtenegger, H., Magnes, W., Nakamura, R., Penz, T., Schwingenschuh, K., Vörös, Z., Zambelli, W., Fornacon, K.-H., Glassmeier, K.-H., Richter, I., Carr, C., Kudela, K., Shi, J. K., Zhao, H., Motschmann, U., & Lebreton, J.-P. 2006, *Planetary and Space Science*, 54, 1336

Highlights of Astronomy, Volume 14
IAU XXVI General Assembly, 14-25 August 2006
Karel A. van der Hucht, ed.

doi:10.1017/S1743921307010903

Mars Express science results and goals for the extended mission

Agustin F. Chicarro

Space Science Department, EsTec, ESA, Noordwijk, the Netherlands
email: agustin.chicarro@esa.int

Abstract. Results of the *Mars Express* mission to Mars are summarized.

Keywords. planets, space vehicles

---

1. Results

The ESA *Mars Express* mission was successfully launched on 02 June 2003 from Baikonur, Kazakhstan, onboard a Russian Soyuz rocket with a Fregat upper stage. The mission comprises an orbiter spacecraft, which has been placed in a polar martian orbit, and the small *Beagle 2* lander, due to land in Isidis Planitia but whose fate remains uncertain.

In addition to global studies of the surface, subsurface and atmosphere of Mars, with an unprecedented spatial and spectral resolution, the unifying theme of the mission is the search for water in its various states everywhere on the planet. A summary of scientific results from all experiments after more than one Martian year in orbit (687 days) is given below.

The High-Resolution Stereo colour imager (HRSC) has shown breathtaking views of the planet, pointing to very young ages for both glacial and volcanic processes, from hundreds of thousands to a few million years old, respectively.

The IR Mineralogical Mapping Spectrometer (OMEGA) has provided unprecedented maps of H_2O ice and CO_2 ice in the polar regions, and determined that the alteration products (phyllosilicates) in the early history of Mars correspond to abundant liquid water, while the post-Noachian products (sulfates) suggest a colder, drier planet with only episodic liquid water on the surface.

The Planetary Fourier Spectrometer (PFS) has confirmed the presence of methane for the first time, which would indicate current volcanic activity and/or biological processes.

The UV and IR Atmospheric Spectrometer (SPICAM) has provided the first complete vertical profile of CO_2 density and temperature, and has discovered the existence of nightglow as well as that of auroras over mid-latitude regions with paleomagnetic signatures.

The Energetic Neutral Atoms Analyser (ASPERA) has identified solar wind scavenging of the upper atmosphere down to 270 km altitude as one of the main culprits of atmospheric degassing.

The Radio Acience Experiment (MaRS) has studied the surface roughness by pointing the spacecraft high-gain antenna to the Martian surface. Also, the martian interior has been probed by studying the gravity anomalies affecting the orbit, and a transient ionospheric layer due to meteors burning in the atmosphere, identified by MaRS.

Finally, results of the subsurface sounding radar (MARSIS) following the late deployment of its antennas due to safety concerns, indicate strong echoes coming from the surface and the subsurface allowing to identify buried craters and tectonic structures. Also, probing of the ionosphere reveals a variety of echoes originating in areas of remnant magnetism.

Mars Express is already hinting at a quantum leap in our understanding of the planet's geological evolution, to be complemented by the ground truth being provided by the American *MER* rovers. The nominal mission lifetime of one Martian year for the orbiter spacecraft has already been extended by another Martian year.

During the extended mission, priority will be given to fulfill the remaining goals of the nominal mission (e.g., gravity measurements and seasonal coverage), to catch up with delayed MARSIS measurements during the nominal mission, to complete global coverage of high-resolution imaging and spectroscopy, as well as subsurface sounding with the radar, to observe atmospheric and variable phenomena, and to revisit areas where discoveries were made.

Also, an effort to enlarge the scope of existing cooperation will be made, in particular with respect to other missions to Mars (such as *MGS*, *MER*, *MRO*) and also missions to other planets carrying the same instruments as *Mars Express* (i.e., *Venus Express*).

2. Mission details

Launch date: 2 June 2003
Arrival at Mars: 25 December 2003
Mass of payload: 116 kg
Primary science instruments:
HRSC (High-Resolution Stereo Camera): high-resolution surface imaging.
ASPERA (Energetic Neutral Atoms Analyser): how the solar wind erodes the Martian atmosphere.
PFS (Planetary Fourier Spectrometer): study of the atmospheric composition and circulation.
OMEGA (visible and infrared Mineralogical Mapping Spectrometer): determination of the surface composition and evolution processes.
MARSIS (sub-surface Sounding Radar Altimeter): search for water in the subsurface.
MaRS (Radio Science experiment): sounding of the internal structure, atmosphere and environment.
SPICAM (ultraviolet and infrared Mars Atmospheric Spectrometer): determination of the composition of the atmosphere of Mars.
Lander (*Beagle 2*): geochemistry and exobiology (lost).

3. In the literature

A detailed description of the Mars Express scientific payload has been published in book form; see Chicarro (2004).

The Mars Express mission has resulted in the publication of several hundred papers and abstracts to date, far too many to cite here. A large number of these results can be found in issues of the journals *Icarus, Journal of Geophysical Research: Planets*, and *Planetary and Space Science* for the years 2005 and 2006.

Reference

A. Chicarro (ed.) 2004, *Mars Express: the Scientific Payload*, ESA SP-1240

Highlights of Astronomy, Volume 14
IAU XXVI General Assembly, 14-25 August 2006
Karel A. van der Hucht, ed.

doi:10.1017/S1743921307010915

Scientific results of the *Mars Exploration Rovers*, *Spirit* and *Opportunity*

W. Bruce Banerdt

Jet Propulsion Laboratory, California Institute of Technology, Pasadena, CA 91109-8099, USA
email: bruce.banerdt@jpl.nasa.gov

Abstract. Results of the *Mars Exploration Rover* mission to Mars are summarized.

Keywords. planets, space vehicles

1. Results

NASA's *Mars Exploration Rover* project launched two robotic geologists, *Spirit* and *Opportunity*, toward Mars in June and July of 2003, reaching Mars the following January. The science objectives for this mission are focused on delineating the geologic history for two locations on Mars, with an emphasis on the history of water. Although they were designed for a 90-day mission, both rovers have lasted more than two years on the surface and each has covered more than four miles while investigating Martian geology.

Spirit was targeted to Gusev Crater, a 300 km diameter impact basin that was suspected to be the site of an ancient lake. Initial investigations of the plains in the vicinity of the landing site found no evidence of such a lake, but were instead consistent with unaltered (by water) basaltic plains. But after a 3 km trek to an adjacent range of hills it found a quite different situation, with abundant chemical and morphological evidence for a complex geological history.

Opportunity has been exploring Meridiani Planum, which was known from orbital data to contain the mineral hematite, which generally forms in the presence of water. The rocks exposed in Meridiani are highly chemically altered, and appear to have been exposed to significant amounts of water. By descending into the 130 m diameter Endurance Crater, *Opportunity* was able to analyze a 10 m vertical section of this rock unit, which showed significant gradations in chemistry and morphology.

2. Mission details

Launch Dates: 10 June 2003 (*Spirit*), and 7 July 2003 (*Opportunity*)
Arrival at Mars: 4 January 2005 (*Spirit*), and 25 January 2005 (*Opportunity*)
Mass of each rover: 180 kg
Primary science instruments:
PANCAM (Panoramic Camera): for determining the mineralogy, texture, and structure of the local terrain.
MINI-TES (Miniature Thermal Emission Spectrometer): for identifying promising rocks and soils for closer examination and for determining the processes that formed Martian rocks. The instrument also looks skyward to provide temperature profiles of the Martian atmosphere.
MB (Mössbauer Spectrometer): for close-up investigations of the mineralogy of iron-bearing rocks and soils.

APXS (ALPHA Particle X-ray Spectrometer): for close-up analysis of the abundances of elements that make up rocks and soils.
Magnets: for collecting magnetic dust particles. The Mössbauer Spectrometer and the Alpha Particle X-ray Spectrometer analyze the particles collected and help determine the ratio of magnetic particles to non-magnetic particles. They also analyze the composition of magnetic minerals in airborne dust and rocks that have been ground by the Rock Abrasion Tool.
MI (Microscopic Imager): for obtaining close-up, high-resolution images of rocks and soils.
RAT (Rock Abrasion Tool): for removing dusty and weathered rock surfaces and exposing fresh material for examination by instruments onboard.

3. In the literature

The scientific results of the *Mars Exploration Rovers* have already been published in many hundreds of papers and abstracts, including special issues of *Science*, on 3 December 2004, and *The Journal of Geophysical Research (Planets)* in February 2006.

A recent overview of the *Spirit Mars Exploration Rover* mission to Gusev Crater, from the landing site to Backstay Rock in the Columbia Hills, can be found at Arvidson *et al.* (2006). A discussion of two years at Meridiani Planum, the results from the *Opportunity Rover* is summarized by Squyres *et al.* (2006).

References

Arvidson, R. E., Squyres, S. W., Anderson, R. C., Bell, J. F., Blaney, D., Brückner, J., Cabrol, N. A., Calvin, W. M., Carr, M. H., Christensen, P. R., Clark, B. C., Crumpler, L., Des Marais, D. J., de Souza, P. A., d'Uston, C., Economou, T., Farmer, J., Farrand, W. H., Folkner, W., Golombek, M., Gorevan, S., Grant, J. A., Greeley, R., Grotzinger, J., Guinness, E., Hahn, B. C., Haskin, L., Herkenhoff, K. E., Hurowitz, J. A., Hviid, S., Johnson, J. R., Klingelhöfer, G., Knoll, A. H., Landis, G., Leff, C., Lemmon, M., Li, R., Madsen, M. B., Malin, M. C., McLennan, S. M., McSween, H. Y., Ming, D. W., Moersch, J., Morris, R. V., Parker, T., Rice, J. W., Richter, L., Rieder, R., Rodionov, D. S., Schröder, C., Sims, M., Smith, M., Smith, P., Soderblom, L. A., Sullivan, R., Thompson, S. D., Tosca, N. J., Wang, A., Wnke, H., Ward, J., Wdowiak, T., Wolff, M., & Yen, A.S. 2006, *Journal Geophys. Res.*, 111, CiteID E02S01

Squyres, S. W., Knoll, A. H., Arvidson, R. E., Clark, B. C., Grotzinger, J. P., Jolliff, B. L., McLennan, S. M., Tosca, N., Bell, J. F., Calvin, W. M., Farrand, W. H., Glotch, T. D., Golombek, M. P., Herkenhoff, K. E., Johnson, J. R., Klingelhöfer, G., McSween, H. Y., & Yen, A. S. 2005, *Science*, 313, 1403

Highlights of Astronomy, Volume 14
IAU XXVI General Assembly, 14-25 August 2006
Karel A. van der Hucht, ed.

doi:10.1017/S1743921307010927

Joint Discussion 11
Pre-solar grains as astrophysical tools

Anja C. Andersen[1] and John C. Lattanzio[2] (eds.)

[1]Dark Cosmology Centre, Niels Bohr Institute, University of Copenhagen,
Juliane Maries Vej 30, DK-2100 Copenhagen, Denmark
email: anja@astro.ku.dk

[2]School of Mathematical Sciences, Monash University,
PO Box 28M, Clayton Campus, Victoria 3800, Australia
email: john.lattanzio@sci.monash.edu.au

Preface

With the first discovery of surviving pre-solar minerals in primitive meteorites in 1987 a new kind of astronomy emerged, based on the study of stellar condensates with all the detailed methods available to modern analytical laboratories. The pre-solar origin of the grains is indicated by considerable isotopic ratio variations compared with Solar System materials, characteristic of nuclear processes in different types of stars.

The astrophysical implication comes from the laboratory studies of such pre-solar grains from meteorites. The studies have provided new and often unique information on galactic chemical evolution, on nucleosynthesis in a variety of stellar objects, on grain formation in stellar outflows, and on the survival of grains in the instellar medium, in the solar nebular, and in meteorite parent bodies.

The full scientific exploitation of pre-solar grains is only made possible by the development of advanced instrumentation for chemical, isotopic, and mineralogical microanalysis of very small samples. Unique scientific information derives primarily from the high precision (in some cases $< 1\%$) of the measured isotopic ratios of various elements in single stardust grains. Known pre-solar phases include diamond, SiC, graphite, Si_3N_4, Al_2O_3, $MgAl_2O_4$, $CaAl_{12}O_{19}$, TiO_2, $Mg(Cr,Al)_2O_4$, and most recently, silicates. Subgrains of refractory carbides (e.g., TiC), and Fe-Ni metal have also been observed within individual pre-solar graphite grains. These grain types represent a wide range of thermal and chemical resistance. Many new breakthroughs are expected in the near future as it is now technically possible to extend isotopic laboratory studies to individual particles down to scales of < 100 nm.

The papers presented here illustrate that the laboratory studies of pre-solar grains provide crucial contributions to several important areas of astrophysics. For example, studying isotopic compositions of grains that condensed from the ejecta of dying stars provide essential boundary conditions for numerical models of stellar nucleosynthesis. The grains disclose information about nucleosynthesis sites of different elements and the relative abundances of different stellar inputs to the Galaxy (e.g., the supernova II/Ia ratio), as well as constraining the degree of mixing of material from diverse stars in the interstellar medium and the types of minerals produced by stars of different metallicity. The grains also probe the conditions of the solar nebula accretion disk during the earliest stages of Solar System formation.

The results from isotopic studies are currently those that bear strongest on other fields of astrophysics. For one, they allow us to pinpoint the grains' stellar sources among which Red Giants stars play a prominent role. In addition, given the precision of the laboratory isotopic analyses, which far exceeds whatever can be hoped to be achieved in remote analyses, they have strong implications for, e.g., the need for an extra mixing process (cool bottom processing) in Red Giants and provide detailed constraints on the operation of the s-process in Asymptotic Giant Branch (AGB) stars. Indeed, they are the only way to provide isotopic abundances for many species, especially for trace elements. A non-standard neutron capture process (neutron burst) may be implied by the small part of the silicon carbide grains which originate from supernovae. The progress in analytical techniques promises more important and exiting results in the near future.

May the next pages be a compass to guide your journey into the facinating discoveries that have come from the studies of pre-solar grains up to now.

Scientific Organizing Committee

Anja C. Andersen (Denmark,co-chair), W. David Arnett (USA), Martin Asplund (Australia), Suchitra C. Balachandran (USA), John P. Bradley (USA), Mounib El Eid (Lebanon), Roberto Gallino (Italia), Sunetra Giridhar (India), John C. Lattanzio (Australia, co-chair), Jacobus Th. van Loon (UK), Nami Mowlavi (Suisse), Antonella Natta (Italy), Takashi Yoshida (Japan), and Ernst Zinner (USA).

Anja C. Andersen and John Lattanzio, co-chairs SOC,
Copenhagen and Monash, November 1, 2006

Highlights of Astronomy, Volume 14
IAU XXVI General Assembly, 14-25 August 2006
Karel A. van der Hucht, ed.

doi:10.1017/S1743921307010939

Pre-solar grains in meteorites and interplanetary dust: an overview

Ulrich Ott and Peter Hoppe

Max-Planck-Institut für Chemie, Joh.-J.-Becherweg 27, D-55128 Mainz, Germany
email: ott,hoppe@mpch-mainz.mpg.de

Abstract. Small amounts of pre-solar grains have survived in the matrices of primitive meteorites and interplanetary dust particles. Their detailed study in the laboratory with modern analytical tools provides highly accurate and detailed information with regard to stellar nucleosynthesis and evolution, grain formation in stellar atmospheres, and Galactic Chemical Evolution. Their survival puts constraints on conditions they were exposed to in the interstellar medium and in the Early Solar System.

Keywords. meteors, meteoroids, stars: AGB and post-AGB, supernovae: general, dust, extinction, ISM: evolution, ISM: general

---

1. Introduction

Almost twenty years have passed since, as a result of the search for host phases of isotopically unusual noble gases, the first discovery in 1987 of surviving pre-solar minerals (diamond and silicon carbide) in primitive meteorites. These were followed by others (graphite, refractory oxides, silicon nitride, and finally silicates) in the years since. Pre-solar grains occur in even higher abundance than in meteorites in interplanetary dust particles (IDPs). The result is a kind of 'new astronomy' based on the study of pre-solar condensates with all the methods available in modern analytical laboratories.

In the 'classical' approach, pioneered in the search for the noble gas carriers diamond, SiC and graphite, pre-solar grains are isolated by dissolving most of the meteorites (consisting mostly of silicate minerals) using strong acids, followed by further chemical and physical separation methods. For overviews at the early stage when only these types were known see reviews by Anders & Zinner (1993) and Ott (1993).

More up-to-date reviews (although only barely including the only recently found pre-solar silicates) are by Zinner (1998), Hoppe & Zinner (2000), Nittler (2003), and Zinner (2004). Refractory oxides and silicon nitride were found in conjunction with the noble gas carrying minerals because of their similar chemical inertness which allowed them to survive the extraction procedure of the noble gas carriers. Identification of pre-solar silicates among the sea of 'normal silicates', however, became possible only with the advent of a new generation of analytical instrumentation, the NanoSIMS (e.g., Hoppe *et al.* 2004) that allowed the imaging search for isotopically anomalous phases in situ, i.e. without the need for chemical / physical extraction.

Central to the identification of pre-solar minerals is the determination of isotopic compositions, which as a rule strongly deviate from the normal (Solar System) composition. Isotopic composition is, as a matter of fact, the main (and only safe) criterion by which they can be identified; hence all our known 'pre-solar grains' are 'circum-stellar condensates' that carry the isotopic signatures of nucleosynthesis processes going on in their parent stars (Fig. 1).

Because pre-solar grains come from different stellar sources, information on individual stars can only be obtained by the study of single grains. This is possible by SIMS (Secondary Ion Mass Spectrometry) for the light to intermediate-mass elements, RIMS (Resonance Ionization Mass Spectrometry) for the heavy elements, and laser heating and gas mass spectrometry for He and Ne.

2. Overview

An overview of the currently known inventory of circumstellar grains in meteorites is presented in Table 1. Abundances of silicates are definitely, those of the other pre-solar grains most likely, higher in IDPs. Some comments follow below, while several specific cases are discussed in detail in other contributions to this volume.

Silicon carbide. All SiC grains in primitive meteorites are of pre-solar origin, and they are the best characterized. This has been helped by their comparably high contents of minor and trace elements. Characteristic for most grains are enhanced ^{13}C, ^{14}N, former presence of ^{26}Al as indicated by overabundances of its daughter ^{26}Mg, neon that is almost pure ^{22}Ne [Ne-E(H)] and heavy elements showing the characteristic isotopic signatures of the s-process. These 'mainstream grains' quite obviously are condensates out of the winds of AGB stars (see contribution by Lugaro & Höfner, this volume). Only a percent or so have a clearly different origin tied to supernovae (the 'X-grains'). They are characterized by high ^{12}C, ^{28}Si, very high former abundances of ^{26}Al as well as ^{44}Ti and not fully understood signature in the heavy trace elements (see contribution by Amari & Lodders, this volume).

Oxides and silicates. Besides diamonds (see below) silicates - not unexpectedly - are the most abundant of the pre-solar grains that have been found. The most characteristic features of oxides and silicates are contained in the oxygen isotopic composition that can

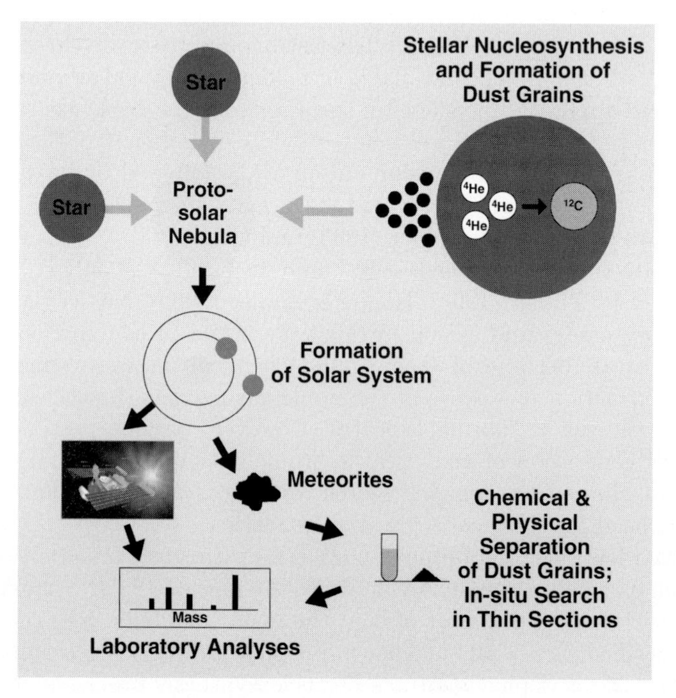

Figure 1. Path of pre-solar grains from their stellar sources to the laboratory.

Table 1. Overview of current knowledge on circum-stellar condensate grains in meteorites.

Mineral	Size [μm] abund. [ppm][1]	Isotopic Signatures	Stellar Sources	Contri-bution[2]
diamond	0.0026 1500	Kr-H, Xe-HL, Te-H	supernovae	?
silicon carbide	0.1 – 10 30	enhanced ^{13}C, ^{14}N, ^{22}Ne, s-process elem. low ^{12}C/^{13}C, often enh. ^{15}N enhanced ^{12}C, ^{15}N, ^{28}Si; extinct ^{26}Al, ^{44}Ti low ^{12}C/^{13}C, low ^{14}N/^{15}N	AGB stars J-type C-stars (?) Supernovae novae	> 90 % < 5 % 1 % 0.1 %
graphite	0.1 – 10 10	enh. ^{12}C, ^{15}N, ^{28}Si; extinct ^{26}Al, ^{41}Ca, ^{44}Ti s-process elements low ^{12}C/^{13}C low ^{12}C/^{13}C; Ne-E(L)	SN (WR?) AGB stars J-type C-stars (?) novae	< 80 % > 10 % < 10 % 2 %
corundum/ spinel/ hibonite	0.1 – 5 50	enhanced ^{17}O, moderately depl. ^{18}O enhanced ^{17}O, strongly depl. ^{18}O enhanced ^{16}O	RGB / AGB AGB stars supernovae	> 70 % 20 % 1 %
silicates	0.1 – 1 140	similar to oxides above		
silicon nitride	1 0.002	enhanced ^{12}C, ^{15}N, ^{28}Si; extinct ^{26}Al	supernovae	100 %

Notes:
[1]For the abund. (in wt. ppm) the reported maximum values from different meteorites are given.
[2]Note uncertainty about actual fraction of diamonds that are pre-solar and for fraction of graphite attributed to SN and AGB stars (see discussion in text).

be used for assigning each grain to one of four groups (Nittler *et al.* 1997). Grains without evidence for the former presence of ^{26}Al are assumed to originate from RGB stars, those with ^{26}Al from AGB stars.

Graphite and silicon nitride. The characteristics of most grains (see Tab. 1) have traditionally led to assume a SN origin (e.g., Zinner 1998; Hoppe & Zinner 2000). However, this percentage may have been overestimated as most high-density graphite grains, although showing enhanced ^{12}C abundances, contain s-process signatures and so are more likely to originate from AGB stars (Croat *et al.* 2005). The rare Si_3N_4 grains show isotopic signatures similar to SiC-X and SN graphite grains and derive probably from supernovae as well.

Nanodiamonds. In several ways these are the most enigmatic. Although discovered first, their pre-solar credentials are based solely on trace elements Te and noble gases that they carry. They are too small for individual analysis each consisting of some 1000 carbon atoms only on average and the carbon isotopic composition of 'bulk samples' (i.e., many diamond grains) is within the range of Solar System materials. What fraction of the diamonds is truly pre-solar is an as yet open question.

3. Implications

Isotopic structures and nucleosynthesis. As isotopic structures are the key for establishing the grains as pre-solar, isotope studies are at the core of investigations that have been performed. Results from isotopic studies in turn are also those that bear strongest on astrophysics. For one, they allow us to pinpoint the grains' stellar sources. In addition, given the precision of the laboratory isotopic analyses, which far exceed whatever can be hoped for in remote analyses, they allow conclusions with regard to details of nucleosynthesis and mixing in the parent stars as well as Galactic Chemical Evolution. They have borne strong on, e.g. the need for an extra mixing process (cool bottom processing) in

Red Giants and provide detailed constraints on the operation of the s-process in AGB stars (e.g., Busso *et al.* 1999). A non-standard neutron capture process ('neutron burst') may be implied by the SiC-X grains from supernovae (Meyer *et al.* 2000) and possibly the trace Xe in the diamonds (e.g., Ott 2002). The progress in analytical techniques promises more important results in the near future.

Grain formation. Chemical composition, sizes, and microstructures of grains constrain conditions during condensation in stellar winds and supernova ejecta. Condensation of SiC apparently occurred under close to the equilibrium conditions (e.g., Lodders & Fegley 1998). Additional constraints are imposed by trace element contents both on average (Yin *et al.* 2006) as well as in individual grains (Amari *et al.* 1995). An important relevant observation is the occurrence of subgrains of primarily TiC within graphite (Croat *et al.* 2005).

The lifecycle of pre-solar grains (and maybe interstellar grains in general). Interstellar grains are expected to be processed and eventually destroyed by sputtering or astration (e.g., Draine 2003), with an as yet unidentified process needed to account for the balance between formation and destruction. Pre-solar grains preserved in meteorites carry, in principle, a record of conditions they have been exposed to, which, however, is difficult to read. Determining an absolute age using long-lived radioisotopes is virtually ruled out by the fact that these systems use decay of rare constituents (e.g., K, Sr, Re, U) decaying into other rare elements with uncertain non-radiogenic composition. However, appearance and microstructures of pristine (i.e. not chemically processed) SiC show little evidence for being processed, indicating either that they were surprisingly young when entering the forming Solar System or that they were protected (Bernatowicz *et al.* 2003); a similar situation is indicated by the lack of detectable spallation Xe produced by exposure to cosmic rays during residence in the ISM (Ott *et al.* 2005). The distribution, finally, among various types of meteorites, provides a measure of processing in the early Solar System.

References

Amari, S., Hoppe, P., Zinner, E., & Lewis R. S. 1995, *Meteoritics*, 30, 490
Anders, E., & Zinner, E. 1993, *Meteoritics*, 28, 490
Bernatowicz, T. J., Messenger, S., Pravdivtseva, O., Swan, P., & Walker, R. M. 2003, *Geochim. Cosmochim. Acta*, 67, 4679
Busso, M., Gallino, R., & Wasserburg, G. J. 1999, *ARAA*, 37, 239
Croat, T. K., Stadermann, F. J., & Bernatowicz, T. J. 2005, *ApJ*, 631, 976
Draine, B. T. 2003, *ARAA*, 41, 241
Hoppe, P., & Zinner, E. 2000, *J. Geophys. Res.*, A105, 10371
Hoppe, P., Ott, U., & Lugmair, G. W. 2004, *New Astron. Revs*, 48, 171
Lodders, K., & Fegley, B. 1998, *Meteorit. Planet. Sci.*, 33, 871
Meyer, B. S., Clayton, D. D., & The, L.-S. 2000, *ApJ* (Letters), 540, L49
Nittler, L. R. 2003, *Earth Planet. Sci. Lett.*, 209, 259
Nittler, L. R., Alexander, C. M. O'. D., Gao, X., Walker, R. M., & Zinner, E. 1997, *ApJ*, 483, 475
Ott, U. 1993, *Nature*, 364, 25
Ott, U. 2002, *New Astron. Revs* 46, 513
Ott, U., Altmaier, M., Herpers, U., Kuhnhenn, J., Merchel, S., Michel, R., & Mohapatra, R. K. 2005, *Meteorit. Planet. Sci.*, 40, 1635
Yin, Q.-Z., Lee, C.-T. A., & Ott, U. 2006, *ApJ*, 647, 676
Zinner, E. 1998, *Ann. Rev. Earth Planet. Sci.*, 26, 147
Zinner, E. 2004, in: K. K. Turekian, H. D. Holland & A. M. Davis (eds.), *Treatise in Geochemistry 1* (Oxford and San Diego: Elsevier), p. 17

Highlights of Astronomy, Volume 14
IAU XXVI General Assembly, 14-25 August 2006
Karel A. van der Hucht, ed.

doi:10.1017/S1743921307010940

What can pre-solar grains tell us about AGB stars?

Maria A. Lugaro[1] and Susanne Höfner[2]

[1]Sterrenkundig Instituut, University of Utrecht,
Postbus 80000, NL-3508TA, Utrecht, the Netherlands
email: m.lugaro@phys.uu.nl

[2]Dept. of Astronomy & Space Physics, Uppsala University,
Box 515, SE-75120 Uppsala, Sweden
email: hoefner@astro.uu.se

Abstract. The vast majority of pre-solar grains recovered to date show the signature of an origin in asymptotic giant branch (AGB) stars. In AGB stars, the abundances of elements lighter than silicon and heavier than iron are largely affected by proton- and neutron-capture processes, respectively, while the compositions of the elements in between also carry the signature of the initial composition of the star. Dust is produced and observed around AGB stars and the strong mass loss experienced by these stars is believed to be driven by radiation pressure on dust grains. We briefly review the main developments that have occurred in the past few years in the study of AGB stars in relation to dust and pre-solar grains. From the nucleosynthesis point of view these include: more stringent constraints on the main neutron source nucleus, ^{13}C, for the *slow* neutron capture process (the *s* process); the possibility of pre-solar grains coming from massive AGB stars; and the unique opportunity to infer the 'isotopic' evolution of the Galaxy by combining pre-solar grain data and AGB model predictions. Concerning the formation of grains in AGB stars, considerable progress has been achieved in modelling. In particular, self-consistent models for atmospheres and winds of C-stars have reached a level of sophistication which allows direct quantitative comparison with observations. In the case of stars with $C/O < 1$, however, recent work points to serious problems with the dust-driven wind scenario. A current trend in atmosphere and wind modelling is to investigate the possible effects of inhomogenieties (e.g., due to giant convection cells) with 2D/3D models.

Keywords. nuclear reactions, nucleosynthesis, abundances, stars: AGB and post-AGB, stars: mass loss, stars: winds, outflows

---

1. Nucleosynthesis in AGB stars and the isotopic compositions of pre-solar grains

Toward the end of their lives stars of masses roughly less than $7\,M_\odot$ go through the asymptotic giant branch (AGB) phase, during which the He- and the H-burning shells are activated alternately on top of a degenerate C-O core and below a large convective envelope. Material in the tiny region between the two shells (intershell) is processed by H-burning and partial He-burning, as well as suffering neutron-capture processes, thus resulting in peculiar nucleosynthesis including the production of ^{12}C, ^{19}F, ^{22}Ne, ^{26}Al as well as of elements heavier than iron such as Zr, Ba and Pb by *slow* neutron captures (the *s* process). This material is mixed to the surface of the star by recurrent episodes of mixing collectively known as the third dredge-up and is thus incorporated into dust grains forming around the star, which are now recovered from primitive meteorites as pre-solar grains.

One of the most important evidences of the origin of pre-solar mainstream SiC grains in AGB stars is the signature of *s*-process nucleosynthesis in all heavy elements analysed to date in these grains, both in bulk, i.e. measurements on a large number of grains, and individually. The main neutron source for the *s* process in the intershell of AGB stars is the $^{13}C(\alpha, n)^{16}O$ reaction. The production of ^{13}C is possible via the ^{12}C+p reaction, if protons from the envelope diffuse into the ^{12}C-rich intershell. However, it is still much debated by which mechanism the proton diffusion occurs and, hence, the amount of ^{13}C nuclei is treated as a free parameter in the models. A detailed analysis of the composition of Sr, Zr, Mo, and Ba in single SiC grains and in AGB models was presented by Lugaro *et al.* (2003). It was shown that a large spread (factor of 24) in the amount of ^{13}C was needed to cover the SiC data, thus confirming the results obtained by Busso *et al.* (2001) by comparing single-star models to spectroscopic observations of the heavy element distribution in AGB stars at different metallicities.

However, one problem with the SiC data is that it is difficult to exclude or evaluate *a priori* the level of contamination of terrestrial material, which would shift isotopic ratios toward solar composition. This is especially true for elements like Sr and Ba, which are more volatile than Zr and Mo, and thus are present in the grains in lower amounts. Barzyk *et al.* (2006) recently measured the composition of more than one heavy element in single SiC grains, thus making it possible to identify contaminated grains. The uncontaminated grain data could be matched by a much smaller spread (a factor of two) than that used by Lugaro *et al.* (2003), confirming recent results obtained by Bonačić Marinović *et al.* (2007) by comparing stellar population models including the *s* process to spectroscopic observations.

From the comparison between grain data for heavy elements and *s*-process models, it is derived that most of the pre-solar grains from AGB stars must have formed around stars of low mass ($M < 3\,M_{\odot}$). However, a unique pre-solar spinel grain, OC2, has been identified to be the first pre-solar grain to possibly show the signature of nucleosynthesis in intermediate-mass (IM) AGB stars ($4\,M_{\odot} < M < 7\,M_{\odot}$), suffering H-burning at the base of the convective envlope (hot bottom burning, HBB). The main hint to such an origin comes from the large observed excesses in the heavy Mg isotopes. Lugaro *et al.* (2007) show that models of IM-AGB stars can reproduce the Mg and O composition of grain OC2 and, within this framework, predictions can be made for the value of the ^{16}O+p and ^{17}O+p reaction rates. Grain OC2, and similar grains that may be recovered in the future, has opened the possibility to derive constraints also on massive AGB models from the study of pre-solar grains.

The compositions of elements such as Si, Ti, Ca and Fe in pre-solar grains from AGB stars are determined by the initial composition of the parent star and by the nucleosynthesis occurring within it. By combining a large amount of Si data from SiC grains of mainstream, Y and Z populations to a variety of detailed models of the nucleosynthesis in AGB stars of different masses and metallicities down to $\sim 1/10\,Z_{\odot}$, Zinner *et al.* (2006) were able to infer the Galactic chemical evolution of the Si isotopes. The evolution based on the grain data predicts much higher ^{29}Si and ^{30}Si abundances at low metallicities than calculated by the models of Timmes & Clayton (1996). This indicates perhaps a low-metallicity source for these isotopes not considered so far, or some other problems with the Galactic chemical evolution model. This kind of work can be extended to other intermediate-mass elements, making pre-solar grains an ideal tool to study the evolution of isotopic abundances in the Galaxy.

2. The role of grains in atmospheres and winds of AGB stars

For a long time, dust has been treated as just another opacity source in models of atmospheres and winds, in order to provide radiation pressure for driving the wind, or fitting observed spectra. Detailed studies of the actual grain formation process in the atmospheres of AGB stars are a rather recent development, especially in the context of self-consistent, time-dependent dynamical models. Such models have become more and more sophisticated during the past decade, and, in some cases, they have reached a level where a direct quantitative comparison with observations becomes possible (see recent reviews, e.g., by Woitke (2003) and Höfner (2005), for an overview of historical developments and a discussion of modelling methods).

The latest generation of models for atmospheres and winds of C-rich AGB stars by Höfner *et al.* (2003), combining a frequency-dependent treatment of radiative transfer with time-dependent hydrodynamics and a detailed description of dust formation, compares nicely with various types of observations, such as low-resolution NIR spectra (Gautschy-Loidl *et al.* 2004) or profiles of CO vibration-rotation lines (Nowotny *et al.* 2005ab). The combination of dynamics, taking into account effects of pulsation and shock waves, and frequency-dependent radiative transfer, accounting for molecules and dust, turns out to be crucial for a reasonably realistic description of AGB atmospheres, and thus for modelling the conditions in the zone where the dust grains are formed. With these non-grey dynamical models it is possible for the first time to simultaneously reproduce the time-dependent behavior of fundamental, first and second overtone vibration-rotation lines of CO, features originating in the outflow, dust formation region, and pulsating atmosphere, respectively, probing the dynamics from the photosphere out into the wind.

Such detailed models, however, rely on fundamental physical and chemical data for dust grain materials measured in the laboratory, and in particular optical properties. Andersen *et al.* (2003) studied the effects of microphysical grain properties in the context of detailed dynamical models for atmospheres and winds of C-stars as described above, using data obtained by Jäger *et al.* (1998). It is commonly assumed that grains in C-stars mostly consists of amorphous carbon, a term that actually covers a variety of materials with varying ratios of sp^2 to sp^3 hybridization of the C atoms. The microphysical structure of the grain material, however, will lead to different optical properties in the near IR, with sp^3-rich material showing a much steeper dependence on wavelength than sp^2-rich material. In non-grey wind models, these differences in the optical properties of the grains influence both the flux-integrated opacity relevant for driving the wind and the grain temperature, and therefore the actual dust formation process.

While recent advances in modelling, in particular the inclusion of frequency-dependent radiative transfer, have lead to a better agreement between models and observations of C-rich AGB stars, the opposite seems to be true for the O-rich case. Jeong *et al.* (2003) presented wind models for M-type stars, combining a detailed description of dust formation with time-dependent dynamics and grey radiative transfer. The combinations of stellar parameters used in this study may be considered as somewhat on the extreme side (high luminosities and low effective temperatures), but their models lead to reasonable wind velocities and mass loss rates. Recent work by Woitke (2006b) and Höfner (2006), including frequency-dependent radiative transfer for gas and dust, however, points to a serious problem with the dust-driven wind scenario. The higher the Fe-content of grains, the steeper the slope of the opacity as a function of wavelength, and thus the higher the radiative equilibrium temperature of the grains. In practice, this means that the Fe-content of silicate grains will be very low, in order to allow for condensation reasonably close to the star. This, in turn, leads to a low flux-integrated opacity of the grains which is not sufficient to drive a stellar wind for typical stellar parameters. At present it seems that alternative wind scenarios for M-type stars should possibly be considered.

Another new development is 2D/3D models of atmospheres and circumstellar envelopes around AGB stars. The computational effort behind such models is considerable, and a number of simplifications have to be introduced in the physics, compared to the detailed spherical models discussed above. In the light of recent interferometric observations which indicate deviations from spherical symmetry, however, it seems necessary to spend some effort testing possible causes and consequences for dust formation and winds. Woitke (2006a) presented 2D (axisymmetric) dust-driven wind models, including time-dependent dust formation and grey radiative transfer, and studied how instabilities in the dust formation process (originally found in earlier spherical models) create intricate patterns in the circumstellar envelope. These models, however, exclude the central star and its pulsation. Freytag & Höfner (2003, 2006), on the other hand, developed complementary 3D RHD 'star-in-a-box' models (including time-dependent dust formation) to investigate the effects of giant convection cells and of the resulting shock waves in the atmosphere on dust formation. The atmospheric patterns created by convective motions are found to be reflected in the circumstellar dust distribution, due to the strong sensitivity of grain formation to temperatures and gas densities.

Acknowledgements

M. L. is supported by NWO under a VENI grant. S. H. acknowledges support from the Swedish Research Council (*Vetenskapsrådet*).

References

Andersen, A. C., Höfner, S., & Gautschy-Loidl, R. 2003, *A&A*, 400, 981

Barzyk, J. G., Savina M. R., Davis, A. M., Gallino, R., Pellin, M. J., Lewis, R. S., Amari, S., & Clayton, R. N. 2006, *Lunar and Planetary Science* XXXVII, abstract 1999

Bonačić Marinović, A., Izzard, R. G., Lugaro, M., & Pols, O. R. 2007, *A&A*, 469, 1013

Busso, M., Gallino, R., Lambert, D. L., Travaglio, C., & Smith, V. V. 2001, *ApJ*, 557, 802

Freytag, B., & Höfner, S. 2003, *AN*, 324, Suppl. Issue 3, 173

Freytag, B., & Höfner, S. 2006, *A&A*, to be submitted

Gautschy-Loidl, R., Höfner, S., Jørgensen, U. G., & Hron, J. 2004, *A&A*, 422, 289

Höfner, S. 2005, in: F. Favata *et al.* (eds.), *Proc. of 13th Cambridge Workshop on Cool Stars, Stellar Systems and the Sun*, ESA SP-560, p. 335

Höfner, S. 2006, in: F. Kerschbaum, C. Charbonnel & R. Wing (eds.), *Why Galaxies Care About AGB Stars, ASP-CS*, 378, in press

Höfner, S., Gautschy-Loidl, R., Aringer, B., & Jørgensen, U. G. 2003, *A&A*, 399, 589

Jäger, C., Mutschke, H., & Henning, Th. 1998, *A&A*, 332, 291

Jeong, K. S., Winters, J. M., Le Bertre, T., & Sedlmayr, E. 2003, *A&A*, 407, 191

Lugaro, M., Davis, A. M., Gallino, R., Pellin, M. J., Straniero, O., & Käppeler, F. 2003, *ApJ*, 593, 486

Lugaro, M., Karakas, A. I., Nittler, L. R., Alexander, C. M. O'. D., Hoppe, P., Iliadis, C., & Lattanzio, J. C. 2007, *A&A* 461, 657

Nowotny, W., Aringer, B., Höfner, S., Gautschy-Loidl, R., & Windsteig, W. 2005a, *A&A*, 437, 273

Nowotny, W., Lebzelter, T., Hron, J., & Höfner, S. 2005b, *A&A*, 437, 285

Timmes, F. X., & Clayton, D. D. 1996, *ApJ*, 472, 723

Woitke, P. 2003, in: N. E. Piskunov, W. W. Weiss & D. F. Gray (eds.), *Modelling of Stellar Atmospheres*, IAU Symposium No. 210 (San Francisco: ASP), p. 387

Woitke, P. 2006a, *A&A* 452, 537

Woitke, P. 2006b, in: F. Kerschbaum, C. Charbonnel & R. Wing (eds.), *Why Galaxies Care About AGB Stars, ASP-CS*, 378, in press

Zinner, E., Nittler, L. R., Gallino, R., Karakas, A. I., Lugaro, M., Straniero, O., & Lattanzio, J. C. 2006, *ApJ* 650, 350

Highlights of Astronomy, Volume 14
IAU XXVI General Assembly, 14-25 August 2006
Karel A. van der Hucht, ed.

doi:10.1017/S1743921307010952

Pre-solar grains from supernovae and novae

Sachiko Amari[1] and Katharina Lodders[2]

[1]Laboratory for Space Sciences & the Physics Department and [2]Department of Earth and Planetary Sciences, Washington University, St. Louis, MO 63130, USA
email: sa@wuphys.wustl.edu, lodders@wustl.edu

Abstract. Pre-solar grains from supernova ejecta – silicon carbide of type X, Si_3N_4 and low-density graphite – are characterized by Si isotopic anomalies (mainly ^{28}Si excesses), low $^{14}N/^{15}N$, high $^{26}Al/^{27}Al$ ratios, and occasionally by excesses in ^{44}Ca (from ^{44}Ti decay). Overall isotopic features of these SiC and graphite grains can be explained by mixing of inner Si-rich zones and the outer C-and He-rich zones, but supernova models require fine tuning to account for $^{14}N/^{15}N$ and $^{29}Si/^{28}Si$ ratios of the grains. Isotopic ratios of Zr, Mo and Ba in SiC X grains may be explained by a neutron burst model. Some of the pre-solar nanodiamonds require a supernova origin to explain measured xenon isotopic ratios. Only a few nova grain candidates, with low $^{12}C/^{13}C$, $^{14}N/^{15}N$, and high $^{26}Al/^{27}Al$ ratios, have been identified.

Keywords. meteors, meteoroids, stars: supernovae: general, stars: novae, cataclysmic variables

---

1. Introduction

Since the recovery of the first pre-solar grains in 1987 a considerable amount of data has been accumulated in this new field of astronomy. Pre-solar minerals currently known include nanodiamond (Lewis *et al.* 1987), SiC (Bernatowicz *et al.* 1987; Tang & Anders 1988), graphite (Amari *et al.* 1990), Al-bearing oxides (Hutcheon *et al.* 1994; Nittler *et al.* 1994), Si_3N_4 (Nittler *et al.* 1995), silicates (Messenger *et al.* 2003; Nguyen & Zinner 2004; Nagashima, Krot & Yurimoto 2004), and refractory inclusions inside graphite (Bernatowicz *et al.* 1991) and SiC grains (Bernatowicz, Amari & Lewis 1992). Isotopic analyses on individual grains by secondary ion mass spectrometry indicate that a small fraction of pre-solar grains originated in supernovae (e.g., Amari & Zinner 1997). There is a handful of grains that show characteristics of novae (Amari *et al.* 2001; Nittler & Hoppe 2005). This paper summarizes information on pre-solar grains of supernova and of probable nova origins. More information can be found in several review papers (Zinner 1998; Clayton & Nittler 2004; Lodders & Amari 2005) and references therein.

2. Pre-solar grains from type II supernovae

2.1. *Oxides, SiC type X, Si_3N_4 and low-density graphite*

The major dust forming elements (e.g., Al, Ca, Mg, Si) are mainly produced and ejected by core-collapse type II supernovae (SNe) and if they condense from supernova (SN) ejecta, there should be abundant grains from SNe among pre-solar grains. However, only a few oxide grains and one olivine aggregate of probable SN origin have been identified (Choi *et al.* 1998; Nittler *et al.* 1998; Messenger, Keller & Lauretta 2005). They show either ^{16}O or ^{18}O excesses (here and elsewhere, an excess is relative to 'normal' isotopic composition). It is odd that SN oxide grains are so rare because ^{16}O is the third most abundant isotope ejected from SNe and overall SN ejecta have $C/O < 1$, so there should be many more SN oxide grains. One explanation could be that SN oxide grains are too small ($\ll 0.1\,\mu$m) to survive for long in the ISM. Such small oxide grains are also unrecoverable from meteorites during chemical pre-solar grain separation procedures.

Strangely, most of the known pre-solar grains from SNe are dominantly carbonaceous: the type X SiC (Amari *et al.* 1992; Hoppe *et al.* 2000) and Si_3N_4 (Nittler *et al.* 1995), and the low-density graphite (Amari, Zinner & Lewis 1995; Travaglio *et al.* 1999) grains. However, only about 1 % of carbonaceous pre-solar grains originated from SNe.

The carbonaceous SN grains are characterized by ^{28}Si excesses (but a few graphite grains show ^{29}Si and/or ^{30}Si excesses), lower ^{14}N/^{15}N ratios than that of air (272), and typically isotopically light C, although their ^{12}C/^{13}C ratios range from 3.4 to 7200 (solar: 89). Many graphite grains show ^{18}O excesses (^{18}O/^{16}O up to 185×solar). Type II supernovae are the main producers of ^{28}Si so excesses of this isotope indicate a SN origin of these grains. Definite proof of their SN origin are excesses in ^{44}Ca from decay of ^{44}Ti ($T_{1/2} = 60$ a) which must have been incorporated when the SiC type X and low-density graphite grains formed (Hoppe *et al.* 1996; Nittler *et al.* 1996). Since ^{44}Ti is produced only in explosive nucleosynthesis, the initial presence of ^{44}Ti in the grains proves supernova origin of these grains. Supernova grains also once contained radioactive ^{26}Al and ^{41}Ca, which are traceable by excesses in their decay products ^{26}Mg and ^{41}K. The initial ^{26}Al/^{27}Al ratios range up to 0.6 in SN grains.

The measured isotopic compositions of the SiC and graphite grains require mixing of different compositional SN zones. For example, high ^{26}Al/^{27}Al ratios require contributions from the He-N zone (where the CNO cycle takes place), and high ^{18}O/^{16}O ratios in low-density graphite grains need contributions from the He-C zone (where the triple α reaction operates). The He-N and He-C zones have to be mixed to account for the observed range of ^{12}C/^{13}C ratios, and the innermost Si-rich zones must contribute e.g., ^{28}Si, and ^{44}Ti. Mixing must also achieve $C/O > 1$ for graphite and SiC condensation. Mixing models by Travaglio *et al.* (1999), using the SN compositions by Woosley & Weaver (1995), reproduce the observed ^{12}C/^{13}C, ^{18}O/^{16}O, and ^{30}Si/^{28}Si ratios, and the inferred ^{26}Al/^{27}Al, ^{41}Ca/^{40}Ca, and ^{44}Ti/^{48}Ti ratios if jets of material from the inner Si-rich zone penetrate the intermediate O-rich zones and mix with matter of the outer C-rich zones. However, the models give less ^{15}N and ^{29}Si than seen in the grains (Nittler *et al.* 1995; Travaglio *et al.* 1999; Hoppe *et al.* 2000), and improvements in stellar structure models and reassessments of reaction rates are necessary. For example, the ^{15}N yield significantly increases when stellar rotation is taken into account, and the ^{26}Mg(α,n)^{29}Si reaction, which is highly temperature dependent, is poorly determined for $T = (1\text{-}4) \times 10^9$ K (e.g., Travaglio *et al.* 1998).

The Zr, Mo, and Ba isotopes in SiC X grains were analyzed with resonant ionization mass spectrometry (Pellin *et al.* 1999; 2000; 2001). Four out of 6 grains have relative ^{95}Mo and ^{97}Mo excesses up to 1.8× solar. Such enrichments are not expected from either the *s*-process (which would enrich mass 96 and 98) nor the *r*-process (which would give the largest excess in mass 100). The excesses in ^{95}Mo and ^{97}Mo can be explained by a neutron burst - akin to a mini *r*-process – occurring in the He-rich zone of an exploding massive star (Meyer, Clayton & The 2000). Excesses in ^{96}Zr and ^{138}Ba in X grains can also be accounted for by this process.

2.2. *Diamond*

Diamond is the most abundant carbonaceous pre-solar mineral, but its origin(s) remain enigmatic. Diamonds are only measurable as aggregates of many nano-size particles that probably came from several different types of sources, including supernovae, as seen from the Xe isotopes. The anomalous Xe (Xe-HL) in diamonds is enriched in both *L*ight (124 and 126) and *H*eavy (134 and 136) isotopes (Lewis *et al.* 1987). These are *p*- and *r*- process only isotopes, respectively, believed to come from supernovae. It remains a puzzle why the excesses in the light and the heavy isotopes are always correlated because the *p*- and *r*-process are assumed to occur in different places within supernovae. Other heavy elements, e.g., Te and Pd also point to a supernova origin of diamonds (Richter, Ott & Begemann 1998; Maas *et al.* 2001). The SN origin of some of the pre-solar diamonds seems to be supported by the mid-IR spectra of SN 1987A, which show a broad feature at 3.40 and 3.53 μm (Meikle *et al.* 1989), consistent with the identification of surface hydrogenated diamond (Guillois, Ledoux & Reynaud 1999).

One the other hand, diamond aggregates have $^{12}C/^{13}C$ of 92 - 93 (e.g., Russell, Arden & Pillinger 1996), close to the solar ratio of 89. The $^{14}N/^{15}N$ ratio of 417-419 (Lewis *et al.* 1987; Russell, Arden & Pillinger 1996) is similar to 435 measured in Jupiter (Owen *et al.* 2001). This may suggest that only a portion of pre-solar diamonds formed in supernovae and prompted the quest for the diamonds' other stellar sources.

3. Pre-solar grains from novae

A handful of pre-solar SiC and graphite grains have low $^{12}C/^{13}C$, low $^{14}N/^{15}N$, and high $^{26}Al/^{27}Al$ ratios (Amari *et al.* 2001), consistent with theoretical predictions for ejecta of both CO and ONe novae (Starrfield, Gehrz & Truran 1997; 1998; José & Hernanz 1998; José *et al.* 2004). All grains have ^{30}Si excesses ($^{30}Si/^{28}Si$ up to $2.1 \times$ solar). The peak temperatures reached in CO novae are not high enough to significantly modify Si isotopic compositions, hence the Si isotopes of the grains indicate they formed in ONe novae. However, to reproduce the grain data it is necessary to mix pure nova ejecta with a huge amount (99%) of close-to-solar material (Amari *et al.* 2001). The apparent lack of known graphite and SiC grains from CO novae agrees with expectations from condensation calculations for nova ejecta (José *et al.* 2004).

Nittler & Hoppe (2005) argued that grains with low $^{12}C/^{13}C$ and $^{14}N/^{15}N$ ratios could also form in supernovae. Isotopic analysis of Ti in the grains could distinguish the two stellar sources because Ti isotopes are not modified in novae. Nittler *et al.* (2006) found a SiC grain with $^{12}C/^{13}C = 1$, which can only be obtained by hot H burning, pointing clearly toward a nova origin.

4. Conclusions

Pre-solar grains from SNe and novae exist. Laboratory analysis of such grains gives us, with unprecedented precision, detailed information about the nucleosynthesis products and grain formation from these stars.

Acknowledgements

This work is supported by NASA grants NNG04GG13G (SA and KL), NNG05GF81G (SA).

References

Amari, S., Anders, E., Virag, A. & Zinner, E. 1990, *Nature*, 345, 238
Amari, S., Gao, X., Nittler, L. R., *et al.* 2001, *ApJ*, 551, 1065
Amari, S., Hoppe, P., Zinner, E., & Lewis, R. S. 1992, *ApJ* (Letters), 394, L43
Amari, S., & Zinner, E. 1997, in: T. J. Bernatowicz & E. Zinner (eds.), *Astrophysical Implications of the Laboratory Study of Pre-solar Materials*, (New York: AIP), p. 287
Amari, S., Zinner, E., & Lewis, R. S. 1995, *ApJ* (Letters), 447, L147

Bernatowicz, T., Fraundorf, G., Ming, T., *et al.* 1987, *Nature*, 330, 728
Bernatowicz, T. J., Amari, S., & Lewis, R. S. 1992, *Lunar Planet. Sci.*, XXIII, 91
Bernatowicz, T. J., Amari, S., Zinner, E. K., & Lewis, R. S. 1991, *ApJ* (Letters), 373, L73
Choi, B.-G., Huss, G. R., Wasserburg, G. J., & Gallino, R. 1998, *Science*, 282, 1284
Clayton, D. D., & Nittler, L. R. 2004, *ARAA*, 42, 39
Guillois, O., Ledoux, G., & Reynaud, C. 1999, *ApJ* (Letters), 521, L133
Hoppe, P., Strebel, R., Eberhardt, P., Amari, S., & Lewis, R. S. 1996, *Science*, 272, 1314
Hoppe, P., Strebel, R., Eberhardt, P., Amari, S., & Lewis, R. S. 2000, *Meteorit. Planet. Sci.*, 35, 1157
Hutcheon, I. D., Huss, G. R., Fahey, A. J., & Wasserburg, G. J. 1994, *ApJ* (Letters), 425, L97
José, J., & Hernanz, M. 1998, *ApJ*, 494, 680
José, J., Hernanz, M., Amari, S., Lodders, K., & Zinner, E. 2004, *ApJ*, 612, 414
Lewis, R. S., Tang, M., Wacker, J. F., Anders, E., & Steel, E. 1987, *Nature*, 326, 160
Lodders, K., & Amari, S. 2005, *Chem. Erde*, 65, 93
Maas, R., Loss, R. D., Rosman, K. J. R., *et al.* 2001, *Meteorit. Planet. Sci.*, 36, 849
Meikle, W. P. S., Allen, D. A., Spyromilio, J., & Varani, G.-F. 1989, *MNRAS*, 238, 193
Messenger, S., Keller, L. P., & Lauretta, D. S. 2005, *Science*, 309, 737
Messenger, S., Keller, L. P., Stadermann, F. J., Walker, R. M., & Zinner, E. 2003, *Science*, 300, 105
Meyer, B. S., Clayton, D. D., & The, L.-S. 2000, *ApJ* (Letters), 540, L49
Nagashima, K., Krot, A. N., & Yurimoto, H. 2004, *Nature*, 428, 921
Nguyen, A. N., & Zinner, E. 2004, *Science*, 303, 1496
Nittler, L. R., Alexander, C. M. O'. D., Gao, X., Walker, R. M., & Zinner, E. K. 1994, *Nature*, 370, 443
Nittler, L. R., Alexander, C. M. O'D., & Nguyen, A. N. 2006, *Meteorit. Planet. Sci.*, 41, A134
Nittler, L. R., Alexander, C. M. O'D., Wang, J., & Gao, X. 1998, *Nature*, 393, 222
Nittler, L. R., Amari, S., Zinner, E., Woosley, S. E., & Lewis, R. S. 1996, *ApJ* (Letters), 462, L31
Nittler, L. R., & Hoppe, P. 2005, *ApJ* (Letters), 631, L89
Nittler, L. R., Hoppe, P., Alexander, C. M. O'D., *et al.* 1995, *ApJ* (Letters), 453, L25
Owen, T., Mahaffy, P. R., Niemann, H. B., Atreya, S., & Wong, M. 2001, *ApJ* (Letters), 553, L77
Pellin, M. J., Davis, A. M., Calaway, W. F., *et al.* 2000, *Lunar Planet. Sci.* XXXI, Abstract #1934
Pellin, M. J., Davis, A. M., Lewis, R. S., Amari, S., & Clayton, R. N. 1999, *Lunar Planet. Sci.* XXX, Abstract #1969
Pellin, M. J., Davis, A. M., Savina, M. R., Kashiv, Y., Clayton, R. N., Lewis, R. S., & Amari, S. 2001, *Lunar Planet. Sci.* XXXII, Abstract #2125
Richter, S., Ott, U., & Begemann, F. 1998, *Nature*, 391, 261
Russell, S. S., Arden, J. W., & Pillinger, C. T. 1996, *Meteorit. Planet. Sci.*, 31, 343
Starrfield, S., Gehrz, R. D., & Truran, J. W. 1997, in: T. J. Bernatowicz & E. Zinner (eds.), *Astrophysical Implications of the Laboratory Study of Pre-solar Materials* (New York: AIP), 203
Starrfield, S., Truran, J. W., Wiescher, M. C., & Sparks, W. M. 1998, *MNRAS*, 296, 502
Tang, M., & Anders, E. 1988, *Geochim. Cosmochim. Acta*, 52, 1235
Travaglio, C., Gallino, R., Amari, S., Zinner, E., Woosley, S., & Lewis, R. S. 1999, *ApJ*, 510, 325
Travaglio, C., Gallino, R., Zinner, E., Amari, S., & Woosley, S. 1998, in: N. Prantzos & S. Harissopulos (eds.), *Nuclei in the Cosmos V* (Paris: Editions Frontires), p. 567
Woosley, S. E., & Weaver, T. A. 1995, *ApJS*, 101, 181
Zinner, E. 1998, *Ann. Rev. Earth Planet. Sci.*, 26, 147

Highlights of Astronomy, Volume 14
IAU XXVI General Assembly, 14-25 August 2006
Karel A. van der Hucht, ed.

doi:10.1017/S1743921307010964

What can pre-solar grains tell us about the solar nebula?

Gary R. Huss[1] and Bruce T. Draine[2]

[1]Hawai'i Institute of Geophysics and Planetology, University of Hawai'i at Mānoa, 1680 East-West Road, Honolulu, HI 96822, USA
email: ghuss@higp.hawaii.edu

[2]Department of Astrophysical Sciences, Princeton University, 108 Peyton Hall, Princeton, NJ 08544-1001, USA
email: draine@astro.princeton.edu

Abstract. Several types of pre-solar grains, grains that existed prior to solar system formation, have been found in the fine-grained components of primitive meteorites, interplanetary dust particles (IDPs), and comet samples. Known pre-solar components have isotopic compositions that reflect formation from the ejecta of evolved stars. Other pre-solar materials may have isotopic compositions very similar to solar system materials, making their identification as pre-solar grains problematic. Pre-solar materials exhibit a range of chemical and thermal resistance, so their relative abundances can be used to probe the conditions in the solar nebula. Detailed information on the relative abundances of pre-solar and solar-system materials can provide information on the temperatures, radiation environment, and degree of radial mixing in the early solar system.

Keywords. pre-solar grains, solar system: formation, ISM: abundances, ISM: dust

1. Introduction

Pre-solar grains are solid objects that existed in interstellar space prior to the formation of the solar system and have survived within meteorites and comets until the present day to be studied in the laboratory. The study of pre-solar grains has been mainly the study of isotopic anomalies, which have historically been defined as isotopic compositions that could not have been produced from the solar system composition, as measured in terrestrial and lunar samples and meteorites, by known processes operating within the solar system. The first anomalies of this type were found in noble gases from primitive meteorites (e.g., Reynolds & Turner 1964; Black & Pepin 1969), although these anomalies were not initially thought of as signatures of pre-solar materials. Additional discoveries of isotopic anomalies in oxygen (Clayton *et al.* 1973; Clayton *et al.* 1977), hydrogen (e.g., Yang & Epstein 1984), carbon (e.g., Swart *et al.* 1983), titanium (e.g., Niederer *et al.* 1980), and other elements eventually led to a consensus that pre-solar materials were not completely homogenized when the solar system formed. The first pre-solar mineral grains to be identified were found after a decades-long search for the carriers of anomalous noble-gas components. They were diamond (carrier of Xe-HL; Lewis *et al.* 1987), silicon carbide (Ne-E(H) and Xe-S; Tang & Anders 1988) and graphite (Ne-E(L); Amari *et al.* 1990). Since these initial identifications, $\sim$20 pre-solar compounds have been identified in meteorites and interplanetary dust particles (Table 1). However, as the abundances clearly show, we have not yet identified a majority of the material that made up the Sun's parent molecular cloud.

Pre-solar materials can be divided into two types, circumstellar condensates, which formed from the ejecta of dying stars and carry the isotopic signature of nucleosynthesis

Table 1. Types and properties of pre-solar materials identified in meteorites and IDPs

Material	Source	Grain Size (μm)	Abundance (ppm)†	Chemical resistance	Thermal resistance
Diamond		~0.002	~1400		
P3 fraction	?			high	low
HL fraction	circumstellar			very high	high
Silicon carbide	circumstellar	0.1-20	13-14	high	high
Graphite	circumstellar	0.1-10	7-10	moderate	low
D-rich organics	interstellar			low to mod.	low to mod.
P1 noble gas carrier	interstellar	*	*	moderate	high
Corundum (Al_2O_3)	circumstellar	0.5-3	0.01	high	very high
Spinel ($MgAl_2O_4$)	circumstellar	0.1-3	1.2	high	very high
Hibonite ($CaAl_{12}O_{19}$)	circumstellar	1-2	0.02	high	very high
Forsterite (Mg_2SiO_4), Enstatite ($MgSiO_3$)	circumstellar	0.2-0.5	10-1800	low to mod.	high
Amorphous silicates	circumstellar	0.2-0.5	20-3600	low	moderate

Notes.
Other pre-solar materials include TiC, MoC, ZrC, RuC, FeC, Si_3N_4, TiO_2, and Fe-Ni metal.
†Abundance in fine-grained fraction (= matrix in primitive chondrites).
* ^{132}Xe-P1 content in carbonaceous residue from primitive chondrites is $\sim(1.2\text{–}1.4)\times10^{-8}$ cc-STP/gram, but the carrier itself is still unknown.
Data: Huss & Lewis (1995); Huss *et al.* (1996); Huss *et al.* (2003); Messenger *et al.* (2006).

in the parent star, and interstellar material, which grew from the gas phase in interstellar space, primarily in molecular clouds. Most pre-solar materials identified in the laboratory are circumstellar condensates, which carry the largest isotopic anomalies relative to solar system materials. Astronomical observations indicate that AGB stars are prolific dust producers. Most of the pre-solar grains studied to date are from AGB stars, although we note that grains below $\sim 0.3\,\mu$m are currently almost impossible to study. Amorphous silicates, Mg-rich crystalline silicates, Al_2O_3, and FeMg oxides are observed in the circumstellar dust shells of O-rich AGB stars (Molster *et al.* 2002; Tielens *et al.* 2005). Silicon carbide is observed in C-rich dust from AGB stars (Speck *et al.* 1997). Amorphous silicates and amorphous C have been reported from observations of the interstellar medium (see, e.g., review by Draine (2003) and references therein). In dark molecular clouds, the grains acquire mantles consisting primarily of H_2O ice but often containing appreciable fractions of CO, CO_2, CH_3OH, and other species.

2. Differential processing of pre-solar materials in the solar nebula

Pre-solar materials exhibit a wide range of resistance to heating and chemical processing (Table 1). This opens the possibility that their relative abundances can be used as probes of conditions in the solar nebula. For example, if a portion of a molecular cloud containing all of the materials in Table 1 (plus many others) is heated to 1000 K, we would expect that the P3 fraction of diamond, graphite, D-rich organics, and some of the poorly crystalline silicates would be destroyed, either through vaporization or through chemical reaction. At higher temperatures, other components would also be destroyed. Heating in a highly oxidizing environment would affect carbonaceous and organic components more than oxides, while reducing environments would increase the survival of carbonaceous and organic components. An initial study of unmetamorphosed chondrites comparing abundances of diamond, silicon-carbide, and graphite in the matrices with the bulk compositions of the meteorites shows a correlation between depletions of fragile pre-solar components and depletions of volatile and moderately volatile elements (Huss *et al.* 2003). This correlation suggests that the pre-solar grains track the thermal processing in the solar nebula that produced the bulk compositions of different classes of meteorites from molecular-cloud material.

3. Formation of organic compounds

Isotopic and chemical compositions of organic materials in meteorites, IDPs, and comet samples may also provide information about conditions in the solar nebula. For example, dense molecular clouds contain a wide variety of relatively complex organic molecules synthesized by radiation-driven chemistry in the gas phase and in icy grain mantles (e.g., Allamandola *et al.* 1988). Some of these molecules, such as those containing the OCN^- ion, are more abundant in the spectra of protostars than in the spectra of background stars (Pendleton *et al.* 1999), suggesting that radiation-driven chemistry may be enhanced in the immediate vicinity of star-formation. Many molecules synthesized in laboratory experiments simulating conditions in molecular clouds are similar or identical to compounds found in primitive meteorites (Bernstein *et al.* 2001). There are clear isotopic signatures (e.g., high D/H) associated with organic molecules produced by radiation chemistry in cold molecular clouds and in dense, cold regions of an accretion disk (e.g., Sandford *et al.* 2001). Interstellar grains may be heavily deuterated in diffuse regions (Linsky *et al.* 2006), with D presumably present in hydrocarbons, possibly including PAHs (Draine 2006). Detailed work on organics in meteorites and interplanetary dust particles is just beginning, as new tools become available. Identifying specific molecules that carry extreme isotope anomalies will help identify the chemical pathways of their creation, which in turn will constrain the environment of formation, either in the molecular cloud or in the early solar nebula.

4. Circulation and mixing in the solar nebula

An *a priori* knowledge of the types of materials in the Sun's parent molecular cloud provides the basis for which to track circulation and mixing in the solar nebula. Late-stage stars produce both amorphous and crystalline silicates in their outflows. The proportion of crystalline silicates may be anywhere from 5 - 10 % up to as much as 40 %, but detection is difficult (e.g., Molster *et al.* 2002; Kemper *et al.* 1999). However, crystalline silicates comprise $< 2\%$ of the silicate material in interstellar space (Kemper *et al.* 2005). Crystalline silicates can be destroyed by interstellar shocks and amorphized by cosmic rays, thus reducing their abundance. On the other hand, observations of young stellar systems show that the abundance of crystalline silicates is much higher in the inner disk than in the outer disk, but that even the outer disks show more crystalline silicates than the interstellar medium (Tielens *et al.* 2005). These observations indicate that crystalline silicates are produced in the inner disk and are mixed outward.

Chemical thermodynamics predicts that pure Mg-end-member olivine (forsterite) and pyroxene (enstatite) grains will be the first major silicate phases to condense from a cooling gas of stellar or solar composition (Grossman 1972; Lattimer *et al.* 1978). In contrast, olivine and pyroxene that crystallize from a melt typically contain Fe and other elements. Nearly pure Mg-end-member forsterite and enstatite are observed along with amorphous silicates and Fe-bearing olivine in the matrices of the most primitive chondrites (e.g., Brearley 1993). Observations of dust released by comets and of porous chondritic IDPs thought to have come from comets also show both crystalline and amorphous silicates (e.g., Wooden *et al.* 2004; Keller & Messenger 2005). These data sets suggest that mixing of high-temperature nebular components from the inner to the outer solar system was significant. Transport of high-temperature material to the outer solar system may have occurred via turbulent mixing (Cuzzi *et al.* 2003), by X-winds (Shu *et al.* 1996), or by other unknown processes. Detailed studies of pre-solar and high-temperature nebula components in primitive meteorites, IDPs, and comet samples will be critical in interpreting the astronomical observations and in understanding mixing processes in the early solar nebula.

Acknowledgements

Supported by NASA grant NNG05GG48G (GRH) and NSF grant AST-0406883 (BTD).

References

Allamandola, L. J., Sandford, S. A., & Valero, G. 1988, *Icarus*, 76, 225
Amari, S., Anders, E., Virag, A., & Zinner, E. 1990, *Nature*, 345, 238
Bernstein, M. P., Dworkin, J. P., Sandford, S. A., & Allamandola, L. J. 2001, *Meteorit. Planet. Sci.*, 36, 351
Black, D. C., & Pepin, R. O. 1969, *Earth Planet. Sci. Lett.*, 6, 395
Brearley, A. J. 1993, *Geochim. Cosmochim. Acta*, 57, 1521
Clayton, R. N., Grossman, L., & Mayeda, T. K. 1973, *Science*, 182, 485
Clayton, R. N., Onuma, N., Grossman, L., & Mayeda, T. K. 1977, *Earth Planet. Sci. Lett.*, 34, 209
Cuzzi, J. N., Davis, S. S., & Dobrovolskis, A. R. 2003, *Icarus*, 166, 385
Draine, B. T. 2003, *ARAA*, 41, 241
Draine, B. T. 2006, in: G. Sonneborn, H. W. Moos, & B.G. Andersson (eds.), *Astrophysics in the Far Ultraviolet, ASP-CS* 348, 58
Grossman, L. 1972, *Geochim. Cosmochim. Acta*, 36, 597
Huss, G. R., & Lewis, R. S. 1995, *Geochim. Cosmochim. Acta*, 59, 115
Huss, G. R., Lewis, R. S., & Hemkin, S. 1996, *Geochim. Cosmochim. Acta*, 60, 3311
Huss, G. R., Meshik, A. P., Smith, J. B., & Hohenberg, C. M. 2003, *Geochim. Cosmochim. Acta*, 67, 4823
Keller, L. P., & Messenger, S. 2005, in: A. N. Krot, E. R. D. Scott & B. Reipurth (eds.), *Chondrites and the Protoplanetary Disk*, *ASP-CS* 341, 657
Kemper, F., Spaans, M., Jansen, D. J., Hogerheijde, M. R., van Dishoeck, E. F., & Tielens, A. G. G. M. 1999, *ApJ*, 515, 649
Kemper, F., Vriend, W. J., & Tielens A. G. G. M. 2005, *ApJ*, 633, 534
Lattimer, J. M., Schramm, D. N., & Grossman, L. 1978, *ApJ*, 219, 230
Lewis, R. S., Tang, M., Wacker, J. F., Anders, E., & Steele, E. 1987, *Nature*, 326, 160
Linsky, J. L., Draine, B. T., Moos, H. W., *et al.* 2006, *ApJ*, 647, 1106
Messenger, S., Sandford, S., & Brownlee, D. 2006, in: D. S. Lauretta & H. Y. McSween (eds.), *Meteorites and the Early Solar System II* (Univ. of Arizona), p. 187
Molster, F. J., Waters L. B. F. M., Tielens, A. G. G. M., *et al.* 2002, *A&A*, 382, 241
Niederer, F. R., Papanastassiou, D. A., & Wasserburg, G. J. 1980, *ApJ* (Letters), 240, L73
Pendleton, Y. J., Tielens, A. G. G. M., Tokunaga, A. T., & Bernstein, M. P. 1999, *ApJ*, 513, 294
Reynolds, J. H., & Turner, G. 1964, *J. Geophys. Res.* 69, 3263
Sanford, S. A., Bernstein, M. P., & Dworkin, J. P. 2001, *Meteorit. Planet. Sci.* 36, 1117
Shu, F. H., Shang, H., & Lee, T. 1996, *Science*, 271, 1545
Speck, A. K., Barlow, M. J., & Skinner, C. J. 1997, *MNRAS*, 288, 431
Swart, P. K., Grady M. M., Pillinger C. T., Lewis R. S., & Anders E. 1983, *Science*, 220, 406
Tang, M., & Anders, E. 1988, *Geochim. Cosmochim. Acta*, 52, 1235
Tielens, A. G. G. M., Waters, L. B. F. M., & Bernatowicz, T. J. 2005, in: A. N. Krot, E. R. D. Scott, & B. Reipurth (eds.), *Chondrites and the Protoplanetary Disk*, *ASP-CS*, 341, 605
Wooden, D. H., Woodward, C. E., & Harker D. E. 2004, *ApJ* (Letters), 612, L77
Yang, J., & Epstein S. 1984, *Nature*, 311, 544

Highlights of Astronomy, Volume 14
IAU XXVI General Assembly, 14-25 August 2006
Karel A. van der Hucht, ed.

doi:10.1017/S1743921307010976

Pre-solar grains: outlook and opportunities for astrophysics

Larry R. Nittler and Conel M. O'D. Alexander

Department of Terrestrial Magnetism, Carnegie Institution of Washington, 5241 Broad Branch Road NW, Washington, D.C. 20015, USA
email: alexande,lrn@dtm.ciw.edu

Abstract. As pristine condensates from a large number of stars, pre-solar grains provide unique information on the sources, types, compositions and processing histories of dust in the Galaxy. However, their promise remains largely unfulfilled. Here we discuss some of the astrophysical problems for which pre-solar grains might provide important insights and constraints in coming years.

Keywords. dust, extinction, circumstellar matter, stars: AGB and post-AGB, supernovae, nuclear reactions, nucleosynthesis, abundances

1. Introduction

Pre-solar grains in meteorites and interplanetary dust particles (IDPs) are relatively pristine samples of circumstellar dust that can be studied in remarkable detail by advanced microanalytical tools in the laboratory (e.g., Nittler 2003). The types of grains that have been found include silicates, oxides, graphite, diamond, SiC and Si_3N_4. They appear to come from red giant branch (RGB) and asymptotic giant branch (AGB) stars, from supernovae and possibly novae. Their analysis can provide elemental and isotopic compositions that are far more precise than is possible astronomically, as well as powerful new information that cannot be obtained astronomically but which complements that obtained from traditional astronomical observations. Many examples are discussed in the other papers from this Joint Discussion. However, the promise of pre-solar grains for astronomy and astrophysics has not been fulfilled to the extent it could be. With several notable exceptions, relatively few studies by traditional astronomers and astrophysicists have taken into account the growing observational data set on pre-solar grains. This is in part due to the fact that for much of the nearly twenty years since pre-solar grains were discovered, most of the studied grains have been atypical, from the astronomical point of view, for circumstellar and interstellar dust. For example, until recently, only grains larger than about $\sim 1\,\mu$m in diameter were amenable to isotopic analysis as single grains, while typical circumstellar and interstellar dust sizes are ~ 100 nm. Similarly, silicates are the dominant type of dust in the Galaxy, but these have been identified in the pre-solar grain population only quite recently (Messenger *et al.* 2003). The situation has greatly improved with recent technological advances. For example, isotopic measurements can now be made on individual grains down to 100 nm in size, and identification of pre-solar silicate grains for detailed study has become fairly routine. Here we discuss some of the astrophysical problems for which pre-solar grains might provide important insights and constraints in coming years.

2. Opportunities for astrophysics

2.1. *Galactic chemical evolution (GCE)*

It is now well-established that the isotopic compositions of some elements, notably Si, Ti, and O, in many pre-solar grains reflect GCE processes (Clayton & Timmes 1997; Alexander & Nittler 1999), but a unified quantitative explanation of the data in terms of GCE is lacking. Unfortunately, at present the limitations of both the grain data set and GCE models make it difficult to judge the relative merits of the disparate ideas put forward to explain the grain data. Perhaps the best hope for progress from the grain front is the acquisition of multi-element data for a large number of grains of different types, for example Si, Ti and O isotopic data for pre-solar silicates, SiC and oxide phases from AGB stars. Such a multi-dimensional dataset would allow the disentanglement of GCE processes from nuclear processes in the parent stars of the grains and could serve as high-precision constraints on future GCE models.

2.2. *Stellar evolution and nucleosynthesis*

Observational data from both stars and pre-solar grains clearly demonstrate the existence of an 'extra mixing' process (often called 'cool bottom processing' or CBP, Wasserburg *et al.* 1995) occurring in low-mass RGB and AGB stars. What is still unknown is the physical-dynamical cause of CBP. In addition to indicating the existence of CBP, pre-solar grains may also shed light on the origin and nature of the mixing. First, comparison of grain data with parameterized models (e.g., Nollett *et al.* 2003) provides constraints on the physical conditions of the mixing (e.g., mixing rate, temperature). Second, the distribution of mixing parameters indicated by the isotopic data, including the lack of mixing in many parent stars, must be reproduced by any reasonable model of CBP.

One of the biggest mysteries of pre-solar grain studies is the origin of the highly ^{13}C-enriched SiC grains, known as Types A and B (Amari *et al.* 2001). J-type and CH-type C-stars are known to be ^{13}C-rich C stars, although their origins are also mysterious. The ratio of A+B grains to normal AGB grains is very similar to the ratio of J-type and N-type C-stars. Thus, it seems likely that most of the A+B grains come from J-type stars. Whether from J-type or CH-type stars, the detailed information that will come from the A+B grains should greatly enhance our knowledge of how these stars form and evolve.

A fraction of pre-solar grains originated in supernova explosions and their compositions provide detailed information about nucleosynthesis and mixing processes in such environments (see e.g., Amari & Lodders, this volume). This topic is likely to expand in coming years with more detailed grain data, with detailed observations of supernova remnants using the *Chandra* and *Spitzer* telescopes, and with multi-dimensional modeling of supernovae, now becoming possible with modern computers.

2.3. *Dust formation around stars*

How dust grains nucleate and grow in stellar environments is not well understood, despite being central to models of dust production in the Galaxy, the evolution of dust in the ISM, and mass loss from stars. Pre-solar grains hold great promise for improving understanding in this field, since formation processes are recorded in the detailed structures and compositions of the grains that can be characterized in minute detail. Thermodynamic equilibrium calculations have been very useful for determining the types of grains that should form and the order in which they condense (Lodders & Fegley 1995). However, astronomical and astrophysical evidence shows that thermodynamic equilibrium is not maintained in stellar outflows. A realistic picture of grain formation will require kinetic models of grain nucleation and growth, coupled to physical models of the conditions that exist in stellar outflows of different kinds (e.g., Gail & Sedlmayr 1999)

Pre-solar grains will provide a wealth of constraints for these models. The size distribution of grains reflects their growth conditions, but there seems to be only modest variations in the range of pre-solar grain compositions with size. So, for instance, different types (masses/metallicities) of AGB stars produce SiC grains with similar size distributions, as do supernovae. Perhaps the similarities in the size distributions of grains from these different objects reflects some common underlying mechanism.

Formation of grains that range in size from 100 nm or less to several microns seems to require a range of gas densities in outflows. In particular, formation of large ($> 1\,\mu$m) AGB grains on reasonable timescales requires much higher densities than estimated average densities in a uniform radially symmetric outflow (e.g., Bernatowicz *et al.* 2005). The density variations recorded by the grains are probably caused by shocks in the outflows that are observed astronomically, and highlight the need for two- and three-dimensional models of grain growth in outflows.

The microstructures of grains also provide clues to the mechanisms of nucleation and growth. Heterogeneous nucleation on pre-existing grains is often invoked to overcome the kinetic barriers to spontaneous nucleation in the gas phase. However, of the grain types that have been examined, only certain types of graphite grain show clear evidence for heterogeneous nucleation (Bernatowicz *et al.* 1996). For thermodynamic and/or kinetic reasons, minerals with the same compositions can have different crystal structure (polytype), depending on the conditions under which they form. Whether AGB Al_2O_3 grains are crystalline or amorphous may depend on the availability of Ti when they grow (Stroud *et al.* 2006). SiC has many polytypes whose stability and preservation depends on the temperature of formation and the rate at which the system cools (Daulton *et al.* 2002).

Dust formation in supernovae (SNe) is of great current interest in astrophysics, both because of observations of copious dust at high redshifts (Bertoldi *et al.* 2003) and because of observations of dust in Galactic SN remnants (Sugerman *et al.* 2006). However, the processes of dust formation in SNe and the types of dust produced are poorly understood. Grains from supernovae have been found in almost every pre-solar grain type and again they provide an opportunity for making progress in this field. For example, in contrast to pre-solar grains from AGB stars, which are typically single euhedral crystals, the SNe-derived SiC, silicates and Si_3N_4 grains studied to date by transmission electron microscopy are polycrystalline aggregates of smaller 10 - 100 nm sized sub-grains (Stroud *et al.* 2004; Messenger *et al.* 2005; Stroud *et al.* 2006), indicating different conditions during grain formation in the different environments. Micron- to ten micron-sized graphitic grains from supernovae contain sub-crystals of Fe-Ni metal and TiC, providing tight constraints on the conditions where the grains formed (Lodders 2006). Isotopic evidence from SiC and diamond grains also provide clues to the timing of grain condensation in supernovae ejecta.

2.4. *Sources and processing of interstellar dust*

Pre-solar grains come from at least tens and, perhaps, hundreds of stars (Alexander 1997) with a considerable range in masses and metallicities. To date, there has been no quantitative explanation for how grains from so many stars became part of the pre-solar molecular cloud. Based on observed encounter rates, direct injection of dust from stars passing through a cloud cannot be the explanation. Presumably, the grains are accumulated by a molecular cloud as it forms, and the number of stellar sources for grains must in some way reflect the volume of the Galaxy sampled by the forming cloud.

Circumstellar grains are thought to be rapidly processed in the ISM by supernova-driven shock waves (Slavin *et al.* 2004). This processing results in the rapid amorphisation of crystalline silicates, sputtering and cratering/shattering in grain-grain collisions. The pre-solar grains appear to be relatively unprocessed, which is surprising given the rapidity of the ISM processing and the number of stars represented. Obviously, a small fraction of circumstellar grains are able to survive in the ISM and those that do survive appear to be largely unscathed. How this might happen in the context of current models of dust evolution in the ISM has yet to be explored.

Given that pre-solar grains come from so many sources and seem to be relatively unprocessed, they also provide constraints on circumstellar dust production rates in the Galaxy. Astronomical estimates of Galactic dust production rates vary widely, particularly in the amount of dust produced by supernovae. The relative abundances of different types of pre-solar grain from various stellar sources are generally roughly what is expected from astronomical estimates of stellar dust production rates that only require a few percent of the dust coming from supernovae (Alexander 1997). The two glaring differences are the low abundances of graphite grains and high abundance of nanodiamonds amongst circumstellar grains in meteorites.

References

Alexander, C. M. O'D. 1997, in: T. J. Bernatowicz & E. K. Zinner (eds.), *Astrophysical Implications of the Laboratory Study of Pre-Solar Materials, AIP-CP*, 402, 567

Alexander, C. M. O'D., & Nittler, L. R. 1999, *ApJ*, 519, 222

Amari, S., Nittler, L. R., Zinner, E., Lodders, K., & Lewis, R. S. 2001, *ApJ*, 559, 463

Bernatowicz, T. J., Cowsik, R., Gibbons, P. C., Lodders, K., Fegley, B., Amari, S., & Lewis, R. S. 1996, *ApJ*, 472, 760

Bernatowicz, T. J., Akande, O. W., Croat, T. K., & Cowsik, R. 2005, *ApJ*, 631, 988

Bertoldi, F., Carilli, C. L., Cox, P., Fan, X., Strauss, M. A., Beelen, A., Omont, A., & Zylka, R. 2003, *A&A* (Letters), 406, L55

Clayton, D. D., & Timmes, F. X. 1997, in: T. J. Bernatowicz & E. K. Zinner (eds.), *Astrophysical Implications of the Laboratory Study of Pre-Solar Materials, AIP-CP*, 402, 237

Daulton, T. L., Bernatowicz, T. J., Lewis, R. S., Messenger, S., Stadermann, F. J., & Amari, S. 2002, *Science*, 296, 1852

Gail, H.-P., & Sedlmayr, E. 1999, *A&A*, 347, 594

Lodders, K., & Fegley, B. 1995, *Meteoritics*, 30, 661

Lodders, K. 2006, *ApJ* (Letters), 647, L37

Messenger, S., Keller, L. P., & Lauretta, D. S. 2005, *Science*, 309, 737

Messenger, S., Keller, L. P., Stadermann, F. J., Walker, R. M., & Zinner, E. 2003, *Science*, 300, 105

Nittler, L. R. 2003, *Earth Planet. Sci. Lett.*, 209, 259

Nollett, K. M., Busso, M., & Wasserburg, G. J. 2003, *ApJ*, 582, 1036

Slavin, J. D., Jones, A. P., & Tielens, A. G. G. M. 2004, *ApJ* 614, 796

Stroud, R. M., Nittler, L. R., & Alexander, C. M. O'D. 2004, *Science* 305, 1455

Stroud, R. M., Nittler, L. R., & Alexander, C. M. O'D. 2006, *Meteorit. Planet. Sci. Supp.*, 41 (Supp.), Abstract #5360

Stroud, R. M., Nittler, L. R., & Hoppe, P. 2004, *Meteorit. Planet. Sci.*, 39 (Supp.), Abstract #5039

Sugerman, B. E. K., Ercolano, B., Barlow, M. J., *et al.* 2006, *Science*, 313, 196

Wasserburg, G. J., Boothroyd, A. I., & Sackmann, I.-J. 1995, *ApJ* (Letters), 447, L37

Highlights of Astronomy, Volume 14
IAU XXVI General Assembly, 14-25 August 2006
Karel A. van der Hucht, ed.

doi:10.1017/S1743921307010988

Joint Discussion 12
Long wavelength astrophysics

T. Joseph W. Lazio and Namir E. Kassim (eds.)
Naval Research Laboratory, 4555 Overlook Ave. SW, Washington, DC 20375, USA
email: joseph.lazio,namir.kassim@nrl.navy.mil

Preface

The greatest discoveries in astronomy have accompanied technological innovations that have opened new windows of the electromagnetic spectrum. One of the last poorly explored regions lies at wavelengths longer than 3 m (100 MHz) to the ionospheric cutoff around 30 m (10 MHz). In the past, variations in the ionosphere have limited ground-based instruments to small (< 5 km) apertures and hence relatively coarse angular resolution and low sensitivity.

Ever-increasing computing power combined with new wide-scale imaging algorithms, self-calibration techniques, and angle-dependent calibration schemes make it possible to overcome this restriction. With these advantages, the Very Large Array's 74 MHz (4 m) observing system and the Indian GMRT have generated many exciting scientific results in the past several years and elegantly demonstrate that connected element interferometry at long wavelengths is no longer limited by phase fluctuations from the ionosphere

With the world-wide activitiy – both the scientific results being generated at the existing facilities and the new facilities that are coming on-line shortly – the IAU XXVI General Assembly seemed timely. Joint Discussion 12 on *Long Wavelength Astrophysics*, was a one-day session at the General Assembly organized to review the current results as well as anticipate those to come shortly.

Reception to JD 12 was enthusiastic, clearly demonstrating the renaissance in this area. A total of 20 oral and 35 poster presentations were made, with scientific topics ranging from the nearby (the Earth's ionosphere) to the most distant (the Universe itself at the epoch of reionization). Nearly a third of the oral presentations described instruments either under development or in construction indicating that the future of long wavelength astrophysics appears quite bright.

We thank all of the presenters for their exciting presentations, a sampling of which is presented here. We also thank our fellow members of the SOC for their excellent suggestions and guidance, and K. Kellermann who assisted at the last minute.

Scientific Organizing Committee

Franklin Briggs (Australia), Gloria Dubner (Argentina), Luigina Feretti (Italy), Nasir E. Kassim (USA, co-chair), T. Joseph Lazio (USA, co-chair), A.V. Megn (Ukraine), Alexander A. Konovalenko (Ukraine), Alain Lecacheux (France), A. Pramesh Rao (India), Huub J.A. Röttgering (Netherlands), and Kurt W. Weiler (USA).

Acknowledgements

Basic research in radio astronomy at the NRL is supported by 6.1 Base funding.

Joseph Lazio and Nasir Kassim,
Washington DC, November 2006

Highlights of Astronomy, Volume 14
IAU XXVI General Assembly, 14-25 August 2006
Karel A. van der Hucht, ed.

doi:10.1017/S174392130701099X

Radio emission from the Sun, planets, and the interplanetary medium

Timothy S. Bastian

National Radio Astronomy Observatory, 520 Edgemont Rd, Charlottessville, VA 22903, USA
email: tbastian@nrao.edu

Abstract. A brief review is given of radio phenomena on the Sun, the planet Jupiter, and the interplanetary medium and its outer boundary. A brief aside is made to draw parallels between radio emission from Jupiter and extrasolar planets.

Keywords. instrumentation: interferometers, instrumentation: spectrographs, radio continuum: solar system, Sun: radio emission, Sun: activity, solar system: planets (Jupiter), interplanetary medium

1. Introduction

Radio emission from the Sun, planets, and the interplanetary medium (IPM) have been studied for decades. Indeed, early solar observations formed the underpinnings of modern radio astronomy. Interest in low frequency radio emission from the solar system remains strong. A lack of space prevents more than a passing mention of several topics of current interest; citations likewise draw attention to only a very small part of the relevant literature.

2. Instrumentation

A large number of both ground and space based low-frequency instruments are available for observations of the Sun, planetary emissions, and the IPM. Ground-based interferometric arrays include the Ukrainian T-shaped Radiotelescope (UTR-2, 10 - 25 MHz; Ukraine), the Nançay Radioheliograph (150 - 450 MHz; France), the Guaribidanur Radioheliograph (40 - 150 MHz; India), the Giant Meter-wavelength Radio Telescope (150, 235, 327 MHz; India), and the VLA (74, 327 MHz; USA). Spectrographs too numerous to list individually are available around the world and provide dynamic spectra of active solar phenomena. The most recent addition (White *et al.* 2006) is the Green Bank Solar Radio Burst Spectrometer (10 - 1050 MHz; USA). In space, both the Ulysses URAP instrument, operating from 1.25 - 940 kHz, and the WIND/WAVES spectrometers, operating from 20 kHz to 13.8 MHz, continue to provide dynamic spectroscopy. In the near future, the Frequency Agile Solar Radiotelescope (50 MHz to 20 GHz), the Miluera Widefield Array (80 - 300 MHz), and the Long Wavelength Array (20 - 80 MHz) will become available on the ground and STEREO/WAVES (10 kHz to 16.1 MHz, plus 50 MHz) will be available in space.

3. Relevant emission mechanisms

There are several emission mechanisms relevant to low frequency radio observations in the solar system that provide unique access to physical processes in the solar, planetary, and interplanetary (IP) environments (e.g., Bastian *et al.* 1998). These are thermal free-free emission (the quiet solar corona and planetary disks), plasma radiation (solar and IP radio bursts), synchrotron radiation (solar flares, coronal mass ejections, and magnetospheric emission), cyclotron maser emission (magnetospheric emissions), as well as other less well-established mechanisms.

4. Radio emission from the Sun and interplanetary medium

Radio emission from the in the meter and decameter wavelength range is particularly rich phenomenologically. It is in this wavelength range that the 'classical' solar radio bursts of Types I - V occur (e.g., McLean & Labrum 1985) in the solar corona. Bursts of Type II/IV and III have received particularly close attention in recent years due to their association with space weather phenomena. Type II bursts result from plasma radiation driven by a fast MHD shock propagating in the solar corona whereas Type IV emission is due to either plasma radiation and/or synchrotron radiation. The shock exciter may be flare ejecta (e.g., Gopalswamy *et al.* 1997), a flare blast wave (e.g., Vršnak *et al.* 2006), or a coronal mass ejection (CME, e.g., Cliver *et al.* 1999). Type III radio bursts, associated with coronal energy release, are due to plasma radiation driven by suprathermal elecron beams and are used as probes of particle acceleration in the IPM (e.g., Krucker *et al.* 1999). Long-duration IP Type IIIs are correlated with solar energetic particle events (Cane *et al.* 2002). Similarly, there are interplanetary analogs to coronal Type II radio bursts that are driven by fast CMEs. Recent theoretical work has emphasize their analogy to planetary bowshocks (e.g., Knock *et al.* 2001, 2003). Progress has been made in detecting thermal free-free (Kathaviran & Ramesh 2005) and synchrotron radiation signatures (Bastian *et al.* 2001; Maia *et al.* 2007) from CMEs in their nascent stages.

5. Radio emission from Jupiter

While all planets are thermal radio emitters, all magnetized planets are nonthermal radio emitters as well, with Jupiter the most prominent source among them. Since its discovery by Burke & Franklin (1955), Jupiter's radio emission has been studied through remote and *in situ* instrumentation. In addition to thermal emission from its planetary disk, there is synchrotron radiation from energetic electrons trapped in its magnetosphere (DIM), and complex, transient radio emissions caused by cyclotron maser emission at decameter wavelengths (DAM), the latter modulated by interactions between the magnetosphere and the Galilean satellites (chiefly Io), and the solar wind (Gurnett *et al.* 2002). Zarka (2004) presents a detailed review.

6. An aside: radio searches for exoplanets

Winglee *et al.* (1986) first pointed out possibility of detecting cyclotron maser emission from magnetized exoplanets and performed a blind search at 327 and 1400 MHz using the VLA. With discovery of *bone fide* planets by Mayor & Queloz (1995), Bastian *et al.* (2000) targeted six known exoplanets and two brown dwarfs at 74, 327, and 1400 MHz with the VLA. On the basis of the 'radiometric Bode's Law' (Desch & Kaiser 1984; Farrell *et al.* 1999), Zarka *et al.* (2001) suggested interaction of 'hot Jupiters' with stellar winds could produce emissions orders of magnitude stronger than Jupiter's (see also Lazio *et al.* 2004). Subsequent attempts to detect exoplanets (e.g., Farrell *et al.* 2003; Lazio *et al.* 2004), have failed. Nevertheless, if detections are eventually made the payoff would be high, for they could be used to measure the planet's rotation period, magnetic field, presence of satellites, their orbital periods, and more.

7. Radio emission from the outer limits

Radio emission from 2 - 3.4 kHz was detected by the Voyager-1 and -2 spacecraft during the period from 1983 - 1984 (Kurth *et al.* 1984), again from 1992 - 1994 (Gurnett *et al.* 1993), and most recently from 2002 - 2003. These emissions have been interpreted as plasma radiation from the heliopause beyond the termination shock of the heliosphere (Cairns & Zank 2002).

8. Concluding remarks

Low frequency observations of solar, planetary, and interplanetary phenomena allow a variety of phenomena – particle acceleration and transport, shock initiation and propagation, emission mechanisms, plasma physics – to be studied in detail. An exciting new generation of instrumentation will be constructed in the coming decade. The future of low frequency radiophysics of the solar system therefore looks extremently promising.

References

Bastian, T. S., Pick, M., Kerdraon, A., Maia, D., & Vourlidas, A. 2001, *ApJ* (Letters), 558, L65
Bastian, T. S., Dulk, G. A., & Leblanc, Y. 2000, *ApJ*, 545, 1058
Bastian, T. S., Benz, A. O., & Gary, D. E. 1998, *ARAA*, 36, 131
Burke, B. F., & Franklin, K. L. 1955, *J. Geophys. Res.*, 60, 213
Cairns, I. H., & Zank, G. P. 2002, *Geophys. Res. Lett.* 29, 47
Cane, H. V., Erickson, W. C., & Prestage, N. P. 2002, *J. Geophys. Res.* (Space Physics), 107, 14
Cliver, E. W., Webb, D. F., & Howard, R. A. 1999, *Solar Phys.*, 187, 89
Desch, M. D., & Kaiser, M. L. 1984, *Nature*, 310, 755
Farrell, W. M., Desch, M. D., & Zarka, P. 1999, *J. Geophys. Res.*, 104, 14025
Farrell, W. M., Desch, M. D., Lazio, T. J., Bastian, T., & Zarka, P. 2003, in: D. Deming & S. Seager (eds.), *Scientific Frontiers in Research on Extrasolar Planets*, *ASP-CS*, 294, 151
Gopalswamy, N., Kundu, M. R., Manoharan, P. K., *et al.* 1997, *ApJ*, 486, 1036
Gurnett, D. A., Kurth, W. S., Hospodarsky, G. B., *et al.* 2002, *Nature*, 415, 985
Gurnett, D. A., Kurth, W. S., Allendorf, S. C., & Poynter, R. L. 1993, *Science*, 262, 199
Kathiravan, C., & Ramesh, R. 2005, *ApJ* (Letters), 627, L77
Knock, S. A., Cairns, I. H., & Robinson, P. A. 2003, *J. Geophys. Res.*, 108, SSH-2
Knock, S. A., Cairns, I. H., Robinson, P. A., & Kuncic, Z. 2001, *J. Geophys. Res.*, 106, 25041
Krucker, S., Larson, D. E., Lin, R. P., & Thompson, B.J. 1999, *ApJ* 519, 864
Kurth, W. S., Gurnett, D. A., Scarf, F. L., & Poynter, R. L. 1984, *Nature*, 312, 27
Lazio, T. J. W., Farrell, W. M., Dietrick, J., *et al.* 2004, *ApJ*, 612, 511
Maia, D. J. F., Gama, R., Mercier, C., *et al.* 2006 *ApJ*, 660, 874
Mayor, M., & Queloz, D. 1995, *Nature*, 378, 355
McLean, D., & Labrum, D. 1985, *Solar Radiophysics* (Cambridge: CUP)
McReady, Pawsey, & Payne-Scott 1947
Vršnak, B., Warmuth, A., Temmer, M., Veronig, A., Magdalenić, J., Hillaris, A., & Karlický, M. 2006, *A&A*, 448, 739
Winglee, R. M., Dulk, G. A., & Bastian, T. S. 1986, *ApJ* (Letters), 309, L59
White, S. M., Bastian, T. S., Bradley, R., Parashare, C., & Wye, L. 2006, in: N. Kassim, M. Perez, W. Junor, & P. Henning (eds.), *From Clark Lake to the Long Wavelength Array: Bill Erickson's Radio Science*, *ASP-CS*, 345, 176
Zarka, P., Treumann, R. A., Ryabov, B. P., & Ryabov, V. B. 2001, *Ap&SS*, 277, 293
Zarka, P. 2004, *Adv. Sp. Res.*, 33, 2045

Highlights of Astronomy, Volume 14
IAU XXVI General Assembly, 14-25 August 2006
Karel A. van der Hucht, ed.

doi:10.1017/S1743921307011003

Sporadic radio emission of the Sun in the decametre range

Valentin N. Melnik[1], **Alexander A. Konovalenko**[1], **Helmut O. Rucker**[2] **and Alain Lecacheux**[3]

[1]Radio Astronomy Institute, 4 Chervonopraporna str., Kharkiv, Ukraine
email: melnik@ira.kharkov.ua

[2]Space Research Institute, Schmiedlstrasse 6, A-8042, Graz, Austria

[3]Department de Radioastronomie CNRS UMR 8644, Observatoire de Paris, France

Abstract. Results of the last observations of solar sporadic radio emission at the UTR-2 radio telescope (Kharkov, Ukraine) at the frequencies 10 - 30 MHz are presented. The use of new backend facilities, the DSP and 60-channel spectrometer, allows us to obtain data with time resolution up to 2 ms and frequency resolution of 12 kHz in the continuous frequency band 12 MHz. Usual Type III bursts, Type IIIb bursts, U- and J-bursts in the decameter range are discussed. Special attention is paid to detection and analysis of Type II bursts and their properties, newly discovered fine time structures of Type III bursts, Type III-like bursts, s-bursts, new observational features of drift pair bursts, and 'absorption' bursts.

Keywords. Sun: radio radiation

Solar sporadic radio emission has been studied in the decameter range with UTR-2 (effective area 150 000 m^2; working frequency band 10 - 30 MHz) since the 1970s. Until the end of the 1990s, the focus was on Type III bursts, Type IIIb bursts, and drift pairs. After putting into operation new backend facilities, DSP (bandwidth 12 MHz; time and frequency resolution 2 ms and 12 kHz, respectively; dynamic range 70 dB; Kleewein *et al.* 2004) and 60-channel spectrometer (single channel bandwidth 3 kHz; time resolution 10 ms; dynamic rage 40 dB), the number of phenomena investigated has increased dramatically. Today, there are opportunities to observe Type II bursts, inverted U- and J-bursts, s-bursts, decameter spikes, Type III-like bursts, and bursts in absorption.

Type III bursts are usually considered to have a smooth time profile. This indicates a uniform electron stream, which generates the bursts. For the first time in 2002, observing with DSP we found great number of Type III bursts with fine time structures. In the most cases the parent Type III burst consists of sub-bursts with 1 s durations, which drift from high frequencies to low frequencies. These cases can be divided into three groups of bursts: (*i*) drift rates equal to or greater that for parent Type III burst; (*ii*) rates about 1 MHz s^{-1}; and (*iii*) rates smaller than 100 kHz s^{-1}. There are some cases when Type III bursts consist of sub-bursts with positive and negative drift rates at the same time.

Decameter Type III-like bursts were registered in the 2002 - 2004 observation campaigns. Their durations were about 1 - 2 s and their drift rates were sometimes that of usual decameter Type III bursts rates. We found that the number of Type III-like bursts is at a maximum when the active region associated with them is situated near the central meridian. The effects of propagation seem to play an important role in the formation of these bursts.

The inverted U- and J-bursts with turning frequency near 10 MHz, which are generated by fast electron streams in high (up to 2 $R_\odot$) coronal loops, were observed for the first time. The spectral and frequency characteristics of ascending branches of these bursts are close to those of usual Type III bursts. Thus, about identical electron streams are propagating in the average corona and into loops. The relations of turning frequencies in observed harmonic J-burst pairs changed in the range 1.5 - 2 with an average value of 1.78.

S-bursts are observed have durations of 0.3 - 0.5 s (Dorovskyy *et al.* 2006). Their drift rates are close to that for drift pairs bursts and are equal to about 1 MHz s^{-1} (Mel'nik *et al.* 2005). In contrast to drift pairs, s-burst drift rates depend upon the frequency. This dependence is approximately similar to well-known dependence for Type III bursts. It seems that in both cases radio emission sources propagate in average corona.

For the first time, we have observed Type II bursts in the decameter range 10 - 30 MHz in 2001 (Mel'nik *et al.* 2004). Such bursts, including Type II bursts with herringbone structure, are registered at UTR-2 regularly, with on average 1 - 2 bursts a month. Our observations show that not only do Type II bursts with herringbone structure consist of sub-bursts, but usual decameter Type II bursts have fine time structure in the form of sub-bursts with positive and negative drift rates. This is evidence for electron acceleration on both sides of the shock. We found a wave-like radio emission of Type II burst backbone, which is interpreted as shock intersection of coronal structures. There are different properties of sub-bursts on both sides of backbone. It gives an opportunity to diagnose plasma before and behind a shock.

The burst in absorption was observed in frequency range 10 - 30 MHz in 2003 (Konovalenko *et al.* 2005). Its drift rate was 120 kHz s^{-1} and duration was about 1 min. The absorption region was very large and it moved with high linear velocity (more than 2000 km s^{-1}). The cause of this phenomenon is still unclear.

Our results show that ground based observations with high frequency and time resolution in the frequency range 10 - 30 MHz promise to give new interesting information about processes and phenomena occurring at heights 0.5 - 2 R_s in the solar corona, which can not be obtained by other methods. So creation of new large radio telescopes (LOFAR, LWA) is extremely urgent.

Acknowledgements

VNM and AAK are supported by INTAS under grant number 03-5727.

References

Dorovskyy, V. V., Mel'nik, V. N., Konovalenko, A. A., Rucker, H. O., Abranin, E. P., & Lecacheux, A. 2006, in: H. O. Rucker, W. S. Kurth & G. Mann (eds.), *Proc. Int. Workshop Planetary Radio Emissions VI* (Vienna: Osterreichischen Akademie der Wissenschaften), p. 383

Kleewein, P., Rosolen, C., & Lecacheux, A. 2004, in: H. O. Rucker, S. J. Bauer & A. Lecacheux (eds.), *Proc. Int. Workshop Planetary Radio Emissions IV* (Vienna: Osterreichischen Akademie der Wissenschaften), p. 349

Konovalenko, A. A., Stanislavsky, A. A., Abranin, E. P., Dorovskyy, V. V., & Mel'nik, V. N. 2005, *Kinematicsand Physics of Celestial Bodies, Suppl. Ser.*, 5, 78

Mel'nik, V. N., Konovalenko, A. A., Rucker, H. O., Stanislavsky, A. A., Abranin, E. P., Lecacheux, A., Mann, G., Warmuth, A., Zaitsev, V. V., Boudjada, M. Y., Dorovskii, V. V., Zaharenko, V. V., Lisachenko, V. N., & Rosolen, C. 2004, *Solar Phys.*, 222, 151

Mel'nik, V. N., Konovalenko, A. A., Dorovskyy, V. V.,Rucker, H. O., Abranin, E. P., Lisachenko, V. N., & Lecacheux, A. 2005, *Solar Phys.*, 231, 143

Highlights of Astronomy, Volume 14
IAU XXVI General Assembly, 14-25 August 2006
Karel A. van der Hucht, ed.

doi:10.1017/S1743921307011015

NRAO 43-m telescope operation at 170 - 1700 MHz: a Bi-Static Radar Collaboration

Glen Langston

National Radio Astronomy Observatory, Green Bank, WV 24944, USA
email: glangsto@nrao.edu

Abstract. The NRAO 43m telescope has been refurbished and begun regular observations in the frequency range 170 - 1700 MHz. The 43 m operations support a Bi-Static Radar Collaboration to measure the Earth's ionospheric turbulence. Researchers from Chalmers University of Technology in Sweden have designed and built a unique design wide-band feed, 150 - 1700 MHz. Lincoln Laboratories/MIT has packaged the feed with room temperature low noise amplifiers. Lincoln Laboratories has installed a high-dynamic range RF system together with a wide-band sampler system. The NRAO operates the 43 m telescope according to schedules authored by Lincoln Laboratories. Currently the 43 m telescope is tracking spacecraft 48 hr a week. The tracking antenna operation is completely automated. A group at MIT/Haystack have installed a second radar experiment at the 43 m as well as an array of 6 'discone' antennas. Their experiment is testing the use of reflected FM radio stations as probes of the ionosphere.

Keywords. atmospheric effects, turbulence

Experiments

Three separate experiments are being implemented at the NRAO 43 m antenna. The main experiment is a "bi-static radar spacecraft tracking experiment" led by R. Sridharan of MIT's Lincoln Laboratory. In this experiment the radar signals are transmitted towards spacecraft and reflections simultaneously sampled at the NRAO 43 m and at the Millstone hill radar station (see documentation at `<http://www.gb.nrao.edu/mlln/>`.)

The second experiment is an 'incoherent scattering radar' in which the ionosphere radar is directly illuminated by a 440 MHz radar at MIT's Haystack observatory. The reflected signals are recorded at the 43 m (see Phil J. Erickson, `<http://www.haystack.mit.edu/~pje>`.

The third experiment is a passive (FM radio) radar prototype experiment (see Frank Lind, `<http://www.haystack.mit.edu/~flind>`). In this experiment commercial FM radio stations are used to illuminate the atmosphere and eventually > 1000 detection stations will sample the signals reflected off of the ionosphere. The collection of these signals will allow measurement of the total electron content (TEC) of the ionosphere.

Highlights of Astronomy, Volume 14
IAU XXVI General Assembly, 14-25 August 2006
Karel A. van der Hucht, ed.

doi:10.1017/S1743921307011027

New Galactic ISM results and progress towards low-frequency spectroscopy

Anne J. Green[1], **Tara Murphy**[1], **Duncan Campbell-Wilson**[1], **Michael Kesteven**[2] **and John Bunton**[2]

[1]School of Physics, University of Sydney, NSW 2006, Australia
email: agreen@physics.usyd.edu.au

[2]CSIRO Australia Telescope National Facility, PO Box 76, Epping, NSW 1710, Australia

Abstract. We dicuss new Galactic ISM results and progress towards low-frequency spectroscopy

Keywords. radio continuum: ISM, stars: formation, ISM: supernova remnants, surveys

1. ISM results with MOST

The Molonglo Observatory Synthesis Telescope (MOST) currently operates as a continuum instrument at 843 MHz with sub-arcminute resolution and excellent brightness sensitivity ($1\,\sigma$ rms, ~ 1 mJy beam^{-1}). An imaging survey of the southern sky (SUMSS; Bock, Large & Sadler 1999) is nearing completion. This includes a second epoch survey of the Galactic plane (Green 2002) from which a catalogue of small diameter sources is in preparation. Correlation of the images and catalogue with 20 GHz observations from the Australia Telescope Compact Array and $8\,\mu$m from the MSX database (following Cohen & Green 2001) will be used to study triggered star formation in the interstellar medium (ISM) by identifying the youngest ultra-compact H II regions and small evolving supernova remnants.

2. Low-frequency spectroscopy with SKAMP

Innovative high-speed digital signal processing techniques will provide a new low-frequency spectral line capability with full polarisation and wide-field imaging, identified as the SKA Molonglo Prototype (SKAMP) project. It will use existing infrastructure and the large collecting area of MOST. Initially, we will target redshifted hydrogen in absorption at $z \simeq 0.7$ through a blind survey over an area of 2000 deg^2. Preliminary testing of the new 96-station, 6000-channel correlator is expected to start in 2007.

Acknowledgements

The MOST is owned and operated by the School of Physics within the University of Sydney, supported by the Australian Research Council and the University of Sydney.

References

Bock, D.-C. J, Large, M. I., & Sadler, E. M. 1999, *AJ*, 117, 1578
Cohen, M., & Green, A. J. 2001, *MNRAS*, 325, 531
Green, A. J. 2002 in: A. P. Rao, G. Swarup & Gopal-Krishna (eds.), *The Universe at Low Radio Frequencies*, Proc. IAU Symposium No. 199 (San Francisco: ASP), p. 259

Highlights of Astronomy, Volume 14
IAU XXVI General Assembly, 14-25 August 2006
Karel A. van der Hucht, ed.

doi:10.1017/S1743921307011039

Diffuse radio sources in clusters of galaxies: models and long-wavelength radio observations

Craig L. Sarazin

Department of Astronomy, University of Virginia, Charlottesville, VA 22903, USA
email: sarazin@virginia.edu

Abstract. Clusters of galaxies contain several types of diffuse radio sources with very steep radio spectra which are associated with the cluster environment, including central radio bubbles, cluster radio relics, and cluster radio halos. Radio halos and relics are found only in merging clusters. Cluster radio relics may be produced by particle acceleration in merger shocks, while radio halos, may result from electron re-acceleration by turbulence produced by mergers. Secondary production of electrons and positrons by hadronic interactions also plays a role. If cluster radio halos and relics are related to mergers, then deep low frequency radio surveys could detect 1000's of clusters. Long-wavelength radio observations have a great potential to help us understand clusters and large scale structure, and can provide a diagnostic of cluster mergers, which affect the use of clusters in cosmological and dark energy studies.

Keywords. shock waves, galaxies: clusters: general, cooling flows, cosmological parameters, intergalactic medium, radio continuum: galaxies, X-rays: galaxies: clusters

1. Introduction

Clusters of galaxies contain several types of radio sources which are associated with the cluster environment. First, the cD galaxies at the centers of cooling core clusters almost always have small, spatially distorted, mainly FR I radio sources; the radio lobes of these sources often appear to be displacing the X-ray emitting cluster gas, producing holes in the X-ray emission surrounded by bright shells (e.g., Blanton *et al.* 2001). Second, there are cluster radio halo sources, which tend to be relatively symmetric and centrally located, are not strongly polarized, and can be ~ Mpc in size (e.g., Feretti *et al.* 2004). Third, there are cluster radio relic sources, which are often elongated, are generally located in the periphery of clusters, are often strongly polarized, and sometimes occur in pairs on opposite sides of the cluster center (e.g., Giovannini & Feretti 2004). Both cluster radio halos and radio relics are not associated with individual radio galaxies, and only occur in merging clusters. Finally, there are centrally located 'radio relics,' which often are filamentary; recent radio and X-ray observations suggest that they may be radio lobes from the central cD galaxies, displaced by buoyancy or motions in the cluster (Fujita *et al.* 2002). The diffuse potions of all of these cluster radio sources tend to have rather steep radio spectra; in part, this is due to the fact that they are confined by the high pressure, X-ray emitting gas in clusters, and thus suffer radiative losses rather than adiabatic expansion.

2. Mergers and radio halos and relics

Clusters of galaxies form hierarchically through the gravitational merger of smaller clusters and groups. Major cluster mergers are the most energetic events in the Universe since the Big Bang, releasing gravitational binding energies of as much as $\sim 10^{64}$ erg. Mergers drive shocks into the intracluster medium, and generate turbulence. The shocks and turbulence may accelerate or re-accelerate relativistic particles; the relativistic electrons will emit synchrotron radio emission in the intracluster magnetic field. The electrons which produce radio emission at most observable wavelengths have energies of many GeV, and have relatively short lifetimes ($\sim 10^8$ yr) in clusters due to synchrotron and inverse Compton (IC) emission. Thus, radio-emitting electrons accelerated in a mergers will tend to survive only for a time scale which is comparable to the duration of the merger. Ions have much longer lifetimes in clusters, and the total input of relativistic hadrons in a cluster from mergers and other sources (e.g., AGNs) will tend to accumulate over the lifetime of the cluster. Thus, another source of relativistic electrons in the intracluster medium is secondary production by hadrons. For example, relativistic protons will collide with thermal protons in the intracluster medium and generate pions and other mesons, $p + p \rightarrow p + p + n\pi$, and the charged pions will decay to produce electrons and positrons, $\pi^{\pm} \rightarrow \mu^{\pm} + \nu_{\mu}(\bar{\nu}_{\mu})$, $\mu^{\pm} \rightarrow e^{\pm} + \nu_e(\bar{\nu}_e) + \bar{\nu}_{\mu}(\nu_{\mu})$. Secondary production will give a steady source of relativistic electrons which should be present in all clusters; the fact that bright radio halos and relics are only found in merging clusters suggests that the bulk of the radio emitting electrons are accelerated during the merging.

Two mechanisms by which cluster mergers can lead to the (re-)acceleration of relativistic particles are shock acceleration and turbulent acceleration. It seems most likely that shock acceleration is the mechanism for producing cluster radio relics. They are elongated, often located at positions which are sensible locations for merger shocks in their host clusters, often have a sharp edge (presumably the side of the relic at the shock), and sometimes have radio spectra which steepen away from the shock edge (as might be expected due to radiative losses by the accelerated electrons) (e.g., Giovannini & Feretti 2004; Clarke & Ensslin 2006). An alternative model for radio relics is that they start as 'radio ghosts', or old radio plasma from a radio galaxy in which the high energy electrons have lost most of their energy. A merger shock could adiabatically compress such a radio ghost, and reactivate it as a visible radio source (Ensslin & Gopal-Krishna 2001).

On the other hand, cluster radio halos are unlikely to be produced directly by merger shock acceleration of electrons. The halos are large and fairly symmetrical, and they extend too far from merger shocks for the radio emitting electrons to have moved without losses. Thus, halos appear to require a distributed source of relativistic electrons. This might be secondary production, or turbulent acceleration following the passage of merger shocks (e.g., Fujita *et al.* 2003; Cassano & Brunetti 2005). For turbulent re-acceleration, one needs a reservoir of low energy, relativistic electrons; these have long lifetimes in clusters, so they may be due to previous mergers and/or AGN activity. One also requires that post-merger regions of merging clusters have turbulence with an energy density of $\sim$ 20% of the thermal energy density of the intracluster gas; this agrees reasonably with the results from numerical simulations (Ricker & Sarazin 2001).

While all the clusters which have so far been observed to contain cluster radio relics or halos are merging clusters, it certainly is not the case that all merging clusters are observed to have diffuse radio emission. However, it is clear that all of the existing surveys for diffuse radio sources in clusters are strongly limited by their surface brightness sensitivity. Because of their steep radio spectra, we need very deep, low frequency radio surveys to detect the bulk of merging clusters (probably 1000's).

3. Clusters and cosmology

Clusters of galaxies are very important cosmological probes, and X-ray or Sunyaev-Zeldovich (SZ) surveys of clusters are likely to be used in the next few years to study cosmology and dark energy (Mohr 2005). All of these surveys require estimates of the masses of the clusters, which are usually based on their X-ray or SZ properties assuming equilibrium. Cluster mergers pose a problem for these surveys, since during a merger the X-ray luminosity and temperature and SZ effect are considerably boosted, and the clusters are not in hydrostatic equilibrium (Randall *et al.* 2002; Wik *et al.* 2006).

In principle, self-calibration (Mohr 2005) may deal with these problems, but it might be better if a simple diagnostic could be found which would allow one to determine which clusters were undergoing mergers, and which would quantify the strength of the merger (e.g., the rate of energy input from merger shocks). Then, merging clusters could either be culled from the cluster samples, or their properties could be corrected for the effects of mergers. This is difficult to do based on either X-ray or SZ observations alone. For example, it is difficult to detect mergers occurring along our line-of-sight. If possible mergers in clusters are detected by using the differences of their X-ray or SZ properties from standard scaling relations, and these clusters are removed from samples, this will tend to bias the results. What would be ideal would be to find a 'Merge-o-Meter' – some diagnostic for mergers that did not rely on their X-ray or SZ properties.

In principle, low frequency, diffuse radio emission (from radio halos and relics) might provide such a diagnostic for cluster mergers for cosmology and dark energy studies. Low frequency radio surveys of clusters could be used to either remove merging clusters from relaxed cluster samples for cosmology, or to attempt to correct the cluster X-ray and/or SZ properties for the boosts produced by mergers. One might expect that the populations of the relativistic electrons generated by mergers through shock or turbulent acceleration would increase in proportion to the energy input for the merger; this connection could provide a physical basis for correcting the X-ray and/or SZ properties for merger boosts.

Acknowledgements

I thank T. Clarke for useful comments. This work was supported by NASA through *XMM-Newton* awards NNG04GO34G, NNG04GO80G, NNG04GP46G, NNG05GA34G, NNG05GO50G, and NNG06GD54G, and through *Chandra* awards GO4-5133X, GO4-5137X,and GO5-6126X.

References

Blanton, E. L., Sarazin, C. L., McNamara, B. R., & Wise, M. W. 2001, *ApJ* (Letters), 558, L15
Cassano, R., & Brunetti, G. 2005, *MNRAS*, 357, 1313
Clarke, T. E., & Ensslin, T. A. 2006, *AJ*, 131, 2900
Ensslin, T. A., & Gopal-Krishna 2001, *A&A*, 366, 26
Feretti, L., Brunetti, G., Giovannini, G., Kassim, N., Orru, E., & Setti, G. 2004, *JKAS*, 37, 315
Fujita, Y., Sarazin, C. L., Kempner, J. C., Rudnick, L., Slee, O. B., Roy, A. L., Andernach, H., & Ehle, M. 2002, *ApJ*, 575, 764
Fujita, Y., Takizawa, M., & Sarazin, C. L. 2003, *ApJ*, 584, 190
Giovannini, G., & Feretti, L. 2004, *JKAS*, 37, 323
Mohr, J. J. 2005, in: S. C. Wolff, & T. R. Lauer (eds.), *Observing Dark Energy*, *ASP-CS*, 339, 140
Randall, S. W., Sarazin, C. L., & Ricker, P. M. 2002, *ApJ*, 577, 579
Ricker, P. M., & Sarazin, C. L. 2001 *ApJ*, 561, 621
Wik, D. R., Sarazin, C. L., Ricker, P. M., & Randall, S. W. 2006, in preparation [see also: *BAAS* 38, 371, #9, #13.33

Highlights of Astronomy, Volume 14
IAU XXVI General Assembly, 14-25 August 2006
Karel A. van der Hucht, ed.

doi:10.1017/S1743921307011040

Particle acceleration processes in the cosmic large-scale structure

Torsten A. Enßlin[1] and Christoph Pfrommer[1,2]

[1]Max-Planck-Institut für Astrophysik,
Karl-Schwarzschild-Straße 1, D85741 Garching, Germany
email: tensslin@mpa-garching.mpg.de

[2]Canadian Institute for Theoretical Astrophysics,
60 St. George Street, Toronto, Ontario, M5S 3H8 Canada

Abstract. The energetic shock waves associated with the violent large-scale structure formation process are able to accelerate relativistic electrons and protons. The induced non-thermal emission, especially at long radio wavelength, provides a fascinating perspective into structure formation, the relativistic Universe, and cosmic magnetic fields.

Keywords. acceleration of particles, large-scale structure of universe

---

The formation of the cosmic large-scale structure is accompanied by shock waves which permeate and enclose the denser regions of the Universe (Figure 1). These shock waves and the induced turbulence can lead to particle acceleration. Four particle acceleration processes may be relevant for the production of relativistic particle populations (cosmic rays, CRs) in the large-scale structure, namely adiabatic compression, diffusive shock acceleration, stochastic acceleration, and particle reactions. All these acceleration processes have to compete against energy losses due to adiabatic expansion, radiative cooling (synchrotron, inverse Compton, bremsstrahlung, hadronic interactions), and non-radiative cooling (Coulomb losses). Different processes will dominate in different cosmic environments, particle energy ranges, and for different particle types (electrons/protons).

Adiabatic compression can revive the observable radio emission of old CR electrons in fossil radio plasma of a former radio galaxy (radio ghost). The electrons gain energy adiabatically, which in combination with the increasing magnetic field strength raises significantly the synchrotron frequency of the electrons at the high energy cooling cutoff. This may explain the small radio relics observed in several galaxy clusters (Ensslin & Gopal-Krishna 2001; Ensslin & Brüggen 2002).

Diffusive shock acceleration at shock waves can accelerate thermal particles, and thereby can become very efficient in generating CRs if the injection spectrum is sufficiently flat, which depends on the shock Mach number. Thus, the higher-Mach number shocks in the outskirts of clusters and the accretion shock waves have a higher CR injection efficiency compared to the shock in cluster centers, as shown in Figure 1. The resulting radio emission probably explains the giant radio relics (radio tsunamis, Ensslin *et al.* 1998; Figure 2).

Stochastic acceleration of CRs by plasma waves can re-accelerate existing CR populations. This may explain the radio halos in galaxy clusters by the re-activation of longer-lived 0.3 GeV electrons to $\sim 10\,\mathrm{GeV}$ (Sarazin 2007, these proceedings; Brunetti 2007, these proceedings).

Particle reactions of long-lived proton populations will induce secondary electrons from the hadronic chain $pp \rightarrow \pi^{\pm} \rightarrow \nu_{\mu}\, \mu^{\pm} \rightarrow \nu_{\mu}\, \nu_{e}\, e^{\pm}$ which may also be able to produce cluster radio halos (Dennison 1980; Figure 2).

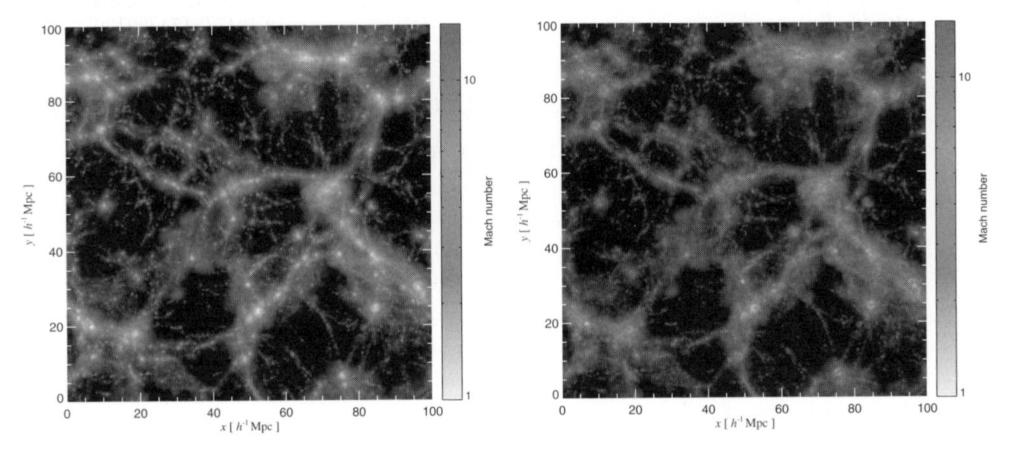

Figure 1. Energy dissipation by shock waves in a cosmological structure formation simulation by Pfrommer *et al.* (2006). The brightness displays the logarithm of the dissipation rate, the color indicates the (dissipation weighted) Mach numbers. *Left*: total dissipation. *Right*: dissipation into cosmic rays.

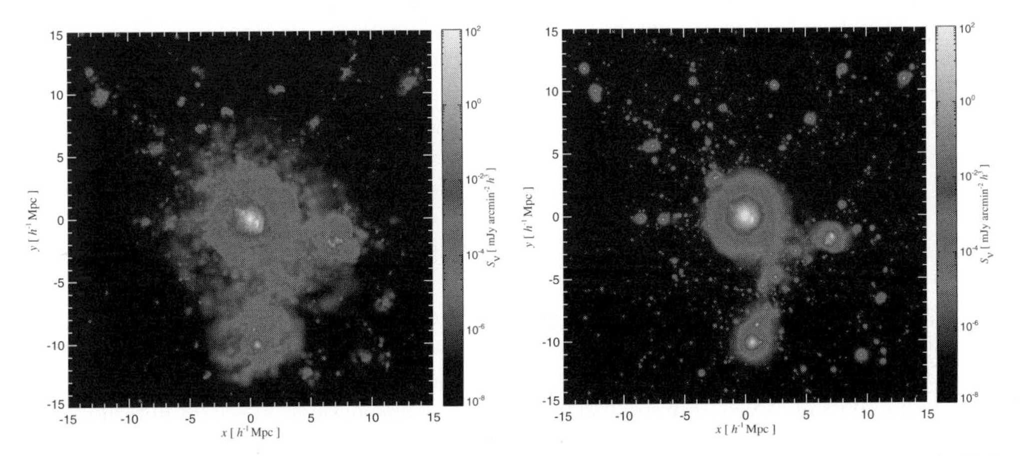

Figure 2. 150 MHz radio emission of a simulated galaxy cluster (Pfrommer *et al.* 2006). *Left*: emission due to shock accelerated electrons (giant radio relics/radio tsunamis). *Right*: emission due to secondary electrons from hadronic interactions of formerly shock accelerated protons (hadronic radio halo).

Long-wavelength radioastronomy is expected to play the leading role in unraveling particle acceleration processes in the cosmic large-scale structure.

Acknowledgements

TAE thanks the SOC for the invitation to the JD12 session and apologizes for the very incomplete presentation of this research field on these two short pages.

References

Dennison, B. 1980, *ApJ* (Letters), 239, L93
Ensslin, T.A., Biermann, P.L., Klein, U., & Kohle, S. 1998, *A&A*, 332, 395
Ensslin, T.A., & Gopal-Krishna 2001, *A&A*, 366, 26
Ensslin, T.A., & Brüggen, M. 2002, *MNRAS*, 331, 1011
Pfrommer, C., Springel, V., Ensslin, T. A., & Jubelgas, M. 2006, *MNRAS*, 367, 113

Highlights of Astronomy, Volume 14
IAU XXVI General Assembly, 14-25 August 2006
Karel A. van der Hucht, ed.

doi:10.1017/S1743921307011052

Particle *re*-acceleration in the ICM and low-frequency observations

Gianfranco Brunetti

INAF – Istituto di Radioastronomia, via P. Gobetti 101, I-40129 Bologna, Italy
email: brunetti@ira.inaf.it

Abstract. The particle reaceleration model is one of the most promising possibilities to explain the Mpc-scale diffuse radio emission detected in a number of galaxy clusters. Ongoing and future radio observations at low frequencies may help in constraining and testing this model.

Keywords. acceleration of particles, turbulence, galaxies: clusters: general

---

The Mpc-scale radio non-thermal emission (Radio Halos, RHs) detected in a growing number of galaxy clusters (GCs) proves the presence of GeV radiating electrons (e.g., Feretti 2005). Relativistic hadrons should accumulate in GCs and direct measurements of the hadronic content may come from future gamma-ray observations (e.g., Blasi 2004).

The origin of the radio emitting electrons in RHs is still a matter of debate: these particles have a lifetime much shorter than that necessary to diffuse over Mpc scales. This *diffusion problem* (Jaffe 1977) poses a fundamental theoretical problem on the origin of the emitting particles. They may be re-accelerated or they may be secondary products of collisions between CR protons and thermal protons in the ICM (e.g., Sarazin 2004).

The injection and confinement of CR hadrons in the ICM has been discussed in several papers (Völk *et al.* 1996; Berezinsky *et al.* 1997; Ensslin *et al.* 1997). These particles actually accumulate in GCs for periods comparable to the age of the clusters. Also, the amplification of the magnetic field in GCs and its decay take cosmological time scales (e.g., Dolag 2006; Subramanian 2006). Thus, in the framework of the secondary model, RHs should be long-lived sources, with lifetimes $\sim$ the ages of clusters. On the other hand, these sources are always found in merging clusters (e.g., Buote 2001) and are quite rare, at least at the level of present radio observations (e.g., Giovannini *et al.* 1999), which means that their typical lifetime is probably relatively short (~ 1 Gyr, Hwang 2004).

A major step forward can be achieved by looking at non-radio emitting clusters; the steep spectrum of RHs makes low frequency observations efficient tools in this research. A pointed survey of a sample of about 50 massive GCs in the redshift bin $z = 0.2 - 0.4$ has been recently carried out at the GMRT at 604 MHz (Venturi *et al.* 2006, 2007). Only about 15 % of these massive GCs host giant RHs, while no diffuse radio emission is detected in the great majority of these GCs at the brightness level of the observations which is ~ 10 times below that of the NVSS at 1.4 GHz. Although the radio power ($P_{\rm R}$) of giant RHs is known to correlate with the X-ray luminosity ($L_{\rm X}$) of the parent GCs (e.g., Feretti 2005), these results suggest that the bulk of GCs does not follow this correlation. The region in the $P_{\rm R}$-$L_{\rm X}$ plane that is filled by the bulk of the massive GCs in the GMRT survey is indeed found to lie about 2 orders of magnitude below (in terms of radio power) that spanned by the $P_{\rm R}$-$L_{\rm X}$ correlation (Brunetti *et al.*, in prep). These findings have important consequences on our present view of the origin of RHs in GCs. They confirm that RHs are transient phenomena while the bulk of GCs is basically non-radio emitting, and this disfavours any scenario in which the emitting electrons are injected in GCs for cosmological (long) time-scales.

Apparently the reacceleration model nicely fits the evidence that RHs are transient and always found in dynamically disturbed GCs. If stochastic reacceleration is driven by turbulence generated in cluster mergers, the cascading time of turbulence from large scales (300 - 500 kpc) sets a natural timescale for the lifetime of RHs which is of the order of 1 Gyr (or even less). This model (e.g., Brunetti *et al.* 2001; Petrosian 2001) is based on the complex issue of stochastic particle re-acceleration whose physical details are still poorly understood and difficult to test. Theoretical attempts to calculate the time-dependent stochastic particle acceleration process in the ICM usually focus on the Alfvén modes (e.g., Brunetti *et al.* 2004), and assume that these modes are injected at small, resonant, scales by some process connected to the presence of turbulence at larger scales. An additional possibility comes from the action of compressible modes injected at large scales in GCs during cluster-cluster mergers. Fast modes and magnetosonic waves couple with relativistic particles via TTD-resonance and non-resonant turbulent compression, and this might provide a contribution to the reacceleration process (Brunetti 2006; Brunetti & Lazarian, in prep).

Attempts to calculate the statistical properties of RHs in the framework of the reacceleration model have been carried out only recently (e.g., Cassano & Brunetti 2005). The probability to develop a RH is found to depend critically on the mass of the hosting GC as the injection of turbulence on Mpc scales is more efficient with increasing cluster mass. This means that there is a threshold in cluster mass below which giant RHs rarely develop, and this drives a low radio power cut-off in the expected luminosity functions of giant RHs (Cassano *et al.* 2006). On the other hand, this mass threshold should depend on observing frequency: at lower frequencies the synchrotron emission comes from lower energy electrons which can be accelerated much easily, and thus RHs in GCs with slightly lower mass come into play. Since the number of GCs per comoving volume increases with decreasing mass, the number of detectable giant RHs in the re-acceleration model is expected to increase at lower observing frequencies. This is a clear expectation of the model which can be tested by future low frequency observations: present calculations show that LOFAR and the LWA might discover more than 1000 new giant RHs (Cassano *et al.* 2006).

References

Blasi, P. 2004, *J. Korean Astron. Soc.* 37, 483
Berezinsky, V. S., Blasi, P., & Ptuskin, V. S. 1997, *ApJ*, 487, 529
Brunetti, G. 2006, *AN*, 327, 615
Brunetti, G., Setti, G., Feretti, L., & Giovannini, G. 2001, *MNRAS*, 320, 365
Brunetti, G., Blasi, P., Cassano, R., & Gabici, S. 2004, *MNRAS*, 350, 1174
Buote, D. A. 2001, *ApJ* (Letters), 553, L15
Cassano, R., & Brunetti, G. 2005, *MNRAS*, 357, 1313
Cassano, R., Brunetti, G., & Setti, G. 2006, *MNRAS*, 369, 1577
Dolag, K. 2006, *AN*, 327, 575
Ensslin, T. A., Biermann, P. L., Kronberg, P. P., & Wu, X-P. 1997, *ApJ*, 477, 560
Feretti, L. 2005, *Adv. Space Res.*, 36, 729
Giovannini, G., Tordi, M., & Feretti, L. 1999, *New Astron.*, 4, 141
Hwang, C.-Y. 2004, *J. Korean Astron. Soc.*, 37, 461
Jaffe, W. J. 1977, *ApJ*, 212, 1
Petrosian, V. 2001, *ApJ*, 557, 560
Sarazin, C. 2004, *J. Korean Astron. Soc.*, 37, 433
Subramanian, K., Shukurov, A., & Haugen, N. E. L. 2006, *MNRAS*, 366, 1437
Venturi, T., Parma, P., de Ruiter, H. R. 2006, *MemSAI Suppl.*, 10, 111
Venturi, T., Giacintucci, S., Brunetti, *et al.* 2007, *A&A*, 463, 937
Völk, H. J., Aharonian F. A., & Breitschwerdt D. 1996, *Space Sci. Revs.*, 75, 279

Highlights of Astronomy, Volume 14
IAU XXVI General Assembly, 14-25 August 2006
Karel A. van der Hucht, ed.

doi:10.1017/S1743921307011064

Long-wavelength spectral studies of giant radio sources

Emanuela Orrù[1,2]

[1]Dipartimento di Fisica, Università degli Studi di Cagliari,
Cittadella Universitaria, I-09042 Monserrato (CA), Italy
email: eorru_s@ira.inaf.it

[2]INAF - Osservatorio Astronomico di Cagliari,
Loc. Poggio dei Pini, Strada 54, I-09012 Capoterra (CA), Italy

Abstract. We study giant radio sources at long wavelengths.

Keywords. giant radio sources

We have performed a long-wavelength spectral study of 3C 35 and 3C 223. These are two giant radio sources selected from a sub-sample of giant radio sources taken from the Lara *et al.* (2001) and Ishwara-*Chandra* (1999) samples.

Giant radio sources (GRS) are defined as those objects with a projected linear size larger than 1 Mpc ($H_0 = 50\,\mathrm{km\,s^{-1}Mpc^{-1}}$, $q_0 = 0.5$).

We produced new Very Large Array images at 74 and 327 MHz. Spectral analyses have been made, including 1.4 GHz images from the literature. For both sources, in this range of frequencies, the spectral shape follows a power law in the hot spot with a slope of $\alpha \simeq 0.5$ - 0.6. In the inner region of the lobe, near the core, the shape of the spectrum shows a curvature in the highest frequency region, where α reaches values of ~ 1.3 - 1.5. Such steepening is in agreement with the synchrotron aging of the emitting relativistic electrons.

The brightness profiles of the lobes of these sources also have been obtained. For 3C 35 the brightness profile shows a constant trend at 74 and 327 MHz while at 1.4 GHz, a decrease of intensity is evident from the hot spot to the inner of the lobe. The brightness profile of 3C 223 is characterized by a decrease of brightness from the hot spot to the inner lobe at all the frequencies analyzed. As radiative losses at long wavelengths are negligible, the different behaviours of the brightness profiles in the two sources might suggest that adiabatic losses are more relevant in 3C 223 than in 3C 35.

A more detailed analysis is in progress (Orrù, PhD Thesis; Orrù *et al.*, in prep.).

References

Lara, L., Márquez, I., Cotton, W. D., Feretti, L., Giovannini, G., Marcaide, J. M., & Venturi, T. 2001, *A&A*, 378, 826

Highlights of Astronomy, Volume 14
IAU XXVI General Assembly, 14-25 August 2006
Karel A. van der Hucht, ed.

doi:10.1017/S1743921307011076

H I 21-cm observations of cosmic re-ionization

Christopher L. Carilli

National Radio Astronomy Observatory, Socorro, NM 87801 USA
email: ccarilli@nrao.edu

Abstract. I review the potential for observing cosmic re-ionization using the H I 21 cm line.

Keywords. atomic processes, line formation, cosmology: observations, cosmology: theory

---

1. Introduction

Cosmic re-ionization corresponds to the transition from a fully neutral intergalactic medium (IGM) to an (almost) fully ionized IGM caused by the UV radiation from the first luminous objects. Re-ionization is a key benchmark in cosmic structure formation, indicating the formation of the first luminous objects. Re-ionization, and the preceding 'dark ages', remain the last of the major phases of cosmic evolution left to explore. Recent observations of the Gunn-Peterson effect, i.e., Ly-α absorption by the neutral IGM, toward the most distant QSOs ($z \simeq 6$), and the large scale polarization of the CMB, have set the first constraints on the epoch of re-ionization. These data, coupled with the study of high-z galaxy populations and other observations, suggest that re-ionization was a complex process, with significant variance in both space and time, starting around $z \simeq 10$, with the last vestiges of the neutral IGM being etched away by $z \simeq 6$ (Fan *et al.* 2006; Ciardi & Ferrara 2005).

The most direct and incisive means of studying cosmic re-ionization is through the 21 cm line of neutral hydrogen (Furlanetto *et al.* 2006). Many programs have been initiated to study the H I 21 cm signal from cosmic re-ionization, including the MWA†, LOFAR‡, PAPER¶, and eventually the SKA‖. Fan *et al.* (2006) and Furlanetto *et al.* (2006) present detailed discussions of the observational challenges.

2. Expected H I 21-cm signals from cosmic re-ionization

The study of H I 21-cm emission from cosmic re-ionization entails the study of large scale structure (LSS), meaning H I masses $> 10^{12}$ M$_\odot$. During this epoch the entire IGM may be neutral, and the LSS in question is not simply mass clustering, but involves a combination of structure in cosmic density, neutral fraction, and H I excitation temperature. Hence, H I 21-cm studies are potentially the 'richest of all cosmological data sets' (Barkana & Loeb 2005).

2.1. *Global signal*

Figure 1 (left panel) shows the latest predictions of the global (all sky) increase in the background temperature due to the H I 21-cm line from the neutral IGM (Gnedin &

† http://web.haystack.mit.edu/arrays/MWA/LFD/
‡ http://www.lofar.org/
¶ D. Backer, in prep
‖ http://www.skatelescope.org/

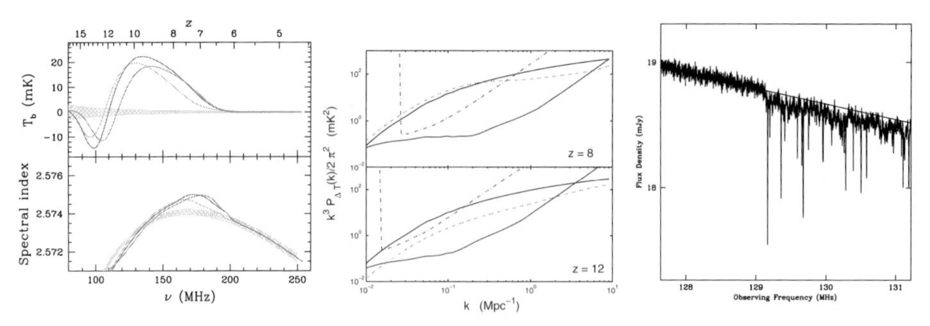

Figure 1. *Left*: Global (all sky) H I signal from re-ionization (Gnedin & Shaver 2003. The shaded region shows the expected thermal noise in a carefully controlled experiment. *Center*: Predicted H I 21-cm brightness temperature power spectrum (in log bins) at redshifts 8 and 12 (Mcquinn *et al.* 2006). The thin black line shows the signal when density fluctuations dominate. The dashed green line shows the predicted signal for $\bar{x}_i = 0.2$ at $z = 12$, and $\bar{x}_i = 0.6$ at $z = 8$, in the Furlanetto *et al.* (2004) semi-analytic model. The thick blue line shows the SKA sensitivity in 1000 hr. The thick red dot-dash show the sensitivity of the pathfinder experiment LOFAR. The cutoff at low k is set by the primary beam. *Right*: The simulated SKA spectrum of a radio continuum source at $z = 10$ (Carilli *et al.* 2002). The straight line is the intrinsic power law (synchrotron) spectrum of the source. The noise curve shows the effect of the 21-cm line in the neutral IGM, including noise expected for the SKA in a 100 hr integration.

Shaver 2003). The predicted H I signal peaks at roughly 20 mK above the foreground at $z \simeq 10$. At higher redshift, prior to IGM warming, but allowing for Ly-α emission from the first luminous objects, the H I is seen in absorption against the CMB. Since this is an all-sky signal, the sensitivity of the experiment is independent of telescope collecting area, and the experiment can be done using small area telescopes at low frequency, with very well controlled frequency response (Subrahmanyan *et al.* 2005, in prep.). Note that the line signal is only $\sim 10^{-4}$ that of the mean foreground continuum emission at ~ 150 MHz.

2.2. *Power spectra*

Figure 1 (middle panel) shows the predicted power spectrum of spatial fluctuations in the sky brightness temperature due to the H I 21-cm line (Mcquinn *et al.* 2006). For power spectral analyses the sensitivity is greatly enhanced relative to direct imaging due to the fact that the Universe is isotropic, and hence one can average the measurements in annuli in the Fourier (u-v) domain, i.e., the statistics of fluctuations along an annulus in the u-v plane are equivalent. Moreover, unlike the CMB, H I line studies provide spatial and redshift information, and hence the power spectral analysis can be performed in three dimensions. The rms fluctuations at $z = 10$ peak at about 10 mK rms on scales $\ell \simeq 5000$.

2.3. *Absorption toward discrete radio sources*

An alternative to emission studies is the possibility of studying smaller scale structure in the neutral IGM by looking for H I 21-cm absorption toward the first radio-loud objects (AGN, star forming galaxies, GRBs, Carilli *et al.* 2002). Figure 1 (right panel) shows the predicted H I 21-cm absorption signal toward a high-redshift radio source due to the 'cosmic web' prior to re-ionization, based on numerical simulations. For a source at $z = 10$, these simulations predict an average optical depth due to 21-cm absorption of about 1 %, corresponding to the 'radio Gunn-Peterson effect', and about five narrow (few km s^{-1}) absorption lines per MHz with optical depths of a few to 10 %. These latter lines are equivalent to the Ly-α forest seen after re-ionization. Furlanetto & Loeb (2002) predict a similar H I 21-cm absorption line density due to gas in minihalos as that expected for the 21-cm forest.

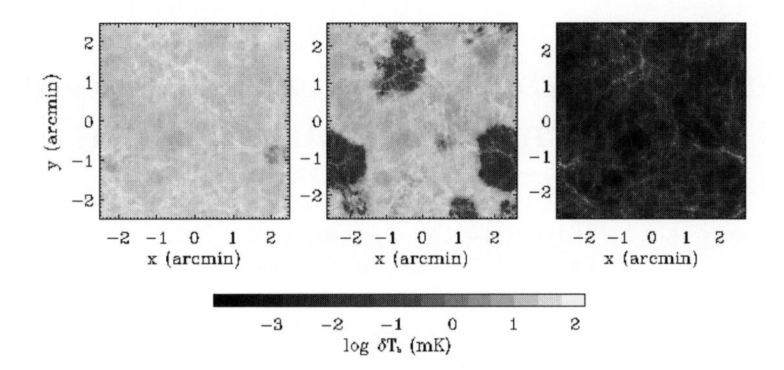

Figure 2. The simulated H I 21-cm brightness temperature distribution during re-ionization at $z = 12$, 9, and 7 (left to right, Zaldariagga *et al.* 2004).

2.4. *Tomography*

Figure 2 shows the expected evolution of the H I 21 cm signal during re-ionization based on numerical simulations (Zaldariagga *et al.* 2004). In this simulation, the H II regions caused by galaxy formation are seen in the redshift range $z \simeq 8$ to 10, reaching scales up to $2'$ (frequency widths ~ 0.3 MHz or physical size ~ 0.5 Mpc). These regions have (negative) brightness temperatures up to 20 mK relative to the mean H I signal. This corresponds to $5\,\mu$Jy beam^{-1} in a $2'$ beam at 140 MHz. Only the full SKA will able to image such structures.

2.5. *Cosmic Strømgren spheres*

While direct detection of the typical structure of H I and H II regions may be out of reach of the near-term EoR 21-cm telescopes, there is a chance that even this first generation of telescopes will be able to detect the rare, very large scale H II regions associated with luminous quasars near the end of re-ionization. The expected signal is $\sim 20\,\text{mK} \times x_{\text{HI}}$ on a scale $\sim 10'$to $15'$, with a line width of ~ 1 to 2 MHz (Wyithe *et al.* 2005). This corresponds to $0.5 \times x_{\text{HI}}$ mJy beam^{-1}, for a $15'$ beam at $z \simeq 6$ to 7, where x_{HI} is the IGM neutral fraction.

Acknowledgements

CC thanks the Max-Planck-Gesellschaft and the Humboldt-Stiftung for support through the Max-Planck-Forschungspreis.

References

Barkana, R., & Loeb, A. 2005, *ApJ* (Letters), 624, L65
Carilli, C., Gnedin, N., & Owen, F. 2002, *ApJ*, 577, 22
Cen, R. 2003, *ApJ*, 591, 12
Ciardi, B., & Ferrara, A. 2005, *Space Sci. Reviews*, 116, 625
Fan, X., Carilli, C., & Keating, B. 2006, *ARAA*, 44, 415
Furlanetto, S. R., & Loeb, A. 2002, *ApJ*, 579, 1
Furlanetto, S. R., Zaldarriaga, M., & Hernquist, L. 2004, *ApJ*, 613, 16
Furlanetto, S. R., Oh, S. P., & Briggs, F. H. 2006, *Phys. Rep.* 433, 181
Gnedin, N., & Shaver, P. 2004, *ApJ*, 608, 611
Mcquinn, M., Zahn, O., Zaldarriaga, M., Hernquist, L., & Furlanetto, S. R. 2006, *ApJ*, 653, 815
Wyithe, J. S., Loeb, A., & Barnes, D. 2005, *ApJ* 634, 715
Zaldarigga, M., Furlanetto, S., & Henquist, L. 2004, *ApJ*, 608, 622

Highlights of Astronomy, Volume 14
IAU XXVI General Assembly, 14-25 August 2006
Karel A. van der Hucht, ed.

doi:10.1017/S1743921307011088

Cosmic re-onization: theoretical modelling and forthcoming observations

Benedetta Ciardi

Max-Planck-Institut fr Astrophysik, Karl-Schwarzschild-Straße 1, D-85741 Garching, Germany
email: bciardi@mpa-garching.mpg.de

Abstract. With the advent in the near future of radio telescopes such as LOFAR and the SKA, a new window on the high-redshift Universe will be opened. In particular, it will be possible, for the first time, to observe the 21-cm signal from the diffuse IGM prior to its complete re-ionization and thus probe the 'dark ages'. I discuss the theoretical modelling of the re-ionization process and its observability through the 21-cm signal and the CMB anisotropies.

Keywords. atomic processes, line formation, cosmology: observations, cosmology: theory

---

In the past few years, the study of the intergalactic medium (IGM) re-ionization has attracted considerable attention and has moved from being a purely theoretical excercise to an observational goal. At present, the only available observations on the re-ionization process are those of high-redshift quasar spectra (which provide information on the abundance of neutral hydrogen during the last stage of the process at $z\, sim\, 6$; e.g., Fan *et al.* 2004) and of anisotropies in the power spectrum of the Cosmic Microwave Background (CMB) radiation (which give an estimate of the global amount of electrons produced by re-ionization; Spergel *et al.* 2007). Despite the most recent progress though, we still lack observations that probe the temporal evolution of the re-ionization process and that answer some still very debated questions: What are the first sources of ionizing radiation? How did re-ionization evolve? Which is its interplay with the galaxy formation process? The perfect tool to start answering these questions is the 21-cm line associated with the hyperfine transition of the ground state of H I. Observations of the 21-cm line at different frequencies can give information on the state of H I at different redshifts. With the advent in the near future of radio telescopes such as LOFAR, MWA, and the SKA, a new window on the high-redshift Universe will be opened and it will be possible, for the first time, to probe the re-ionization history and the 'dark ages'.

From a theoretical point of view, the ingredients needed to model the re-ionization process are: (*i*) simulations of galaxy formation; (*ii*) properties of the ionization sources; and (*iii*) radiative transfer (RT) of ionizing photons in the IGM. The latter point is particularly tricky given the complexity of the equations involved and the present impossibility of solving them. For this reason, several groups are working on the development of RT codes based on different assumptions and approximations (for a comparison between the RT codes present in the cosmological community, see Iliev *et al.* 2006).

We have run numerical simulations of the re-ionization process in a box of comoving length $20\,(10)\,h^{-1}$ Mpc representing a 'mean' ('proto-cluster') region of the Universe. The sources are stars with variable properties (Ciardi *et al.* 2003a; Ciardi *et al.* 2003b) and the radiative transfer has been followed using the code CRASH (Ciardi *et al.* 2001; Maselli *et al.* 2003). The simulations have been run also with the addition of sub-grid physics to take into account the effect of mini-halos (MHs, Ciardi *et al.* 2006). Figure 1 shows an example. The main results can be summarized as follows: (*i*) the environment can have a deep impact on re-ionization; (*ii*) the presence of MHs can delay re-ionization up

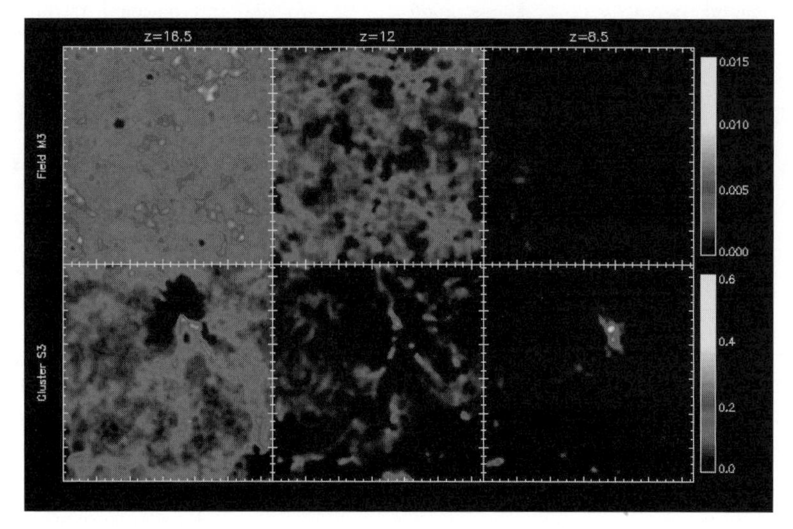

Figure 1. Slices cut through the simulation boxes. The panels show the H I number density for the 'mean' region of the Universe (upper) and the proto-cluster (lower) at $z = 16.5$, 12, and 8.5.

to $\Delta z \simeq 2$; and (*iii*) a suitable choice of the parameters involved in the calculations can reproduce the Thomson scattering optical depth measured by WMAP1 and WMAP3 (Spergel *et al.* 2007). The latter point suggests that to better contrain the theoretical modelling, additional observations are needed.

Ideal observations will be those of the 21-cm line from H I. If the IGM has been heated, the line is observed in emission and its intensity (described by the brightness temperature, $T_{\rm b}$) is directly proportional to the abundance of H I. Thus, it is straightforward to produce maps of 21-cm line emission from those of H I obtained from the above simulations. From the analysis of such maps we can conclude that (Ciardi & Madau 2003; Salvaterra *et al.* 2005; Valdes *et al.* 2006): (*i*) fluctuations of $T_{\rm b}$ show a behaviour almost independent of the re-ionization history, with a peak at ~ 10 mK (earlier re-ionization peaks at frequencies lower than a later re-ionization); (*ii*) the expected emission is within the sensitivity range of the next generation of radio telescopes; (*iii*) 21-cm observations can be used in combination with other measurements (i.e., CMB) to maximize the extracted information; and (*iv*) the greatest challenge will be dealing with foreground contamination of both Galactic and extragalactic origin.

Acknowledgements

I thank my collaborators in these projects.

References

Ciardi, B., Ferrara, A., Marri, S., & Raimondo, G. 2001, *MNRAS*, 324, 381
Ciardi, B., Ferrara, A., & White, S. D. M. 2003, *MNRAS* (Letters), 344, L7
Ciardi, B., & Madau, P. 2003, *ApJ*, 596, 1
Ciardi, B., Scannapieco, E., Stoehr, F., *et al.* 2006, *MNRAS*, 366, 689
Ciardi, B., Stoehr, F., & White, S. D. M. 2003a, *MNRAS*, 343, 1101
Fan, X., Hennawi, J. F., Richards, G. T., *et al.* 2004, *AJ*, 128, 515
Iliev, I., Ciardi, B., Alvarez, M. A., *et al.* 2006, *MNRAS*, 371, 1057
Maselli, A., Ferrara, A., & Ciardi, B. 2003, *MNRAS*,345, 379
Salvaterra, R., Ciardi, B., Ferrara, A., & Baccigalupi, C. 2005, *MNRAS*, 360, 1063
Spergel D.N., Bean, R., Dor, O.,*et al.* 2007, *ApJS*, 170, 377
Valdes, M., Ciardi, B., Ferrara, A., Johnston-Hollitt, M., & Röttgering, H. 2006, *MNRAS* (Letters), 369, L66

Highlights of Astronomy, Volume 14
IAU XXVI General Assembly, 14-25 August 2006
Karel A. van der Hucht, ed.

doi:10.1017/S174392130701109X

High-redshift radio galaxies: the most massive galaxies at every epoch

Carlos De Breuck

European Southern Observatory, Karl Schwarzschild Straße 2, D-85748 Garching, Germany
email: cdebreuc@eso.org

Abstract. Radio galaxies are identified with the most massive host galaxies known out to $z = 5$ and can put strong constraints on galaxy evolution models, provided their space density is accurately determined. Here, I present the important role low-frequency radio surveys will have in this by selecting the highest redshift radio galaxies using the steepness of their radio spectra.

Keywords. galaxies: high-redshift, radio continuum: galaxies

1. The massive hosts of radio galaxies

The best studied type 2 AGN at high redshift are radio galaxies. Their radio selection allows the construction of samples which are unbiased against dust properties. The very wide area of the parent radio surveys (up to 75 % of the sky for NVSS; Condon *et al.* 1998) also allows the selection of objects with a very low space density.

The host galaxies of nearby radio galaxies are almost uniquely identified with massive elliptical galaxies (Matthews *et al.* 1964). Over the past two decades, there has been mounting evidence that the host galaxies of higher redshift powerful radio sources are also amongst the most massive at each epoch. Much of this evidence is based on the Hubble K - z diagram (e.g., Lilly & Longair 1984; Eales *et al.* 1997; van Breugel *et al.* 1998; De Breuck et al. 2002, Rocca-Volmerange *et al.* 2004). However, the interpretation of the K-z diagram in terms of stellar masses has been complicated by (*i*) strong band-shifting effects: the observed K-band at $z = 4$ corresponds to rest-frame B band; and (*ii*) the uncertain contributions of the obscured AGN. To remedy these problems, we have started a major project to determine the stellar masses of a sample of 70 radio galaxies from $z = 1$ to 5.2 covering a range in redshift and 3 GHz radio power. Our results confirm that radio galaxies have stellar masses between 10^{11} and $10^{12}\,M_{\odot}$ (Seymour *et al.* this volume; Seymour *et al.* in preparation). They thus trace the very upper end of the stellar mass function, even at high redshift. As such, they can provide some of the strongest constraints on galaxy formation models, provided that their space density is well determined. It is therefore important to construct large and complete samples of high redshift radio galaxies. This is an area where future low-frequency radio surveys will be able to make an important impact.

2. Ultra-steep spectrum selection techniques

The main difficulty in deriving samples with complete redshift information is the often extreme optical faintness of radio sources. Often, the redshift determination is only possible thanks to bright optical emission lines photo-ionized by the AGN (e.g., the pioneering work on the 3CR sample by Spinrad *et al.* 1985). However, this emission-line luminosity is correlated with the radio power (Willott *et al.* 1999), meaning that it is very difficult to obtain redshifts of progressively fainter radio sources. Optical redshift determinations of the highest redshift radio galaxies often require integrations of more than 1 hr on 8 - 10 m class telescopes.

As present-day radio surveys contain up to 2 million sources, a drastic culling of the catalogues is needed to allow follow-up redshift determinations. In particular, one is mostly interested in the highest redshift radio galaxies. To select these, the most successful technique has been the selection of radio sources with ultra steep radio spectra (USS, $\alpha < -1.0$ for $S_\nu \propto \nu^\alpha$). This technique is based on the empirical correlation between spectral index and redshift (De Breuck *et al.* 2002).

It has been by far the most successful filter to find the highest redshift radio galaxies to the extent that all known $z > 3.5$ radio galaxies have been found from such USS samples. The physical interpretation of this z-α correlation has generally been sought in a band-shifting of the steeper high-frequency part of the radio spectrum into the observed low-frequency window. However, Klamer *et al.* (2006) have recently challenged this interpretation based on the observation of straight radio spectra in 92 % of the observed USS radio galaxies. Such an absence of high-frequency steepening has been previously reported by Mangalam & Gopal-Krishna (1995). The alternative explanation is that the highest redshift radio galaxies are located in increasingly denser media, which would lead to steeper radio spectra, as seen in low-redshift radio sources in dense environments (Athreya & Kapahi 1998).

If the radio spectra are indeed straight at rest frequencies of $\sim$1 GHz, one thus has to move to lower frequencies to find the spectral downturn due to synchrotron self-absorption. Selecting USS at rest frequencies of a few hundred MHz is thus expected to be a very efficient selection technique of $z > 5$ radio galaxies, as they should stand out even more compared to low-redshift sources having much flatter spectra. The wide-field capabilities of the new low-frequency observatories such as LOFAR, LWA, and SKA will likely open a completely new window for USS searches of high redshift radio galaxies. In addition, the broadband spectral capabilities may allow us to directly obtain redshifts in the radio regime using the H I 21-cm line and/or the 21-cm forest (Carilli 2006, these proceedings).

References

Athreya, R., & Kapahi, V. 1998, *JAA*, 19, 63
Condon, J., Cotton, W. D., Greisen, E. W., *et al.* 1998, *AJ*, 115, 1693
De Breuck, C., van Breugel, W., Stanford, S. A., *et al.* 2002, *AJ*, 123, 637
Eales, S., Rawlings, S., Law-Green, D., Cotter, G., & Lacy, M. 1997, *MNRAS*, 291, 593
Klamer, I. J., Ekers, R. D., Bryant, J. J., *et al.* 2006, *MNRAS*, 371, 852
Lilly, S., & Longair, M. 1984, *MNRAS*, 211, 833
Mangalam, A., & Gopal-Krishna 1995, *MNRAS*, 275, 976
Matthews, T., Morgan, W., & Schmidt, M. 1964, *ApJ*, 140, 35
McCarthy, P. J., Kapahi, V. K., van Breugel, W., *et al.* 1996, *ApJS*, 107, 19
Rocca-Volmerange, B., Le Borgne, D., De Breuck, C., Fioc, M., & Moy, E. 2004, *A&A*, 415, 931
Spinrad, H., Marr, J., Aguilar, L., & Djorgovski, S. 1985, *PASP*, 97, 932
van Breugel, W., Stanford, S. A., Spinrad, H., Stern, D., & Graham, J. R. 1998, *ApJ*, 502, 614
Willott, C., Rawlings, S., Blundell, K., & Lacy, M. 1999, *MNRAS*, 309, 1017

Highlights of Astronomy, Volume 14
IAU XXVI General Assembly, 14-25 August 2006
Karel A. van der Hucht, ed.

doi:10.1017/S1743921307011106

Deep GMRT images of the *Spitzer* extragalactic First Look Survey at 610 MHz

David A. Green and Timothy S. Garn

Cavendish Laboratory, 19 J.J. Thomson Avenue, Cambridge, CB4 3PT, UK
email: dag@mrao.cam.ac.uk, tsg25@cam.ac.uk

We have observed the *Spitzer* extragalactic First Look Survey (xFLS) region with the Giant Meterwave Radio Telescope (GMRT), near Pune, India, at 610 MHz. Seven pointings were observed with the GMRT, one central and six in a surrounding hexagon. Each field was observed with an integration time of $\sim$200 min., in a series of scans spread over a range of LST to improve u-v coverage. Two sidebands, each of 16 MHz with 128 channels – to allow narrow band interference to be excised efficiently – were observed. The synthesised beam of the images is $\sim 5''$, and as the primary beam of the GMRT at 610 MHz is $\sim 43'$, these seven pointings cover most of the xFLS region.

Initial comparison of sources detected in the overlap regions between adjacent pointings, and with existing surveys of the region, revealed two problems. (*i*) The coordinates of the GMRT images were slightly distorted, which was found to be due to slight errors in the timestamps, and hence (u, v, w) coordinates, of the visibilities. (*ii*) The flux densities of sources seen in adjacent fields varied systematically with position, implying an offset to the effective primary beam of the telescope. After correcting for these problems, the final images have an r.m.s. noise of typically 30 μJy beam^{-1}, before primary beam correction, although dynamic range issues limit the quality of the images near bright sources.

These results will be combined with existing radio data and IR data from the *Spitzer Space Telescope*, to investigate the far-infrared – radio correlation up to redshifts of about $z = 2$.

Highlights of Astronomy, Volume 14
IAU XXVI General Assembly, 14-25 August 2006
Karel A. van der Hucht, ed.

doi:10.1017/S1743921307011118

A steradian of the southern sky at 151.5 MHz using the Mauritius Radio Telescope

V. N. Pandey and N. Udaya Shankar

Raman Research Institute, Bangalore 560 080, India
email: vnpandey,uday@rri.res.in

Abstract. We present a few results of a low-frequency southern sky survey carried out using the Mauritius Radio Telescope (MRT).

Keywords. catalogs, galaxies: clusters: general, surveys, supernova remnants

---

We present and discuss wide field images covering ~ 1.25 sr ($18^{\rm h} \leqslant \alpha \leqslant 24^{\rm h}$, $-75^\circ \leqslant \delta \leqslant -10^\circ$) of the sky at 151.5 MHz with an angular resolution of $4' \times 4\farcm6$ and an rms noise of 260 mJy beam^{-1}. A source catalog of ~ 2800 sources and its comparative study including cross identification with the Molonglo Reference Catalogue at 408 MHz and Culgoora catalogue at 160 MHz is described. The comparison reveals that there are no systematics in the final source list and the positional accuracies agree typically to about 0.07 times the MRT beam width. The procedures developed for flux calibration, which involves scaling different images to a common level, estimation of the primary beam shape of the helix, and recovery of the amplitude information of the signal lost in a 2-bit, 3-level correlator with an AGC are discussed. Our true flux density estimates agree to within 6.3 % with the known flux densities of unresolved sources in the MS 4Jy sample.

Due to low-frequency and availability of short spacings, our images (surface brightness sensitivity $\sim 2.1 \times 10^{-21}$ W m^{-2} Hz^{-1} sr^{-1}) are sensitive to extended sources like giant radio sources, relics and supernova remnants (SNRs). Recently observed sub-structures in X-ray temperature maps support the existence of extended shock waves at locations of several known cluster relics. We show a specific example of such extended radio emission from these proposed shocks around the X-ray cluster Abell 3667 in the MRT image. These are steep spectrum sources, and we expect to detect quite a few such sources in the MRT survey.

An additional 0.6 sr of the sky ($15^{\rm h} \leqslant \alpha \leqslant 18^{\rm h}$, $-75^\circ \leqslant \delta \leqslant -10^\circ$) has also been imaged and its comparison with existing low frequency images to search for new SNR candidates is under progress. As an example of interesting sources in the images, we briefly discuss a few steep spectrum sources, giant radio sources, double sources, fossil galaxies and report the structure of a few resolved SNRs.

References

Pandey, V. N., Oozeer, N., Udaya Shankar, N., & Somanah, R. 2002, *Bull. Astron. Soc. India*, 30, 773
Pandey, V. N., Udaya Shankar, N., & Somanah, R. 2002, *Ap&SS*, 282, 29
Pandey, V. N., & Udaya Shankar, N. 2005, *URSI General Assembly*, Proc. Session J

Highlights of Astronomy, Volume 14
IAU XXVI General Assembly, 14-25 August 2006
Karel A. van der Hucht, ed.

doi:10.1017/S174392130701112X

A very brief description of LOFAR – the Low Frequency Array

Heino D. Falcke[1,2,†], Michiel P. van Haarlem[1], A. Ger de Bruyn[1,3], Robert Braun[1], Huub J.A. Röttgering[4], Benjamin Stappers[1,5], Wilfried H.W.M. Boland[4], Harvey R. Butcher[1], Eugène J. de Geus[1], Leon V. Koopmans[3], Robert P. Fender[5,6], H. Jan M.E. Kuijpers[2], George K. Miley[4], Richard T. Schilizzi[7,4], Corina Vogt[1], Ralph A.M.J. Wijers[5], Michael W. Wise[5], Willem N. Brouw[3], Johan P. Hamaker[1], Jan E. Noordam[1], Thomas Oosterloo[1], Lars Bähren[1,2], Michiel A. Brentjens[1,3], Stefan J. Wijnholds[1], Jaap D. Bregman[1], Wim A. van Cappellen[1], André W. Gunst[1], G.W. (Dion) Kant[1], Jan Reitsma[1], Kjeld van der Schaaf[1], and Cornelis M. de Vos[1]

[1]ASTRON, Postbus 2, NL-7990AA Dwingeloo, the Netherlands
email: falcke@astron.nl

[2]Department of Astronomy, Radboud University, Nijmegen, the Netherlands

[3]Kapteyn Astronomical Institute, Groningen University, Groningen, the Netherlands

[4]Leiden Observatory, Leiden University, Leiden, the Netherlands

[5]Astronomical Institute Anton Pannekoek, University of Amsterdam, the Netherlands

[6]School of Physics & Astronomy, University of Southampton, Southampton, SO17 1BJ, UK

[7]International SKA Project Office, c/o ASTRON, Dwingeloo, the Netherlands

Abstract. LOFAR (Low Frequency Array) is an innovative radio telescope optimized for the frequency range 30 - 240 MHz. The telescope is realized as a phased aperture array without any moving parts. Digital beam forming allows the telescope to point to any part of the sky within a second. Transient buffering makes retrospective imaging of explosive short-term events possible. The scientific focus of LOFAR will initially be on four key science projects (KSPs): (*i*) Detection of the formation of the very first stars and galaxies in the universe during the so-called epoch of reionization by measuring the power spectrum of the neutral hydrogen 21-cm line (Shaver *et al.* 1999) on the $\sim 5'$ scale; (*ii*) Low-frequency surveys of the sky with of order 10^8 expected new sources; (*iii*) All-sky monitoring and detection of transient radio sources such as γ-ray bursts, X-ray binaries, and exo-planets (Farrell *et al.* 2004); and (*iv*) Radio detection of ultra-high energy cosmic rays and neutrinos (Falcke & Gorham 2003) allowing for the first time access to particles beyond 10^{21} eV (Scholten *et al.* 2006). Apart from the KSPs open access for smaller projects is also planned. Here we give a brief description of the telescope.

Keywords. instrumentation: interferometers, telescopes, cosmology: observations, cosmic rays

---

1. LOFAR – how it works

In its first phase LOFAR will consist of 77 *stations* distributed within a ring of ~ 100 km diameter; 32 stations will be clustered in a central core of ~ 2 km diameter located in the northeastern Netherlands near the village of Exloo. Each station has two antenna systems: the Low-Band and High-Band Antennas (LBA, HBA). The LBA system operates primarily in the frequency range 30 - 80 MHz with a switch to observe over a 10 - 80 MHz

† Visiting Miller Professor, Astronomy Department, University of California at Berkeley

band as well. The HBA is optimized for the range 110 - 240 MHz with a possibility to observe up to 270 MHz with lower sensitivity. The LBA field is 60 m in diameter and contains 96 inverted-V crossed dipoles oriented NE-SW and SE-NW (i.e., dual polarization) in a randomized distribution with a slight exponential fall-off in density with radius. The HBA field consists of 96 tiles distributed in an as yet undetermined manner over roughly 50 m. Each tile consists of a 4×4 array of bowtie-shaped crossed dipoles with an analog 5-bit beam former using true time delays. Radio waves are sampled with a 12-bit A/D-converter – to be able to cope with expected interference levels – operating at either 160 or 200 MHz in the first, second or third Nyquist zone (i.e., 0 - 100, 100 - 200, or 200 - 300 MHz band respectively for 200 MHz sampling). The data from the receptors is filtered in 512×195 kHz sub-bands (156 kHz subbands for 160 MHz sampling) of which a total of 3 MHz bandwidth (164 channels) can be used at any time. Data from each receptor can be buffered in a transient buffer board (TBB) for as long as $\sim 10\,\mathrm{min}/(\Delta\nu/196\,\mathrm{kHz})$. Subbands from all antennas are combined on a station-level in a digital beamformer allowing eight independently steerable beams which are sent to the central processor via a glas fibre link that handles a $0.7\,\mathrm{Tbit\,s^{-1}}$ data rate. The beams from all stations are further filtered into 1-kHz channels, cross-correlated and integrated on typical timescales of 1 - 10 s. The integrated visibilities are then calibrated on 10-s intervals to remove the effects of the ionosphere and images are produced. Channels with disturbing radio frequency interference (RFI) are dropped. For the correlation we use four of the six racks of an IBM Blue Gene/L machine in Groningen with a total of $\sim 12\,000$ processors. We expect a typical input rate of $\sim 0.5\,\mathrm{Tbit\,s^{-1}}$. Unix clusters are used as input and output nodes for pre- and postprocessing.

The expected $3\,\sigma$ point source sensitivities of LOFAR for one hour integration over 4 MHz bandwidth dual-polarization are 2 mJy, 1.3 mJy, 70 μJy, and 60 μJy at, respectively, 30, 75, 120, and 200 MHz. The resolution will be $25''$, $10''$, $6''$, and $3.5''$ for the same frequencies. The field of view is 3° at 150 MHz (HBA) and 7.5° at 50 MHz (LBA).

The project is currently in discussions with consortia in Germany, U.K., France, Italy, and Sweden to expand the baseline and increase the resolution by up to a factor of ten.

National funding of LOFAR has been obtained at a level of ~ 75 MEuro plus various in-kind contributions. Construction of the first station was completed in 2006 September. Further stations are expected to be rolled out in the course of 2007, so that commissioning and start of operation is foreseen in 2008.

2. Outlook and conclusions

With its new concept (Bregman 2000) of a broad-band aperture array and digital beamforming LOFAR is expected to pave the way for a new generation of telescopes and to be an important pathfinder for the Square Kilometre Array. LOFAR will improve the resolution and sensitivity of previous telescopes for continuum observations by roughly two orders of magnitude over a wide frequency range. It will also provide instantaneous access to a large fraction of the sky at low frequencies at once, making serious and regular all-sky radio monitoring possible for the first time. With these unusual properties LOFAR promises a wealth of new discoveries.

References

Bregman, J. D. 2000, in: H.R. Butcher (ed.), *Radio Telescopes*, Proc. SPIE 4015, 19
Falcke, H., & Gorham, P. 2003, *Astropart. Phys.*, 19, 477
Farrell, W. M., Lazio, T. J. W., Zarka, P., *et al.* 2004, *Planet. Space Sci.*, 52, 1469
Scholten, O., Bacelar, J., Braun, R., *et al.* 2006, *Astropart. Phys.*, 26, 219
Shaver, P. A., Windhorst, R. A., Madau, P., & de Bruyn, A. G. 1999, *A&A*, 345, 380

Highlights of Astronomy, Volume 14
IAU XXVI General Assembly, 14-25 August 2006
Karel A. van der Hucht, ed.

doi:10.1017/S1743921307011131

The Long Wavelength Array

Gregory B. Taylor †

Department of Physics and Astronomy,
University of New Mexico, Albuquerque, NM 87131, USA
email: gbtaylor@unm.edu

Abstract. The Long Wavelength Array (LWA) will be a new, open, user-oriented astronomical instrument operating in the relatively unexplored window from 20 - 80 MHz near arcsecond angular resolution and milliJansky sensitivity. Operated by the University of New Mexico on behalf of the Southwest Consortium (SWC) the LWA will provide a unique training ground for the next generation of radio astronomers. Students may also put skills learned on the LWA to work in computer science, electrical engineering, and the communications industry, among others. The development of the LWA will follow a phased build which benefits from lessons learned at each phase. Four university-based Scientific Testing and Evaluation (ST&E) teams with different areas of concentration: (*i*) high-resolution imaging and particle acceleration; (*ii*) wide-field imaging and large scale structures; (*iii*) ionospheric physics; and (*iv*) radio frequency interference (RFI) suppression and transient detection will provide the feedback needed to assure that science objectives are met as the build develops. Currently in its first year of construction funding, the LWA team is working on the design for the first station (see also Ray *et al.* 2006).

Keywords. instrumentation: interferometers, radio continuum: general, radio lines: general

---

1. Introduction

We are entering into a new era of exploration of the Universe at long wavelengths by a variety of new instruments including the Long Wavelength Array (LWA), LOFAR (Falcke, these proceedings), MWA (Webster *et al.*, these proceedings), FASR (Gary *et al.*, these proceedings), and others. Key science drivers that are motivating these advances include: (*i*) acceleration, propagation, and turbulence in the interstellar medium (ISM), including the space distribution and spectrum of Galactic cosmic rays, supernova remnants, and pulsars; (*ii*) the high-redshift universe, including the most distant radio galaxies and clusters – tools for understanding the earliest black holes and the cosmological evolution of dark matter and dark energy; (*iii*) planetary, solar, and space science, including space weather prediction and extrasolar planet searches; and (*iv*) the radio transient universe, including the known (e.g., SNe, GRBs) and the unknown.

Because the LWA will explore one of the last and least investigated regions of the spectrum, the potential for new discoveries, including new classes of physical phenomena, is high, and there is a strong synergy with exciting new X- and γ-ray measurements, e.g., for cosmic ray acceleration, transients, and galaxy clusters. Further discussion of the scientific goals of the LWA can be found in Kassim *et al.* (2005) or at the LWA web pages `<http://lwa.unm.edu>`.

† On behalf of the Southwest Consortium consisting of the University of New Mexico, the Naval Research Laboratory, the Applied Research Laboratory at the University of Texas, and the Los Alamos National Laboratory.

2. Current status

Construction and testing of the Long Wavelength Demonstrator Array (LWDA) is currently underway. The LWDA consists of 16 pairs of dipoles and is located a few kilometers from the center of the VLA. Lessons learned from the LWDA will feed into the design work on the first two LWA stations. The very first LWA station will eventually supplant the LWDA, and will consist of 256 dipole pairs. The site for the second LWA station is currently under study, and is likely to be located near one of the arms of the VLA. Early science should be possible with the first two stations in combination with the 74 MHz system on the VLA. The full LWA is expected to consist of 52 stations on baselines ranging from 400 m to 400 km. This will provide arcsecond level resolution and mJy level sensitivity.

An RFI survey has begun at the LWDA site. Preliminary results are encouraging, both for the environmental conditions, and for the level of self-generated RFI. While FM stations are clearly present at all times, the spectrum from 20 - 85 MHz is generally clean, with only narrowband signals present the majority of the time.

Acknowledgements

We acknowledge support for the LWA project from the Office of Naval Research.

References

Kassim, N. E., Polisensky, E. J., Clarke, T. E., Hicks, B. C., Crane, P. C., Stewart, K. P., Ray, P. S., Weiler, K. W., Rickard, L. J., Lazio, T. J. W., Lane, W. M., Cohen, A. S., Nord, M. E., Erickson, W. C., & Perley, R. A. 2005, in: N. Kassim, M. Perez, M. Junor & P.Henning (eds.), *From Clark Lake to the Long Wavelength Array: Bill Erickson's Radio Science*, *ASP-CS*, 345, 392

Ray, P. S., Ellingson, S., Fisher, R., Kassim, N. E., Rickard, L. J., & Clarke, T. 2006, LWA Memo Series No. 35, `<http://www.ece.vt.edu/swe/lwa/>`

Highlights of Astronomy, Volume 14
IAU XXVI General Assembly, 14-25 August 2006
Karel A. van der Hucht, ed.

doi:10.1017/S1743921307011143

The Square Kilometre Array (SKA)

T. Joseph W. Lazio[1] and Bryan M. Gaensler[2]

[1]Naval Research Laboratory, 4555 Overlook Ave. SW, Washington, DC 20375, USA
email: joseph.lazio@nrl.navy.mil

[2]School of Physics A29, University of Sydney, Sydney, NSW 2006, Australia
email: bgaensler@usyd.edu.au

Abstract. The Square Kilometre Array is intended to be the centimeter- and meter-wavelength telescope for the 21st century. At long wavelengths, the SKA's key science projects include the search for highly redshifted hydrogen, including the signal from the epoch of re-ionization, and the search for the first supermassive black holes.

Keywords. atomic processes, line formation, instrumentation: interferometers, galaxies: active, cosmology: observations

---

1. Introduction

The Square Kilometre Array (SKA)† will be one of a suite of new, large astrophysics facilities for the 21st century, probing fundamental physics, the origin and evolution of the Universe, the structure of the Milky Way Galaxy, and the formation and distribution of planets. The SKA will be at least 50 times more sensitive than any other centimeter- to meter-wavelength telescope ever built. In addition to answering fundamental scientific questions, the vast increase in sensitivity provided by the SKA will also almost certainly lead to the discovery of new and totally unexpected celestial phenomena.

2. Key science

Five Key Science Projects for the SKA have been identified by the international community (Carilli & Rawlings 2004). These are: (*i*) the Cradle of Life and astrobiology (Lazio *et al.* (2004; (*ii*) Strong Field Tests of Gravity Using Pulsars and Black Holes (Kramer *et al.* 2004); (*iii*) the Origin and Evolution of Cosmic Magnetism (Gaensler *et al.* 2004); (*iv*) Galaxy Evolution, Cosmology, and Dark Energy (Rawlings *et al.* 2004); and (*v*) Probing the Dark Ages (Carilli *et al.* 2004). These Key Projects all represent unanswered questions in fundamental physics and astrophysics. Furthermore, each of these projects has been selected using the criterion that it represents science which is either unique to the SKA, or is a topic which is complementary to other data sets, but in which the SKA plays a key role (Gaensler 2004).

Observations will be conducted at wavelengths longer than 1 m in support of both the *Galaxy Evolution, Cosmology, and Dark Energy* and *Probing the Dark Ages* projects. A common goal of both of these projects is observing hydrogen at large redshifts ($z > 3$). For the galaxy evolution project, the goal would be to probe the assembly of the first structures, primarily by conducting an *unbiased* survey for damped Ly-α absorbers. Observations at these wavelengths are unaffected by dust obscuration, thus these observations can search in an unbiased manner for these H I-rich structures. At even larger redshifts, the SKA should have the sensitivity to image the formation of the structures during the Epoch of Reionization as the first luminous objects in the Universe heated the surrounding H I gas (Carilli 2006, these proceedings; Ciardi 2006, these proceedings).

† http://www.skatelescope.org/

As a secondary goal, the SKA will also conduct a survey for the first supermassive black holes. With the generally higher densities in the earlier Universe, as the first supermassive black holes accrete, any jets that they form are likely to result in radio-loud active galactic nuclei. A deep continuum survey should find radio-loud supermassive black holes out to their formation epoch.

3. SKA Reference Design

Over the past several years, initial design and prototyping efforts have focussed the design (Hall 2005) and have led to the recently adopted Reference Design. The Reference Design comprises both a set of specifications, informed by the Key Science Projects, as well as a concept for realizing those specifications.

The Reference Design emphasizes that, contrary to the popular notion of a telescope being the collecting area (a mirror or a reflecting parabolic dish), the most complex aspect of the SKA will be the data transmission and central processing facility. Radio signals will be fed to this central processing facility by one of three 'front-end' collectors, in a manner analogous to that of an optical telescope in which the mirror feeds visible-wavelength photons to one of a number of 'back-end' instruments. The three front-end collectors are: (*i*) an array of small-diameter antennas with 'smart feeds': the diameter is of the order of 10 m, and the feeds comprise phased arrays in the focal planes of the antennas for frequencies between 0.3 and 3 GHz, and wide-band feeds at higher frequencies up to 25 GHz; (*ii*) aperture array tiles: this innovative technology provides a 'radio fish-eye lens' for all-sky monitoring in the frequency range 0.3 up to 1 GHz and multiple independent observations; and (*iii*) an Epoch of Re-onization array: operating in the 0.1 to 0.3 GHz range, it will will make use of broad-band dipoles similar to those developed for the Low Frequency Array (LOFAR), the Milurea Wide-field Array (MWA), and the Long Wavelength Array (LWA). One of the motivations for this multiple front-end design is that the technology required to cover the full 0.1 to 25 GHz frequency range cannot be obtained by a single collector.

The Reference Design also describes the distribution of the collecting area: 20 % should be within a 1 km diameter region, 50 % should be within 5 km of the core site, 75 % within 150 km of the core, with the remainder on maximum baselines of about 3000 km.

The current timeline for the SKA calls for construction to begin in the early part of the next decade, with a facility having approximately 10 % of the collecting area of the full telescope, probably a relatively compact array operating at frequencies around 1 GHz. Over the rest of the decade, the telescope would be expanded to longer baselines and higher frequencies.

Acknowledgements

Basic research at the NRL is supported by 6.1 Base funding.

References

Carilli, C. L., Furlanetto, S., Briggs, F., *et al.* 2004, *New Astron. Revs.*, 48, 1029

Carilli, C. L., & Rawlings, S. (eds.) 2004, *Science with the Square Kilometre Array, New Astron. Rev.*, 48

Gaensler, B. M. 2004, *Key Science Projects for the SKA, SKA Memo Series*, No. 44

Gaensler, B. M., Beck, R., & Feretti, L. 2004, *New Astron. Revs.* 48, 1003

Hall, P. 2005, *The Square Kilometre Array: An Engineering Perspective* (Springer-Verlag: Berlin)

Kramer, M., Cordes, J. M., Backer, D. C., *et al.* 2004, *New Astron. Revs.* 48, 993

Lazio, T. J. W., Tarter, J. C., & Wilner, D. J. 2004, *New Astron. Revs.*, 48, 985

Rawlings, S., Abdalla, F. B., Bridle, S. L., *et al.* 2004, *New Astron. Revs.*, 48, 1013

Highlights of Astronomy, Volume 14
IAU XXVI General Assembly, 14-25 August 2006
Karel A. van der Hucht, ed.

doi:10.1017/S1743921307011155

Posters presented at Joint Discussion 12

A. A. Konovalenko, A. A. Stanislavsky, E. P. Abranin, V. V. Dorovskyy, V. N. Melnick, M. L. Kaiser, A. Lecacheux, & H. O. Rucker
Comparative analysis of solar observations to a strong absorption on background of sporadic radio emission from the Sun.

S. L. Rashkovskiy, & V. A. Shepelyev
Influence of space plasma and ionosphere on interferometer measurements at decametre wavelengths.

M. R. Olyak
On the possibility of study of the external solar wind thin structure in decameter radio waves.

I. S. Falkovich, A. A. Konovalenko, N. N. Kalinichenko, M. R. Olyak, A. A. Gridin, I. N. Bubnov, A. Lecacheux, & H. O. Rucker
Variations of parameters of the solar wind stream structure at the distances more than 1 AU in 2003 - 2004.

V. N. Melnik, A. A. Konovalenko, B. P. Rutkevych, H. O. Rucker, V. V. Dorovskyy, E. P. Abranin, A. Lecacheux, & A. I. Brazhenko
Decameter Type III-like bursts.

V. N. Melnik, A. A. Konovalenko, N. V. Shevchuk, H. O. Rucker, E. P. Abranin, V. V. Dorovskyy, & A. Lecacheux
Properties of solar spikes at decameter wavelengths.

G. V. Lytvynenko, A. Lecacheux, H. O. Rucker, A. A. Konovalenko, V. V. Vinogradov, V. E. Shaposhnikov, & U. Taubenschuss
High sensitive investigations of the sporadic Jovian radio emission.

A. A. Konovalenko, D. V. Mukha, & S. V. Stepkin
Detection of carbon recombination lines in the direction of Galactic plane at decametric wavelengths.

O. M. Ulyanov, V. V. Zakharenko, A. A. Konovalenko, A. Lecacheux, C. Rosolen, & H. O. Rucker
Detection of individual pulses of the pulsars B0809+74, B0943+10, B0950+08, B1133+16 at decametre wavelength.

M. V. Popov, A. D. Kuzmin, O. M. Ulyanov, A. A. Deshpande, A. A. Ershov, V. V. Zakharenko, V. I. Kondratev, S. V. Kostyuk, B. Ya. Losovski, & V. A. Soglasnov
Instantaneous radio spectra of giant pulses from the Crab pulsar from decimeter to decameter wavelengths.

S. V. Stepkin, A. A. Konovalenko, N. G. Kantharia, & N. Udaya Shankar
Investigations of the interstellar medium by observations of radio recombination lines at decametric wavelength – the largest bound atoms in space.

A. P. Miroshnichenko
The jet structure and parameters of radio emission of quasars and galaxies.

A. P. Miroshnichenko
The jet velocity at kiloparsec scale.

A. I. Brazhenko, V. N. Melnick, A. A. Konovalenko, E. P. Abranin, V. V. Dorovskyy, R. V. Vashchishin, A. V. Frantsuzenko, H. O. Rucker, & A. Lecacheux
Polarization of drifting pairs at decameter waves.

Ya. M. Sobolev
Towards synchrotron radiation theory in curved magnetic field lines.

K. Niinuma, K. Takefuji, S. Kida, A. Takeuchi, R. Nakamura, T. Tanaka, S. Suzuki, K. Asuma, M. Kuniyoshi, N. Matsumura, T. Daishido
A bursting transient was detected at high Galactic latitude in Waseda Nasu Pulsar Observatory.

T. J. W. Lazio, P. S. Rey, S. Ellingson, S. Close, P. Crane, S. D. Hyman, B. A. Jacoby, W. Junor, N. E. Kassim, S. R. Kulkarni, Y. M. Pihlström, G. B. Taylor, & D. Werthimer
The Long Wavelength Array and the radio transient sky.

M. A. Sidorchuk, & E. A. Abramenkov
Observations of the supernova remnants HB 3, IC 443, Cygnus Loop, and some others in the direction $\ell = 65°$ at UTR-2 radio telescope.

N. N. Kalinichenko, I. S. Falkovich
A search for compact decametric radio sources in supernova remnants using interplanetary scintillation method.

A. I. Brazhenko, G. A. Inyutin, V. V. Koshovyy, A. B. Lozinskyy, O. A. Lytvinenko, A. V. Megn, S. L. Rashkovskiy, V. A. Shepelyev, & R. V. Vaschishin
Angular structure of the radio sources at decameter wavelengths.

S. Ya. Braude, K. M. Sidorchuk, M. A. Sidorchuk, S. L. Rashkovsky, A. P. Miroshnichenko, & S. M. Zakharenko
Decameter discrete sources survey of the northern sky using the UTR-2 radio telescope.

N. M. Vasilenko, M. A. Sidorchuk, D. V. Mukha, & S. M. Zakharenko
Very low-frequency continuum survey of the Northern sky.

R. Cassano, G. Brunetti, T. Venturi, G. Setti, S. Giacintucci, D. Dallacasa, & S. Bardelli
Statistics of giant radio halos: expectations, and recent (GMRT) and future (LOFAR, LWA) observations.

A. Omar
Radio sources at 333 MHz at 1 mJy – GMRT results.

D. A. Roshi, S. K. Sethi, U.-L. Pen, J. Peterson, R. Subrahmanyan, T.-C. Chang, C. M. Hirata, J. Roy, & Y. Gupta
HI signal from the Epoch of Re-ionization: a pilot observation with the GMRT.

G. I. Shanin, & A. S. Hojaev
Progress on Suffa Large Radiotelescope Project.

A. A. Konovalenko, H. O. Rucker, A. Lecacheux, V. N. Melnick, I. S. Falkovich, S. L. Rashkovskij, A. I. Brazhenko, & V. V. Koshevoj
Current status of long-wavelength radio astronomy in Ukraine.

A. S. Belov, A. S. Ivanov, A. B. Lozinskyy, S. L. Rashkovskiy, & V. A. Shepelyev
The new wide-band equipment for the URAN interferometers.

N. E. Kassim, T. E. Clarke, A. S. Cohen, P. C. Crane, T. Gaussiran, C. Gross, P. A. Henning, B. C. Hicks, W. Junor, W. M. Lane, T. J. W. Lazio, N. Paravatsu, Y. M. Pihlström, E. J. Polisensky, P. S. Ray, K. P. Stewart, G. B. Taylor, & K. W. Weiler
Exploring the last electromagnetic frontier with the Long Wavelength Array (LWA).

J. R. Dickel, P. C. Crane, W. H. Gerstle, & E. Aguilera, Y. M. Pihlström, J. York, A. Kerkhoff, J. Copeland, C. Slack, & D. Munton
The Long Wavelength Demonstrator Array.

P. S. Ray, S. W. Ellingson, J. R. Fisher, N. E. Kassim, L. J. Rickard, & T. E. Clarke
A baseline design for the Long Wavelength Array stations.

H. R. Dickel, Y. M. Pihlström, T. L. Gaussiran, P. A. Henning, A. Kerkhoff, W. Junor, N. E. Kassim, & G. B. Taylor
The Long Wavelength Array (LWA): a multi-disciplinary educational opportunity.

M. Kuniyoshi, N. Matsumura, K. Takefuji, K. Niinuma, S. Kida, A. Takeuchi, R. Nakamura, S. Suzuki, K. Asuma, & T. Daisido
The automatic radio burst search system at Nasu Observatory.

R. D. Dagkesamanskiy, V. M. Malofeev, I. A. Alekseev, V. I. Kostromin, S. M. Kutuzov, & S. V. Logvinenko
The Multi-Beam Meter Wavelengths Array.

J. Lazio, R. J. MacDowall, K. W. Weiler, D. L. Jones, S. D. Bale, L. D. Demaio, & J. C. Kasper
Astrophysics with a Lunar radio telescope.

Highlights of Astronomy, Volume 14
IAU XXVI General Assembly, 14-25 August 2006
Karel A. van der Hucht, ed.

doi:10.1017/S1743921307011167

Joint Discussion 13
Exploiting large surveys for Galactic astronomy

Christopher J. Corbally[1], **Coryn A.L. Bailer-Jones**[2], **Sunetra Giridhar**[3], **and Thomas H. Lloyd Evans**[4] **(eds.)**

[1]Vatican Observatory Group, University of Arizona, Tucson AZ 85721, USA
email:corbally@as.arizona.edu

[2]Max-Planck-Institut für Astronomie, Königstuhl 17, D-69117 Heidelberg, Germany
email: calj@mpia.de

[3]Indian Institute of Astrophysics, Koramangala, Bangalore 560034, India
email: giridhar@iiap.res.in

[4]SUPA, School of Physics and Astronomy, University of St Andrews,
North Haugh, St Andrews, Fife KY16 9SS, UK
email: thhle@st-andrews.ac.uk

Abstract. This summary of JD13, *Exploiting large surveys for Galactic astronomy*, is based on the talks that were given during the meeting and/or submitted to the full proceedings which appeared in *MemSAI* 77 no. 4, 2006. The electronic version of this *MemSAI* volume also has the abstracts and images of the posters that were displayed and discussed during JD13. Here can only be listed their titles and authors. A panel discussion followed the talks, and a summary of the topics covered is added. Finally, there come the concluding remarks.

Keywords. surveys, Galaxy: fundamental parameters, stars: fundamental parameters, techniques: miscellaneous

---

Preface

While surveys have always been the lifeblood of astronomy, the current era has seen an explosion in both quantity and quality of survey data, made possible by digital instrumentation and electronic databases. Since these surveys are using immense resources, it seemed important to pose the question of how to get the best out of them. This IAU Joint Discussion 13 concentrated particularly on those surveys whose targets are stars and so bear on astrophysics within the Milky Way and Local Group. Time limitations dictated emphasis on the optical and near-infrared surveys, but without excluding the higher and lower energy surveys.

The JD13 started by reviewing the major surveys to find out what these were telling us about the formation and evolution of our Galaxy, the model of an average late-type spiral galaxy. It looked carefully at the techniques of photometric and spectroscopic classification used to identify stars of different types, at stellar variability for distances, and at the complementary kinematic data from radial velocities and proper motions. The relationship of these data with theoretical models of stellar structure and atmospheres was then considered, for these models are critical to age determinations and parameterization.

Our meeting highlighted the remarkable achievements of large surveys, but it also provided a forum in which participants could discuss and reflect on how far the analysis methods were achieving what was hoped from these surveys and so how best to exploit future surveys. The opportunity given at the IAU General Assembly XXVI in Prague for a gathering of astronomers from many fields proved ideal, if short, for promoting this fruitful exchange of expertise and ideas.

Individual contributions from the authors of the eighteen invited talks and the thirty-nine posters, the panel discussion, and the concluding remarks, all of which composed our day and a half meeting, were beyond the page limit set for JD13 in these Highlights of Astronomy. Instead, you will find in the following pages summaries. These were written by a few of those closely involved in the conception and execution of this meeting's topic and were based on the submitted papers. For the full papers of JD13, authored by the individual speakers, and for viewing the very attractive posters, with updated abstracts, please consult the *Memorie della Società Astronomica Italiana*, vol. 77, no. 4, at `<http://sait.oats.inaf.it>`

It is our great pleasure to acknowledge the IAU Secretariat and Executive Committee for grants that made the attendance of some speakers and key participants possible. We also warmly thank the IAU-GA XXVI's most attentive Local Organizing Committee.

Scientific Organizing Committee

Further, this JD13 would not have been possible without the expertise and active involvement of all the Organizing Committee members:
Coryn A.L. Bailer-Jones (Germany, co-chair), Christopher J. Corbally (Vatican City, co-chair), Laurent Eyer (Switzerland), Sunetra Giridhar (India, co-chair), Thomas H. Lloyd Evans (U.K.), Dante Minniti (Chile), Heather L. Morrison (USA), Birgitta Nordström (Denmark), Imants Platais (USA), and Patricia A. Whitelock (South Africa).

It was a pleasure to work with them all and together achieve a meeting that was enjoyable at the time and will hopefully prove of lasting benefit to Galactic astronomy.

We look forward to the future surveys and to our further reflections on them.

Christopher Corbally, Coryn Bailer-Jones, and Sunetra Giridhar, co-chairs SOC,
and Thomas Lloyd Evans, co-author of proceedings.
Tucson, Arizona, USA, October 31, 2006

1. Photometric, spectroscopic and kinematic Surveys

Reported by Thomas H. Lloyd Evans

1.1. *Rosemary F.G. Wyse: Lessons from surveys of the Galaxy*

Wide-field imagers and multi-object spectroscopy, from the ground and increasingly from space, permit large surveys of stars in galaxies of the Local Group. These may be compared to the predictions of Galaxy formation and to the results of high-redshift surveys on the properties of galaxies in the early Universe. Early surveys, which comprised mainly photographic photometry in a few pass-bands, sufficed to give an outline of the galactic halo and the thick and thin disks. Spectroscopy of relatively small samples gave an indication of the metal contents of these subsystems and permitted the estimation of some dynamical properties. Distance estimates and proper motions together then enable samples to be plotted in phase space. Much subsequent work was guided by the ELS model of the formation of the Galaxy. Stars of high proper motion were used to trace the halo and thick disk. The work of Tinsley stressed the need to understand the evolution of the Milky Way, a typical galaxy, as a precondition for understanding the Hubble diagram.

The current paradigm of galaxy formation by mergers, instead of the collapse of a single cloud postulated by ELS, introduces new considerations. The orbital energy of the merging galaxies is converted into internal degrees of freedom. The low density outer regions of the smaller systems are removed by tides and the thin disks are heated, a permanent consequence for the stars though not for the gas. The outer parts gain angular momentum at the expense of the inner parts, so gas and stars are driven towards the centre and any disk formed later will have a short scale-length. Predictions include the late formation of the extended thin disk, after merging is largely completed, and the formation of the stellar halo from disrupted satellites. The extent of change depends on the relative masses involved: collision with a galaxy with less than 20 % of the mass of the disk will heat the thin disk to create a thick disk and add material to the bulge, while a more significant merger may transform a disk galaxy to an S0 or an elliptical galaxy. The disk may be re-created by subsequent accretion of gas and possibly stars. Many stars in the solar neighbourhood formed at redshifts greater than 2, or an age $\geqslant 10$ Gyr, and retain memories of conditions at these early times, thus enabling a local approach to cosmology.

The thin disk has an exponential structure with scale-length of 3 kpc and scale-height of older stars of 300 pc, with a mass around $6 \times 10^{10}\,M_{\odot}$. Observations of nearby stars indicate an approximately constant star formation rate with low-amplitude bursts every few Gyr. The mean metallicity is close to solar and there is not a close correlation of age and metal content. The presence of very old thin disk stars at the solar circle, or 3 scale-lengths, is hard to explain in the ΛCDM model, hence the suggestion that they have been accreted. Many galaxies with prominent extended disks have been observed at redshifts of 2 and greater. The old stars in the local thin disk formed at a redshift $z > 1.5$.

The thick disk has a scale length of 3 kpc and a scale-height of 1 kpc and accounts for about 5 % of the local density. The mean metallicity is −0.6 dex and the age is similar to those of globular clusters of similar metallicity, 10 - 12 Gyr. The elemental abundance pattern is distinctive. The thick disk lags the thin disk by 30 - 50 km/s and has a vertical dispersion of 45 km/s, which is hotter than would be expected from heating by disk perturbations such as giant molecular clouds and spiral arms. The predominantly old age argues against extended heating of the thin disk, while if it was merger-induced, the last significant merger was surprisingly long ago, at redshift about 2. The high metallicity is difficult to account for if the bulk of the thick disk is from accreted satellites. The Milky Way is not unusual in possessing a thick disk.

The Bulge has exponential scale-length of 500 pc, scale-height of 300 pc and is mildly triaxial with a bar; the mass is 10^{10} $M_\odot$. The mean metallicity is below solar and the dominant age is old. The few elemental abundances which have been measured suggest enrichment by Type II supernovae, which suggests a short duration of star formation. The IMF of low mass stars is the same as that seen in metal-poor globular clusters, the local disk and in the UMi dwarf spheroidal (dSph) galaxy.

The stellar halo within 15 kpc of the Galactic Centre has a total mass of 10^9 $M_\odot$ and is flattened by c/a of about 0.5 at the solar distance. It comprises mainly old, metal-poor stars in orbits of low angular momentum. These are very different from those in dwarf galaxies now, so late accretion has not been important. There is remarkably little scatter in elemental abundance ratios, with a flat 'Type II plateau' at all metallicities: this is unlike the situation in dwarf galaxies, and provides no evidence for changes in the massive star IMF.

The outer halo, with dynamical timescales in excess of 1 Gyr, is the best place to find structure, and several streams have been found in both coordinate space and kinematics. The Sagittarius dwarf galaxy accounts for much of this, with streams around the sky detected in 2MASS photometry. SDSS data have recently revealed a narrow 'orphan' stream, as well as a Monoceros stream which may indicate disk accretion. There is a stream associated with the globular cluster Pal 5, but this results from star loss by the cluster and not accretion by the Galaxy. A component of the thick disk, identified in faint F/G stars at several kpc above the Galactic Plane, is characterised by low angular momentum and metallicity and may represent a shredded satellite galaxy. The same data reveal a possible retrograde stream. Moving groups and stellar streams within the thin disk may reflect past disturbances by spiral arms, rather than stars accreted from dwarf galaxies. The new surveys have sufficient depth to include stars in the dwarf galaxies which surround the Milky Way. SDSS data have revealed additional dSph galaxies in Boo and CVn; radial velocities confirm the reality and delineate the extent of the Boo dSph.

Large surveys are needed to quantify both small-scale and large-scale structures. Excellent photometry, positions and proper motions are required to analyse spatial structure in colour space. Stars need to be selected carefully for spectroscopy of medium resolution, to give kinematics and metallicity, while a subset must be observed at high resolution to provide elemental abundances. Several large scale spectroscopic surveys are under way.

1.2. *Heidi J. Newberg: Galactic structure from photometric surveys*

Photometric surveys are the most effective way to study large scale Galactic structure, at least until radial velocity surveys on a similar scale are available. Photometric criteria permit the selection of specific types of star: examples are BHB stars, RR Lyrae stars, F-type turnoff stars, M- and K-type giants. The luminosities of these stars are sufficiently well known to permit their use as tracers of galactic structure. Results for the smooth components of the Galactic spheroid and disk structures are not in good agreement with one another, perhaps because discrete structure within each component has not been accounted for. Published results do not agree to within the respective errors: K-type giants selected from 2MASS gave scale heights of the thin and thick disks of 269 ± 13 pc and 1062 ± 52 pc, respectively; main sequence stars with SDSS data gave 330 ± 3 pc and 580 - 750 pc in one investigation and 280 pc and 1200 pc in another. These discrepancies are unlikely to be attributable to uncertain luminosity estimates, as the results for the thin and thick disks are not in the same ratio in the different studies. It may be that the situation is confused by structure additional to the large scale structure envisaged.

Seventeen of the twenty dwarf galaxies known to surround the Milky Way have been found by photometry. Most recent discoveries have been made with SDSS data and one, the possible example in Canis Major, from 2MASS. The latest discoveries are all of very low surface brightness and it is difficult to distinguish between dwarf galaxies and star clusters. These objects are detected as statistically significant excesses of star counts at a two-dimensional location in the sky, and locality in distance is determined from a colour-magnitude diagram in which features such as the main sequence, horizontal branch and giant branch may be recognised.

At least six tidal debris streams, which extend over tens of degrees of sky, have been identified, five of them in SDSS data. Three may be associated with dwarf galaxies and three are probably associated with globular clusters. Five of them were discovered in the SDSS data by convolving a template colour-magnitude density profile with the data. The Sgr dwarf galaxy was detected using carbon stars as tracers, but the associated stream was detected spatially with A star tracers from the SDSS, and later shown to extend all the way around the sky using stellar tracers from the SDSS and from 2MASS. These dominate photometric surveys of much of the sky, so that no photometric survey of a small area could measure the parameters of the spheroid reliably.

The spheroid is not axially symmetric but may be fitted by a triaxial profile. An overdensity in Virgo, apparent in studies of A and F stars, makes a large contribution to this asymmetry, which is therefore hard to establish. The Virgo overdensity itself overlaps both the leading and the trailing Sgr tidal tails, but this structure is receding and is unlikely to be associated with the infalling leading tidal tail of Sgr, while the Virgo stars are redder than those typical of the Sgr tail. It is widely separated from the trailing Sgr debris and in any case it appears to have a centre, unlike the tidal tails. Some of the BHB stars in the photometric survey have measured radial velocities. The resulting picture is complex, but there is a significant positive peak in velocities near the centre of the Virgo overdensity, consistent with the finding from RR Lyrae stars.

1.3. *Željko Ivezić: The SDSS spectroscopic survey of stars*

The Milky Way is usually modeled by three components: the thin disk, with $\sigma_z \simeq 20$ km/s and a scale height of ~ 300 pc, the thick disk with $\sigma_z \simeq 40$ km/s and a scale height of ~ 1 kpc and a lower average metallicity ($[Z/Z_\odot] \simeq -0.6$), while the halo has low metallicity ($[Z/Z_\odot] < -1.5$) and little or no net rotation. A full description of the Milky Way requires knowledge of the distributions of the three spatial coordinates, three velocity components and metallicity.

The SDSS provides photometry and spectroscopy for stars in a quarter of the celestial sphere in the North Galactic Cap. Flux densities are measured in u, with effective wavelength 3540 Å and limiting magnitude 22.1, g (4760 Å, 22.4), r (6280 Å, 22.1), i (7690 Å, 21.2) and z (9250 Å, 20.3). A total of 100 million stars in 10 000 deg^2 will be observed, with completeness dropping from 99.3 % at the bright end to 95 % at the faint limits quoted. The RMS photometric accuracy is 0.02 mag at the bright end and the absolute zero point calibration is accurate to within about 0.02 mag. The astrometric positions are accurate to about 0.1 arcsec in each co-ordinate for stars brighter than $r \simeq 20.5$ mag, and star-galaxy separation is reliable to $r \simeq 21.5$ mag.

Spectra for over 150 000 stars in the North Galactic Cap have been obtained with a spectroscopic resolution of 2000 over the range 3800 - 9200 Å. Spectral types and radial velocities are determined by matching the measured spectrum to a set of templates, which are calibrated using the ELODIE stellar library. Random errors in the velocities depend on spectral type but are usually $< 5\,\mathrm{km/s}$ for stars brighter than $g \simeq 18$, rising to $\sim 25\,\mathrm{km/s}$ for stars with $g = 20$. Multiple observations suggest that these errors are underestimated by a factor of ~ 1.5.

Sixteen stars of spectral type near F8 were selected on each plate as spectroscopic standards to calibrate the spectrophotometry. The spectra were used to estimate the effective temperature, gravity and metallicity of the stars observed. Linear relationships were established between effective temperature and g-r and between metallicity and u-g, in a limited range of colour in each case. Errors of 100 - 200 K in T_{eff} for stars in the $-0.3 < (g-r) < 1.0$ colour range and 0.3 dex in metallicity for stars at the blue tip of the stellar locus, $(u-g) < 1$, have been attained. These relationships were then used to determine effective temperature and metallicity for the much larger number of stars for which only photometric data were obtained.

Metallicity estimates were then made for a sample of 10 000 blue main-sequence stars with $14.5 < g < 19.5$, $0.7 < (u-g) < 2.0$, and $0.25 < (g-r) < 0.35$. The last condition selects stars of temperature 6000 - 6500 K, with an additional restriction on distance from the locus in the $(g-r)$ *vs.* $(u-g)$ plane to confine the sample to the main stellar locus. The metallicity distribution of stars a few kpc from the galactic plane is bimodal, with a local minimum near [Fe/H] = −1.3. The stars of higher metallicity have a much larger density gradient in the z direction than in the R direction, whereas the stars of lower metallicity show little dependence on position within 4 kpc of the Sun. The high metallicity stars presumably represent the thin and thick disks, but there is no indication of the existence of two distinct groups in these data. Radial velocity data similarly show no such distinction. The stars of low metallicity have a velocity dispersion which is about 2.5 times higher than that of those of high metallicity, extending this well-known correlation to a much larger volume of space. Stars of high metallicity in an anticentre field show a higher velocity dispersion and an anomalous rotational velocity, possibly indicating the presence of a stellar stream.

1.4. *Johan Holmberg: Revisiting the Geneva-Copenhagen Survey*

The Geneva-Copenhagen Survey (GCS) provides metallicities, ages, kinematics and Galactic orbits for 14 000 F- and G-type dwarfs with $V \leqslant 8.3$. These were based on *uvby* photometry, *Hipparcos/Tycho-2* parallaxes and proper motions, and new accurate radial velocities. The GCS provides large samples of stars over a full range of age, metallicity and abundance, while it is largely free of selection by kinematics or metallicity. The transformations from observational data to astrophysical parameters may introduce systematic errors and this paper re-investigates the underlying calibrations.

The GCS obtains effective temperatures from a calibration of $b-y$, m_1 and c_1. The availability of the 2MASS photometry permits a calibration based instead on $V-K$, which appears to give more consistent results. There is less systematic error, but the observational error of the $b-y$ index is less than that in $V-K$, so the $V-K$ temperature scale is used to provide a new calibration of $\mathrm{T_{eff}}$ as a function of $b-y$. The resulting calibration recovers the correct effective temperature of the Sun, and has a dispersion of 60 K.

The GCS used a composite metallicity scale which was based on several separate calibrations. Nearly 600 stars, which have recent abundance determinations based on a correct temperature scale, have been used to relate the *uvby* indices to metallicity. The redder and bluer stars are still calibrated with the earlier relations. The Hyades presents a particular problem, because the published photometry is not on the same system as that used for the other stars, and because the cluster has an unusual He/Fe ratio. It cannot therefore be used to check the age and metallicity scales for field stars. Ages and masses depend on isochrones in the CM diagram, and these need to be revised using the new temperature and metallicity calibrations. Wide physical binaries provide a check on the calibrations, and differences in age and metallicity of the components are reduced with the new calibrations. The new calibration of age is systematically different from that of the GCS: the greatest ages are reduced by about 10 %. Incorrect results are still obtained for a small number of stars in the 'hook' region of the HR diagram, where a star could be on the main sequence or on the subgiant branch.

The age-metallicity relation (AMR) may be used to study the chemical evolution of the Milky Way. The GCS comprises about 90 % disk, 8 % thick disk and 2 % halo, and we assemble a simulated catalogue with this same mix and apply the same selection criteria and calibrations as were used in the real case. Hence the overall slope in the AMR, with the mean metallicity increasing to younger age, is shown to be real, but an apparent deficiency of metal poor stars at the youngest ages results from the blue cutoff to the sample. This latter point argues against a closed-box model for galactic evolution.

1.5. *Matthias Steinmetz: Radial velocity surveys*

We require radial velocities to provide phase-space information in order to understand how the Galaxy was formed, but by 2000 only about 40 000 radial velocities had been determined for stars in the Galaxy, although a million galaxies have measured redshifts. The Geneva-Copenhagen survey provided radial velocities for 20 000 stars selected from the Hipparcos catalogue. Two major projects are already under way. The SDSS/SEGUE project aims to obtain radial velocities and metallicities for 250 000 stars of magnitude $14.5 < g < 20$, while RAVE will provide radial velocities and chemical abundances for a million stars, to a limiting magnitude $I = 12$, by 2010. SEGUE will observe stars in strips across the Galactic plane in the northern hemisphere, with resolution 2000 over the wavelength range 3800 - 9100 Å. RAVE offers more complete coverage of 15 000 square degrees of the southern sky, with a resolution of 7500 over a narrow spectral region, 8460 - 8740 Å, covering the Ca II triplet. Abundances of many elements will be measured.

SEGUE spectroscopy is supported by accurate and homogeneous multicolour photometry. It will reach at least 100 000 pc in dust-free directions and will delineate substructure in the distant halo, the galactic anticentre and the outer disk. The spectra, with 3 Å resolution, will provide radial velocities accurate to 7 km/s at $g = 18.2$. The expected uncertainties in atmospheric parameters are 150 K in $T_{\rm eff}$, 0.3 dex in [Fe/H] and 0.5 dex in $\log g$. External checks are being made against stars which have high resolution spectroscopic data, especially in globular and open star clusters. A total of 90 000 spectra in 75 fields had been obtained by August 2006; errors of 7 km/s were obtained for main sequence turn off stars of the thick disk, with $g - r > 0.45$, and 11 km/s for metal poor main sequence turn off and blue horizontal branch stars with $g - r < 0.45$.

RAVE observes stars with $9 < I < 12$ with a fixed exposure time of an hour, so the SNR is magnitude dependent. Half the stars observed have errors under 2.0 km/s and re-observations indicate that stability is maintained, while external sources of high precision velocities confirm the zero point and scale of the velocities. A total of 131 500 stars had been measured by August 2006. Initial results show that errors are generally less than 0.1 dex in [Fe/H], 0.15 dex in $\log g$, 2 % in $T_{\rm eff}$ and 6 km/s in $v_{\rm rot}$. The intensity of the prominent diffuse interstellar band near 8660 Å shows a linear correlation with $E(B-V)$. A preliminary estimate of the mass of the Galaxy is $1.45 \times 10^{12}\,M_\odot$.

1.6. *Catherine Turon: astrometric surveys*

Astrometric surveys provide trigonometric parallaxes and proper motions, which together with radial velocities give the full phase space information. Global astrometric surveys, covering the whole sky and linked to a reference frame, provide the reference frame for photometric and spectroscopic surveys as well as for small field observations with large telescopes. Homogeneous data sets with clearly defined limits are essential, both in order to control selection effects and to identify the sources of error.

Ground-based surveys, accurate to the mas level, are still limited by atmospheric motions, by the deformation of the telescope under gravity, and by the small field of the telescopes. Attempts to find all the nearest stars used photometric and spectroscopic as well as astrometric data, and some of the findings were contradicted by the accurate *Hipparcos* parallaxes. This work continues, extended to fainter magnitudes in order to find brown dwarfs, and many more stars have been found near the Sun.

The USNO CCD Astrographic Catalog, UCAC, is an astrometric all-sky catalogue made with the USNO Twin Astrograph. The preliminary UCAC2 contains data for 48 million stars with $R = 8$ - 16 from $-90°$ declination to at least $+40°$ but lacks stars with $R \leqslant 8$, multiple stars with separations up to $6''$ and stars without good proper motions. The final catalogue, UCAC3, will include 60 million stars with positional errors of 15 mas at best and systematic errors smaller than 10 mas, as well as being more complete. The USNO-B1.0 catalogue is based on digital scans of all the major sky surveys from Palomar, AAO and ESO. It contains magnitudes and star/galaxy estimators for more than a billion objects to $V = 21$. The J2000 positions have an accuracy of $0.2''$ and relative proper motions are provided. Other important ground-based astrometric surveys are the Carlsberg Meridian Catalogue 14 (CMC14), the Bordeaux Proper Motion Catalogue (PM2000) and NOMAD, a compilation of the best astrometric data. 2MASS and SDSS combine astrometric positions with photometric/spectroscopic data, while USNO-B has been combined with SDSS to give a proper motion catalogue. Major future projects under way or under consideration are URAT, Pan-STARRS and LSST.

Observations in space eliminate the effects of atmospheric seeing and gravitational distortion of the telescope. The instrument has access to the whole sky, so that all observations can be linked. The only survey to date was the Hipparcos mission, which provided proper motions and parallaxes with a median accuracy of about 0.8 mas/yr and 1 mas respectively, and a link to the International Reference System to 0.6 mas, for 118 000 pre-selected stars. Multi-epoch photometry to 0.0001 - 0.0005 mag was obtained for them all. The data from the star mappers was used to provide data for 2.5 million stars to $V = 11.0$, as the *Tycho-2* catalogue.

1.7. *Javier López-Santiago: Young stellar populations in the Solar neighbourhood*

Young stellar populations are best enumerated by stellar X-ray surveys. These enable the age distribution of main sequence stars of low mass to be determined, as the X-ray luminosity of such stars change by 3 - 4 orders of magnitude even though there is negligible change in optical luminosity. It follows that young stars may be detected at much greater distances than old stars in X-ray flux-limited surveys. The smaller scale heights of younger stars means that old stars will be the dominant population in deep high-latitude stellar X-ray samples, while young stars will dominate in shallow samples. An excess of young stars in the solar neighbourhood was detected by the *Einstein*-EMSS, *EXOSAT* and *ROSAT* missions, excluding the possibility that the stellar birthrate has decreased in the last billion years. The excess stars detected in the EMSS survey were yellow stars, whereas the *Chandra* Deep Field-North survey shows a lack of F-, G- and K-type stars but many dM-type stars. This difference may be related to the dominance of old disk stars in the CDF-N, while the EMSS survey is dominated by young stars.

Two new shallow surveys, the Bright Source Sample (BSS) of the *XMM-Newton* Bright Serendipitous Survey and the stellar sample of the *ROSAT* North Ecliptic Pole (NEP)

Survey, may be used to study stellar populations near the Sun. The BSS provides X-ray spectroscopy for 58 stellar sources with count rates of at least 0.01 cnt/s in the energy range 0.5 - 4.5 keV at galactic latitude $|b| > 20$ deg. The NEP contains 151 sources detected in the energy range 0.1 - 2.4 keV in a $9° \times 9°$ field centred on the NGP. The spectra found in the BSS may be fitted with two components with similar contributions and of temperatures around 0.31 keV and 0.98 keV, each with a narrow Gaussian half-width near 0.17 keV. Similar spectra are known for the coronae of slowly-rotating solar-type stars in the Pleiades. This suggests that the BSS is dominated by young main sequence stars. Their infrared counterparts lie on the main sequence in the *JHK* two-colour diagram from 2MASS data. The BSS and NEP samples both comprise mainly main sequence stars of type G - M.

Comparison of the spectral type distribution of the stars observed in the BSS and NEP surveys with predictions from the X-ray galactic model XCOUNT shows that there is an excess of FGK stars, which would be relatively young, although the X-ray spectra and the position of the stars in the *JHK* two-colour diagram suggest the stars are of intermediate or old age. A possible solution is that the excess FGK stars are binaries with a yellow primary and an M dwarf secondary.

1.8. *Daisuke Ishihara: The mid-infrared all sky survey with AKARI*

The first mid-infrared sky survey was completed by *IRAS* in 1983: almost the entire sky was covered in four broad wavebands centred on 12, 25, 60 and 100 μm. The later *Midcourse Space Experiment* (*MSX*) repeated crowded regions such as the Galactic plane and the Magellanic Clouds, as well as the small area which had not been covered before the helium coolant in IRAS was exhausted. Four broad bands, at 8.28, 12.13, 14.65 and 21.34 μm, were observed with higher sensitivity and spatial resolution.

The *AKARI* satellite, formerly known as *Astro-F*, was launched from Japan into a Sun-synchronous orbit in February; observations began in May 2006. The Ritchey-Chretien telescope has a primary mirror of 685 mm diameter, cooled to 6 K by superfluid liquid helium and by mechanical coolers. It will conduct an all-sky survey in the mid infrared using the InfraRed Camera (IRC). This has broad bands centred on 9 μm and 18 μm, with a spatial resolution of 10 arcsec. This relatively high resolution, which is nearly ten times better than that of *IRAS*, will provide a much better depiction of crowded Milky Way fields and star formation regions. The sensitivity is better than 50 and 120 mJy, respectively, nearly ten times better than *IRAS*, so it will be able to reach a T Tauri star at a distance of 100 pc. The *AKARI*-IRC will also undertake imaging and low resolution spectroscopy in pointed mode. Absolute calibration is carried out with a network of 615 bright K- and M-type giants, which will place the data on the system established by *MSX*, and 249 fainter calibrators, which were set up to calibrate the IRAC camera on the *Spitzer* satellite. *AKARI*-IRC observations are thus especially well suited to the study of young stellar objects and debris disk systems, including those in regions of active star formation. The all-sky survey will be important for the study of mass loss in the later stages of stellar evolution.

The Far Infrared Surveyor experiment (FIS) on *AKARI* will also observe the entire sky in four broad bands between 50 and 200 μm. The sensitivity and angular resolution of FIS also improve on the performance of *IRAS* by an order of magnitude.

The various instruments are fed from different parts of the focal plane, so that the various wavebands are not observed simultaneously in a specific direction. The two mid-infrared bands of the *AKARI*-IRC are observed in separate fields, and the *AKARI*-FIS in a third. Each of the IRC detectors covers a field of approximately $10' \times 10'$ for pointed observations, while the all-sky survey observations use only two of 256 pixels in the direction of scan. The virtual pixel size is 4×4 of the intrinsic pixel size, by virtue of a readout time of a quarter of the dwell time of a point source on a pixel in the scan direction, and by binning four pixels in the cross-scan direction. Each channel of the IRC – the two mid-infrared cameras and the one devoted to the near-infrared – has three broad band filters and two spectroscopic dispersers for low-resolution spectroscopy.

The first *AKARI* point source catalogues of the IRC and the FIS all-sky surveys are expected to be generally available two years after the end of the survey. A faint source catalogue and a small scale structure catalogue are planned to follow later, as well as image data.

2. Impact on Galactic astronomy

Reported by Coryn A.L. Bailer-Jones

2.1. *Joss Bland-Hawthorn: Galactic history – formation and evolution*

The primary motivation for large surveys, both past/ongoing (*Hipparcos*, SDSS, SEGUE, RAVE) and future (pan-STARRS, *Gaia*, WFMOS), is, arguably, exploring the formation and evolution of galaxies. This topic will dominate observational cosmology and galactic studies for decades to come, mostly because it is difficult to define a unique model for galaxy formation (assuming one exists) which fits the diversity of observations. It is not sufficient to simply identify the building blocks of galaxies at high redshift and to demonstrate that numerical simulations can explain their properties and those of galaxies at later ages. We must also construct a detailed picture which self-consistently includes the underlying physics.

Observations indicate quite conclusively that galaxies form through the gradual build-up of dark matter and baryons. Both the Milky Way and M 31 show (sub)structure in most of their components. A major development over the past decade started with the discovery of a dwarf spheroidal galaxy in Sagittarius that was seen to be falling into the Galactic disk, plus a tidally disrupted stream associated with this. Numerous other streams have since been discovered in the halo of the Milkey Way. (Some argue that most or all of these Milkey Way streams are either associated with the Sgr dSph or are part of the outer disk.) *Hipparcos* found evidence for substructure in the thin disk: the distinct ages of these clumps suggest that they are dynamical in origin rather than a result of patchy star formation. A large tidal stream has also been discovered associated with M 31. Indeed, the (metal poor) halo of M 31 is now believed to extend to at least 140 kpc, some 20°on the sky. M 33, a satellite to M 31, has been found to have a halo with a similar metallicity to those of both M 31 and the Milkey Way.

It can be argued that establishing a theory of galaxy formation is primarily about understanding the processes which formed disks in the early universe. The thick disk of the Milkey Way is about 10 - 12 Gyr old and, compared to the thin disk, is dynamically 'hotter' and more metal poor (also with different alpha element abundances, as discussed below), possibly with a distinct abundance gradient and longer radial scale length. One interpretation is that the thick disk was formed through dynamical heating from the merger of another galaxy. Other scenarios for forming the thick disk include dissipational collapse of gas or that it is the dispersed remnants of unbound star clusters in the early universe. Recent observations suggest that thick disks are common in galaxies, so we are probably not dealing with a rare event for which an ad hoc explanation suffices.

The Local Group is a relatively low density cluster. As there are indications that galaxy formation depends on environment, future surveys must extend out to the nearest dense clusters, such as Virgo. This will be possible with *JWST*, but optical and UV observations are also necessary to determine metallicities, perhaps from ground-based 30 - 40 m class telescopes. Of particular interest is the origin of galactic bulges and halos. For example, while the Milkey Way and M 31 are comparable in mass, M 31 has a much more prominent bulge. What caused this? To answer this, high spatial resolution infrared surveys are essential. Ancient stars are important for understanding old halo populations: these very low metallicity stars are probably some of the first stars to have formed in the universe. Their exact abundances and distributions will tell us about the earliest phases of star formation in proto-galaxies.

Extragalactic surveys provide support for the ΛCDM paradigm for galaxy formation, and computer simulations yield structures which look a lot like real galaxies and galaxy clusters. But there are problems. For example, these models predict higher densities of dark matter in the cores of galaxies than are inferred from observations. Self-consistent models of the Galactic bulge also suggest that baryons dominate the total mass within the Solar Circle, in contradiction with CDM simulations. To improve our picture of galaxy formation, future surveys will need to combine information on both the kinematic phase space (positions and velocities) and the chemical phase space (abundances of various elements with different product paths).

2.2. *Sofia Feltzing: Abundance structure of the Galactic disk*

Stars in the solar neighbourhood can be tagged to different stellar populations (e.g., thin disk, thick disk, halo) based on their kinematics, abundances and ages. Stars in the Galactic disk with distinct kinematical signatures have distinctly different elemental abundances. Specifically, for stars at a given [Fe/H], those with thick disk kinematics have higher abundances of the alpha elements (O, Ne, Mg, Si, S, A, Ca, synthesized and released predominantly by Type II, or core collapse, supernovae) than do stars with thin disk kinematics. Oxygen in particular is important because it is produced almost exclusively in SN II, the progenitors of which are massive stars with lifetimes of tens to one hundred million years. Other elements are produced both by SN Ia and SN II. Hence the [O/Fe] abundance can be used a clock to trace the relative contributions of SN Ia and SN II. The trends in the [O/Fe] *vs.* [Fe/H] plot (Fig. 1) can be interpreted to indicate that at early times (clocked by $[Fe/H] \simeq -0.5$), the thick disk (filled circles) underwent a period of intense star formation. Only SN II would have been operational at this time, and so the [O/Fe] ratio remained constant. Later, SN Ia start to contribute, thus increasing the ratio of Fe compared to O, and so [O/Fe] decreases as [Fe/H] increases. The plot for the thin disk stars is slightly different and implies a lower star formation rate with more equal contributions from SN Ia and SN II and therefore a smoother trend across the whole plot.

Analyses of other elements reveal the following. [C/Fe] shows similar trends amongst thin and thick disk stars. (This is based on carbon abundances derived from a forbidden transition and so is robust to departures from LTE in the stellar atmosphere.) This implies that the sources which produce carbon operated on the same timescale as those which produced iron (which is SN Ia). If we could nail down the production of carbon we could infer the evolution of the SN Ia rate. Unfortunately, the formation of carbon is not yet well understood, with no unique observational picture emerging. A survey of 95 kinematically-selected dwarfs shows that the stars with thin disk kinematics have enhanced levels of manganese relative to oxygen ([Mn/O]) compared to the stars with thick disk kinematics. Furthermore, for thick disk stars [Mn/Fe] steadily increases with increasing [Fe/H], whereas for thin disk stars it is flat below [Fe/H]=0 and increases above that value. Plotting instead against [O/H] (which makes more sense as oxygen is only produced in SN II) we see that for thick disk (and halo) stars, [Mn/O] is flat below

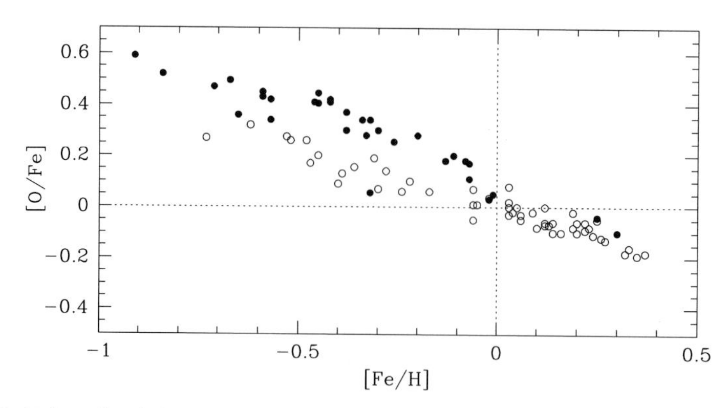

Figure 1. [O/Fe] *vs.* [Fe/H] for two samples of stars. • denote stars that have kinematics typical of the thick disk whilst ∘ are stars with kinematics typical of the thin disk. From Feltzing (2006) in the proceedings of the JD13 published in *MemSAI.*

[O/H] = −0.5. This indicates a common or synchronized Mn and O production. Above [O/H] = −0.5 the [Mn/O] abundance rises, which is interpreted as metallicity-dependent yields in SN II.

Upcoming surveys will make a significant impact in this field. Most spectroscopic abundance surveys have been confined to the solar neighbourhood. Deep photometric surveys will provide targets at larger distances, in particular dwarfs (which are easier to model) which can then be observed spectroscopically on large telescopes to measure individual elemental abundances. Geometric distances (i.e., parallaxes) are required to then map the abundance structure of the Galactic disk, and accurate proper motions allow us to extend abundance – velocity analyses to larger volumes. Parallaxes are also necessary to determine the evolutionary phase or surface gravity (which cannot always be deduced unambiguously from line broadening) which is vital for determining the elemental abundances. Finally, we also need age estimates. These are notoriously difficult and/or inaccurate to estimate for individual field stars, and can only be obtained with any accuracy for MS turn-off stars or for young stars where we believe chromospheric activity proxies. While thick disk stars appear, on average, to be older than thin disk stars, there is still a debate concerning how well separated their age distributions are. Larger samples will help.

2.3. *Amina Helmi: Dwarf spheroidal satellites of the Galaxy*

Most of the satellite galaxies of the Milky Way are very low luminosity systems, the dwarf spheroidals (dSph). While they show a large variety of star formation and chemical evolution histories, all contain a population of old, metal-poor stars (with metallicities similar to stars in the Galactic halo). Because the metallicity of a galaxy increases with time (due to enrichment by successive generations of stars of the interstellar medium, ISM), the metal poor stars are presumably some of the earliest to have formed in the universe. The red giant branch stars in these dSphs are believed to contain in their atmospheres a representative sample (unpolluted by dredge-up) of the chemical elements of the ISM from which they formed.

The DART team has carried out a spectroscopic study of the RGB stars in four dSphs: Sculptor, Fornax, Sextans and Carina. Their CM diagram for Sculptor is shown in Fig. 2. The metallicites are determined from the EW of the individual Ca II triplet lines and radial velocities via cross correlation. Members are then selected via radial

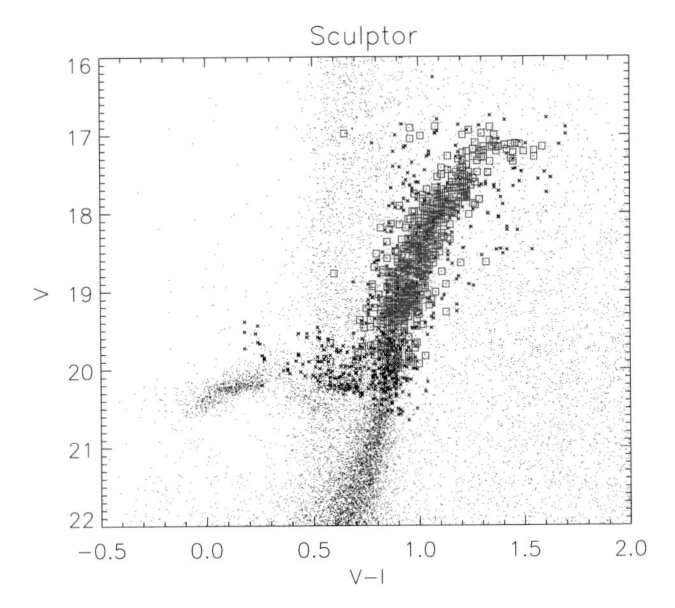

Figure 2. Colour-magnitude diagram of Sculptor. Light squares denote kinematic members and the dark asterisks probable foreground stars, as identified spectroscopically. From Helmi (2006) in the proceedings of the JD13 published in *MemSAI*.

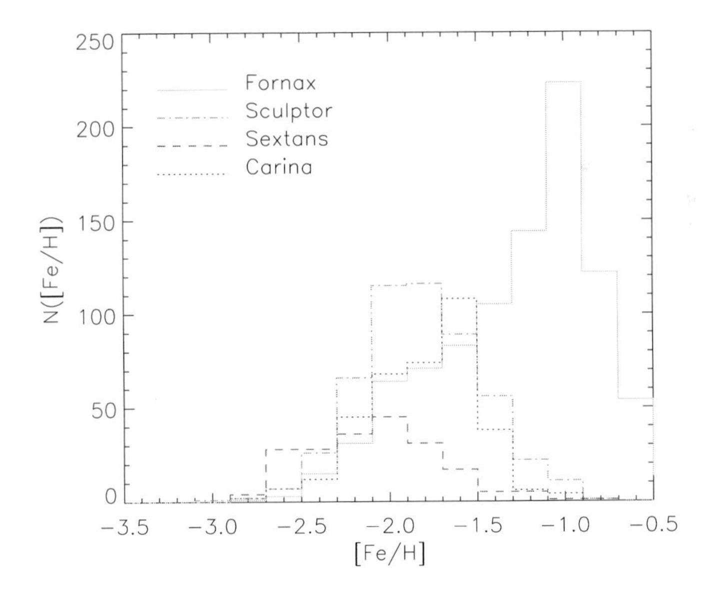

Figure 3. The metallicity distribution for the stars in four dSph. There is an evident lack of objects more metal-poor than [Fe/H] < -3 dex. From Helmi (2006) in the proceedings of the JD13 published in *MemSAI*.

velocity and spectral type. The frequency distribution over metallicity for the RGB stars differ between the four dSph (Fig. 3), yet all have one thing in common: the lack of stars with [Fe/H] < -3.0. Using a closed-box model of chemical evolution with initial enrichment shows that the gas in all four of the dwarf galaxies must have been enriched to $[Fe/H] \simeq -3.0$ prior to the earliest epoch of star formation which led to the present observed population. The fact that this is the same in four otherwise very different dSph galaxies implies that the pre-stellar gas was uniformly enriched over a large volume ($\sim 1\,Mpc^3$, the volume of the Local Group).

In contrast, the low metallicity tail of the Galactic halo extends well below $[Fe/H] = -3.0$. Models indicate the Milky Way halo must have had a very low initial enrichment of $[Fe/H] = -4.5$ or less, even consistent with no initial enrichment ($[Fe/H] = -\infty$).

One idea for the hierarchical build-up of galaxies, and our Galaxy in particular, is that they formed via the accretion of smaller satellites. What is the relation between these putative satellites and the dSph galaxies? To answer this we must compare the metal poor (i.e., old) Galactic halo stars not with the present-day properties of the dSph galaxies but rather with their properties at the time they would have been accreted. The building blocks which accreted onto the Galaxy many billions of years ago had little time to evolve as independent entities, unlike the dSph galaxies we see now (which, of course, were not accreted). Therefore, to make the comparison with the metal poor halo stars, we must focus on the metal poor stars (only) in the dSph galaxies, as these too presumably correspond to their first stellar populations. The conclusion: The fact that the Galactic halo has stars of significantly lower metallicity indicates that it cannot have formed via accretion of the progenitors of the present-day dSph galaxies. Statistical tests (bootstrap and Kolmogorov-Smirov) suggest that this conclusion is not due to the relatively small number of stars in the comparison.

A possible explanation of this conclusion is that whereas the Galaxy formed from a high-σ density fluctuation in the early universe (in the ΛCDM paradigm), the lower mass dSph galaxies formed from the low-σ fluctuations. The latter are predicted to collapse (on average) much later. Before the dSph galaxies formed (not before 12 Gyr and so after reionization) the intergalactic medium could have been pre-enriched. Alternatively the initial mass function could have differed in the early dSph and in the Galactic building blocks, thus altering their chemical evolution.

3. Survey data: gains, limitations and challenges

Reported by Sunetra Giridhar

The new and ongoing surveys are providing a large volume of data comprised of positions, kinematics, multi-band photometry and in some cases spectroscopy at moderate resolutions, reaching out farther in distances, with extended wavelength coverage and with better precision than before. The stellar parameters derived from such extended data sets have the potential of making a large impact on our understanding of the structure and formation of our Galaxy, as well as that of the Local Group, by providing observational input as well as constraints on the models of galaxy evolution. However, careful analyses of these datasets from various surveys and attempts at normalization/standardization of these datasets by experts have also exposed certain gray areas about which the users are being cautioned. A versatile approach based on combining carefully chosen multi-band photometry with spectroscopy could lead to more precise stellar parameters for objects belonging to all parts of the HR diagram. Due to their objectivity and capacity of handling large datasets, automated methods of classifications and parameterization such as those based on Neural Network are extremely important and would become a necessity rather than a choice for the datasets from future surveys. The population synthesis approach is known to be useful for the analysis of large datasets. The consolidated stellar information from these datasets, such as evolutionary stages, kinematics and metallicity and 3D extinction distribution, would be used to test global scenarios of formation and evolution of the Galaxy, including kinematical, dynamical and chemical evolutionary constraints. However, developing accurate and efficient methods for data handling would be a challenging task and should be given high priority if we want to get maximum benefit from the present and future surveys.

3.1. *Ian S. Glass: Classic and new photometric systems*

Many large surveys are based on a new system of color bands. These new multi-band surveys offer some sort of classification and help in identifying unusual objects. A good example is the filters used in search of brown dwarfs of type T that show hot star colors in $J-H$ *vs.* $H-K$ two-color diagrams due to extraordinary molecular features in their spectra, e.g., H_2O, CH_4 and H_2 bands. By adding Y (1.02 μm) and Z (0.89 μm) to the conventional JHK system, these cool dwarfs can be better discriminated in the $Y-J$ *vs.* $J-H$ diagram. However, use of these uncommon filters also requires new observations of standard stars whose magnitudes are within the range of the new system, new tables of intrinsic colors, new models, new corrections of interstellar absorption, etc., if high precision is to be achieved. It should also be remembered that transformations and color equations are valid only for smooth spectra without deep absorptions and emission lines. Hence the colors of late type stars or QSOs having strong and variable emission lines cannot be transformed.

Some of the new frontiers such as planetary transits, studies of star-spots and high modes of stellar oscillations require very high precision photometry. A precision of one part in 10^4 can be achieved only with very stable instrument response and seeing conditions, and only when the comparison star is present in the same CCD frame.

In studies of the galactic disk, the extinction $A(\lambda)$ may vary significantly across the pass-bands of a broadband filter. The problem becomes more acute in IR where the functional form of $A(\lambda)$ beyond 2.5 μm changes from a simple power law and appears to flatten out until the region of the SiO band beyond 8 μm. Better determination of $A(\lambda)$ can be found in Nishiyama *et al.* (2006).

3.2. *Richard O. Gray: Parameterization of stars*

The Strömgren $uvby\beta$ system has been successfully used in parameterization of stars over the spectral type range B through G. A photometric system comprising 5 broad and 14 medium pass-bands, as envisaged for *Gaia* (but later replaced by low resolution spectra), had versatility to characterize stars over much of the HR diagram as explained in Jordi *et al.* (2006). The proposed low-resolution option for *Gaia* may offer opportunities to overcome the problems inherent to photometry. The photometric strategies involving construction of a spectral energy distribution (SED) (or at-least sizable part of it), by converting available magnitudes for a star into fluxes and comparing them with those predicted by model atmospheres, may be adequate for certain stellar parameters such as the temperatures. The method is not sensitive to other parameters like metallicity. Better photometric strategies of stellar parameterization employ carefully calibrated photometric indices, e.g., the $T_{\rm eff}$ calibration of Alonso *et al.* (1996), the reddening correction of Olsen (1988), the [Fe/H] calibration of Nordström *et al.* (2004). But all these calibrations have a strict limit of validity and hence cannot be applied to surveys dealing with stars with a vast range in stellar parameters. The photometric systems do suffer from local and global degeneracies. An illustrative example is the global degeneracy between $T_{\rm eff}$ and A_v, made more difficult to handle by the fact that the extinction law varies from place to place in the Galaxy.

A better method based on artificial neural network in being explored by Bailer-Jones and his colleagues. This system maps the observation space into parameter space and provides 'super' calibration. However, such a system needs to be trained initially with photometric indices derived from stellar spectra. The difficulty with this approach is that synthetic spectra (in some parts of the HR diagram) do not adequately reproduce actual spectra. Since we train the neural network using the data of normal stars, the system will have difficulty in recognizing truly unusual and astrophysically important objects, which would certainly be present in new surveys covering hundreds of million objects.

Since *Gaia* spacecraft design has substituted the multi-color photometric system with low resolution prism spectra, the SEDs generated with these flux calibrated spectra may be used to derive stellar parameters. But instead of comparing these fluxes to those obtained from models directly (which would be equivalent to doing an unweighted fit to the data), Bailer-Jones attempted to optimally sample the SED using filter parameters as free parameters that could be determined by optimizing how well the filter system discriminates between stellar types and avoids degeneracies. The resulting filter system had broad overlapping bands, and these broad bands would be compatible with the low-resolution SEDs to be obtained by *Gaia*. Using this finding one may contemplate designing optimal pass-bands or weighting for different spectral types. The analysis pipeline may first determine a rough stellar type and then apply the pass-bands optimized for that particular stellar type to determine physical parameters. This approach may be able to overcome many degeneracies of traditional fixed photometric systems.

The strategy of adding spectra to photometry has been adopted in the Nearby Stars (NStars) program wherein all dwarfs and giants earlier than M0 within 40 pc of the Sun, as determined by the *Hipparcos* mission, were characterized. Here the Strömgren *uvby* system is augmented with ultraviolet fluxes for the hotter stars and with Johnson-Cousins *RI* photometry and medium resolution spectra in the blue-violet region. The stellar parameters are determined by simultaneously fitting synthetic spectra and fluxes to the blue-violet spectra and fluxes from photometric sources. This strategy resulted in well-constrained fits over a wide range in stellar parameters and the parameters could be determined precisely with fewer spurious solutions. The blue-violet spectral region used in NStars would not be suitable for surveys covering objects with large extinction in the Galactic plane. *Gaia* and RAVE have chosen the 840 - 880 nm region due to its absence of telluric lines and since it contains important lines like the Ca II infrared triplet (sensitive to $T_{\rm eff}$ and gravity as well as to chromospheric activity), diffuse interstellar lines to estimate extinction, lines of α-elements and Paschen series lines of hydrogen; but this spectral region is not really suitable for parameterization of early-type stars. Secondly, even for F-G type stars lines in 840 - 880 nm are a slowly varying function of temperature, and a metal-poor early-G star looks very similar to a mid-F type star. This degeneracy may lead to a poor determination of temperature and/or metallicity. A well designed photometric system, combined with spectroscopy, and in a carefully chosen region of the spectrum, is an ideal approach, and one that opens the survey to measuring much more than basic physical parameters since the spectra and photometry work in complementary ways, where one makes up for the weakness of the other.

3.3. *Sunetra Giridhar: Automated classification and stellar parameterization*

Many large telescopes are now equipped with multi-object spectrometers enabling coverage of a large number of objects per frame for stellar systems like globular clusters. On-going and future surveys and space missions would collect a large number of spectra for stars belonging to different components of our Galaxy. Such a large volume of data can be handled only with automatic procedures, which would also have the advantage of being objective and providing a homogeneous data set most suited for Galactic structure and evolutionary studies. Another outcome would be the detection of stellar variability and finding peculiar objects. Although Minimum Distance Method (MDM) and Gaussian Probability Method (GPM) have been popular, automated spectral classification using Principal Component Analysis (PCA) and Artificial Neural Network (ANN), would have wider applications now and in the future.

PCA is a method of representing a set of N-dimensional data by means of their projection onto a set of optimally defined axes. Since these axes (Principal components) form an orthogonal set, a linear transformation of the data is achieved. Components that represent large variance are important, while those that represent least variance can be ignored and the data set can be replaced by significant components alone, thereby resulting in a reduction of the data size. These compressed data sets are used as input for neural networks. Bailer-Jones *et al.*(1998) had demonstrated that precise calibration could be done using these compressed spectra and that the optimal compression also results in noise removal. Neural Network is a computational method which can provide non-linear parameterized mapping between an input vector (a spectrum for example) and one or more outputs like SpT and LC, or T_{eff}, $\log g$ and [M/H]. The method is generally supervised; this means that for the network to give the required input-output mapping, it must be trained with the help of representative data patterns. These are stellar spectra for which classification or stellar parameters are well determined. The training proceeds by optimizing the network parameters (weights) to give minimum classification error. Once the network is trained and the weights are fixed, the network can be used to produce output SpT and LC, or T_{eff}, $\log g$ and [M/H] for an unclassified spectrum. A summary of the application of ANN can be found in Bailer-Jones (2002).

At IAAP we have made a modest effort to use ANN for parameterization of a sample of stars in the temperature range 4500 to 8000 K using medium resolution spectra ($R \simeq 1000$) obtained with the 2.3 m Vainu Bappu Telescope at VBO, Kavalur, India. The spectra have a wavelength coverage of 3800 - 6000 Å. The stars from the list of Allende *et al.* (1999) and Snider *et al.* (2001) were included to develop a library of stars with known temperatures, gravities and metallicities. These spectra were used for training and testing the network. The preliminary results based on 680:11:3 architecture gave stellar parameters with RMS errors of 200 K for T_{eff}, 0.3 dex for [Fe/H] and 0.4 dex for $\log g$.

It is very important to envisage an approach that would give quick, reliable spectral classifications (or stellar parameters) for stars falling in all regions of the HR diagram. A single ANN architecture may not give the same desired accuracy over the full range of spectral types and luminosity classes. A pilot program, using the photometric inputs in visual and UV, and using special photometric indices measuring the strengths of molecular bands for late-type stars, could serve as preprocessor and help in identifying a set of specialist networks which would lead to classification of the desired accuracy. A specialist system also needs to be evolved for A-type stars to give quick identification of chemically peculiar, magnetic or emission line stars. A special network needs to be developed for objects displaying complex spectra such as symbiotic stars, novae and supernovae. Here the network must be trained on flux calibrated spectra, and it must use emission line strength as well as shape and structure of the continuum (a composite for symbiotic stars and novae) for classification purposes.

3.4. *Annie C. Robin: Data sets and population synthesis*

The population synthesis approach aims at assembling current scenarios of galaxy formation and evolution, theories of stellar formation and evolution, models of stellar atmospheres and dynamical constraints, in order to make a consistent picture explaining currently available observations of different types (photometry, astrometry, spectroscopy) at different wavelengths. The Galactic model generally describes a smooth Galaxy, while inhomogeneities are observed in the disk as well as in the halo. It is intended to produce a useful tool to compute the probable stellar content of large data sets and therefore to test the usefulness of such data to answer a given question in relation to Galactic structure and evolution.

The originality of the Besançon population synthesis models lies in its dynamical self-consistency. The Boltzmann equation allows the scale height of an isothermal and relaxed population to be constrained by its velocity dispersion and the Galactic potential. The use of this dynamical constraint avoids a set of free parameters quite difficult to determine: the scale height of the thin disk at different ages. It gives the model an improved physical credibility. The main scheme of the model is to reproduce the stellar content of the Galaxy, using some physical assumptions and a scenario of formation and evolution. We essentially assume that stars belong to four main populations: the thin disk, the thick disk, the stellar halo (or spheroid), and the outer bulge. The modeling of each population is based on a set of evolutionary tracks, assumptions on density distributions, constrained either by dynamical considerations or by empirical data, and guided by a scenario of formation and evolution.

The thick disk formation scenario has been studied using photometric and astrometric star counts in many directions, which also provided its velocity ellipsoid, local density, scale height, and mean metallicity. These physical constraints led to a demonstration that the probable origin of the thick disk is an accretion event early in the history of the Galaxy. The 3D extinction model has been inferred from star counts and color distribution from the 2MASS survey by Marshall *et al.* (2006). It furnishes an accurate description of the large scale structure of the disk of dust. It shows a scale height of 98 ± 21 pc. A big hole almost free of dust is identified around the Galactic center with a radius of $\sim 3.2 \pm 0.5$ kpc, but it also contains an elongated feature which resembles a dust lane with an inclination with regard to the Sun-Galactic center direction of $\sim 32^\circ \pm 10^\circ$. This feature may trace the dust falling into the center along the stellar bar. This 3D extinction map can be used to make detailed prediction of the star density and luminosity in the Galactic plane.

The microlensing survey also provides very important constraints. The microlensing optical depth, that is the instantaneous number of ongoing microlensing events per source star, is a key measurable for these surveys and provides an important constraint on the bulge surface mass density. Its dependency upon direction provides, in principle, a unique and powerful probe of the three-dimensional geometry of the bulge stellar mass distribution.

The data sets from different surveys need to be combined to enrich as well as refine the global scenario of formation and evolution of the Galaxy. An accurate and efficient method of data fitting for such large data sets is a primary requirement.

4. Future strategies

Reported by Coryn A.L. Bailer-Jones

4.1. *Coryn A.L. Bailer-Jones: Prospects for Gaia and other space-based surveys*

Space-based surveys are essential for various types of observations and naturally complement ground-based surveys. A space platform is required for the UV, mid- and far-infrared and X-ray, and provides a lower background in almost all wavebands. It permits diffraction limited observations and allows us to overcome temporal and spatial variations of the atmosphere's refractivity, which limit the precision of wide-field astrometry to about 1 mas.

Several upcoming missions with a significant Galactic component take advantage of this environment. *Herschel* is a far infrared and sub-mm ESA observatory due for launch in 2008. It comprises several imaging and spectroscopic instruments operating between 60 and 670 μm. Its key science objective is the formation of stars and galaxies via deep imaging surveys. It will investigate the formation and evolution of galaxy bulges and elliptical galaxies during the first third of the present age of the universe, determining how the galaxy luminosity function and star formation rate has evolved with time. *JWST*, in contrast, will observe between 0.6 and 27μm and is optimized for performing deep 'pencil beam' surveys. It has four instruments operating over different wavelength ranges, three of which have spectroscopic modes with resolving powers between 100 and 7000. One of the main science objectives of *JWST* is to study the early universe, in particular the epoch of the first stars and the formation of the first galaxies. *JWST*'s current launch date is 2013. *SIM PlanetQuest* is a pointed astrometric mission. Several key programmes on Galactic structure have been approved, including calibration of the stellar Mass-Luminosity relation, measuring the distances and ages of globular clusters and measuring the Galactic potential via stellar proper motions. At the time of writing (August 2006), *SIM* faces a very uncertain future due to the shifted priorities of NASA. A launch cannot be expected (if at all) before 2015.

Gaia is an all-sky astrometric and photometric survey due for launch in 2011. It will measure accurate parallaxes and proper motions for everything brighter than $G = 20$ ($V = 20$ - 22; ca. 10^9 stars). Its primary objective is to study the composition, origin and evolution of our Galaxy from the 3D structure, 3D velocities, abundances and ages of its stars. *Gaia* will achieve an astrometric accuracy of 12 - 25 μas at $G = 15$ (providing a distance accuracy of 1 - 2 % at 1 kpc) and 100 - 300 μas at $G = 20$. (These numbers are also the approximate parallax accuracy in μas and the proper motion accuracy in μas/year.) It will measure distances to better than 1 % for about 11 million stars, compared to about 200 now.

One of the main contributions of *Gaia* will be mapping structure across the Galaxy in position, velocity and chemical abundance. This includes small scale structure which could be the remnants of past mergers, which are predicted (and to some extent already observed) to be a major mechanism of galaxy formation. *Gaia* will also map the dark matter in our Galaxy via two derivatives of its survey. First, from the 3D stellar kinematics we can map the total gravitational potential (dark and bright). Second, using its parallaxes and photometry we can make a detailed and accurate measurement of the stellar luminosity function over a large volume. This is converted into a mass function with an M-L relation which itself will be determined to higher accuracy than is presently possible (and over a wide mass range) using *Gaia* binaries. Subtracting the stellar mass distribution from the total mass distribution yields the dark matter distribution. *Gaia* will address many other areas of astrophysics, including calibrating the extragalactic distance ladder, identifying Near-Earth Objects and discovery and characterization of exoplanetary systems and their host stars. In the area of stellar astrophysics, accurate stellar luminosities for stars in open and globular clusters can be derived from their parallaxes and line-of-sight extinctions (determined from the onboard spectrophotometry). From this we can derive the bulk Helium abundance and investigate phenomena such as convective overshooting and diffusion.

The data processing for *Gaia* is complex and challenging, unlike anything yet undertaken in astronomy. *Gaia* continuously sweeps the sky for five years. The basic data product is several image strips more than two million degrees long (7000 great circle scans), with each of the one billion objects appearing about 100 times in each strip. The tiny relative displacements within the strip allow us to determine the parallax (distance) and proper motions. As *Gaia* is self-calibrating (i.e., the science data are also the calibration data), the data reduction involves a simultaneous iterative solution of hundreds of millions of source and calibration parameters. On top of this are many other tasks: treatment of binary stars; General Relativistic effects; photometric and CCD calibration; spectroscopic extraction; source classification; variability detection; just to mention a few.

The next significant advance in understanding the formation, structure and evolution of galaxies will come about from three lines of pursuit. The first is detailed astrometric and chemical abundance surveys of our own Galaxy. This is addressed primarily by Gaia, but also by *SIM* and *Jasmine* if they fly. While our Galaxy retains fossils of its evolution, these will only ever tell us part of the story, and then only for one galaxy. The second line of pursuit, therefore, is the observation of galaxies at different stages of their life, i.e., at a range of redshifts. Several ground- and space-based surveys are already addressing this, but the next generation of satellites, in particular *JWST* and *Herschel*, will focus much more on the earliest epoch of galaxy formation (and star formation) in the high-redshift universe. Together, these two lines of pursuit will significantly advance our understanding of galaxy formation, dark matter, chemical evolution and stellar structure and evolution (to know galaxies we must know stars). The third line is the development of powerful models and data analysis tools. These are essential for processing and then exploiting the *Gaia* data, but will also be required to draw together knowledge obtained from our 'near-field' cosmological studies (our Galaxy) with that from high-redshift galaxies. Effort must be invested into developing models and data analysis techniques with as much zeal as the instrumentation and space platforms.

4.2. *Nick Kaiser: Pan-STARRS*

Pan-STARRS is a deep imaging survey of the sky. It is split into two phases. The first, PS1, comprises a single 1.8m telescope with a seven square degree field-of-view imaging in five filters: g,r,i,z,y. Its major survey mode is to observe the 3π steradian north of $\delta = -30°$(31 000 square degrees) twice every month. The PS1 science observations will start in March 2007 and take 3.5 years. At the end of this, every point in the survey area will have been observed 12 times, and the predicted 5σ point source sensitivity is 24.6 mag in g, 24.0 in i and 21.5 in y. This is 1.3 mag deeper than SDSS (in g) and covers almost four times the area on the sky.

PS1 has numerous science goals, including the identification of potentially hazardous Near-Earth Objects, constraints on dark energy and dark matter (e.g. via weak lensing), detection of microlensing events and stellar/exo-planet transits, the stellar and substellar mass function and, of particular relevance to this JD, Galactic structure. PS1 will contribute a deep photometric and astrometric survey of 3/4 of the sky. The final catalogue should achieve an absolute photometric accuracy of about 0.01 mag and a relative astrometric accuracy of 10 mas (absolute 100 mas) for source-noise limited observations. Given 3.5 years of observations, relative proper motions precise to a few mas/yr (and parallaxes to a few mas) should be possible, limited by the Earth's atmosphere. Although not as accurate as Gaia, it goes much deeper. This catalogue will permit numerous studies in Galactic structure, such as the search for streams in/around the Milky Way and local group galaxies, luminosity function, identification of bound structures, etc.

PS1 is also intended as a test-bed for the larger PS4 survey, which will have four times the collecting area. This will follow once PS1 is complete and address similar science goals, but to fainter magnitudes and higher precisions.

5. A topic-ordered list of poster papers

This ordering of the posters presented at the Joint Discussion 13, *Exploiting Large Surveys for Galactic Astronomy*, follows that of Birgitta Nordström, with due thanks. Most of the actual posters can be found, with their abstracts, at the full, electronic version of the proceedings in *Memorie della Società Astronomica Italiana*, volume 77, n.4 at <http://sait.oat.ts.astro.it/>
Their original abstracts are also available at the IAU XXVIth GA site, <http://www.astronomy2006.com/list-of-registered-abstracts.php?event=jd13>.

5.1. *New and recent surveys*

Radio observations of the HII region complex RCW 95
U. Barres de Almeida, Z. Abraham and A. Roman-Lopes
The Digitized First Byurakan Survey (DFBS): a unique database for proper motion,variability studies, and object classification
A.M. Mickaelian, K.S. Gigoyan, R. Nesci, C. Rossi
A 2dF Survey for Omega Centauri Members at and beyond the Tidal Radius
G.S Da Costa, M.G Coleman
Proper motion sky survey of 2.7 million stars with the Bordeaux automated CCD meridian circle
C. Ducourant, J.F. Le Campion, M. Rapaport and 9 co-authors
Survey of open star cluster. Optical monitoring and photometry
A.S. Hojaev
Near infrared survey of the nuclear region of the Milky Way
U.C. Joshi, S. Ganesh, K.S. Baliyan, I.S. Glass and T. Nagata
The search for Post-AGB stars with dusty discs
T. Lloyd Evans, A.M. Smith, J. McCombie, P.J. Sarre
Near infrared survey of the nuclear region of the Milky Way
G.J. Madsen, L.M. Haffner, R.J. Reynolds
AST/RO sub-mm survey of the galactic center
C.L. Martin, W.M. Walsh, K. Xiao and 5 co-authors
Preliminary results from an open clusters polarimetric survey
A.M. Orsatti, M.M. Vergne, C. Feinstein, R.E. Martinez
The Guide Star Catalog II. Properties of the GSC 2.3 release
A. Spagna, M.G. Lattanzi, B. McLean and 9 co-authors
First brown dwarfs from the UKIRT Infrared Deep Sky Survey (UKIDSS)
R. Tata, E.L. Martin, T. Kendall and 3 co-authors
Kinematics of nearby K-M dwarfs: first results
A.R. Upgren, R.P. Boyle, J. Sperauskas, S. Bartašiūtė

5.2. *New calibrations and models*

Estimating interstellar extinction toward to elliptical galaxies and star clusters
E.B. de Amôres and J.R.D. Lépine
Using the genetic algorithms to study the galactic structure
E.B. de Amôres, A.C. Robin
DSS-I value added catalog of stellar parameters and the SEGUE pipeline
T.C. Beers, Y. Lee, T. Sivarani, C. Allende Prieto, R. Wilhelm, P. Re Fiorentin, C. Bailer-Jones, J.E. Norris, and the SEGUE Calibration Team
All-sky counts in Gaia's unfiltered passband
R. Drimmel, A. Spagna, B. Bucciarelli, M. Lattanzi, R. Smart

Chemical abundances in the ancient Milky Way: G-type SDSS stars Automated determination of T_{eff}, $\log g$, [Fe/H] and $[\alpha/Fe]$.
P. Girard, C. Allende Prieto, C. Soubiran

Search for and investigation of new open clusters using the data from huge astronomical catalogues
S. Koposov, E. Glushkova

Classification of eclipsing binaries in large surveys
O. Malkov, E. Oblak

The Strömvil photometric system: classifying faint stars
A.G.D. Philip, R.P. Boyle

A library of synthetic galaxy spectra for GAIA: comparison with SDSS
P. Tsalmantza, M. Kontizas, R. Korakitis, and 8 co-authors

5.3. *Exploiting existing surveys*

Using the Geneva-Copenhagen survey to study the nature of the Hyades stream
B. Famaey, F. Pont, X. Luri, S. Udry, M. Mayor, A. Jorissen

VIMOS@VLT photometric and spectroscopic survey of the Sagittarius dwarf spheroidal galaxy
G. Giuffrida, S. Zaggia, C. Izzo and 6 co-authors

Investigation of star clusters detected automatically in 2MASS Point Source Catalogue
E. Glushkova, S. Koposov

Searching for White Dwarfs in Surveys
A. Kawka, S. Vennes

Short timescale variability in the faint sky variability survey
L. Morales-Rueda, P.J. Groot, T. Augusteijn, G. Nelemans, P.M. Vreeswijk, E.J.M. van den Besselaar

The White Dwarf Population in IPHAS
L. Morales-Rueda, P.J. Groot, R. Napiwotzki, J.E. Drew

Variable stars in the MOA database
L. Skuljan, I.A. Bond

The RAVE survey: Using the local escape velocity to determine the mass of the Milky Way
M.C. Smith, G.R. Ruchti, A. Helmi, R.F.G. Wyse, and the RAVE collaboration

Kinematics of nearby disk stars from Hipparcos database
R. Teixeira, R.E. de Souza

5.4. *Future survey projects*

JASMINE-astrometric map of the galactic bulge
N. Gouda, Y. Kobayashi, Y. Yamada and 12 co-authors

Vista variables in the Via Lactea (VVV)
D. Minniti, P. Lucas, A. Ahumada and 55 co-authors

LOBSTER telescopes as X-ray All Sky Monitors
R. Hudec, M. Skulinova, L. Pina, L. Sveda

A very small astrometry satellite mission: Nano-JASMINE
Y. Kobayashi, N. Gouda, T. Tsujimoto

Development of a very small telescope for space astrometry surveyer
M. Suganuma, Y. Kobayashi, N. Gouda and 4 co-authors

A near-infrared high-resolution spectroscopic survey of Galactic bulge stars
T. Tsujimoto, N. Kobayashi, Y. Ikeda and 5 co-authors
New method for astrometric measurements in Space Mission, JASMINE
T. Yano, N. Gouda, Y. Yamada
JASMINE simulator
Y. Yamada, N. Gouda, T. Yano and 7 co-authors

6. Discussion and conclusions

Reported by Christopher J. Corbally

6.1. *The panel discussion*

About an hour was set aside to ensure that this meeting truly was a *joint discussion.* With the speakers seated in front, questions and comments on large surveys in the Galaxy were invited.

A first question concerned how detailed should be the information on abundances. It was concluded that first the overall α-element abundances should be determined but that individual element abundances, since they have different nucleosynthesis origins, should certainly be investigated subsequently.

This invoked a reminder that surveys were useful for finding not only the very low metallicity stars, but also stars with higher than solar abundances. The larger the surveys, the more likely to find such stars. Large surveys would also have the best chance of finding the very rare types of peculiar stars, but to do this effectively auto-classification tools needed further refinement.

The discussion then turned to the advantages forthcoming in the larger surveys, those of the order of 100s of millions of stars, with radial velocity as well as full photometric and spectroscopic data, about which the participants had been hearing earlier in the morning (Session IV). These promised improved calibrations, and so improvement in derived parameters and the understanding of stellar atmospheres and convective modeling. However, if these surveys were to fulfill their promise, the adequate modeling of such data was needed, as is already recognized. It was agreed that critical components for such modeling were the Galactic bulge and arms. There was a prediction that, when we have data on a billion stars, we shall realize that the Milky Way is a dynamical, not an equilibrium system.

It was important, when thinking of the large surveys, to recognize niches that they could not fill. One such was a survey with high sensitivity to low metallicities, i.e., the SkyMapper project at Siding Spring Observatory, and another was the analysis of proper motion survey catalogs for data on wide binaries, which can be cleanly classified as to Galactic population.

One aspect of Galactic astronomy could not be well covered in the time available for the Joint Discussion; this was radio data, and so the role of molecular clouds and gas dynamics within the Milky Way. Clearly a further, longer meeting on large surveys and the Galaxy would be needed in the not-too-distant future.

So, the panel discussion served to solidify concepts presented by the speakers during the JD13, to broaden horizons, and to enhance the sense of common enterprise and excitement as all looked forward to exploiting the really large databases for research into Galactic astronomy.

6.2. *Tim de Zeeuw: Concluding Remarks*

In understanding the formation of galaxies, the Milky Way has the unique advantage of having a 'fossil record' that can be read in detail, since we have access to billions of individual stars. The Milky Way is also an average galaxy, and so provides 'near-field' cosmology and a testing ground for the paradigm of formation through hierarchical merging of many small building blocks. So specific questions need to be asked of the Milky Way: (*i*) when its stars formed; (*ii*) when and how it was assembled; and (*iii*) how its dark matter is distributed. These are indeed the questions which many ongoing and planned surveys address and which have been reviewed in this Joint Discussion.

Ongoing surveys, including 2MASS and SDSS/SEGUE, reveal many coherent structures in the Galactic halo, such as tidal tails. These are targets for photometric surveys coming on line (Pan-STARRS, *AKARI*, VST/VISTA) and those planned, if not yet funded (Pan-STARRS 4 and LSST). From thousands to multiple millions of radial velocities have issued from present surveys (Geneva/Copenhagen, SDSS/SEGUE, RAVE) and are anticipated in the future (*Gaia*), while the impact of the astrometric work of *Hipparcos* has stimulated both work from the ground and the planning of future missions (*Gaia*, and Japanese and US missions). Such photometric, kinematic and spectroscopic surveys are providing the all-important stellar parameters that input into Galactic structure information.

There is mounting evidence for different kinematic groups with very homogeneous elemental abundances within the group, but distinct difference between groups. Such evidence is found not only in the stellar halo but now also in the thick disk, and it is very suggestive of successive merging of separate building blocks. These building blocks appear not to be the same as the precursors of the dwarf spheroidals that we observe around the Milky Way today.

The large data sets from ongoing and future surveys promise to unravel at least part of not only the Milky Way formation history but also of the Local Group history, if in less detail. However, these surveys face some significant challenges. Photometric calibration is critical and the derivation of stellar parameters for all classes of stars has to be approached cautiously. Correcting for extinction will be important in some directions and can be ignored in others, while astrometric parallaxes must be of sufficient precision as to give independent distances. The additional physical information provided by variable stars needs multi-epoch observations in a well-planned cadence, something more easily achieved from the ground than from spinning satellites. The input of all these data into the dynamical modeling of the Galaxy will have massive discriminating power between the various formation scenarios, but only if the existing computational machinery for this is developed considerably to deal with the immense data sets which are foreseen.

If there are challenges for surveys, there are also opportunities. Ground-based surveys, e.g., RAVE, can find much complementarity with space-based surveys, e.g., the all-sky *Gaia*. The complementarity is both in providing a pilot project to pave the way for a fainter, space probe, and also in eventually going fainter than the space mission limit. The space-based photometry can also help calibrate the ground-based measurements. Japanese space astrometry plans focus on an infrared study of the Bulge, certainly a complement to the reach of *Gaia* which is confined there to the low-extinction windows. High-resolution spectroscopic follow-ups, e.g., with the proposed WFMOS for Subaru, are going to be essential complements to space-based data, and perhaps even more such projects should be developed.

In conclusion, there is much to look forward to in the next decade for our understanding of the formation of the Milky Way and Local Group.

Acknowledgements

While the Scientific Organizing Committee's preparations were essential to this meeting, its obvious success owes much to the considerable effort and enthusiasm contributed by the speakers, by the poster presenters, and by those who attended. We thank them and their various granting agencies wholeheartedly.

References

Allende Prieto, C., & Lambert, D. L. 1999, *A&A*, 352, 555
Alonso, A., Arribas, S. & Martinez-Roger, C. 1996, *A&A*, 313, 873
Bailer-Jones, C. A. L. 2002, in: R. Gupta, H. P. Singh & C. A. L. Bailer-Jones (eds.), *Automated Data Analysis in Astronomy* (New Delhi, London: Narosa Pub. House), p. 83
Bailer-Jones, C. A. L., Irwin, M., & von Hippel, T. 1998, *MNRAS*, 298, 361
Jordi, C., Høg, E., Brown, A. G. A., *et al.* 2006, *MNRAS*, 367, 290
Marshall, D. J., Robin, A. C., Reylé, C., Schultheis, M., Picaud, S. 2006, *A&A*, 453, 635
Nishiyama, S., Nagata, T., Kusakabe, N., *et al.* 2006, *ApJ*, 638, 839
Nordström, B., Mayor, M., Andersen, J., *et al.* 2004, *A&A*, 418, 989
Olsen, E. H. 1988, *A&A*, 189, 173
Snider S., Allende Prieto, C., von Hippel, T., *et al.* 2001, *ApJ*, 562, 528

Highlights of Astronomy, Volume 14
IAU XXVI General Assembly, 14-25 August 2006
Karel A. van der Hucht, ed.

doi:10.1017/S1743921307011179

Joint Discussion 14
Modeling dense stellar systems

Alison I. Sills[1], Ladislav Subr[2], and Simon F. Portegies Zwart[3] (eds.)

[1]Department of Physics and Astronomy, McMaster University, Hamilton, Canada
email: asills@mcmaster.ca

[2]Astronomical Institute, Charles University, Praha 8, Czech Republic
email: subr@sirrah.troja.mff.cuni.cz

[3]Astronomical Institute Anton Pannekoek, University of Amsterdam, the Netherlands
email: spz@science.uva.nl

Preface

Joint Discussion 14 was held at the General Assembly of the International Astronomical Union from August 17 until 23 in the beautiful Bohemian capital, Prague. The blueprints for this meeting were laid out during the MODEST-5 workshop, held in the Canadian city of Hamilton, Ontario in August 2004. We were sitting in a nice cafe with local brew and food, discussing the future of the MODEST community when we posed the idea for this Joint Discussion at the General Assembly. The meeting was then coined MODEST-7.

MODEST stands for MOdeling DEnse STellar systems, which is an initiative between scientists with a common interest in, well..., you guess. One interesting aspect of the MODEST community is that it brings together scientists who feel that the best way to make progress in their respective sub-domains in computational astrophysics and observational astronomy is by collaboration with scientists in neighboring disciplines. By close collaboration, exchanging information via workshops, conferences and via the MODEST website (`<http://www.manybody.org>`) great progrees has been made recently on the understanding and modeling of dense stellar systems like young star formation regions, massive star clusters, galactic star clusters, globular clusters and the nuclei of galaxies.

One of the interesting aspects of MODEST is that it brings together friend and foe, collaborator and competitor in a friendly environment. Luckily the MODEST community consists of people who do not fear competition and therefore science is discussed in the lecture room in the heat of the moment. The scientific output from this 'gang of scientists' is measured in tens of papers for which the basis has been laid at some MODEST event, most likely while appreciating some local gastronomic speciality.

In this proceedings you will find many examples of the recent progress in this open scientific community, including the summary of Douglas Heggie. For the meeting, and the open way the MODEST community operates I would like to introduce the term 'open science', to emphasis the openness in which we communicate, compete and share our findings. I hope that we will have many more occasions in which we can enjoy this open scientific community.

Scientific Organizing Committee

Christian M. Boily (France), Melvyn B. Davies (Sweden), Douglas C. Heggie (UK), Piet Hut (USA), Ralf Klessen (Germany), Junichiro Makino (Japan), Rosemary A. Mardling (Australia), Steve L. W. McMillan (USA), Georges Meylan (Switzerland), Giampaolo Piotto (Italy), Simon F. Portegies Zwart (the Netherlands), Alison I. Sills (Canada, co-chair), Rainer Spurzem (Germany), and Ladislav Subr (Czech Republic, co-chair).

Simon F. Portegies Zwart, for the SOC,
Amsterdam, 26 October 2006

Highlights of Astronomy, Volume 14
IAU XXVI General Assembly, 14-25 August 2006
Karel A. van der Hucht, ed.

doi:10.1017/S1743921307011180

Modeling dense stellar systems: background

Piet Hut
Institute for Advanced Study, Princeton, NJ 08540, USA
email: piet@ias.edu

Abstract. I provide some background about recent efforts made in modeling dense stellar systems, within the context of the MODEST initiative. During the last four years, we have seen more than fifteen MODEST workshops, with an attendance between twenty and a hundred participants, and topics ranging from very specialized discussions to rather general overviews.

---

1. Dense stellar systems

The study of star clusters, and of dense stellar systems in general, has recently seen great progress, through observations as well as simulations, as is evident from the papers in the proceedings of this meeting, JD14. The label 'dense' is given when stars are close enough that significant interactions between them occur on a time scale short compared to the age of the stellar system.

A star forming region is dense in this sense, because the contracting protostellar clouds have a high probability to interact with each other during star formation. Old open clusters are called dense when their age exceeds their half-mass relaxation time. Most globular clusters in our galaxy are dense for that reason, and in addition, many globulars have a central density high enough for physical collisions between stars to occur frequently.

The most spectacular type of dense stellar system is that found in the nucleus of most galaxies. Our own Milky Way galaxy is no exception: in the central parsec around the central supermassive black hole, there are frequent collisions between the stars, which have a total mass of a few million solar masses.

2. Multi-scale simulations

Twenty-five years ago, stellar dynamics was split up in a number of different subfields that could be studied independently. Planetary dynamics, simulations of star forming regions, star cluster dynamics, modeling of galactic nuclei, the study of interacting galaxies, and cosmological simulations formed six different areas of research that had rather little in common.

In contrast, all six areas are now firmly integrated. In many cases, it makes little sense to study only of these in isolation. Starting at the smallest scales, a detailed simulation of planet formation may have to take into account the influence of neighboring stars within the same star forming region. Or looking from the largest scales, that of cosmological simulations, the most detailed modeling efforts resolve the encounters between individual galaxies; and in turn, a detailed simulation of such an encounter shows how new dense star clusters are formed in the process.

As a result, detailed simulations now routinely span multiple scales, on which the same physical laws show rather different emerging properties. In stellar dynamics, relaxation effects between stars can be ignored on galactic scales, yet are essential in the more dense areas of star clusters and galactic nuclei. And in star forming regions, hydrodynamics shows quite different behaviors on different scales.

3. Multi-physics simulations

The steady increase in computer power (Makino 2006) has made it possible to simulate multiple aspects of the physics of a single system. In a dense star cluster, we now can model the stellar dynamical history of a modest cluster, as well as the stellar evolution of each star. In addition, we can also model the hydrodynamical interactions that play a role when two or more stars have a close encounter, possibly resulting in a collision. Given that computer speed has increased by a factor of a million in the last thirty years, we can now follow the evolution of a million stars as quickly as we could calculate the track of a single star in the mid seventies.

As a consequence, the main bottlenecks in performing multi-scale, multi-physics simulations are no longer related to hardware, but rather to software limitations. For example, none of the existing legacy codes for stellar evolution can pass through all stages of stellar evolution in a robust way, without human intervention. One priority is to develop simpler and more robust versions of all three types of codes needed in the study of dense stellar systems, modeling the stellar dynamics, evolution, and hydrodynamics. Another priority is to find ways to combine these codes in easy and realible ways.

4. MODEST

These two priorities have been the main aims of the MODEST initiative, short for MOdeling DEnse STellar systems (`<http://www.manybody.org/modest.html>`). Starting four years ago with the first workshop in the American Museum for Natural History in New York city (Hut *et al.* 2002), we now hold a main workshop each year, as well as a number of satellite meetings, typically once every few months. Most meetings have a few dozen participants, though some of the yearly meetings have attracted a hundred participants or more. The topics of the workshops range from very specialized discussions to rather general overviews.

5. Frameworks

Having robust versions of stellar dynamics, stellar evolution, and hydrodynamics codes is not enough, if they cannot be coupled in modular and flexible ways. What is really needed is an umbrella software environment, a *framework* that contains the 'glue' to connect different codes with each other. In addition, such a framework should contain user-friendly tools for visualization, archiving, and comparisons with observations.

During the MODEST-6d workshop in Amsterdam in March 2006, we started a first prototype framework for dense stellar systems, MUSE. More information can be found on the 'projects' page of the MODEST home page, mentioned above. On the 'workshops' page, you can find the schedule of future frameworks-related workshops. We plan to hold several such workshops each year, to coordinate the ongoing development efforts.

References

Hut, P., Shara, M. M., Aarseth, S. J., Klessen, R. S., Lombardi, J. C., Makino, J., McMillan, S., Pols, O. R., Teuben, P. J., & Webbink, R. F. 2003, *New Astronomy*, 8, 337

Makino, J. 2006, these proceedings p. 424

Highlights of Astronomy, Volume 14
IAU XXVI General Assembly, 14-25 August 2006
Karel A. van der Hucht, ed.

doi:10.1017/S1743921307011192

Special-purpose computing for dense stellar systems

Junichiro Makino

Division of Theoretical Astophysics, National Astronomical Observatory of Japan,
2-21-1 Ohsawa, Mitaka, Tokyo 181-8588, Japan
email: makino@astron.s.u-tokyo.ac.jp

Abstract. I'll describe the current status of the GRAPE-DR project. The GRAPE-DR is the next-generation hardware for N-body simulation. Unlike the previous GRAPE hardwares, it is programmable SIMD machine with a large number of simple processors integrated into a single chip. The GRAPE-DR chip consists of 512 simple processors and operates at the clock speed of 500 MHz, delivering the theoretical peak speed of 512/226 Gflops (single/double precision). As of August 2006, the first prototype board with the sample chip successfully passed the test we prepared. The full GRAPE-DR system will consist of 4096 chips, reaching the theoretical peak speed of 2 Pflops.

Keywords. celestial mechanics, methods: n-body simulations, globular clusters.

1. Introduction

Since 1989, we have developed several GRAPE (GRAvity PipE) hardwares for astrophysical N-body simulations (Sugimoto *et al.* 1990; Makino & Taiji 1998). The first hardware, GRAPE-1 completed in 1989, had the speed of 240 Mflops. The present generation GRAPE-6 has the peak speed of 64 Tflops, and was the fastest computer at the time of its completion, 2002. We believe GRAPE project so far has been fairly successful, and we hope to continue to develop new hardwares.

A practical problem with a follow-on project for GRAPE-6 is that the initial cost of a custom chip has become too high. Initial cost of GRAPE-4 chip was USD 250 K. That of GRAPE-6 was around 1.5 M. A new chip will cost 5-20 M, depending on which company you talk to. The development cost of GRAPE-4 was already pretty large for a research project within the community of theoretical astrophysics of Japan. For GRAPE-6, we were very lucky to be selected as one project within a national program for 'Computational Science'. However, even with such a national program, USD 10 M just for the initial development of the chip which can be used only for astrophysical N-body simulation is way too much.

Therefore, we had to change our basic strategy of making highly specialized custom LSI chips for astrophysical N-body simulation. One option is to use FPGA (Field programmable Gate Array) chips to reduce the initial cost. This approach is quite effective if the required accuracy is not too high. However, even with the most advanced FPGA chips available today, the performance of high-accuracy force calculation is not much more than that of the 7-yr-old GRAPE-6 chip.

2. GRAPE-DR

Figure 1 shows the basic architecture of GRAPE-DR, the next-generation GRAPE hardware. It consists of a number of processing elements (PEs), each of which consists of

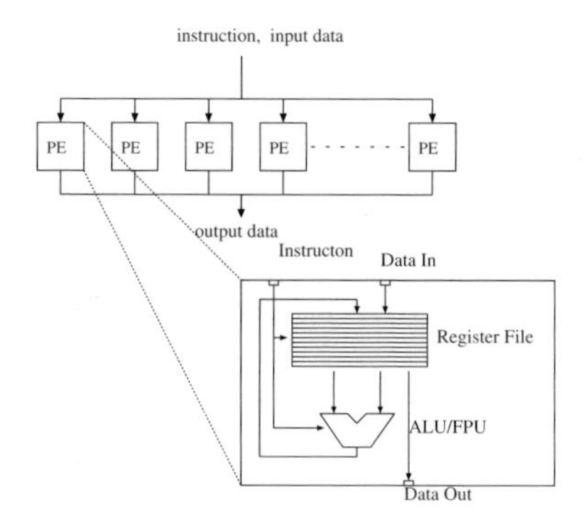

Figure 1. Basic structure of an SIMD processor

an FPU and a register file. They all receive the same instruction from outside the chip, and perform the same operation. This architecture is similar to that of classical SIMD machines such as Illiac-IV or TMC CM-2. Main difference is with GRAPE-DR we limit the size of the local memory of PEs so that we can fit a large number of PEs into a single chip.

With this architecture, each PE 'emulates' GRAPE pipeline by software. The absolute performance of such chip is around a factor of five less than full-custom GRAPE chip made with the same technology, because of additional transistors needed for memory and other control logics. However, with this approach we can vastly widen the application area, since PEs of GRAPE-DR are programmable by software. We hoped this wider application area would justify the large initial cost. We decided to call this architecture Greatly Reduced Array of Processor Elements, or GRAPE.

One practical problem with this new GRAPE architecture is that the number of PEs in one chip is too large. If each PE calculates the force on its own particle, a 512-PE GRAPE chip needs at least 512 particles to share the same time to attain good performance, when used with individual timestep algorithm. To reduce the number of PEs visible from the application program, we added a binary-tree reduction network which can add the results calculated on multiple PEs (Hence the name GRAPE-DR, with DR stands for Data Reduction). This tree has 16 inputs, and 512 processors are divided to 16 groups each with 32 PEs. These 32 PEs calculate force on different particles, and different groups calculate forces from different particles. This reduction tree turned to be useful for many other applications, including LINPACK. Sample chips arrived in May 2006 and operated successfully with the designed value of 500MHz clock on our prototype board. We are currently working on control logic on board, to run real applications.

Acknowledgements

This research is partially supported by the Special Coordination Fund for Promoting Science and Technology (GRAPE-DR project), Ministry of Education, Culture, Sports, Science and Technology, Japan.

References

Makino, J., & Taiji, M. 1998, *Scientific Simulations with Special-Purpose Computers – The GRAPE Systems* (John Wiley and Sons)

Sugimoto, D., Chikada, Y., Makino, J., Ito, T., Ebisuzaki, T., & Umemura, M. 1990, *Nature*, 345, 33

Highlights of Astronomy, Volume 14
IAU XXVI General Assembly, 14-25 August 2006
Karel A. van der Hucht, ed.

doi:10.1017/S1743921307011209

How to build and use special purpose PC clusters in stellar dynamics

Rainer Spurzem

Astronomisches Rechen-Institut, Zentrum Astronomie Universität Heidelberg,
Mönchhofstr. 12-14, D-69120 Heidelberg, Germany
email: spurzem@ari.uni-heidelberg.de

1. The challenge

Large scale, direct particle-particle, brute force N-body simulations are required to accurately resolve numerically transport processes of energy and angular momentum due to two-body relaxation, and interactions between supermassive black holes and other particles having a much smaller mass. Direct accurate N-body codes are the widely used tool for such simulations, e.g., NBODY4 or NBODY6 (Aarseth 1999, 2003), see also Harfst *et al.* (2007) for a less complex code variant, used for benchmarks in this paper. Makino (2002) has presented another direct N-body summation code, which is optimized for a quadratic layout of processor (p required to be a square number).

2. The cluster hardware at RIT and ARI

Computing clusters incorporating the micro-GRAPE-6A boards (see Makino, this volume, for GRAPE) have been installed at the Rochester Institute of Technology (RIT, 'gravitySimulator') and the Astronomisches Rechen-institut (ARI) at the University of Heidelberg (project GRACE = GRAPE + MPRACE). Both clusters consist of 32 compute nodes plus one head node. In addition to a standard Gbit-ethernet, the nodes are connected via a low-latency Infiniband network with a transfer rate of 10 Gbit/s (ARI: 20 Gbit/s). The total, theoretical peak performance is approximately 4 Tflop/s, given by the combined speed of the 32 micro-GRAPE-6A boards. In addition to that the ARI cluster is equipped with reconfigurable FPGA cards (called MPRACE) in addition to GRAPE. MPRACE is optimized to compute neighbour forces and other types of forces between particles such as used for SPH or e.g. molecular dynamics.

It is a relatively new approach to address the challenges of high performance computing by the use of reconfigurable logic, i.e., architectures based on FPGAs (Field Programmable Gate Array). FPGAs mainly consist of a matrix of programmable logic elements plus an interconnecting routing network. Both of them can be configured in a fraction of a second by software, which is minimal compared to typical computing times. We think that the use of reconfigurable logic, presently a relatively new approach to high-performance computing and its power should increase dramatically in the future (Hamada *et al.* 2005; Nakasato *et al.* 2007).

3. Discussion, summary and outlook

Here we show the result of benchmark simulations for a Plummer model on the two clusters. Different models and detailed discussion can be found in Harfst *et al.* (2007). We have varied processor number and particle number. It is a robust result that at one million particles and 32 nodes used, our clusters achieve a sustained performance of 50 %

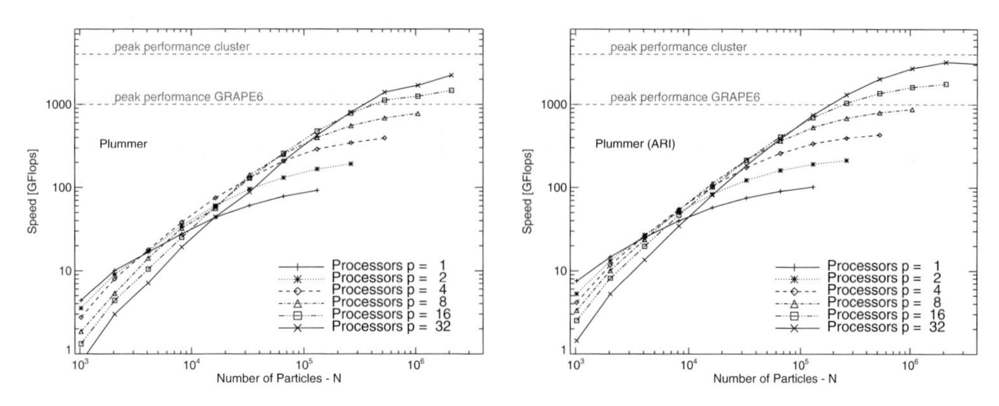

Figure 1. Speed *vs.* particle numbers N for different numbers of processors p. The plots show the results for a Plummer model on the RIT (*left*) and the ARI cluster (*right*).

to 100 % of the peak performance. The total computing time required, and the achieved fraction of the peak performance depend on the details of the astrophysical model, e.g., whether the system is more or less cuspy in central density. The latter will influence the time step structure, and this in turn will affect the parallel run performance. As our complex and communication intensive code typically reaches only a few per cent of peak performance in general purpose parallel computers (e.g., IBM Jump at FZ Jülich), it means that our clusters at RIT and ARI compete in delivered, sustained performance for our application with the top of the list of fastest computers in the world (presently at 280 Tflop/s) at a small fraction of the cost.

Acknowledgements

This work and the GRACE cluster in Heidelberg were financed from the Volkswagen Foundation under grant No. I80 041-043, and the Ministry of Science, Education and Arts of the state of Baden-Württemberg, Germany.

References

Aarseth, S. J. 1999, *PASP*, 111, 1333
Aarseth, S. J. 2003, *Gravitational N-Body Simulations* (Cambridge, UK: CUP)
Hamada, T., Fukushige, T., & Makino, J. 2005, *PASJ*, 57, 799
Harfst, S., Gualandris, A., Merritt, D., Spurzem, R., Portegies Zwart, S., & Berczik, P. 2007, *New Astronomy*, 12, 357
Makino, J. 2002, *New Astronomy*, 7, 373
Nakasato, N., Hamada, T., & Fukushige, T. 2007, *PASJ*, submitted [astro-ph/0604295]
Spurzem, R. 1999, *Jl. Comp. Appl. Math.*, 109, 407

Highlights of Astronomy, Volume 14
IAU XXVI General Assembly, 14-25 August 2006
Karel A. van der Hucht, ed.

doi:10.1017/S1743921307011210

N-body codes

Sverre J. Aarseth

Institute of Astronomy, University of Cambridge, Madingley Road, Cambridge CB3 0HA, UK
email: sverre@ast.cam.ac.uk

Abstract. We review some advances relating to direct N-body codes. In particular, there has been significant progress in dealing with large-N systems containing a few dominant members. The simulation of massive black holes also requires treatment of relativistic effects for strongly bound two-body orbits. Although somewhat costly, the addition of post-Newtonian terms is still straightforward when used in connection with regularization methods. Several versions of multiple regularization are especially well suited to studying black hole problems. We also report on a new stability criterion for the general three-body problem which will provide a robust test in systems where hierarchies are a troublesome feature, as in the case of star cluster simulations with primordial binaries.

Keywords. stellar dynamics, methods: n-body simulations, celestial mechanics

1. Introduction

Many technical challenges arise in the quest for realistic simulations of globular clusters and galactic nuclei. Already special-purpose GRAPE and parallel supercomputers enable the study of systems with up to 10^6 members over significant times. This development has focused attention on problems involving massive objects in the form of black holes, which are of great astrophysical interest. However, until recently, a careful treatment of dominant two-body motions was mostly missing and the effect of relativistic terms rarely included. In this short review we concentrate mainly on these aspects which are vital for further progress. Since compact subsystems also play an important role in general star cluster models, we summarize a new development which gives rise to more reliable stability criteria for hierarchical systems.

2. Large-N simulations

Several large N-body simulations have included massive bodies representing black holes without the introduction of relativistic terms. Agreement with theory was found for the growth of a density cusp around a massive central object and up to 2.5×10^5 particles using GRAPE hardware (Preto *et al.* 2004). This simulation employed chain regularization for the central subsystem. Simulations of three massive objects in systems with $N \leqslant 128\,\mathrm{K}$ have also been carried out with gravitational radiation added (Iwasawa *et al.* 2005). This work employed the standard Hermite scheme with different softening for the Newtonian interactions and produced conditions leading to coalescence for relatively large binary masses and short times, partly aided by large eccentricities. The long-term evolution of rotating systems containing two massive bodies was studied on two GRAPE clusters for models based on direct summation with $N \leqslant 4 \times 10^5$ and relatively small softening (Berczik *et al.* 2006). Here the hardening rate was independent of N, albeit for a modest range in shrinkage of the semi-major axis.

3. Multiple regularizations

Given the large range in length scale required for achieving GR coalescence, it is desirable to explore regularization methods which are able to deal with small separations. This still leaves the serious problem of small time-scales even if only a few particles are involved. Chain regularization has been tried with some success for modest binary shrinkage in combination with the GRAPE-6 (Szell *et al.* 2005). However, the time-transformed leapfrog method (Mikkola & Aarseth 2002) is more accurate for studying significant orbital shrinkage in the massive binary problem. An attractive alternative method has also been developed recently based entirely on time transformations combined with the leapfrog integrator (Mikkola & Merritt 2006). In the so-called algorithmic regularization, a time-symmetric leapfrog scheme can be constructed even when the forces depend on velocities, as in the post-Newtonian case. Provisional tests of the stand-alone code are very promising and indicate that a full-scale simulation code would be effective for studying extreme configurations, albeit at some cost.

The search for powerful methods has lead to an implementation of the so-called wheel-spoke regularization which was developed to deal with one massive object surrounded by strongly bound particles (Zare 1974). Here the basic idea is to regularize all the members in a compact subsystem with respect to the dominant body and include a small softening for the other internal interactions. Recent tests show that substantial binary shrinkage can be treated, including coalescence due to post-Newtonian terms (Aarseth 2006).

4. Hierarchical stability

Following Wisdom (1980), a new stability criterion for the general three-body problem has been constructed using the concept of resonance overlap (Mardling 2006). Resonances between the inner and outer orbits of a hierarchical triple can be identified via a Fourier expansion of the disturbing function. A given configuration is defined to be resonant if at least one Fourier argument, say ϕ, *librates*, that is, $\phi_{\rm min} < \phi < \phi_{\rm max}$, where $|\phi_{\rm max} - \phi_{\rm min}| < 2\pi$. If two or more Fourier arguments librate with different libration frequencies, the system is said to be in a state of resonance overlap and hence chaotic. Almost all chaotic triples are unstable to the escape of one of the bodies: the resonance overlap stability criterion provides a powerful way to identify unstable configurations.

The new stability criterion employs a simple analytic formula expressed in terms of the orbital parameters of the system to determine its resonant (or otherwise) state. A single number is calculated for several 'neighbouring' Fourier components; if two or more of these numbers are negative the system is identified as unstable to the escape of one of the bodies.

References

Aarseth, S. J. 2006, in preparation
Berczik, P., Merritt, D., Spurzem, R., & Bischof, H. 2006, *ApJ* (Letters), 642, L21
Iwasawa, M., Funato, Y., & Makino, J. 2006, *ApJ*, 651, 1059
Mardling, R. 2006, in preparation
Mikkola, S., & Aarseth, S. J. 2002 *Celes. Mech. Dyn. Ast.*, 84, 343
Mikkola, S., & Merritt, D. 2006, *MNRAS*, 372, 219
Preto, M., Merritt, D., & Spurzem, R. 2004, *ApJ* (Letters), 613, L109
Szell, A., Merritt, D., & Mikkola, S. 2005, *Ann. NY Acad. Sci.*, 1045, 225
Wisdom, J. 1980, *AJ*, 85, 1122
Zare, K. 1974, *Celes. Mech.*, 10, 207

Highlights of Astronomy, Volume 14
IAU XXVI General Assembly, 14-25 August 2006
Karel A. van der Hucht, ed.

doi:10.1017/S1743921307011222

Stellar evolution and feedback connections to stellar dynamics

Francesca D'Antona

INAF - Osservatorio di Roma, Via Frascati 33, Monteporzio, I-00040, Italy
email: dantona@oa-roma.inaf.it

Abstract. Until a few years ago, the common paradigm for the formation of Globular Clusters (GCs) was that they constitute a 'simple stellar population' in which all the stars were formed from a chemically homogeneous cluster medium within a relatively short interval of time, at the beginning of the galactic life. In recent years, the spectroscopic information on the low luminosity (turnoff) cluster stars have extended to the unevolved stars the recognition that chemical anomalies are a common feature of GCs and not an exception. This has provoked a revolution in the simple view of GC formation, and requires an adequate dynamical modelling including gas dynamics. It is by now well accepted that at least two different stellar components are common in most GCs. These are almost unequivocally identified with (*i*) a first stellar generation, which gave origin to stars of all masses; and (*ii*)) a second generation, born from the ejecta of the most massive asymptotic giant branch stars of the first generation, in the first 100 - 200 Myr from the first burst of star formation. A 'third' population is present only in some GCs, and is more difficult to be understood. It is characterized by stars having a huge helium content ($Y \simeq 0.4$, if stellar modelling is reasonable) and extreme chemical anomalies in the proton capture elements (Na, O, Al). The status of understanding of the GC properties, based on our most recent models of stellar evolution, is discussed.

Keywords. stars: abundances, stars: AGB and post-AGB, stars: horizontal-branch, stars: Population II, globular clusters: general, globular clusters: individual (NGC 2808, NGC 6441, M 13), stars: formation

---

1. Introduction

Globular Cluster (GC) are the oldest objects in the Galaxy, generally regarded as important tests of stellar evolution. Rewarding aims in the study of their stellar content were a direct stellar measurement of a lower limit to the age of the Universe, the measurement of the initial helium content emerging from the Big Bang, the derivation of the initial mass function of the oldest stellar systems down to the mass limit for hydrogen ignition. Under this work was hidden the very important hypothesis that GC stars constitute the best example of a 'simple stellar population', that is stars born all at the same time with the same initial element abundances. In this case, indeed, the only problems which related stellar evolution and cluster dynamics were to look for explanation of anomalous features such as the presence of rich populations of blue stragglers, or the embarrassing 'second parameter' problem and, most recently, the presence of a fraction of interacting binaries containing neutron stars much larger than in the galactic field. Contrary to the basic assumption, hints were known 25 years ago that star to star abundance anomalies in the elements involved in the hot CNO cycle are present in many GCs. When these anomalies were discovered to be present in the turnoff star (notably by Gratton *et al.* 2001), this stopped the debate on whether they should be attributed to deep, non-canonical mixing in giants or to some kind of 'self-enrichment', which was basically attributed to massive AGBs (Ventura *et al.* 2001). Later on, a new explanation of the old 'second parameter'

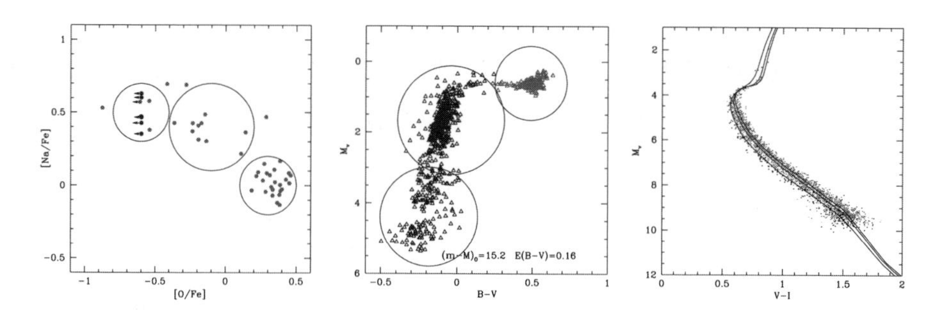

Figure 1. The three different aspects pointing to three different populations in NGC 2808: Na-O anti-correlation (*left*), HB morphology (*center*) and color distribution in the MS (*right*).

problem was proposed by D'Antona *et al.* (2002), namely that the peculiar morphologies of the horizontal branch (HB) of clusters otherwise similar for age and chemistry was due to a population of stars enriched in helium *in all the structure*. This hypothesis can explain the extreme blue tails of some HB, the extended blue HB of the metal rich clusters NGC 6388 and NGC 6441, and the dichotomy of the HB in NGC 2808 (D'Antona & Caloi 2004), later on confirmed by the color distribution of the main sequence stars in this same cluster by D'Antona *et al.* (2005). The correspondence between the HB morphology, the MS colors and the distribution of abundances of sodium and oxygen in the stars of this cluster (Carretta *et al.* 2006) remain the most complete example of the complex star formation history which this 'cluster' has suffered (see Fig. 1).

The clusters which show a small fraction of stars with very high helium content ($Y \simeq 0.40$), such as NGC 2808, Ω Cen, and the much more metal rich cluster NGC 6441 (Caloi & D'Antona 2007 are indeed the most difficult to explain, as such a high helium abundance is not a natural remnant of the AGB evolution, not even in the most massive intermediate mass stars.

It is then possible that, after the first burst of star formation, in these cluster there is star formation from an as yet not well identified stellar generation (small mass supernovae, as suggested by Piotto *et al.* (2005), high mass AGBs, as suggested by D'Antona *et al.* (2005), Wolf Rayet winds, as suggested by Prantzos & Charbonnel (2006), followed by a prolonged phase of star formation emerging directly from the 'normal' AGBs of mass $M \gtrsim 4\,M_{\odot}$. The mass budget required for this 'third' stellar generation is very high indeed (see, e.g., D'Antona & Caloi 2004) and requires a detailed study of the star formation process in clusters.

References

Caloi, V., & D'Antona, F. 2007, *A&A*, 463, 949
Carretta, E., Bragaglia, A., Gratton, R. G., *et al.* 2006, *A&A*, 450, 523
D'Antona, F., & Caloi, V. 2004, *ApJ*, 611, 871
D'Antona, F., Caloi, V., Montalbán, J., Ventura, P., & Gratton, R. 2002, *A&A*, 395, 69
D'Antona, F., Bellazzini, M., Caloi, V., *et al.* 2005, *ApJ*, 631, 868
Gratton, R. G., Bonifacio, P., Bragaglia, A., *et al.* 2001, *A&A*, 369, 87
Piotto, G., Villanova, S., Bedin, L. R., *et al.* 2005, *ApJ*, 621, 777
Prantzos, N., & Charbonnel, C. 2006, *A&A*, 458, 135
Ventura, P., D'Antona, F., Mazzitelli, I., & Gratton, R. 2001, *ApJ* (Letters), 550, L65

Highlights of Astronomy, Volume 14
IAU XXVI General Assembly, 14-25 August 2006
Karel A. van der Hucht, ed.

doi:10.1017/S1743921307011234

Stellar dynamics and feedback connections to stellar evolution

Nataliya M. Ivanova

Canadian Institute for Theoretical Astrophysics, University of Toronto,
60 St. George Street, Toronto, ON M5S 3H8, Canada
email: nata@cita.utoronto.ca

Abstract. In dense stellar systems, dynamical interactions between objects inevitably lead to frequent formation of exotic stellar objects and multiple systems, thereby imposing new questions for the stellar evolution theory. The evolutionary path of such systems could be different from that of the unperturbed objects, therefore, we must re-evaluate their evolutionary treatment to clarify their consequent dynamical evolution. We review briefly the classes of important dynamical encounters and discuss several post-encounter outcomes that may require more detailed attention or development of a new treatment in stellar evolution: evolution of complex merger products, spun-up stars, binaries with stripped giants and triples.

Keywords. globular clusters: general, stars: evolution

Let us examine encounters between objects in the order of increasing distance of closest approach. In the smallest limit, an encounter leads to a head-on collision. The result of such event depends on the relative velocity at infinity v_∞ and on the binding energies of the participating objects. It may lead either to a complete merger of the two objects, to a significant mass-loss or, in the case of collisions with giants, to a binary formation. As the distance increases, participants of the encounters still can be destroyed due to strong tidal forces. When the pericenter of an encounter is far enough for stars not to be destroyed through tidal interactions, but tidal forces are still very strong, binary formation may occur – a so-called tidal capture. However, if v_∞ is large, the total kinetic energy that is needed to be dumped in stars by tidal interactions to form a bound system, becomes comparable to the binding energy of the participants, hence the tidal capture regime becomes unviable, and an encounter may lead only to the envelope spin up.

Among all the varieties of possible physical collisions and their outcomes, most well studied are collisions between two main sequence (MS) stars. They are popular as they lead to the formation of blue stragglers, a class of objects well distinguished observationally. Recent progress in the subject includes the automatizing of the evolutionary calculation of the merger product (Glebbeek & Pols, in prep). However, this does not yet take into account the rotation of the merger product, and the rotation affects strongly the appearance (luminosity) and the evolutionary life-time of a blue straggler (Sills *et al.* 2001). Compared to collisions of MS stars, very little is done on what occurs in the collisions of more evolved stars. These are not rare: we find that at least 5 % of all red giants present in the globular clusters (GCs) cores at ~ 10 Gyr have experienced collisions that resulted either in a significant increase of the envelope mass, or in the formation of hybrid core (where, in comparison to a normal AGB core, very massive He shell of 0.1 - 0.4 $M_\odot$ sits on the CO core). The evolution of such objects has not been studied within the standard stellar evolution.

Another outcome of a physical collision with a giant is the formation of an eccentric binary. The stripped giant core retains bound remnant envelope up to 0.1 $M_\odot$ (Lombardi *et al.* 2006). An envelope of such mass rebounds to its pre-collision size on a thermal time-scale. In addition, the mass of the envelope is not large enough to tidally circularize binary on this time-scale. As a result, the mass transfer may start in a non-circularized binary with rapidly expanding giant envelope.

When a strong tidal interaction results in a binary formation, it is important to clarify where and how much energy is deposited in the star, as the depth and the quantity affect the star expansion and the time-scales on which star will expand and contract (the latter is about of a thermal time-scale of the expanded star). Detailed stellar evolution with proper energy input should show whether it is possible to form a tight binary, or star will remain expanded too long compared to the time-scale of the tidal orbital decay. Special case is a fly-by encounter when neither a bound binary was formed nor destruction nor significant mass loss has occurred, but significant amount of the angular momentum was transferred to stars. The efficiency of the spin-up increases with v_∞, and therefore any effect associated with extra mixing possible in fast rotating giants (see, e.g., Sweigart 1997; Denissenkov *et al.* 2006) will manifest stronger in clusters with bigger velocity dispersion.

Last, we want to emphasize the importance of triples evolution in GCs. Observations of field population reveal that significant fraction of stars are members of hierarchical systems of three and higher multiplicity Tokovinin 2001), and the fraction of triples is increasing when one considers short-periodic binaries (Pribulla & Rucinski 2006) or massive stars (Zinnecker 2006 Zinnecker 2006). In addition to primordial triple population, binary-binary encounters frequently lead to the formation of triple hierarchical systems. The rate of their formation at 10-11 Gyrs is so high that up to 5 % of all binaries have had participated in the triple formation during 1 Gyr (Ivanova 2006). Triples formation and evolution should lead to the modification of the binary population. Indeed, we find that the Kozai Mechanism (KM, Kozai 1962) will act in at least 30 % of all dynamically formed triples. KM leads to the angular momentum transfer between the outer orbit to the inner one, resulting in the oscillations in the eccentricity of the inner orbit and the inclination angle between the two orbits. If an affected by KM triple has the characteristic time-scale between two successful collisions longer than its Kozai cycle, the inner binary can merge or set on the mass transfer before next encounter (Eggleton & Kiseleva-Eggleton 2001, 2006), therefore triples should affect, e.g., the formation and evolution of X-ray binaries in GCs. This suggestion is also supported by results of our simulations where we find that the relative fraction of compact objects in the population of inner binaries of dynamically formed triples is two times larger than in the population of all the binaries.

References

Denissenkov, P. A., & Chaboyer, B., Li, K. 2006, *ApJ*, 641, 1087

Eggleton, P. P., & Kiseleva-Eggleton, L. 2006, in: S. Hubrig & A. Tokovinin (eds.), *Multiple Stars across the H-R Diagram*, ESO Proc. (in press)

Eggleton, P. P., & Kiseleva-Eggleton, L. 2001, *ApJ* 562, 1012

Ivanova, N. 2006 in: S. Hubrig & A. Tokovinin (eds.), *Multiple Stars across the H-R Diagram*, ESO Proc., in press

Kozai, Y. 1962, *AJ*, 67, 591

Lombardi, J. C., Proulx, Z. F., Dooley, K. L., *et al.* 2006, *ApJ* 640, 441

Pribulla, T., & Ruciknski S. 2006, *AJ*, 131, 2986

Sills, A. I., Faber, J. A., Lombardi, J. C., Rasio, F. A., & Warren, A. R 2001, *ApJ*, 548, 323

Sweigart, A. V. 1997 *ApJ* (Letters), 474, L23

Tokovinin, A. A. 2001, in: H. Zinnecker & R. D. Mathieu (eds.), *The Formation of Binary Stars*, Proc. IAU Symp. No. 200 (San Francisco: ASP), p. 84

Zinnecker, H. 2006, in: S. Hubrig & A. Tokovinin (eds.), *Multiple Stars across the H-R Diagram*, ESO Proc., in press

Highlights of Astronomy, Volume 14
IAU XXVI General Assembly, 14-25 August 2006
Karel A. van der Hucht, ed.

doi:10.1017/S1743921307011246

Multiple populations in globular clusters

Giampaolo Piotto

Dipartimento di Astronomia, Università di Padova,
Vicolo dell'Osservatorio 3, I-35122 Padua, Italy
email: giampaolo.piotto@unipd.it

Abstract. I briefly present the most relevant observational facts supporting the idea of one or more star formation episodes in globular clusters.

Keywords. globular clusters: general, stars: AGB and post AGB, stars: formation, Hertzsprung-Russel diagram

1. Introduction

Globular clusters (GC) have been always presented in our textbooks as ideal laboratories for the study of the evolution of 'simple' stellar populations, with neither metal content nor age dispersion. However, it has become harder and harder to fit into this simple picture two continuously increasing sets of observational facts: (*i*) The presence of abundance inhomogeneities, found in all evolutionary sequences, and (*ii*) the presence of anomalous HBs, which generated the so called second parameter problem, i.e., the evidence that some parameters other than the metal content is ruling the properties of core He-burning stars.

The discovery that abundance inhomogeneities are present also on virtually unevolved stars (see Gratton *et al.* 2004) was an important piece of evidence that inhomogeneity should have already been present at the moment of their formation, and the suspect that at least a fraction of GC stars could have formed from material somehow polluted by older stars became almost a certainty. D'Antona *et al.* (2002) proposed that pollution from an early generation of intermediate mass (4 - 6 $M_\odot$) AGB stars could explain the observed abundance inhomogeneities. This polluting material must also be He-rich (Ventura *et al.* 2002), and He enhancement could also explain the anomalously extended HB observed in some GCs, as shown, e.g., by D'Antona & Caloi (2004).

However, we were missing the smoking gun, i.e., a 'direct' evidence of the fact that more than one stellar generation is present in GCs.

2. Direct evidence of multiple stellar populations in GCs

The first direct evidence of a double generation of stars in a GC was found only in 2004. From a set of *HST*-WFPC2 and *HST*-ACS images of the Galactic GC ω Centauri, Bedin *et al.* (2004) definitely showed that its MS was splitted into two distinct sequences, confirming earlier suspects of a double MS (DMS) raised by J. Anderson in his 1997 PhD thesis. It became immediately evident that the DMS discovery was bringing with itself another puzzling fact: the bluest sequence (bMS) was containing a small fraction ($\sim 20\,\%$) of the cluster stars, at variance with what expected from the metallicity distribution of the RGB stars, which shows that the bulk of ω Centauri stars must be metal poor. An immediate spectroscopic followup with VLT-FLAMES confirmed that the bMS was indeed *more* metal rich than the reddest one (rMS), contrary to any expectation from canonical stellar models. The only way to reproduce the bMS was to assume an anomalously high He content ($Y = 0.38$), as indeed proposed on the basis of simple theoretical arguments by Norris (2004), immediately after the publication of Bedin *et al.* (2004) results. These observational facts must be considered as the first empirical evidence of a double generation of stars in the same GC, with the second generation polluted by material ejected by an earlier generation. It must be clear, however, that the sequence identified by Bedin *et al.* (2004) corresponds to an extreme pollution, which cannot be easily explained in terms of normal intermediate mass AGB ejecta, but more likely, by low mass (10 -12 $M_\odot$) SNe events. *If and how* SNe ejecta can be retained within a GC remains still to be explained. Clearly, the bMS population discovered in ω Centauri calls for a generation of stars different from the mildly He enhanced stars proposed by D'Antona *et al.* (2002), which, on the other hand, are apparently sufficient to explain even rather peculiar HBs, as the HB of NGC 2808 (D'Antona & Caloi 2004).

3. Other GCs with multiple populations

Surely, ω Cen is a special object among Galactic GCs. And, as shown in Villanova *et al.* (2007), its stellar population is even more complex than what we described above, with the presence of a third MS, and at least five distinct groups of stars in the SGB, showing the presence of multiple star formation episodes which lasted for almost half of the cluster lifetime. There might be the suspect that ω Cen is an unique case, because of its mass (2.5×10^6 $M_\odot$ at the present time), which could have helped to retain inside the cluster the polluting material.

A rising number of new results seems to demonstrate the contrary. D'Antona *et al.* (2005) have shown that also NGC 2808, a cluster about 2.5 times less massive than ω Cen has a secondary (20 % of its stars), bluer MS which can be explained only assuming a strongly He-enriched ($Y = 0.40$) stellar population. Further (indirect) evidence is provided by Busso *et al.* (2006), who have shown that also the two metal rich GCs NGC 6388 and NGC 6441, with 1/2 and 1/3 the mass of ω Cen must have a population with a He-enrichment $Y \simeq 0.40$ (for about 13 % of the stars in NGC 6388) and $Y \simeq 0.38$ (for about 8 % of the stars in NGC 6441), in order to explain their anomalously blue HB. In addition, Caloi & D'Antona (2007) show that the entire HB of NGC 6441 can be explained only by including a third generation of stars with $0.27 < Y < 0.35$.

Prompted by these results, we recently started a broader investigation, based on proprietary data from the *HST* GO10922 (PI Piotto), and *HST* GO10775 (Pi Sarajedini) to search for extended or multiple MSs in other GCs. The preliminary results are very encouraging. We confirm the presence of a DMS in the already mentioned clusters, and find evidence of a DMS in a dozen of clusters, and a suspected extended sequence in an additional 15, out of the 60 investigated objects. Though these results must be considered as very preliminary, it is interesting to note that, on average, the clusters with the presence of a DMS have an order of magnitude larger masses than the others. Though, some massive clusters, like 47 Tuc, does not seem to have a clear evidence of a DMS.

References

Bedin, L. R., Piotto, G., Anderson, J., *et al.* 2004, *ApJ* 605, 125
Busso, G., Piotto, G., Cassisi, S., *et al.* 2006, *A&A*, submitted
D'Antona, F., Bellazzini, M., Caloi, V., *et al.* 2005, *ApJ*, 631, 868
D'Antona, F., Caloi, V., Montalban, J., Ventura, P. & Gratton, R. 2002, *A&A*, 395, 69
D'Antona, F., & Caloi, V. 2004, *ApJ*, 611, 871
Caloi, V., & D'Antona, F. 2007, *A&A*, 463, 949
Gratton, R., Sneden, C., & Carretta, E. 2004, *ARAA*, 42, 385
Piotto, G., Villanova, S., & Gratton, R., *et al.* 2005, *ApJ*, 621, 777
Villanova, S., Piotto, G., King, I. R., Anderson, J., *et al.* 2007, *A&A*, 663, 296

Highlights of Astronomy, Volume 14
IAU XXVI General Assembly, 14-25 August 2006
Karel A. van der Hucht, ed.

doi:10.1017/S1743921307011258

Dynamical implications of multiple stellar populations

Alison I. Sills and Jonathan M. Downing

Department of Physics and Astronomy, McMaster University,
1280 Main Street West, Hamilton, ON, L8S 4M1, Canada
email: asills@mcmaster.ca

Abstract. We investigate some implications of having two star formation episodes in globular clusters, rather than the traditional single-burst approximation. Evidence for more than one stellar generation is accumulating in observations of abundances of elements lighter than iron in globular cluster stars, and is thought to imply some self-enrichment of the globular cluster gas. In particular, we explore models based on the assumption that the self-enrichment comes from an early generation of asymptotic giant branch (AGB) stars.

Keywords. stars: AGB and post-AGB, globular clusters: general, globular clusters: individual (NGC 2808)

---

1. Introduction

Globular clusters are usually thought of as a simple population of stars that were formed in a single burst from a cloud of uniform composition. However, such a model is inconsistent with the detailed observations of abundances in globular cluster stars. Almost all clusters seem to have a constant value of iron or iron peak elements in their stars. However, every cluster that has been carefully studied shows star-to-star variation of lighter metals (C, N, O, Al, Mg, etc.) and possibly also helium for stars in all regions of the HR diagram. One possible mechanism to explain these variation is the self-enrichment scenario. This scenario takes a variety of forms, but always involves more than one star formation episode in the cluster. Dynamical models of star clusters have, to date, always assumed that reasonable 'initial conditions' are a coeval population of stars in dynamical equilibrium. These models tend to reproduce cluster observations well, but are inconsistent with the self-enrichment scenario. In this paper, we look at the dynamical implication of a second stellar generation. We follow other work (e.g., D'Antona & Caloi 2004) in assuming that the first generation is weighted towards stars which have a significant AGB phase.

2. Dynamical models

We use STARLAB (Portegies Zwart *et al.* 2001) and a micro-GRAPE (Fukushige *et al.* 2005) to perform a direct N-body integration of our two-generation cluster. Our initial conditions consist of a Plummer distribution of the first generation of stars, which has an initial mass function (IMF) that is weighted such that there is an over-abundance of $3-5\,M_\odot$ stars. These are stars which will reach the AGB phase in about 200 Myr, and are the ones which produce the helium-enhanced material from which the second generation forms. We used the mass function from D'Antona & Caloi (2004) which produced the smallest total mass of the cluster. This initial cluster was placed inside a static, analytic Plummer potential which represents the gas which will form the second generation. The mass of this potential was scaled so that it matched the mass lost from the first generation

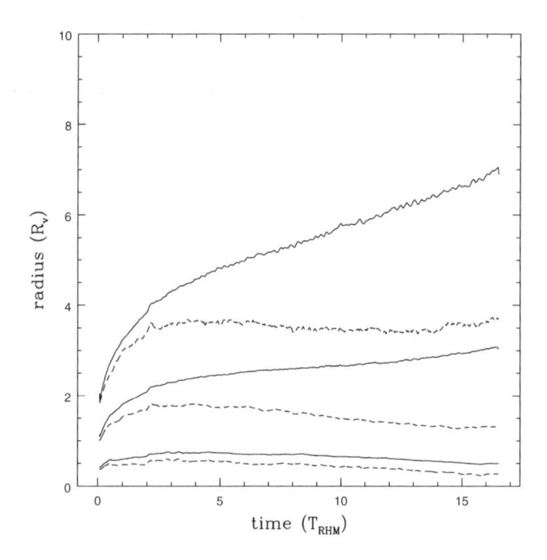

Figure 1. Time evolution of the 10 %, 50 % and 75 % Lagrangian radii for a two-generation cluster, starting after the initial 200 Myr evolution. The *solid* line shows the Lagrangian radii for all the stars in the cluster; the *dotted* line shows only the 10 % most massive cluster members.

in 200 Myr. For simplicity, we assumed that there were no primordial binaries, there is no initial anisotropy, and that the cluster was not affected by the Galactic tidal field. Our initial virial radius for the cluster was taken to be 5 pc.

This initial cluster was evolved for 200 Myr. The simulation was then stopped. The Plummer potential representing the gas was removed, and a second generation of stars was added. This second generation had a Salpeter IMF, and was initially distributed in a Plummer sphere with the same scale radius as the first. The simulation was then restarted with the two generations, and allowed to evolve for at least 12 Gyr. Figure 1 shows the evolution of the 10 %, 50 % and 75 % Lagrangian radii as a function of initial half mass relaxation times in the cluster. It looks completely normal, and dynamically the same as a single generation cluster. The only observable difference between the single and multiple generation clusters is that the two-generation clusters are more abundant in white dwarfs than would be expected from a normal stellar population. The velocity dispersions of the main sequence stars are indistinguishable from those of a normal Salpeter IMF cluster, and the mass-to-light ratio is only slightly higher (~ 2 instead of ~ 1.5).

3. Conclusions

Dynamically, we find that clusters which have two rapid bursts of star formation are almost indistinguishable from the canonical single generation clusters. The possible observational hints of a second generation come from the overabundance of white dwarfs compared to main sequence stars, and from the mass-to-light ratio. For more details of these simulations, see Downing & Sills (2007).

References

D'Antona, F., & Caloi, V. 2004, *ApJ*, 611, 871
Downing, J. M. B., & Sills, A. I. 2007, *ApJ*, 662, 341
Fukushige, T., Makino, J., & Kawai, A. 2005, *PASJ*, 57, 1009
Portegies Zwart, S. F., McMillan, S. L. W., Hut, P., & Makino, J. 2001, *MNRAS*, 321, 199

Highlights of Astronomy, Volume 14
IAU XXVI General Assembly, 14-25 August 2006
Karel A. van der Hucht, ed.

doi:10.1017/S174392130701126X

Observations of blue straggler stars in globular clusters

Francesco R. Ferraro[1] **and Barbara Lanzoni**[2]

[1]Astronomy Department, Bologna University, Via Ranzani 1, I-40127 Bologna , Italy
email: francesco.ferraro3@unibo.it

[2]INAF - Osservatorio Astronomico di Bologna, Via Ranzani 1, I-40127 Bologna , Italy
email: barbara.lanzoni@bo.astro.it

Abstract. Blue stragglers stars (BSS) define a sparsely populated sequence extending to higher luminosity than the turnoff point of normal main sequence stars in the color magnitude diagrams of stellar aggregates, thus mimicking a rejuvenated (more massive) stellar population. The nature of these stars has been a puzzle for many years and their formation mechanism is not completely understood, yet. Two mechanisms have been proposed to produce BSS: (*i*) the mass transfer in binary systems; and ((*ii*) the merger of two stars induced by stellar interactions. In this contribution we schematically report on the main properties of BSS in globular clusters (GCs) in the light of the most recent photometric and spectroscopic observations. These results, combined with dynamical simulations, indicate that both the proposed formation mechanisms play an important role in the production of BSS in GCs.

Keywords. stars: blue stragglers, stars: Population II, stars:evolution, techniques: photometric, techniques: spectroscopic

1. Introduction

Globular clusters (GCs) are important astrophysical laboratories for studying the evolution of single stars as well as binary systems. In particular, the evolution and the dynamical interactions of binary systems in high-density environment can generate objects that cannot be explained by standard stellar evolution (like X-ray binaries, millisecond pulsars, etc.). In this respect the most common by-product of binary evolution are the so-called blue straggler stars (BSS). They are commonly defined as those stars brighter and bluer (hotter) than the main sequence (MS) turnoff (TO) stars. BSS lie along an extrapolation of the MS, and thus mimic a rejuvenated stellar population. First discovered by Sandage (1953) in M 3, BSS are more massive than the normal MS stars, thus indicating that some process which increases the initial mass of single stars must be at work. Such effects could be related either to mass transfer (MT) between binary companions, the coalescence of a binary system or the merger of two single or binary stars driven by stellar collisions. Thus, BSS represent the link between classical stellar evolution and dynamical processes. The realization that BSS are the ideal diagnostic tool for a quantitative evaluation of the dynamical effects inside star clusters has led to a remarkable burst of searches and systematic studies, using UV (see Ferraro *et al.* 2003), optical broad-band photometry (see Piotto *et al.* 2004) and high-resolution spectroscopy (Ferraro *et al.* 2006), that provided a number of interesting results:

2. The radial distribution of BSS

In at least five GCs (namely M 3, 47 Tuc, NGC 6752, M 55 and M 5), the radial distribution of the BSS specific frequency has been found to be bimodal: highly peaked in the cluster center, rapidly decreasing at intermediate radii (the so-called zone of avoidance) and rising again outward. Though the number of the surveyed clusters is still low, these discoveries suggest that this *could be the 'natural' radial distribution of BSS*. Moreover, first results from dynamical simulations (Mapelli *et al.* 2004) indicate that the position of a given BSS in the GC may represent a strong dynamical clue on its formation mechanism: if it is located outside the zone of avoidance, the BSS almost certainly results from MT in primordial binaries (PBs), whereas the BSS found close to the cluster core have most likely a collisional origin.

3. Specific frequency and cluster mass

Piotto *et al.* (2004) noted an anticorrelation between BSS specific frequency and the cluster absolute magnitude (mass). Based on these results, Davies *et al.* (2004) suggested that BSS in low mass systems ($M_V > -8$) arise mostly from MT in PB. In more massive systems stellar interactions produce mergers of the primordial binaries early in the cluster history, hence BSS resulting from these mergers are already evolved away. Then, the BSS that we are currently observing in the cores of the most massive systems ($M_V < -9$) are mostly collisional BSS. However detailed cluster-to-cluster comparison has shown that the emerging scenario is much more complex than this, since the dynamical history of each cluster apparently plays a significant role in determining the origin and radial distribution BSS content (Ferraro *et al.* 2003).

4. The chemical signature of the BSS formation process

Indication about the origin of the BSS can be obtained from high resolution spectroscopy. Indeed the chemical signature of the MT-BSS formation process has been recently detected in 47 Tuc (Ferraro *et al.* 2006), where a sub-population of BSS showing a significant depletion of carbon and oxygen with respect to the dominant population has been discovered. This evidence suggest the presence of CNO burning products on the BSS surface coming from a deeply peeled parent star, as expected in the case of MT process. This is the first detection of a chemical signature clearly pointing to a specific BSS formation process in a GC. The C-O depleted BSS seem to share the same radial distribution of 'normal' BSS. A few of them have been identified as W UMa systems (i.e., shrinking binary systems which are losing orbital momentum because of magnetic braking and that would finally merge into a single star). Most of the observed BSS are found to be slow rotators, with velocities compatible with those measured in unperturbed TO stars. However, it is interesting to note that the few fast rotator BSS, the C-O depleted BSS and the W UMa stars are all located in a narrow strip in the low-luminosity region of the BSS distribution, peraphs suggesting that they are the most recently born. The cooler, older BSS rotate more slowly and have 'normal' C-O abundances. This evidence would suggest that, as the rotation starts to slow, mixing might be induced diluiting the surface C-O under-abundance, pushing C-O back toward 'normalcy'. The acquisition of similar sets of data in clusters with different structural parameters and/or in different regions of the same cluster will provide an unprecedented tool to finally address the BSS formation processes and their complex interplay with the dynamical evolution of the cluster.

References

Davies, M. B., Piotto, G., & de Angeli, F. 2004, *MNRAS*, 349, 129
Ferraro, F. R., Sills, A., Rood, R. T., Paltrinieri, B., & Buonanno, R. 2003, *ApJ*, 588, 464
Ferraro, F. R., Sabbi, E., Gratton, R., *et al.* 2006, *ApJ* (letters), 647, L53
Mapelli, M., Sigurdsson, S., Colpi, M., *et al.* 2004, *ApJ* (Letters), 605, L29
Piotto, G., De Angeli, F., King, I. R., *et al.* 2004, *ApJ* (Letters), 604, L109
Sandage, A. R. 1953, *AJ*, 58, 61

Highlights of Astronomy, Volume 14
IAU XXVI General Assembly, 14-25 August 2006
Karel A. van der Hucht, ed.

doi:10.1017/S1743921307011271

Observational evidence for origin of stellar exotica in globular clusters

Franciscus W.M. Verbunt

Astronomical Institute, Utrecht University, the Netherlands
email: f.w.m.verbunt@astro.uu.nl

Abstract. The formation of special binaries in a globular cluster is regulated by the total encounter rate Γ in the cluster, but their life expectancy by the number of encounters γ that one system experiences. The orbital periods indicate whether a neutron star or white dwarf entered a binary via direct collision, via tidal capture, or via exchange encounter. The numbers of X-ray binaries with a neutron star scales with Γ. Magnetically active binaries (including blue stragglers) are formed via evolution of primordial binaries, and their numbers scale with the cluster mass. Cataclysmic variables are formed by stellar encounters or via evolution of a primordial binary in clusters with high and low central density, respectively.

Keywords. globular clusters, binaries, X-rays: binaries

The strong over-abundance of bright X-ray binaries in globular clusters is the consequence of stellar encounters that allow a neutron star to become member of a binary. Direct collisions with a giant can lead to a binary with an ultrashort-period, of 5 - 20 min. The observed orbital periods indicate this origin for half of the systems. The other periods range from hours to a day, indicating that a neutron star catches a main-sequence star by tidal capture, or takes its place in a binary via an exchange encounter. All these mechanisms scale with the total encounter rate in a cluster Γ:

$$\Gamma \propto \int n_c n A v \mathrm{d}V \propto \int \frac{n_c n R}{v} \mathrm{d}V \propto \frac{\rho_o^2 r_c^3}{v} \propto \rho_o^{1.5} r_c^2 \tag{0.1}$$

where n_c and n are the number densities of the neutron stars and of the objects (giants, single stars, binaries) with which they interact, respectively, A is the encounter cross section, v the velocity dispersion, and the integral is over the cluster volume. The second proportionality follows because the cross section scales as $A \propto R/v^2$, with R the size of the object; the third assumes that the encounter rate is dominated by the core (radius r_c, mass density ρ_o) and that $n_c \propto n \propto \rho_o$; and the last equality uses the virial theorem: $v \propto \sqrt{\rho_o} r_c$. Another important number is the encounter rate for a single binary γ:

$$\gamma = nAv \propto \frac{\rho_0}{v} R \propto \frac{{\rho_o}^{0.5}}{r_c} \tag{0.2}$$

This rate must be low if a binary has to live long for an observed state to be reached (e.g., to leasurely spin up a neutron star to millisecond period, or to evolve from a wide initial binary into a cataclysmic variable).

If we superpose lines of constant Γ on the positions of clusters in a diagram showing central density of globular clusters versus core radius (Fig. 1 left), we find that a high value for Γ is a good predictor for the presence of a bright X-ray source. If we superpose lines of constant γ, we find that the two long-period binaries with pulsars are indeed in clusters where they can survive, with low γ.

Chandra has really opened up our research into the low-luminosity X-ray sources. As predicted, low-luminosity sources include low-mass X-ray binaries with low accretion

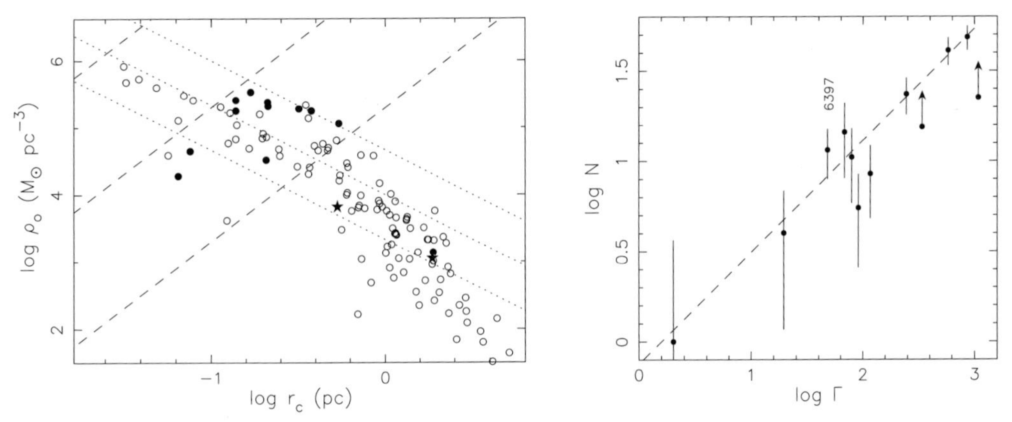

Figure 1. *Left*: Central density and core radius for globular clusters, with lines of constant Γ and γ. Clusters containing bright (above 10^{35} erg/s) X-ray sources are indicated with •, containing pulsars in long-period binaries with a star (after Verbunt 2003). *Right*: Number of low-luminosity sources as a function of Γ (Eq. 0.1), with fit $N \propto \Gamma^{0.62}$ (cf. Pooley *et al.* 2003).

rate (qLMXBs), cataclysmic variables (CVs), millisecond pulsars, and active binaries, as reviewed by Verbunt & Lewin (2006). Various efforts have been made to separate these classes, and investigate their numbers and the scaling with Γ. The number of qLMXBs appears to scale with Γ, but the number of CVs has a more shallow dependence, $N_{\rm CV} \propto \Gamma^{\alpha}$, with $\alpha \simeq 0.6 - 0.7$ (Heinke *et al.* 2003, 2006; Pooley & Hut 2006), possibly because the CV numbers are contaminated by active binaries. The low-density cluster NGC 288 has primordial cataclysmic variables (Kong *et al.* 2006). A dependence of N on metallicity, found for bright sources in globular clusters with other galaxies, is not found for the low-luminosity sources. NGC 6397 warns against premature conclusions on the exceptionality of individual clusters, being normal in Fig. 1 (*right*), but being away from the trend when Γ is computed by integrating a King model. Comparison of clusters through Γ doesn't take into account differences in mass segregation and in neutron star retention. It is therefore important to consider simulations as well (Ivanova, this volume).

An origin via stellar encounters is indicated by the study of several individual systems, such as a blue straggler triple and two sub-subgiants in the old open cluster M 67 (van den Berg *et al.* 2001; Mathieu *et al.* 2002), and the sub-subgiant companion to a pulsar in NGC 6397 (Orosz & van Kerkwijk 2003). The pulsar in the far outskirts of NGC 6752 now has an accurate mass, but may not belong to the cluster after all (Bassa *et al.* 2006).

References

Bassa, C. G., van Kerkwijk, M. H., Koester, D., & Verbunt, F. 2006, *A&A*, 456, 304
Heinke, C. O, Grindlay J. E., Lugger, P. M., *et al.* 2003, *ApJ*, 598, 501
Heinke, C. O, Wijnands, R., Cohn, H. N, *et al.* 2006, *ApJ*, 651, 1098
Kong, A. K. H., Bassa, C., Pooley, D., *et al.* 2006, *ApJ*, 647, 1065
Mathieu, R. D., van den Berg, M., Torres, G., *et al.* 2002, *AJ* 125, 246
Orosz, J., & van Kerkwijk, M. 2003, *A&A*, 397, 237
Pooley, D., & Hut, P. 2006, *ApJ* (Letters), 646, L143
Pooley, D., Lewin, W. H. G., Anderson, S. F., *et al.* 2003, *ApJ* (Letters), 591, L131
van den Berg, M., Orosz, J., Verbunt, F., & Stassun, K. 2001, *A&A*, 375, 375
Verbunt, F. 2003, in: G. Piotto, G. Meylan, G. Djorgovski & M. Riello (eds.), *New Horizons in Globular Cluster Astronomy, ASP-CS*, 296, 245
Verbunt, F., & Lewin, W. H. G. 2006, in: W. H. G. Lewin & M. van der Klis (eds.), *Compact stellar X-ray sources* (Cambridge: CUP), p. 341

Highlights of Astronomy, Volume 14
IAU XXVI General Assembly, 14-25 August 2006
Karel A. van der Hucht, ed.

doi:10.1017/S1743921307011283

Models of M 67

Jarrod R. Hurley

Centre for Astrophysics and Supercomputing, Swinburne University of Technology, PO Box 218, Hawthorn, VIC 3122, Australia
email: jhurley@astro.swin.edu.au

Abstract. The old open cluster M 67 is an ideal test case for current star cluster evolution models because of its dynamically evolved structure and rich stellar populations that show clear signs of interaction between stellar, binary and cluster evolution. Here we discuss a direct N-body model of M 67. This model of 12,000 single stars and 12,000 binaries is evolved from zero-age and takes full account of cluster dynamics as well as stellar and binary evolution. At an age of 4 Gyr the model cluster matches the mass and structure of M 67 as constrained by observations. We discuss the role of the primordial binary population and the cluster environment in shaping the nature of the stellar populations of M 67, with a focus on X-ray binaries and blue stragglers.

Keywords. stellar dynamics, methods: n-body simulations, binaries: close, blue stragglers

---

Despite advances in computing power made over the last decade or so, the globular clusters of our Galaxy remain out of reach of the direct N-body method. However, old open clusters such as M 67 and NGC 188 with ages in the range 4 - 8 Gyr offer dynamically evolved systems where the number, N, of stars required in the starting model is less than 100 000 and within current limitations (which also depend on the primordial binary fraction). A large number (29) of blue stragglers (BSs) are observed in M 67 and there are a number of indicators that the cluster environment, in addition to close binary evolution, has played a role in shaping this population. These include: (*i*) the BSs are concentrated towards the centre of M 67; (*ii*) BSs are found in eccentric binaries (with an orbital period of $P \sim 4$ d in one case: see Latham, these proceedings); and, (*iii*) the ratio of BSs to main-sequence stars is excessive for an open cluster and greater than can be produced by standard binary population synthesis (see Hurley *et al.* 2001). An N-body model of M 67 was presented by Hurley *et al.* (2001) and showed that binary evolution in combination with 3- and 4-body interactions can create the variety of BSs and BS-binaries observed. However, this model was semi-direct in that the dynamical evolution was not modelled directly for the first 2.5 Gyr of the cluster lifetime. Thus it was not ideal.

A preferred model of M 67 was presented by Hurley *et al.* (2005). This started with 12 000 single stars and 12 000 binaries. Stellar masses were chosen from the initial mass function of Kroupa *et al.* (1993) between the limits of 0.1 - 50 $M_\odot$ to give a total mass of 18 700 $M_\odot$. The cluster was placed on a circular orbit at 8 kpc from the Galactic centre – the time averaged semimajor axis for M 67 which has a slightly eccentric orbit – with an orbital speed of 220 km s^{-1}. This gave an initial tidal radius of 32 pc. A Plummer density profile was assumed for the starting model with the stars in virial equilibrium. Solar metallicity was assumed.

The model was evolved to an age of 5 Gyr using NBODY4 (Aarseth 1999) – a Hermite integration code utilising GRAPE-6 hardware with stellar and binary evolution included as described in Hurley *et al.* (2001). At an age of 4 Gyr the model cluster has a mass of $\sim 2000\, M_\odot$ within a tidal radius of 15 pc, a half-mass radius of 2.7 pc and a binary frequency of 50 %. This provides an excellent match to the properties of M 67 (see (Fan

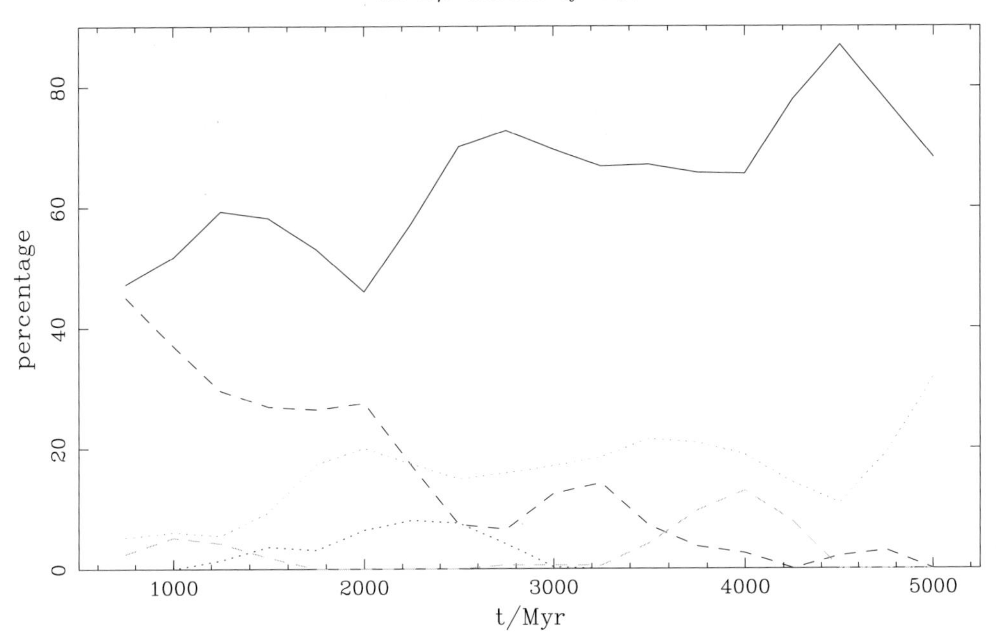

Figure 1. Blue straggler configurations as a function of time for the M 67 N-body model. Shown are single BSs (*solid* line) and BSs in binaries with $P < 1000$ d (*dashed* lines) or longer periods (*dotted* lines – bold/grey indicates circular/eccentric orbits).

et al. 1996, for example). We note that the core binary fraction rises from 0.5 to 0.8 during the simulation and is thus not depleted even if neglected soft binaries are counted.

The model at 4 Gyr contains 20 BSs with 8 in eccentric binaries. The half-mass radius of the BSs is 1.1 pc. Figure 1 summarizes the respective BS configurations as the model evolves – single BSs dominate but the proportion of BSs in eccentric binaries grows with time. Indications for M 67 are that 50 % of the BSs are single with 20 % in short-period binaries (all eccentric) and the remainder in long-period binaries (~ 20 % eccentric, ~ 10 % circular). The 12 000 primordial binaries in the model had periods drawn from a flat distribution of log (P). Evolving these with a binary evolution algorithm (no dynamics) predicts 25 BSs at 4 Gyr with 75 % single and 25 % in circular binaries. So the cluster environment creates a more realistic configuration spread but also destroys potential BSs via hardening of close binaries. Repeating the N-body model with initial periods drawn from the distribution of Kroupa (1995) – which gives a 50 % reduction in short-period binaries – produces only one BS at 4 Gyr. This also leads to a reduced number of X-ray active BY Draconis binaries (see Hurley *et al.* 2005 for a full explanation and references).

Acknowledgements

JRH would like to thank the IAU for assistance in the form of a travel grant.

References

Aarseth, S. J. 1999, *PASP*, 111, 1333

Fan, X., Burstein, D., Chen, J.-S., *et al.* 1996, *AJ*, 112, 628

Hurley, J. R., Tout, C. A., Aarseth, S. J., & Pols, O. R. 2001, *MNRAS*, 323, 630

Hurley, J. R., Pols, O. R., Aarseth, S. J., & Tout, C. A. 2005, *MNRAS*, 363, 293

Kroupa, P. 1995, *MNRAS*, 277, 1507

Kroupa, P., Tout, C. A., & Gilmore, G. 1993, *MNRAS*, 262, 545

Highlights of Astronomy, Volume 14
IAU XXVI General Assembly, 14-25 August 2006
Karel A. van der Hucht, ed.

doi:10.1017/S1743921307011295

Spectroscopic binaries in M 67†

David W. Latham

Harvard-Smithsonian Center for Astrophysics, 60 Garden Street, Cambridge, MA 02138, USA
email: dlatham@cfa.harvard.edu

Abstract. We summarize the characteristics of 85 spectroscopic orbits derived from more than two decades of radial-velocity monitoring of stars in the old open cluster M 67, with special emphasis on the blue stragglers and other members that do not fall on the evolutionary tracks expected for isolated single stars.

Keywords. stars: binaries: spectroscopic, Galaxy: open clusters and associations: individual (M 67)

In November 1979 one of the CfA Digital Speedometers (Latham1992) was used for the first time to obtain a spectrum of a star in the old open cluster M 67. Not long after that first exposure, Bob Mathieu arrived at CfA and convinced us to undertake a serious radial-velocity survey of the brightest cluster members, with the goals of confirming cluster membership, and identifying spectroscopic binaries and deriving their orbits. The initial project was successful (Mathieu & Latham 1986), and the survey was extended to fainter objects in stages, eventually reaching $V = 15.5$ mag. Altogether 6921 spectra have now been obtained of 411 proper-motion members. For 39 of the targets the mean radial velocities are inconsistent with cluster membership. Among the confirmed members we have identified more than 100 stars with variable velocity and have derived 85 spectroscopic orbits that are of publication quality. Twenty-five of our orbits are double-lined, while eight of our orbits are for members of triple systems.

From the beginning we had a special interest in the 13 classical blue stragglers, and we soon confirmed Armin Deutch's unpublished short-period eccentric orbit for S 1284 = F 190 with $P = 4.18$ d and $e = 0.24$ (Milone & Latham 1992). Four of the hottest and brightest blue stragglers (S 977 = F 81, S 1066 = F 156, S 1280 = F 184, and S 1434 = F 280) rotate too rapidly for the CfA Digital Speedometers to provide reliable velocities. Another two hot blue stragglers (S 968 = F 153 and S 1263 = F185) rotate slowly enough to give good velocities but show no velocity variation. Five of the cooler blue stragglers (S 752 = F 55, S 975 = F 90, S 997 = F 124, S 1195 = F 207, and S 1267 = F 238) show long-period orbits with moderate or low eccentricity. These results were the subject of Ale Milone's PhD thesis (Latham & Milone 1996). Subsequently Sandquist *et al.* (2003) showed that S 1082 = F 131 is a triple composed of an eclipsing binary with $P = 1.06$ d and a third star with $P = 1189$ d and $e = 0.57$. It is tempting to suppose that the system is a bound hierarchical triple, but this has not yet been demonstrated conclusively.

Two short-period binaries (S 1063 and S 1113) lie well below the subgiant branch. Both appear to be cluster members based on their proper motions and radial velocities (Mathieu *et al.* 2003). The evolutionary history of these binaries is not clear and may require dynamical interactions. In addition there are other binaries, such as S 1072, which occupy puzzling positions in the color-magnitude diagram for the cluster and may also require dynamical interactions to understand their formation and evolution.

† Some of the observations reported here were obtained with the Multiple Mirror Telescope, operated jointly by the Smithsonian Institution and the University of Arizona.

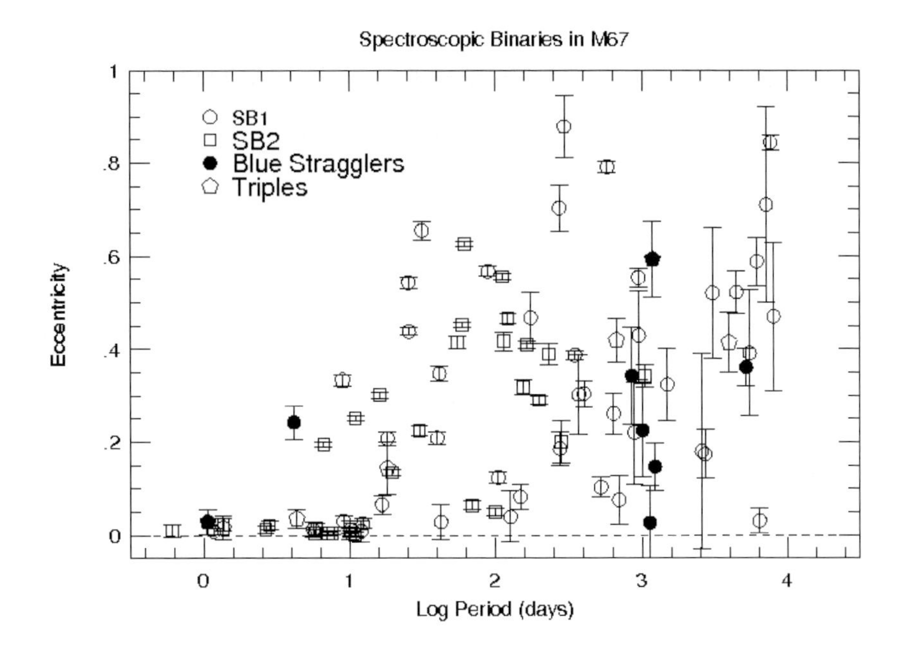

Figure 1. Eccentricity e *versus* period $\log(P)$ for 85 spectroscopic binaries in M 67.

Acknowledgements

Our survey of the spectroscopic binaries in M 67 has been carried out in collaboration with Bob Mathieu and Ale Milone. We thank the many observers who obtained spectra for us with the CfA Digital Speedometers, with special thanks to Bob Davis for operating our data reduction pipeline and maintaining our data archives.

References

Latham, D. W. 1992, in: H. McAlister & W. Hartkopf (eds.), *Complementary Approaches to Binary and Multiple Star Research*, Proc. IAU Coll. No. 135, *ASP-CS*, 32, 110,

Latham, D. W., & Milone, A. A. E. 1996, in: E. F. Milone & J.-C. Mermilliod (eds.), *The Origins, Evolution, and Destinies of Binary Stars in Clusters*, *ASP-CS*, 90, 385

Mathieu, R. D., & Latham, D. W. 1986, *AJ*, 92, 1364

Mathieu, R. D., van den Berg, M., Torres, G., Latham, D. W., Verbunt, F., & Stassun, K. 2003, *AJ*, 125, 246

Milone, A. A. E., & Latham, D. W. 1992, in: Y. Kondo, R. F. Sistero, & R. S. Polidan (eds.), *Evolutionary Processes in Interacting Binary Stars*, Proc. IAU Symp. No. 151 (Dordrecht: Kluwer Academic Publishers), p. 475

Sandquist, E. L., Latham, D. W., Shetrone, M. D., & Milone, A. A. E. 2003, *AJ*, 125, 810

Highlights of Astronomy, Volume 14
IAU XXVI General Assembly, 14-25 August 2006
Karel A. van der Hucht, ed.

doi:10.1017/S1743921307011301

Observations of internal dynamics of globular clusters

Luigi 'Rolly' Bedin

European Southern Observatory, Karl-Schwarzschild-Straße 2, D-85748, Garching, Germany
email: lbedin@eso.org

Abstract. In preparation for the first simulation of a real globular cluster with a few 10^5 particles and several Gyr of evolution, which will be made possible by the advent of the new GRAPE-8, the MODEST community encharged our working group (WG-9) to provide all of the needed observational constrains. The selected clusters for this experiment were NGC 6121 (M 4) and NGC 6397. We present the status of the project.

Keywords. Galaxy: globular clusters: individual (NGC 6121, NGC 6397)

1. Cluster properties

The requested observational inputs are: cluster membership, internal proper motion, internal line of sight velocities, geometrical distance, mass function, mass segregation, anisotropy, binary fraction (and, possibly, binary properties), spatial distribution of stars, white dwarf (WD) counts, tidal tail distribution, and absolute proper motion and radial velocities (in order to characterize the orbit).

In Fig. 1 we show in a schematic way all the observational quantities that needs to be known in order to constrain MODEST simulations.

2. Instruments and techniques

To obtain these observational quantities, we started an extensive observational campaign in order to get high precision photometry and astrometry with both *HST* – from space – and with wide-field-imagers – from ground. In addition to this, for the third component of the motions (along the line of sight) we got time at the ESO VLT-FLAMES multi-fiber spectroscope.

Furthermore, our exquisite photometry from the tip of the red giant branch to the bottom of the WD cooling sequence allows: (*i*) to obtain local present day mass functions (down to the hydrogen burning limit) at different distances from the cluster center, and cleaned from contamination of foreground/background objects; (*ii*) to study the WDs down to the cooling sequence end, (*iii*) estimate the photometric binaries. Comparison of internal proper motions with line of sight velocities for thousands of stars provide geometric determination of the cluster distance with uncertainties of the order of a few percent (see Bedin 2003a for a short description of the method). The internal proper motions and radial velocities provide information on the stellar kinematics inside the cluster.

A project born inside MODEST to determine the spectroscopic binary fraction in M 4 has been recently approved, and observations are carried out in these days at VLT-FLAMES.

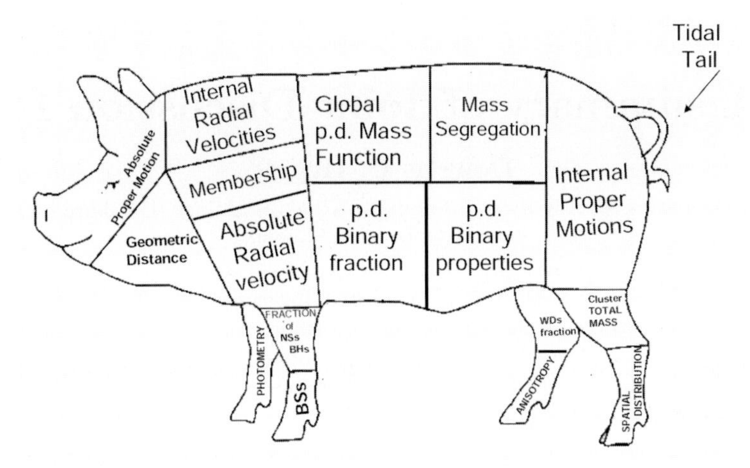

Figure 1. Typical Galactic globular cluster with indication of its main observational quantities and parameters useful to constrain MODEST simulations. (p.d.: present day.)

High-precision astrometry and photometry also allow to disentangle blends from real binaries, for which we can now provide an estimate on their (present day) number, and radial distribution.

In Table 1 we summarized the status of the project.

Table 1. Summary of the status of the project. *Notes*: '-': not yet available; *a*: Bedin *et al.* (2003b); *b*: Milone *et al.* (2006); *c*: Bedin *et al.* (2001); *d*: King *et al.* (1998); *e*: Anderson *et al.* (2006).

Observational quantities	Proposed	Acquired	Reduced	Published
absolute proper motions	M 4/NGC 6397	M 4/NGC 6397	M 4/NGC 6397	M 4[a]/NGC 6397[b]
absolute radial velocity	M 4/NGC 6397	M 4/NGC 6397	M 4/NGC 6397	–/NGC 6397[b]
internal proper motions	M 4/NGC 6397	M 4/NGC 6397	M 4/NGC 6397	–/–
internal radial velocity	M 4/NGC 6397	M 4/NGC 6397	M 4/NGC 6397	–/–
geometrical distance	M 4/NGC 6397	M 4/NGC 6397	M 4/NGC 6397	–/–
orbital parameters	M 4/NGC 6397	M 4/NGC 6397	M 4/NGC 6397	–/NGC 6397[b]
deep *HST* photometry	M 4/NGC 6397	M 4/NGC 6397	M 4/NGC 6397	M 4[c]/NGC 6397[d]
wide field photometry	M 4/NGC 6397	M 4/NGC 6397	M 4/NGC 6397	M 4[e]/NGC 6397[e]
membership	M 4/NGC 6397	M 4/NGC 6397	M 4/NGC 6397	M 4[a,c,e]/NGC 6397[b,d,e]
p.d. mass function	M 4/NGC 6397	M 4/NGC 6397	M 4/NGC 6397	M 4[c]/NGC 6397[d]
p.d. mass segregation	M 4/NGC 6397	M 4/NGC 6397	M 4/NGC 6397	–/–
p.d. binary fraction	M 4/–	M 4/–	–/–	–/–
p.d. binary properties	M 4/–	–/–	–/–	–/–

References

Anderson, J., Bedin, L. R., Piotto, G., Yadav, R. S., & Bellini, A. 2006, *A&A*, 454, 1029

Bedin, L. R., Anderson, J., King, I. R., & Piotto, G. 2001, *ApJ* (Letters), 560, L75

Bedin, L. R., Piotto, G., Anderson, J., & King, I. R. 2003a, in: G. Meylan, S. G. Djorgovski & M. Riello (eds.), *New Horizons in Globular Cluster Astronomy*, *ASP-CS*, 296, 360

Bedin, L. R., Piotto, G., King, I. R., & Anderson, J. 2003b, *AJ*, 126, 247

King, I. R., Anderson, J., Cool, A. M., & Piotto, G. 1998, *ApJ* (Letters), 492, L37

Milone, A., Villanova, S., Bedin, L. R., Piotto, G., Carraro, G., Anderson, J., King, I. R., & Zaggia, S. 2006, *A&A*, 456, 517

Highlights of Astronomy, Volume 14
IAU XXVI General Assembly, 14-25 August 2006
Karel A. van der Hucht, ed.

doi:10.1017/S1743921307011313

A summary of Joint Discussion 14

Douglas C. Heggie

School of Mathematics, University of Edinburgh, King's Buildings, Edinburgh EH9 3JZ, UK
email: d.c.heggie@ed.ac.uk

Abstract. Modelling rich star clusters is at an exciting time, now that detailed star-by-star modelling of all open star clusters has become possible. We should renew attention to the modelling of globular star clusters, to enable us to exploit the flood of excellent observational data now available. At the same time, new ideas in star cluster astrophysics require us to broaden the realism of our modelling in various directions.

Keywords. methods: n-body simulations, binaries: general, stars: evolution, pulsars: general, globular clusters: general, open clusters and associations: general.

N-body modelling has reached a major landmark with the modelling of the open cluster M 67 (Hurley†). It is the hardest of all, because it is so old, and must have started life with far more stars than it has now. I hope that the modelling of open star clusters will now become a major industry, as there is lots to be done, even on M 67 (Latham; Verbunt).

Beyond the realm of the open star clusters, N-body techniques are pushing to larger N through both software and hardware advances (Aarseth; Makino), and their application to the study of tidal tails (Lee *et al.*), mass segregation (Vesperini *et al.*) and rotation (Vesperini & Zepf) is constantly being developed. And yet, *there is no star-by-star N-body model of any globular cluster*. This JD 14, as well as JD 6, has exposed several exciting problems which cry out for such modelling, including the constraints provided by the newly discovered abundance of pulsars in Ter 5 (Ransome), the many open dynamical questions on the binary populations in globular clusters (Richer; Piotto), the dynamical estimation of the white dwarf population in globulars (Fahlman), and so on. The very existence of exquisite observational data on increasing numbers of clusters (Piotto & Bedin; Dieball et al.; Ferraro *et al.*) demands a complementary effort in modelling.

Unfortunately, N-body techniques are still too slow. Furthermore, the pessimistic view is often expressed that there are just too many uncertainties in star-by-star modelling to make the effort worth while, such as in the evolution of collision remnants‡, stellar rotation/magnetism, tidal effects, etc. But we do not need to know everything about the physics of a problem to start modelling it. After all Kepler, who published his theory of the motion of Mars in Prague about 400 years ago, thought that the planets were pushed around by magnetic forces; but he managed to produce an excellent model which has stood the test of time, thanks to the wonderful observations of Tycho. Surely, we can do the same, given our better understanding of stellar dynamics and evolution, thanks to the wonderful and abundant observations of globular clusters.

JD 6 was better in giving appropriate weight to a viable alternative to N-body modelling: the Monte Carlo technique (Fregeau). Even the gas model can be used to understand many processes quantitatively (Boily), and in fact much can be understood by

† Where an author's name appears without a year, the reference is to a speaker or poster or panel contribution at either this Joint Discussion or JD6.

‡ Nevertheless, Glebbeek & Pols reported ongoing progress at this JD 14.

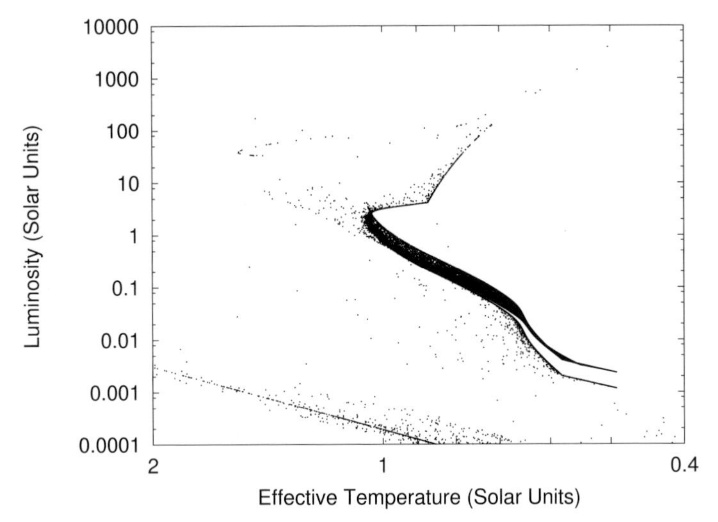

Figure 1. Preliminary Monte Carlo model of the globular cluster M4. The figure is a scatter plot of luminosity and effective temperature. The initial model has initially 75000 single stars and the same number of binaries. The curious bifurcation in the lower main sequence is an artefact of the way in which the initial masses of binary components are chosen, which had a lower limit of 0.1 $M_\odot$. Other choices for initial binary parameters and initial mass function were similar to those of Hurley *et al.*(2005). The initial model is a King model with $W_0 = 7$ and a tidal radius of 60 pc. At 12 Gyr the mass is close to that of M 4 but the radius is too large. The simulation uses a Monte Carlo code (Giersz 2006) combined with stellar evolution packages (Hurley *et al.* 2000, 2002) by means of the McScatter interface (Heggie *et al.* 2006).

modelling the dynamics in even simpler ways (Ivanova; Orlov et al). All that is needed is a concerted effort to bring the realism of some of these techniques up to the level of N-body methods. Indeed the author demonstrated ongoing work in collaboration with M. Giersz to do just this for the Warsaw Monte Carlo code (Fig. 1). In a general way, modelling of globular clusters has been undertaken for decades, but the modelling of *individual* clusters, with all the sophistication required by both the observations and the present state of knowledge of how populations evolve, needs fresh impetus from within the community.

This JD 14 has been very interesting for exposing the kind of issues which are usually neglected in modelling but which are perfectly feasible for inclusion now: multiple generations of stars (D'Antona; Piotto; Sills), spiral arms (Gieles et al), 'live' stellar evolution (Justham et al), planetary systems (Fregeau et al), massive black holes (Gebhardt), primordial triples (Eggleton) and rotation (Spurzem). These are themes which constantly arise in the regular meetings of the MODEST consortium, of which these two Joint Discussions have been one. With this momentum, our subject is in a rapid state of development.

References

Giersz, M. 2006, *MNRAS*, 371, 484

Heggie, D. C., Portegies Zwart, S., & Hurley, J. 2006, *New Astron.*, 12, 20

Hurley, J. R., Pols, O. R., & Tout, C. A. 2000, *MNRAS*, 315, 543

Hurley, J. R., Tout, C. A., & Pols, O. R. 2002, *MNRAS*, 329, 897

Hurley, J. R., Pols, O. R., Aarseth, S. J., & Tout, C. A. 2005, *MNRAS*, 363, 293

Highlights of Astronomy, Volume 14
IAU XXVI General Assembly, 14-25 August 2006
Karel A. van der Hucht, ed.

doi:10.1017/S1743921307011325

Joint Discussion 15 New cosmology results from the *Spitzer Space Telescope*

George Helou and David T. Frayer (eds.)
Spitzer Science Center, California Institute of Technology, Pasadena, CA 91125, USA
email: gxh@ipac.caltech.edu

Abstract. We present and discuss recent new results from the *Spitzer Space Telescope* and their impact on cosmology.

Keywords. cosmology: observations, large-scale structure of universe

1. Introduction

Joint Discussion 15 reviewed results from the *Spitzer Space Telescope* and their impact on cosmology, and sought to stimulate a synthesis between them and corresponding recent results from increasingly powerful X-ray, radio, optical, near-infrared and γ-ray observatories. By bringing together experts from all these fields we hoped to formulate new approaches to the following questions.

- How do the currently known galaxy populations emerge and evolve?
- What is the relationship of the infrared populations at high redshift to the populations in the local universe?
- How does the ultraviolet/optical-based star formation history compare to the history derived from infrared studies?
- How is the rising abundance of heavy chemical elements over time related to the history of galaxy formation and evolution?
- What is the contribution of Active Galactic Nuclei (AGN) to the luminosity of these populations, and how does that contribution evolve?
- What is the role of a starburst phase or a heavy accretion phase in the life cycle of individual galaxies?
- What is the nature of the sources detectable in the X-Rays and infrared, but not the ultraviolet/optical, and what other new populations of objects emerge from the new surveys?

During the introductory session, Tom Soifer, Director of the *Spitzer* Science Center reported on the status of the *Spitzer* Observatory, and gave an overview of the *Spitzer* Science Program related to Cosmology.

2. The modern universe

Seb Oliver presented a multi-wavelength view of the universe at $z<1$, providing the scientific motivation for relatively shallow surveys of large areas on the sky, such as the *Spitzer* Wide-area Infrared Extragalactic Survey (SWIRE, Lonsdale *et al.* 2004). Such surveys result in a wide variety of objects, challenging the population models to account for all of them (Lagache *et al.* 2004; Rowan-Robinson *et al.* 2005). Davoodi *et al.* (2006) compared SWIRE detections to Sloan Digital Sky Survey (SDSS) detections, looking in particular for the infrared properties of galaxies in the context of their blue sequence or red sequence classification. They found that 18 % of their objects (195 of 1114) have red optical colors and 24 μm excess, reflecting either AGN or enhanced star formation activity, suggesting this activity is superposed on a bulge-dominated system.

Rob Kennicutt reported on progress in the study of galaxies in the local universe, highlighting a better characterization of the integrated spectral energy distributions from the ultraviolet to the infrared (Dale *et al.* 2007), combining data from *GALEX* (*Galaxy Evolution Explorer*) and *Spitzer*; new robust star formation indicators combining Hα and 24 μm emission from galaxies (Calzetti *et al.* 2007; Kennicutt *et al.* 2007); and a careful analysis of variability in the Aromatic Feature emission in the range 5 μm $< \lambda <$ 20 μm (Smith *et al.* 2007).

Vassilis Charmandaris and *Jim Houck* summarized the state of knowledge on dust properties in low metallicity environments, best advanced by the study of nearby dwarf galaxies. The existence of a threshold in metallicity below which Aromatic Feature emission is clearly suppressed is now well established (Engelbracht *et al.* 2005; Smith *et al.* 2007). However, the underlying causes for this deficiency and other contributing factors are still open to discussion (Rosenberg *et al.* 2006; Draine *et al.* 2007).

Eckhard Sturm reviewed new results on Ultra-Luminous Infrared Galaxies (ULIRG) and AGN in the local universe, stressing the diversity in their spectra and showing that these classes display evidence for emission associated with both star formation and black hole heating, often modified by strong extinction in dense interstellar medium surrounding the power source (Armus *et al.* 2007; Weedman *et al.* 2006); this high extinction often reveals crystals and ices on very cold grains (Spoon *et al.* 2006). Silicate emission has now been clearly detected for the first time in AGN. It is also becoming clear from the study of galaxies with both AGN and star formation heating diagnostics like ratios of Aromatic Feature emission or fine-structure line emission can be modified by factors other than the presence of AGN, and should therefore be interpreted with caution (Sturm *et al.* 2006; Dale *et al.* 2006).

3. The evolving universe

Emanuele Daddi considered what we are learning from *Spitzer* about stars and star formation in the universe at a redshift $z \simeq 2$. Substantial insight into the properties of stellar populations of galaxies at these distances are yielded by studies in the 3 to 8 μm range using the IRAC instrument on *Spitzer*. Maraston *et al.* (2006) demonstrate the importance of including all phases of stellar evolution in the population models, in particular the thermally pulsing asymptotic giant branch, in order to reproduce the *Spitzer*-IRAC data. One result of this stellar population investigation is that most massive galaxies ($\sim 10^{11}$ M$_\odot$) observed at $z \simeq 2$ are also bright in the infrared and qualify as Ultra-Luminous Infrared Galaxies, whereas massive but relatively inactive galaxies are rare (Daddi *et al.* 2005).

Ranga-Ram Chary discussed dusty galaxies as seen by *Spitzer* in the deepest 24 μm surveys, echoing the result that red, massive galaxies at redshifts $1 < z < 3.5$ have a specific star formation rate an order of magnitude greater than their analogs at redshifts $0.3 < z < 0.75$, suggesting that the bulk of star formation in massive galaxies occurs at early cosmic epochs and is largely complete by $z \simeq 1.5$ (Papovich *et al.* 2006). The spectral energy distributions of galaxies at high redshifts appear consistent with those of galaxies in the local universe at similar luminosity given available data; with that assumption, the cosmic infrared background can be reproduced if one includes all sources down to $f_\nu(24\mu\text{m}) = 24\,\mu$Jy, with AGN contributing a small fraction of the total background.

David Frayer reviewed results from deep and ultra-deep far-infrared surveys with *Spitzer* at 70 and 160 μm. The infrared-radio correlation seems to hold for galaxies out to redshifts $z \simeq 1$ (e.g., Appleton *et al.* 2004). Source counts have been measured down to $f_\nu(70\mu\text{m}) = 1.2$ mJy, accounting for about 60% of the cosmic infrared background (Frayer *et al.* 2006). The 70 - 24 μm colors suggest that galaxies are on average cooler at $z \simeq 1$ than expected from the luminosity distribution at that epoch.

Guilaine Lagache spoke to results obtained from spectroscopic studies of high-redshift infrared-luminous galaxies using *Spitzer*. While the interpretation of these spectra is still evolving rapidly, it is pretty clear that trends identified in the local universe continue to apply at $z \simeq 1$ - 2, such as the greater contribution of AGN to powering sources as the source luminosity rises (Yan *et al.* 2007). However, the potential existence of galaxies at $L_{\rm IR} \simeq 10^{13} L_\odot$ which are energetically dominated by star formation is unique to high redshifts.

Pieter van Dokkum addressed the nature of massive galaxies at redshifts 2 to 3, with an emphasis on the multiplicity of techniques to identifying such objects, and approaches to constructing samples that avoid the biases inherent to each technique. With the near-infrared bump shifting to longer wavelengths, *Spitzer* is clearly at an advantage in identifying such objects, though inconsistencies remain among various authors regarding the fraction of massive galaxies at $z \simeq 2$ - 3 which appear completely quiescent (e.g., Kriek *et al.* 2006).

4. The distant universe

Tommy Wiklind examined how *Spitzer* has helped identify and study objects whose optical emission is redshifted into the mid-infrared, and in particular galaxies at $z > 5$. Building on the early detection of a massive post-starburst galaxy at $z \simeq 6.5$ (Mobasher *et al.* 2005), the GOODS (Great Observatories Origins Deep Survey) team has identified 18 candidates for similar objects at $z > 5$, using *Spitzer* data as the key. The objects average $\sim 2 \times 10^{11} M_\odot$ and run between 0.2 and 1 Gyr in age, with little or no on-going star formation, and modest or low extinction. They must have formed at $z > 6$ and possibly as high as $z \simeq 25$, and are observed with a density 4 - 5×10^{-5} Mpc^{-3} that starts to challenge the predictions of ΛCDM models of galaxy formation (Somerville 2005).

Xiaohui Fan reviewed studies of quasars at high redshifts, where *Spitzer* can detect and study accretion disk and dust emission, helping to find $z > 5$ quasars with a frequency of 1 per 1.5 sq.deg, greater than optical ground-based surveys (Cool *et al.* 2006). These quasars are also among the least luminous known at these redshifts, and appear similar in their spectral energy distribution to lower redshift objects, adding to the evidence for a lack of strong evolution in the properties of quasars. While $z \simeq 6$ quasars tend to show prominent hot dust emission, the origin of that dust remains uncertain, especially that some objects probably have unusual dust properties (Jiang *et al.* 2006).

Richard Ellis summarized searches for the sources responsible for cosmic re-ionization, where *Spitzer* plays a key role because of its ability to fill in the spectrum beyond the longest wavelength available from the ground. The follow-up of i'-band drop-out sources for instance reveals that $z \simeq 6$ sources have dominant stellar populations with ages > 100 Myr, and a history of vigorous star formation at $z > 7$. The progenitors of such systems must have allowed star formation to play a prominent role in reionizing the universe (Eyles *et al.* 2005).

Anton Koekemoer presented joint *Spitzer*/*HST*/*Chandra* results on Extreme X-ray / Optical sources (EXOs), defined by their large ratio of X-ray to visible flux. EXOs are readily detected in the near-infrared and by *Spitzer* in the mid-infrared. While many EXOs reside at $2 < z < 5$ and exhibit high optical depth or low optical depths and active star formation, at least one candidate remains best explained as a $z > 7$ AGN with a power law spectrum and excess emission beyond 10μm (Koekemoer *et al.* 2005). There are no obvious local universe counterparts to EXOs.

5. Integrated perspectives

Marijn Franx offered an overview of the evolution of galaxies measured by *Spitzer* and multi-wavelength surveys, emphasizing the two powerful tools offered by *Spitzer*: imaging and photometry at 3 - 8 μm to trace the evolution of stellar mass over cosmic time scales, and at 24 μm to trace the star formation history. While clear patterns of decline in the cosmic star formation rate are detected (e.g., Bell *et al.* 2005), a remaining limitation is the translation of 24 μm measurements into bolometric luminosities, critical for estimating the star formation rate. Studies combining *Spitzer* data with submillimeter 850 μm and radio 20 cm data suggest a downward revision of the amount of star formation at higher redshifts, and half the total star formation occurring since $z \simeq 1.4$ (Wang *et al.* 2006).

Mark Dickinson discussed the state of our understanding of the global history of star formation in the *Spitzer* era, again underscoring the essential need for multi-wavelength studies, and the importance not only of estimating properly bolometric luminosities for individual objects, but also of integrating over populations when only the most luminous objects are detected. Multi-wavelength studies are unveiling new populations, such as AGN undetected at X-rays but evident in radio and infrared data (e.g., Donley *et al.* 2005), or Ultra-Luminous Infrared Galaxies with cooler far-infrared colors than is common in the local universe (e.g., Pope *et al.* 2006).

Carlos Frenk described the state of the art in interpreting *Spitzer* data in the context of Cold Dark Matter numerical simulations of cosmic history. While the models still manage to explain *Spitzer* data, they are strained in their ability to represent Aromatic Feature emission properties, and they require modifications to the Initial Mass Function of stars, direct evidence for which is sparse (e.g., Granato *et al.* 2004).

Jean-Loup Puget reviewed the relation between the extragalactic background light and the star formation history, pointing out the critical role *Spitzer* plays in elucidating the rise of dust, which converts at least half the visible photons into infrared photons. Of particular interest is the rise of Aromatic compounds, which require cycling of material from stellar outflows through dense molecular clouds then out into photo-dissociation regions, going from simple carbon compounds to aliphatic then aromatic compounds. The infrared background is clearly consistent with source counts in the mid-infrared including 24 μm, but only indirectly related to source counts at far-infrared to submillimeter bands (e.g., Dole *et al.* 2006).

Sylvain Veilleux discussed Ultraluminous Infrared Galaxies in the context of cosmic evolution, stressing in particular their likely role as progenitors of spheroids, and their likely central role in black hole growth. In addition, these objects probably dominate the enrichment of the intergalactic medium, because of their over-abundance at epochs $z > 2$

and their tendency to generate super-winds that disperse supernova ejecta well beyond their halos.

Thorsten Naab reported on recent progress in understanding galaxy mergers, a key process in the evolution of the universe. Current models reproduce quite well the mass interactions, but more work is needed to represent more accurately the star formation activity triggered in close encounters and mergers.

6. Conclusions

In his concluding remarks, *Simon Lilly* described the impact of *Spitzer* on our understanding of the high redshift universe as enormous, fulfilling the most optimistic expectations. He highlighted the value of Legacy Science projects, pioneered by *Spitzer*, and their key contribution to a timely diffusion and assimilation of the scientific returns from the observatory. Looking ahead, he advocated the need for 'a billion galaxy survey', achievable by going to $I_{\rm AB} \simeq 25$ mag over a large area of the sky, perhaps 10^4 square degrees. A critical component of such an undertaking would be a survey at 3.6 and 4.5μm during the *Spitzer* warm mission reaching a commensurate depth, and requiring by itself several years of the warm mission.

Scientific Organizing Committee

Jacqueline A. Bergeron (France), Catherine J. Cesarsky (Germany), Vassilis Charmandaris (Greece), Thierry J-L. Courvoisier (Switzerland), George Helou (USA, chair), Myungshin Im (South Korea), Leopoldo Infante (Chile), Robert J. Ivison (UK), Haruyuki Okuda (Japan), Jan Palous (Czech Republic), Charles C. Steidel (USA), and Rasjid A. Sunyaev (Russian Federation).

Acknowledgements

We would like to acknowledge the members of the Science Organizing Committee. NASA, through the *Spitzer* Science Center, provided travel support for several of the speakers. The *Spitzer Space Telescope* is operated by the Jet Propulsion Laboratory, California Institute of Technology under contract with NASA.

References

Appleton, P. N., Fadda, D. T., Marleau, F. R., *et al.* 2004, *ApJS*, 154, 147
Armus, L., Charmandaris, V., Bernard-Salas, J., *et al.* 2007, *ApJ*, 656, 148
Bell, E. F., Papovich, C., Wolf, C., *et al.* 2005, *ApJ*, 625, 23
Calzetti, D., Kennicutt, R. C., Engelbracht, C. W., *et al.* 2007, *ApJ*, 666, 870
Cool, R. J., Kochanek, C. S.; Eisenstein, D., *et al.* 2006, *AJ*, 132, 823
Daddi, E., Dickinson, M., Chary, R., *et al.* 2005, *ApJ* (Letters), 631, L13
Dale, D. A., Smith, J. D. T., Armus, L., *et al.* 2006, *ApJ*, 646, 161
Dale, D. A., Gil de Paz, A., Gordon, K. D., *et al.* 2007, *ApJ*, 655, 863
Davoodi, P., Pozzi, F., Oliver, S., *et al.* 2006, *MNRAS*, 371, 1113
Dole, H., Lagache, G., Puget, J.-L., *et al.* 2006, *A&A* 451, 417
Donley, J. L., Rieke, G. H., Rigby, J. R., & Pèrez-González, P. G. 2005, *ApJ*, 634, 169
Draine, B. T., Dale, D. A., Bendo, G., *et al.* 2007, *ApJ*, 663, 866
Engelbracht, C. W., Gordon, K. D., Rieke, G. H., *et al.* 2005, *ApJ*, 628, 29
Eyles, L. P., Bunker, A. J., Stanway, E. R., *et al.* 2005, *MNRAS*, 364, 443
Frayer, D. T., Huynh, M. T., Chary, R., *et al.* 2006, *ApJ* (Letters), 647, L9
Granato, G. L., De Zotti, G., Silva, L., *et al.* 2004, *ApJ*, 600, 580

Jiang, L., Fan, X., Hines, D. C., *et al.* 2006, *AJ*, 132, 2127
Kennicutt, R. C., Calzetti, D., Walter, F., *et al.* 2007, *ApJ*, in press [2007arXiv0708.0922K]
Koekemoer, A. M., Alexander, D. M., Bauer, F. E., *et al.* 2005, in: A. Wilson (ed), *The Dusty and Molecular Universe: A Prelude to Herschel and ALMA*, ESA SP-577, p. 111
Kriek, M., van Dokkum, P. G., Franx, M., *et al.* 2006, *ApJ* (Letters), 649, L71
Lagache, G., Dole, H., Puget, J.-L., *et al.* 2004, *ApJS*, 154, 112
Lonsdale, C., Polletta, M.d.C., Surace, J., *et al.* 2004, *ApJS*, 154, 54
Maraston, C., Daddi, E., Renzini, A., *et al.* 2006, *ApJ*, 652, 85
Mobasher, B., Dickinson, M., Ferguson, H. C., *et al.* 2005, *ApJ*, 635, 832
Papovich, C., Moustakas, L. A., Dickinson, M., *et al.* 2006, *ApJ*, 640, 92
Pope, A., Scott, D., Dickinson, M., *et al.* 2006, *MNRAS* 370, 1185
Rowan-Robinson, M., Babbedge, T., Surace, J., *et al.* 2005, *AJ*, 129, 1183
Rosenberg, J. L., Ashby, M. L. N., Salzer, J. J., Huang, J.-S. 2006, *ApJ* 636, 742
Smith, J. D. T., Draine, B. T., Dale, D. A., *et al.* 2007, *ApJ*, 656, 770
Somerville, R. S. 2005, in: A. Renzini, & R. Bender (eds.), *Multiwavelength Mapping of Galaxy Formation and Evolution*, Proc. ESO Workshop, Venice, Italy, 13-16 October 2003 (Berlin: Springer), p. 131
Spoon, H. W. W., Tielens, A. G. G. M., Armus, L., *et al.* 2006, *ApJ*, 638, 759
Sturm, E., Hasinger, G., Lehmann, I., *et al.* 2006, *ApJ*, 642, 81
Wang, W.-H., Cowie, L. L., & Barger, A. J. 2006, *ApJ*, 647, 74
Weedman, D. W., Soifer, B. T., Hao, L., *et al.* 2006, *ApJ*, 651, 101
Yan, L., Sajina, A., Fadda, D., Choi, P., Armus, L., Helou, G., *et al.* 2007, *ApJ*, 658, 778

Highlights of Astronomy, Volume 14
IAU XXVI General Assembly, 14-25 August 2006
Karel A. van der Hucht, ed.

doi:10.1017/S1743921307011337

Joint Discussion 16
Nomenclature, precession and new models in fundamental astronomy
Applications and scientific contribution to astronomy

Nicole Capitaine[1], Jan Vondrák [2] and James L. Hilton[3] (eds.)

[1]SYRTE, Observatoire de Paris, 61 Avenue de l'Observatoire, F-75014 Paris, France
email: nicole.capitaine@obspm.fr

[2]Astronomical Institute, Czech Academy of Sciences,
Boční II, 1401, CZ-141 31, Praha 4, Czech Republic
email: vondrak@ig.cas.cz

[3]Astronomical Applications Department, US Naval Observatory (USNO),
1274 Hwy 238, Jacksonville, OR 97530, USA
email: jhilton@usno.navy.mil

Preface

The IAU Joint Discussion 16 was held at the IAU XXVI General Assembly in Prague, in August 2006. The title of the meeting was *Nomenclature, Precession and new models in Fundamental Astronomy. Applications and scientific contribution to astronomy.* It was organized by IAU Division I (*Fundamental Astronomy*) and Commission 19 (*Earth Rotation*), with the participation of IAU Division X (*Radio Astronomy*) and all the Division I Commissions, as well as with the support of the International Association of Geodesy (IAG). The Scientific Organizing Committee was made up of the three organizers and the representatives of these scientific bodies.

This JD 16 was of 1.5-day duration and was composed of the six following sessions:
1. State of the art of the implementation of the *IAU Resolutions* and the ICRS
2. Precession and the ecliptic
3. High accuracy models for reducing astronomical observations
4. New terminology in fundamental astronomy, time and relativity
5. Scientific applications of high-accuracy astronomy
6. General discussion and educational efforts

Each session consisted of two invited talks, one to four oral presentations and a number of posters.

Presentations covered both completed works and prospects for the future. The meeting was also an opportunity to present to a large audience the proposals for new IAU *resolutions* related to the topic of the meeting, which were submitted to be voted on during the second session of the 26th IAU GA.

The main purpose of JD 16 was to discuss recent and future IAU *resolutions* on reference systems. The International Celestial Reference System (ICRS) and its realization the International Celestial Reference Frame (ICRF) were adopted by the IAU XXIII General Assembly in 1997. At the IAU XXIV General Assembly in 2000, a number of additional *resolutions* were passed concerning the definition of the celestial and terrestrial reference systems and transformations between them. These *resolutions* contain several new concepts. Implementation of these resolutions requires a consistent and well defined terminology that is recognized and adopted by the astronomical community. The

Working Group on *Nomenclature for Fundamental Astronomy* was to make related educational efforts for addressing the issue to the larger community of scientists. Two *resolutions* on new terminology and an improved definition of Barycentric Dynamical Time (*Resolutions 2 and 3*) submitted to the IAU XXVI General Assembly in 2006 were discussed.

Discussion of the IAU 2000A precession-nutation at the IAU XXV GA in 2003, revealed a requirement for a new precession model that was consistent with both dynamical theories and the IAU 2000A nutation model; it also revealed the need for an improved definition for the ecliptic. The Division I Working Group on *Precession and the Ecliptic* was created to address these requirements. This WG has selected a new, high-accuracy precession model to replace the IAU 2000 precession. A proposal to adopt this precession model has been submitted to the IAU 2006 GA (*Resolution 1*). This *resolution* has been presented and discussed along with proposals for next generation models.

The adopted reference systems and the high-accuracy models which have been recently developed make possible various scientific applications in a number of fields of astronomy. Another objective of JD 16 was the presentation of scientific applications of high accuracy astrometric observations, models and accurate realizations of reference systems for ephemerides, celestial mechanics, astrometry, Earth rotation, time and radio-astronomy.

Improvements in astrometric models and catalogues were discussed. Effects such as Earth rotation, nutation, light deflection, and relativistic transformations, with potential for various scientific applications were presented, emphazing the recent progress in models and observations (Earth dynamics, spacecraft observations and planetary ephemerides, time synchronization and navigation in deep space). Presentations about future space astrometric missions, like *Gaia* and *SIM*, were also discussed. The final discussion emphasized the necessary educational effort to be done in order that a sufficient number of students all over the world be educated with the knowledge and skills in astrometry in order to support existing and future projects.

The scientific programme of the meeting included 12 invited papers, 20 oral and 45 poster presentations, 90 % of which have been provided for the proceedings. The proceedings are divided into six parts corresponding to the sessions of the meeting, each one including one-page, one-half page and one-fourth page abstracts corresponding to the invited papers, the oral papers and the posters, respectively.

We express our thanks to the Scientific Organizing Committee for its valuable help in preparing the scientific programme and chairing the sessions, and to all the authors of presentations for their very interesting contributions. We thank the Local Organizing Committee for the very efficient help during the meeting, especially the Chair of the SOC, Cyril Ron.

Scientific Organizing Committee

Aleksander Brzezinski (Polska), Mark R. Calabretta (Australia), Nicole Capitaine (France, co-chair), Veronique Dehant (Belgique), Toshio Fukushima (Japan), James L. Hilton (USA, co-chair), Kenneth J. Johnston (USA), Irina I. Kumkova (Russia), Andrea Milani (Italia), Robert A. Nelson (USA), Kenneth P. Seidelmann (USA), Michael Soffel (Germany), and Jan Vondrák (Czech Republic, co-chair).

Nicole Capitaine, James Hilton, Jan Vondrák, co-chairs of the JD 16 SOC,
October 31, 2006

Highlights of Astronomy, Volume 14
IAU XXVI General Assembly, 14-25 August 2006
Karel A. van der Hucht, ed.

doi:10.1017/S1743921307011349

Present status of the celestial reference system and frame

Chopo Ma
NASA Goddard Space Flight Center, Greenbelt, MD 20771, USA
email: cma@gemini.gsfc.nasa.gov

Abstract. We discuss the present status of the celestial reference system and frame.

Keywords. astrometry, catalogs, reference systems.

From 1 January 1998 the IAU adopted the International Celestial Reference System (ICRS) oriented by distant extragalactic objects in the framework of general relativity with the underlying assumption that the ensemble of such objects has no global rotation. The initial fundamental realization of the ICRS is the ICRF (International Celestial Reference Frame) based on positions of 212 defining radio sources measured with VLBI along with 396 other sources with consistent positions (Ma *et al.* 1998). The position uncertainty floor is $\sim$0.25 mas with the frame axes determined at the $\sim$0.03 mas level. VLBI data acquired from 1995 to 2002 have been used to determine positions of 109 additional sources in ICRF-Ext.2 (Fey *et al.* 2004) as well as to refine the positions of the non-defining sources. Considering also the 23 sessions of the VLBA Calibrator Survey (VCS), in which new sources were observed in only one session (Petrov *et al.* 2006), there are $\sim$3500 sources with astrometric positions better than a few mas.

There have been several important developments since 1995 when the initial ICRF analysis was completed. The number of 24-hr sessions has increased by $\sim$65 % and the number of observations by $\sim$200 % with more robust observing networks, but data acquisition is still primarily for geodetic monitoring. More systematic methods for identifying astrometrically stable sources have been developed. A permanent IVS (International VLBI Service for Geodesy and Astrometry) program for monitoring at least once every six months $\sim$300 stable, potentially stable and defining sources not used in geodetic observing was started in 2004, and astrometric observing in the southern hemisphere has been enhanced. Observations have begun at K- and Q-bands, where source structure is generally more compact. Better geophysical modeling should permit simultaneous adjustment of the VLBI celestial and terrestrial reference frames without distortion.

With the accumulated improvements in data, modeling and analysis a new ICRF is desirable. The challenges are selecting revised defining sources, handling source position variation in the analysis, and preparing for the optical frame of *Gaia*. Because the quasars that are expected to be most accurately measured by *Gaia* are much brighter optically than most radio sources currently used in astrometric VLBI, the identification and observation of the best radio-optical tie objects must be pursued. In any case, even though the precision of the optical frame is anticipated to be an order of magnitude better than the uncertainty of current VLBI frame, the VLBI frame will continue to be essential for measuring variations in Earth orientation parameters.

References

Fey, A. L., Ma, C., Arias, E. F., *et al.* 2004, *AJ*, 127, 3587
Ma, C., Arias, E. F., Eubanks, T. M., *et al.* 1998, *AJ*, 116, 516
Petrov, L., Kovalev, Y. Y., Fomalont, E. B., & Gordon, D. 2006, *AJ*, 131, 1872

Highlights of Astronomy, Volume 14
IAU XXVI General Assembly, 14-25 August 2006
Karel A. van der Hucht, ed.

doi:10.1017/S1743921307011350

Tools for implementing the recent IAU resolutions

George H. Kaplan and John A. Bangert

Astronomical Applications Department, U.S. Naval Observatory (USNO),
3450 Massachusetts Avenue NW, Washington, DC 20392-5420, USA
email: gkaplan@usno.navy.mil

Abstract. We discuss tools for implementing the recent IAU resolutions, notably the USNO Circular 179 and the NOVAS software package.

Keywords. reference systems, astrometry, ephemerides, time, standards

---

The resolutions on positional astronomy adopted at the 1997 and 2000 IAU General Assemblies are far-reaching in scope, affecting both the details of various computations and the basic concepts upon which they are built. For many scientists and engineers, applying these recommendations to practical problems is thus doubly challenging. Because the U.S. Naval Observatory (USNO) serves a broad base of users, we have provided two different tools to aid in implementing the resolutions, both of which are intended for the person who is knowledgeable but not necessarily expert in positional astronomy. These tools complement the new material that has been added to *The Astronomical Almanac*, which is prepared jointly by Her Majesty's Nautical Almanac Office (HMNAO) in England and the USNO Astronomical Applications Department (AA). See the paper by Hohenkerk in this Joint Discussion.

USNO Circular 179 is a 118-page book that introduces the recent IAU resolutions to non-specialists. It includes extensive narratives describing the background and basic concepts as well as compilations of the equations necessary to apply the recommendations. The resolutions have been logically grouped into six main chapters: Relativity, Time Scales, The Fundamental Celestial Reference System, Ephemerides of the Major Solar System Bodies, Precession and Nutation, and Modeling the Earth's Rotation. The Circular is available as a hard-cover book or as a PDF file that can be downloaded from the USNO/AA web site `<http://aa.usno.navy.mil/ publications/docs/Circular_179.html>`. It is also available from `<arXiv.org>` as [astro-ph/0602086].

NOVAS (Naval Observatory Vector Astrometry Subroutines) is a source-code library available in both Fortran and C. It is a long established package with a wide user base that has recently been extensively revised (in version 3.0) to implement the recent IAU resolutions. However, use of NOVAS does not require detailed knowledge of the resolutions, since commonly requested data – for example, topocentric positions of stars or planets – are provided in a single call to one of the high-level subroutines or functions in the package. There are also low-level routines for individual calculations such as precession, nutation, aberration, sidereal time, etc. NOVAS can be downloaded from the USNO/AA web site `<http://aa.usno.navy.mil/software/novas/>`.

The two tools are linked, since Circular 179 provides documentation for many of the algorithms used in NOVAS; conversely, NOVAS provides software that implements the formulas given in Circular 179. Both Circular 179 and NOVAS version 3.0 encompass the recommendations of the 2003 - 2006 IAU working groups on precession and nomenclature that were adopted as Resolutions 1 - 3 of the IAU XXVI IAU General Assembly, 2006.

Highlights of Astronomy, Volume 14
IAU XXVI General Assembly, 14-25 August 2006
Karel A. van der Hucht, ed.

doi:10.1017/S1743921307011362

Selecting highly-compact radio sources for the definition of the celestial reference frame

Patrick Charlot[1], Alan L. Fey[2], Roopesh Ojha[2], David A. Boboltz[2], J. I. B. Camargo[3] and A. Collioud[1]

[1]Observatoire de Bordeaux (OASU) - CNRS/UMR 5804, BP 89, F-33270 Floriac, France
email: charlot@obs.u-bordeaux1.fr

[2]Earth Orientation Department, U.S. Naval Observatory (USNO),
3450 Massachusetts Avenue NW, Washington, DC 20392-5420, USA
email: afey,rojha,dboboltz@usno.navy.mil

[3]Observatório do Valongo, UFRJ, Lad. Pedro Antonio 43, Rio de Janeiro, RJ 20080-090, Brasil

Abstract. We discuss the issue of selecting highly-compact radio sources for the definition of the celestial reference frame.

Keywords. astrometry, reference systems, radio continuum: general, techniques: high angular resolution, techniques: interferometric, galaxies: active

---

The intrinsic radio structure of the extragalactic sources is one of the limiting errors in the definition of the International Celestial Reference Frame (ICRF). Based on multi-epoch VLBI images obtained with the Very Long Baseline Array and other VLBI telescopes around the world between 1994 and 2005, we evaluate this effect for 560 ICRF sources (about 80 % of the current frame) and calculate a so-called 'structure index' to define the astrometric suitability of the sources.

The structure index ranges from 1 for the most compact sources to 4 for the most extended sources. The number of epochs for which the structure index is available for a given source varies from 1 for the least-observed sources to 20 for the intensively-observed sources. From this calculation, we identify a subset of 221 ICRF sources which have very good or good astrometric suitability (i.e., a structure index of either 1 or 2) at any of the available epochs.

We argue that these compact sources are potential candidates for defining the celestial frame with the highest accuracy when a future realization of the ICRF is made.

Highlights of Astronomy, Volume 14
IAU XXVI General Assembly, 14-25 August 2006
Karel A. van der Hucht, ed.

doi:10.1017/S1743921307011374

Véron & Véron-based optical extragalactic reference frame – progress report

Alexander H. Andrei[1,2], D. N. da Silva Neto[2], Marcelo Assafin[2], Norbert Zacharias[3], Roberto Vieira Martins[1], J. I. B. de Camargo[2], Jean Souchay[4] and Anna P. B. de Melo[1]

[1]Observatório Nacional, MCT,
R. Gal. Jose Cristino 77, São Cristóvão, Rio de Janeiro, RJ 20921-400, Brasil
email: oat1@on.br

[2]Observatório do Valongo, UFRJ,
Lad. Pedro Antonio 43, Rio de Janeiro, RJ 20080-090, Brasil
email: massaf@ov.ufrj.br

[3]U.S. Naval Observatory (USNO),
3450 Massachusetts Avenue NW, Washington, DC 20392-5420, USA
email: nz@usno.navy.mil

[4]SYRTE, Observatoire de Paris, 61 Avenue de l'Observatoire, F-75014, Paris, France
email: jean.souchay@obspm.fr

Abstract. We present a Véron & Véron based optical extragalactic reference frame progress report.

Keywords. astrometry, surveys, catalogs, methods: data analysis, quasars: general

---

Taking as input the quasars from the V&V list, in this project the aim is to build a dense, optical extagalactic reference frame on the ICRS, directly aligned by the ICRF, and bridging the magnitude gap to the *Hipparcos* frame.

Using the UCAC2 as reference star catalog this enables an accurate astrometry independent of the USNO B1.0 catalog. The preliminary B1.0 positions, for the V&V sources, are locally corrected using UCAC2 stars. Here, the northernmost portion is also corrected, by using preliminary extracts of the UCAC2. With this project the sample of quasars with precise radio position was expanded to 4,400 objects (30% increase). Preliminary reductions using harmonic functions relate the VLBI obtained positions to the Hipparcos frame (through the UCAC2).

The final positions are precise to better than 100 mas, while the resulting frame adheres to the ICRF at the formal level of 3 mas.

Highlights of Astronomy, Volume 14
IAU XXVI General Assembly, 14-25 August 2006
Karel A. van der Hucht, ed.

doi:10.1017/S1743921307011386

Problems of the reference radio source selection

Oleg A. Titov

Geoscience Australia, GPO Box 378, Canberra, ACT 2601, Australia
email: oleg.titov@ga.gov.au

Abstract. We discuss problems of the reference radio source selection.

Keywords. data analysis, quasars: general, reference systems

The International Celestial Reference Frame (ICRF) is realised by high precision coordinates of the extragalactic radio sources observed by VLBI. Only radio sources with stable positions should be used in the ICRF catalogue to maintain long-term stability of the reference system. However, some radio sources previously treated as astrometrically stable, show significant variations in position on different time scales. In this paper the positional instabilities of selected radio sources are discussed. It is shown that apparent proper motions for frequently observed objects can reach 1 mas/yr over several years. Therefore, more careful procedure of the reference radio source selection should be applied.

Several global solutions based on different sets of reference radio sources have also been obtained using OCCAM software by the least squares collocation method. It was shown that the existing lists of reference radio sources produce some controversial results. These solution statistics and resulted astrometric position catalogues were investigated to draw conclusion about more effective scheme for the reference radio source selection.

Highlights of Astronomy, Volume 14
IAU XXVI General Assembly, 14-25 August 2006
Karel A. van der Hucht, ed.

doi:10.1017/S1743921307011398

Review of the Division I Working Group on Precession and the Ecliptic

James L. Hilton

Astronomical Applications Department, U.S. Naval Observatory,
3450 Massachusetts Ave. NW, Washington, DC 20392, USA
email: jhilton@aa.usno.navy.mil

Abstract. We present a review of the Division I Working Group on *Precession and the Ecliptic.*

Keywords. astrometry, reference systems

During the 2003-2006 triennium the members of the IAU Division I Working Group on *Precession and the Ecliptic* consisted of J.L. Hilton (Chair)(U.S. Naval Observatory), N. Capitaine (Systèmes de Référence Temps-Espace), J. Chapront (Systèmes de Référence Temps-Espace), J.M. Ferrandiz (U. de Alicante), A. Fienga (Institut de Mécanique Céleste), T. Fukushima (National Astronomical Observatory Japan, Tokyo, Japan), J. Getino (U. de Valladolid), P. Mathews (U. of Madras), J.-L. Simon (Institut de Mécanique Céleste), M. Soffel(T. U. Dresden), J. Vondrak, (Czech Acad. Sci.), P. Wallace (Her Majesty's Nautical Almanac Office), and J. Williams (Jet Propulsion Laboratory). The report of their work is published in Hilton *et al.* (2006).

This work also culminated in *Resolution 1* presented to the IAU XXVI General Assembly. The recommendations of *Resolution 1* are:

1. that the terms *lunisolar precession* and *planetary precession* be replaced by *precession of the equator* and *precession of the ecliptic*, respectively,
2. that, beginning on 1 January 2009, the precession component of the IAU 2000A precession-nutation model be replaced by the P03 precession theory, of Capitaine *et al.* (2003) for the precession of the equator (Eqs. 37) and the precession of the ecliptic (Eqs. 38); the same paper provides the polynomial developments for the P03 primary angles and a number of derived quantities for use in both the equinox based and *Celestial Intermediate Origin* based paradigms,
3. that the choice of precession parameters be left to the user, and
4. that the ecliptic pole should be explicitly defined by the mean heliocentric orbital angular momentum vector of the Earth-Moon barycenter in the Barycentric Celestial Reference Frame (BCRS), and this definition should be explicitly stated to avoid confusion with other, older definitions.

Acknowledgements

I would like to gratefully acknowledge the hard work of the members of the IAU Division I Working Group on precession and the ecliptic. Without their hard work this paper would not have been possible.

References

Capitaine, N., Wallace, P. T., & Chapront, J. 2003, *A&A*, 412, 567

Hilton, J. L., Capitaine, N., Chapront, J., Ferrandiz, J. M., Fienga, A., Fukushima, T., Getino, J., Mathews, P., Simon, J.-L., Soffel, M., Vondrák, J., Wallace, P., & Williams, J. 2006, *Celest. Mech.*, 94, 351

Highlights of Astronomy, Volume 14
IAU XXVI General Assembly, 14-25 August 2006
Karel A. van der Hucht, ed.

doi:10.1017/S1743921307011404

Long-term solution for the insolation quantities of the Earth

Jacques Laskar

IMCCE, Observatoire de Paris, 77 Avenue Denfert-Rochereau, Bat. A, F-75014 Paris, France

Abstract. We discuss a long-term solution for the insolation quantities of the Earth.

Keywords. solar system: general, celestial mechanics, ephemerides, Earth

---

A precise solution for the motion of the Earth axis (precession and nutation) is required for the reduction of astronomical observations over a few centuries. On the other hand, over long time scales, the short term variations of the Earth axis can be neglected and only the long term variations of the obliquity and precession angle become dominant. These solutions are used for the analysis of the paleoclimate signal present in the sedimentary records over several millions of years, according to Milankovitch theory of paleoclimates.

The agreement between the computed insolation signal (that depends on the secular evolution of the Earth's orbit and spin axis) and the sedimentary records is now so well established that in the recently published geological timescale GTS2004 (Gradstein, Ogg & Smith, eds., 2004), the whole Neogene period (0 - 23 Ma) has been calibrated using the astronomical solution of Laskar *et al.* (2004). In the continuation of this work, there is now an international effort for the astronomical calibration of the full Paleogene period (65 Ma). This goal is a difficult challenge for the computation of the Earth parameter evolution. Indeed, due to the chaotic evolution of its orbit (Laskar 1989, 1990), the uncertainty on the solution diverges exponentially by a factor of 10 every 10 Myr. The present orbital solution of Laskar *et al.* (2004) is estimated to be valid over about 40 Myr. To extend this solution over 65 Myr will require to improve the accuracy of the model by more than 2 orders of magnitude. The situation of the solution for the evolution of the Earth's axis is even worse, due to the uncertainty of the past evolution of the tidal dissipation in the Earth-Moon system. A first step towards the construction of a new generation of long term insolation solutions that will attempt to meet this challenge has been achieve recently in our group with the development of a new high accurate planetary ephemeris (INPOP06) fitted over all available planetary observations (see the presentation of A. Fienga in this volume).

References

Fienga, A., Manche, H., Laskar, J., & Gastineau, M. 2006, preprint, see: www.inpop.eu

Gradstein, F.M, Ogg, J.G., & Smith, A.G. (eds.) 2004, *A Geologic Time Scale 2004* (Cambridge: CUP).

Laskar, J. 1989, *Nature*, 338, 237

Laskar, J., 1990, *Icarus*, 88, 266

Laskar J., Robutel, P., Joutel, F., Gastineau, M., Correia, A.C.M., & Levrard, B. 2004, *A&A*, 428, 261

web site for the long term Laskar *et al.* 2004 solutions, see: http://www.imcce.fr/Equipes/ASD/insola/earth/earth.html

web site for the short term INPOP06 solutions, see: http://www.inpop.eu

web site for the Geological Time Scale 2004 (GTS2004), see: http://www.stratigraphy.org/

Highlights of Astronomy, Volume 14
IAU XXVI General Assembly, 14-25 August 2006
Karel A. van der Hucht, ed.

doi:10.1017/S1743921307011416

Using the P03 precession model

Patrick T. Wallace[1] and Nicole Capitaine[2]

[1]CCLRC/, Rutherford Appleton Laboratory, Didcot, UK
email: ptw@star.rl.ac.uk

[2]SYRTE/UMR8630, Observatoire de Paris, 61 Avenue de l'Observatoire, F-75014 Paris, France
e-mail: n.capitaine@obspm.fr

Abstract. We discuss aspects of using the P03 precession model.

Keywords. reference systems, ephemerides, astrometry

The IAU 2000 precession model comprised the existing Lieske *et al.* (1977) model plus rate corrections.

Though a good fit to existing VLBI observations, the IUA 2000 model model is not consistent with dynamical theory, and so the IAU Working Group on precession and the ecliptic recommended (Hilton *et al.* 2006) that it be replaced by the 'P03' model of Capitaine *et al.* (2003).

P03 provides improved models for both the equator and the ecliptic, and also includes parameterized provision for future adjustment to match new determinations of properties of the non-rigid Earth.

Practical use of the new model involves choices, and various ways have been studied (Capitaine & Wallace 2006) of generating the directions of the celestial intermediate pole and origin (CIP, CIO), from which the usual rotation matrices can be obtained.

From a wide range of possible procedures we have selected two that target different classes of application, typified by the SOFA software and the IERS Conventions respectively.

These procedures achieve a high standard of consistency, both internal and mutual, as well as being efficient and versatile. One is based on the Fukushima-Williams precession-nutation angles, the other on series for the CIP coordinates.

Both use the CIO locator s, and both deliver the full range of products, supporting classical equinox/GST methods in addition to the CIO/ERA 'new paradigm'.

References

Capitaine, N., Wallace, P. T., & Chapront, J. 2003, *A&A*, 412, 567

Capitaine, N., & Wallace, P. T. 2006, *A&A*, 450, 855

Hilton, J. L., Capitaine, N., Chapront, J., Ferrandiz, J. M., Fienga, A., Fukushima, T., Getino, J., Mathews, P., Simon, J.-L., Soffel, M., Vondrák, J., Wallace, P., & Williams, J. 2006, *Celest. Mech.*, 94, 351

Lieske, J. H., Lederle, T., Fricke, W., Morando, B. 1977, *A&A*, 58, 1

Highlights of Astronomy, Volume 14
IAU XXVI General Assembly, 14-25 August 2006
Karel A. van der Hucht, ed.

doi:10.1017/S1743921307011428

Long-periodic precession parametrization

Jan Vondrák

Astronomical Institute, Academy of Sciences of the Czech Republic,
Boční II, 141 31 Prague 4, Czech Republic
email: vondrak@ig.cas.cz

Abstract. We discuss aspects of long-periodic precession parametrization.

Keywords. methods: analytical, astrometry, reference systems, celestial mechanics, ephemerides

---

Both old (IAU1976) and newly adopted (IAU2006) parametrization of precession is based on polynomial developments of the associated angles. Such developments are designed to achieve very high precision for several centuries around the central epoch, J2000. However, when extrapolated to more distant epochs, they start to rapidly diverge from a more realistic long-term predictions coming from a numerical integration of the motion of the rotating Earth and the bodies of the solar system.

The aim of this presentation is to propose another parametrization of precession, based on long-periodic functions of time rather than polynomials, which would yield results with precision approaching that of IAU2006 for epochs close to J2000 (up to a few thousand years), and a good fit to numerical integration for much longer time intervals (up to several hundred centuries).

It is demonstrated that such solution is possible with relatively low number of periodic terms, both for the precession of the ecliptic (six periodic terms with periods ranging from 500 to 2300 centuries) and the general precession / obliquity (ten periodic terms with periods ranging from 200 to 4000 centuries).

This study is a preparatory phase for the prepared work by N. Capitaine, P.T. Wallace and J. Vondrák that should lead to a next generation of precession model.

Highlights of Astronomy, Volume 14
IAU XXVI General Assembly, 14-25 August 2006
Karel A. van der Hucht, ed.

doi:10.1017/S174392130701143X

Models for high-precision spacecraft and planetary and lunar ephemerides

E. Myles Standish and James G. Williams

JPL, California Institute of Technology, JPL 301-150, Pasadena, CA 91109, USA
email: ems,james.williams@jpl.nasa.gov

Abstract. We discuss models for high-precision spacecraft and planetary and lunar ephemerides.

Keywords. reference system, time, standards, ephemerides

The accuracies of the observational data fit by ephemerides are expected to increase by a full order of magnitude in the near future. Spacecraft ranging should improve from the present 1 m level to perhaps 10 cm; directional measurements (VLBI, VLBA, etc.) will be accurate to 0.1 mas or less; and Lunar Laser Ranging measurements will be taken near the 1 mm level.

For such measurements to be fit by the ephemerides, a number of modeling improvements will be required for the ephemeris creation process. For the planetary ephemerides, it will be necessary to consider that many planets have distinct satellites (as opposed to being modeled with their barycenter); the perturbations of more than the just the present 300 asteroids must be considered, as well as some of the largest Kuiper belt objects; and the effect of the media through which the electromagnetic signals travel must be more accurately calibrated, possibly using dual or even triple frequency ranging.

For the lunar ephemeris, many physical and observational features must be considered and further refined: thermal expansion of the retroreflectors, a possible lunar inner core, decrease of the solar mass, refined movements of the telescopes and retroreflectors. These are in addition to the presently accounted-for features: computation of the lunar librations; nonspherical gravitational fields of the moon, earth, and sun; earth and moon tidal effects; separate modeling of the rotating lunar mantle and fluid core; atmospheric time-delays depending on pressure, temperature, and humidity at the telescope; and relativistic effects upon each station's clock, position, and light-time.

The lunar and planetary integration program is necessarily done now in quadruple precision. Relativity for the point-mass motions is complete through order $1/(c^2)$; the need for the next order must be studied. Once the modeling does justice to the accuracies of the upcoming observations, a number of interesting tests will be vastly improved over the present-day status; PPN beta and gamma, dG/dt, dAU/dt, strong and weak equivalence principles, dark matter, etc. will be vastly improved over the present-day status: PPN beta and gamma, dG/dt, dAU/dt, strong and weak equivalence principles, dark matter, etc.

Highlights of Astronomy, Volume 14
IAU XXVI General Assembly, 14-25 August 2006
Karel A. van der Hucht, ed.

doi:10.1017/S1743921307011441

Relativistic aspects of Earth's rotation

Michael Soffel and Sergei A. Klioner
Lohrmann Observatory, Dresden Technical University, D-01062 Dresden, Germany
email: soffel,klioner@rcs.urz.tu-dresden.de

Abstract. We discuss relativistic aspects of Earth's roration

Keywords. astrometry, reference systems, relativity, Earth

Various relativistic aspects related with Earth's rotation are reviewed (see also Soffel & Klioner 2005).

First the problem of reference systems is discussed, where the BCRS (barycentric celestial reference system) and GCRS (geocentric celestial reference system) act as basic systems. For the near future the first post-Newtonian approximation to Einstein's theory of gravity will be sufficient.

This does not only apply to the problem of reference systems but also to the construction of global quantities such as angular velocity, moments of inertia etc., and to the corresponding dynamical equations of motion. In a series of papers, Damour, Soffel & Xu (1991, 1992, 1993) have constructed an improved post-Newtonian celestial mechanics with the BCRS and GCRS as basis. The dynamical equation for rotational motion of the Earth is given explicitly with full post-Newtonian expression for the torque in the GCRS. For practical applications a model of rigidly rotating multipoles has been suggested (Klioner *et al.* 2001). This approach is presently pursued to derive an improved consistent post-Newtonian nutation series.

In a series of papers Xu and coworkers laid the foundation for a relativistic description of elastic deformable rotating astronomical bodies (Xu *et al.* 2001, 2003, 2005). This local approach still poses fundamental problems since the relation of the formalism with observed quantities related with Earth's rotation is still unclear.

References

Damour, T., Soffel, M., & Xu, C. 1991, *Phys. Rev.*, D 43, 3273
Damour, T., Soffel, M., & Xu, C. 1992, *Phys. Rev.*, D 45, 1017
Damour, T., Soffel, M., & Xu, C. 1993, *Phys. Rev.*, D 47, 3124
Klioner, S., *et al.* 2001, in: N. Capitaine (ed.), Proc. Journées 2001 (Paris: Observatoire)
Soffel, M., & Klioner, S. 2005, in: N. Capitaine (ed.), Proc. Journées 2004 (Paris: Observatoire)
Xu, C., Wu, X., & Soffel, M. 2001, *Phys. Rev.*, D 63, 043002
Xu, C., Wu, X., Soffel, M., & Klioner, S. A. 2003, *Phys. Rev.*, D 68, 064009
Xu,C., Wu, X., & Soffel, M. 2005, *Phys. Rev.*, D 71, 024030

Highlights of Astronomy, Volume 14
IAU XXVI General Assembly, 14-25 August 2006
Karel A. van der Hucht, ed.

doi:10.1017/S1743921307011453

The dynamical model of the planet motions and EPM ephemerides

Elena V. Pitjeva

Institute of Applied Astronomy of RAS, Nab Kutuzova 10, RU-191187 St Petersburg, Russia
email: evp@ipa.nw.ru

Abstract. We discuss the dynamical model of the planet motions and EPM ephemerides.

Keywords. celestial mechanics, astrometry, ephemerides, reference systems, time, solar system

---

The dynamical model of the motion of planets used for construction of high-precision numerical ephemerides EPM (Ephemerides of Planets and the Moon) at the Institute of Applied Astronomy RAS is presented. The model of EPM2006 ephemerides includes mutual perturbations from the nine planets, the Sun, the Moon, lunar physical libration, perturbations from 301 biggest asteroids, as well as perturbations from the solar oblateness, the massive asteroid ring, and the Kuiper belt objects. The total shift of the barycenter of the solar system due to 19 largest trans-Neptunian objects is 6140 m within the lifetime of *Gaia* (2011 - 2020).

Ephemerides of the planets and the Moon have been produced by numerical integration in the PPN metric over a 170-yr time interval (1880 - 2050) and were oriented onto ICRF by using the ICRF-base VLBI measurements of spacecraft.

The EPM2006 ephemerides have resulted from a least square adjustment to observational data totaling more than 440 000 position observations (1913 - 2005) of different types including radiometric observations of planets and spacecraft, CCD astrometric observations of the outer planets and their satellites, meridian transits and photographic observations. In addition to the TDB-base primary ephemerides, a version of the EPM ephemerides was also constructed in the TCB time scale.

The EPM ephemerides has formed the basis for the Russian 'Astronomical Yearbook' since 2006 and are available at: ftp://quasar.ipa.nw.ru/incoming/EPM2004.

Highlights of Astronomy, Volume 14
IAU XXVI General Assembly, 14-25 August 2006
Karel A. van der Hucht, ed.

doi:10.1017/S1743921307011465

INPOP06: a new planetary ephemeris

Agnes Fienga[1,2], H. Manche[1], Jacques Laskar[2] and Michael Gastineau[1]

[1]IMCCE - CNRS UMR8028, Observatoire de Paris,
77 Avenue Denfert-Rochereau, Bat. A, F-75014 Paris, France
email: fienga@imcce.fr

[2]Observatoire de Besançon - CNRS UMR6091, Besançon, France
email: agnes@obs-besancon.fr

Abstract. We present INPOP06: a new planetary ephemeris.

Keywords. celestial mechanics, astrometry, ephemerides, reference systems, time, solar system

---

The new numerical planetary ephemeris developped at IMCCE - Observatoire de Paris, named INPOP06 (*Intégrateur numérique planétaire de l'Observatoire de Paris*), is presented. Dynamical models are shown as well as observation datasets used to adjust the solutions. Determinations of physical parameters like asteroid masses, densities, Sun oblatness and PPN β and γ are also presented. Comparisons to new *Mars Express* observations are also presented.

INPOP06 is a numerical integration of the motion of the nine planets and the Moon (Moyer 1971) fitted to the most accurate available observations. It also integrates the motion of 300 perturbing main belt asteroids, the Earth's rotation and Moon libration. Interactions between non-spherical objects and point-mass objects are also taken into account.

We used more then 45 000 observations including the last tracking data of the *MGS* and *Mars Odyssey* missions. The accuracy obtained with INPOP06 is comparable to the last versions of the JPL DE solutions and of the EPM solutions. Good estimations of physical parameters are given and compared with others found in literature. Comparisons to new observations not used in any fit (MEX) are also presented, which shows the good extrapolation capabilities of INPOP06. Two versions of INPOP exists: one using the TDB time scale, the other based on TCB.

Reference

Moyer, T. D. 1971, *JPL Technical report*, 32, 1157

Highlights of Astronomy, Volume 14
IAU XXVI General Assembly, 14-25 August 2006
Karel A. van der Hucht, ed.

doi:10.1017/S1743921307011477

Accurate harmonic development of Lunar ephemeris LE-405/406

Sergey M. Kudryavtsev

Sternberg Astronomical Institute, Moscow State University,
Universitetsky Pr. 13, RU-119992 Moscow, Russia
email: ksm@sai.msu.ru

Abstract. We discuss an accurate harmonic development of Lunar ephemeris LE-405/406

Keywords. Moon, ephemerides, methods: analytical

By using a new method of spectral analysis of an arbitrary tabulated function of Sun/Moon/planets coordinates to Poisson series (Kudryavtsev 2004) we made accurate harmonic development of the long-term numerical lunar ephemeris LE-405/406 (Standish 1998). Spherical lunar coordinates r (geocentric distance), V (ecliptic longitude) and U (ecliptic latitude) are represented by Poisson series in the form used by analytical theories of lunar motion (Chapront-Touzé & Chapront 1983; Chapront & Francou 2003).

The complete solution LEA-406a includes 42 270 terms of minimal amplitude equivalent to 1 cm and is valid over 1500 - 2500. The simplified solution LEA-406b includes 7952 terms of minimal amplitude equivalent to 1 m and is valid over 3000 BC - 3000 AD. The maximum difference in lunar coordinates r, V, U calculated by means of the analytical development LEA-406a and numerical ephemeris LE-405/406 is respectively 1.7 m, $0.''0038$, $0.''0013$ over 1900 - 2100, and 3.2 m, $0.''0056$, $0.''0018$ over 1500 - 2500.

Over 3000 BC - 3000 AD the maximum difference in coordinates r, V, U calculated by means of LEA-406b and LE-406 is respectively 0.20 km, $0.''42$ and $0.''33$. It is better than accuracy of the modern analytical theory of lunar motion ELP/MPP02 by Chapront & Francou (2003) (in particular, over 1500 - 2500 the gain in accuracy is from a factor of 9 to a factor of 70 depending on the coordinate). The number of terms in series LEA-406a,-b is less than that in ELP/MPP02.

The Poisson series LEA-406a,-b analytically representing the lunar ephemeris LE-405/406 are available at: `<http://lnfm1.sai.msu.ru/neb/ksm/moon/LEA-406.zip>`.

Acknowledgements

An IAU grant provided to the author is sincerely acknowledged. The work is supported in part by the Russian Foundation for Basic Research under grant number 05-02-16436.

References

Chapront, J., & Francou, G. 2003, *A&A*, 404, 735
Chapront-Touzé, M., & Chapront, J. 1983, *A&A*, 124, 50
Kudryavtsev, S. M. 2004, *J. Geodesy*, 77, 829
Standish, E. M. 1998, *A&A*, 336, 381

Highlights of Astronomy, Volume 14
IAU XXVI General Assembly, 14-25 August 2006
Karel A. van der Hucht, ed.

doi:10.1017/S1743921307011489

Solar quadrupole moment from planetary ephemerides: present state of the art

Sophie Pireaux[1], E. Myles Standish[2], Elena V. Pitjeva[3] and Jean-Pierre Rozelot[4]

[1]ARTEMIS, Observatoire de la Côte d'Azur, Avenue N. Copernic, F-06130 Grasse, France
email: sophie.pireaux@obs-azur.fr

[2]JPL, Caltech, 301-150, Pasadena, CA 91109, USA
email: ems@jpl.nasa.gov

[3]Institute of Applied Astronomy of RAS, Nab Kutuzova 10, RU-191187 St Petersburg, Russia
email: evp@ipa.nw.ru

[4]GEMINI, Observatoire de la Côte d'Azur, Avenue N. Copernic, F-06130 Grasse, France
email: jean-pierre.rozelot@obs-azur.fr

Abstract. We discuss the present state of the art of the solar quadrupole moment from planetary ephemerides.

Keywords. gravitation, relativity, ephemerides, Sun: fundamental parameters

---

Even though the order of magnitude of the Solar quadrupole moment, J_2, is known to be 10^{-7}, its precise value is still discussed. Furthermore, stellar equations combined with a differential rotation model, the Theory of Figures of the Sun, as well as inversion techniques applied to helioseismology, are methods which are solar model dependent, i.e., implying solar density and rotation laws. Hence the need for dynamical estimates of the solar quadrupole moment, based on the motion of spacecrafts, celestial bodies or light in the gravitational field of the Sun.

We present an attempt to estimate the solar quadrupole moment via JPL and IAA planetary ephemerides, along with the other ephemeris parameters, through a single step fit to observations. Even though, in principle, it would be possible to extract J_2 from planetary ephemerides, we observe that it is significantly correlated with other solution parameters (semi-major axis of Mercury or Venus, mass of asteroid ring, ...) in the present ephemerides fitted to the now available observation data. We shall focus on the J_2 correlations with the Post-Newtonian parameters, characterizing alternative theories of gravitation, according the set of observations considered and the corresponding weight.

The situation shall improve with new data sets (additional VLBI data, additional spacecraft measurements with the ongoing missions, new space missions), the increasing precision in (ranging) observations and the development of new ephemerides.

Highlights of Astronomy, Volume 14
IAU XXVI General Assembly, 14-25 August 2006
Karel A. van der Hucht, ed.

doi:10.1017/S1743921307011490

Proposed terminology in fundamental astronomy based on IAU 2000 resolutions

Nicole Capitaine[1], Alexandre H. Andrei[2], Mark R. Calabretta[3], Véronique Dehant[4], Toshio Fukushima[5], Bernard R. Guinot[1], Catherine Y. Hohenkerk[6], George H. Kaplan[7], Sergei A. Klioner[8], Jean Kovalevsky[9], Irina I. Kumkova[10], Chopo Ma[11], Dennis D. McCarthy[7], Kenneth P. Seidelmann[12] and Patrick T. Wallace[13]

[1]SYRTE, Observatoire de Paris, 61 Avenue de l'Observatoire, F-75014 Paris, France
email: n.capitaine@obspm.fr

[2]Observatorio Nacional, R. Gal. Jose Cristino 77, São Cristóvão, Rio de Janeiro, RJ 20921-400, Rio de Janeiro, Brazil

[3]CSIRO / Australia Telescope National Facility, PO Box 76, Epping, NSW 1710, Australia

[4]Royal Observatory Belgium, Avenue Circulaire 3, B-1180 Brussels, Belgium

[5]National Astronomical Obesrevatory, 2-21-1 Osawa, Mitaka-shi, Tokyo 181-8588, Japan

[6]HM Nautical Almanac Office, UK Hydrographic Office, Admiralty Way, Taunton, Taunton TA1 2DN, UK

[7]Astronomical Applications, US Naval Observervatory, 3450 Massachusetts Ave NW, Washington DC 20392-5420, USA

[8]Lohrmann Observatory, Technical University, D-01062 Dresden, Germany

[9]ARTEMIS, Observatoire de la Côte d'Azur, Avenue N. Copernic, F-06130 Grasse, France

[10]Sobolev Astronomical Institute, St Petersburg State University, Dvortsovaya emb 18, RU-191186 St Petersburg, Russia

[11]NASA Goddard Space Flight Center, Code 698, Greenbelt, MD 20771, USA

[12]Virginia University, 129 Fontana Ct, Charlottesville, VA 22911-3531, USA

[13]Department of Space Science & Technology, STFC, Clilton, Didcot OX11 0QX, UK

Abstract. We present the proposals of the IAU Division I Working Group on *Nomenclature for Fundamental Astronomy* (NFA) that was formed at the IAU XXV General Assembly in 2003.

Keywords. reference system, time, ephemerides, standards, Earth

---

The NFA WG has worked on selecting a consistent and well defined terminology for all the quantities based on the IAU 2000 Resolutions on reference systems in order that it will be understood, recognized and adopted by the astronomical community.

We first recall the main nomenclature issues associated with the implementation of the IAU 2000 resolutions and especially those related to new concepts. We then report on the final NFA WG recommendations on terminology choices and guidelines that have been supported by explanatory documents. We finally discuss the resolution proposals to the IAU XXVI General Assembly, 2006, that have been prepared by the WG.

The IAU 2000 Resolutions recommended the use of an improved precession-nutation model, the use of a new definition of Universal Time and the adoption of a new origin on the equator as a replacement of the equinox. The NFA Recommendations related to these issues consist for example of specifying the terminology associated with the new paradigm, defining the celestial and terrestrial 'intermediate systems', keeping the classical terminology for 'true equator and equinox' and giving the name 'equation of the origins' to the distance between the Celestial intermediate origin (CIO) and the equinox along the intermediate equator. The NFA WG has also proposed a re-definition of Barycentric Dynamical Time (TDB) through a linear transformation of Barycentric Coordinate Time (TCB). These recommendations have resulted from a detailed discussion within the WG on issues described in a number of Newsletters and documents issued by the WG and posted on the NFA web site at ¡http://syrte.obspm.fr/iauWGnfa/¿.

A special page of the NFA web site makes available documents with educational purposes relevant to the NFA issue.

The NFA explanatory document supplies information on the terminology and guidelines recommended by the WG. The NFA Glossary provides a set of detailed definitions that best explain all the terms required for implementing the IAU 2000 resolutions. The other sections provide complementary and supporting material to facilitate the understanding and implementation of the IAU resolutions, as well as illustrating the Glossary.

The NFA WG has submitted two resolution proposals to the IAU 2006 General Assembly; one is a 'Supplement to the IAU 2000 Resolutions on reference systems' for harmonizing the name of the pole and origin to 'intermediate' and fixing the default orientation of the BCRS and GCRS; the other one is a re-definition of TDB.

Highlights of Astronomy, Volume 14
IAU XXVI General Assembly, 14-25 August 2006
Karel A. van der Hucht, ed.

doi:10.1017/S1743921307011507

Implementation of the new nomenclature in The Astronomical Almanac

Catherine Y. Hohenkerk

HM Nautical Almanac Office, UK Hydrographic Office, Admiralty Way, Taunton, Taunton TA1 2DN, UK
email: catherine.hohenkerk@ukho.gov.uk

Abstract. This talk charts the implementation of the resolutions of the IAU XXIV General Assembly in 2000 (IAU 2000) in The Astronomical Almanac (AsA).

Keywords. reference system, time, standards.

The AsA is a joint publication of the Nautical Almanac Office of the US Naval Observatory (USNO) and the HM Nautical Almanac Office (HMNAO) at the UK Hydrographic Office. All updates must be made in the context of the document into which they are being introduced. The AsA is an annual reference product. It is not a text book but it must maintain standards and be up to date and reliable. It is not at the leading edge, but it must be useful to users and users require continuity. The aspects of the resolutions that affect Section B - Times Scales and Coordinate Systems of the AsA are the implementation of IAU 2000 precession-nutation theory, the recommendation to use what is now called the Celestial Intermediate Origin (CIO) as the origin for right ascension and the Earth rotation angle (ERA). The final version of the IAU2000 precession-nutation software became available in December 2002, and the IAU-SOFA software, the underlying software being used by HMNAO, both equinox and CIO-based, was released in April 2003. For the AsA production this meant implementation in the 2006 edition, with the IAU2000 precession-nutation being used throughout the almanac. It is only when new concepts are fully considered, the software tools developed and an established nomenclature exists, that a new algorithm may successfully be introduced. The areas considered were new tables, new explanations, adding new material into existing pages and finally updating the software, both for the calculations and the new layout.

All the existing quantities are retained including GMST / GAST / GAST−GMST the quantities from the Universal and Sidereal Times pages and the precession and nutation matrix (NPB). The new quantities are the Earth rotation angle (ERA), the equation of the origins (ERA − GAST), the celestial intermediate pole (CIP) and location of the CIO (X, Y, s), and the matrix transformation (C) from the GCRS (Geocentric Celestial Reference System) to the celestial intermediate reference system. No matter how easy it may seem to produce tables of numbers, it is essential to show how the new and existing quantities are used. The explanation of the calculation of intermediate places (CIO-based) and the formation of hour angles has been introduced alongside the equinox-based algorithm for calculating apparent places, which has been extended to calculate hour angles. Although, these processes have different routes, they are described in parallel, and the end results are the same.

The AsA for 2006 was our first almanac to implement the IAU 2000 resolutions, while the 2007 edition is now published and the 2008 edition has just gone to press. USNO Circular 179 by G. Kaplan, at http://aa.usno.navy.mil/publications/docs/Circular 179.html, should be consulted for more detailed explanatory material.

Highlights of Astronomy, Volume 14
IAU XXVI General Assembly, 14-25 August 2006
Karel A. van der Hucht, ed.

doi:10.1017/S1743921307011519

TDB or TCB: does it make a difference?

Sergei A. Klioner

Lohrmann Observatory, Dresden Technical University,
Mommsenstraße 13, D-01062 Dresden, Germany
email: klioner@rcs.urz.tu-dresden.de

Abstract. We argue that with the refined definition of TDB to be adopted by the IAU, the transformation between TDB and TCB becomes trivial.

Keywords. time, relativity, standards

Although the relativistic framework as adopted by the IAU in 1991 and refined in 2000 recommends the coordinate time TCB to be used for the modeling of any physical phenomena not localized to the vicinity of the Earth, an older time scale TDB is often used instead. In this presentation the current status of TDB, its relation to TCB as well as its advantages and disadvantages for practical problem will be discussed in detail.

It is argued that with the refined definition of TDB to be adopted by the IAU, the transformation between TDB and TCB becomes trivial. Special attention will be paid to the use of TCB and TDB for developing new solar system ephemerides. A clear algorithm to make the time argument of future ephemerides fully consistent with TCB will be presented. A few open issues, where additional relativity-related conventions are still required will also be discussed.

Highlights of Astronomy, Volume 14
IAU XXVI General Assembly, 14-25 August 2006
Karel A. van der Hucht, ed.

doi:10.1017/S1743921307011520

From atomic clocks to coordinate times

Gérard Petit

Bureau Intenational des Poids et des Mesures,
Pavillon de Breteuil, F-92312 Sèvres Cedex, France
email: gpetit@bipm.org

Abstract. With the present level of accuracy it may be necessary to use space borne atomic clocks to directly generate TCG, and to start addressing the limitations in the coordinate time transformations.

Keywords. time, relativity, standards

The IAU1991 Resolution A4, complemented by IAU2000 Resolution B1.3-4, provide rigorous definitions for barycentric and geocentric reference systems in a relativistic framework and define the coordinate times of these systems as TCB and TCG, respectively. Other coordinate times in use are TT, defined from TCG through IAU2000 Resolution B1.9, and TDB, whose rigorous definition from TCB has been adopted in 2006.

For practical use, these coordinate times must be realized and the proper time provided by atomic clocks (Atomic time AT) generates all coordinate times. The present sequence is $\mathrm{AT} => \mathrm{TT} \equiv> \mathrm{TCG} \rightarrow \mathrm{TCB} \equiv> \mathrm{TDB}$, where $\equiv>$ is an exact transformation and where $=>$ indicates the complex series of operations involved in generating International atomic time TAI.

The present uncertainty of realization of TAI and of TT(BIPM), another realization of TT, is about 1×10^{-15} in frequency with good prospects to reach 1×10^{-16} in the coming years. The uncertainty brought by the transformation $\mathrm{TCG} \rightarrow \mathrm{TCB}$ is below 1×10^{-17}. Future evolutions of atomic clocks, e.g., using new optical transitions, could allow reaching this level and will bring new performances to the generation of coordinate time scales.

At this level of accuracy it may be necessary to use space borne atomic clocks to directly generate TCG, and to start addressing the limitations in the coordinate time transformations.

Highlights of Astronomy, Volume 14
IAU XXVI General Assembly, 14-25 August 2006
Karel A. van der Hucht, ed.

doi:10.1017/S1743921307011532

Pulsar as barycenter coordinate clock

Yuri P. Ilyasov[1], **S. M. Kopeikin**[2], **Mikhail V. Sazhin**[3], **and Vladimir E. Zharov**[3]

[1]Pushchino Radio Astronomical Observatory, Lebedev Physical Institute,
Leninskii prosp 53, RU-142290 Puschino, Russian Federation
email: ilyasov@prao.psn.ru

[2]Department of Physics and Astronomy, University of Missouri,
322 Physics Building, Columbia, MO 65211, USA

[3]Sternberg Astronomical Institut, Moscow State University,
Universitetskij pr., 13, RU-119992 Moscow, Russian Federation
email: snn,zharov@sai.msu.ru

Abstract. Pulsars can be considered as very precise clocks if they are observed from the barycenter of the Solar system. Pulsar Timing Array (PTA) can be used to establish new astronomical reference frame which describes both space and time properties. Among these arrays most remarkable are the Parks Pulsar Timing Array (PPTA) and Kalyazin Pulsar Timing Array (KPTA).

Keywords. pulsars: general, time, standards

The pulsar motion in space has to be taken into account. The pulsar proper motion is determined by timing technique, and transversal velocity is observed and measured by VLBI. It is impossible to measure pulsar radial velocity because there are not known spectral lines in pulsar spectra. The radial component of velocity of pulsar can not be determined by timing as well due to fact that radial velocity, radial acceleration are renormalized in pulsar period and its derivatives due to the Doppler shift. The Shapiro time delay, which is generated by average gravitational field of our Galaxy, affects the pulsar clock, as well.

Stability of the set of pulsar clocks can be satisfactory when the mutual motion of a pulsar and the observer (referred to the barycenter of the Solar system) can be considered as linear. It is evident, that pulsars having large acceleration can be excluded from PTA, for instance, pulsars belonging to Globular clusters. It is why we have to know the relative acceleration pulsar versus observer.

From ten years of observations of the pulsars at Kalyazin one can say that intrinsic and extrinsic (interstellar medium) instabilities for the best millisecond reference pulsars are about 10^{-14}. One can expect that instability of pulsar time scale during 50 - 100 years will be about 10^{-18}, and secular aberration of the Solar system should be taken into account on the level -10^{-20}/yr.

Each of effects which affect pulsar long-term stability will be very important astronomical discovery.

Highlights of Astronomy, Volume 14
IAU XXVI General Assembly, 14-25 August 2006
Karel A. van der Hucht, ed.

doi:10.1017/S1743921307011544

Accurate optical reference catalogs

Norbert Zacharias
U.S. Naval Observatory (USNO),
3450 Massachusetts Avenue NW, Washington, DC 20392-5420, USA
email: nz@usno.navy.mil

Abstract. Current and near future all-sky astrometric catalogs on the ICRF are reviewed with the emphasis on reference star data at optical wavelengths for user applications.

Keywords. astrometry, catalogs, surveys, reference systems, stars: kinematics, telescopes, instrumentation: detectors

---

The standard error of a *Hipparcos* Catalogue star position is now about 15 mas per coordinate. For the *Tycho-2* data it is typically 20 to 100 mas, depending on magnitude.

The USNO CCD Astrograph Catalog (UCAC) observing program was completed in 2004 and reductions toward the final UCAC3 release are in progress; for updates see <ad.usno.navy.mil/ucac> . This all-sky reference catalog will have positional errors of 15 to 70 mas for stars in the 10 to 16 mag range, with a high degree of completeness, including positions of double stars from blended images. Proper motions for the about 60 million UCAC stars will be derived by combining UCAC astrometry with available early epoch data, including yet unpublished measures (StarScan) of the complete set of AGK2, Hamburg Zone astrograph and USNO Black Birch programs, reaching up to 14th magnitude from about 1900, 2300, and 700 plates, respectively. Dedicated deep CCD imaging of about 600 QSOs obtained between 1996 and 2004 is part of the UCAC program to allow a direct tie to the ICRF independent of *Hipparcos*.

Other accurate optical astrometric catalogs are the 14th Carlsberg Meridian Catalog (CMC14) going deeper ($V = 17$) than UCAC and covering $-30°$ to $+50°$ declination, the Bordeaux zone PM2000 catalog with block-adjustment-type reductions of early (AC, Carte du Ciel) and new (scanning CCD) observations of about 2.7 million stars in the $+11°$ to $+18°$ zone (excellent proper motions), and the ongoing Southern Proper Motion (SPM) survey (Yale, San Juan), providing absolute proper motions in large areas of the southern sky from ties to background galaxies, similar to the already completed Northern Proper Motion (NPM, Lick Observatory) program.

Accurate positional and proper motion data are combined in the Naval Observatory Merged Astrometric Dataset (NOMAD) which includes *Hipparcos*, *Tycho-2*, UCAC2, USNO-B1, NPM + SPM plate scan data for astrometry, and is supplemented by multi-band optical photometry as well as 2MASS near infrared photometry, covering all stars down to about 20th magnitude. Accurate positions of fainter stars in selected areas are available in the Sloan Digital Sky Survey (SDSS) and the Deep Astrometric Standards (DAS) project using 4-meter class telescopes on 4 fields of 10 sq.deg. size each to 24th magnitude.

The optical design of the USNO Robotic Astrometric Telescope (URAT) is complete and the world largest, monlithic CCD chip (111 million pixel, 95 mm square) was fabricated in June 2006. The URAT project aims at 5 mas positions for 14 to 18th mag stars, including proper motions and parallaxes with a limiting magnitude of $R = 21$.

Highlights of Astronomy, Volume 14
IAU XXVI General Assembly, 14-25 August 2006
Karel A. van der Hucht, ed.

doi:10.1017/S1743921307011556

Scientific potential of the future space astrometric missions

Lennart Lindegren
Lund Observatory, Lund University, Box 43, SE-22100 Lund, Sweden
email: lennart@astro.lu.se

Abstract. We discuss the scientific potential of the future space astrometric missions.

Keywords. space vehicles: instruments; astrometry; reference systems; minor planets, asteroids; stars: fundamental parameters; Galaxy: structure

---

Visual and infrared astrometric observations from space have enormous potential for many branches of stellar and galactic astrophysics, solar-system research, cosmology and fundamental physics. The technical requirements for space astrometry are reviewed in very general terms. Considerations of diffraction and photon noise suggest that micro-arcsec astrometry is readily achieved with metre-size instruments, while nano-arcsec astrometry requires extremely challenging interferometry with 100 - 1000 m baselines. The 'parallax horizon' for trigonometric distances was some 100 pc for *Hipparcos*, and will be some 10 kpc for *Gaia* and 100 kpc for *SIM PlanetQuest*. This trend cannot be further extrapolated by a large factor even when nano-arcsecond astrometry becomes feasible, since the parallax method is less useful at cosmological distances, due to the paucity of sources that are both small (< 1 AU) and luminous enough (thus extremely hot).

Gaia is a fully funded ESA mission, now in the detailed design and implementation phase (B2/C/D) with a planned launch in late 2011. During its 5 - 6 year lifetime, it will observe most point sources brighter than 20th magnitude, including ~ 1000 million stars, half a million of quasars and very large numbers of asteroids, extragalactic supernovae, etc. The expected accuracy of the positions at mean epoch, parallaxes and annual proper motions is about 7 μarcsec for bright (< 12 mag) objects, 12 - 25 μarcsec at 15th mag, and 150 - 300 μarcsec at 20th mag. Quasi-simultaneous spectrophotometry will be obtained for all objects, using dispersed images at resolution $R \simeq 13$ - 30 in two wavelength bands. Radial velocities at 1 - 10 km/s precision will be determined for objects brighter than 16 - 17 mag using slitless spectra of the near-infrared Ca II triplet region.

The main impact of *Gaia* is expected in galactic and stellar research. Galactic studies will benefit from the flux-limited survey of all six components of phase space, including spectrophotometric classification. This will probe the distribution stars, dust and dark matter in the Galaxy, and identify dynamical processes such as density waves, bars, warps, and past merger events. In stellar astrophysics, the massive numbers of accurate trigonometric distances (e.g., $\sim 10^5$ better than 0.1 %, $\sim 10^7$ better than 1 %) for a wide range of stellar masses and evolutionary stages will provide new calibrators for stellar parameters and put new constraints on theoretical models. A detailed census of binaries and extrasolar systems within a few hundred pc will result, detecting Jupiter-size or larger companions to stars of all spectral types and providing unambiguous orbit and mass determinations. Half a million solar-system objects will obtain very accurate orbits allowing detailed dynamical analysis including many new mass-determinations.

The *Gaia* reference frame will be linked to ICRS through quasars (for a non-rotating frame) and optical counterparts of VLBI sources (for frame orientation). The expected accuracy is 0.5 μarcsec/yr for the rotation and 30 μarcsec for the orientation.

Highlights of Astronomy, Volume 14
IAU XXVI General Assembly, 14-25 August 2006
Karel A. van der Hucht, ed.

doi:10.1017/S1743921307011568

Space astrometry with the *Milli-Arcsecond Pathfinder Survey*: mission overview and science possibilities

Ralph A. Gaume[1], Bryan Dorland[1], Valeri V. Makarov[2], Norbert Zacharias[1], Kenneth J. Johnston[1], and Gregory S. Hennessy[1]

[1]Astrometry Depertment, US Naval Observatory (USNO),
3450 Massachusetts Avenue NW, Washington DC 20392-5420, USA
email: rgaume,bdorland,nz,kjj,gsh@usno.navy.mil

[2]Michelson Science Center, California Institute of Technology,
MS 301 486, 4800 Oak Grove Drive, Pasadena, CA 91109, USA
email: valeri.v.makarov-119624@jpl.nasa.gov

Abstract. We present a mission overview and science possibilities of space astrometry with the *Milli-Arcsecond Pathfinder Survey* (*MAPS*).

Keywords. astrometry, surveys, reference systems, stars: kinematics, instrumentation

---

The *Milli-Arcsecond Pathfinder Survey* (*MAPS*) mission is a space-based, all-sky astrometric and photometric survey from 2nd through 15th magnitude with a 2010 launch date goal. The primary mission goal for MAPS is the generation of a 1 mas all-sky astrometric catalog for the 2010 epoch.

The instrument consists of a 15 cm telescope and large (64 megapixel) active pixel sensor focal plane with associated processing electronics carried aboard a microsatellite bus in a 900 km sun-synchronous low Earth orbit.

MAPS technology, including the very large format detector, the onboard processing electronics, and next generation space-based GPS-receiver, will serve as a pathfinder in support of future space missions.

A 1 mas (or better) all-sky survey through 15th magnitude will have a tremendous impact on our current understanding of the galaxy and stellar astrophysics. *MAPS* science topics include:

(*i*) a kinematic and photometric exploration of the nearest star forming regions and associations; an understanding of the dynamics and membership of nearby open clusters;

(*ii*) a survey of nearby stars that addresses the 130 missing systems within 10 pc;

(*iii*) recalibration of the cosmic distance scale via distances to nearby clusters, and the period-luminosity relationship using high accuracy proper motion (*Hipparcos* and *MAPS* positions and a twenty year baseline) and parallax measurements;

(*iv*) discovery of giant planets and brown dwarfs orbiting nearby stars; kinematic detection of galactic cannibalism and mergers in the Milky Way; and

(*v*) discovery of low-mass black holes and neutron stars in astrometric binaries.

Highlights of Astronomy, Volume 14
IAU XXVI General Assembly, 14-25 August 2006
Karel A. van der Hucht, ed.

doi:10.1017/S174392130701157X

Does the magnetic field in the fluid core contribute a lot to Earth nutation?

Cheng-Li Huang[1], Véronique Dehant[2], Xin-Hao Liao[1], Olivier de Viron[2], and Tim van Hoolst[2]

[1]Shanghai Astronomical Observatory, CAS, 80 Nandan Road, Shanghai 200030, P.R. China
email: clhuang,xhliao@shao.ac.cn

[2]Royal Observatory of Belgium, Avenue Circulaire 3, B-1180 Brussels, Belgium
email: veronique.dehant,o.deviron,timvh@oma.be

Abstract. We discuss the influence of the magnetic field in the fluid core to the Earth's nutation.

Keywords. Earth, reference systems, magnetic fields

The existence of relative nutational motions between the liquid core and its surrounding solid parts induces a shearing of the magnetic field. An incremental magnetic field is then created, which perturbs the nutations themselves. This problem has already been addressed within a nutation model by Buffett (1992, Paper 1) and Buffett *et al.* (2002, Paper 2). These authors used an angular momentum budget approach, and their results are taken into account in the MHB nutation model.

In this work, the magnetic field influence is incorporated in the numerical integration method used in precise nutation theory. New equations for that problem and new boundary conditions inside the Earth are developed, and a new strategy to compute nutations is established. The Coriolis force, which is ignored in Paper 1 is included in our work. Our results show that the change of the free core nutation period is very consistent with the change in the main nutation (18.6 yr and retro-annual) terms.

Comparisons of these results with Paper 1 and 2 are made, and discussions on the contribution of the Coriolis force and the magnetic field itself on the coupling constants are also presented.

References

Buffett, B. A. 1992, *J. Geophys. Res.*, 97, 19,581 - 19,597
Buffett, B. A., Mathews, P. M., & Herring, T. 2002, *J. Geophys. Res.*, 107, B4, 10.1029 / 2001JB000056

Highlights of Astronomy, Volume 14
IAU XXVI General Assembly, 14-25 August 2006
Karel A. van der Hucht, ed.

doi:10.1017/S1743921307011581

Retrieving diurnal and semi-diurnal signals in Earth rotation from VLBI observations

Aleksander Brzeziński[1] **and Sergei Bolotin**[2]

[1]Space Research Centre, Polish Academy of Sciences,
Bartycka 18 A, PL-00 716 Warsaw, Poland
email: alek@cbk.waw.pl

[2]Main Astronomical Observatory, National Academy of Sciences,
27 acad. Zabolotnoho st., Kiev 03680, Ukraine
email: bolotin@mao.kiev.ua

Abstract. We discuss the issue of retrieving diurnal and semi-diurnal signals in Earth rotation from VLBI observations.

Keywords. Earth, reference systems, astrometry

Polar motion and UT1 contain physical signals within the diurnal and semi-diurnal frequency bands. The dominant part (< 1.0 mas) is due to the gravitationally forced ocean tides. There is also a small variation (< 0.1 mas) caused by the direct influence of the tidal gravitation on the triaxial structure of the Earth. The remaining part (< 0.1 mas) comprises the atmospheric and nontidal oceanic influences driven by the daily cycle in the solar heating.

The modeling efforts and observations concern mostly the purely harmonic tidal variation which is the dominant effect in the diurnal and semi-diurnal bands. The high frequency geophysical signals, which are either irregular or quasi-harmonic, have rather poor observational evidence. Here we discuss how the diurnal and semi-diurnal signals in polar motion and UT1 can be estimated from the routine VLBI observations with one session in 3 to 5 days. The method relies upon the so-called complex demodulation technique.

We demonstrate its application to real data by using the VLBI analysis software Steel-Breeze. Spectral analysis of the demodulated time series reveals significant corrections to the conventional model of the ocean tide variations as well as the broad band variability with excess of power near the frequencies of the tidal lines S1 and S2. These series are suitable for the time domain comparisons with the available sub-diurnal estimates of the atmospheric and oceanic excitation.

Highlights of Astronomy, Volume 14
IAU XXVI General Assembly, 14-25 August 2006
Karel A. van der Hucht, ed.

doi:10.1017/S1743921307011593

Education in astrometry

William F. van Altena[1] and Magda G. Stavinschi[2]

[1]Physics Department, Yale University, PO Box 208121, New Haven, CT 06520, USA
email: vanalten@astro.yale.edu

[2]Astronomical Institute, Romanian Academy of Sciences,
Cutitul de Argint 5, RO-040557 Bucharest, Romania
email: magda@aira.astro.ro

Abstract. We discuss education in astrometry.

Keywords. reference systems, astrometry, ephemerides, time, standards

---

The potential for studies of the structure, kinematics and dynamics of our Galaxy, the physical nature of stars and the cosmological distance scale is without equal in the history of astronomy. Cutting-edge areas of research are now accessible due to the dramatic increases in accuracy provided by *Hipparcos*, *HST* and the next generation of astrometric space missions such as *SIM* and *Gaia*. The decadal committee in the US and the multinational ESA continuously place *SIM* and *Gaia* at the top of their scientific priority list and they have committed to spending on the order of a billion dollars on each mission. Major improvements in detector technology, such as the orthogonal-transfer arrays, have significantly improved our measurement precision and large ground-based facilities will enable us to probe more deeply into the universe. These opportunities oblige us to assume responsibilities for ensuring the success astrometric missions and facilities as well as educating astronomers to use them creatively and analyze the data with rigor.

Most countries outside the United States have vital educational programs in astrometry, but in nearly all cases optimism about the future of education in astrometry is low, except in France and Russia where extensive educational programs exist. The only full-term course in astrometry regularly taught in the US during the past decade has been at Yale, while at other institutions only a few lectures in a course on some aspect of observational astronomy are devoted to astrometric techniques.

We face a very troubling future for astronomy. Too few students are being educated with the knowledge and skills in astrometry to support existing and future projects. While outside the US astrometrists are regularly hired into teaching positions, that is not the case in the US. To minimize the impact of this problem, the *SIM* Michelson Science Center held a summer school in 2005 dealing with the detection of extrasolar planets by astrometric methods and Yale held a summer mini-course on basic astrometric methods.

We acknowledge our many colleagues who provided us with information on the status of astrometric education in their institutions and countries, especially N. Capitaine and V. Vityazev. In addition, we thank the NSF, NASA, ESA, and ESO, who have provided us with outstanding facilities for astrometric research.

This investigation was supported in part by the NSF.

Reference

van Altena, W. F., & Stavinschi, M. G. 2005, in: P. K. Seidelmann & A. K. B. Monet (eds.), *Astrometry in the Age of the Next Generation of Large Telescopes*, *ASP-CS* 338, 311

Highlights of Astronomy, Volume 14
IAU XXVI General Assembly, 14-25 August 2006
Karel A. van der Hucht, ed.

doi:10.1017/S174392130701160X

The 3D representation of the new transformation from the terrestrial to the celestial system

Véronique Dehant[1], **Olivier de Viron**[1], **and Nicole Capitaine**[2]

[1]Royal Observatory of Belgium, Avenue Circulaire 3, B-1180 Brussels, Belgium
email: veronique.dehant,o.deviron@oma.be

[2]SYRTE, Observatoire de Paris, 61 Avenue de l'Observatoire, F-75014 Paris, France
email: n.capitaine@obspm.fr

Abstract. We offer a 3D representation of the new transformation from the terrestrial to the celestial system.

Keywords. astrometry, reference systems, Earth

In order to study the sky from the Earth or to use navigation satellites, we need two reference systems, a celestial reference system (CRS) without no intrinsic rotation with respect to space and a terrestrial reference system (TRS) rotating with the Earth. Additionally, we need a way to go from one reference system to the other. This transformation involves the Earth rotation rate, polar motion, and precession-nutation. It is done using an intermediate system obtained by transforming the CRS with taking into account precession-nutation. Previously, one used an intermediate system related to the equinox; the new paradigm involved a point, denoted the Celestial Intermediate Origin (CIO), which, due to its kinematical property of 'Non Rotating Origin' (NRO), allows better describing the Earth rotation angle. The use of a NRO for the point related to the TRS has also been introduced; the corresponding name of this origin is the Terrestrial Intermediate Origin (TIO). Using or not using the CIO and the TIO only affects the intermediate system used in the transformation between the TRS and the CRS. The use of the CIO allows a perfect decontamination between Earth rotation and precession nutation. The use of the TIO allows a perfect decontamination between Earth rotation and polar motion.

Several 3D animations have been performed in order to explain the definition of the NRO, the transformation between the TRS and CRS, and the conceptual advantage with using the CIO. The new paradigm for performing the transformation between TRS and CRS based on the CIO and TIO uses the small quantities s and s'. As a first step, the CIP motion in the TRS is accounted for in order to move the z-axis of the TRS to the CIP. Then, the rotation s' around the CIP is applied in order for the new x-axis to be brought on to the non-rotating origin ϖ corresponding to the TIO. The third step consists in a rotation around the CIP axis of the so-called Earth Rotation Angle (ERA) denoted θ in order to move ϖ on the celestial counterpart of the non-rotating origin, σ, corresponding to the CIO. The rotation s is then applied, still around the CIP, and finally, the celestial motion of the CIP is taken into account in order to reach the CRS (rotation around the CIO, σ, to move the CIP on to the Z-axis of the CRS). The small quantities s and s' are for contributions associated with polar motion and nutation, respectively and are easily solved for kinematically from expressions for the polar motion and precession-nutation.

Highlights of Astronomy, Volume 14
IAU XXVI General Assembly, 14-25 August 2006
Karel A. van der Hucht, ed.

doi:10.1017/S1743921307011611

The small telescopes still useful for astrometry

Magda G. Stavinschi

Astronomical Institute, Romanian Academy of Sciences,
Cutitul de Argint 5, RO-040557 Bucharest, Romania
email: magda@aira.astro.ro

Abstract. We make a case for the usefulness of small telescopes astrometry.

Keywords. astrometry, solar system, telescopes

The IAU XXIV General Assembly approved the foundation of a special working-group on the *Future development of the ground-based astrometry*, as a Working Group of IAU Division I. It worked for six years.

The contribution of the small telescopes raises great questions, especially when we think of astrometry which is made today by means of the space missions or of the great telescopes. Nevertheless, there are still a series of programs for which they should continue to work, not in competition but in order to complete or prepare their activity.

There are some of its objectives still valid for the next few years:

- astrometric observations of some bodies of the solar system bodies;
- monitoring selected asteroids approaching the Earth;
- observations of artificial objects and space events and other natural phenomena generating hazards in the vicinity of the Earth;
- improving double star orbits;
- astrometric observations of the areas around extragalactic radiosources to extend Hipparcos system to the faint stars;
- rediscovering of recently discovered asteroids with the help of digital plate archive;
- the program 'Before *Gaia*'
- the observation of the mutual phenomena of Uranus' satellites within the campaign PHEURA 07.

Last but not least, there are the educational efforts for the training of a new generation of astrometrists, now before the launching of new space specialized missions and the processing of a huge amount of data collected so far.

Highlights of Astronomy, Volume 14
IAU XXVI General Assembly, 14-25 August 2006
Karel A. van der Hucht, ed.

doi:10.1017/S1743921307011623

Joint Discussion 16. Poster papers

1. State of the art of the implementation of the IAU resolutions and the ICRS

Searching candidate radio sources for the link with the future *Gaia* frame
P. Charlot, J.-F. Le Campion, & G. Bourda

New data of linking optical-radio reference frames
Z. Aslan, I. Khamitov, R.Gumerov W. Jin, Z. Tang, & S. Wang

Systematic errors and combination of the individual CRF solutions in the framework of the international project for the next ICRF
J. Sokolova, & Z. Malkin

Kinematic control of the inertiality of the system of *Tycho-2* and UCAC2 stellar proper motions
V.V. Bobylev, & M.Yu. Khovritchev

VLA radio star measurement of the rotation of the *Hipparcos* frame with respect to the ICRF
D.A. Boboltz, A.L. Fey, W.K. Puatua, N. Zacharias, M.J. Claussen, et al.

Deep Astrometric Standards
I. Platais, S.G. Djorgovski, C. Ducourant, A. Fey, S. Frey, Z. Ivezic, et al.

Better accuracy of *Hipparcos* proper motions in declination for stars observed with 10 Photographic Zenith Tubes
G. Damljanović, & J. Vondrák

Torino Observatory Parallax Program in the Catalog of Nearby Stars
R.L. Smart, M.G. Lattanzi, H. Jahreiß, B. Bucciarelli, & G. Massone

The influence of choice of fundamental catalogue on calculated apparent places of stars
M. Sekowski

Connection between ICRS and ITRS consistent with IAU 2000 resolutions
I.I. Kumkova, & M.V. Stepashkin

Earth orientation catalogue EOC-3: an improved optical reference frame
V. Štefka, & J. Vondrák

Catalogue of reference stars for observation of extragalactic radio sources of the Northern Sky
V. Ryl'kov, N. Narizhnaja, A. Dement'eva, G. Pinigin, N. Maigurova, et al.

Estimation of CRF and TRF from VLBI observations by the Least Squares Collocation method
S.L. Kurdubov

Computation of the veritable inclination between FK5 and *Hipparcos* equators: a critical discussion
M.J. Martínez, F.J. Marco, & J.A. López

The Russian astronomical yearbooks and IAU 2000 resolutions
N.I. Glebova, M.V. Lukashova, G.A. Netsvetaeva, & M.L. Sveshnikov

2. Precession and the Ecliptic

Precession-nutation solution consistent with the general planetary theory
V.A. Brumberg, & T.V. Ivanova

New expressions for the celestial coordinates of the CIP
M. Folgueira, N. Capitaine, & J. Souchay
Comparison of the nutation theories with the VLBI observations
V.E. Zharov, & S.L. Pasynok
Comparison of nutation series from GPS and VLBI observations using IAU80 and IAU2000 nutation models
K. Snajdrová, S. Englich, R. Weber, & H. Schuh

3. High accuracy models for reducing astronomical observations

Limitations on some physical parameters from position observations of planets
E.V. Pitjeva
Model of atmospheric radiation for large field radio astronomical data reduction
S. Ryś, & M. Urbanik
Relativistic ray tracing applied to a rotating optical system
Anglada-Escudé, G., Klioner, S.A., Soffel, M., & Torra, J.
VLBI antenna thermal deformation
E.A. Skurikhina

4. New terminology in fundamental astronomy, time and relativity

About the reference axis of the rotation of the Earth
R.O. Vicente
Weak microlensing effect and stability of pulsar time scale
M.S. Pshirkov, & M.V. Sazhin
Relativistic angular distance using Synge's world function
C. Le Poncin-Lafitte, & P. Teyssandier
Positioning systems and relativity
J.-F. Pascual-Sánchez

5. Scientific applications of high accuracy astronomy

Ten years timing of millisecond pulsars at Kalyazin
Yu.P. Ilyasov, & V.V. Oreshko
The Parkes Pulsar Timing Array Project
R.N. Manchester
Preliminary results in asteroid mass determination
Z. Aslan, I. Khamitov, R. Gumerov, L. Hudkova, A. Ivantsov, & G. Pinigin
Evidences of correlations between masses and minor planets elements. Analysis of the sources of errors
F.J. Marco, M.J. Martínez, & J.A. López
On geophysical excitation of prograde diurnal Polar Motion
M.V. Kudryashova, A. Brzezinski, & S.D. Petrov
Atmospheric excitation of UT1 variations during CONT05 campaign
Y. Masaki
High frequency variability in Earth rotation break from VLBI and GNSS data
S. Englich, K. Snajdrova, R. Weber, & H. Schuh
Optimum parameterization in estimating sub-daily earth rotation parameters with Very Long Baseline Interferometry
P.J. Mendes Cerveira, J. Böhm, S. Englich, R. Weber R., & H. Schuh

Highlights of Astronomy, Volume 14
IAU XXVI General Assembly, 14-25 August 2006
Karel A. van der Hucht, ed.

doi:10.1017/S1743921307011635

Joint Discussion 17 Highlights of recent progress in the seismology of the Sun and Sun-like stars

Timothy R. Bedding[1], Allan S. Brun[2], Jørgen Christensen-Dalsgaard[3], Ashley Crouch[4], Peter De Cat[5], Raphael A. García[2], Laurent Gizon[6], Frank Hill[7], Hans Kjeldsen[3], John W. Leibacher[7,8,9] (ed.), Jean-Pierre Maillard[10], S. Mathis[5], M. Cristina Rabello-Soares[11], Jean-Pierre Rozelot[12], Matthias Rempel[13], Ian W. Roxburgh[14], Réza Samadi[9], Suzanne Talon[15], and Michael J. Thompson[16]

[1]School of Physics, University of Sydney, NSW 2006, Australia
email: bedding@physics.usyd.edu.au

[2]DSM/DAPNIA/Service d'Astrophysique & UMR AIM 7158, CEA Saclay, F-91191 Gif-sur-Yvette, France
email: sacha.brun,rafael.garcia,stephane.mathis@cea.fr

[3]Institut for Fysik og Astronomi, Aarhus Universitet, DK-8000 Aarhus C, Denmark
email: jcd,hans@phys.au.dk

[4]Canadian Space Agency, Longueuil, QC, J3Y 8Y9, Canada
email: ash@astro.umontreal.ca

[5]Royal Observatory Belgium, Avenue Circulaire 3, B-1180 Brussels, Belgium
email: peter@oma.be

[6]Max-Planck-Institut für Sonnensystemforschung, D-37191 Katlenburg-Lindau, Germany
email: gizon@linmpi.mpg.de

[7]National Solar Observatory, NOAO, 950 N Cherry Ave, Tucson, AZ 85719-4933, USA
email: fhill,jleibacher@nso.edu

[8]Institut d'Astrophysique Spatiale, Centre Universitaire d'Orsay, F-91405 Orsay, France

[9]LESIA, Observatoire de Paris, CNRS UMR 8109, F-92195 Meudon, France
email: reza.samadi@obspm.fr

[10]Institut d'Astrophysique de Paris, 98bis Bd Arago, F-75014 Paris, France
email: maillard@iap.fr

[11]Hansen Experimental Physics Laboratory, Stanford University, CA 94305, USA
email: csoares@sun.stanford.edu

[12]Observatoire de la Côte d'Azur, GEMINI, F-06130Grasse, France
email: jean-pierre.rozelot@obs-azur.fr

[13]High Altitude Observatory, NCAR, PO Box 3000, Boulder, CO 80307-3000, USA
email: rempel@hao.ucar.edu

[14]Astronomy Unit, Queen Mary, University of London, London E1 4NS, UK
email: i.w.roxburgh@qmul.ac.uk

[15]Département de Physique, Université de Montréal, QC H3C 3J7, Canada
email: talon@astro.umontreal.ca

[16]SP2RC, Department of Applied Mathematics, University of Sheffield, Sheffield S3 7RH, UK
email: michael.thompson@sheffield.ac.uk

Abstract. The seismology and physics of localized structures beneath the surface of the Sun takes on a special significance with the completion in 2006 of a solar cycle of observations by the ground-based Global Oscillation Network Group (GONG) and by the instruments on board the *Solar and Heliospheric Observatory* (*SOHO*). Of course, the spatially unresolved Birmingham Solar Oscillation Network (BiSON) has been observing for even longer. At the same time, the testing of models of stellar structure moves into high gear with the extension of deep probes from the Sun to other solar-like stars and other multi-mode pulsators, with ever-improving observations made from the ground, the success of the *MOST* satellite, and the recently launched *CoRoT* satellite. Here we report the current state of the two closely related and rapidly developing fields of helio- and asteroseimology.

Keywords. Sun: interior, Sun: helioseismology, Sun: rotation, stars: interiors, stars: oscillations (including pulsations), stars: rotation, space vehicles: instruments

1. Global helioseismology and models

Global helioseismology, with the study of global modes of oscillation of the Sun, has existed for some thirty years. It continues to produce new and challenging results. Of particular current interest are the temporal variations of the modes and the possible detection of gravity (g) modes, both reviewed here. An aspect that still requires improvement is the accurate determination of the mode parameters used in helioseismology: efforts for the high-degree modes are described below. Three-dimensional modelling of the solar internal dynamics are making substantial strides, as A.S. Brun described at the meeting; and as J. Toomre described, there is a fruitful interplay between the observational results of helioseismology and the results from the numerical simulations. Another aspect of state-of-the-art modelling addresses mixing in the Sun by, for example, gravity waves. Such efforts may provide at least part of the answer to why new abundance measurements, which were described by M. Asplund, make agreement between helioseismology and standard solar models worse rather than better. Dynamo modelling is also gaining inspiration and guidance from helioseismic results, as described further below.

Solar Cycle 23 is the first activity cycle to be completely and continuously covered by modern imaging helioseismology observations, in particular those from the Global Oscillation Network Group (GONG) program, and the Michelson-Doppler Imager (MDI) on the *Solar and Heliospheric Observatory* (*SOHO*) spacecraft. In addition, the integrated-light Birmingham Solar Oscillation Network (BiSON) has been continuously operational since 1992. This unprecedented coverage allows us to study a number of aspects of the relationship between the oscillation parameters and the activity, as well as changes in the global properties of the dynamics of the solar rotation and convection zone.

Solar oscillations are observed typically either through Doppler velocity shifts of lines formed in the lower atmosphere of the Sun, or through the associated intensity variations. The oscillations are set up by waves that are turbulently excited by subphotospheric convection. If these waves survive long enough as they propagate inside the Sun, they set up resonant global modes. These modes are the objects of study of global helioseismology. Because the Sun is nearly spherical, the horizontal structure of each mode is described by a spherical harmonic of some degree (ℓ) and some azimuthal order (m). Being three dimensional, the modes also have structure in the radial direction, which can be labelled by a third integer, the radial order (n). Thus each mode, and its particular frequency (ω), can be labelled by the three integers (n, ℓ, m).

1.1. *Temporal variation of mode parameters*

It has been known for more than two decades that the oscillation frequencies change with the activity cycle. Using ACRIM data, Woodard & Noyes (1985) found that the frequencies increase as the number of sunspots rises. The shift was seen to increase with frequency, reaching a value of $+0.4\,\mu$Hz at about 4 mHz. Further work showed that the shift is highly correlated with various solar activity indices, in particular the global surface magnetic field. Later observations (e.g., Libbrecht & Woodard 1990) extended the results to higher frequencies and showed that the shift has a maximum of $+0.5\,\mu$Hz at the photospheric acoustic cutoff value of 5.6 mHz, then changes sign and rapidly drops to $-2\,\mu$Hz within 0.2 mHz of the cutoff. Modern observations with large-scale ring diagrams (How *et al.* 2006a) have reached even higher frequencies and reveal an even more complicated structure – the shift has a local minimum of $\sim 4\,\mu$Hz at 6 mHz and then sharply decreases in absolute magnitude once again. These observations suggest different mechanisms for the interaction between the surface magnetic field, the trapped subsurface p modes below 5.6 mHz, and the higher-frequency propagating atmospheric waves. The p modes sense magnetic changes to the acoustic cutoff frequency, while the atmospheric waves at higher frequencies interact with the surface field and excite various MHD waves.

Other mode parameters also vary with activity. Both the amplitudes and the life times of the modes are seen to decrease with increasing activity (Salabert & Jiménez-Reyes 2006; Komm *et al.* 2000; Chaplin *et al.* 2003), in agreement with observations that active regions suppress the convection which excites the oscillations. This mechanism is also supported by the observation that the energy supply rate to the modes does not change with increasing surface activity but the damping increases (Salabert *et al.* 2007). The spatial and temporal pattern of the mode parameter shifts can be projected back onto the solar surface (Komm *et al.* 2002; How *et al.* 2002) and is seen to be highly correlated with the activity bands.

The continuous coverage also allows us to study changes in the frequencies on short time scales. Tripathy *et al.* (2007) analyzed all of the GONG and *SOHO*-MDI data in nine-day segments, substantially shorter than the standard 108-day GONG or 72-day MDI analyses. They found that the nine-day frequencies are also highly correlated with activity indices, with the correlation at low and high frequencies decreasing more rapidly than in the longer analyses. The low-frequency decrease is likely due to increased noise and poorer frequency resolution in the short time series which degrades the precision of the measurements. The high-frequency decrease may arise from the short lifetimes of the modes, which do not survive long enough to interact with the surface activity. Tripathy *et al.* (2007) also found that the sensitivity of the short-time scale frequency shifts to the magnetic field index is significantly higher in the ascending phase of the cycle compared to the descending phase. This may be related to the hysteresis seen in other solar activity indices.

1.2. *3-D MHD global simulations of solar internal dynamics*

The Sun's internal dynamics is very complex, involving nonlinear interactions between convection, turbulence, instabilities, rotation, shear, and magnetism. Being able to understand the intricate interplay between all of these processes is crucial if one wants to progress in our knowledge of the Sun and the stars for which the Sun is our only spatially resolved example. Numerical simulations are thus becoming more and more useful for tackling this challenging problem. We briefly report here on the current status of 3-D global MHD simulation of solar internal magnetohydrodynamics obtained with the 3-D MHD ASH code (Clune *et al.* 1999; Miesch *et al.* 2000; Brun *et al.* 2004). In particular we wish to summarize the physical processes behind differential rotation, meridional circulation, and the solar magnetic activity and cycle.

The differential rotation of the Sun's convective envelope is peculiar since it is almost constant along a radial line at mid-latitude, a property which is counter intuitive, since it has been known for a long time that rotating fluids usually possess a cylindrical rotation profile (Taylor-Proudman theorem and columns, see Brun & Toomre 2002). What can

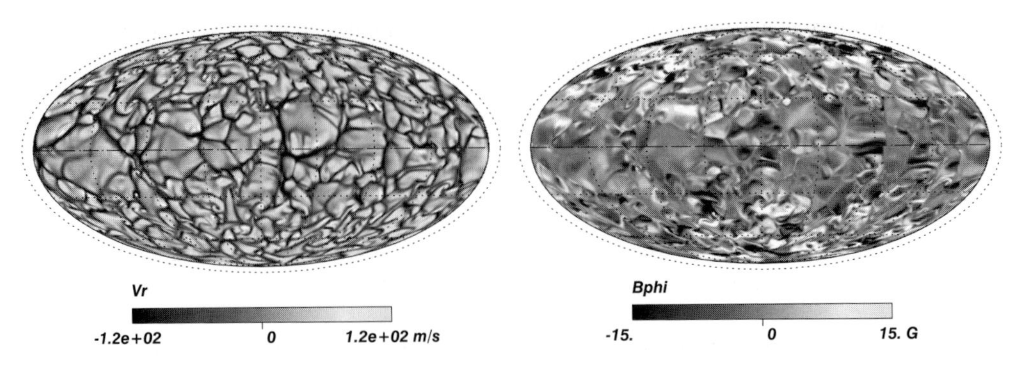

Figure 1. Snapshot using a Mollweide projection, of the radial velocity and toroidal field, near the top of the domain for the purely unstable magnetized convection model with a small magnetic Prandlt number $Pm = 0.8$. Typical field strengths are indicated, with dark tones corresponding to downward velocities and negative polarities. The dashed line indicates the equator.

3-D global models tell us about the processes leading to such a special rotation profile? A careful study of the redistribution of the angular momentum in a turbulent convection shell indicates that Reynolds stresses, i.e., non-linear correlations between the radial and latitudinal velocities with the azimuthal one) are the key players in transporting the angular momentum from the polar region toward the equator thus accelerating the equatorial region. The ASH simulations also show that a strong thermal wind is associated with this pronounced differential rotation, with associated temperature contrast of order few K. The conical shape is the result of a subtle latitudinal heat and angular momentum transport that favor a fast equator and slow poles and a weak pole to equator temperature contrast. A small change of the amplitude of this temperature contrast is found to have a drastic effect on the resulting profile of angular velocity, at 3 K the profile is cylindrical, at 13 K it is almost horizontal, with a value of 10 K giving the best agreement with helioseismic inversion (Miesch *et al.* 2006). Such a very delicate nonlinear balance between heat and angular momentum transport, is likely to be at the origin of the current observed differential rotation. As a consequence, small change in the parameters of convection models lead to a large range of differential rotation profiles (Brun & Toomre 2002). It thus remains to be seen, as the level of turbulence of the simulations are increased to become more realistic, how anisotropic nonlinear heat transport could finally setup a temperature contrast of 10 K found to reproduce reasonably well the observations.

The meridional circulation driven by the convection is found to be much weaker than the differential rotation, since it can be considered as a small departure from geostrophic balance. In the current 3-D global simulations, meridional flows are found to be multi-cellular both in latitude and radius, even in the case possessing a conical solar-like profile (Miesch *et al.* 2006). A recent study by Jouve & Brun (2007) indicates that such multi-cellular meridional flows will modify the butterfly diagram and decrease the cycle period in 2-D mean field models. If such complex meridional flows were to persist over many years in the Sun, they will either lead to a modified butterfly diagram or push the solar dynamo community to reconsider standard Babcock-Leighton solar dynamo models (Dikpati *et al.* 2004), as in the mid-1980's the inversion of the angular velocity lead the community to modify the standard α-ω model. Clearly, inverting the meridional flow down to the tachocline is rather urgent if one wants to progress in our understanding of the internal dynamics of the Sun.

Concerning 3-D modeling of solar magnetism, today the most accepted scenario to explain the Sun's large variety of dynamical phenomena rests on the operation of an internal dynamo in the highly-turbulent-convective solar envelope and at its base, in the tachocline (Parker 1993; Dikpati *et al.* 2004; Brun *et al.* 2004; Jouve & Brun 2007). This dynamo is thus operating at different location and physical scales and organize the large-scale field such as to make it cyclic. High-resolution computations performed with the ASH code (Brun *et al.* 2004), confirmed that dynamo action in a rotating turbulent convection zone generates magnetic fields at all scales present in the velocity field and that the field filled up the sphere leading to the formation of an $\ell = 1$ dipolar component. This dipolar field contains about 1% of the total magnetic energy generated in the convection zone which is mostly found in non axisymmetric field, and its amplitude is about 10 G. It is found to oscillate irregularly with a pseudo-period of about 400 days, almost a factor of ten faster than the real solar 11-yr cycle. Further we do not find in our purely unstable simulations strong toroidal mean magnetic field, as the cyclic emergence of sunspot on the solar surface suggest. A simple explanation is that the highly dynamical convection zone is not able to generate strong flux tube like structures locally. The fast, irregular reversal and the weak, toroidal mean field together point to the necessity or having a stable layer, where both the dynamical time scale is slower and the winding up of a poloidal field into strong toroidal structure is made easier. This has thus led to the incorporation of the role of the tachocline in a 3-D MHD simulation of the global solar magnetism (Browning *et al.* 2006). The results are encouraging since having a tachocline generates much stronger mean toroidal fields than previously found (by a factor found to be between 10 to 100), that possess the antisymmetry observed in sunspots between the two hemispheres (Hale's law). Also the presence of a strong field in the stable layer seems to have a stabilizing influence of the reversal of the weak poloidal field generated above in the turbulent convection zone, suggesting that the solar activity can be controlled by processes occurring rather deep in our star. A recent work by Brun & Zahn (2006)as also demonstrated that if a deep, fossil magnetic field exists in the solar radiative interior, this field is most likely in a mixed poloidal/toroidal configuration rather than just in a simple dipolar shape. Indeed they confirmed the pioneering work of Tayler in the mid-1970's that showed that a purely poloidal or toroidal magnetic field will become unstable respectively to high m and $m = 1$ instabilities. This complex topology and evolution of the inner field, and its coupling with the dynamo field of the tachocline can potentially lead to interesting modulation of the solar magnetic cycle. It is thus crucial to consider the Sun has a whole rather than to split it into its radiative, convective, and external parts. The time has come to develop a fully integrated model of the solar dynamics that coupled as much as possible all the fascinating dynamics occurring in its interior, transition layers and at its surface (Brun 2007).

1.3. *Solar interior dynamics and their temporal variation*

Turning to the internal solar dynamics, Howe *et al.* (2000) found a 1.3-yr oscillation in the rotation rate in the tachocline. While Basu & Antia (2001) were unable to reproduce the result, it remains intriguing and further work by How *et al.* (2006b) have shown it may be intermittent, since it apparently ceased after solar cycle 23 maximum in 2002. Continued observations will be needed to determine if it reappears in the ascending phase of Cycle 24, which is just now starting in August 2006.

Global helioseismology has also revealed that the so-called torsional oscillation seen on the surface for the last 30 years extends almost all of the way through the convection zone (How *et al.* 2005; Vorontsov *et al.* 2002). There is evidence that this zonal flow originates deep in the convection zone and then rises to the surface over the course of the cycle (How *et al.* 2005). The current temporal evolution of the flow can be described by the sum of an 11-yr and an 11/2-yr period sinusoid, and the near-surface helioseismic inferences of the flow are very highly correlated with the surface observations (How *et al.* 2006a). Finally, the helioseismic observations show the zonal flow associated with Cycle 24 began to migrate from the poles towards the equator in 2002, some four years before the first Cycle 24 active region was seen. This supports the idea of an extended solar cycle, as observed in coronal green-line observations (Altrock & Howe 2004).

1.4. *A flux-transport dynamo model*

Recent work has aimed at producing a non-kinematic flux-transport dynamo model to address the non-linear saturation through Lorentz force feedback as well as cycle variations of differential rotation and meridional flow. To this end one combines the differential rotation and meridional flow model developed recently by Rempel (2005) with a flux-transport dynamo similar to the models of Dikpati & Charbonneau (1999) and Dikpati & Gilman (2001).

The differential rotation model utilizes a meanfield Reynolds-stress approach that parametrizes the turbulent angular momentum transport (Λ-effect, Kitchatinov & Rüdiger 1993) leading to the observed equatorial acceleration. The tachocline is forced through a uniform rotation at the lower boundary of the computational domain. A meridional circulation, as required for a flux-transport dynamo, follows self-consistently through the Coriolis force resulting from the differential rotation. The computed differential rotation and meridional flow are used to advance the magnetic field in the flux-transport dynamo model, while the magnetic field is allowed to feed back through the macroscopic Lorentz force $\langle \boldsymbol{J} \rangle \times \langle \boldsymbol{B} \rangle$ (the contribution of the fluctuating part $\langle \boldsymbol{J}' \times \boldsymbol{B}' \rangle$ is not well known and neglected here).

It is found that the dynamo saturates through a reduction of the mean differential rotation when the toroidal field strength at the base of the convection zone reaches values around 1 to 1.5 T. The feedback on the meridional flow remains weak for these values, meaning that the equatorward transport of toroidal field at the base of the convection zone (crucial ingredient in a flux-transport dynamo) is only marginally affected.

The feedback of the macroscopic Lorentz force on differential rotation leads to a poleward propagating branch of torsional oscillations in agreement with observations in terms of amplitude and phase; however, the equatorward propagating low-latitude branch requires additional physics. Parameterizing the idea proposed by Spruit (2003) that the low-latitude torsional oscillation is a geostrophic flow caused by increased radiative loss in the active region belt (due to small scale magnetic flux) leads in a model to a surface oscillations pattern in good agreement with observations (see Fig. 2). As a side effect the cooling produces close to the surface an inflow into the active region belt of a few $\mathrm{m\,s^{-1}}$ in amplitude, which has been observed through surface Doppler measurements by Komm at al. (1993); Komm (1994), and with local helioseimology by, e.g., Gizon (2004) and Zhao & Kosovichev (2004). Around 50 Mm depth we find an outflow with an amplitude reduced by one order of magnitude (due to the increase in density). Recently Gizon & Rempel (2006) found observational support for such an outflow on a qualitative level.

Further details and properties of the solution of the coupled-differential-rotation dynamo model can be found in Rempel (2005) and Rempel (2006).

1.5. *High-degree mode parameter determination*

An aspect of helioseismic data analysis that still requires improvement is the accurate determination of the mode parameters used in helioseismology, particularly at high degree. The inclusion of high-degree modes (i.e., ℓ up to 1000) has the potential to improve dramatically the inference of the sound speed and the adiabatic exponent (Γ_1) in the outermost 2 to 3 % of the solar radius (Rabello-Soares *et al.* 2007), a region of great

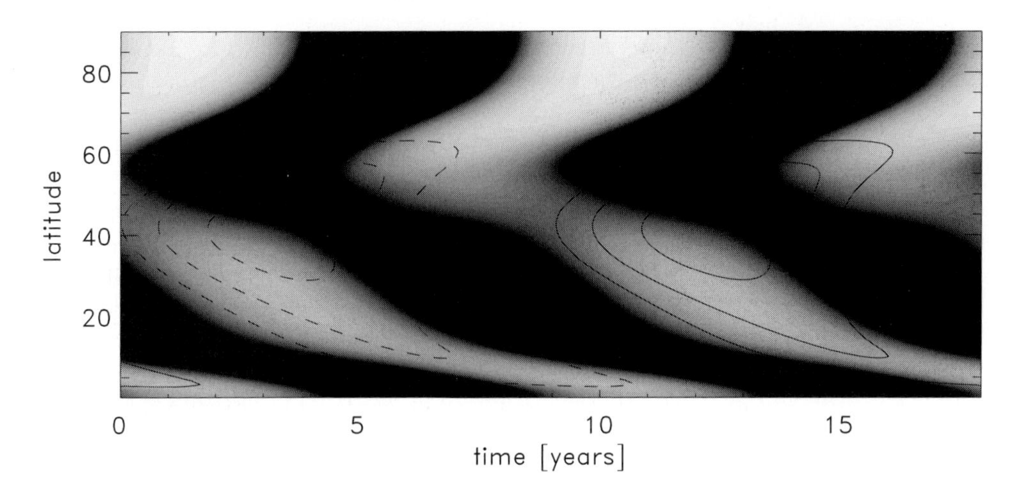

Figure 2. Surface pattern of torsional oscillations in a model considering macroscopic Lorentz force feedback and cooling of the active region belt. The contour lines indicate the toroidal field at the base of the convection zone. Bright colors refer to faster rotation, dark colors to slower rotation.

interest. However, the spherical harmonic spatial filtering is not orthonormal over less than the full Sun and the resulting spatial leaks get closer in frequency as the degree of a mode increases. As a result, the high-degree individual modes blend into ridges which will mask the true underlying mode parameters. This has so far prevented the estimation of unbiased mode parameters at high degrees using global helioseismology. As discussed by M.C. Rabello-Soares, in order to recover the mode characteristics, a very good model of the relative amplitude of all the modes that contribute to the ridge is needed, which requires a very good knowledge of the instrumental properties. Such a model of the relative amplitudes is called a 'leakage matrix'.

Rabello-Soares and collaborators have incorporated most of the observational and instrumental effects of the *SOHO*-MDI instrument important to high-degree analysis into a new analysis. Compared to their previous work (Korzennik *et al.* 2004), they include the following effects: correct instantaneous plate scale, radial image distortion, CCD tilt ($\sim 2°$) with respect to the optical axis, effective P angle ($0.2°$) and Carrington elements correction of $0.1°$ (Giles 1999). They have analysed variations of the central frequencies for $100 \leqslant \ell \leqslant 1000$ caused by the solar activity cycle. They showed that the high-degree changes during solar cycle 23 are in good agreement with the medium-degree results, except for years when the instrument was highly defocused (1996 to 1998) which were excluded from the analysis.

Although the correlation of the frequency changes with the solar cycle has been known for some time, its physical origin has been a matter of debate. The analysis of its behaviour will hopefully help in its understanding. Rabello-Soares *et al.* observed that the p modes and f (fundamental) modes behave differently, which suggests different physical effects. Their results confirm that the p-mode frequency shift scaled by the relative mode inertia is a function of frequency alone (see Rabello-Soares *et al.* 2007). The scaled frequency shift increases until 4.8 mHz. At higher frequencies, the frequency variation does not appear correlated with solar activity.

1.6. *The search for solar g-modes*

A detailed knowledge of the interior of the Sun has been obtained by means of precise measurements of its acoustic eigenmodes and the subsequent structural inversions, providing the stratification of crucial variables like the sound speed down to $0.05\,R_\odot$ (Turck-Chièze *et al.* 2001). The radial profile difference of the observed sound speed with the one extracted from the solar evolution model depends strongly on the physics used. The seismic model which has been built to reduce this difference to practically zero in the radiative zone enables to predict neutrino fluxes in excellent agreement with their detection (Couvidat *et al.* 2003a). However, the acoustic modes are less sensitive to other structural variables such as the density. In this case, the agreement with the models is still poor in the radiative zone. Moreover, the dynamical properties of the nuclear core are practically unknown. Effectively, due to the limitation in the number of acoustic modes penetrating in this region, the solar rotation profile is very uncertain below $0.2\,R_\odot$ (Chaplin *et al.* 1998; Couvidat *et al.* 2003a; García *et al.* 2004).

To improve our knowledge of the deepest layers of the Sun, the detection of gravity modes is required. These modes propagate in the radiative zone but are evanescent in the convective region and their expected surface amplitudes could be very small (Andersen 1996; Kumar *et al.* 1996). These modes have been intensively looked for since the 1980s. In the last ten years, two complementary approaches have been developed: to look for individual modes (relevant spikes in the Fourier spectrum of the observed signal) above a given statistical threshold (typically 90 % confidence level) at frequencies above 15 μHz; and to look for their predicted asymptotical properties at lower frequencies. The individual mode search has provided several *g*-mode candidates in *SOHO*-GOLF observations with more than 90% confidence level. In this study, multiplets instead of single peaks have been looked for (Turck-Chièze *et al.* 2004). The frequencies of such modes are extremely sensitive to the structural properties of the radiative zone (Mathur 2006) and their splittings provide information on the solar core rotation. But, today, there remains some ambiguity on the proper identification of the detected components in terms of their ℓ, m, n properties. Some scenarios have been proposed with consequences on the core dynamics (Turck-Chièze 2006a). The search for the asymptotic *g*-mode properties has been with us from the earliest days of helioseismology (summarized in Pallé 1991 and Hill *et al.* 1991) but using low-quality observational data sets (pre-space and pre-networks era). The asymptotic approach (where $n \gg \ell$) predicts a constant separation between the periods of adjacent (n, n+1) gravity modes. Consequently, periodicities in the form of broad peaks are expected in the power spectrum (PS) of the power spectral density (PSD) expressed in period corresponding to the observed velocity time series. This year, a peak structure has been detected (see Fig. 3), with more than 99.86 % confidence level (García *et al.* 2006), thanks to the high quality datasets available from *SOHO*-GOLF (García *et al.* 2005). The detailed analysis of such structure provides access to the dynamics of the inner core, in particular its rotation rate. At the present stage of this investigation, it would favour a core rotating faster than the rest of the radiative zone. If the detection of these *g*-mode signatures is confirmed, a very exciting future will be ahead of us to improve our knowledge of the dynamics of the deepest layers inside the Sun. Both analyses are fully complementary and we are encouraged to apply them to other instruments aboard *SOHO* and to improve once more our capability of detection. A new generation of instruments is necessary to reveal the complete dynamics of the core including its magnetic field. We are hoping to be able to improve the capability of intensity measurement at the limb (*SODISM*-PICARD) and to optimize the signal-to-noise ratio at these very low frequencies by reducing the solar noise (GOLFNG). This

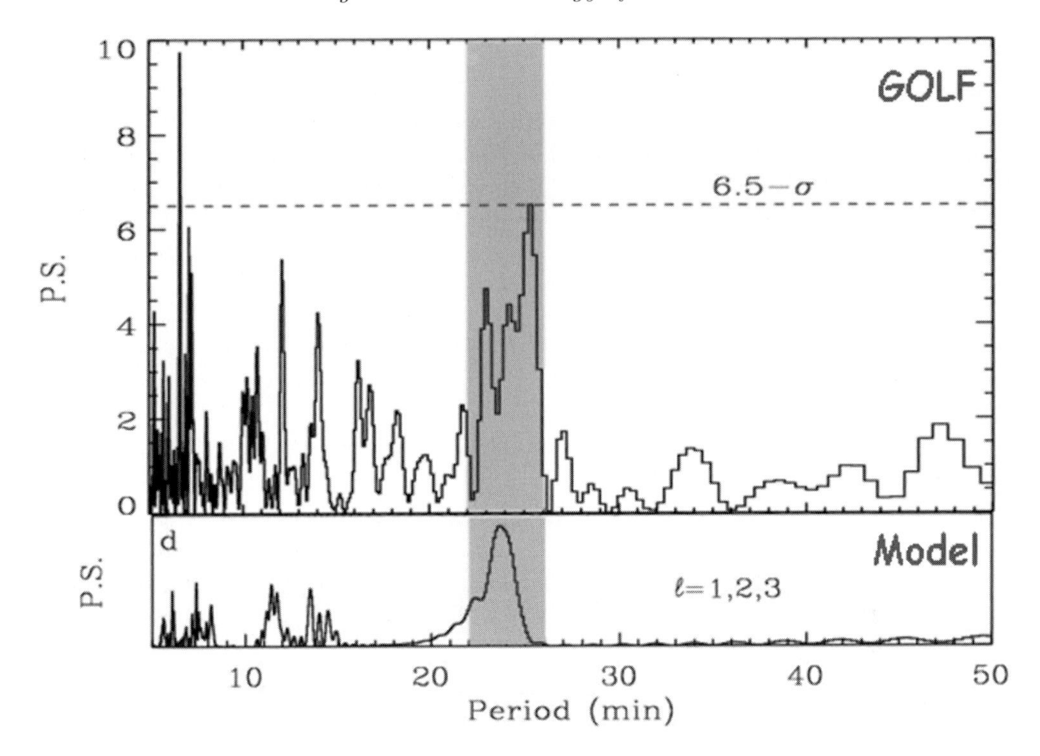

Figure 3. *Top*: Power spectrum of the power spectral density expressed in period computed from ten years of *SOHO*-GOLF velocity time series. *Bottom*: PS of the PSD from a model containing theoretical g-modes computed from the seismic model and using a rigid core rotation.

is the motivation of the *DynaMICS* (*Dynamics and Magnetism from the Inner Core to the Chromosphere (& Corona) of the Sun*) mission (Turck-Chièze *et al.* 2006c), which aims to make use of a small satellite with the GOLFNG instrument (Turck-Chièze *et al.* 2006b) while *Solar Dynamics Observatory* (*SDO*) and *SODISM*-PICARD are operating, as well as possibly as part of a bigger platform within the ESA Cosmic Vision 2015 – 2025 programme (Turck-Chièze *et al.* 2005).

1.7. *Solar interior dynamics: angular momentum transport by gravity waves*

It is now well established that the interior of the Sun is rotating more or less at the same rate as the average rotation rate of the surface convection zone. We also know that, in the past, the Sun was rotating much faster; the interaction of the Sun's lost mass with the outer magnetic fields explains the spin-down of the convection zone, but there remains the need for a very efficient mechanism to carry angular momentum from the radiative core to the surface convection zone.

Internal gravity waves excited at the base of the convection zone may be an important source of angular momentum redistribution, since they take momentum from the region where they are excited and deposit it where they are damped. These waves have low frequencies (around 1 μHz) and are the propagating counterparts of the higher-frequency g-modes that form standing waves. Talon *et al.* (2002) showed how differential filtering exerted by an asymmetrical-shear-layer oscillation (itself related to the damping of high-degree internal gravity waves) favors the penetration of low-degree, low-frequency *retrograde* waves that may deposit their negative angular momentum in the deep interior, causing the spin-down of the core.

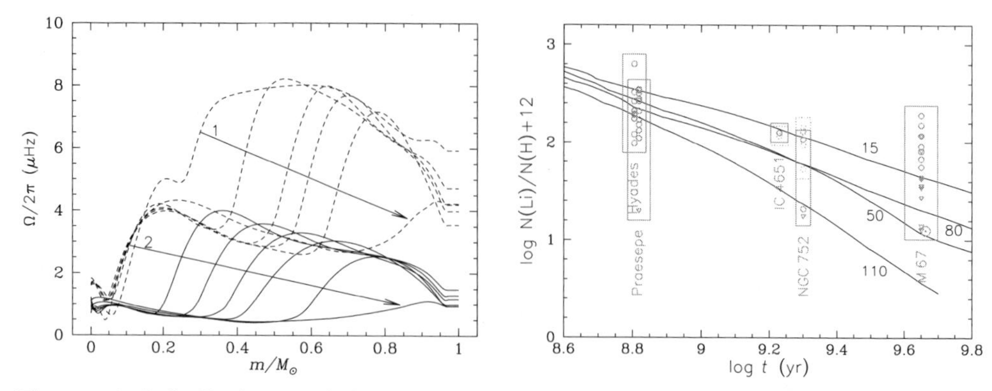

Figure 4. *Left*: Evolution of the solar rotation profile under the effect of surface braking, meridional circualtion, shear turbulence, and internal gravity waves (ages: 0.2, 0.21, 0.22, 0.23, 0.25, 0.27, 0.5, 0.7, 1, 1.5, 3, 4.6 Gy). *Right*: Evolution of the surface lithium abundance for various initial rotation rates (black lines) compared to measurements of Li in solar-mass stars in clusters of various ages (grey points and boxes). Adapted from Charbonnel & Talon (2005).

Charbonnel & Talon (2005) applied that model to an evolving solar-mass model in which the surface convection zone is slowly spun-down with time. Internal gravity waves then produce a front of 'slowness', that propagates from the core to the surface, resulting in a much reduced differential rotation at the solar age when compared with the one obtained using 'classical' rotational-mixing models (Fig. 4). The time-evolution of lithium destruction obtained in those models is consistent with Li measurements in solar-mass stars of various ages and in the Sun, using free parameters that are calibrated on massive stars.

1.8. *Solar asphericities*

As a different tool for analysis, solar-shape coefficients (c_n), also referred to as asphericities, reflect the internal non-homogeneous mass distribution and non-uniform angular velocity (Fig. 5, *left*). A careful inspection of the curve $c_n(r)$ reveals two breaks, of which one is the signature of the tachocline. The second, located around 0.995 $R_\odot$ was dynamically studied through the f modes, which probe the physical changes just below the photosphere (Lefebvre & Kosovichev 2005).

The study of asphericities, directly linked with solar gravitational moments is not only crucial for solar physics, but also for astrometry (when computing light deflection in the vicinity of the Sun), celestial mechanics (relativistic precession of planets, planetary orbit inclination and spin-orbit couplings) and for future tests of alternative theories of gravitation (correlation of J_2 with Post-Newtonian parameters).

According to the temporal variation of the f-mode frequencies, it is found that the very near solar surface is stratified in a thin double layer, interfacing the convective zone and the surface. This 'leptocline' is the seat of many phenomena: an oscillation phase of the seismic radius, together with a non-monotonic expansion of this radius with depth (Fig. 5, *right*), a change in the turbulent pressure (Fig. 6), likely an inversion in the radial gradient of the rotation velocity rate at about 50° in latitude, and the cradle of hydrogen and helium ionisation processes (Lefebvre *et al.* 2006).

2. Local helioseismology, magnetic activity and mode physics

Local helioseismology has developed rapidly in the last few years, offering a three-dimensional view of the Sun. Thanks to high-resolution, uninterrupted observations par-

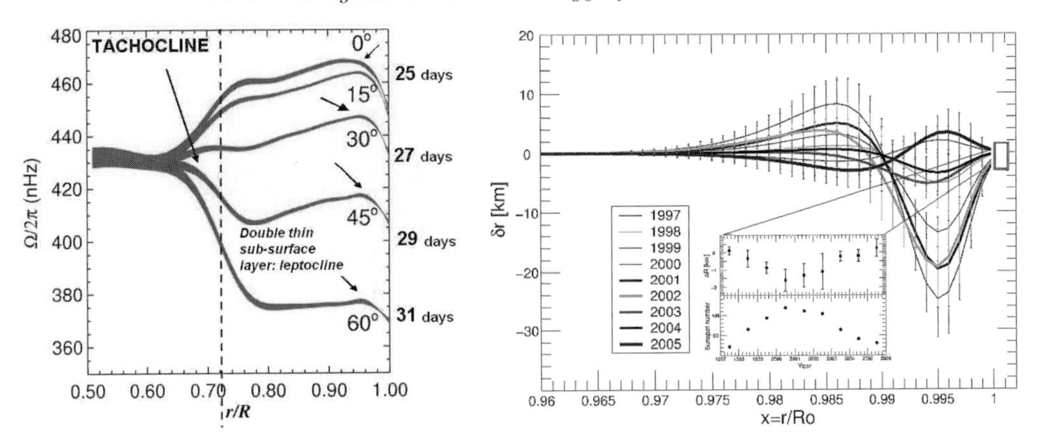

Figure 5. *Left*: The velocity rotation rate indicates a break near the surface, from Howe (2003). *Right*: *f*-mode analysis show a non-monotonic expansion of the solar radius with depth and a phase change with activity. From Lefebvre & Kosovichev (2005).

ticularly from MDI onboard *SOHO* and GONG++, it has become possible to image convective structures, sunspots, and active regions below the solar surface. Progress was reviewed by L. Gizon and A.G. Kosovichev. A particular challenge for the correct interpretation of local-helioseismic data in active regions, for example, is the theoretical modelling of the interaction of waves with the magnetic field (see below). Another important area of advance is in numerical simulations pertinent to understanding the flows of the upper convection zone and the interaction of acoustic waves with these flows. R. Stein reviewed work on a realistic solar surface convection simulation on supergranulation scales (48 Mm wide by 20 Mm deep), whose duration is currently 48 hr. There is a rich spectrum of *p*-modes excited in the simulation, and the data set is available for studying solar oscillations and local helioseismic inversion techniques.

2.1. *Time-varying component of the meridional circulation*

At the Sun's surface, the average motion in meridional planes is from the equator to the poles with a maximum amplitude of 10 - 20 m s^{-1} near 25° latitude. Temporal variations in the longitudinal average of the meridional flow have been reported by several authors using direct Doppler measurements or correlation tracking of small photospheric magnetic features. Only local helioseismology can provide measurements of the subsurface meridional circulation (Giles *et al.* 1997). It is known that the time-varying component of the meridional flow with respect to a long-term average does not exceed ± 5 m s^{-1} and is consistent with a small near-surface inflow toward active latitudes (Basu & Antia 2003; Gizon 2004; Zhao & Kosovichev 2004; González Hernández *et al.* (2006); Komm *et al.* 2006) and an outflow from active latitudes at depths greater than 20 Mm (Chou & Dai 2001; Beck *et al.* 2002; Chou & Ladenkov 2005).

Recently, Gizon & Rempel (2006) obtained independent measurements of the temporal variations of the meridional circulation near the solar surface and at a depth of about 60 Mm. These measurements confirm the previous observations listed above. In addition, the observations were compared with Rempel's theoretical model (Section 1.4) based on a flux-transport dynamo combined with a geostrophic flow caused by increased radiative loss in the active region belt, according to Spruit's (2003) original idea. The model is qualitatively consistent with the observations, in particular concerning the phase of the solar-cycle variations of the flows near the surface and 60 Mm below. Near the surface, the model is agreement with the data: the relative amplitudes of torsional oscillation

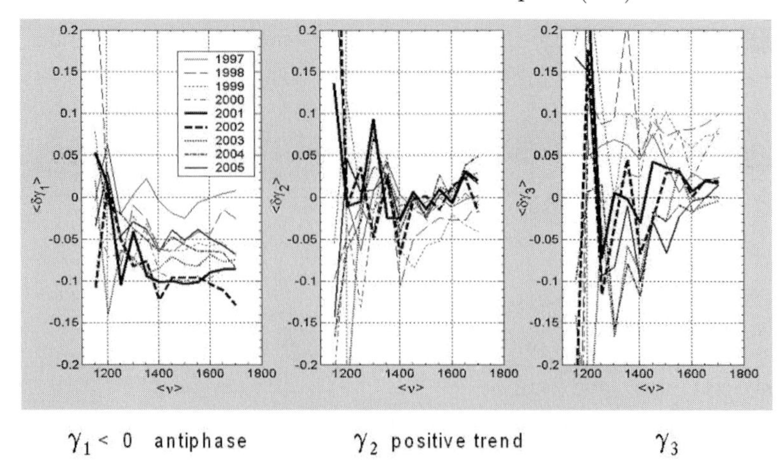

Figure 6. Three first asphericities parameters γ, i.e., even-a coefficients of f-modes. The layer around 0.995 $R_\odot$ is the seat of numerous physical changes. From Lefebvre *et al.* (2006).

and the time-varying component of the meridional flow and their relative phase are well reproduced. Deeper in the interior, it appears that the model underestimates the amplitude of the time variations of the meridional flow by nearly an order of magnitude; however, the flow variation is in antiphase to the surface flow as seen in the data. Overall, it is fair to say that the model is encouraging. We note that a local treatment of the regions of strong magnetic-field concentrations (sunspots and active regions) might be necessary to obtain a better match between the model and the data (see Gizon 2004).

2.2. *Magnetohelioseismology*

It is well known that sunspots absorb energy from and increase the phase speed of f- and p-modes incident upon them (Braun 1995). When this was first observed it was hoped that this information could be used to probe the subsurface and internal structure of sunspots. However, interpretation of the observations was difficult. It was only recently that the likely mechanism causing each of these effects was identified (although the idea was originally put forward some time ago by Spruit 1991). Models based on the conversion of acoustic oscillations to magnetohydrodynamic waves within sunspots, although simplistic in several respects, can now explain the absorption and phase speed changes simultaneously (Cally *et al.* 2003; Crouch *et al.* 2005). This confirms that mode conversion is the most promising absorption mechanism with wave energy channeled both up and down the magnetic field inside sunspots. It was also discovered that field inclination is a vital ingredient for mode conversion to work efficiently (Crouch & Cally 2003, 2005).

Ray conversion theory, as developed by Cally (2006) and Schunker & Cally (2006), has provided several new insights into the problem of how solar oscillations interact with magnetic field. Under that formalism, mode conversion is evident as ray splitting. Ray conversion theory has shown that fast ray travel times can be reduced by several minutes in comparison to the nonmagnetic acoustic ray in the same solar model, with obvious helioseismic implications. In addition, it has also shown that fast to slow mode conversion is very sensitive to the angle of attack that the wavevector makes with the magnetic field at the equipartition layer, where the sound speed and Alfvén speed coincide. A major consequence of this insight is that the magnetic field acts as a filter, preferentially allowing through acoustic signal from a narrow range of incident directions. This directional filtering effect has been confirmed by three dimensional wave mechanical calculations, which show that the upward flux of acoustic energy is indeed greatest when the propagation direction is aligned with the magnetic field (which is analogous to a fine attack angle).

With the most recent wave mechanical models now accounting for wave propagation in both the solar interior and atmosphere, we can predict how strong magnetic fields influence the polarisation of the velocity signal at heights where it is typically observed (such as at the formation height of the 676.8 nm Ni I spectral line used by *SOHO*-MDI and GONG). It turns out that the eccentricity and inclination of the principal axis of the velocity ellipse are very sensitive to the ratio of sound and Alfvén speeds at a given height. At great height in the atmosphere, where the Alfvén speed greatly exceeds the sound speed, the fast and Alfvén waves are evanescent; whereas the slow mode (which dominates the velocity signal at these levels) is a field guided travelling acoustic wave and causes the velocity ellipse to be highly eccentric and field aligned. At lower levels, where observations are typically made (and the sound and Alfvén speeds are comparable), the fast and Alfvén waves also contribute to the velocity signal and the slow mode is less dominant, resulting in a more complex pattern. In the near future we aim to compare these model predictions with observations of the surface velocity (e.g., Schunker *et al.* 2005).

2.3. *Seismic diagnostics inferred from amplitudes of stochastically-excited modes*

Solar-like oscillations in main-sequence stars, as well as in some red giant stars, are excited by turbulent convection in the strongly superadiabatic outermost convective-zone boundary layer. From the measurement of the amplitude and line-width of resonant modes, it is possible to infer the power supplied to the mode by turbulent convection, which in turn provides a unique probe of turbulent convection in stars. Such measurements have been used in helioseismology for some time now (e.g., Goldreich & Keeley 1977; Balmforth 1992; Samadi & Goupil 2001; Chaplin *et al.* 2005), and recently stellar measurements have become available from α Cen A from the ground (Bedding *et al.* 2004; Kjeldsen *et al.* 2005) and from space (*WIRE*) (Schou & Buzasi 2001; Fletcher *et al.* 2006). Recent developments have been reviewed by Houdek (2006) and Samadi *et al.* (2007).)

The model of Samadi & Goupil (2001), with the improvements proposed by Belkacem *et al.* (2006b), was presented in depth. Such a model requires a prescription of the temporal-correlation between the turbulent elements, which is obtained from a 3D simulation of the upper part of the solar convective zone, e.g., Samadi *et al.* (2003a). At large scales, this seems to be better fitted by a lorentzian than a gaussian, and the lorentzian fit appears to also agree better with the excitation rates inferred from helioseismic observations by Chaplin *et al.* (1998). The question remains as to whether this result remains valid for other stars. The model also requires a prescription for the fourth-order moments involving the entropy fluctuations and the turbulent velocity. The quasi-normal approximation, which consists in splitting the fourth-order moments into the product of two second-order moments, provides a convenient starting point, but it is significantly biased in the solar convective zone (Belkacem *et al.* 2006a) leading to an under-estimate of the p-mode excitation rate. More sophisticated closure models can be used, for example the two-scale mass-flux model (Gryanik & Hartmann 2002) which takes the asymmetries in the medium into account, but it is only applicable for quasi-laminar flows. Belkacem *et al.* (2006a) have generalized the approach by taking the turbulent properties of the medium into account, introducing plumes dynamics following Rieutord & Zahn (1995). As shown by Belkacem *et al.* (2006b), calculations based on this new closure model increase the contribution of the Reynolds stress to the excitation rates of the solar modes, and when the additional contribution due to entropy fluctuations is included, the new theoretical calculations fit rather well the maximum in the solar mode excitation rates derived by Baudin *et al.* (2005). The mode amplitudes and lifetimes from a significant sampling of stars that should be forthcoming from *CoRoT* will enormously improve the observational contraints on these models, and help us better understand the exitation of solar modes as well.

3. A new era of asteroseismology

Asteroseismology is at a particular exciting juncture with excellent new ground-based observations and with the recent launch of the satellite mission *CoRoT*. One mission already successfully launched and producing exciting results is *MOST*, which was reviewed by the mission's PI, J. Matthews. Much of what was presented at the meeting is summarized below. Solar-like oscillations have now been detected in quite a few main-sequence stars and sub-giants, as well as in red giants. One goal of asteroseismology is to study the internal rotation of other stars, as has been achieved for the Sun by helioseismology. M.-J. Goupil reviewed the challenges of seismic detection of rotation in other stars, and illustrated these with the results from a Hare-and-Hounds excercise carried out by the *CoRoT* team.

3.1. *An introduction to asteroseismology*

Helioseismology has provided very detailed and precise information about one specific star, the Sun. This has served as a crucial test of the theory of stellar internal structure and dynamics and has raised new challenging questions, such as the evolution of the rotation rate in the solar radiative interior, the origin of the differential rotation in the solar convection zone and, recently, the consequences of the new determinations of the solar surface abundances. However, studies of just one star, regardless of their quality, are evidently insufficient as a basis for understanding the broad range of phenomena occurring in stars of all masses over their eventful lives. Fortunately, stars throughout the HR diagram, from red supergiants to white dwarfs, show pulsations and hence offer at least some possibilities of asteroseismic studies.

In many cases, particularly near and below the main sequence, a sufficient number of modes are observed to allow, at least in principle, detailed constraints on stellar properties. Accurate determinations of stellar masses, radii and evolutionary states are of obvious interest; however, a more fundamental goal is to obtain data of sufficient quality to allow tests of the physical properties of stellar interiors and indications of the required improvements to our understanding of these properties.

Two general classes of modes are observed in stars. In p-modes, the dominant restoring force is pressure and the modes have the nature of standing acoustic waves. Such modes dominate the observed spectrum of solar oscillations. In g-modes, which are essentially standing internal gravity waves, buoyancy dominates the restoring force; this obviously requires departures from spherical symmetry and hence g-modes are only found with non-zero spherical harmonic degree (ℓ). In unevolved stars there is a clear separation between these two classes. However, in evolved stars with strong internal gradients in the chemical composition or large gravitational acceleration in a compact core, the characteristic buoyancy frequency of gravity waves becomes so high that the modes may have a mixed nature, behaving like a g-mode in the deep interior of the star and a p-mode in the external layers. Such mixed modes have very considerable diagnostic value.

For the foreseeable future almost all observations of stellar oscillations will take place in light integrated over the stellar disk. This strongly suppresses modes of degree ℓ exceeding three and hence the data are dominated by low-degree modes; this is particularly

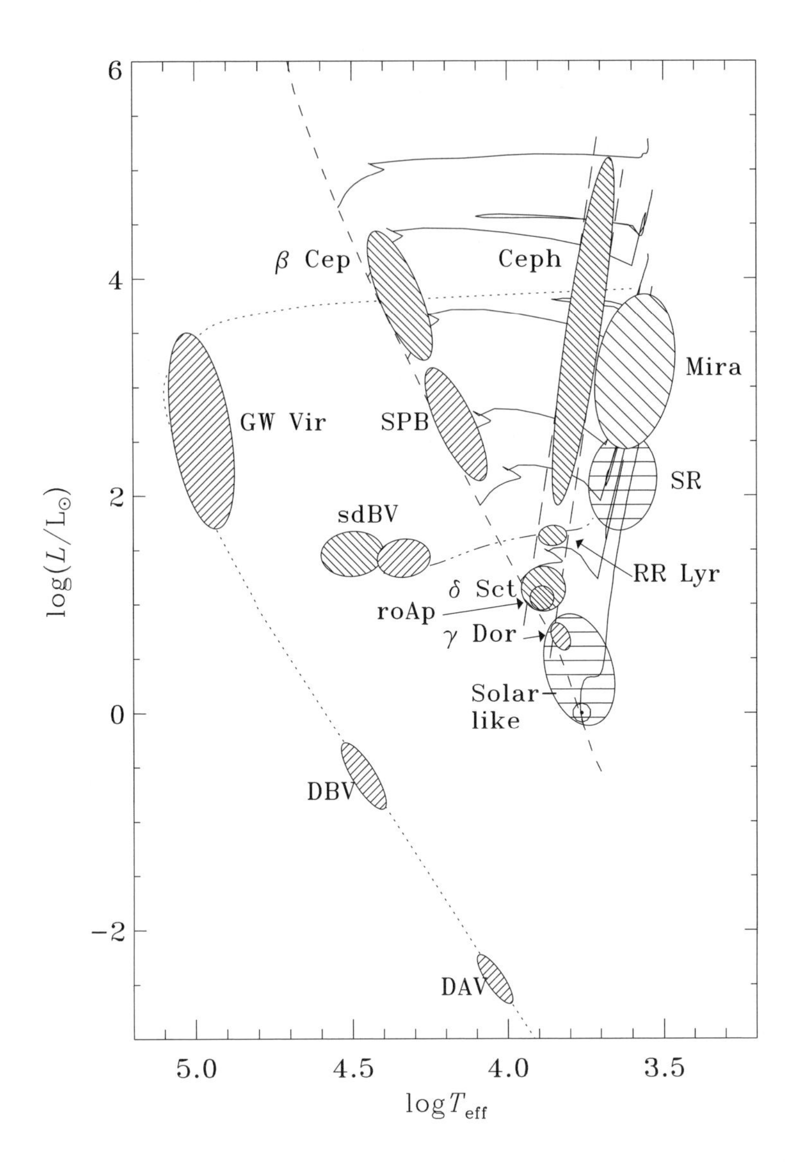

Figure 7. Hertsprung-Russell diagram showing the location of the principal types of asteroseismic variables stars discussed below.

true for solar-like oscillations where the stochastic excitation leads to true amplitudes that on average are largely independent of ℓ. Compared with the extensive solar data this is obviously a strong restriction; however, fortunately, amongst the p-modes it is precisely the low-degree modes which penetrate most deeply into the star and hence provide information about its core properties.

Which modes are actually excited to observable amplitudes in a given star obviously depends on the relevant excitation mechanisms. A mode can either be self-excited or intrinsically damped but excited by forcing external to the mode. In the former case a critical region of the star operates as a heat engine, being heated in the compressional phase of the oscillation and cooled in the expansion phase. Such a region is typically associated with specific features in the opacity (κ). To cause overall excitation the region has to be placed at the appropriate depth within the star. This constraint leads to relatively well-defined instability areas in the HR diagram. In the Cepheid instability belt, which extends to the δ Scuti stars near the main sequence, the opacity feature is associated with the second ionization of helium. An opacity feature at higher temperature, related to iron-group elements, causes instability in B-type stars near the main sequence and in subdwarf B stars on the extreme blue end of the horizontal branch. Convection may also cause heat-engine instability, typically at long periods; this is the case in the γ Doradus stars and some of the pulsating white dwarfs.

An unstable mode grows exponentially with time, according to linear theory. The final amplitude is determined by nonlinear effects. For large-amplitude pulsators such as Cepheids, which typically show a single mode, amplitude limitation probably occurs through a saturation of the driving mechanism when the amplitude becomes sufficiently high. In stars near the main sequence, where many modes are typically excited, the distribution of observed amplitudes amongst the unstable modes is generally highly irregular, and no definite mechanism determining this distribution has so far been identified. Also, the irregular selection of observable modes often greatly complicates the identification of the observed modes with modes of stellar models.

In most stars on the cool side of the Cepheid instability strip, including the Sun, the modes appear to be linearly stable, and the observed oscillations are caused by stochastic excitation, driven by near-surface convection of such stable modes. In this case the mode amplitudes are determined on average by the balance between the energy input from convection and the damping rate. The result is a typical distribution of amplitude, such as observed in the Sun where the largest amplitudes are found near periods of five minutes. Within this overall amplitude envelope most modes are excited, although with varying amplitudes due to the stochastic nature of the excitation. As a result it is generally possible to identify the observed modes, using also the typically regular structure of the frequency spectrum. Together with the generally substantial number of observed modes this makes solar-like oscillations of great interest to asteroseismology.

3.2. *Observational results for solar-type stars*

Observations of solar-like oscillations are accumulating rapidly, and measurements have now been reported for several main-sequence, subgiant, and giant stars. The following list includes the most recent observations and is ordered according to decreasing stellar density (which also corresponds to a decreasing large frequency separation (between modes of different radial order n):

- τ Cet (G8 V): T.C. Teixeira *et al.* (in prep.)
- 70 Oph, a visual binary: Carrier & Eggenberger (2006)
- α Cen A and B (see below)
- μ Ara, a planet-hosting star: Bouchy *et al.* (2005)
- HD 49933, a potential *CoRoT* target: Mosser *et al.* (2005)
- β Vir (F9): Martić *et al.* (2004b); Carrier *et al.* (2005b)
- Procyon A (see below)
- β Hyi, a G2 subgiant: Bedding *et al.* (2007)
- δ Eri, a K0 subgiant: Carrier *et al.* (2003)
- η Boo (see below)
- ν Ind, a metal-poor subgiant: Bedding *et al.* (2006); Carrier *et al.* (2007)
- η Ser (K0 III): Barban *et al.* (2004)
- ξ Hya (G7 III): Frandsen *et al.* (2002); Stello *et al.* (2006)
- ϵ Oph (G9 III): De Ridder *et al.* (2006)

On the main sequence, the most spectacular results have been obtained for the α Cen system. In the A component, more than 40 modes have been measured, with angular degrees of $\ell = 0$ to 3 – see Bouchy & Carrier (2002) and Bedding *et al.* (2004). The mode lifetime is about 2 - 4 days and there is now evidence of rotational splitting from photometry with the *WIRE* satellite analysed by Fletcher *et al.* (2006) and also from ground-based spectroscopy (M. Bazot *et al.*, in prep.). Meanwhile, nearly 40 modes have been measured in – see Carrier & Bourban (2003) and Kjeldsen *et al.* (2005).

Oscillations in the bright G0 subgiant η Boo were measured by two independent groups using ground-based spectroscopy: Kjeldsen *et al.* (2003) and Carrier *et al.* (2005a). Some of those frequencies were seen in spaced-based photometry by the *MOST* satellite, as reported by Guenther *et al.* (2005), but those authors also claimed to detect oscillations at very low frequencies, the reality of which remains controversial.

The star Procyon A also generated controversy when *MOST* data reported by Matthews *et al.* (2004) failed to reveal oscillations that were claimed from ground-based data. For the latter, see Martić *et al.* (2004a) , Bouchy *et al.* (2004), and Claudi *et al.* (2005) for the most recent examples. However, Bedding *et al.* (2005) argued that the *MOST* non-detection was consistent with the ground-based data.

In the future, we expect further ground-based observations using Doppler techniques (for example, a multi-site campaign on Procyon has been organized for January 2007). The new spectrograph SOPHIE at l'Observatoire de Haute-Provence in France should be operating very soon (`<http://www.obs-hp.fr/>`). From space, the *WIRE* and *MOST* satellites continue to return data and we look forward with excitement to the results from *CoRoT* (launched December 2006) and *Kepler* (to be launched in 2008).

Although remarkable, the results already obtained on solar-like oscillations are not sufficient to allow the desired detailed study of the internal properties of the stars. Simulations have shown that a frequency precision of the order of 0.1 μHz is required for detailed investigations of stellar cores, e.g., through inversion. A similar precision is required to extract the signatures of the base of the convective envelope and hence determine its depth, as well as to infer the envelope helium abundance from the effect of helium ionization on the sound speed. Also, such a frequency resolution is required to reliably measure the rotational splitting caused by slow rotation as in the Sun. To achieve this precision and resolution observations extending over at least several months are required. Mode identification itself remains a significant challenge as well. Furthermore, it is of great interest to extend the observations towards lower frequencies; here solar-like oscillations have long lifetimes, allowing high precision in the frequency determination, and are more sensitive to features such as the helium ionization. On the other hand, the amplitudes of such modes tend to be low, placing stringent constraints on the allowable noise level, both observational and stellar.

3.3. *Theory developments: dynamics of stellar radiative zones*

With the dramatic advances of helioseismology and of asteroseismology, a coherent picture of the evolution of rotating stars and of their internal dynamical processes is needed. In fact, rotation, and more precisely differential rotation, has a major impact on the internal dynamics of stars. First, as is known from the theory of rotating stars, rotation induces some-large scale circulations, both in radiation and convection zones, which act to transport simultaneously the angular momentum, the chemicals but also the magnetic field by advection. In radiation zones, the large-scale circulation, which is called the meridional circulation, is due to the differential rotation, to the transport of angular momentum and to the action of the perturbing forces, namely the centrifugal force and the Lorentz force (cf., Busse 1982; Zahn 1992; Maeder & Zahn 1998; Garaud 2002a; Rieutord 2006). Next, the differential rotation induces hydrodynamical turbulence in radiative regions through various instabilities: the secular and the dynamical shear instabilities, the baroclinic and the multidiffusive instabilities. In the same way that the atmospheric turbulence in the terrestrial atmosphere, it acts to reduce the gradients of angular velocity and of chemical composition and thus, it is modelled as a diffusive process (cf., Talon & Zahn 1997; Garaud 2001; Maeder 2003). On the other hand, rotation has a strong impact on the stellar magnetism. For example, it interacts with turbulent convection in convective envelopes of solar-type stars (cf., Brun *et al.* 2004) to lead to a dynamo mechanism and, as it is expected from observations, to a cyclic magnetism. In radiation regions, it interacts with fossil magnetic fields where the secular torque of the Lorentz force and the magneto-hydrodynamical instabilities such as the Tayler-Spruit instability and the multidiffusive magnetic instabilities have a strong impact on the transport of angular momentum and of chemicals (cf., Charbonneau & MacGregor 1993; Garaud 2002a; Spruit 1999; Spruit 2002; Menou *et al.* 2004; Maeder & Meynet 2004; Eggenberger *et al.* 2005; Braithwaite & Spruit 2005; Braithwaite 2006; Brun & Zahn 2006). Finally, waves constitute the last transport process in single stars where they are also interacting with rotation. Internal waves, which are excited at the borders with convective zones, propagate inside radiation zones where they extract or deposit angular momentum where they are damped leading to a modification of the angular velocity profile and thus of the chemicals distribution (cf., Goldreich & Nicholson 1989; Talon *et al.* 2002 ; Talon & Charbonnel 2003, 2004, 2005; Rogers & Glatzmaier 2005). Note also that rotation modifies stellar winds and mass losses (cf., Maeder 1999).

On the other hand, in closed binary systems, where the companion could be a star as well as a planet, there are transfers of angular momentum between the star, its companion and the orbit due to the dissipation acting on flows induced by the tidal potential; that could be the equilibrium tide (cf., Zahn 1966) due to the hydrostatic adjustement of the star or the dynamical tide which is due to the tidal excitation of internal waves (cf., Zahn 1975. This dynamical evolution modifies the internal rotation of each component that have consequences on the properties of their internal transport.

To conclude, all the processes, with which rotation interacts, transport angular momentum and matter that modifies the internal angular velocity, the chemical composition and the nucleosynthesis. Therefore, differential rotation has imperatively to be taken into account to get a coherent picture of the internal dynamics and the evolution of the stars. To achieve this goal, advances in the secular magnetohydrodynamics of stellar radiation zones have been made.

First, rotational transport of type I – where the angular momentum and the chemical species are transported by the meridional circulation and by the hydrodynamical turbulence due to shear instabilities – has been studied. Its present modelling has been generalized to treat simultaneously the bulk of radiation zones and their interfaces with convective zones, the tachoclines (cf., Mathis & Zahn 2004). Next, a new prescription for the horizontal turbulent transport has been obtained. It has been derived from Couette-Taylor laboratory experiments that enable the study turbulence in differentially rotating flows (cf., Mathis *et al.* 2004). However, the introduction of these two hydrodynamical mechanisms in stellar models leads to results which fail to reproduce the observations of solar-type stars, because these have been slowed down by the wind during their evolution and hence the rotational processes are less efficient. Therefore, rotational transport of type II has been introduced. The chemical species are still transported by meridional circulation and turbulence, but angular momentum is carried by an another process: the two candidates being the magnetic field and internal waves. First, the effects of a fossil magnetic field have been introduced in a consistent way. The action of turbulence,

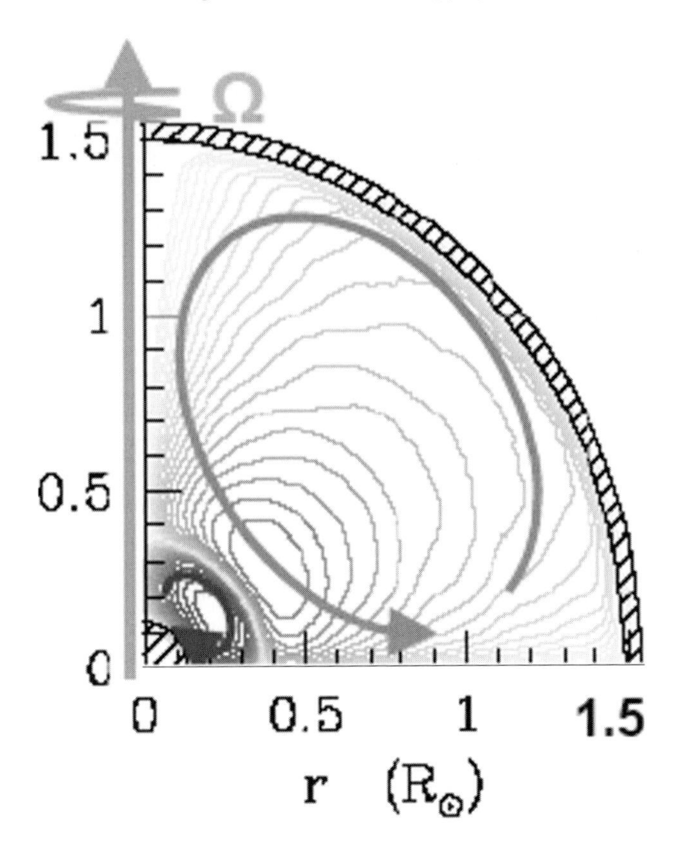

Figure 8. Meridional circulation currents in a 1.5 $M_\odot$ star with a solar metallicity and $v_{\rm ini} = 100\,{\rm km\,s^{-1}}$. The age is 7.604×10^8 years with a central hydrogen mass fraction $X_c = 0.57$. In this model, the outer cell is turning counterclockwise allowing the equatorial extraction of angular momentum by the wind.

differential rotation, and meridional circulation on the field, and also its feed-back on momentum and heat transport, are taken into account (cf., Mathis & Zahn 2005). Next, the effects of the Coriolis force have been introduced in the modelling of internal waves (cf., Mathis & Zahn 2005). This allows the inclusion of gravito-inertial waves in the description of the angular momentum transport. Finally, a coherent treatment of tidal processes has been derived.

On the other hand, numerical simulations, based on those theoretical results which allow the following in 2D of the secular hydrodynamics of rotating stars, have been developed with the associated diagnosis tools to identify the dominant processes in angular momentum transport, meridional circulation, and chemical species mixing. The first step – where we assume that the anisotropic turbulence in stellar radiation zones enforces a shellular rotation law $\overline{\Omega}(r)$ – has been achieved. Work is now in progress to implement in stellar evolution codes the differential rotation in latitude and the transport by the magnetic field and the gravito-inertial waves, those two last processes being crucial to understanding the internal angular momentum transport in the Sun and the properties of low-mass stars. Moreover, theoretical work is now underway on the description of MHD instabilities and of the internal-wave excitation to improve our description of stellar radiation zones in support of helioseismology and of asteroseismology.

3.4. *A new era of asteroseismology: hot stars*

The theoretical g-mode frequency spectrum is by far the most dense and their periods are the longest, making g-mode pulsators more difficult to study than p-mode pulsators. During the last few years, big steps forward have been made for the 'hot' stars, i.e., pulsating stars situated along the main sequence which are hotter than solar type stars.

The β Cep and δ Sct stars are early-B and mid-A to early-F type stars, respectively, pulsating in low order p- and g-modes with periods of the order of hours. The modes are excited by the κ mechanism acting on a partial ionization zone of metals (Fe) and helium (He), respectively. Currently, more than 100 β Cep and 1200 δ Sct stars are known. The multi-periodic ones have beat-periods of weeks up to months. Recently, a magnetic field of a few hundred Gauss has been detected for three β Cep stars (Henrichs *et al.* 2000; Neiner et al 2003b; Hubrig et al 2006). The best studied β Cep star is ν Eri (HD 29248), for which large scale multi-site campaigns have been organized in 2002 - 2003 (Handler *et al.* 2004; Aerts et al 2004; De Ridder *et al.* 2004) and 2003 - 2004 (Jerzykiewicz *et al.* 2005). A total of 14 independent frequencies are found, including the radial fundamental ($\ell = 0$, p_1), three rotationally-split triplets ($\ell = 1$, g_1, p_1, and p_2), and two unidentified high-order g-modes. Asteroseismic modeling of β Cep stars based on a few identified modes (including the radial fundamental or first overtone) has led to evidence for convective core overshoot and in two cases for non-rigid rotation (the core rotates faster than the surface layers). For the observed modes of ν Eri to be unstable in the theoretical models, a metal enhancement in the whole interior (Ausseloos *et al.* 2004) or in the driving region only (Pamyatnykh *et al.* 2004) has been assumed.

Currently, FG Virginis is the best studied δ Sct star, for which multi-site campaigns have been organized within the Delta Scuti Network between 1992 and 2004. So far, 67 independent frequencies have been detected in the photometric data, which confirms the theoretical prediction of a large number of unstable modes (Breger *et al.* 2005). This includes a lot of high frequencies with an amplitude below 0.2 mmag that cannot be explained with unstable low ℓ modes (Daszyńska-Daszkiewicz *et al.* 2006) and close frequency pairs around the expected positions of radial modes (Breger & Pamyatnykh 2006). Thirteen of these frequencies have been confirmed in spectroscopic data. The mode identification reveals: (*i*) a lot of axisymmetric modes, including at least two radial ones; (*ii*) a rotationally-split $\ell = 1$ mode; and (*iii*) evidence for insufficient convection (Viskum *et al.* 1998; Breger *et al.* 1999; Mantegazza & Poretti 2002; Daszyńska-Daszkiewicz *et al.* 2005; Zima *et al.* 2006;). These results are used as input for asteroseismic modeling, which is ongoing at the moment.

The slowly pulsating B (SPB) and γ Dor stars are mid-B to late-B and early-F type stars, respectively, pulsating in high-order g-modes with periods of the order of days. The modes are excited by the κ mechanism acting on the partial ionization zone of metals (Fe) and by a flux blocking mechanism at the base of the convective envelope, respectively. For both classes, more than one hundred (candidate) members are known up to now. The multi-periodic ones have beat-periods of months up to years. Recently, a magnetic field of a few hundred Gauss has been detected for 14 SPB stars (Neiner *et al.* 2003a; Hubrig *et al.* 2006). HD 160124 holds the record of the largest number of detected frequencies (8) for an SPB star based on ground-based observations (Waelkens 1991), while 21 frequencies were detected in the 37 days of white light observations obtained with the *MOST* satellite for the new SPB star HD 163830 (Aerts *et al.* 2006). The highest frequency is a known artefact of the satellite and the two lowest frequencies could be associated with the star's rotational frequency, but the remaining 18 frequencies can easily be interpreted in terms of unstable low-degree g-modes. An accurate mode identification is lacking which prevents us from further digging into the seismic models.

For the γ Dor stars, we are still in the inventory stage. Large-scale ground-based observation campaigns have been organized both in the northern (Kaye *et al.* 1999; Henry *et al.* 2001; Henry & Fekel 2002; Henry & Fekel 2003; Fekel *et al.* 2003; Mathias *et al.* 2004; Henry & Fekel 2005; Henry *et al.* 2005) and southern (Eyer *et al.* 2002; De Cat *et al.* 2006) hemisphere to search for members of this class. Large steps forward are achieved by the inclusion of time-dependent convection in the models which enables to calculate the instability of the modes and the theoretical amplitude ratios and phase shifts that can be used for mode identification (Grigahcène *et al.* 2005; Dupret *et al.* 2005). The frequency ratio method has led to promising results for γ Dor stars. It is based on the first-order asymptotic g-mode expression and enables the selection of seismic models that are compatible with the observations as soon as three frequencies are observed (Moya *et al.* 2005). In case of 9 Aur, a unique model has been found (Moya *et al.* 2006).

The potential of seismology of 'hot' stars is excellent, and for the hybrid stars it is even better, since both p- and g-modes are excited. The currently known (candidate) β Cep/SPB hybrids are ι Her (Chapellier *et al.* 2000), 53 Psc (Le Contel *et al.* 2001), ν Eri (Jerzykiewicz *et al.* 2005), γ Peg (Chapellier *et al.* 2006), and HD 13745 & HD 19374 (De Cat *et al.* 2006), while HD 209295 (Handler *et al.* 2002) and HD 8801 (Henry & Fekel 2005) are δ Sct/γ Dor hybrids. For the future, we need: (*i*) accurate mode identifications, especially for g-modes; (*ii*) the inclusion of magnetic fields, diffusion, and rotation in theoretical models; and (*iii*) data from multi-site campaigns and/or from satellite missions like *MOST*, *CoRoT*, and *Kepler*.

4. Future activities

The future for the fields of helioseismology and asteroseismology is bright. Numerous new initiatives and forthcoming activities and missions were discussed at the meeting. Several of these, including NASA's *Solar Dynamics Observatory* (expected launch data 2008) and the proposed *DynaMICS* mission in the field of helioseismology, and *CoRoT* (launched December 2006) in the area of asteroseismology, have been mentioned already above.

Solar Orbiter, ESA's next mission to study the Sun, is scheduled to be launched in May 2015 according to the current baseline (December 2006). The extended mission will be completed in January 2024. The most interesting aspects of the mission for helioseismology reside in the unique vantage points from which the Sun will be viewed (Gizon 2007 Gizon 2007; Woch & Gizon 2007). The spacecraft will use multiple gravity-assist manoeuvres at Venus and Earth to reach its science orbit after a cruise phase of about 3.4 years. The orbit design will include two main characteristics, both of which offer novel perspectives for helioseismology. First, *Solar Orbiter* will make observations away from the ecliptic plane to provide views of the Sun's polar regions. The inclination of the spacecraft's orbit to the Ecliptic will incrementally increase at each Venus swing-by manoeuvre to reach at least 30° toward the end of the mission. Second, *Solar Orbiter* will cover a large range of spacecraft-Sun-Earth angles. In combination with data collected from the ground or near-Earth orbit, *Solar Orbiter* will thus mark the advent of steroscopic helioseismology. One important goal is to gain a better understanding of solar activity and variability by probing the solar interior to higher latitudes and larger depths, beyond what can be achieved with Earth-side observations alone.

The SIAMOIS fourier tachometer (Seismic Interferometer Aiming to Measure Oscillations in the Interior of Stars; <http://siamois.obspm.fr/>) is planned for Dome C in

Antarctica . The Antarctic site presents the unique advantage of requiring only a single instrument for the continuous observation of a target over several months. This post-CoRoT instrument, based on an interferometric method to measure the Doppler shifts induced by the p-modes, should be put in operation during the 2011 austral winter with this objective. Two bright stars could be observed simultaneously with two robotic 40-cm telescopes feeding the interferometer through a fiber. There are about ten southern, circumpolar solar-type stars observable, with among them α Cen A and B, with such small telescopes. With future developments on the site, this type of instrument would find its place behind a larger telescope to conduct similar long observations on fainter stars.

The NASA *Kepler* mission, planned for launch in November 2008, will provide extensive asteroseismic data. *Kepler* will survey of the order of 100 000 stars in a 105-square-degree field in Cygnus, nearly continuously for more than four years. The goal is to detect extra-solar planetary systems through the observation of slight decreases in the luminosity of the stars caused by planetary transits. The bulk of the stars will be observed with a cadence of 30 minutes. However, 512 selected stars will at any given time be observed with a one-minute cadence, hence for example allowing the study of solar-like oscillations. Each such star will be observed for at least three months, and in some cases observations of a given star continuously or repeatedly throughout the mission may be warranted, e.g., to search for frequency variations associated with cyclic activity. For stars of magnitude between 9 and 11 it is estimated that detailed asteroseismic investigations will be possible; for fainter stars simulations indicate that it will be possible to determine the large and small frequency separations, hence providing measures of stellar radii, and evolutionary states. Also, even the longer-cadence data are of great asteroseismic interest for the study of solar-like oscillations in giant stars or, for example, long-period oscillations in γ Doradus or slowly pulsating B-type stars.

The *CoRoT* and *Kepler* missions, as well as the formerly proposed *Eddington* mission, highlight the benefits of combining observations for asteroseismology and exo-planets in a single mission. The *CoRoT* mission will spend five months on each of five fields of 10 square degrees. The asteroseismology component will monitor the light curves of a total of 50 target starst with magnitudes in the range 5 to 9 with a cadence of 32 seconds, yielding a precision on the frequencies of order 0.1 μHz. The planet search component will monitor of the order of 10,000 stars in each field with magnitudes in the range 11 to 16; 1000 with a cadence of 32 seconds the rest with a cadence of 512 seconds. The planet search data will also be available for asteroseismology studies.

Looking to the future, *PLATO* is a new mission concept being developed by a European consortium to be submitted to the European Space Agency for inclusion in the new Cosmic Vision programme. The current concept is to have an aligned cluster of small telescopes with very wide fields of view which will continuously monitor of the order of 100,000 stars for both asteroseismology and planet finding. The concept is currently undergoing industrial studies.

Finally, there are ambitious plans to build SONG (Stellar Oscillations Network Group) (<http://astro.phys.au.dk/SONG>), which will be a global network of small telescopes equipped with high-resolution spectrographs and dedicated to asteroseismology and planet searches.

References

Aerts, C., De Cat, P., Handler, G., *et al.* 2004, *MNRAS*, 347, 463
Aerts, C., De Cat, P., Kuschnig, R., *et al.* 2006, *ApJ* (Letters), 642, L165
Altrock, R., & Howe, R. 2004, *BAAS*, 36, 1411

Andersen, B. 1996, *A&A*, 312, 610
Ausseloos, M., Scuflaire, R., Thoul, A., & Aerts, C. 2004, *MNRAS*, 355, 352
Balmforth, N. J. 1992, *MNRAS*, 255, 639
Barban, C., De Ridder, J., Mazumdar, A., *et al.*, 2004, in: D. Danesy (ed.), Proc. SOHO 14/GONG 2004 Workshop, *Helio- and Asteroseismology: Towards a Golden Future*, ESA SP-559, p. 113
Basu, S., & Antia, H. M. 2001, *MNRAS*, 324, 498
Basu, S., & Antia, H. M. 2003, *ApJ*, 585, 553
Baudin, F., Samadi, R., Goupil, M.-J., *et al.* 2005, *A&A*, 433, 349
Beck, J. G., Gizon, L., & Duvall, T. L. 2002, *ApJ* (Letters), 575, L47
Bedding, T. R., Kjeldsen, H., Arentoft, T., *et al.* 2007, *ApJ*, 663, 1315
Bedding, T. R., Butler, R. P., Carrier, F., *et al.* 2006, *ApJ*, 647, 558
Bedding, T. R., Kjeldsen, H., Bouchy, F., *et al.* 2005, *A&A* (Letters), 432, L43
Bedding, T. R., Kjeldsen, H., Butler, R. P., *et al.* 2004, *ApJ*, 614, 380
Belkacem, K., Samadi, R., Goupil, M., & Kupka, F. 2006a, *A&A*, 460, 173
Belkacem, K., Samadi, R., Goupil, M., Kupka, F., & Baudin, F. 2006b, *A&A*, 460, 183
Bouchy, F., Bazot, M., Santos, N. C., Vauclair, S., & Sosnowska, D. 2005, *A&A*, 440, 609
Bouchy, F., & Carrier, F. 2002, *A&A*, 390, 205
Bouchy, F., Maeder, A., Mayor, M., *et al.* 2004, *Nature*, 432, 7015
Braithwaite, J. 2006, *A&A*, 449, 451
Braithwaite, J., & Spruit, H. 2005, *Nature*, 431, 819
Braun, D. C. 1995, *ApJ*, 451, 859
Breger, M., Lenz, P., Antoci, V., *et al.* 2005, *A&A*, 435, 955
Breger, M., & Pamyatnykh, A. A. 2006, *MNRAS*, 368, 571
Breger, M., Pamyatnykh, A. A., Pikall, H., & Garrido, R. 1999, *A&A*, 341, 151
Browning, M., Miesch, M. S., Brun, A. S., & Toomre, J. 2006, *ApJ* (Letters), 648, 157
Brun, A. S. 2004, *Solar Phys.*, 220, 33
Brun, A. S. 2007, *AN* 328, 329
Brun, A. S., Miesch, M., & Toomre, J. 2004, *ApJ*, 1014, 1073
Brun, A. S., & Toomre, J. 2002, *ApJ*, 570, 865
Brun, A. S., & Zahn, J.-P. 2006, *A&A* 457, 665
Busse, F. H. 1982, *ApJ*, 259, 759
Cally, P. S. 2006, *Phil. Trans. Roy. Soc. Lond. A*, 364, 333
Cally, P. S., Crouch, A. D., & Braun, D. C. 2003, *MNRAS*, 346, 381
Carrier, F., & Bourban, G. 2003, *A&A* (Letters), 406, L23
Carrier, F., Bouchy, F., & Eggenberger, P. 2003, in: M. J. Thompson, M. S. Cunha & M. J. P. F. G. Monteiro (eds.), *Asteroseismology Across the HR Diagram* (Dordrecht:Kluwer), p. 311
Carrier, F., & Eggenberger, P. 2006, *A&A*, 450, 695
Carrier, F., Eggenberger, P., & Bouchy, F. 2005a, *A&A*, 434, 1085
Carrier, F., Eggenberger, P., D'Alessandro, A., & Weber, L. 2005b, *New Astron.*, 10, 315
Carrier, F., Kjeldsen, H., Bedding, T. R., *et al.* 2007, *A&A* 470, 1059
Chapellier, E., Le Contel, D., Le Contel, J. M., Mathias, P., *et al.* 2006, *A&A*, 448, 697
Chapellier, E., Mathias, P., Le Contel, J., *et al.* 2000, *A&A*, 362, 189
Chaplin, W. J., Elsworth, Y., Isaak, G. R., *et al.* 1998, *MNRAS* (Letters), 298, L7
Chaplin, W. J., Elsworth, Y., Isaak, G. R., *et al.* 2003, *ApJ* (Letters), 582, L115
Chaplin, W. J., Houdek, G., Elsworth, Y., *et al.* 2005, *MNRAS*, 360, 859
Charbonneau, P., & MacGregor, K. B. 1993, *ApJ* 417, 762
Charbonnel, C., & Talon, S. 2005, *Science*, 309, 2189
Chou, D.-Y., & Dai, D. 2001, *ApJ* (Letters), 559, L175
Chou, D.-Y., & Ladenkov, O. 2005, *ApJ*, 630, 1206
Claudi, R. U., Bonanno, A., Leccia, S., *et al.* 2005, *A&A* (Letters), 429, L17
Clune, T. L., Elliott, J. R., Glatzmaier, G. A., *et al.* 1999, *Parallel Comput.* 25, 361
Couvidat, S., Turck-Chièze, S., & Kosovichev, A. 2003a, *ApJ*, 599, 1434
Couvidat, S., Garcìa, R. A., Turck-Chièze, S., *et al.* 2003b, *ApJ* (Letters), 597, L77

Crouch, A. D., & Cally, P. S. 2003, *Solar Phys.* 214, 201
Crouch, A. D., & Cally, P. S. 2005, *Solar Phys.* 227, 1
Crouch, A. D., Cally, P. S., Charbonneau, P., *et al.* 2005, *MNRAS*, 363, 1188
Daszyńska-Daszkiewicz, J., Dziembowski, W. A., & Pamyatnykh, A. A. 2006, *Mem. S.A.It.*, 77, 113
Daszyńska-Daszkiewicz, J., Dziembowski, W. A., Pamyatnykh, A. A., *et al.* 2005, *A&A*, 438, 653
De Cat, P., Eyer, L., Cuypers, J., *et al.* 2006, *A&A*, 449, 281
De Ridder, J., Barban, C., Carrier, F., *et al.* 2006, *A&A*, 448, 689
De Ridder, J., Telting, J. H., Balona, L. A., *et al.* 2004, *MNRAS*, 351, 324
Dikpati, M., & Charbonneau, P. 1999, *ApJ*, 518, 508
Dikpati, M., & Gilman, P. A. 2001, *ApJ*, 559, 428
Dikpati, M., Cally, P. S., & Gilman, P. A. 2004, *ApJ*, 610, 597
Dupret, M.-A., Grigahcène, A., Garrido, R., Gabriel, M., & Scuflaire, R. 2005, *A&A*, 435, 927
Eggenberger, P., Maeder, A., & Meynet, G. 2005, *A&A* (Letters), 440, L9
Eyer, L., Aerts, C., van Loon, M., Bouckaert, F., & Cuypers, J. 2002, in: C. Sterken & D. W. Kurtz (eds.), *Observational Aspects of Pulsating B- and A Stars*, *ASP-CS* 256, 203
Fekel, F. C., Warner, P. B. & Kaye, A. B. 2003, *AJ*, 125, 2196
Fletcher, S. T., Chaplin, W. J., Elsworth, Y., Schou, J., & Buzasi, D. 2006, *MNRAS*, 371, 935
Frandsen, S., Carrier, F., Aerts, C., *et al.* 2002, *A&A* (Letters), 394, L5
Garaud, P. 2001, *MNRAS* 324, 68
Garaud, P. 2002, *MNRAS*, 329, 1
Garaud, P. 2002, *MNRAS*, 335, 707
García, R., Corbard, T., Chaplin, W. J., *et al.* 2004, *Solar Phys.*, 220, 269
García, R., Turck-Chize, S., Boumier, P., *et al.* 2005, *A&A*, 442, 385
García, R., Turck-Chièze, S., Jiménez-Reyes, S. J., *et al.* 2006, in: K. Fletcher & M. J. Thompson (eds.), *SOHO 18/ GONG 2006/ HELAS I, Beyond the Spherical Sun: Towards a Golden Future*, ESA SP-624, 23.1
Giles, P. M. 1999, PhD thesis, Stanford University, USA
Giles, P. M., Duvall, T. L., Scherrer, P. H., & Bogart, R. S. 1997, *Narture*, 390, 52
Gizon, L. 2004, *Solar Phys.*, 224, 217
Gizon, L. 2007, in: *The Second Solar Orbiter Workshop*, ESA SP-641, in press
Gizon, L., & Rempel, M. 2006, in: K. Fletcher & M. J. Thompson, (eds.), *SOHO 18/ GONG 2006/ HELAS I, Beyond the Spherical Sun: Towards a Golden Future*, ESA SP-624, 129.1
Goldreich, P., & Keeley, D. A. 1977, *ApJ*, 212, 243
Goldreich, P., & Nicholson, P. D. 1989, *ApJ*, 342, 1079
González Hernández, I., Komm, R., Hill, F., *et al.* 2006, *ApJ*, 638, 576
Grigahcène, A., Dupret, M.-A., Gabriel, M., Garrido, R., & Scuflaire, R. 2005, *A&A*, 434, 1055
Gryanik, V., & Hartmann, J. 2002, *J. Atmos. Sci.*, 59, 2729
Guenther, D. B., Kallinger, T., Reegen, P., *et al.* 2005, *ApJ* 635, 547
Handler, G., Balona, L. A., Shobbrook, R. R., *et al.* 2002, *MNRAS*, 333, 262
Handler, G., Shobbrook, R. R., Jerzykiewicz, M., *et al.* 2004, *MNRAS*, 347, 454
Henrichs, H. F., de Jong, J. A., Donati, D.-F., *et al.* 2000, in: Yu.V. Glagolevskij & I. I. Romanyuk (eds.), *Magnetic Fields of Chemically Peculiar and Related Stars*, Proc. Intern. Meeting Special Astrophysical Observatory, September 23-27, 1999, p.57
Henry, G. W., & Fekel, F. C. 2002, *PASP*, 114, 988
Henry, G. W., & Fekel, F. C. 2003, *AJ* 126, 3058
Henry, G. W., & Fekel, F. C. 2005, *AJ*, 129, 2026
Henry, G. W., Fekel, F. C., & Henry, S. M. 2005, *AJ*, 129, 2815
Henry, G. W., Fekel, F. C., Kaye, A. B. & Kaul, A. 2001, *AJ*, 122, 3383
Hill, H., Frohlich, C., Gabriel, M. & Kotov, V. A. 1991, in: A. N. Cox, W. C. Livingston & M. S. Matthews (eds.), *Solar Interior and Atmosphere* (Tucson: Univ. of Arizona Press), p. 562
Houdek, G. 2006, in: K. Fletcher & M. J. Thompson, (eds.), *SOHO 18/ GONG 2006/ HELAS I, Beyond the Spherical Sun: Towards a Golden Future*, ESA SP-624, p. 28.1
Howe, R. 2003, in: H. Sawaya-Lacoste (ed.), *Proceedings of SOHO 12 / GONG+ 2002. Local and Global Helioseismology: the Present and Future*, ESA SP-517, p. 81

Howe, R., Christensen-Dalsgaard, J., Hill, F., *et al.* 2000, *Science*, 287, 2456
Howe, R., Christensen-Dalsgaard, J., Hill, F. 2005, *ApJ*, 634, 1405
Howe, R., González Hernández, I., Hill, F. & Komm, R. 2006a, in: K. Fletcher (ed.), *SOHO 18/GONG 2006/HELAS I, Beyond the Spherical Sun. A New Era of Helio- and Asteroseismology*, ESA SP-624, 68.1
Howe, R., Komm, R., & Hill, F. 2002, *ApJ*, 580, 1172
Howe, R., Hill, F., Komm, R., *et al.* 2006b, paper presented at HMI Team meeting, Monterey, California, February 2006
Howe, R., Komm, R., Hill, F., *et al.* 2006c, *Solar Phys.*, 235, 1
Hubrig, S., Briquet, M., Schöller, M., *et al.* 2006, *MNRAS* (Letters), 369, L61
Jerzykiewicz, M., Handler, G., Shobbrook, R. R., *et al.* 2005, *MNRAS*, 360, 619
Jouve, L., & Brun, A. S. 2007, *A&A*, 474, 239
Kaye, A. B., Henry, G. W., Fekel, F. C., *et al.* 1999, *AJ* 118, 2997
Kitchatinov, L. L., & Rüdiger, G. 1993, *A&A*, 276, 96
Kjeldsen, H., Bedding, T. R., Baldry, I. K., *et al.* 2003, *AJ*, 126, 1483
Kjeldsen, H., Bedding, T. R., Butler, R. P., *et al.* 2005, *ApJ*, 635, 1281
Komm, R. W. 1994, *Solar Phys.*, 149, 417
Komm, R. W., Howard, R. F., & Harvey, J. W. 1993, *Solar Phys.*, 147, 207
Komm, R. W., Howe, R., & Hill, F. 2000, *ApJ*, 543, 472
Komm, R. W., Howe, R., & Hill, F. 2002, *ApJ*, 572, 663
Komm, R. W., Howe, R., & Hill, F. 2006, *Adv. Space Res.*, 38, 845
Korzennik, S. G., Rabello-Soares, M. C., & Schou, J. 2004, *ApJ*, 602, 481
Kumar, P., Quataert, E. J. & Bahcall, J. N. 1996, *ApJ* (Letters), 458, L83
Le Contel, J.-M., Mathias, P., Chapellier, E., & Valtier, J.-C. 2001, *A&A*, 380, 277
Lefebvre, S., & Kosovichev, A. G. 2005, *ApJ* (Letters), 633, L149
Lefebvre, S., Kosovichev, A. G., Nghiem, P., *et al.*, 2006, in: K. Fletcher & M. J. Thompson (eds.), *SOHO 18/ GONG 2006/ HELAS I, Beyond the Spherical Sun: Towards a Golden Future*, ESA SP-624, 9.1
Libbrecht, K. G., & Woodard, M. F. 1990, *Nature*, 345, 779
Maeder, A. 1999, *A&A*, 347, 185
Maeder, A. 2003, *A&A*, 399, 263
Maeder, A., & Meynet, G. 2004, *A&A*, 422, 225
Maeder, A., & Zahn, J.-P. 1998, *A&A*, 334, 1000
Mantegazza, L., & Poretti, E. 2002, *A&A*, 396, 911
Martić, M., Lebrun, J.-C., Appourchaux, T., & Korzennik, S. G. 2004a, *A&A*, 418, 295
Martić, M., Lebrun, J. C., Appourchaux, T., & Schmitt, J. 2004b, in: D. Danesy (ed.), Proc. SOHO 14/GONG 2004 Workshop, *Helio- and Asteroseismology: Towards a Golden Future*, ESA SP-559, p. 563
Mathias, P., Le Contel, J.-M., Chapellier, E., *et al.* 2004, *A&A*, 417, 189
Mathis, S., Palacios, A., & Zahn, J.-P. 2004, *A&A*, 425, 243
Mathis, S., & Zahn, J.-P. 2004, *A&A*, 425, 229
Mathis, S., & Zahn, J.-P. 2005, *A&A*, 440, 653
Mathis, S., & Zahn, J.-P. 2005, in: F. Casoli, T. Contini, J. M. Hameury, & L. Pagani (eds.) Proc. Semaine de l'Astrophysique Francaise 2A-2005, *EdP-Sciences CS*, 319
Mathur, S.,Turck-Chièze, S., Couvidat, S., García, R. 2006, in: K. Fletcher & M. J. Thompson, (eds.), *SOHO 18/ GONG 2006/ HELAS I, Beyond the Spherical Sun: Towards a Golden Future*, ESA SP-624, p. 95.1
Matthews, J. M., Kuschnig, R., Guenther, D. B., *et al. Nature*, 430, 51. Erratum: 430, 921
Menou, K., Balbus, S. A., Spruit, H. C. 2004, *ApJ*, 607, 564
Miesch, M. S., Brun, A. S., & Toomre, J. 2006, *ApJ*, 641, 618
Miesch, M. S., Elliott, J. R., Toomre, J., *et al.* 2000, *ApJ*, 532, 593
Mosser, B., Bouchy, F., Catala, C., *et al.* 2005, *A&A* (Letters), 431, L13
Moya, A., Grigahcene, A., Suárez, J. C., *et al.* 2006, *Mem. S.A.It.*, 77, 466
Moya, A., Suárez, J. C., Amado, P. J., Martin-Ruíz, S., & Garrido, R. 2005, *A&A*, 432, 189
Neiner, C., Geers, V. C., Henrichs, H. F., *et al.* 2003a, *A&A*, 406, 1019
Neiner, C., Henrichs, H. F., Floquet, M., *et al.* 2003b, *A&A*, 411, 565

Pallé, P. L. 1991, *Adv. Space Res.*, 11 (4), 29
Pamyatnykh, A. A., Handler, G., & Dziembowski, W. A. 2004, *MNRAS*, 350, 1022
Parker, E. N. 1993, *ApJ*, 408, 707
Rabello-Soares, M. C., Basu, S., Christensen-Dalsgaard, J., *et al.* 2000, *Sol. Phys.*, 193, 345
Rabello-Soares, M. C. Korzennik, S. G., & Schou, J. 2007, *Adv. Space Res.*, E22-0005-06, in press
Rempel, M. 2005, *ApJ*, 622, 1320
Rempel, M. 2006, *ApJ*, 647, 662
Rieutord, M. 2006, in: M. Rieutord & B. Dubrulle (eds.), *Stellar Fluid Dynamics and Numerical Simulations: From the Sun to Neutron Stars, EAS-PS*, 21, 275
Rieutord, M., & Zahn, J.-P. 1995, *A&A*, 296, 127
Rogers, T., & Glatzmaier, G. 2005, *MNRAS*, 364, 1135
Salabert, D., & Jiménez-Reyes, S. J. 2006, in: K. Fletcher & M. J. Thompson, (eds.), *SOHO 18/ GONG 2006/ HELAS I, Beyond the Spherical Sun: Towards a Golden Future*, ESA SP-624, 90.1
Salabert, D., Chaplin, W. J., Elsworth, Y., New, R., & Verner, G. A. 2007, *A&A*, 463, 1181
Samadi, R., Belkacem, K., Goupil, *et al.* 2007, in: F. Kupka, I. W. Roxburgh & K. L. Chan (eds.) *Convection in Astrophysics*, Proc. IAU Symp. No. 239 (Cambridge: CUPP), p. 119
Samadi, R., & Goupil, M. 2001, *A&A* 370, 136
Samadi, R., Nordlund, Å, Stein, R. F., Goupil, M. J., & Roxburgh, I. 2003a, *A&A* 404, 1129
Schou, J., & Buzasi, D. L. 2001, in: A. Wilson & P. L. Pallé (eds.), *SOHO 10/GONG 2000: Helio- and Asteroseismology at the Dawn of the Millennium*, ESA SP-464, 391
Schunker, H., Braun, D. C., Cally, P. S., & Lindsey, C. 2005, *ApJ* (Letters), 621, L149
Schunker, H., & Cally, P. S. 2006, *MNRAS*, 372, 551
Spruit, H. C. 1999, *A&A* 349, 189
Spruit, H. C. 1991, in: J. Toomre & D. O.Gough (eds), *Challenges to Theories of the Structure of Moderate Mass Stars, Lecture Notes in Physics*, 388, 121
Spruit, H. C. 2002, *A&A*, 381, 923
Spruit, H. C. 2003, *Solar Phys.*, 213, 1
Stello, D., Kjeldsen, H., Bedding, T. R. & Buzasi, D. 2006, *A&A*, 448, 709
Talon, S., & Charbonnel C. 2003, *A&A*, 405, 1025
Talon, S., & Charbonnel C. 2004, *A&A*, 418, 1051
Talon, S., & Charbonnel C. 2005, *A&A*, 440, 981
Talon, S., Kumar P., & Zahn J.-P. 2002, *ApJ* (Letters) 574, L175
Talon, S., & Zahn, J.-P. 1997, *A&A*, 317, 749
Tripathy, S. C., Hill, F., Jain, K., & Leibacher, J. W. 2007, *Solar Phys.*, submitted
Turck-Chièze, S., Couvidat, S., Kosovichev, A. G., *et al.* 2001, *ApJ* (Letters), 555, L69
Turck-Chièze, S., García, R. A., Couvidat, S., *et al.* 2004, *ApJ*, 604, 455
Turck-Chièze, S., Appourchaux, T., & Ballot, J. 2005, in: F. Favata, J. Sanz-Forcada, A. Giménez & B. Battrick (eds.), *Trends in Space Science and Cosmic Vision 2020*, Proc. 39th ESLAB Symposium, ESA SP-588, 193
Turck-Chièze, S. 2006a, *Adv. Space Res.*, 37, 1569
Turck-Chièze, S., Carton, P.-H., Ballot, J., *et al.* 2006b, *Adv. Space Res.*, 38, 1812
Turck-Chièze, S., *et al.* 2006c, in: K. Fletcher & M. J. Thompson, (eds.), *SOHO 18/ GONG 2006/ HELAS I, Beyond the Spherical Sun: Towards a Golden Future*, ESA SP-624, 24.1
Viskum, M., Kjeldsen, H., Bedding, T. R., *et al.* 1998, *A&A*, 335, 549
Vorontsov, S. V., Christensen-Dalsgaard, J., Schou, J., *et al.* 2002, *Science*, 296, 101
Waelkens, C. 1991, *A&A*, 246, 453
Woch, J., & Gizon, L. 2007, *AN*, 328, 362
Woodard, M. F., & Noyes, R. W. 1985, *Nature*, 318, 445
Zahn, J.-P. 1966, *Ann. d'Astrophys.* 29, 313 (I), 489 (II)
Zahn, J.-P. 1975, *A&A*, 41, 329
Zahn, J.-P. 1992, *A&A*, 265, 115
Zhao, J., & Kosovichev, A. G. 2004, *ApJ*, 603, 776
Zima, W., Wright, D., Bentley, J., *et al.* 2006, *A&A*, 455, 235

III. Special Sessions

Highlights of Astronomy, Volume 14
IAU XXVI General Assembly, 14-25 August 2006
Karel A. van der Hucht, ed.

doi:10.1017/S1743921307011647

Special Session 1 Large astronomical facilities of the next decade

Gerard F. Gilmore[1] **and Richard T. Schilizzi**[2] **(eds.)**

[1]Institute of Astronomy, University of Cambridge, Madingley Road, Cambridge CB3 0HA, UK
email: gil@ast.cam.ac.uk

[2]Square Kilometre Array, ASTRON, PO Box 2, NL-7990AA Dwingeloo, the Netherlands
email: schilizzi@skatelescope.org

Preface

The questions addressed in this large and popular meeting were: What mature proposals for the new large astronomical facilities of the next decade are available? When can they be funded? What international planning is underway?

It is understood that astronomy requires major facilities across and beyond the electromagnetic spectrum. It is agreed those built should be innovative, and where possible complementary, to generate a scientific optimum. Some complementary pairs of facilities generate enormous synergy – *HST* imaging with spectra from the large ground telescopes being a good example.

Very many excellent proposals are available. Funding and approving these projects in the chronological order of their origin is not obviously ideal: some sort of a rank is needed. This raises the primary question of interest: what is the 'best' way to set a priority order to spend the available finite funding in a global context? This decision and funding process must take place in a context in which the global astronomy community is building mid-price (US$10M to US$100M) new facilities at a rate higher than one per year, all of which require operations and upgrade support well into the future. Cumulative operations costs compete for funding with new capital costs.

The list of facilties which are in operation, under development, or proposed for near-term development which was outlined at the discussion includes the following:

UHE: ICE-CUBE, Amanda, Baikal, Fiona, km3Net, Super-K, ...

VHE: FlysEye, Auger, Hegra, Magic, HESS, Tel-Array, Magic-2, 5X5 ...

HE: *Integral*, *Batse*, *Swift*, *XTE*, *Chandra*, *XMM-Newton*, *Suzaku*, *FUSE*, *Con-X*, *HXMT*, *Spec-X*, *XEUS*, ...

mid-IR: *Herschel*, *Spitzer*, *Akari*, Sofia, *SPICA*, ...

mm-wave: Carma, MMA, GMT, CC-LMT, ALMA, ...

cm-, m-wave: LOFAR, Allen, KAT, MWA, Madrid, Sardinia, GBT, E-VLA, e-Merlin, FAST, GMRT, VSOP-2, X-NTD, SKA, ...

optical-IR: Keck-I,II, Gemini-N,S, VLT ($\times$ 4), Magellan ($\times$ 2), SALT, HET, LBT, Subaru, Lamost, LSST, JELT, CFGT, GMT, E-ELT, TMT, ...

planetary missions: *CoRot*, *Kepler*, *Gaia*, *Darwin*, *TPF*, (*SIM*), ...

virtual observatories: AVO, JVO, CVO, ..., IVOA

This list is certainly incomplete. Each of these requires construction funds, together with operation and upgrade costs which typically exceed the capital construction cost by a factor of two over an operational lifetime.

No simple solution was forthcoming. An interesting future awaits.

Highlights of Astronomy, Volume 14
IAU XXVI General Assembly, 14-25 August 2006
Karel A. van der Hucht, ed.

doi:10.1017/S1743921307011659

Atacama Large Millimeter/Submillimeter Array (ALMA)

Jean L. Turner[1] **and H. Alwyn Wootten**[2]†

[1]Department of Physics and Astronomy, UCLA, Los Angeles, CA 90095-1547, USA
email: turner@astro.ucla.edu

[2]NRAO, 520 Edgemont Road, Charlottesville, VA 22903-2475, USA
email: awootten@nrao.edu

Abstract. The Atacama Large Millimeter/Submillimeter Array (ALMA) is an international effort to construct an instrument capable of matching the exquisite imaging properties of optical space telescopes at millimeter and submillimeter wavelengths. ALMA science will transform our vision of the cold, dusty, and gaseous universe, from extrasolar planets to the youngest galaxies.

Keywords. telescopes, submillimeter, techniques: high angular resolution, techniques: interferometric

---

1. Introduction to ALMA

The Atacama Large Millimeter/Submillimeter Array, ALMA, is the largest new facility for long wavelength astronomy. It will be a large array of telescopes on a high, dry plain in the Atacama desert of northern Chile. ALMA is a joint venture of the European Southern Observatory (ESO) and Spain, the National Radio Astronomy Observatory (NRAO) of Associated Universities Inc., funded by the U.S. National Science Foundation with support from the National Research Council of Canada, and the National Institutes of Natural Sciences of Japan with the Academia Sinica of Taiwan, in cooperation with the Republic of Chile. The total cost of ALMA, based on the rebaseline of the project completed in late 2005, is $\sim$1 billion current euros. ALMA will consist of up to 64 12-m antennas, an additional 12 7-m antennas for mapping extended emission (the Atacama Compact Array), and 4 enhanced 12-m antennas capable of total power observing. ALMA will ultimately cover all atmospheric windows at wavelengths from 3 mm to 350 μm.

ALMA is currently under construction at Llano de Chajnantor, near the historic town of San Pedro de Atacama. It shares the site with several existing and future millimeter and submillimeter facilities. Construction highlights of the past year include the completion of a road from the Operations Support Facility at 2900 m to the telescope site that wide enough to transport the antennas. The Array Operations Site Technical Building is nearly complete at the 5000 meter level (16,570 ft), and is the largest steel-frame building in the world at this altitude. The ALMA Test Facility, an interferometer composed of two antenna prototypes, is being tested at the VLA site. Antenna contracts have been awarded by the partners, and the first 12-m antenna is due in Chile in 2007.

2. ALMA science

ALMA will be a mighty leap forward in capability, with an order of magnitude gain in sensitivity over current millimeter arrays in the northern hemisphere, and the potential

† Present address: ALMA, 40 El Golf, Piso 18, Las Condes, Santiago, Chile.

for two orders of magnitude gain in imaging resolution and fidelity, at a high, dry site suited to submillimeter observing. The ALMA Level-One design goals exemplify the types of projects that will drive this transformational instrument. These goals are: (*i*) to detect CO in an L* galaxy at $z = 3$; (*ii*) to image molecular lines in a protoplanetary disk with a resolution of 1 AU out to a distance of 150 pc; and (*iii*) to obtain high fidelity imaging at $0''.1$ to match *HST*, *JWST*, or AO imaging. ALMA will allow us to study the gas in 'normal' galaxies at redshifts at which galaxies should be significantly different from the present, in the 'Gas Ages'. Closer to home, ALMA will enable study of the birth of planetary systems, allowing imaging of the nearest protoplanetary disks with 1 AU spatial resolution and sufficient spectral resolution to resolve motions on planet scales within a gaseous protoplanetary disk.

The science capabilities of ALMA span all fields of astronomy.

• **ALMA and planet formation.** At its highest resolution of 5 mas, ALMA will easily detect thermal dust continuum emission from protoplanetary or debris disks out to the nearest star clusters at distances of 150 pc. It will be able to image the thermal emission from young Jupiter-sized protoplanets out to 100 pc.

• **ALMA and molecular gas.** ALMA will image emission from lines of the over 140 molecules currently detected in space, a list that includes organic molecules of biological interest, including amino acids. In addition to astrochemistry, studies of molecular abundances and excitation in interstellar clouds will provide diagnostics for interstellar physics, from shocks to gas dynamics, allowing the study of star formation and the feedback of star formation on molecular clouds.

• **ALMA and nearby galaxies.** ALMA will image molecular clouds in galaxies to give us a new view of galactic structure, galaxy interactions and galaxy mergers, and star formation on galactic scales. ALMA will probe the gas dynamics within a few parsecs of supermassive black holes in the centers of galaxies. At its highest resolution, ALMA can directly resolve structures of 50 AU ($\sim$600 Schwarzschild radii) in Sgr A*.

• **ALMA and the solar system.** The Submillimeter Array in Hawaii recently detected the thermal emission from Pluto and Charon: while Charon's surface is the expected 50 K for this distance from the sun, Pluto's temperature is a chilly 38 K; sublimation of surface ice is suspected. With its unprecedented spatial resolution and sensitivity, ALMA will be able to image surface features on the outer planets, in addition to asteroids, dwarf planets and icy Kuiper belt objects. ALMA will easily detect and perhaps even resolve the most distant object known in the solar system, UB313. The composition, structure, and kinematics of planetary atmospheres will also be within ALMA's reach.

• **ALMA and the distant universe.** ALMA will be a unique eye on the early universe. Actively star-forming galaxies, such as Arp 220, can be detected easily in CO to $z = 10$. Images of gas and dust emission will allow us to study many properties of primeval galaxies, such as structure, dynamical masses, and gas masses. While the Hubble Deep field contains many nearby galaxies and few distant ones, the ALMA Deep Field is the opposite: to a given brightness, simulations show that galaxies beyond $z = 1.5$ are a hundred times more numerous than nearby galaxies.

3. When can I use ALMA?

After the first antenna arrives in Chile in 2007, there will be an extended period of system integration, verification, and commissioning. 'Early Science', when ALMA is first opened to the community, is scheduled to begin in 2010, when the array has 12 - 16 antennas. ALMA reaches full capability, with all antennas and receiver bands, in 2012. More ALMA information and an ALMA Cam are available at `<http://www.alma.cl>`.

Highlights of Astronomy, Volume 14
IAU XXVI General Assembly, 14-25 August 2006
Karel A. van der Hucht, ed.

doi:10.1017/S1743921307011660

James Webb Space Telescope

Hervey S. Stockman

Space Telescope Science Institute, 3700 San Martin Drive, Baltimore, MD 21218, USA
email: stockman@stsci.edu

Abstract. The *James Webb Space Telescope* (*JWST*) is the scientific successor to the *Hubble* and *Spitzer* missions. Its wavelength range (1 - 28μm) and sensitivity (1 nJy - 1 μJy) complement the submillimeter facilities of the coming decade, *Herschel* and ALMA. The *JWST* development is on schedule for a June 2013 launch to L2 on an Ariane 5.

Keywords. space vehicles, space vehicles: instruments, telescopes

1. Introduction

In 2000, the Astronomy and Astrophysics Survey Committee of the US National Academy of Sciences recommended the *James Webb Space Telescope* (*JWST*, *Webb*) as the highest priority new space facility for the decade and a Giant Segmented Mirror Telescope (GSMT) as the highest priority ground based facility. These ambitious projects are the scientific successors to the very successful *Hubble Space Telescope* and the 8 - 10 m ground-based telescopes developed in the 1990s. While the emphasis in both new initiatives is for superb imaging and sensitivity in the near infrared (1 - 5 μm), the equally exciting prospect for *Webb* was to extend the 'discovery space' of the *Spitzer* mission in the mid-infrared (5 - 28μm). The 'discovery space' improvement that *Webb* offers over GSMT and *Spitzer* is shown in Fig. 1.

The superb sensitivity of *Webb* at long wavelengths is shown in Fig. 2. along with similar estimates for the *Herschel* mission to be launched in 2008 and the Atacoma Large Milllimeter Array (ALMA). Also shown are sample spectra of two sources with spectra that peak in the 60 - 200 μm: a low-mass Class 0 protostar and a high-redshift luminous infrared galaxy (starburst or AGN). This figure shows the complementarity of the three facilities in the mid-IR and sub-millimeter range.

2. Science goals

The *Webb* science goals are described in *The James Webb Space Telescope* (Gardner *et al.* 2006). They fall into four major areas listed below and are illustrative of the scientific power of the *Webb*. Recently, the discovery of dozens of eclipsing exo-planets by photometric surveys of stars both near the Sun and toward the Galactic Center has raised the potential of *Webb* studying the atmospheric properties of planets, down to Earth-like masses for nearby systems.

- First Light: The first luminous objects and the epoch of re-ionization
- Galaxy Assembly: the origins and growth of galactic structures and cosmo-chemistry
- The Birthplaces of Stars: The environments of star and planet formation
- Planets and Life: The evolution of planetary systems and the ways they could support life

3. Project status

The *Webb* development is a collaborative effort involving NASA, the European Space Agency and the Canadian Space Agency. Goddard Space Flight Center (NASA) is the

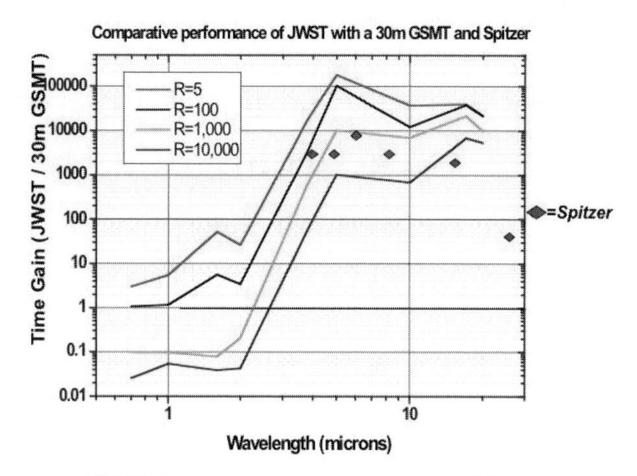

Figure 1. '*Webb* discovery space compared to GSMT and *Spitzer*

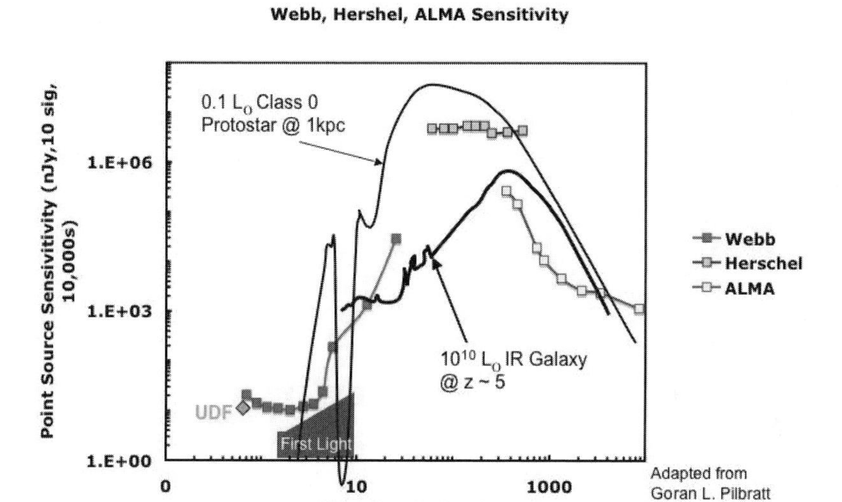

Figure 2. *Webb*, *Herschel*, and ALMA sensitivities in the mid-IR and sub-millimeter region. $10\,\sigma$, $10^4\,s$ observations.

lead organization and Northrop Grumman Space Technologies (NGST) is the prime contractor. The Project is on schedule for a June 2013 launch and will have demonstrated the flight worthiness of the key enabling technologies by January 2007. For a list of the SWG members and current information, see `<www.jwst.gsfc.nasa.gov>` and `<www.stsci.edu/jwst>`.

References

Gardner, J. P., Mather, J. C., Clampin, M., Doyon, R., Greenhouse, M. A., Hammel, H. B., Hutchings, J. B., Jakobsen, P., Lilly, S. J., Long, K. S., Lunine, J. I., McCaughrean, M. J., Mountain, M., Nella, J., Rieke, G. H., Rieke, M. J., Rix, H.-W., Smith, E. P., Sonneborn, G., Staivelli, M., Stockman, H. S., Windhorst, R. A., Wright, G. S., 2006, *Space Sci. Rev.*, 123, 485

Highlights of Astronomy, Volume 14
IAU XXVI General Assembly, 14-25 August 2006
Karel A. van der Hucht, ed.

doi:10.1017/S1743921307011672

Status of the European Extremely Large Telescope (E-ELT)

Guy J. Monnet

European Southern Observatory, Karl-Schwarzschild-Straße 2, D-85748 Garching, Germany
email: gmonnet@eso.org

Abstract. The European initiative for an Extremely Large Telescope is presented. During the past year the transition from the 100 m OWL concept to a 30 m to 42 m diameter telescope has been made. The history and the current status of this development is described.

Keywords. telescopes

---

1. The prehistory

Conceptual studies for a new generation of Extremely Large Telescopes, or 'ELT' in short, started quite early in Europe. A university consortium led by Lund University produced the EURO-50 concept in the 1998 to 2003 period, while ESO conducted the OWL conceptual development from 1997 to 2005. At 50 m diameter, EURO-50 had about the absolute maximal size that a 'classical' telescope, based on a segmented aspheric primary, could attain, while OWL at a 60 m to 100 m size with segmented spherical primary and secondary mirrors has explored the absolute maximal possible size allowed by mass production. Both have been conceived as general-use facilities, combining large flux collection and exquisite image quality down to the diffraction limit and, consequently, incorporate intrinsic Adaptive Optics correction capabilities.

A European-wide technological programme regrouping close to ninety research institutes and industrial companies has been established to develop and validate the many enabling technologies that are needed to produce such a challenging facility with its very large collecting area and advanced instrumentation. It combines basic R&D conducted under the aegis of OPTICON (`<http://www.astro-opticon.org/>`), the FP6 EC-funded Optical Infrared Coordination Network for Astronomy, and the FP6 EC-funded ELT Design Study (`<http://www.eso.org/projects/elt-ds/>`) for the validation of the many crucial subsystems of any ELT.

2. The E-ELT scientific case

In parallel and with close coordination with the conceptual studies, more than one hundred astronomers have built a scientific case for Extremely Large Telescopes under the aegis of OPTICON (`<http://www-astro.physics.ox.ac.uk/~imh/ELT/>`). Synergy with future ground and space based capabilities, like ALMA and *JWST*, was taken into account. Main ELT scientific challenges are found in the areas of planetary systems detection and characterisation, study of resolved stellar populations up to the Virgo cluster, study of high-redshift galaxies and detection of the first lights, and use of astronomical objects to probe the frontiers of physics (black holes, quantum gravity effects ...). The science case is continuously being refined and its latest version, geared towards the European ELT, can be found at `<http://www.eso.org/projects/e-elt/publications.html>`.

Figure 1. Possible mechanical design for an E-ELT

3. The E-ELT genesis

As mandated by its Council in December 2004, ESO has embarked into the Basic Reference Design (BRD) for the European ELT (<http://www.eso.org/projects/e-elt/>). The effort is fully coordinated with the continuing science definition (OPTICON) and technological development (OPTICON & ELT Design Study). It is conducted with the help of working groups with extensive help from community astronomers. The goal is to get a high-performance science-optimised general-use facility for the European community. The telescope size is envisioned to be in the 30 m to 42 m range and will be set by the cost/risk versus performance ratio. The BRD will be presented to the community at large in Marseilles at the end of November (<http://www.elt2006.org>) and early December to the ESO Council for decision on launching the Project.

4. Conclusion

Within this ambitious European programme, there is ample need for cooperation with the equally ambitious developments occurring on the other side of the Atlantic. E-ELT, GMT and TMT share most of the same enabling technology needs and coordination in this area shall be maintained and ideally much augmented. There is also potential for useful complementarities between these future facilities and optimising their combined science output does look like a promising approach, which deserves to be seriously explored.

Highlights of Astronomy, Volume 14
IAU XXVI General Assembly, 14-25 August 2006
Karel A. van der Hucht, ed.

doi:10.1017/S1743921307011684

Current and future status of gravitational wave astronomy - gravitational wave facilities

Sheila Rowan

SUPA, Institute for Gravitational Research, Department of Physics and Astronomy, University of Glasgow, Glasgow G12 8QQ, UK
email: s.rowan@physics.gla.ac.uk

Abstract. Currently a network of interferometric gravitational wave detectors is in operation around the globe, in parallel with existing acoustic bar-type detectors. Searches are underway aimed at the first direct detection of gravitational radiation from astrophysical sources. This paper briefly summarizes the current status of operating gravitational wave facilities, plans for future detector upgrades, and the status of the planned space-based gravitational wave detector *LISA*.

Keywords. gravitational waves

---

1. Introduction

Gravitational waves are strains in space-time caused by the acceleration of mass, as predicted by Einstein's General Theory of Relativity. Sources of astrophysical interest include black hole and neutron star coalescences, low-mass X-ray binaries, supernova explosions and rotating asymmetric neutron stars (Hough *et al.* 2005). Predicted astrophysical signals are at a level where detectors with strain sensitivities of 10^{-23} or better are required.

2. Status of gravitational wave detectors

2.1. *Current detectors*

Currently a gravitational wave detector network exists based on sensing the changes, induced by gravitational waves, in the relative arm lengths of large interferometers. The US LIGO project (Waldman *et al.* 2006) comprises two 4 km long detectors, one in Hanford, WA, and one in Livingston, LA. A 2 km detector exists inside the same evacuated enclosure at Hanford. The 3 km long French/Italian VIRGO detector (Acernese *et al.* 2006) near Pisa is close to completing commissioning and the 300 m long Japanese TAMA 300 detector (Takahashi *et al.* 2004) is operating at the Tokyo Astronomical Observatory. The German/British detector, GEO600 (Hild *et al.* 2006), through use of novel technologies is expected to reach a sensitivity at frequencies above a few hundred Hz close to those of VIRGO and LIGO in their initial operation. Complementing the interferometers is a network of cryogenic bar detectors, running continuously with amplitude spectral densities of less than 10^{-21}. Links to projects can be found at: `<http://gwic.gravity.psu.edu/>`.

Four science data taking runs have been completed with the LIGO detectors. Three of these involved the GEO and TAMA detectors. New 'upper limits' have been set on the strength of gravitational waves from a range of sources, see for example, publications available at `<http://www.ligo.org/results/>`. The 5th LIGO science run started on 4th Nov 2005 with GEO having joined in January 2006. Ongoing searches for gravitational wave signals are at a level where it is plausible that a detection could be made.

3. Planned upgrades to detectors

However, detection with initial systems is not guaranteed; a ten times improvement in sensitivity is needed to reach levels where many signals are expected. Thus plans for an upgraded LIGO, 'Advanced LIGO', are already mature (Fritschel 2003). The upgrade is approved by the National Science Board with the start of US construction funds from the NSF account for large projects (MREFC) anticipated in 2008, construction expected to start in 2010, and with initial operation by 2014. Contributions from the UK (PPARC) and Germany (MPG) are already approved. Around the same time, an upgrade to VIRGO is planned along with the rebuilding of GEO as a detector with high sensitivity at kHz frequencies, and the building of an underground detector, LCGT (Kazuaki *et al.* 2006), in Japan. In Australia, the ACIGA consortium operates an 80 m interferometric testbed and has plans for a future full-scale interferometer. The advanced technology planned for the 'DUAL' acoustic-type detector could allow it to have a sensitivity equal to that of interferometers at high frequencies.

Ongoing laboratory research will complement the proposal in Europe of a design study for a 3rd generation interferometric detector, aimed at the European Community Framework 7 funding call.

4. *LISA*

LISA is a space-based interferometric gravitational wave detector proposed jointly by a US/European team (Danzmann & Rüdiger 2003), aimed at signals in the region of 10^{-4} Hz to 10^{-1} Hz. A demonstrator mission '*LISA Pathfinder*' is in phase C/D, preparing for launch in 2009. A US study to prioritize the launches of the *LISA*, *Con-X*, and *JDEM* missions is imminent. On the ESA side, final commitment to *LISA*s implementation will be influenced by the success of LPF. However, work is underway before *LPF* launch to define the *LISA* mission and prepare the invitation to tender for the implementation phase. With NASA's selection in FY 2009 and ESA's final commitment, *LISA* is expected to enter the implementation phase in 2011, with launch in the 2015 to 2016 timeframe.

Acknowledgements

The author would like to thank PPARC and the University of Glasgow for support, and colleagues in the field. In addition thanks go to the funding agencies, projects and Universities around the World involved in supporting the field of gravitational wave detection.

References

Acernese, F., *et al.* 2006, *Class. Quantum Grav.*, 23, S63
Danzmann, K., & Rüdiger, R. 2003, *Class. Quant. Grav.*, 20, S1
Fritschel, P. 2003 in: M. Cruise & P. Saulson (eds.), *Proceedings of SPIE Vol. 4856 'Gravitational Wave Detection'* (SPIE, Bellingham, WA,) p. 282
Hild, S., for the LIGO Science Collaboration, 2006, *Class. Quantum Grav.*, 23, S643
Hough, J., Rowan, S., & Sathyaprakash, B. S. 2005, *J. Phys. B: At. Mol. Opt. Phys.*, 38, S497
Kazuaki K., for the LCGT Collaboration, 2006, *Class. Quantum Grav.*, 23, S215
Takahashi, R., for the TAMA Collaboration, 2004, *Class. Quantum Grav.*, 21, S403
Waldman, S. J., for the LIGO Science Collaboration, 2006, *Class. Quantum Grav.*, 23, S653

Highlights of Astronomy, Volume 14
IAU XXVI General Assembly, 14-25 August 2006
Karel A. van der Hucht, ed.

doi:10.1017/S1743921307011696

International Virtual Observatory Alliance

Masatoshi Ohishi

Astronomy Data Center, National Astronomical Observatory of Japan,
2-21-1, Osawa, Mitaka, Tokyo, 181-8588, Japan
email: masatoshi.ohishi@nao.ac.jp

Abstract. The International Virtual Observatory Alliance is briefly introduced as a concensus-based group to construct International Virtual Observatory – a new, planet-wide research infrastructure for the 21st century astronomy. Standardized protocols by the IVOA were used to interconnect more than 10 astronomical obsrvatories and data centers to provide astronomers with multiwavelength astronomical data. The priority areas for technical development and planned developments are described.

Keywords. standardized protocol, astronomical database, international collaboration

---

In recent years many large telescopes are operating, under construction, and planned to produce Peta-byte scale data. It is crucial to develop a mechanisim to utilize and integrate such data to accerelate astronomical researches. The International Virtual Observatory Alliance (IVOA: `<http://www.ivoa.net>`) represents 16 international projects, as of August 2006, working in coordination to realize the essential technologies and interoperability standards necessary to create a new, planet-wide research infrastructure for 21st century astronomy. This international Virtual Observatory will allow astronomers to interrogate multiple data centres in a seamless and transparent way, will provide new powerful analysis and visualisation tools within that system, and will give data centres a standard framework for publishing and delivering services using their data.

The first step for the IVOA projects is to develop the standardised framework that will allow such creative diversity. Since its inception in June 2002, the IVOA has already fostered the creation of a new international and widely accepted, astronomical data format (VOTable) and has set up technical working groups devoted to defining essential standards for service registries, unified content descriptions (UCDs), data access (Images, Spectra, Catalogs, etc.), data models and query languages to access distributed databases following developments in the grid community. These standards are still evolving, and readers are suggested to visit the IVOA web site, as well as links to each national project, to get the most recent information. Fig. 1 shows a schematic diagram regarding relationship of each standards.

These new standards and technologies were used to build science prototypes, demonstrations, and applications. As of 2006, more than 10 observatories and data centers in Canada, Europe, Japan and the United States of America have been interconnetced to provide astronomers around the world with large scale, multi-wavelength data. More observatories and data centers are expected to join this international framework, and some VO projects have moved to their operations phase.

The ultimate goal of the International Virtual Observatory will be to provide not only a seamless and transparent way to access data in the world, but new powerful analysis and visualisation tools within that system. IVOA has been working to standardize workflow (pipeline) mechanism to perform data access, data analysis and visualization as a single job. It is required to define workflow description language, workflow execution mechanism,

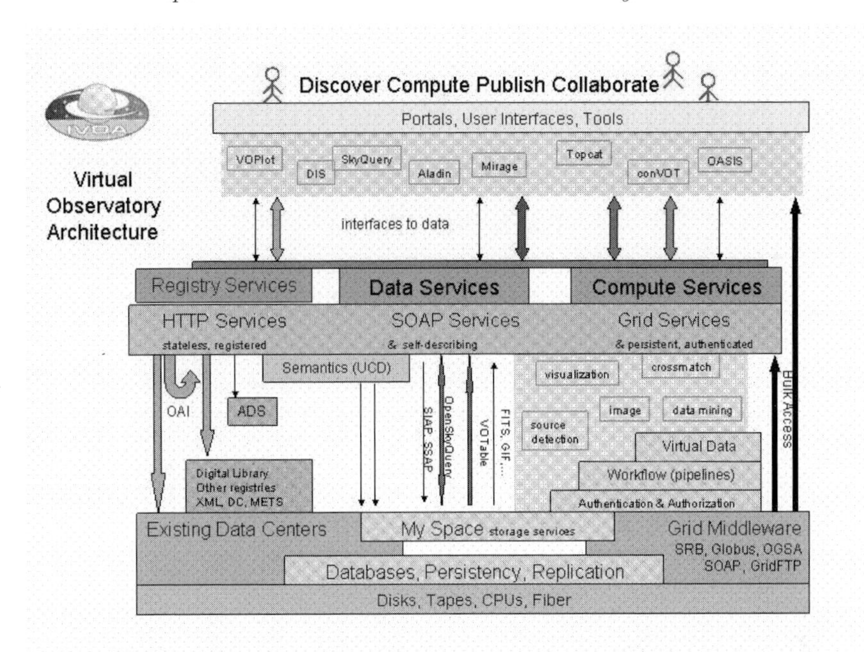

Figure 1. Relationship of IVOA standards and computing resources for astronomy.

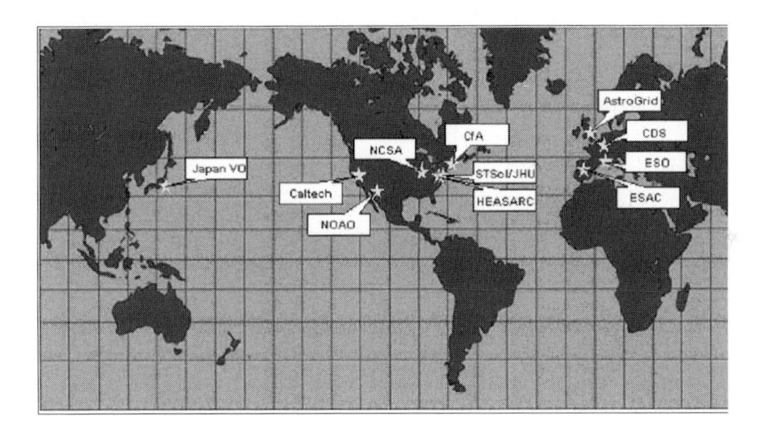

Figure 2. Observatories and data centers around the World, as of August 2006, that are being inter-operated as International Virtual Observatory.

metadata definition to describe applications, and single-sign-on mechanism to seamlessly access several sites that are operated under different access policies.

Finally, it should be noted that the IAU has set up the Working Group *Virtual Observatories, Data Centers nad Networks*, under Division XII / Commission 5 on *Documentation and Astronomical Data*, to authorize the IVOA standards.

Acknowledgements

I would like to acknowledge all member projects in the International Virtual Observatory Alliance and the staff of the National Astronomical Observatory of Japan in supporting the development of the International Virtual Observatory. This work was also supported by the JSPS Core-to-Core program, and by the MEXT KAHENHI (18049074).

Highlights of Astronomy, Volume 14
IAU XXVI General Assembly, 14-25 August 2006
Karel A. van der Hucht, ed.

doi:10.1017/S1743921307011702

ESA Space Science Programme, *Cosmic Vision 2015-2025*, for astrophysics

Catherine Turon

GEPI - UMR CNRS 8111, Observatoire de Paris,
Section de Meudon, F-92195 Meudon cedex, France
email: catherine.turon@obspm.fr

Abstract. After a brief description of the ESA Science Programme, the long-term plan for Astrophysics is described, as well as possible strategies for its implementation.

Keywords. astrophysics, space, ESA

1. Introduction

The European Space Agency, ESA, is an inter-governmental organisation with a mission to provide and promote – for exclusively peaceful purposes – the exploitation of space science, research and technology, and space applications. The Science Programme is the only mandatory programme of ESA. With a budget of about 400 M€ per year, i.e., about 12% of the total ESA budget, it funds satellites, some (part of) payloads, satellite operation, data scientific validation and access to the data. The programme is chosen by the community, with long-term planning renewed every about 10 years. For over 30 years, ESA's space science projects have shown the scientific benefits of multi-nation cooperation.

Starting from April 2004, a new long term planning exercise was initiated by the ESA Science Programme. This plan is the third step in a decadal series. In 1983-1984, the first similar exercise, *Horizon 2000* (Bonnet & Bleeker 1984), replaced the previous *à la carte* style of mission selection. It is on the basis of this ambitious programme that Europe has developed *SOHO* and *Cluster*, landed on Titan with the *Huygens* probe early in 2005, built *XMM-Newton* and *Integral*, launched *Mars Express*, *Rosetta* and *Venus Express*. Two missions are still to come: *Planck* and *Herschel*. In 1994-95, the *Horizon 2000-Plus* plan (Bonnet & Woltjer 1995) resulted in a series of high-profile missions which are now in preparation: *Gaia*, *BepiColombo*, and *JWST*, and also *LISA* in cooperation with NASA, and *Solar Orbiter*.

The numerous and enthousiast answers to the Call for Themes for Cosmic Vision 2015-2025 issued in April 2004 by the ESA Directorate of Science show the richness of new ideas coming from European scientists, and the potential for innovation open to industry.

On the basis of the many ideas proposed by the community, three main themes were identified by each of the ESA Science Working Groups. The three main themes identified by the Astronomy Working Group (Turon *et al.* 2005) were:

(*a*) other worlds and life in the Universe: detection, census and characterisation of exoplanets, search for extraterrestrial life, formation of stars and planetary systems;

(*b*) the Early Universe: investigating Dark Energy, probing inflation, observing the Universe taking shape; and

(*c*) the evolving violent Universe: matter under extreme conditions, black holes and galaxy evolution, supernovae and the life cycle of matter.

These themes were debated during two meetings open to the community: a workshop in Paris in September 2004, and the 39th ESLAB Symposium, *Trends in Space Science and Cosmic Vision 2020* (39th ESLAB Symposium), held at ESTEC in April 2005. In looking for synergies and coherence, four themes were finally selected by the Space Science Advisory Committee. They address four basic questions concerning the Universe and our place in it:

(*a*) What are the conditions for planetary formation and the emergence of life? Place the Solar System into the overall context of planetary formation, aiming at comparative planetology. The astrophysical aspects of this theme are two-fold: (*i*) Map the birth of stars and planets by peering into the highly obscured cocoons where they form. Investigate which properties of the host stars are more favourable to the formation of planets;and (*ii*) Search for and image planets around stars other than the Sun, looking for biomarkers in their atmospheres. Direct detection of Earth-like planets. Physical and chemical characterization of their atmospheres for identification of unique biomarkers. Systematic census of terrestrial planets.Ultimate goal: image terrestrial planets.

(*b*) How does the Solar System work?

(*c*) What are the fundamental physical laws of the Universe?

(*d*) How did the Universe originate and what is it made of? This theme is essentially astrophysics, with strong connections to theme (*c*) centred on the fundamental laws explaining the phenomena. This theme is divided in three parts: (*i*) The early Universe: investigate the physical processes that lead to the inflationary phase in the early Universe during which a drastic expansion took place. Investigate the nature and origin of the Dark Energy that currently drives our Universe apart; (*ii*) The Universe taking shape: find the very first gravitationally bound structures assembled in the Universe (precursors to todays galaxies and clusters of galaxies) and trace their evolution to today; and (*iii*) The evolving violent Universe: formation and evolution of the super-massive black holes at galaxy centres, in relation to galaxy and star formation. Life cycle of matter in the Universe along its cosmic history.

These four themes are described in detail in *Cosmic Vision 2015-2025, Space Science for Europe, 2015-2025* (Bignami *et al.* 2005).

2. Implementation strategy

To implement the major objectives of Cosmic Vision 2015-2025 while keeping some flexibility in the overall planning, successive slices of between 1 and 1.3 B€ each will be considered. The first Call for Mission Proposals is expected to be issued in early 2007. The Call is planned to be fully open, i.e., with no *a priori* size restriction, but with clear different categories of cost and length of development for the proposed missions. Two categories are considered: proposals for medium size missions, with an overall cost to ESA smaller than 300 M€, and a first launch opportunity in the 2015-2018 timeframe; proposals for concepts of large size missions with an overall cost to ESA smaller than 650 M€. The concepts of missions selected in this category will serve for long term technological developments, with a first possible launch after 2020.

References

Bignami, G., Cargill, P., Schütz, B., & Turon, C. 2005, *ESA-BR* 247, October 2005, ed. A. Wilson

Bonnet, R.-M., & Bleeker, J. 1984, *ESA-SP* 1070, December 1984, eds. N. Longdon & H. Olthof

Bonnet, R.-M., & Woltjer, L. 1994, *ESA-SP* 1180, November 1994, August 1995, ed. B. Battrick

39th ESLAB Symposium 2005, *ESA-SP* 588, eds. F. Favata, *et al.*

Turon, C., Done, C., Quirrenbach, A., *et al.* 2005, *ESA-SP* 588, eds. F. Favata, *et al.*, p. 53

Highlights of Astronomy, Volume 14
IAU XXVI General Assembly, 14-25 August 2006
Karel A. van der Hucht, ed.

doi:10.1017/S1743921307011714

Japan's optical/infrared astronomy plan

Masanori Iye

National Astronomical Observatory, Mitaka, Tokyo 181-8588, Japan
email: iye@optik.mtk.nao.ac.jp

Abstract. The status of Japan's planning of the optical/infrared astronomy projects for the 2010's is briefly reviewed. The road map shows a 30 m class extremely large ground based telescope project with advanced adaptive optics capability, JELT, and a mid-infrared optimized 3.5 m space telescope project, *SPICA*, as the top-priority major project to be accomplished probably on international collaboration basis.

Keywords. Akari, ELT, SPICA, Subaru

---

1. Status of the projects

The optical/infrared astronomical community of Japan conducted a two-year study and produced a report booklet in 2005 to recommend major optical/infrared astronomy missions to be accomplished in the 2010 decade. The report expects to complete JELT, an extremely large telescope of about 30 m in diameter with innovative adaptive optics capability, to make advances in observational capabilities to investigate various newly emerging science problems as well as ever standing key questions. It also identifies *SPICA*, a 3.5 m astronomical space telescope project optimized for far-infrared regime, as the key space mission for Japan's optical/infrared astronomy community to accomplish. *SPICA* will bridge the wavelength gap between those covered by ALMA and *JWST/Herschel*.

JELT and *SPICA* are natural missions advancing the expertise cultivated around 8.2 m Subaru Telescope and 60 cm *Akari* satellite, respectively. There are communities heritage on engineering and science around these leading projects.

As for the design and development efforts on JELT, the core group is making some generic researches on: (*i*) new zero-expansion ceramic material to be used for next generation mirrors; (*ii*) high-precision grinding method to figure out aspheric surface to shape error less than a wavelength; and (*iii*) developments for new methodology sensors. There is a concrete plan to construct a 3.5 m telescope with circularly segmented mirrors to prove these new technologies. National Astronomical Observatory is especially interested in promoting the JELT on Mauna Kea to make good use of Subaru heritage in people and facilities.

SPICA would achieve the highest sensitivity among proposed missions in the wavelength range 15 - 130 μm and could be a mission complementary to *Herschel* and *JWST*. *SPICA* will be placed at an L2 orbit to avoid thermal background from the Sun and the Earth. Research and development for (*i*) light-weight mirrors, (*ii*) cryogenic system with radiative cooling, and (*iii*) science instruments, are under investigation.

Additional space missions under conceptual phase are *JASMINE*, a space telescope for astrometry to map out Galactic bulge structure and *JTPF*, a mission to study extra-solar planets.

National funding policy on basic science is not in highly encouraging situation and costly projects are facing financial difficulties despite well received publicity of astronomical achievements.

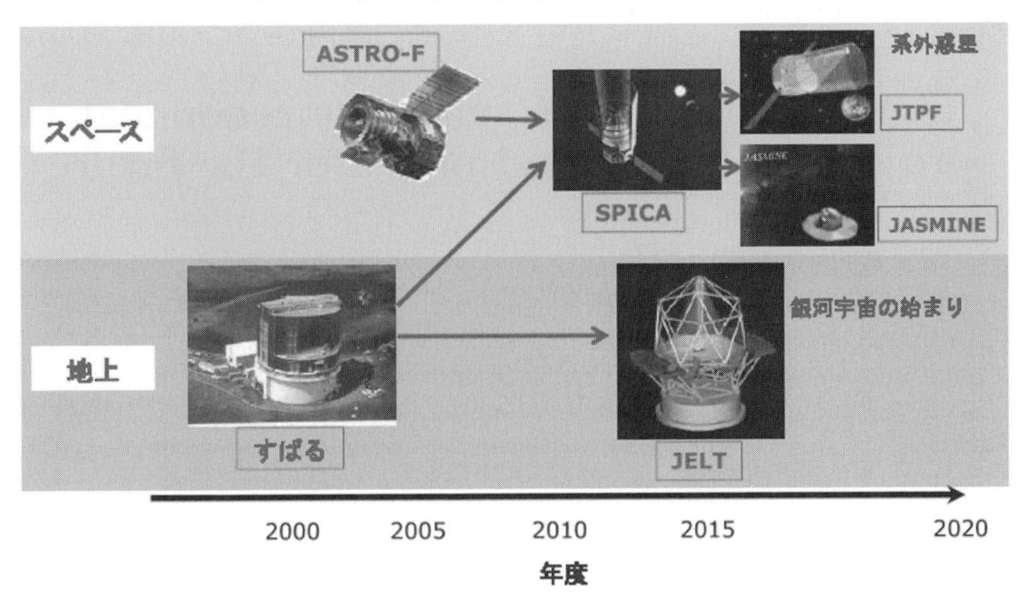

Figure 1. Road map of Japan's optical/infrared astronomy of 2010s.

Since the required budgetary scale of these ambitious missions is likely to be beyond the level any single nation can fully support, organizing an international consortium to accomplish these science missions will be extremely important. In this regard, positive discussion toward international coordination for common goals is essential and the present Special Session in the IAU XXVI General Assembly, for instance, is highly useful for those involved groups.

The relevant URLs for further information are:
JELT home page: `<http://jelt.mtk.nao.ac.jp/index_e.html>`, and
SPICA mission: `<http://www.ir.isas.jaxa.jp/SPICA/index.html>`.

References

Iye, M. 2006, in: P. A. Whitelock, M. Dennefeld & B. Leibundgut (eds.), *The Scientific Requirements for Extremely Large Telescopes*, Proc. IAU Symp. No. 232 (Cambridge: CUP), p. 381

Onaka, T., & Nakagawa, T. 2005, *Adv. Space Res.* 36, 1123

Highlights of Astronomy, Volume 14
IAU XXVI General Assembly, 14-25 August 2006
Karel A. van der Hucht, ed.

doi:10.1017/S1743921307011726

An introduction to major astronomical projects in China and plans for the future

Xiangqun Cui

National Astronomical Observatories,
Nanjing Institute of Astronomical Optics and Technology,
CAS, 188 Bancang Street, Nanjing 210042, P.R. China
email: xcui@niaot.ac.cn

Abstract. This report is a general introduction to Chinese major astronomical projects. It includes the ongoing project 'Large Sky Area Multi-Object Fiber Spectroscopic Telescope' (LAMOST), and three major projects which have finished their feasibility study and development of key technologies: Five-hundred-meter Aperture Spherical (radio) Telescope (FAST); *Space Solar Telescope* (*SST*); *Hard X-ray Modulation Telescope* (*HXMT*). Among them, FAST and *HXMT* have been approved by government in 2006, and *SST* is pending for the next five years plan. Besides these major projects, a site survey in the west of China, a plan for developing Antarctic Dome A for astronomy, and a preliminary study of Chinese future giant optical/infrared telescopes are also briefly introduced.

Keywords. large ground based telescope, space telescope, site survey, antarctic astronomy

1. Large Sky Area Multi-Object Fiber Spectroscopic Telescope

LAMOST is a meridian reflecting Schmidt telescope with an average clear aperture of 4-meter, and a field of view of 5°. There are 4000 optical fibers on its focal surface to transfer light into the 16 spectrographs. Its main scientific goals are: (1) extragalactic spectroscopic survey large-scale structure of the universe and physics of galaxies; (2) stellar spectroscopic survey for structure of the Galaxy and stellar astrophysics; and (3) cross identification of multi-wave band surveys. The main technical challenges in LAMOST include: (a) active optics for segmented deformable mirrors; and (b) parallel controllable fiber positioning. The site of LAMOST is at Xinglong station, about 180 km north of Beijing. The observing sky area is $-10°$ to $+90°$ (total 24,000 square degrees). The spectral resolution is 1000 - 2000 for low-, and 5000 - 10000 for medium-resolution spectroscopy. With a spectral resolution of 1 nm, the survey maximum magnitude capability is 20.5 mag in about 1.5 hr integration time. The total budget for LAMOST was US$30M in 1997. Before the end of 2005, the key technologies such as the active optics and the fiber positioning technology have been tested successfully. Integration of mechanical parts has been completed and installed on their concrete pillars. Manufacturing of 4000 optical fibers has started. Now, most mirror polishing is finished and all 61 mirrors will be ready for assembly before the end of 2006. The enclosure and building will be completed in September of 2006. The first light for a 'small LAMOST' with the partial optical aperture (2 m), 250 optical fibers, one spectrograph and two CCD cameras will be in 2007, and the first light with full aperture and 4000 optical fibers will be in 2008.

For further details, see: `<http://www.lamost.org/en/>`.

2. Five-hundred-meter Aperture Spherical (radio) Telescope (FAST)

The Five hundred meter Aperture Spherical Telescope is an Arecibo type large radio telescope. It is also a Chinese concept for SKA. FAST is proposed to be built in the unique karst area of southwest China. It will be over twice as large as Arecibo coupled with much wider sky coverage. Technically, FAST is not simply a copy of the existing Arecibo telescope but has rather a number of innovations: (1) the proposed main spherical reflector, by conforming to a paraboloid of revolution in real time through actuated active control, enables the realization of both wide bandwidth and full polarization capability while using standard feed design; (2) a feed support system which integrates optical, mechanical and electronic technologies will effectively reduce the cost of the support structure and control system. The reflector is a dish with a diameter of 500 m and a radius of curvature 300m. The illuminated aperture is 300 m. The opening angle is 120°. For the sky coverage, the maximum observing zenith angle is 50°. Its working frequencies are 0.1 - 2 GHz, up to C-band, X-band. The pointing accuracy is 4′, and the slewing speed is 10° per minute. The prototype for the feed, the active panels and a 50 m prototype of FAST have been finished, and some experiments have been successfully done during recent years. The FAST project has passed review by the Chinese Academy of Sciences in 2005, and was approved by the Chinese government in 2006. The total budget of FAST is US$85M, and is planned to be completed in 2013.

For further details, see: <http://www.bao.ac.cn/bao/LT/>.

3. Space Solar Telescope (SST)

The *Space Solar Telescope* is proposed to take the advantage of rocketry and abundant research work in solar magnetic field and velocity field in China. *SST* is a 1 m solar space telescope with diffraction limited optical images, equipped with an innovative 2-d spectrograph, a $0''.5$ resolution soft X-ray telescope, and wide band spectrometer. It will be used in the wide band, continuing evolution observation. With its main character of high resolution, and concentrated in detecting the magnetic elements, the study and observation in the instant and stable magnetic hydrokinetics will take place. *SST* is waiting for getting funds in the next five years plan (2011 - 2015).

For further details, see: <http://www.bao.ac.cn/bao/SST>.

4. Hard X-ray Modulation Telescope (HXMT)

The *Hard X-ray Modulation Telescope* has been selected recently as the first Chinese space astronomy mission for launch in 2010. The total budget is about US$80M. Its main scientific goals are deep hard X-ray all-sky survey between 20 - 250 keV to discover about 1000 new supermassive black holes, pointed observations of faint objects: high-sensitivity timing studies, and high-sensitivity detection of other hard X-ray source. The characteristics are:

- main detector NaI(Tl)/CsI(Na) Phoswich
- total detector area 5000 cm^2
- energy range 20 - 250 keV
- energy resolution 22 % (60 keV)
- continuum sensitivity 3.0×10^{-7} ph cm^{-2} s^{-1} keV^{-1} (3 sec, 100 keV, 10^5 s)
- field of view $5^\circ.7 \times 5^\circ.7$ (FWHM)
- source location 1′ (20 s)
- angular resolution 5′ (20 s)

- mass $\sim$1100 kg, payload $\sim$700 kg
- dimensions $1.7 \times 1.7 \times 1.2$ m
- nominal mission lifetime 2 years
- orbit altitude 550 km; inclination 43°
- three-axis stabilized
- control precision $0^\circ.25$
- stability $0^\circ.005$ /s
- measurement accuracy $< 0^\circ.01$

A laboratory prototype has been successfully developed and tested over the last several years. Possible secondary instruments through international collaborations could be a soft X-ray telescope and a wide field X-ray monitor.

For further details, see: <http://www.hxmt.org/english>.

5. Site survey in west of China

Western China is a vast area. Most places of this area are at high altitude, with clean air, low temperature, dry environment, less cloudy and dark sky. Targeting on the sites for ELT (optical-IR-submm), National Astronomical Observatories, CAS has started a site survey in the western of China (Tibet, Xinjiang, Qinghai, Yunnan) and set up several stations in 2005.

6. Plan for Antarctic Dome A

A Chinese expedition team arrived at Dome A on Jan. 10, 2005, starting Chinese astronomical activities involved in PANDA - a Chinese key international program for IPY. Some efforts have been done together with the international astronomical community for preparing a site survey at Dome A and studying several wide field telescope plans.

7. Preliminary study for Extremely Large Telescopes

Several configurations for extremely large optical/infrared telescope (CFGT) have been studied (Su *et al.* SPIE Vol.4004, 2000; Su *et al.* SPIE Vol. 5489, 2004; Su *et al.* CAA, 28, 2004), and two configurations for extremely large wide field telescopes have been put forward. For CFGT: (1) one primary focus with FOV 100 % energy concentrated in diameter (d100) in $0''.36$ in FOV $20'$, 4 Nasmyth foci with d100 $0''.086$ in FOV $8'$ (diffraction limited in FOV $2'.83$), one Cassegrain focus with d100 $0''.11$ in FOV $8'$ (diffraction limited in FOV $2'.17$), one Coudé focus with FOV $29''.59$ (diffraction limited); (2) all with R-C system; (3) aspherical primary mirror with F/1.2 and 1122 partial annular sub-mirrors; (4) one secondary mirror with diameter 2.476 m for all optical systems. For the extremely large wide field telescope, two configurations have been studied: (1) LAMOST type extremely large wide field telescope; and (2) a corrector array extremely large wide field telescope. For both cases, all mirrors are plano and spherical.

Highlights of Astronomy, Volume 14
IAU XXVI General Assembly, 14-25 August 2006
Karel A. van der Hucht, ed.

doi:10.1017/S1743921307011738

ASTRONET: strategic planning for European astronomy

Anne-Marie Lagrange[1] **and Johannes Andersen**[2]

[1]ASTRONET Coordinator, INSU/CNRS, France
anne-marie.lagrange@cnrs-dir.fr

[2]Chair ASTRONET Board, Nordic Optical Telescope, Spain
ja@astro.ku.dk

Abstract. ASTRONET is an ERA-NET led by a group of European funding agencies, including ESO and ESA. Its aim is to establish a comprehensive, long-term planning process for all of European astronomy – at all wavelengths, from the ground and from space, and for all of Europe. By addressing both long-term scientific goals, infrastructure needs, and resource management procedures, ASTRONET seeks to consolidate the continued development of European astronomy as a front-line player in the field.

Keywords. planning and coordination,future large facilities, terrestrial intelligence

---

1. Introduction

As its most visible effort, ASTRONET will develop a strategic plan for European astronomy for the next $\sim$25 years - much longer than the 7-year span of Framework Programme 7 (FP7). The 'Decadal Surveys' conducted regularly in the USA are a model for the process, but ASTRONET's broader goals will pursued in three steps:

2. The Science Vision

First, a *Science Vision* will identify the key research priorities for European astronomy over the period, making maximum use of existing national and regional plans. This effort is led by a Working Group, assisted by expert panels addressing the following four overarching questions:

- Do we understand the extremes of the Universe?
- How do galaxies form and evolve?
- How do stars and planets form?
- How do we fit in?

The group and panel members have been selected for their scientific eminence, not as lobbyists for any discipline or project. A draft of their report will be published in November 2006, and comments from the community are invited via the web and in an open discussion at a large Symposium in January 2007, after which the report will be finalised and published.

3. The Infrastructure Roadmap

Taking over from the *Science Vision*, an *Infrastructure Roadmap* will be developed to chart out the tools needed to reach these scientific goals. It will comprise not only the next generation of observational facilities (ALMA, ELT, SKA, space missions, etc.), but also computing facilities and networks, virtual observatories, and human resources. An optimum implementation plan will be drafted later.

Close coordination will be maintained with the discipline oriented initiatives (e.g., OPTICON, RadioNet, ASPERA, ELT and SKA Design Studies, etc.) as well with the European Strategy Forum for Research Infrastructures (ESFRI), the European Science Foundation, the European Research Council, and the European Astronomical Society. A similarly open procedure as adopted for the *Science Vision* will be followed in the preparation of the final report on the *Infrastructure Roadmap*.

4. Involving all of Europe

Engaging all intellectual resources in Europe is crucial for the success of these initiatives which, we hope, will help to shape a strong and competitive future for European astronomy. ASTRONET is therefore contacting the astronomical communities and funding agencies in all EU Member and Associated States to involve them in developing the *Science Vision* and *Infrastructure Roadmap* and but enlist their participation in making them reality. Only then will we truly succeed.

5. Resource management

At present, no reliable statistics exist on the overall human and financial resources deployed in European astronomy. Similarly, national strategic plans - if they exist at all - are different in scope, form, and schedule, making rational long-term planning at the European level difficult at best.

ASTRONET is therefore conducting a survey of the organisation, funding, and national and regional strategic plans for astronomy in Europe. On this basis, proposals will be made for better-coordinated and more effective resource management procedures for European astronomy. A call for a jointly funded research project will then be issued as a pilot project.

6. The big picture

To be sure, ESA and ESO have developed long-term plans for their future investments; ESFRI is collecting plans for large European infrastructures for all the sciences; and the OECD Global Science Forum does the same on the global scene. So what will ASTRONET do that is not already being done?

It is crucial to note here that the budget for FP7 can fund only a tiny fraction of the new research infrastructures that Europe wants; the rest must come from the national funding agencies including the budgets of ESO and ESA, etc., whose plans cover just their own fields. But the funding agencies know they will in the end pay for it all and want to see how it all fits together, and ESFRI is not set up to recommend priorities or schedules.

On this background, ASTRONET is the chance offered by the funding agencies to European astronomy to prove that we can define our own priorities and present a common long-term plan that will convince them that their money will be well spent by us. If we succeed, a few more of our hopes may be realised. If we fail, we will all be worse off.

But there is an even wider perspective: Astronomy's greatest ambitions will no doubt only be achieved by truly global planning and cooperation. For Europe to play a role that befits its scientific, technical, and financial potential, we must sort out our own plans and priorities first. Only then can we aspire to more than playing second or third fiddle in the world of tomorrow.

For updated detailed information, see <http://www.astronet-eu.org>.

Highlights of Astronomy, Volume 14
IAU XXVI General Assembly, 14-25 August 2006
Karel A. van der Hucht, ed.

doi:10.1017/S174392130701174X

The Square Kilometre Array

Richard T. Schilizzi

International SKA Project Office,
c/o ASTRON, Postbus 2, NL-7990AA Dwingeloo, the Netherlands
email: schilizzi@skatelescope.org; url: www.skatelescope.org

Abstract. The Square Kilometre Array (SKA) is a global project to design and build a new generation radio telescope at metre to centimetre wavelengths.

---

1. Introduction

The Square Kilometre Array is under development by more than 50 institutes in 17 countries as one of the major observatories serving the global astronomical community for the coming decades. It will operate at metre and centimetre radio wavelengths with enormous sensitivity and wide field of view, and will have unprecedented surveying power. The range of key science to be tackled by the SKA (Carilli & Rawlings 2004) covers the epoch of re-ionization, galaxy evolution, dark energy, cosmic magnetism, strong field tests of gravity, gravitational wave detection, transients, proto-planetary disks, and the search for extra-terrestrial life. The major increase in performance compared to existing telescopes and the flexibility inherent in the telescope design allows us to predict that unexpected discoveries will be made with the SKA.

2. Telescope concept

The SKA is to be a radio telescope with

• the sensitivity to detect and image hydrogen in the early universe through its enormous collecting area of about 1 million square metres. This will make it about 50 times more sensitive than the EVLA, and able to reach an rms noise level of 10 nano-Jy in an 8 hr integration for a continuum observation;

• 50 % of the collecting area concentrated in the central 5 km diameter for optimal detection of hydrogen, pulsars, and magnetic fields;

• a fast surveying capability over the whole sky through its sensitivity and very large angle field of view of several tens of square degrees. The SKA will be 10000 times faster than the EVLA in surveying the sky;

• the capability for detailed imaging of compact objects like active galactic nuclei through its large physical extent of at least 3000 km;

• a frequency range from $\leqslant 100$ MHz to 25 GHz;

• data transport to the central data processor via very wide-band (terabit/sec) fibre links. Pflops/sec capacity is required for the central processor.

This combination of an enormous increase in sensitivity across the frequency range, a wide field of view, and the capability to sample the spatial, frequency, and time domains with high resolution will revolutionize many fields of astronomy.

The Reference Design for the SKA is an interferometer array capable of imaging the radio sky at frequencies from $\leqslant 100$ MHz to 25 GHz, and providing an all-sky monitoring capability at frequencies below 1 GHz. It covers the frequency band with three different

Figure 1. Artist's impression of the central 5 km of the SKA

kinds of receptors: (*i*) dipole arrays for $\leqslant$ 100 - 300 MHz to observe the Epoch of Reionization; (*ii*) a small parabolic dish array, with phased focal plane arrays in the 300 MHz to 3 GHz range and broad-band single-pixel feeds above 3 GHz; and (*iii*) aperture array tiles in the core of the array for all-sky monitoring in the frequency range 0.3 - 1 GHz and multiple independent field observations. These three receptor components all make use of the same data transport, processing, and software infrastructure. An artist's impression of the central core of the array is shown in the figure.

Much of the required technology is being developed in the course of the construction of several 1 % SKA Pathfinder instruments. In Europe, a Design Study (SKADS) is underway, funded in part by the European Commission. The Low Frequency Array (LOFAR) is under construction in the Netherlands and Germany. The extended New Technology Demonstrator (xNTD) and the Mileura Wide-field Array-Low Frequency Demonstrator (MWA-LFD), are under construction in Western Australia, as is the Karoo Array Telescope (KAT) in South Africa, and the Allen Telescope Array (ATA) in the USA.

3. Location

Following an evaluation of proposals for four potential locations for the telescope by the International SKA Steering Committee (ISSC) and its advisory committees, the ISSC decided in August 2006 that the short-list of acceptable sites would comprise Australia and Southern Africa. A final choice is expected to be made late in this decade.

4. Timeline and cost

The major milestones foreseen in the project are the following:

2008	external review of concept design
2009-2011	costed system design completed, final site selection, agreement on funding
2012	start construction of the SKA
2014	science with 10 % SKA
2020	construction completed, full science operations

The target cost for the SKA is 1 billion Euro.

Reference

Carilli, C., & Rawlings, S. 2004, *New Astronomy Reviews*, 48, 979

Highlights of Astronomy, Volume 14
IAU XXVI General Assembly, 14-25 August 2006
Karel A. van der Hucht, ed.

doi:10.1017/S1743921307011751

Special Session 2
Innovation in teaching and learning astronomy

Rosa M. Ros[1] and Jay M. Pasachoff[2] (eds.)

[1]Applied Mathematics IV, Universitat Politècnica de Catalunya, ES-08034 Barcelona, Spain
email: ros@mat.upc.es

[2]Hopkins Observatory, Williams College, Williamstown, MA 01267, USA
email: jay.m.pasachoff@williams.edu

Abstract. On August 17 and 18, 2006, Commission 46 on Astronomy Education and Development held a Special Session at the IAU XXVI General Assembly in Prague. The session, on *Innovation in Teaching/Learning Astronomy*, was organized around four themes: (*i*) general strategies for effective teaching, (*ii*) connecting astronomy with the public, (*iii*) effective use of instruction and information technology, and (*iv*) practical issues connected with the implementation of the 2003 IAU Resolution that recommended including astronomy in school curricula, assisting schoolteachers in their training and backup, and informing them about available resources. Approximately 40 papers were presented orally; in addition, 60 poster papers were displayed.

1. General strategies for effective teaching

The 20 papers constituting the first day's proceedings were devoted to the first theme. After welcoming remarks, commission president Jay M. Pasachoff and Rosa M. Ros, who later became the IAU Commission 46 vice-president, briefly summarized the main objectives of the Special Session, mentioning a variety of new methods of information dissemination (e.g., World Wide Web, Astronomy Picture of the Day, podcasts), the role astronomy can play in attracting young people to careers in science and technology, and the usefulness of technology both to observers and to teachers.

Former Commission 46 president *John Percy* of Canada next spoke about 'Learning Astronomy by Doing Astronomy'. Percy contrasted learning astronomy facts from lectures and textbooks and 'doing astronomy' in a more intellectually engaging fashion, emulating the actual scientific process, than called for in standard activities culminating in a predetermined result. While recognizing the value of 'hands-on' activities, such as making scale models of the solar system, Percy argued that 'minds-on' activities are more valuable, such as involving students in meaningful ways in their teachers professional research (even if the result is not publishable). He pointed out that in laboratories, which should mirror actual research, students can manipulate actual data, using the same computer languages and software used by real researchers. In these ways they can grasp that astronomy facts do not emerge fullblown from textbooks but are figured out by astronomers based on ongoing research. Percy spoke also of the value of having students themselves assume the role of astronomy communicators by tutoring peers or younger students or making classroom or public presentations. Percy referred to conference participant Richard Gelderman's assertion that students should they be exposed to recreational science, such as science fairs, 'for curiosity, interest, and ... for fun, fellowship, and ... mental well-being,' just as they are encouraged from their earliest years to participate in recreational sports for physical and mental health. He noted, too, that even urban students can learn to understand and make the observations that underlie the religious observances of major world faiths. He concluded with the thought that 'the ultimate goal of astronomy education' is to reach every student. While only a fraction will become professional astronomers, every student may become an amateur astronomer, with a lifetime passion for astronomy.

Like Percy, *Roger Ferlet* of France underscored the importance of engaging pupils in 'observing, arguing, sharing, discussing and interpreting real astronomical data, in order to enhance autonomy and reasoning; in brief, learning science by doing science'. Ferlet discussed the European Union's Hands-on Universe project as a tool to reverse the 'clear disaffection for scientific studies at universities' by convincing middle and high school students that scientific understanding 'can be a source of pleasure'. The project, a partnership of eight European countries under the auspices of the French University Pierre and Marie Curie in Paris, invites these younger students 'to manipulate and measure images' in class, using 'real observations acquired through an internet-based network of robotic optical and radio telescopes or with didactical tools such as Webcam,' assisted by scientific experts and by a select group of teachers who are trained in special workshops. The newly trained teachers go on to become 'resource agents' for their own countries, with a mandate to train other educators. This European undertaking is important not only to the international scientific community but to society as a whole, since not only does a 'sustainable economy' depend on innovations by 'a critical number of scientists and engineers,' but also societies that undervalue science 'regress to more primitive and much less attractive' conditions. An additional social benefit of the project is that it encourages communication among students from different countries. The positive reaction to Hands-on Universe has aroused hopes that similar European initiatives will result in 'Hands-on life' for biology, 'Hands-on Earth' for geology/ecology, etc'.

Former Commission 46 president *Edvard V. Kononovich* of Russia described a manual about the Sun, the Earth, and their interactions assembled by the Sternberg Astronomical Institute of Russia for older high school students and for 'students of natural faculties of universities and teachers colleges' with an interest in solar problems. The manual is also a useful teaching aid for courses covering solar physics and solarterrestrial relations. The manual, which makes use of both groundbased observations and results from *SOHO*, *Yohkoh*, *Ulysses*, *TRACE*, *CORONAS*, and other space telescopes, is divided into three sections. The first considers the Sun as a star, solar activity, and helioseismology. The second describes the Earth in space, its structure, its atmosphere, its magnetic field, its weather and climate, and active phenomena on the terrestrial surface. The third section considers such solar-terrestrial relationships as ionospheric disturbances, solar cosmic rays, and auroras.

Bill MacIntyre of New Zealand presented 'A Model of Teaching Astronomy to Pre-Service Teachers that Allows for Creativity in Communicating Students' Understanding of Seasons'. (In MacIntyre's paper, 'students' are teacher trainees, not the young people they will go on to teach.) MacIntyre began by distinguishing among mental models ('cognitive notions held by individuals'), expressed models (mental models that have been communicated to others), consensus models (expressed models valued by a social group and widely used by it), scientific models (consensus models that are used by scientists for further scientific developments), and teaching models (used to provide opportunities for teachers-in-training to develop their understanding of basic astronomy along with pedagogical skills). A goal of the 'investigating with models' approach is to have students understand 'the limitations and strengths' not only of their own models but also of scientists' models. Since expression of a scientifically inaccurate mental model has the potential to embarrass the student, teachers must make sure to have the revelations take place 'in a small group' and 'in a nonthreatening way'. The exercise requires students to collect evidence that will either support or lead them to change their mental models, a requirement that guarantees they will practice 'aspects of the nature of science (observations, inferences, creativity and empirically based knowledge) that relate to the systematic nature of investigating'. MacIntyre described in detail the process by which two teachersintraining creatively developed new expressed models after identifying specific aspects of their original expressed model that other members of their group had difficulty comprehending. Why is such an exercise for teacher trainees useful? 'If we expect classroom teachers to cater for the creative-productive gifted students during astronomy teaching in primary and secondary schools, then pre-service teacher training must model the appropriate classroom environment that allows it to occur'.

In 'How to Teach, Learn About, and Enjoy Astronomy', *Rosa M. Ros* of Spain described 'what I learnt after 10 years of the [European Association for Astronomy Education] Summer School,' drawing on the questionnaire responses of approximately 600 opinions of teachers of secondary school students from over 20 countries. To make students – in this case, the teachers – 'feel like actors in the teaching-learning process', she, like other SpS2 session participants, advocates 'learning by doing'. During the summer sessions the teachers are exposed to a variety of approaches to the teaching of astronomy, including modelmaking, drawing, playground activity. Just as a classroom teacher must be prepared to answer spontaneous questions from pupils about an astronomy topic that interests them, the summer school directors had to modify their lectures to accommodate 'the topics, matters, and methods' about which the teacher participants wanted to know more, bearing in mind always that astronomy concepts must be presented in some context, not in isolation. Instead of presenting a body of facts for students to memorize, 'It is important to connect the concepts with the personalities related to the topics, with the scientific situation in the past or maybe the social implications of the subjects'. One goal of the summer school is to encourage more inspired and more passionate teaching of math and physics: 'If the teachers enjoy teaching, the students will also enjoy their classes'. Ros lamented the fact that science museums tend to mount exhibits about science that stress the 'spectacular and funny', leaving to schools 'the boring science area'. Nonetheless, she noted that 'Not everything can be fun at school', arguing that teachers must also 'introduce the culture of making an effort to students'. Even underfunded schools with limited resources can include creative astronomy activities in the curriculum: 'All schools have a sky over their buildings. It must be used to observe and take measurements. If the school does not have tools and devices for making observations, we can encourage the students to produce their own instruments'. Whatever is lost in precision by doing so is more than compensated for in student commitment.

Based on his experience as director of astronomical laboratories at the University of Colorado in the United States, *Douglas Duncan* advocated the use of 'clickers' wireless student response systems as 'the easiest interactive engagement tool' in teaching large lecture classes. Studies show that students enjoy using clickers, which transform students from 'passive listeners' to 'active participants' in the learning process. The use of clickers also enables teachers to determine their level of comprehension without waiting for testtime. Studies show that the average student in a large physics lecture course, regardless of the effectiveness of the lecturer, truly comprehends at most 30 % of newly introduced concepts. To master a new concept, students 'must think about the idea and its implications, fit it into what they already know, and use it'. They must dislodge the misconceptions with which they enter the lecture hall. While professional scientists bounce ideas back and forth among themselves and within their own minds, students often believe 'that taking notes, memorizing, and repeating material on an exam is all there is to learning'. Duncan also advocated using clickers to facilitate peer instruction, since studies show that 'when comparable numbers of students start with right and wrong conceptions, peer instruction usually results in students agreeing on the correct answer, not the wrong one'. He cautioned, however, that 'like any technology, clickers can be misused, and it is important to practice and to explain their use to students before starting'.

Gilles Theureau reported on his experience with co-author *Karl L. Klein* teaching a two-semester course for students with varied backgrounds and interests at the University of Orleans (France) on the 'history and epistemology of the concepts of stars and galaxies' from antiquity to the early twentieth century. The cross-disciplinary approach is unusual in France, 'where pupils start to be specialized' from the age of 15. The course opens with a study of world systems from the pre-Socratics to the philosophers of the Middle Ages, moves on to concepts of mechanics and planetary motion from Aristotle to Newton (and a little beyond), proceeding then to a discussion of spectroscopy and the nature of stars, and concluding with descriptions of the Milky Way and the nature of nebulae. The course emphasizes 'mechanisms of knowledge,' including observation, experiment, and theory, as well as mythology, theology, philosophy, metaphysics, physics, mathematics, and instrumentation, in order to demonstrate that human ideas of the universe evolved 'as a part of human history and culture,' that 'science belongs to the patrimony of humanity and that it has no frontiers,' and that astronomy draws on both the humanities and the sciences. The course makes use of original documents that are considered in their historical context, including Aristotle on meteorology (350 B.C.E.), Nicolas Oresme's challenge to Ptolemy's view of celestial motion (c. 1380), William Herschel's 'On the Construction of the Heavens' (1785), and Agnes M. Clerke's Problems in Astrophysics (1903). The class also paid a visit to the Nancay Radio Observatory Astronomy Museum and Visitor Center; for the majority of students, who have 'never looked at the sky,' this visit provided 'an impression of the questions of interest to past generations of philosophers and astronomers'. Although in such a course it would have been easier to evaluate students through exams exclusively, the professors opted for the more difficult choice of assigning individual written projects in the first semester and a comprehensive exam in the second semester. The professors broadened their own cultural outlook by teaching the course, but found the presentation of material to and evaluation of such a heterogeneous group of students very challenging.

In 'Educational Opportunities in Pro-Am Collaboration', *Richard Fienberg*, editor of *Sky and Telescope* magazine, echoed other symposium participants in asserting that 'the Best way to learn science is to do science,' and called, as Michael Bennett would do in a later paper, for collaborations among amateurs, professionals, and educators. 'Amateurs will benefit from mentoring by expert professionals, pros will benefit from observations and data processing by increasingly knowledgeable amateurs, and educators will benefit from a larger pool of skilled talent to help them carry out astronomy-education initiatives'. Noting the important contributions amateur astronomers have historically made to the field, the loss of access of professional astronomers to mid-sized meter-class telescopes, and the need to follow up on 'countless interesting objects' being discovered by automated all-sky surveys, Fienberg recommended serious amateurs-many of whom have access to digital imagers on computer-controlled mounts-be given the opportunity to do some of this monitoring. He identified the American Astronomical Society Working Group for Professional-Amateur Collaboration as a 'forum for collaboration between amateur and professional astronomers'. Fienberg noted that amateurs continue to make important contributions to astronomy in areas including occultations; variable stars; meteor showers; CCD photometry and astrometry; and the search for and discovery of novae, supernovae, GRB counterparts, comets, and asteroids. In addition to this important work, amateurs can also help with astronomy outreach within their local communities and over the Internet with other researchers and educators around the world.

Jose Maza reported on his two-decades of experience 'Teaching History of Astronomy to Second-Year Engineering Students at the University of Chile'. The course partly fills the twocourse 'Humanistic Studies' requirement that each engineering graduate must complete. As a result of the course, men and an increasing number of women who will go on to work at and often become senior executives at major Chilean companies are exposed to the basics of astronomy and to its development over history. The first part of the course, 'a tour to the scientific revolution,' begins with the ancient civilizations, leads up to Newtons construction of modern science, and ends with the contributions of Euler, Clairaut, Lagrange, D'Alembert, and Laplace to celestial mechanics. The second part begins with William Herschel and the discovery of the Milky Way and proceeds over several weeks to a discussion of the big bang, the cosmic background radiation, and dark energy, before culminating with a lecture on the history of astronomy in Chile. Maza would be happy to exchange ideas with other astronomy educators.

Jay M. Pasachoff spoke about 'Education Efforts of the International Astronomical Union'. He described how the work of the commission, which resulted from a merger of the commissions on education and on astronomy in developing countries, is carried out in ten program groups. These groups include the world-wide development of astronomy, which sends some of its members to visit countries prospective for advances in carrying out astronomy and perhaps even becoming members of the International Astronomical Union; teaching for astronomy development, which provides visiting experts or lecturers to help advance a country's astronomical education; exchange of astronomers, which arranges international visits of several months or longer for people from developing countries to visit major research institutions; the IAU International School for Young Astronomers that is held every non-General-Assembly year for some dozens of new astronomers or graduate students; a semiannual newsletter; a group charged with coordinating with international institutions such as UNESCO, and that will now work with the *International Year of Astronomy* scheduled for *2009*; a group involved in international exchanges of journals that could aid developing countries; and a group related to taking advantage of public interest at the times of solar eclipses to spread astronomical knowledge, including but not limited to the eclipse itself. All these groups, the newsletters, and other related activities are accessible through the Commission 46 Website at `<http://www.astronomyeducation.org>`.

Magda G. Stavinschi of Romania, who became Commission 46 president at the end of the Prague IAU XXVI General Assembly, argued that astronomy is an integral part of human culture, in the evolution of which is often played a significant role. The discoveries of archaeoastronomers have proven that prehistorical civilizations pondered cosmological questions and wondered about the place of humankind in the universe. From the beginnings of history people across cultures have recorded significant events through markers that include not only human events, such as wars and the births and deaths of leaders, but also cosmic events, notably comets and eclipses. Stavinschi called attention to the often overlooked relationship between politics and astronomy, noting the post-World War I confirmation of Einsteins general theory of relativity by the Englishman Sir Arthur Eddington. During the war, Germany, Einsteins native land and the country that employed him, was a bitter enemy of England. This scientific collaboration not only served as 'a perfect proof of scientific internationalism' but also helped reincorporate German scientists into the scientific community after the war. After briefly mentioning the links between astronomy and geography, mathematics, physics, chemistry, meteorology, technology, medicine, and pharmacology, Stavinschi pointed out some famous examples of the incorporation of astronomy in art, music, heraldry, folklore, and literature. She spoke of the importance of astronomy education: those with an appreciation of the universe understand the need to protect Earth from manmade devastation 'much before its natural end'. While the mass media have succeeded in educating the public through coverage of space missions, and television programs featuring scientists like Carl Sagan have popularized astronomy, the media also are responsible for disseminating misinformation about astronomy. After briefly reviewing the history of astrology, she pointed to the astronomers' 'moral duty' to 'prove the quackery of astrology'. She concluded by arguing that 'there is no conflict between science and religion,' since they are 'two different ways of considering the world,' and dismissed the usefulness of arguing the relationship between astronomy and philosophy, since 'all that defines philosophy intimately contains the Universe and especially man in the Universe'. Astronomy can continue to play a role in the development of culture in pointing humanity in the direction of 'what it has to do from now on'.

Margarita Metaxa of Greece, where she teaches at the Arsakeio High School in Athens, spoke about 'Light Pollution: A Tool for Astronomy Education', which can help motivate not only students but also 'the public, government officials and staff, and lighting professionals'. Like Duncan, Metaxa noted that students 'hold misconceptions about the physical world that actually inhibit the learning of scientific concepts' and that 'students can remember less' than their teachers often assume. She called attention to a two-year program on light pollution sponsored by the Greek Ministry of Education and Religion, with backing from the EU; to the Internet Forum on Light Pollution, sponsored by the netd@ys Europe project, a European Commission initiative in the area of education, culture and youth for the promotion of new media; and to a UNESCO-backed conference on 'Youth and Light Pollution' held in Athens in autumn 2003. Outside of Europe, Chile has played a significant role in educating students about 'the effects of light pollution on the visibility of stars in the night sky'. She concluded by emphasizing that bringing both astronomy and light pollution to the world's attention in order 'to protect the prime astronomical places and the 'dark skies' as a world heritage' represents a significant challenge. It is one, however, that astronomers can meet by working 'together with interested organizations'.

In 'Student Gains in Understanding the Process of Scientific Research', *Travis A. Rector et al.* described 'Research-Based Science Education' as 'a method of instruction that models the processes of scientific inquiry and exploration used by scientists to discover new knowledge'. As an example of 'self-guided, cooperative groups' of undergraduates tackling 'a real research project', they summarized a student search for novae in the Local Group of galaxies by blinking images from the Kitt Peak WIYN (Wisconsin-Indiana-Yale-National Radio Optical Astronomy Observatory) 0.9 m telescope and then generating light curves and measuring decay rates through use of aperture photometry. Each student then selected a question to explore, 'such as comparing the location of novae in the galaxy and their rates of decay', presenting their conclusions both in a written research paper and in an oral class presentation. A comparison of student 'concept maps on the topic of scientific research', one completed before they undertook their research and one after, showed an overall deepening of their understanding of ten concepts inherent to scientific research. 'On average, students increased the number of the ten understood concepts ... from 2.8 before the class to 5.4 afterwards'.

In 'Effects of Collaborative Learning on Students' Achievements in Introductory Astronomy', *Myung-Hyun Rhee et al.* summarized their efforts over several years with non-science-major students at Yonsei University in Seoul, Korea. Students were divided into four groups, with one group participating in nine Collaborative Learning sessions, a second group in five, a third group in only two, and a control group participating in no such sessions at all. Students who participated in the greatest number of Collaborative Learning sessions experienced much less 'Communication Apprehension' than their peers, with the effect enduring even after six months had elapsed. The students with nine Collaborative Learning Sessions under their belts also rated higher in assessments of Academic Achievement and Class Satisfaction than students in the other three groups.

On behalf of a group of collaborators, *Stewart P.S. Eyres* of the UK described 'World-wide On-line Distance Learning' from astronomy courses prepared by the University of Central Lancashire. The student subscribers to these online courses range in age from 16, though most are over 21. They might include a retired industrial professional with a doctorate in chemical engineering, an English teacher with a deep interest in astronomy, an employee of an examinations board responsible for school astronomy curricula, a high school student preparing for university entrance exams, a primary school classroom assistant working toward a degree, among others. While all students share an interest in astronomy, they differ in what they hope to achieve through participating in the on-line distance learning program. Although distance learning in the UK is often associated primarily with the Open University, the University of Central Lancashire (UCLan) has been offering adult education courses in astronomy for about a decade. It is now possible to earn an honors bachelor of science degree in astronomy through UCLan. Eyres explained why the traditional 'teacher focused' astronomy education, in which an 'expert in the subject decides what the student needs to know at the end of the course, and works backwards from there to determine where they must start,' is inappropriate to distance learning. The framework of modules in the UCLan program enables students to determine if they are capable of higher-level work before they sign up for a lengthy course of study. Students are able to study modules provided by other institutions and use them for UCLan credit. They may also receive credit for skills they may already have in such fields, for example, as IT or math. Students correspond with their tutors both through email and through 'discussion and chat tools on UCLan's virtual learning environment'. With the introduction of competitive tuition fees at UK brick-and-mortar universities, the honors bachelor of science degree available through UCLan has the potential 'to attract students from the traditional 18-to-21 year-old UK degree market'. Even students who can pay the tuition fees at traditional universities may choose to take distance courses as a way of drawing attention to their qualifications and making themselves stand out from other applicants for admission. UCLan is also thinking of entering the teacher training market.

Donald Lubowich of Hofstra University, on New York's Long Island, USA, demonstrated how he has successfully used 'Edible Astronomy Demonstrations' to motivate students of all ages and to enhance their understanding of such varied concepts as differentiation, plate tectonics, convection, mud flows on Mars, formation of the Galactic Disk, formation of spiral arms, curvature of space, expansion of the Universe, and radioactivity and radioactive dating. His materials have included chocolate, marshmallows, candy pieces, nuts, popcorn, cookies, and brownies. Echoing other participants' comments that passionate and joyful teachers are effective teachers, he urged symposium participants to be 'creative, create your own edible demonstrations, and have fun teaching astronomy'.

In 'Amateur Astronomers as Public Outreach Partners', *Michael A. Bennett*, executive director of the Astronomical Society of the Pacific (ASP), which is based in San Francisco, USA, identified 'a huge, largely untapped source of energy and enthusiasm to help astronomers reach the general public' as volunteer science educators and urged astronomers and astronomy educators around the world 'to consider more formal cooperation with amateurs'. The ASP has estimated that, if one defines 'amateur astronomer' as one who has joined a club of like-minded people, there are over 50,000 'affiliated amateur astronomers' in the US alone. It has also estimated that US amateur astronomers 'reach some 500,000 members of the general public every year' through public star parties, classroom visits, community fairs, and museum/science center events. In March 2004 the ASP, with funding from the Navigator Public Engagement Program at NASA/JPL, launched its NASA Night Sky Network (NSN) to provide amateurs 'with tested Outreach ToolKits on specific topics that can be used in a wide variety of ways with many different types of audiences,' as well as training in how to use these resources. Amateur clubs must meet certain criteria in order to become members of NSN, and approximately 200 clubs had joined by summer 2006, representing approximately 20,000 amateur astronomers who have participated in over 4500 public outreach events. NASA funding is expected to continue for this effective program. Bennett urged astronomers around the world to identify ways to engage 'outreach amateur astronomers' in their own countries.

Underlying the paper of former Commission 46 president *Syuzo Isobe* of Japan about 'Does the Sun Rotate Around the Earth or Does the Earth Rotate Around the Earth? An Important Aspect of Science Education' is the conviction that effective astronomy teaching must begin with the consideration of four variables: the pupil's class year, ability, level of interest, and future career. While of course it is more accurate to teach that the Sun does not rotate around the Earth, it is not quite correct to teach that the Earth rotates around the Sun, since 'solar system bodies rotate around a gravity center different from the center of the Sun'. While for most students the assertion that the Earth is a sphere is adequate, students who have a higher interest level and students who may go on in scientific professions should know that the Earth 'is an ellipsoid or a geoid'.

Fernando J. Ballesteros and his colleague *Bartolome Luque*, both of Spain, made a case for 'Using Sounds and Sonification' -and not merely impressive astronomy photographs- 'for Astronomy Outreach'. The authors have a successful weekend radio program, 'The Sounds of Science', broadcast on the national radio station of Spain. Sometimes, in fact, as with pulsars, 'the images are not very spectacular but the sounds are strangely attractive'. Astronomical sounds are also available to blind people in a way that images are not. [SpS 2 co-editors' note: At least three books of astronomy images are available to the blind: Noreen Lawson Grice's *Touch the Stars, Touch the Universe: A NASA Braille Book of Astronomy* and *Touch the Sun: A NASA Braille Book*. Pasachoff reviewed them in the U.S. college honor society Phi Beta Kappa's *The Key Reporter*, spring 2006, pp. 15-16, downloadable at <http://www.pbk.org>.] Ballesteros identified a number of Internet resources for astronomy sounds, the computer software 'Sounds of Space' available for both PCs and Macs, and the possibility for professional astronomers to 'sonificate' their own data, by passing them 'to an audible format'. Addressing the issue that there is no sound in the vacuum of space, Ballesteros notes that this fact represents a teachable moment in itself, since 'in many cases the sounds will be radio signals passed to sound', as is the case with both pulsars and aurorae. Similarly, black holes, lightning storms on Saturn, and ionization tracks from shooting stars also emit radio signals. Ballesteros noted that in some cases there are real sounds, such as 'when a shooting star crosses the sound barrier,' and in others the sound may be inaudible but can be indirectly reconstructed, as in the case of 'sound waves crossing the solar surface', which the vacuum of space prevents from reaching Earth, but which *SOHO* instruments can record indirectly and reconstitute after the fact.

Basing their argument on successful activities offered for school students at the Sydney (Australia) Observatory, *Nicholas R. Lomb* and *Toner M. Stevenson* contend in 'Teaching

Astronomy and the Crisis in Science Education' that the trend in some countries to shun careers in math and science can be overcome by using astronomy 'as a tool to stimulate students scientific interest'. The matter is of some importance, since if the trend is not offset, 'there may not be enough people with Science, Engineering & Technology (SET) skills to satisfy the demand from research and industry'. Not only will it be necessary to replace retirees from the 'baby boomer' generation but also burgeoning industries including nanotechnology, biotechnology, and information technology will require workers with SET skills. In Australia, however, studies show a decline in the number of high school students studying advanced mathematics and both the life and physical sciences. Data from other countries are similarly dispiriting. Unless their parents have a positive attitude toward science, students often shun the physical sciences because they think of them as boring and irrelevant. Many students perceive science to be so difficult that only highly gifted students can succeed at them. Even those with high aptitude sometimes enroll in science courses only to improve their chances at excelling in university entrance exams. Lomb and Stevenson assert that planetaria and public observatories can improve student attitudes to science by engaging students' interest in a personal way.

In 'Astronomy for All as Part of a General Education', *John E.F. Baruch et al.* discussed the pluses and minuses of using www.telescope.org/, a Web-based education program in basic astronomy available free of charge to anyone with Internet access. The authors explained the advantages of truly autonomous robotic telescopes, such as the Bradford Robotic telescope, which can 'deliver the initial levels of astronomy education to all school students in the UK', over remotely driven telescopes that reach only 'a tiny percentage ... of students'. They also discussed 'practical solutions' for assisting teachers lacking not only a deep knowledge of basic astronomy but also confidence in working with information technology.

The 20 oral papers presented on Friday, August 18, related to the remaining three themes of the special session: connecting astronomy with the public, effective use of instruction and information technology, and practical issues connected with the implementation of the 2003 IAU Resolution.

2. Connecting astronomy with the public

The second day of the Special Session began with a status report delivered by *Dennis Crabtree* on behalf of the IAU Division XII Working Group on *Communicating Astronomy with the Public* (WG-CAP), which was established in late 2003. After giving the URL <http:// www.communicatingastronomy.org> for the WG-CAP's 'effective website ... describing its activities', he singled out one of those activities ('to promulgate adoption of the Washington Charter by various professional and amateur astronomy socities, funding agencies, and observatories'), mentioned that WG-CAP held 'a very successful 2005 meeting on Communicating Astronomy with the Public' at ESO and had begun planning for another such meeting in 2007, and noted that at the current General Assembly the Working Group had been converted to Commission 55 under Division XII.

After a status report from the Division XII Working Group, *Julieta Fierro* of Mexico won the prize for the liveliest presentation of the session, conveying 'Outreach Using Media' with the 'passion and ... joy' that she hopes all astronomers will bring to their outreach activities. While her presentation involved her leaping onto tables, throwing books out into the audience, and getting normally staid astronomers up on their feet to swing dance according to her instructions, her message was far from frivolous. She recommended that astronomers hire professional fundraisers to help find funding for astronomy outreach activities, which she believes are an ideal way to excite the general public about science in general. The point of Fierro's 'dance lesson' was to model some characteristics of the type of informal learning she believes astronomy educators can provide: the teaching is done incrementally to eager participants, who are given a lot of time to practice, are encouraged to learn from their mistakes, and to accept help from their peers in their quest for mastery. The point of her 'book toss' was to encourage other astronomers to follow her example in writing illustrated popular astronomy books to capture the public's attention. Book-writing is not Fierro's only outreach activity; for almost a decade she has hosted, together with a chemist co-host, a 45-minute weekly radio program, aired during the evening rush hour. Even if every astronomer does not host a regular show, everyone can be available for interviews. She also recommended television for public outreach. Noting that many media personalities lack a strong science background, she encouraged astronomers to give guest lectures at journalism schools. Fierro reported on a graduate program for public outreach, offering both a master's and a doctorate, offered by Mexico's national university, with students majoring in at least one branch of the media and one scientific field. Since Fierro's presentation was mainly directed at astronomers in developing nations, she concluded by urging public policy makers to fund scientific outreach activities, since only a scientifically literate society can flourish. Noting that 'If women are not prejudiced against science, their children will perform better at school,' she also called for special programs for women.

In 'Hands-on Science Communication', *Lars Lindberg Christensen* named several astronomy-related 'fundamental issues with a great popular appeal', including 'How was the world created? How did life arise? Are we alone? How does it all end?' He noted the growing importance of communicating science to the public, not only to attract future scientists to the field but also to create support for public funding of science. He noted that university statutes in some countries are being redrafted 'to include communication with the public as the third mandatory function besides research and education'. Christensen then identified several 'interesting 'lessons learned' from the daily work at the Education and Public Outreach (EPO) of the ESA/Hubble ST-ECF'. These included the most effective flow of communication from scientist to public, the criteria for a successful press release, the benefits to an EPO office of a 'commercial approach', the appropriate skills base in a modern EPO office, and the more efficient use of modern technology for communicating science.

Silvia Torres-Peimbert presented a 'Critical Evaluation of the New Hall of Astronomy for the Science Museum' of the University of Mexico, which opened to the public in December 2004. The Science Museum as a whole, which opened in December 1992, covers a wide range of topics, ranging from mathematics to agriculture, and also includes a library, a 3-D theater, auditoriums, and outdoor displays. The original Hall of Astronomy's displays were primarily focused on the Solar System and paid little attention to the Universe beyond. The current renovation 'comprises 60% of the space assigned to astronomy'; in a later stage the Solar System displays will be renovated. Divided into three sections (the Sun, stars, and interstellar matter; galaxies, clusters, and the Universe as a whole; astronomical tools), the primary displays of the new Hall of Astronomy convey the idea that the Universe and all its components 'are undergoing continuous evolution'. Additional displays include 'a representation of the vastness of space [through a powers of 10 display of pictures], a time line from the Big Bang to the present epoch [laid out in a long strip on the floor], and some video clips from local astronomers [answering frequently asked questions and explaining their own research],' as well as a section on the history of astronomy [in a cartoon display, including the pre-Columbian astronomers]. The exhibit's goals were to 'attract young students to science, ... present modern day astronomical results and show that astronomy is an active science, ... show that we can interpret cosmic phenomena by means of the laws of physics, ... show modern-day concepts of the structure of the Universe and its constituents, ... show that modern technology has played a major role in increasing our knowledge of the universe'. Torres and her coauthor carried out their evaluation of the new Hall of Astronomy by observing the behavior of a sample of 50 visitors and by interpreting a sample of 100 visitors' responses to a questionnaire they distributed. From their observation of visitor behavior, the authors concluded that 'The favorite displays were those that are larger and interactive'. They attribute the lack of interest in the time line 'to the fact that it is not well advertised'. From analyzing the questionnaire, the authors concluded that 'many topics are too complicated' for visitors under 15, but that older visitors 'found interesting new information'. While 'not everybody is attracted to the most challenging displays,' interested visitors enjoyed the opportunity to delve deeply into these subjects. The authors also concluded that the new hall 'can be of assistance to science teachers'.

A special lecture on astronomy education research by *Timothy F. Slater* of the Conceptual Astronomy and Physics Education Research (CAPER) Team at the University of Arizona, USA, was the final presentation on Theme 2. The main thrust of 'Revitalizing Astronomy Teaching Through Research on Student Understanding' is that the lecture-tutorial model for teaching introductory astronomy is more effective for a majority of students (excluding, interestingly enough, 'those most likely to become faculty themselves') than the traditional lecture-only model. Not only do students learn more, but the opportunity to engage in Socratic dialogue during tutorial also leads students 'to reason critically about difficult concepts in astronomy and astrobiology'. The lecture-tutorial model 'does not require any outside equipment or drastic course revision' on the part of the instructor, and is more 'learner-centered' than the lecture-only model, transforming passive listeners to active participants in the learning process. Slater made the interesting point that although the course is called 'introductory astronomy', for many of the more than 200,000 students who take the course annually 'it is their terminal course in astronomy, and in fact marks the end of their formal education in science'. For that reason, introductory astronomy 'represents an opportunity to engender the excitement of scientific inquiry in students who have chosen to avoid science courses throughout their academic career'. Since many future schoolteachers take this course, it is also an opportunity to model 'effective instructional strategies' for them. Like other participants in the session, Slater noted that professors tend to believe that their students learn more than they actually do, since students not only lack the basic vocabulary of astronomy to begin with but also come in with misconceptions that impede their grasp of basic concepts. In noting the 'small cognitive steps' students can make in the lecture-tutorial model and the method of 'having students work collaboratively in pairs in order to capitalize on the benefits of social interactions', Slater echoed points made earlier by Fierro. During the collaborative-learning tutorials, which take place in the regular lecture hall, the professor steps out of the lecturer role and into the role of facilitator, 'circulating among the student groups, interacting with students, posing guiding questions when needed, and keeping students on task'. In end-of-semester course evaluations, students 'frequently commented positively on the lecture-tutorials, even without being prompted', generally noting how much better they understand material after hashing out their difficulties with classmates.

3. Effective use of instruction and information technology

The first of five presentations relating to Theme 3 was made by *Douglas Pierce-Price*, representing a group of collaborators from Garching, Germany, and Santiago, Chile. In 'ESO's Astronomy Education Programme,' he described several educational activities run by the Educational Office of the European Organization for Astronomical Research in the

Southern Hemisphere (ESO), some of which are run collaboratively with the European Association for Astronomy Education (EAAE). Among the recent activities of the ESO, which is headquartered in Garching with three astronomical observatories in Chile, were the coordination of over 1500 teams of international observers at the time of the 2004 Transit of Venus and the celebration of the World Year of Physics in 2005 by providing equipment for measuring solar radiation levels to students in schools throughout Chile. For several years ESO and EAAE have jointly sponsored 'Catch a Star!,' an international competition aimed at developing 'interest in science and astronomy through investigation and teamwork'. Student teams from around the world do research on an astronomical object or theme, 'and discuss how large telescopes such as those of ESO can play a part in studying it'. Although some of the contributions are judged by an international jury (with awards including travel to ESO's VLT facility in Chile or to observatories in Europe), some prizes are also awarded by lottery in order 'to avoid a sense of elitism'. In autumn 2005, ESO and its partners in the EIROforum (a partnership of Europe's seven largest intergovernmental research organizations) sponsored the first 'Science on Stage,' a science education festival. The next such festival is scheduled for spring 2007 in Grenoble. ESO also provides astronomy-related articles for 'Science in School' (a new European science education journal for teachers, scientists, and others) and produces 'Journey Across the Solar System' (a series of informational sheets) and a series of astronomy exercises 'based on real data from the VLT or HST, the former in collaboration with EAAE and the latter in collaboration with ESA. A new undertaking is ALMA ITP, where ALMA stands for the Atacama Large Millimeter Arraya new astronomical facility under construction in Chile's Atacama desert by ESO 'as part of a global collaboration' -and ITP stands for Interdisciplinary Teaching Project. The teaching material 'will highlight the links between 21st-century astronomy and the topics in engineering, earth sciences, biology, medicine, history and culture' related to ALMA's location in the Earth's driest spot. The goal of ALMA ITP is 'to introduce scientific topics to students as part of other school subjects, and also to put scientific research into a wider context'. Like other participants in the session, Pierce-Price spoke about the importance of such educational activities 'to ensure that future citizens, whatever their careers, have the scientific literacy they need to make informed decisions about issues related to science'.

Mary Ann Kadooka from the Institute for Astronomy at the University of Hawaii spoke about the special challenges and rewards of running an astronomy educational outreach program in Hawaii, which is very culturally diverse. In 'Astronomy Remote Observing Research Projects of USA High School Students,' Kadooka explained how the NSF's Toward Other Planetary Systems (TOPS) teacher enhancement workshops, held over 18-day periods between 1999 and 2003, were the first major effort to introduce math and science teachers from Hawaii and the Pacific Islands to astronomy. The participants, some of whom returned for several years, became 'a master teacher cadre to serve as the backbone of our student project efforts today'. Some of these continue to mentor former students, encouraging them to attend star parties, public lectures in astronomy, and other science-oriented events. Among the ongoing partnerships resulting from the TOPS program are those with the Bishop Museum, the Hawaii Astronomical Society, the American Association of Variable Star Observers, the Faulkes Telescope North (now part of the Las Cumbres Observatory), and NASA's Deep Impact Mission. Demonstrating that the PRO-AM relationship advocated by Fienberg and Bennett on the first day of the special session is already in effect in some locations, Kadooka described the contributions to astronomy education in Hawaii made by amateur astronomers. Hawaii's TOPS-trained teachers have mentored many student participants in science fairs in Hawaii and in Oregon (where one of Hawaii's master teachers now teaches in a Portland private school, where she offers an electivescience research course). A NASA IDEAS (Initiative to Develop Education through Astronomy and Space Science) grant awarded in May 2006 is making possible outreach to a new interisland target group of students from grades 7-10, including Native Hawaiians and 'at-risk, rural students'. To that end, mini-workshops were offered in autumn 2006 to interested and committed teachers on the islands of Maui and Molokai. To maximize the chances of entering 'exemplary student astronomy research projects' in the 2008 Hawaii State Science Fair, and with the hopes of having some of these make the cut for the 2008 Intel International Science Fair, students committed to doing research will be recruited for a summer 2007 week-long workshop on astrobiology.

Like other special session participants, *Richard Gelderman* of Western Kentucky University, USA, believes that 'our primary job as teachers is to prepare tomorrow's citizens, rather than to prepare tomorrow's scientists' and that 'Astronomy is perhaps the field of science where the biggest contribution can be made toward the creation of a scientifically literate society'. A proponent of 'hands-on, minds-on astronomy experiences', he advocates giving students 'greater access to astronomical telescopes'. In 'Global Network of Autonomous Observatories Dedicated to Student Research', he described the Bradford Robotic Telescope and 'the beginnings of global networks of autonomous observatories'. The Bradford Robotic Telescope, inaugurated in 1993, has had a positive impact on science education by providing for 'the general public's requests for queue-scheduled service observations as well as remote, manual operation', thus 'offering access to an autonomous observatory coupled with welldesigned projects and guided activities'. The opportunities that should become available from the proposed networks of autonomous observatories will doubtless enhance 'hands-on astronomical education', but 'other successful ways of engaging and teaching young people' should not be overlooked. Gelderman described some of these other educational possibilities, such as having students build their own 'simple, low-cost' telescopes, exposing them to WebCam technology, and-echoing Ferlet's earlier paper-engaging them in the Hands-On Universe program.

David H. McKinnon and co-author *Lena Danaia*, from Charles Sturt University, Bathurst, Australia, described in 'Remote Telescopes in Education: an Australian Study', how 'the use of remote telescopes can be harnessed to impact in positive ways the attitudes of students'. Echoing the earlier paper by Lomb *et al.*, McKinnon lamented the decline in science enrollments 'during the post-compulsory years of education' in Australia and elsewhere. In the mid-1990s McKinnon built the Charles Sturt University (CSU) Remote Telescope, which enabled students to get their own images of celestial objects. Impressed by 'the motivational impact that the control aspect of the CSU Remote Telescope had on primary age students', the Federal Government of Australia commissioned a study that developed educational materials for students and their teachers and assessed some outcomes in those who used them. Students who were given access to remote control of the CSU telescope not only acquired 'a significantly greater ability to explain astronomical phenomena', but also 'increased their astronomical knowledge significantly' and significantly improved their 'attitudes towards science in general and astronomy in particular'. McKinnon concluded by expressing the hope that access for students to the Las Cumbres Observatory's global telescope network will confirm 'the hypothesis that a love of astronomy can be engendered in more students'.

The final paper related to the 'effective use of instruction and information technology' was 'Visualizing Large Astronomical Data Holdings', by *Carol Ann Christian et al.* As 'huge quantities of observed or simulated data' become available, it is important both for educators and researchers to optimize the visual display of the astronomical information. Christian described several tools available for scientific visualization, including the

Sloan Digital Sky Survey Navigate Tool, which enables users of the SkyServer website to browse, create finding charts, and display portions of the sky in the form of catalog data; World Wind, a NASA Learning Technologies project, which allows exploration of Mars and the sky 'as represented by a number of all-sky surveys'; Digital Universe Atlas, from the American Museum of Natural Historys Hayden Planetarium, which enables users to browse the sky, find brown dwarfs, and carry out a number of educational activities; Science on a Sphere, developed by NOAA, which already makes it possible to examine data and phenomena on Earth, Mars, and the Moon, and soon will be adapted for all-sky survey data; and Google Earth, which currently enables the display of astronomical images, and should eventually 'allow interfaces to data archives, press release archives, and possibly the National Virtual Observatory'. McKinnon concluded by expressing the belief that 'with the emergence of large multi-wavelength all-sky survey data and collections of data ... visualization tools can be important for public understanding of science and education both in formal classroom and informal science center settings'.

4. Practical issues connected with the implementation of the 2003 IAU Resolution

The responses of a number of different countries to the IAU Resolution were discussed in this session, beginning with *Edvard V. Kononovich*'s 'Stellar Evolution for Students of Moscow University'. Advanced students specializing in astrophysics and with a strong background in theoretical astrophysics are assigned 'a special practicum work' that requires them to solve five problems using a PC program based on Paczynski code and supported by the Web interface, enabling them to use the Internet. Problem 1, which deals with zeroage main sequence (ZAMS) stellar models, requires students 'to calculate ZAMS models for three different star masses and two variants of the chemical composition'; problem 2, dealing with main sequence stellar models, requires student 'to calculate the evolutionary tracks for three stars with different masses during the time of hydrogen burning in the star core'; problem 3, dealing with 'the evolutional peculiarity of stars with different masses and different chemical composition,' requires them 'to calculate the evolutional tracks for three stars with the same composition and different masses,' and in one case 'to change the chemical composition'; problem 4, dealing with 'structure of red giants and supergiants,' asks students 'to calculate an evolutionary star track up to the supergiant branch'; and problem 5, relating to the evolutionary model of the present sun, asks students 'to compute the standard solar model ... using as free parameters that of the convection zone and of the chemical composition'.

Moving from the particular to the universal, *Maria C. Pineda de Carias* presented 'Astronomy for Everybody: An Approach from the CASAO', the acronym for Central America Suyapa Astronomical Observatory, part of the National Autonomous University of Honduras, where currently all professional astronomers in Honduras are employed, along with regional and foreign colleagues. After noting that 'Honduras is a country of very young people' and that 'most students are at elementary school level' with fewest students in university, she explained that the astronomical observatory of Honduras was inaugurated in 1997. In the ensuing decade not only did the observatory receive 'regional accreditation' but also initiated 'a regional program in astronomy and astrophysics at both undergraduate and graduate levels' for Central Americans'. She described three different outreach programs, targeting different audiences. One program, aimed at 'elementary and secondary school students, teachers, college students, parents, and media communicators', involves two-hour visits by groups of 20 to 100 to CASAO, where an astronomer delivers a lecture 'organized and adapted to the interest of the participants', followed by practical activities aimed at familiarizing the visitors with the use of 'small telescopes to observe the sun and planets' and at demonstrating 'how the Maya of Central America used stelae' to measure solar time and design solar calendars. The goals of this program include motivating 'the study of science, mathematics, and space exploration,' arousing 'curiosity for learning about what exists' beyond 'our own environment,' and introducing new resources for teaching and learning science. A second program involves Friday-night visits by the general public to the observatory. The goals of the 'Astronomical Nights Program' are to familiarize visitors with the night sky and the visible universe and give them the opportunity to make naked-eye observations under the guidance of a professional astronomer. Recognizing that most of the beneficiaries of these two programs come from the capital city, Tegucigalpa, the observatory is making an effort to reach children elsewhere in the country. A third program, 'Introduction to Astronomy @ Internet', is an online course written in Spanish, offered as an elective 'to all career students' in Honduras 'as part of their education' but particularly recommended for elementary and secondary school teachers, as well as to students and teachers elsewhere in Central and Latin America. The course 'represents an opportunity ... to be part of a scientifically literate generation, trained by professional astronomers'.

In 'The Epistemological Background of Our Strategies', *Mirel I. Birlan et al.* focused on the importance of identifying the fundamental concepts that astronomers wish to convey in educating the public. Only by being aware of 'the importance of the epistemological background' will astronomers succeed in 'defining the strategy of education in (through) astronomy', and in determining the role of astronomy should play as a required course within a general education curriculum, as it interacts with other scientific disciplines.

Turkey's response to the 2003 IAU Resolution, 'Towards a New Program in Astronomy Education in Secondary Schools in Turkey,' was presented by *Zeki Aslan*. Although before 1974, astronomy was taught on its own at the secondary level in Turkey, since then it has been incorporated into secondary-school physics courses as well as into elementary science and geography courses on the primary level. Most teachers on those levels, however, have had little formal preparation in astronomy, so that the teaching has not been 'very effective'. In addition, topics in astronomy 'are generally scheduled at the end of a particular course,' when there is little time left to devote to them. To remedy the situation, in 2005 the ITAK National Observatory (TUG) proposed to the Ministry of Education that a national meeting on 'teaching of astronomy and using astronomy to teach physics' be held for teachers of physics and astronomy teachers at the time of the 29 March 2006 total solar eclipse. The approximately 120 hand-picked schoolteachers and 10 schoolchildren 'had a beautiful sky to see the eclipse and to carry out ... experiments'. According to Aslan, the meeting, which involved 'astronomers and physicists from Turkish universities and educators from the Ministry of Education plus three educators from abroad', was 'very successful'. TUG subsequently submitted to the Ministry of Education a number of proposals, suggesting, among other things, that the Ministry of Education and TUG should collaborate in providing educational materials and in holding summer schools and in-service training for primary and secondary school teachers. Independently, in 2005 ME published a draft for a new primary-school science and technology course, in which astronomy topics, previously taught on the primary level as parts of other subjects and now presented under the rubric 'The Earth and the Universe', will make up nearly 10% of the entire course. A similar program for secondary schools is being prepared.

In 'Astronomy in the Russian Scientific-Educational Project', *Alexander V. Gusev* and *Irina Kitiashvili* decribed the International Center of the Sciences and Internet Technologies 'GeoNa' at Kazan University in the Republic of Tatarstan. Historically known as the home institution of mathematician Nikolai Lobachevsky, father of non-

Euclidian geometry, Kazan University is now also the home of Center GeoNa ('Geometry of Nature'), 'a modern complex of conference halls including the Center for Internet Technologies, a 3D planetarium, ... an active museum of natural sciences, an oceanarium, and a training complex ...'. Through Center GeoNa, scientists and educators at Russian universities will be able to share their 'advanced achievements in science and information technologies' with foreign colleagues in 'scientific centers around the world'. In addition to hosting 'conferences, congresses, fundamental scientific research sessions' on lunar research, Center GeoNa also hopes to initiate a 'more intense program of exchange between scientific centers and organizations for a better knowledge and planning of their astronomical curricula and the introduction of the teaching of astronomy'.

Cecilia Scorza, on behalf of colleagues at ESO and the universities of Heidelberg and Leiden, presented 'Universe Awareness for Young Children' (UNAWE). This international program, 'motivated by the premises that access to the simple knowledge about the Universe is a birthright and that the formative ages of 4 to 10 years play an important role in the development of a human value system,' targets 'economically disadvantaged young children' in this age group and exposes them to 'the inspirational aspects of modern astronomy'. The program should be operational by 2009, the *International Year of Astronomy*, with goals including production of 'entertaining material in several languages and cultures,' organization of training courses for those who will present the program, and provision of a network for exchange of ideas and experiences. In 2006 pilot projects were carried out in Venezuela and Tunisia to examine UNAWE's feasibility. More information relating to UNAWE is available at <http://www.UniverseAwareness.org/>.

Ahmed A. Hady spoke about 'Education in Egypt and the Egyptian Response to Eclipses'. He began by summarizing the history of modern astronomy education at the university in Egypt, which began in 1936. The University of Cairo offers the bachelor of science degree in astronomy and physics, in astronomy, and in space science; a masters degree in astrophysics, in theoretical physics, and in space science; and a doctorate. Professional astronomical research is conducted at Cairo University and at Helwan Observatory in a variety of fields. Egyptian scientists participated in international observations of total solar eclipses in Egypt on 25 February 1952 and 29 March 2006. The more recent observations are being coordinated with the ESA/NASA *Solar and Heliospheric Observatory* (*SOHO*) and NASA's *Transition Region and Coronal Explorer* (*TRACE*) observations to show the magnetic structure of the corona.

Paul Baki described 'Spreading Astronomy Education Through Africa'. After explaining how different African societies 'practice astronomy largely for understanding and predicting the weather and climatic changes for seasons', he described 'some traditional tools that are used by some ethnic communities in East Africa to interpret astronomical phenomena for solving their local problems'. He noted that these communities 'combine the knowledge of plant and animal behavior changes together with sky knowledge to predict the weather and climate for the coming season'. Baki argued that from the African perspective, 'it seems that the best way to spread knowledge in astronomy is to begin by appreciating its cultural value'. He suggested incorporating the traditional practices into the standard astronomy curriculum, and predicted that by leading to environmental conservation and increased crop yields, as well as to an increase of tourism, such an approach would 'get recognition and possible funding from the various African governments'.

Located on the *Pampa Amarilla* in western Argentina, the Pierre Auger Cosmic Ray Observatory not only studies the highest energy particles in the universe but also participates, according to Beatriz Garcia, in 'a wide range of outreach efforts that link schools and the public with the Auger scientists and the science of cosmic rays, particle rays, particle physics, astrophysics, technologies'. In 'Education at the Pierre Auger Observatory:

The Cinema as a Tool in Science Education', Garcia described the use of educational videos for children between 6 and 11 and for general audiences, as well as the use of animation in science teaching and learning. She identified scientific outreach as a means of encouraging 'scientific vocations', particularly in countries where they are not accorded a high social status. Garcia asserted that 'if we want to help the student-public to think and be able to solve problems, the audio-visual language must be characterized by its originality and the search of new forms of expression that stimulate the imagination'.

The session concluded with *Mary Kay M. Hemenway*'s 'Freshman Seminars: Interdisciplinary Engagements in Astronomy'. To facilitate the transition of the diverse population of students entering the University of Texas at Austin to college academic and social life, the university offers freshman seminars limited to 15. Instructors invited to participate in the program may design the course of their choosing. The only stipulations are that students must complete a certain number of certain types of writing assignments, and must also attend 'sessions on time-management and using the library'; students whose seminars involve only two hours of class time each week must also attend an additional hour-long weekly event. Hemenway reported on two seminars rooted in astronomy. For a seminar focused on the life of Galileo, students modeled 'rotation and revolution of solar system objects with their own bodies'; experimented with lenses to master the concepts of focal length, field of view, and refracting telescopes; compared positions of Jovian satellites as Galileo drew them in *Sidereus Nuncius* with those calculated by the *Starry Night* computer program for the dates and location corresponding to Galileos depictions; and a dramatic reading of an English translation of Bertolt Brechts play 'Galileo'. These classroom activities were enriched by visits to the Blanton Museum of Art to view Italian art of the period, as a springboard to discussing the influence of religious struggles on contemporaneous artists; and to the Harry Ransom Center for humanities research, to expose students to original seminal works in the history of astronomy, including, among others, Ptolemy's *Almagest*, Copernicus's *De Revolutionibus*, and several by Galileo himself. A highlight of this seminar for many students is the mock trial 'in which Galileo has the benefit of something he lacked in reality-a defense team'. Students not assigned to either the prosecution or defense team participate in the judgment phase. A second seminar, 'Astronomy and the Humanities', students are exposed to science fiction, which they are asked to contrast with science facts known now and at the time of the writing; to poetry and literature with some astronomical connection; to a range of music with an astronomical theme; and to art work relating to astronomy. Hemenway concluded by noting that while the seminars' goal differs from that of introductory astronomy courses, Astronomy 101 instructors might do well to pick and choose from among the broad humanistic connections the seminars highlight to enrich the teaching of those standard academic offerings.

5. Abstracts of oral contributions

5.1. *Main Objectives for this IAU Special Session on Innovation in Teaching and Learning Astronomy. By Rosa M. Ros (Spain) and Jay M. Pasachoff (USA)*

In the IAU resolution on the Value of Astronomy Education, passed by the IAU XXV General Assembly in 2003, it was recommended: to include astronomy in school curricula, to assist schoolteachers in their training and backup, and to inform teachers about available resources.

The aim of this Special Session 2 on 'Innovation in Teaching/Learning Astronomy' is to contribute to the implementation of these recommendations, introducing innovative points of view regarding methods of teaching and learning. Astronomers from all countries – developed or developing – will be equally interested.

Astronomy attracts many young people to education in important fields in science and technology. But in many countries, astronomy is not part of the standard curriculum, and teachers do not receive adequate education and support. Still, many scientific and educational societies and government agencies have produced materials and educational resources in astronomy for all educational levels. Technology is used in astronomy both for obtaining observations and for teaching. In any case, it is useful to take this special opportunity to learn about the situation in different countries, to exchange opinions, and to collect information in order to continue, over at least the next triennium, the activities related to promoting astronomy throughout the world.

In particular, we would like to invite all participants to explain their positive original experiences so they can be adapted for other regions. Everyone is invited to exchange their initiatives and to try to involve other countries in common projects. All of us are in the same boat.

5.2. *Learning astronomy by doing astronomy. By John R. Percy (Canada)*

In the modern science curriculum, students should learn science knowledge or 'facts'; they should develop science skills, strategies, and habits of mind; they should understand the applications of science to technology, society, and the environment; and they should cultivate appropriate attitudes toward science. While science knowledge may be taught through traditional lecture-and-textbook methods, theories of learning (and extensive experience) show that other aspects of the curriculum are best taught by doing science – not just hands-on activities, but 'minds-on' engagement. That means more than the usual 'cookbook' activities in which students use a predetermined procedure to achieve a predetermined result. The activities should be 'authentic'; they should mirror the actual scientific process.

In this presentation, I will describe several ways to include science processes within astronomy courses at the middle school, high school, and introductory university level. Among other things, I will discuss: topics that reflect cultural diversity and 'the nature of science'; strategies for developing science process skills through projects and other practical work; activities based on those developed and carried out by amateur astronomers; topics and activities suitable for technical-level courses (we refer to them as 'applied' in my province); projects for astronomy clubs and science fairs; and topics that expose students to astronomy research within lecture courses.

5.3. *Hands-on Universe – Europe. By Roger Ferlet (France)*

The EU-HOU project aims at re-awakening the interest for science through astronomy and new technologies, by challenging middle and high schools pupils. It relies on real observations acquired through an internet-based network of robotic optical and radio telescopes or with didactical tools such as Webcam. Pupils manipulate and measure images in the classroom environment, using the specifically designed software SalsaJ, within pedagogical trans-disciplinary resources constructed in close collaboration between researchers and teachers. Gathering eight European countries coordinated in France, EU-HOU is partly funded by the European Union. All its outputs are freely available on the Web, in English and the other languages involved. A European network of teachers is being developed through training sessions.

5.4. *Life of the Earth in the solar atmosphere. By Edvard V. Kononovich, Olga B. Smirnova, T.V. Matveychuk, G.V. Jakunina, and S.A. Krasotkin (Russia)*

The theory of stellar interior is a very stimulating tool of the physical and astrophysical curricula. To support the corresponding lecture courses, a practical work was proposed and elaborated upon in 1991 for advanced students of physics specializing in astrophysics at Moscow State University. The work is recommended for 5th year students and requires significant knowledge in theoretical astrophysics. The main purpose of the work is to calculate the evolutionary set of stellar models, including those for the Sun. The PC program is based on the B. Paczynski set of routines and supported by the Web interface. This allows working via Internet. The results of the work may be presented both in a table and graph form.

5.5. *A model of teaching astronomy to pre-service teachers that allows for creativity in communicating students' understanding of seasons. By W.R. MacIntyre (New Zealand)*

This paper details a model of teacher development for astronomy concepts that allows students to demonstrate their understanding of basic astronomy concepts as well as communicating that understanding in creative ways. Several key features of the model is the inclusion of starting from students' initial understanding about astronomical concepts, providing the time for students to assess their mental models with 3-D models, and individual student assessment of their astronomical understanding using 3-D models. It appears that the three features collectively provide an appropriate creative environment for students. Two students created two new 3-D models in order to communicate specific aspects of seasons – different solar inputs and varying lengths of day/night throughout the year. The students are interviewed highlighting the rationale for creating the new 3-D models. The uptake by other students, during the modelling assessment task, demonstrated their usefulness in communicating astronomical understanding of seasons. The model of teacher development illustrates how teacher educators can teach for astronomy understanding as well as allow for creative ways to communicate that understanding to others an essential disposition to being an effective astronomy educator in the classroom.

5.6. *How to teach, learning, doing and enjoying astronomy. By Rosa M. Ros (Spain)*

This contribution deals with the author's experience organising a summer school for European teachers over ten years and the parallels with the everyday school for students.

The main interests for teachers are similar to students. It is necessary to give them:

- answers to their questions;

- practical activities: learning by doing;

- study astronomy using different approaches: making models, cutting, drawings, playing in the playground and, in general, make them feel like actors in the teaching/learning process;

- astronomical activities can help teachers/students to teach/learn mathematics or physics in a more appropriate way to attract young people to science;

- simple and clear language. It is good to reduce the specialized language and try to play with the proximity to the student;

- methods which promote rationality, curiosity and creativity. All schools have a sky over their buildings, it must be used to observe and take measurements;

- a contextualized approach to astronomy. Do not present the concepts in an isolated way. The school must be connected with the place where students are living;

In summary, students should feel a positive passion related to some astronomical experiences; then they will add a positive connotation to astronomy.

This presentation will mix some concrete examples of all these ideas.

5.7. *Clickers: a new teaching tool of exceptional promise. By Douglas K. Duncan (USA)*

Wireless student response systems – 'clickers' – address two of the oldest and most fundamental challenges in teaching: how to engage students, and how to determine if they are learning what you are teaching. Clickers are relatively low cost and easy to use, and their use is spreading remarkably fast throughout the US, with many universities using thousands. Astronomy textbooks may be ordered with coupons for clickers in them.

The way the system works is that each student's clicker has buttons *a*, *b*, *c*, *d*, and *e*. Any time the instructor wants feedback, he or she asks a multiple choice question. Student responses go to a receiver that plugs into a computer (e.g., the instructor's) showing responses as a bar chart. The chart is often shown to the class via an LCD projector. Clickers give immediate feedback about what each student is thinking. The instructor can decide whether to proceed or to spend more time on a particular topic. Equally valuable, the student learns immediately whether he or she understands the concept the teacher is presenting, without waiting for a test or raising their hand to ask a question.

I will present extensive research data that show that when clickers are used well in large lecture classes, they increase the engagement of students and improve their learning by a significant amount. Students overwhelmingly like using clickers and believe they increase their learning. They also increase class attendance, typically by 20 %. Like any technology, it is possible to misuse clickers. Common mistakes made by new clicker users and how to avoid them will be described, as well as 'best clicker uses' such as peer instruction. (Peer instruction means that when the class answers are split, students have to debate with their neighbor who is right, rather than the instructor telling them.)

This presentation will feature a demonstration of the clickers, with one given to each person in attendance.

5.8. *Teaching the evolution of stellar and Milky Way concepts through the ages: a tool for the construction of a scientific culture using astrophysics. By Gilles Theureau (France)*

I will report on a two-semester experience at the Orlans University (France) of a course of history and epistemology of the concepts of stars and galaxies from Antiquity to early XXth century. The framework is a 'module d'ouverture' of the new Licence-Master-Doctorate reform of French University, i.e., a transversal course aiming at providing a scientific culture to a mixed set of students from various fields (law, languages, sport, physics, etc.). Due to the number of students and to their wide heterogeneity, the form chosen has been a 22 hours lecture distributed in 10 lessons plus one planetarium session. Special attention was made to regularly refer to and read through original historical texts. The final evaluation was centered on collecting reading notes and commentaries on an original (full) text book. The text was chosen among a list of various references covering the whole period of interest, each student or group presenting his own report.

5.9. *Educational opportunities in pre-Am collaboration. By Richard T. Fienberg (USA)*

While many backyard stargazers take up the hobby just for fun, many others are attracted to it because of their keen interest in learning more about the universe. The best way to learn science is to do science. Happily, the technology available to today's amateur astronomers – including computer-controlled telescopes, CCD cameras, powerful astronomical software, and the Internet – gives them the potential to make real contributions to scientific research and to help support local educational objectives.

Meanwhile, professional astronomers are losing access to small telescopes as funding is shifted to larger projects, including survey programs that will soon discover countless interesting objects needing follow-up observations. Clearly the field is ripe with opportunities for amateurs, professionals, and educators to collaborate. Amateurs will benefit from mentoring by expert professionals, pros will benefit from observations and data processing by increasingly knowledgeable amateurs, and educators will benefit from a larger pool of skilled talent to help them carry out astronomy-education initiatives.

We will look at some successful pro-am collaborations that have already borne fruit and examine areas where the need and/or potential for new partnerships is especially large. In keeping with the theme of this special session, we will focus on how pro-am collaborations in astronomy can contribute to science education both inside and outside the classroom, not only for students of school age but also for adults who may not have enjoyed particularly good science education when they were younger. Because nighttime observations with sophisticated equipment are not always possible in formal educational settings, we will also mention other types of pro-am partnerships, including those involving remote observing, data mining, and/or distributed computing.

5.10. *Education efforts of the International Astronomical Union. By Jay M. Pasachoff (USA)*

I describe the education activities of the IAU, particularly the work of Commission 46 on Education and Development. We are most interested in education in schools and for general university education rather than for pre-professional training or graduate schools. We have over 75 National Liaisons, mostly from member countries of the IAU but some from nonmembers or regional groupings. We operate through 10 program groups, which are described at our website at <http://www.astronomyeducation.org>. We also organize Special Sessions at IAU General Assemblies, such as this Special Session 2 on Innovation in Teaching/Learning Astronomy Methods, organized by Rosa M. Ros and me, and Special Session 5 on Astronomy for the Developing World, organized by John B. Hearnshaw. A modified version of our Special Session from the 2003 IAU XXV General Assembly in Sydney was published as *Teaching and Learning Astronomy: Effective Strategies for Educators Worldwide* (Jay M. Pasachoff & John R. Percy, eds., 2005, Cambridge: CUP). Michèle Gerbaldi and Edward F. Guinan run the IAU International Schools for Young Astronomers. James C. White heads the IAU Program Group on Teaching Astronomy for Development. John B. Hearnshaw runs the IAU Program Group for the Worldwide Development of Astronomy. Charles R. Tolbert and John R. Percy run the IAU Exchange of Astronomers program with a limited number of grants for stays of over three months between astronomers in developing countries and established astronomical institutions. Barrie W. Jones, as vice-president, aided by Tracey J. Moore, runs the Newsletter and keeps track of the National Liaisons list. I run the Program group of Public Education at the Times of Solar Eclipses.

5.11. *Astronomy and culture. By Magda G. Stavinschi (Romania)*

Astronomy is, by definition, the sum of the material and spiritual values created by mankind and of the institutions necessary to communicate these values. Consequently, astronomy belongs to the culture of each society and its scientific progress does nothing but underline its role in culture. It is interesting that there is even a European society which bears this name 'Astronomy for Culture' (SEAC). Its main goal is 'the study of calendric and astronomical aspects of culture'. Owning ancient evidence of astronomical knowledge, dating from the dawn of the first millennium, Romania is interested in this topic. But astronomy has a much deeper role in culture and civilization. There are many aspects that deserve to be discussed. Examples? The progress of astronomy in a certain society, in connection with its evolution; the place held by astronomy in literature and, generally, in art; the role of the SF in the epoch of super-mediatization; astronomy and belief; astronomy and astrology in the modern society, and so forth. These are problems that can be of interest for IAU; but, the most important one could be her educational role, in the formation of the culture of the new generation, in the education of the population for the protection of our planet, and in the ensuring of a high level of spiritual development of the society in the present epoch.

5.12. *Light pollution a tool for astronomy education. By Margarita Metaxa (Greece)*

The problem of light pollution exists most everywhere and is still growing rapidly. The maintenance of dark skies at a prime astronomical location, and elsewhere as well, depends very much on the awareness of the public, and particularly with key decision makers responsible for developments, including lighting engineers. It is necessary to continually promote awareness of light pollution and its effects. Thus, the preservation of the astronomical environment is strongly connected and requires effective education. We will present the educational project that the newly formed Commission for the Prevention of Light Pollution, which the Hellenic Astronomical Society will support based on innovating teaching of astronomy. The framework of the project will be to collaborate through our National Pedagogical Institute with all possible school networks so to efficiently introduce the topic to schools and to relate it with our national curriculum. The help of astronomers and lighting engineers through the respective Commission will facilitate and provide the natural environment for this educational project. The duration will be two years, and through the project we expect the students-teachers to act as 'reporters' for this serious problem.

5.13. *Student gains in understanding the process of scientific research. By Travis A. Rector, Catherine A. Pilachowski, and Melina J. Young (USA)*

Research-Based Science Education is a method of instruction that models the processes of scientific inquiry and exploration used by scientists to discover new knowledge. It is 'research based' in the sense that students work together in self-guided, cooperative groups on a real research project. In other words, in order to learn science, students are given the opportunity to actually do science. Here we present the results of a study of undergraduate students that were given the opportunity to work on a research project underway to search for novae in Local Group galaxies. Students analyzed images obtained regularly from the WIYN 0.9 m telescope on Kitt Peak. Novae were found by blinking these images. Aperture photometry was used to generate light curves and measure decay rates. Students then explored individually chosen questions, such as comparing the location of novae in the galaxy and their rates of decay. Students then wrote research papers and gave oral presentations to the class. To assess their development in the understanding of science as a process, students completed pre- and post-concept maps on the topic of 'scientific research'. Each map was assessed for an understanding of the following ten concepts. Scientific research is: a process (i.e., a series of many steps over time); based upon prior knowledge or previous research; based on a hypothesis/question; uses experimentation; data collection; data representation (e.g., charts, tables and graphs); requires equipment; analysis/interpretation; generates results/conclusions; and results that link back to modify the initial hypothesis iteratively. Overall, students made significant gains on the concept maps, showing greater depth in the number of concepts and their relationships. On average, students increased the number of the ten understood concepts listed above from 2.8 before the class to 5.4 afterwards.

5.14. *Effects of collaborative learning on students' achievements in introductory astronomy.*
By Myung-Hyun Rhee, S.-W. Kim, E.-J. Kim, J. Kim (Republic of Korea)

For the last few years, we have performed various Collaborative Learning (CL) sessions in the classes of the Introductory Astronomy course for non-science majors at Yonsei University, Seoul, Korea. We present some results from these experiments, mainly focusing on the effects of Collaborative Learning (CL) on university students' Communication Apprehension (CA), Class Satisfaction and Academic Achievement.

The main results we found are as follows:

(1) The amount of CA reduction is proportional to the number of CL sessions; the amount of CA reduction of the nine CL students was much higher than that of the zero CL (control group), two CL and five CL students.

(2) The amount of CA reduction was greater with the higher CA students.

(3) CA reduction effect was intact after a half year later.

(4) Academic Achievement of the nine CL students was higher than that of the two CL, five CL and control group students.

(5) Students' Class Satisfaction also showed more or less the same results with Academic Achievement.

5.15. *Worldwide on-line distance learning university astronomy. By Stewart P.S. Eyres, B.J.M. Hassal, I. Butchart, and Gordon E. Bromage (United Kingdom)*

The University of Central Lancashire operates a suite of distance learning courses in Astronomy, available both on-line and via CD-ROM. The courses are available worldwide, and emphasize flexibility of study. To this end, students can study anything from a single module (1/6th of a full year at degree level) all the way up to an entire degree entirely by distance learning. Study rates vary from one to four modules each year, and students can move on to Level 2 modules (equivalent to second year level in a UK degree) before completing the full set of Level 1 modules. Over 1000 awards have been made to date. The core syllabus is Astronomy and Cosmology at Level 1, alongside skills in literature research, using computers, and basic observing. We also offer a basic history of European astronomy. At Level 2, we look at the astrophysics of the Sun, the stars, and galaxies including the Milky Way. By Level 3, students are expected to engage in a large individual project, and a collaborative investigation with other students, alongside high-level courses in cosmology, relativity, extreme states of matter and the origins of the elements, life and astronomical objects. While many students are retired people looking to exercise their brains, keen amateurs or professionals with disposable incomes, a significant fraction are teachers seeking to improve their subject knowledge or high school students gaining an edge in the UK University entrance competition. Via our involvement with SALT, we offer our courses to members of previously disadvantaged communities. This leads to an incredibly diverse and lively student body.

5.16. *Edible astronomy demonstrations. By Donald A. Lubowich (USA)*

By using astronomy demonstrations with edible ingredients, I have been able to increase student interest and knowledge of astronomical concepts. This approach has been successful with all age groups from elementary school through college students. I will present some of the edible demonstrations I have created including using popcorn to simulate radioactivity; using chocolate, nuts, and marshmallows to illustrate density and differentiation during the formation of the planets; and making big-bang brownies or chocolate-chip cookies to illustrate the expansion of the universe. Sometimes the students eat the results of the astronomical demonstrations. These demonstrations are an effective teaching tool and the students remember these demonstrations after they are presented.

5.17. *Amateur astronomers as public outreach partners. By Michael A. Bennett (USA)*

Amateur astronomers involved in public outreach represent a huge, largely untapped source of energy and enthusiasm to help astronomers reach the general public. Even though many astronomy educators already work with amateur astronomers, the potential educational impact of amateur astronomers as public outreach ambassadors remains largely unrealized.

Surveys and other work by the ASP in the US show that more than 20 % of astronomy club members routinely participate in public engagement and educational events, such as public star parties, classroom visits, work with youth and community groups, etc. Amateur astronomers who participate in public outreach events are knowledgeable about astronomy and passionate about sharing their hobby with other people. They are very willing to work with astronomers and astronomy educators. They want useful materials, support, and training. In the USA, the ASP operates 'The Night Sky Network', (funded by NASA). We have developed specialized materials and training, tested by and used by amateur astronomers. This project works with nearly 200 local astronomy clubs in 50 states to help them conduct more effective public outreach events. It has resulted in nearly 3,600 outreach events (reaching nearly 300,000 people) in just two years. In this presentation we examine key success factors, lessons learned, and suggest how astronomers outside the US can recruit and work with 'outreach amateur astronomers' in their own countries.

5.18. *Does the Sun rotate around the Earth or roes the Earth rotate around the Sun? – An important aspect of science education. By Syuzo Isobe (Japan)*

Sciences are continuously developing. This is a good situation for the sciences, but when one tries to teach scientific results, it is hard to decide which levels of science should be taught in schools. The point to evaluate is not only the quality of scientific accuracy, but also the method with which school students of different scientific abilities study scientific results. In astronomy, an important question, which is 'Does the Sun rotate around the Earth or does the Earth rotate around the Sun?' can be used to evaluate student abilities. Scientifically, it is obvious that the latter choice is the better answer, but it is not so obvious for the lower-grade students and also for the lower-ability students even in the higher grades. If one sees daily the sky without scientific knowledge, one has an impression of 'the Sun rotates around the Earth,' and for his rest of his life he will not see any problem. If one wants to be a scientist, though, he should know that 'the Earth rotates around the Sun' before reaching university level. If he will become a physical scientist, he should understand that it is not correct to say 'the Earth rotates around the Sun,' but he should know that the Earth rotates around the center of gravity of the solar system. A similar type of question is 'has the Earth the shape of a sphere, or a pear, or a geoid?'

There are many teachers with varying ranges of students who do not understand the proper level of science instruction. When students of lower capacity are instructed to understand concepts with the higher degrees of sophistication, they can easily lose their interest in the sciences. This happens in many countries, especially in Japan, where there are many different types of people with different jobs. We, as educators, should appreciate that the students can be interested in any given scientific idea, no matter what level of sophistication it is.

5.19. *Using sounds and sonifications for astronomy outreach. By Fernando J. Ballesteros (Spain)*

It is well known that good astronomy pictures play a great role in astronomy outreach, triggering curiosity and interest, as in the case of *HST* pictures. But this same aim can also be obtained by means of sounds. Here I present the use of astronomy-related sounds and data sonifications to be used for astronomy outreach. Examples of these sounds are the case of the *Cassini* probe passing through Saturn's rings, radio signals from pulsars, black holes, aurorae, or signals from space missions, among many others. These are sounds that people, usually, will never hear and are a good tool for provoking an interest when teaching astronomy. In our case, sounds are successfully used in a weekend science-spreading program of the Spanish National Broadcast RNE, called 'The sounds of science'. But teachers can also make use of them in the classroom easily, as sounds only require a simple cassette player

5.20. *Teaching astronomy and the crisis in science education. By Nicholas R. Lomb, T.M. Stevenson, M.W.B. Anderson, and G.G. Wyatt (Australia)*

In Australia, as in many other countries, the fraction of high school students voluntarily choosing to study the core sciences such as physics and chemistry has dropped in recent decades. There seems to be a number of reasons for this worrying trend, including the perception that they are difficult subjects that lack relevance to the lives of the students. Family influence to choose courses that are believed to be more likely to lead to highly paid careers is also a major factor.

Astronomy has a broad public appeal and escapes much of the negative feelings associated with most other scientific fields. Anecdotally and logically, this allows astronomy to be used as a tool to stimulate students' scientific interest. While this is most evident at college level in the USA and at Australian universities, informal education centres can play an important role. Investment in public facilities and the provision of resources for astronomy outreach can be highly beneficial by engaging the imagination of the public. We will discuss activities offered at Sydney Observatory where public attendance have more than doubled in the last decade. These include a regular schools program and preliminary results from a survey of teachers' experiences and attitudes to their class visit will be given.

5.21. *Astronomy for all as part of a general education. By John E.F. Baruch, D.G. Hedges, J. Machell, C.J. Tallon, and K. Norris (United Kingdom)*

This paper evaluates a new initiative in support of the aim of Commission 46 of the IAU to develop and improve astronomy education at all levels throughout the world. This paper describes a free facility to support education programmes which include basic astronomy and are delivered to students who have access to the internet on <www.telescope.org/>.

This paper discusses the role of robotic telescopes in supporting both students and their teachers and shows that, although robotic telescopes have been around for some time, almost all of them are designed to cater to an elite, a tiny percentage, group of students. These telescopes are generally not true autonomous robots but remotely driven telescopes. This paper shows how truly autonomous robots offer the possibility of delivering a learning experience for all students in their general education. The experience of the Bradford Robotic telescope is discussed. This telescope is on track to deliver the initial levels of astronomy education to all school students in the UK.

The problems of delivering a web-based education programme to very large numbers of students are discussed. Traditionally innovative astronomy programmes have been delivered through enthusiastic teachers with considerable expertise in IT and astronomy.

This paper looks at the problems of delivering such a programme with teachers who have little confidence working with IT and little knowledge of basic astronomy and discusses practical solutions. The facility is available free of charge and it is intended to continue to be so.

5.22. *A status report from the IAU Division XII Working Group on Communicating Astronomy with the Public. By Dennis R. Crabtree (Canada), E. Ian Robson (UK), and Lars Lindberg Christensen (Germany)*

This Division XII Working Group was created in 2004 following a conference entitled 'Communicating Astronomy to the Public' held at the US National Academy of Sciences in Washington, DC, in early October, 2003. The Working Group's Mission Statement is as follows:

Mission statement:

To encourage and enable a much larger fraction of the astronomical community to take an active role in explaining what we do (and why) to our fellow citizens.

To act as an international, impartial coordinating entity that furthers the recognition of outreach and public communication on all levels in astronomy.

To encourage international collaborations on outreach and public communication.

To endorse standards, best practices and requirements for public communication.

This paper will report on the achievements and progress made since the working group's formation and present our plans for the next three years.

5.23. *Outreach using media. By Julieta Fierro (Mexico)*

Outreach is the best way to carry out informal astronomy education during a person's life. Astronomy is such an attractive and evolving discipline that people of all age levels are usually attracted to it in spite of feeling threatened by the apparent difficulty of science. During my presentation I shall address the importance of astronomical popularization. I shall mention fundraising for its diffusion employing radio and television programs. Simple demonstrations will also be included. I will also remind members of Commission 46 that 2009 might become the *Year of Astronomy* and mass media are an ideal way for publicizing astronomical outreach projects.

5.24. *Hands-on science communication. By Lars Lindberg Christensen (Germany)*

Many of the most important questions studied in science touch on fundamental issues with a great popular appeal, such as: How was the world created? How did life arise? Are we alone? How does it all end?

Communication of science to the public is important and will play an even greater role in the coming years. The communication of achieved results is more and more often seen as a natural and mandatory activity to inform the public, attract funding, and attract science students. In some countries, university statutes are even being rewritten in these years to include communication with the public as the third mandatory function besides research and education.

A number of interesting 'lessons learned' from the daily work at the Education and Outreach (EPO) office of the European Space Agency's Hubble Space Telescope will be presented. The topics include conventional as well as unconventional issues such as:

* How does the flow of communication from scientist to public work, which actors are involved, and which pitfalls are present in their interaction? How can possible problems be avoided?

* What are the criteria that determine whether press releases 'make it' or not?

* How can a commercial approach benefit an EPO office?

* What is the right skills base in a modern EPO office?
* How can modern technology be used to communicate science more efficiently?

5.25. *Critical evaluation of the New Hall of Astronomy for the Science Museum.*
By Silvia Torres-Peimbert (Mexico)

In December 2004, a new astronomy exhibit was opened at 'Universum', the Science Museum at the University of Mexico. The displays are presented in several sections: (*i*) Sun, stars and matter between the stars; (*ii*) clusters, galaxies and the universe as a whole; and (*iii*) the tools of astronomers.

The main concept is not limited to the description of each component, but also incorporates the idea that all components, including the universe, are subject to continuous evolution. In addition, a representation of the vastness of space, a timeline from the Big Bang to the present epoch, and some video clips from local astronomers are included. As a complement to the exhibit, a section on the history of astronomy is also included. We are now in the process of assessing the impact of the different elements of this exhibit among the visitors. The results of this evaluation will be presented.

5.26. *Revitalizing astronomy teaching through research on student understanding.*
By Timothy F. Slater (USA)

Over the years, considerable rhetoric exists which instructional strategies induce the largest conceptual and attitude gains in non-science majoring, undergraduate university students. To determine the effectiveness of lecture-based approaches in astronomy and astrobiology, we found that student scores on a 68-item pre-test/post-test concept inventory showed a statistically significant increase from 30 % to 52 % correct. In contrast, students evaluated after the use of Lecture-Tutorials for Introductory Astronomy increased to 72 %. The Lecture-Tutorials for Introductory Astronomy are intended for use during lecture by small student groups and compliment existing courses with conventional lectures. Based on extensive research on student understanding, Lecture-Tutorials for Introductory Astronomy offer professors an effective, learner-centered, classroom-ready alternative to lecture that does not require any outside equipment or drastic course revision for implementation. Each 15-minute Lecture-Tutorial for Introductory Astronomy poses a carefully crafted sequence of conceptually challenging, Socratic-dialogue driven questions, along with graphs and data tables, all designed to encourage students to reason critically about difficult concepts in astronomy and astrobiology.

5.27. *ESO's astronomy education programme.*
By Douglas P.I. Pierce-Price, Henri Boffin, and Claus Madsen (Germany)

ESO, the European Organisation for Astronomical Research in the Southern Hemisphere, has operated a programme of astronomy education for some years, with a dedicated Educational Office established in 2001. We organise a range of activities, which we will highlight and discuss in this presentation. Many are run in collaboration with the European Association for Astronomy Education (EAAE), such as the 'Catch a Star!' competition for schools, now in its fourth year.

A new endeavour is the ALMA Interdisciplinary Teaching Project (ITP). In conjunction with the EAAE, we are creating a set of interdisciplinary teaching materials based around the Atacama Large Millimeter Array project. The unprecedented astronomical observations planned with ALMA, as well as the uniqueness of its site high in the Atacama Desert, offer excellent opportunities for interdisciplinary teaching that also encompass physics, engineering, earth sciences, life sciences, and culture.

Another ongoing project in which ESO takes part is the 'Science on Stage' European science education festival, organised by the EIROforum – the group of seven major European Intergovernmental Research Organisations, of which ESO is a member. This is part of the European Science Teaching Initiative, along with Science in School, a newly launched European journal for science educators.

Overviews of these projects will be given, including results and lessons learnt. We will also discuss possibilities for a future European Astronomy Day project, as a new initiative for European-wide public education.

5.28. *Astronomy remote observing research projects of USA high school students. By Mary Ann Kadooka (HI, USA)*

In order to address the challenging climate for promoting astronomy education in the high schools, we have used astronomy projects to give students authentic research experiences in order to encourage their pursuit of science and technology careers. Initially, we conducted teacher workshops to develop a cadre of teachers who have been instrumental in recruiting students to work on projects. Once identified, these students have been motivated to conduct astronomy research projects with appropriate guidance. Some have worked on these projects during non-school hours and others through a research course. The goal has been for students to meet the objectives of inquiry-based learning, a major US National Science Standard. Case studies will be described using event-based learning with the NASA Deep Impact mission. Hawaii students became active participants investigating comet properties through the NASA *Deep Impact* mission. The Deep Impact Education and Public Outreach group developed materials which were used by our students. After learning how to use image processing software, these students obtained Comet 9P/Tempel 1 images in real time from the remote observing Faulkes Telescope North located on Haleakala, Maui, for their projects. Besides conducting event-based projects which are time critical, Oregon students have worked on galaxies and sunspots projects. For variable star research, they used images obtained from the remote observing offline mode of Lowell Telescope, located in Flagstaff, Arizona. Essential to these projects has been consistent follow-up required for honing skills in observing, image processing, analysis, and communication of project results through Science Fair entries. Key to our success has been the network of professional and amateur astronomers and educators collaborating in a multiplicity of ways to mentor our students. This work-in-progress and process will be shared on how to inspire students to pursue careers in science and technology with these projects.

5.29. *Global network of autonomous observatories dedicated to student research. By Richard Gelderman (USA)*

We will demonstrate operation of one or more meter-class telescopes devoted to student-initiated astronomical research projects. For multiple decades, astronomers have promised each other the development of global networks of telescopes. For the last decade, without ever fulfilling the initial promise, we have upped the ante and promised global networks of robotic telescopes. Sometimes the network is to be composed of 20- to 40-cm aperture telescopes; other times the network will include meter-class telescopes. Sometimes the network is exclusive to a select, small group of users; other times the dream is open to any interested parties. Western Kentucky University, the Hands-On Universe project, and NASA's Kepler mission have achieved the first components of a network of telescopes established for educational programs. We will discuss the process used by teachers and students to make use of a substantial fraction of the network's observing time, and to access most of the archived data. Examples of student projects will be shared, along with immediate plans for expanding the network.

5.30. *Remote telescopes in education: an australian study.* *By David H. McKinnon, and L. Danaia (Australia)*

In 2004, the Australian Federal Department of Education, Science and Training funded a study into the impact of using remote telescopes in education in four educational jurisdictions: The Australian Capital Territory, New South Wales, Queensland, and Victoria. A total of 101 science teachers and 2033 students in grades 7 - 9 provided pre-intervention data. Students were assessed on their astronomical knowledge, alternative conceptions held, and ability to explain astronomical phenomena. They also provided information about the ways in which science is taught and their attitudes towards the subject. Teachers provided information about the ways in which they teach science. Students (N = 1463) and teachers (N = 35), provided the same data after the intervention was completed. The return rate for students and teachers was 71 % and 34 % respectively. This represents the largest study undertaken involving the use of remote telescopes in education. The intervention comprised a set of educational materials developed at Charles Sturt University (CSU) and access to the CSU Remote Telescope housed at the Bathurst Campus, NSW. Outcomes showed that students had increased their astronomical knowledge significantly.

5.31. *Visualizing large astronomical data holdings.* *By Carol A. Christian, A. Connolly, A. Conti, N. Gaffney, S. Krughoff, B. McClendon, A. Moore, and R. Scranton (USA)*

Scientific visualization involves the presentation of interactive or animated digital images for interpretation of potentially huge quantities of observed or simulated data. Astronomy visualization has been a tool used to convey astrophysical concepts and the data obtained to probe the cosmos. With the emergence of large astrophysical data archives, improvements in computational power and new technologies for the desktop, visualization of astronomical data is being considered as a new tool for exploration of data archives with a goal to enhance education and public understanding of science.

We will present results of some exploratory work in the use of visualization technologies from the perspective of education and outreach. These tools are also being developed to facilitate scientists2̆019 use of such large data repositories as well.

5.32. *Stellar evolution for students of Moscow University.* *By Edvard V. Kononovich (Russia)*

Theory of stellar interior is a very stimulating tool of the physical and astrophysical curricula. To support the corresponding lecture courses, a practical work was proposed and elaborated upon in 1991 for advanced students of the physics department who specialized in astrophysics at Moscow State University. The work is recommended for 5th year students and requires significant knowledge of theoretical astrophysics. The main purpose of the work is to calculate the evolutionary set of stellar models, including those for the Sun. The PC program is based on the B. Paczynski set of routines and supported by the WEB interface. This allows working via Internet. The results of the work may be presented both in table and graph forms.

5.33. *Astronomy for everybody: an approach from the CASAO.* *By Maria C. Pineda de Carias (Honduras)*

Astronomy is a science that attracts the attention of all ages of people from a variety of views and interests. At the Central America Suyapa Astronomical Observatory of the National Autonomous University of Honduras (CASAO/NAUH), the formal general course of Introduction to Astronomy (AN-111) for all-careers students and the regular courses for a Master in Astronomy and Astrophysics students, three different academic outreach programs have become of importance, after less than a decade of experience. A 'Visiting the CASAO/NAUH Program' is aimed at elementary and secondary schools, where astronomers three times per week present to groups from fifteen up to one hundred students and their teachers. Conferences on selected topics of astronomy, illustrated with real sky and astronomical objects and images, give the opportunity to observe the sun, the moon and planets using a small telescope. They explain how astronomers today perform their observations and, also, how the Mayas that inhabited Central America did their observations during their time. The 'Astronomical Nights Program', intended for the general public, children, youth and adults who attend on Friday nights at the Astronomical Observatory, learn about astronomical bodies' properties, the sky of the week, and the differences in making observations using small telescopes and with a naked eye. 'Intro_Astro@Internet' is an online course program designed for school teachers and is also used by college and university students of Central America willing to learn more systematically on their own using new technologies about the sky, the solar system, the stars, and the universe. In this paper we present a complete description of these programs and the ways they are currently developed at CASAO/NAUH, and a discussion of how these programs contribute to the implementation of the IAU Resolution on the Value of Astronomy Education.

5.34. *The epistemological background of our strategies.*
By Mirel I. Birlan (France), Gheorghe Vass, and Constantin Teleanu (Romania)

The major objectives of the long-term strategy in promoting a scientific discipline into the education sphere may be the following:
- A permanent presence of the discipline inside the general education;
- A good harmonisation with other disciplines (at least the scientific ones); and
- A maximal efficiency of its presence as a required discipline.

Before any practical approaches (social, administrative, or any other), it is required (necessary) that the promoters of any discipline formulate and underline its cognitive and formative contribution. This can be done only by starting with the epistemological statute of the discipline and its pedagogical implications.

Defining the epistemological statute, as well as the requirements of an optimal communication among the options concerning the orientation (re-orientation) of educational strategies, brings forth the necessity of analysing some fundamental concepts, with epistemological characters.

Even if it seems to be known by everyone, usually the concepts are not taken into account with their real importance. The fundamental concept we refer to herein constitute a real 'conceptual network' who, *volens-nolens*, is the epistemological background for any strategy of education relative to science.

Being conscious of the importance of the epistemological background will permit the modification of the beginning and the references used in defining the strategy of education in (through) astronomy.

5.35. *Towards a new program in astronomy education in secondary schools in Turkey.*
By Zeki Aslan and Zeynel Tunca (Turkey)

It has been of great concern for Turkish astronomers that the teaching of astronomy, which is a part of the physics course in secondary schools, is not very effective, or not taught at all, mainly because the majority of the physics teachers have had no formal education in astronomy. Knowing this, TUBITAK National Observatory (TUG) proposed to the Ministry of Education in 2005 that a national meeting for physics and astronomy teachers be held during the opportune time of the total solar eclipse of 29 March, 2006, with the subject matter 'teaching of astronomy and using astronomy to teach physics'. The meeting, with participation of teachers from all over Turkey, has been very successful. The speakers were astronomers and physicists from Turkish universities and educators from the Ministry, plus three educators from abroad. A text containing minutes of the meeting has been submitted to the Ministry of Education. The details and their relevance to the 2003 IAU Resolution and the role TUG has undertaken will be presented.

5.36. *Astronomy in the Russian scientific-educational project KazanGeoNa2010. By Alexander V. Gusev and Irina Kitiashvili (Russia)*

The European Union promotes the Sixth Framework Programme. One of the goals of the EU Programme is opening national research and training programs. A special role in the history of the Kazan University was played by the great mathematician Nikolai Lobachevsky – the founder of non-Euclidean geometry (1826). Historically, the thousand-year old city of Kazan and the two-hundred-year old Kazan University carry out the role of the scientific, organizational, and cultural educational center of the Volga region. For the continued successful development of educational and scientific-educational activity of the Russian Federation, Kazan (in the Republic of Tatarstan) was offered the national project: the International Center of the Sciences and Internet Technologies 'GeoNa'. Geometry of Nature - GeoNa - wisdom, enthusiasm, pride, grandeur. This is a modern complex of conference halls including the Center for Internet Technologies, a 3D Planetarium - development of the Moon, PhysicsLand, an active museum of natural sciences, an oceanarium, and a training complex 'Spheres of Knowledge'. Center GeoNa promotes the direct and effective channel of cooperation with scientific centers around the world. GeoNa will host conferences, congresses, fundamental scientific research sessions of the Moon and planets, and scientific-educational actions: presentation of the international scientific programs on lunar research and modern lunar databases. A more intense program of exchange between scientific centers and organizations for a better knowledge and planning of their astronomical curricula and the introduction of the teaching of astronomy are proposed. Center GeoNa will enable scientists and teachers of the Russian universities with advanced achievements in science and information technologies to join together to establish scientific communications with foreign colleagues in the sphere of the high technology and educational projects with world scientific centers.

5.37. *Universe awareness for young children. By Cecilia Scorza, George K. Miley, Carolina Ödman, and Claus Madsen (the Netherlands, Germany)*

Universe Awareness (UNAWE) is an international programme that will expose economically disadvantaged young children aged between 4 and 10 years to the inspirational aspects of modern astronomy. The programme is motivated by the premise that access to simple knowledge about the universe is a basic birthright of everybody. These formative ages are crucial in the development of a human value system. This is also the age range in which children can learn to develop a 'feeling' for the vastness of the universe. Exposing young children to such material is likely to broaden their minds and stimulate their world-view. The goals of Universe Awareness are in accordance with two of the United Nations Millennium goals, endorsed by all 191 UN member states, namely (*i*) the achievement of universal primary education and (*ii*) the promotion of gender equality in schools.

We propose to commence Universe Awareness with a pilot project that will target disadvantaged regions in about 4 European countries (possibly Spain, France, Germany and The Netherlands) and several non-EU countries (possibly Chile, Colombia, India, Tunisia, South Africa and Venezuela). There will be two distinct elements in the development of the UNAWE program: (*i*) Creation and production of suitable UNAWE material and delivery techniques, (*ii*) Training of educators who will coordinate UNAWE in each of the target countries. In addition to the programme, an international network of astronomy outreach will be organised.

We present the first results of a pilot project developed in Venezuela, where 670 children from different social environments, their teachers and members of an indigenous tribe called Yekuana from the Amazon region took part in a wonderful astronomical and cultural exchange that is now being promoted by the Venezuelan ministry of Education at the national level.

5.38. *Education in Egypt and Egyptian response to solar eclipses. By Ahmed A. Hady (Egypt)*

Astronomy and space science education started in Egypt at the university level since 1939 at the Department of Astronomy and Meteorology, Cairo University. Undergraduate and graduate education in Egypt will be discussed in this work. About 15 students yearly obtain their PhD degrees in Astronomy from the Egyptian universities. Seven international groups under my supervision have done the total solar eclipse observations that took place on 29 March, 2006, in El-Saloum (Egypt). The results of observations and photos will be discussed. An Egyptian-French group have done the total solar eclipse observations that took place on 25 February, 1952, in Khartoum by using a Worthington Camera. The research groups of astrophysics in Cairo University and Helwan Observatory are interested in the fields of solar physics, binary stars, celestial mechanics, interstellar matter and galaxies. Most of the researches have been published in national scientific journals, and some of them were published in international journals.

5.39. *Spreading astronomy education through Africa. By Paul Baki (Kenya)*

Although astronomy has been an important vehicle for effectively passing a wide range of scientific knowledge, teaching the basic skills of scientific reasoning, and for communicating the excitement of science to the public, its inclusion in the teaching curricula of most institutions of higher learning in Africa is rare. This is partly due to the fact that astronomy appears to be only good at fascinating people but not providing paid jobs. It is also due to the lack of trained instructors, teaching materials, and a clear vision of the role of astronomy and basic space science within the broader context of education in the physical and applied sciences. In this paper we survey some of the problems bedeviling the spread of astronomy in Africa and discuss some interdisciplinary traditional weather indicators. These indicators have been used over the years to monitor the appearance of constellations. For example, the Orions are closely intertwined with cultures of some ethnic African societies and could be incorporated in the standard astronomy curriculum as away of making the subject more 'home grown' and to be able to reach out to the wider populace in popularizing astronomy and basic sciences. We also discuss some of the other measures that ought to be taken to effectively create an enabling environment for sustainable teaching and spread of astronomy through Africa.

5.40. *Education at the Pierre Auger Observatory: the cinema as a tool in science education. By Beatriz Garcia and C. Raschia (Argentina)*

The Auger collaboration's broad mission in education, outreach and public relations is coordinated in a separate task. Its goals are to encourage and support a wide range of outreach efforts that link schools and the public with the Auger scientists and the science of cosmic rays, particle physics, astrophysics in general, and associated technologies. This report focuses on recent activities and future initiatives and, especially, on a very recent professional production of two educative videos for children between 6 and 11 years: 'Messengers of Space' (18 min), and for general audiences: 'An Adventure of the Mind' (20 min). The use of new resources, as 2D- and 3D-animation, to teach and learn in sciences is also discussed.

5.41. *Freshman seminars: interdisciplinary engagements in astronomy. By Mary Kay M. Hemenway (USA)*

The Freshman Seminar program at the University of Texas is designed to allow groups of fifteen students an engaging introduction to the university. The seminars introduce students to the resources of the university and allow them to identify interesting subjects for further research or future careers. An emphasis on oral and written communication by the students provides these first-year students a transition to college-level writing and thinking. Seminar activities include field trips to an art museum, a research library, and the Humanities Research Center rare book collection. This paper will report on two seminars, each fifteen weeks in length. In 'The Galileo Scandal', students examine Galileo's struggle with the church (including a mock trial). They perform activities that connect his use of the telescope and observations to astronomical concepts. In 'Astronomy and the Humanities', students analyze various forms of human expression that have astronomical connections (art, drama, literature, music, poetry, and science fiction); they perform hands-on activities to reinforce the related astronomy concepts. Evaluation of the seminars indicates student engagement and improvement in communication skills. Many of the activities could be used independently to engage students enrolled in standard introductory astronomy classes.

6. List of posters

Approximately 60 posters were presented by participants from over 20 countries. The titles of the poster papers, their authors names, and the countries of their affiliation follow:

- An educational CD-ROM based on the making of Guide Star Catalog II, *R.L. Smart*, Italy
- An astronomer in the classroom: Observatoire de Paris' partnership between teachers and astronomers, *A. Doressundiram, C. Barban*, France
- An effective distance mode of teaching astronomy, *V.B. Bhatia*, India
- Astrobiology and extrasolar planets - a new lecture course at Potsdam University, *S.A. Franck, W. von Bloh, Ch. Bounama*, Germany
- `<Astronomia.pl>` portal as a partner for projects aimed at students or public, *K. Czart, J. Pomierny*, Poland
- Astronomical black holes as an exciting tool and object for teaching relativistic physics, *V. Karas*, Czech Republic
- Astronomy and space sciences in Portugal: communication & education, *P. Russo, A. Pedrosa, M. Barrosa*, Portugal

• Astronomy education in the Republic of Macedonia, *O. Galbova, G. Apostolovska*, Macedonia

• Astronomy education in Ukraine, the school surriculum, and a lecture course at Kyiv Planetarium, *N.S. Kovalenko, K.I. Churyumov*, Ukraine

• Astronomy education with movement and music, *C.A. Morrow*, US

• Astronomy in the laboratory, *B. Suzuki*, Japan

• Astronomy in the training of teachers and the tole of practical rationality in sky observation, *P.S. Bretones, M. Compiani*, Brazil

• Astronomy, the Australian school curriculum, and the role of the Sir Thomas Brisbane Planetarium, *A. Axam, M. Rigby, W. Orchiston*, Australia

• Challenges of astronomy: classification of eclipses, *Sonja Vidojevic*, Serbia

• Cosmic deuterium and social networking software, *J.M. Pasachoff, D.A. Lubowich, T.-A. Suer, T. Glaisyer*, USA

• Cosmology and globalization, *D.K. Perkins*, USA

• Crayoncolored planets: using childrens drawings as guides for improving astronomy Teaching, *A.B. De Mello, E.A.M. Gonzalez, B.C.G. De Lima, D.H. Epitácio Pereira, R.V. De Nader*, Brazil

• Critical evaluation of the New Hall of Astronomy for the Science Museum at the University of Mexico, *S. Torres-Peimbert, C. Doddoli*, Mexico

• Daytime utilization of a University Observatory for laboratory instruction, *J.R. Mattox*, USA

• e-SpaceCam : development of a remote cooperative observation system for telescopes with P2P (Peer-to-Peer) agent network using location, *T. Okamoto*, Japan

• Education and public outreach for eGY: virtual observatories that connect teachers with authentic science data, *P.A. Fox*, USA

• Education at the Pierre Auger Observatory: The cinema as a tool in science education, *B. Garcia, C. Raschia*, Argentina

• Educational opportunities in pro-am collaboration, *R.T. Fienberg, R.E. Stencel*, USA

• Elementary astronomy, *J. Fierro Gossman*, Mexico

• Experiences in the Sky Classroom, *A.T. Gallego*, Spain

• Gemini Observatory outreach, *A. Garcia*, Chile

• Gemini Observatory's innovative education and outreach for 2006 and beyond, *J. Harvey*, USA

• History of the teaching of astronomy in serbian schools, *Sonja Vidojevic, S. Segan*, Serbia

• History of Ukrainian culture and science in astronomical toponymy, *I.B. Vavilova*, Ukraine

• Identification and support of outstanding astronomy students, *A.D. Stoev, E.S. Bozhurova*, Bulgaria

• Image subtraction using a space-varying convolution kernel, *J.P. Miller, C.R. Pennypacker, G.L. White*, USA

• Light pollution: a tool for astronomy education, *M. Metaxa*, Greece

• Malargüe light pollution: a study carried out by measuring real cases, *B. Garcia, A. Risi, M. Santander, A. Cicero, A. Pattini, M.A. Cantón, L. Córica, C. Martínez, M. Endrizzi, L. Ferrón*, Argentina

• Making and using astronomical fairy-tales on DVD in planetarium, *V.G. Goncharova*, Russia

• Modern facilities in astronomy education, *H.A. Harutyunian, A.M. Mickaelian*, Armenia

- News from the cosmos: daily astronomical news web page in Spanish, *A. Ortiz*, Spain
- Outreach activities of the National Astronomical Observatory of Japan, *T. Ono, J. Watanabe, H. Agata*, Japan
- Physics education: a significant backbone of sustainable development in developing countries, *A.R. Akin*
- Podcast, blogs, and new media outreach techniques, *A. Price, P. Gay, T. Searle*, USA
- Popularization of astronomy through cooperation between students and educators in Japan: the TENPLA project (1), *M. Hiramatsu*, Japan; (2) *K. Kamegai*, Japan
- Reproduction of William Herschel's metallic mirror telescope, *N. Okamura, S. Hirabayashi, A. Isida, A. Komori, M. Nishitani*, Japan
- Research thinking development by teaching archaeo-astronomy, *P.V. Muglova, A.D. Stoeva*, Bulgaria
- Role of creative competitions and mass media in the astronomy education of school students, *E.Yu. Aleshkina*, Russia
- Sendai Astronomical Observatory: its renewal and history as an observatory for the general public, *J. Watanabe*, Japan
- Simple, joyful, instructive: make a unique telescope of your own and explore the Universe, *Y. Hanaoka*, Japan
- 'Solar System: Practical Exercises' and 'Astronomy: Practical Works' for secondary scholars, *A. Tomic*, Serbia
- Star Week: a successful campaign in Japan, *J. Watanabe*, Japan
- Successful innovative methods in introducing astronomy courses, *T.K.C. Chatterjee*, Mexico
- The 2005 annular eclipse: a classroom activity at EPLA, *H. Filgaira*, Spain
- The Armagh Observatory human orrery, *M.E. Bailey, D.J. Asher, A.A. Christou*, Northern Ireland, UK
- The constellations of the zodiac: astronomy for low vision and blind people, *B. Garcia, A. Cicero, M. Farrando, P. Bruno*, Argentina
- The distance-learning part-time masters and doctoral internet programs in astronomy at James Cook University, Australia, *G.L. White, A. Hons, W. Orchiston, D. Blank*, Australia
- The first two years of the Latin-American Journal of Astronomy Education (RELEA), *P.S. Bretones, L.C. Jafelice, J.E. Horvath*, Brazil
- The Pomona College undergraduate 1-meter telescope, astronomy laboratory, and remote observing program, *B.E. Penprase*, USA
- The recent globe at night initiative involving schoolchildren and families from 96 countries, *C.E. Walker*, USA
- The Universe: helping to promote sstronomy, *R.M. Ros, F.J. Moldón*, Spain
- Use of modern technologies in improving astronomy education in Tanzania, *N. Jiwaji*, Tanzania
- Visualization of the astronomy domain: a mapping strategy in teaching and learning astronomy, *S. Gulyaev*, New Zealand
- Weaving the cosmic web: frontiers of astronomy education on the internet, *D.K. Perkins*, USA
- What mathematics is hidden behind the astronomical clock of Prague? *M. Krizek, A. Solcová, L. Somer*, Czech Republic
- With weekly astronomy tips against the weekly papers' astrology humbug, *G.A. Szécsényi-Nagy*, Hungary

Scientific Organizing Committee

Michael A. Bennett (USA), Julieta Fierro (Mexico), Michele Gerbaldi (France), Petr Heinzel (Czech Republic), Bambang Hidayat (Indonesia), Syuzo Isobe (Japan), Edvard V. Kononovich (Russia), Margarita Metaxa (Greece), Jay M. Pasachoff (USA, co-chair), John R. Percy (Canada), Rosa M. Ros (Spain, co-chair), Magda G. Stavinschi (Romania), Richard M. West (Germany) and Lars Lindberg Christensen (ESO, webmaster).

Summarizer: Naomi Pasachoff. Assistants: Javier Moldon and Madeline Kennedy.

Epilogue

A book based on the proceedings of this Special Session will be published by Cambridge University Press, editors Jay M. Pasachoff, Rosa M. Ros, and Naomi Pasachoff. Its publication will be announced at `<http://www.astronomyeducation.org>`.

Highlights of Astronomy, Volume 14
IAU XXVI General Assembly, 14-25 August 2006
Karel A. van der Hucht, ed.

doi:10.1017/S1743921307011763

Special Session 3 The Virtual Observatory in action: new science, new technology, and next generation facilities

Nicholas A. Walton[1], Andrew Lawrence[2], and Roy Williams[3] (eds.)

[1]Institute of Astronomy, University of Cambridge, Madingley Road, Cambridge CB3 0HA, UK
email: naw@ast.cam.ac.uk

[2]University of Edinburgh, Royal Observatory, Blackford Hill, Edinburgh EH9 3HJ, UK
email: al@roe.ac.uk

[3]Center for Advanced Computer Research, California Institute of Technology, MC 158-79, 1200 E. California Boulevard, Pasadena, CA 91125, USA
email: roy@cacr.caltech.edu

Preface

The vision of the Virtual Observatory (VO) is to make access to astronomical databases as seamless and transparent as browsing the World Wide Web is today. It will federate the data flows from current and future facilities and large scale surveys, and the computational resources and new tools necessary to fully exploit them. This requires both technological developments and an international commitment to standardisation and working culture. Increasingly, it will alter the way that astronomers do science, and the way that future facilities and projects plan for their data management, and the scientific exploitation of their data. It will make an impact on a wide variety of astronomical topics, but especially those using very large databases, and those needing a multiwavelength approach, or more generally the use of multiple archives.

To date, there are fifteen VO projects worldwide, who co-ordinate their efforts through an International Virtual Observatory Alliance (IVOA). This body evolves and agrees technical standards as well as sharing best practice and software. The various VO projects have laid the foundations for the VO: international standards, fundamental infrastructure, early demonstrations, and the first published science papers using VO tools. The VO is now becoming an operational reality. The next stage involves (*i*) deployment of the new infrastructure at data centres; (*ii*) science results enabled via the use of VO services; (*iii*) making links to existing and planned facilities; and (*iv*) much more ambitious data mining analysis services.

Following endorsement and sponsoring by IAU Divisions and Commissions and subsequent approval by the IAU Executive Committee, this IAU Special Session on *'The Virtual Observatory in Action'* allowed for the latest developments in the area of the Virtual Observatory to be presented. Talks and posters covered an exciting range covering new science results, plans for major facilities, and discussion of technical advances in data mining and the VO.

The presentations given here allow for a brief summary of the range of topics covered. We have also made the slides for all the oral presentations given at this Special Session available at the SPS 3 conference website:
`<http://www.ivoa.net/pub/VOScienceIAUPrague/programme/index.html>`.

The smooth running of the VO Special Session, organised as one of the constituent sessions of the IAU XXVI General Assembly in Prague, August 2006, greatly benefited from the excellent local organisation. We greatly acknowledge the help of Prof. Meszaros and the many helpful local students and staff, along with the session chairs, in ensuring the smooth running of this Special Session.

Scientific Organizing Committee

Giuseppina Fabbiano (USA), Françoise Genova (France), Robert J. Hanisch (USA), Ajit K. Kembhavi (India), Andrew Lawrence (UK, chair), Oleg Yu. Malkov (Russia), Atilla Meszaros (Czech Republic), Raymond P. Norris (Australia), Masatoshi Ohishi (Japan), Peter Quinn (Germany), Isabelle F. Scholl (France), Enrique Solano (Spain), Alexander Szalay (USA), Nicholas A. Walton (UK), Roy Williams (USA), and Yongheng Zhao (China).

Andrew Lawrence, chair SOC,
Nicholas Walton & Roy Williams, editors,
Edinburgh, Cambridge, Pasadena, November 30, 2006

Highlights of Astronomy, Volume 14
IAU XXVI General Assembly, 14-25 August 2006
Karel A. van der Hucht, ed.

doi:10.1017/S1743921307011775

The Virtual Observatory: what it is and where it came from

Andrew Lawrence

Institute for Astronomy, University of Edinburgh,
Royal Observatory, Blackford Hill, Edinburgh EH9 3HJ, UK
email: al@roe.ac.uk

The Virtual Observatory (VO) is driven partly by new science goals, partly by external technological changes, and partly by the flood of data coming our way. More astronomy is done on line through organised 'science ready' archives, and with processed data (source catalogues etc) from large survey projects. Users assume on-line availability of anything useful. The normal methods to date are to download files and analyse them at home, but very likely in the future data will be analysed *in situ* and analysis software will be as standardised as data reduction software is now. Many science goals require combining data from multiple archives (e.g. crossmatching different wavelengths), and often demand the processing of huge amounts of data, e.g. rare object searches, computing correlation functions, etc. All users want to be 'power users'.

Two technical bottlenecks make this hard - firstly CPU-disk I/O, which has grown much slower than Moore's law, and secondly end-user network bandwidth (the infamous 'last mile problem'). These bottlenecks mean that jobs like searching and analysing very large databases are best performed as *services* provided by the data centre, or by collaborating data centres, joined by fat pipes. Such 'on tap' services will need to be standardised, professional and reliable. New internet software technologies (especially XML, SOAP, and Grid software) mean that computers can exchange data and/or share processing power in a transparent way, thus these kind of standardised services are plausible. The final driver for the VO is data expansion. Many archives now grow at TB/yr rates, several will be growing at 100TB/yr over the next few years, and soon we will have facilities like ALMA and SKA which will deliver PB/yr. Dealing with these rates is one of logistics. Modern facilities need organised data management, and methods for serving science ready data to users. From the users point of view, the problem is not the number of bits, but the number of archives. If they all speak different languages, we are sunk. The two key requirements are therefore *archive interoperability* and *well funded data centres*.

The aim of the VO is that using astronomical data from all over the world should feel just as transparent as using the Web does today. Like the Web, the VO is not a monolithic system – it is a way of life. It is an agreed set of standards, a community of interoperable data collections, and – perhaps not so obviously – a collection of interoperable software modules. If pieces of software have the same screw-threads as it were, we do not have design and build some grand system – it will grow organically. The VO is not an enormous warehouse; it is not a hierarchy, like the LHC Grid; it is not a peer-to-peer system, like Napster. Rather, it is a small set of *service centres* and large population of *end users*. Its not so different from internet shopping.

To achieve this vision we need global standards; well funded data centres; working data services; infrastructure software; VO aware client tools; and VO aware data mining services. Considerable progress has been made on standards - data access protocols, service metadata, table column semantics, software interfaces - by the formation of the *International Virtual Observatory Alliance (IVOA)*.

Highlights of Astronomy, Volume 14
IAU XXVI General Assembly, 14-25 August 2006
Karel A. van der Hucht, ed.

doi:10.1017/S1743921307011787

Large surveys and the Virtual Observatory

Konrad Kuijken

Leiden Observatory, Leiden University, PO Box 9513, NL-2300RA Leiden, the Netherlands
email: kuijken@strw.leidenuniv.nl

Increasingly, large surveys of the sky in multiple wavelength bands are becoming an important part of how astronomy is being done. These surveys produce huge homogeneous datasets, and as such are prime material for dissemination and analysis through the VO. As an example, the KIDS survey (see Fig. 1), which will map 1500 square degrees of sky in *ugriZYJHK* using ESO's VST and VISTA telescopes, and which is expected to start in 2007, will generate some 15 TB of pixel data, as well as several TB of derived source parameters. More than 100 TB of raw data will have to be processed.

Analysis of such massive datasets requires that the end users are empowered to interact with their data flexibly. Given the huge data volumes, it will often be much more efficient to turn calculations into complex queries on the database, particularly so if other data sets (X-ray or radio maps, or earlier epochs from older surveys such as SDSS, for example) are to be incorporated.

Surveys help the VO by populating it with well-qualified data, while the VO helps the community to extract the science from this vast resource.

A critical application is to measure accurate colours for extended objects (galaxies), in order to derive photometric redshifts. Since different wavebands will typically be observed in different seeing conditions, or even at different telescopes with different pixel scales, this is not trivial. The development of techniques that correct the photometric catalogues for different image quality is vital if colours can be measured efficiently; the traditional technique of convolving pairs of images to a common seeing is simply too laborious when half a dozen different surveys are considered. Details on KIDS survey can be found at `<http://www.strw.leidenuniv.nl/~kuijken/KIDS/>`.

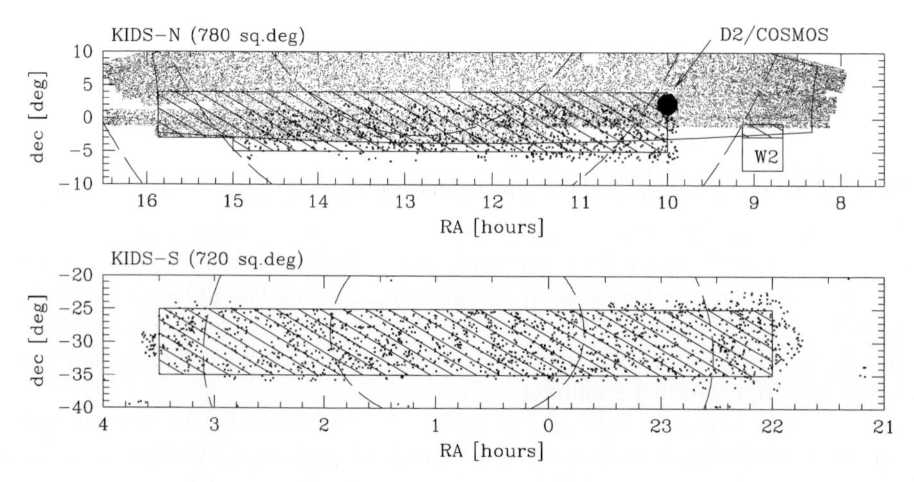

Figure 1. Layout of the KIDS survey fields, with SDSS (*dots*) and 2dF (*crosses*).

Highlights of Astronomy, Volume 14
IAU XXVI General Assembly, 14-25 August 2006
Karel A. van der Hucht, ed.

doi:10.1017/S1743921307011799

A VO study of SDSS AGNs with X-ray emission from *ROSAT*-PSPC pointed observations

Robert J. Hanisch[1], **Anatoly A. Suchkov**[2], **Timothy M. Heckman**[2], **and Wolfgang H. Voges**[3]

[1]Space Telescope Science Institute, 3700 San Martin Drive,Baltimore, MD 21218, USA
email: hanisch@stsci.edu

[2]Department of Physics and Astronomy, The Johns Hopkins University,
3400 N. Charles St., Baltimore, MD 21218, USA

[3]Max-Planck-Institut für extraterrestrische Physik,
Giessenbachstraße 1, D-85748 Garching, Germany

We use VO facilities to study AGNs with X-ray emission. We present a sample of 1744 of Type 1 AGNs from the Sloan Digital Sky Survey Data Release 4 (SDSS DR4) spectroscopic catalog with X-ray counterparts in the White-Giommi-Angelini catalogue (WGACAT) of *ROSAT*-PSPC pointed observations. Of 1744 X-ray sources, 1410 (80.9 %) are new AGN identifications. Of 4,574 SDSS DR4 AGNs for which we found radio matches in the catalogue of radio sources from the Faint Images of the Radio Sky at Twenty (FIRST) cm survey, 224 turned up in our sample of SDSS X-ray AGN.

We illustrate the content of our catalogue and its potential for AGN studies by providing statistical relationships for the catalogue data. The potential of the morphological information is emphasized by confronting the statistics of optically resolved, mostly low-redshift AGNs with unresolved AGNs that occupy a much wider redshift range. The immediate properties of the catalogue objects include significant correlation of X-ray and optical fluxes, which is consistent with expectations.

Also expected is the decrease of X-ray flux toward higher redshifts. The X-ray-to-optical flux ratio for the unresolved AGNs exhibits a decline toward higher redshifts, in agreement with previous results. The resolved AGNs, however, display the opposite trend. The X-ray hardness ratio shows a downward trend with increasing low-energy X-ray flux and no obvious dependence on redshift. At a given optical brightness, X-ray fluxes of radio-loud AGNs are on average higher than those of radio-quiet AGNs by a factor of 2.

We conclude that:

- VO-enabled comparisons and classifications of multi-wavelength source catalogs provide insights into nature of AGNs.
- Statistical analyses confirm several previously known correlations, but with much larger samples.
- The relationship between x-ray flux and optical flux differs for low-z resolved AGN and unresolved AGN.
- The sample of X-ray AGNs can be extended to $i > 19.1$ by using AGN candidates with only photometric data.

The full version of this talk is available at the SPS 3 conference website:
`<http://www.ivoa.net/pub/VOScienceIAUPrague/programme/sess2-hanisch.ppt>`.

Highlights of Astronomy, Volume 14
IAU XXVI General Assembly, 14-25 August 2006
Karel A. van der Hucht, ed.

doi:10.1017/S1743921307011805

A VO-based solution to the origin of soft X-ray emission in obscured AGN

Matteo Guainazzi[1], I. Barbarisi[1], Stefano Bianchi[1], Miguel Cerviño[2], Pedro Osuna[1], and Jesus Salgado[1]

[1]ESA European Space Astronomy Center, Madrid, Spain
email: Matteo.Guainazzi@sciops.esa.int

[2]Istituto de Astrofísica de Andalucia, Granada, Spain

Recent high-resolution imaging observations with *Chandra* have unveiled a striking similarity between the Narrow Line Region (NLR) and the diffuse soft (0.1 - keV) emission-line dominated X-ray morphologies in a number of obscured Active Galactic Nuclei (AGN) on scales as large as 0.1 - 2 kpc (Bianchi *et al.*, 2006, A&A, 448, 499). This discovery suggests a strong link between these components, which points toward a common physical origin. AGN photo-ionization is a natural explanation, consistent with soft X-ray high-resolution spectroscopy (Guainazzi & Bianchi, 2006, submitted). However, the possibility that the bulk of this component is generated by 'local' photo-ionization induced by shocks in the interaction between a radio jet and the interstellar matter, or by mechanical heating in regions of intense star formation cannot be ruled out yet.

In this paper, we discuss a Virtual Observatory (VO) based project, which aims at determining whether this diffused high-energy component can be produced in regions of intense star formation. This approach is based on the comparison between stellar synthesis models and observed Spectral Energy Distributions (SEDs) in a suitable sample of nearby obscured AGN.

The project makes use of the stellar synthesis model SED@ (Cerviño *et al.* 2002, A&A, 392, 19). This is the only stellar synthesis model currently available, which attempts at predicting the X-ray emission in starbursts. In this model it is assumed that the bulk of high-energy photons are produced by gas mechanical heating, i.e., conversion of gas kinetic energy into radiation according to a given efficiency parameter – one of the free parameters in the model) and supernova remnants. SED@ is implemented in a *Theoretical Simple Access Protocol* compatible server, and can, therefore, be accessed by any VO tools able to interpret this protocol. We have applied SED@ to a small (six objects) sample of Compton-thick AGN of the Guainazzi *et al.* (2005, MNRAS, 356, 295) sample. UV to X-rays (EPIC-pn) SEDs were generated from *Simple Spectra Access Protocol* compatible servers through the spectra visualization and analysis tool `VOSpec`, and compared with the SED@ models which best reproduce the observed UV flux density.

In all cases the SED@ predictions lay well below the soft X-ray photometric measurements. In three out of the six cases studied (NGC 1068, NGC 2273, NGC 3393) the starburst contribution is constrained to be $\leqslant 30\,\%$. In NGC 1068 this result is in excellent agreement with the interpretation of high-resolution soft X-rays spectra (Kinkhabwala *et al.*, 2002, ApJ, 575, 732). In the other three cases the constraints are looser, and still a substantial fraction, as high as 70 - 80 % of the X-ray emission, can be attributed to starburst. Nonetheless, the outcomes of the independent research streams outlined above seem all to converge towards the same explanation: the X-ray emitting NLR gas is primarily photoionized by the AGN, although with a strong contribution by resonant scattering (Guainazzi & Bianchi 2006).

Highlights of Astronomy, Volume 14
IAU XXVI General Assembly, 14-25 August 2006
Karel A. van der Hucht, ed.

doi:10.1017/S1743921307011817

Determination of radio spectra from catalogues and identification of gigahertz peaked sources using the Virtual Observatory

Bernd Vollmer[1], Sebastian Derriere[1], Thomas P. Krichbaum[2], Thomas Boch[1], Brice Gassmann[1], Emmanuel Davoust[3], Pascal Dubois[1], Françoise Genova[1], François Ochsenbein[1] and Wim van Driel[4]

[1]CDS, Observatoire astronomique de Strasbourg, UMR 7550,
11 rue de l'Université, F-67000 Strasbourg, France,
email: bvollmer@astro.u-strasbg.fr

[2]Max-Planck-Institut fr Radioastronomie, Auf dem Hügel 69, D-53121 Bonn, Germany

[3]UMR 5572, Observatoire Midi-Pyrénées, 14 avenue E. Belin, F-31400 Toulouse, France

[4]Observatoire de Paris, Section de Meudon, GEPI, CNRS UMR 8111 and Université de Paris,
5 place Jules Janssen, F-92195 Meudon, France

We have used the 20 largest radio continuum catalogues contained in VizieR (CDS) to determine radio continuum spectra between wavelengths of 2 cm and 1 m. For 67,000 out of the 3.5 million catalogued sources we could extract radio spectra with measurements at at least three independent frequencies (Vollmer *et al.* 2005, A&A, 431, 1177). These have been validated by comparison with existing spectral indices from the literature. This work allowed us to investigated the compatibility between the 20 radio continuum catalogues (Vollmer *et al.* 2005, A&A, 436, 757). Our radio spectra data base was searched for Gigahertz peaked source candidates, which we then observed quasi-simultaneously with the Effelsberg 100-m radio telescope at 6 cm (4.85 GHz), 2.8 cm (10.45 GHz), and 9 mm (32 GHz). This represents an efficient procedure to discover new Gigahertz peaked sources, which are believed to be AGNs at the beginning of their radio evolution. In our sample of more than 200 sources we find more than 50 % bona fide GPS sources. In addition, we can estimate the percentage of variable sources in our multi-epoch sample of radio sources which show an inverted spectrum. We are generalizing the method by using VO capabilities to: (*i*) identify pertinent radio catalogues in the VO registry using Uniform Content Descriptions (UCDs); all catalogues containing a user defined set of UCDs (e.g., PHOT_FLUX_RADIO* for a radio flux, POS_EQ_RA and POS_EQ_DEC for the position) are located in the VO registry and listed for further queries; (*ii*) extract relevant data, the user can easily assign a row of a given catalogue to a row of a previously defined output catalogue; and (*iii*) normalize these for the determination of radio spectra; units can be converted, aconymes can be created, flags can be created, etc. This procedure allows to homogenize the information retrieved from a heterogenuous set of catalogues. For this purpose software allowing semi-automated information retrieval is being developed at the CDS within the framework of the European VO-TECH project. The potential usage of all available radio catalogues will strongly increase the number of independent radio source cross-correlations and radio spectra. Our aim is to include more than 100 radio catalogues into the radio spectra determination software. The results are progressively being made available to the community through the CDS services.

Highlights of Astronomy, Volume 14
IAU XXVI General Assembly, 14-25 August 2006
Karel A. van der Hucht, ed.

doi:10.1017/S1743921307011829

Study on the environment around QSOs with redshift of 1–3 using the JVO System

Yuji Shirasaki[1], Masahiro Tanaka[1], Satoshi Honda[1], Satoshi Kawanomoto[1], Yoshihiko Mizumoto[1], Masatoshi Ohishi[1], Naoki Yasuda[2], Yoshifumi Masunaga[3], Yasuhide Ishihara[4], Jumpei Tsutsumi[4], Hiroyuki Nakamoto[5], Yuusuke Kobayashi[5] and Michito Sakamoto[5]

[1]National Astronomical Observatory of Japan, 2-21-1 Osawa, Mitaka Tokyo, 181-8588 Japan

[2]University of Tokyo, 5-1-5 Kashiwa-no-Ha, Kashiwa Chiba 277-8582, Japan

[3]Ochanomizu Univerisity, 2-1-1 Otsuka Bunkyo-ku, Tokyo 112-8610, Japan

[4]Fujitsu Ltd., 4-1-1 Kamikodanaka Nakahara-ku, Kawasaki 211-8588, Japan

[5]Systems Engineering Consultants Co. Ltd., 22-4 Sakuraoka-cho Shibuya-ku, Tokyo 150-0031, Japan

Virtual Observatory (VO) is an emerging astronomical infrastructure for sharing the astronomical data set in the world. National Astronomical Observatory of Japan (NAOJ) started its VO project (Japanese Virtual Observatory – JVO) in 2002, and developed JVO portal prototypes. We have carried out several science use cases, such as cosmic string searches and QSO environment studies, by using the prototype system to examine the functionality of the system. This paper describes a preliminary result of the latter science use case.

QSOs trace the most massive structures, so they can be used as a probe of clusters of galaxies. Since a survey using the QSOs as a lighthouse is not biased by galaxy types, it provides averaged feature of galaxy structure than that conducted by a survey that utilizes a narrow band filter to pick up an emission line for a specific redshift. In addition, it is also possible to explore the redshift desert around $z = 1.5 - 3.0$ by taking multi-color images between optical and near-infrared region.

We have constructed databases of QSOs and of Subaru SuprimeCam, and put a SkyNode interface on them. SkyNode is a VO standard data access interface based on SQL. JVO portal can federate the distributed SkyNode services, so it is possible to execute queries which associate the two different databases.

Using the JVO portal, we searched for SuprimeCam images which contains catalogued QSOs, then found 700 QSOs were observered by the SuprimCam with at least three filters and total exposure of longer than 1000 sec.

From these images, we created catalogues of detected objects around QSOs, estimated redshifts, and calculated a correlation between a QSO and the objects. We analyzed 100 out of the 700 QSOs. We calculated spatial QSO-Galaxy cross-correlation amplitude $B_{\rm QG}$ for each redshift interval. The result is: $\rm B_{QG} = 53.6 \pm 13.3$, 122.2 ± 29.3, 164.0 ± 80.9, 185.8 ± 108.2, and 538.5 ± 1072.1 for redshift ranges of, respectively, $z = 0.5 - 1.0$, $1.0 - 1.5$, $1.5 - 2.0$, $2.0 - 3.0$, and $3.0 - 4.0$, respectively.

Although the result still has large statistical errors, there can be seen a marginal clustering around QSOs. We expect that this study will be improved by addition of available data in the VO and by follow-up observations in the near-infrared bands.

Highlights of Astronomy, Volume 14
IAU XXVI General Assembly, 14-25 August 2006
Karel A. van der Hucht, ed.

doi:10.1017/S1743921307011830

Towards a VO compliant ESO science archive

Paolo Padovani, on behalf of the Virtual Observatory Systems department at ESO

European Southern Observatory, Karl-Schwarzschild-Straße 2,
D-85748 Garching bei München, Germany
email: ppadovan@eso.org

Data centres have a major role in the Virtual Observatory (VO), as they are the primary source of astronomical data. The VO cannot (and does not) dictate how a data centre handles its own archive. However, 'VO-layer' is needed to 'translate' any locally defined parameter to the standard (i.e., International Virtual Observatory Alliance compliant) ones. The longer term vision of the VO is also to hide away any observatory/telescope/instrument specific detail and work in astronomical units, for example, 'wavelength range' and not grism or filter name. Data providers are then advised to systematically collect metadata ('data about data') about the curation process, assign unique identifiers, describe the general content (e.g., physical coverage) of a collection, and provide interface and capability parameters of public services. Finally, the VO will work at its best with high-level ('science-ready') data, so that the VO user is spared as much as possible any complex and time consuming data reduction. Data centres should then make an effort to provide such data.

All these issues affect also the European Southern Observatory, which operates one of the largest largest astronomical archives in the world. This currently holds more than 100 Terabytes of data but is predicted to increase its size ~ 16 times in the next 6 years. The Data Management and Operations Division has then created the Virtual Observatory Systems (VOS) Department on November 1, 2004. The role of VOS is to manage ESO's involvement in VO activities, but above all to make the Science Archive Facility (SAF) into a powerful scientific resource for the ESO community. The new department includes, at present, the Virtual Observatory Technology (VOT) and the Advanced Data Products (ADP) groups and is made up of fifteen people.

Current/recent SAF/VO activities include: an improved archive interface, which represents a first step towards a VO-compliant interface(`<http://archive.eso.org>`); science-driven requirements already implemented into the ESO metadata repository and data models to serve both the real and virtual observatory; the creation of ADPs, that is 'science-ready' data; and preparation for ADP ingestion and publication. ADPs currently scattered around the community are also being collected, validated and published into the SAF (according to detailed guidelines). ESO, in fact, now requests Large Programs and the upcoming VST and Vista surveys to provide their reduced data at the time of publication. This represents a paradigm change and underscores the fact that the investigators of an observing program are in the best position to analyze and provide properly reduced data. Science-ready imaging ADPs are also being produced in VOS by the ADP/MVM data reduction system. Based on the EIS/MVM software, this is aimed at ISAAC, WFI, SOFI and other VLT imaging instruments ADP production and was used for the reduction of the ISAAC/GOODS ADPs, released on September 30, 2005, which cover the deepest large near-IR field carried out to date.

Our VO plans for the next two years include: more ADPs (internal and above all from ESO PIs), a brand new VO-compliant archive interface (based on metadata registry), VO tools to enhance the archive interface, and the ingestion of Quality Control products.

Highlights of Astronomy, Volume 14
IAU XXVI General Assembly, 14-25 August 2006
Karel A. van der Hucht, ed.

doi:10.1017/S1743921307011842

Storing and accessing the largest astronomical catalogues with the SAI CAS project

Sergey E. Koposov[1,2,3], **Oleg Bartunov**[2] **and Sergey Karpov**[4]

[1]Max Planck Institute for Astronomy, Königstuhl 17, Heidelberg, D-69117, Germany

[2]Sternberg Astronomical Institute, Universitetskiy pr. 13, Moscow, 119992, Russia

[3]Institute of Astronomy , Madingley Road, Cambridge CB3 0HA, UK

[4]Special Astrophysical Observatory, pos. Nijniy Arkhyz, Russia
email: math@sai.msu.ru

One of the main goals of the Virtual Observatory activities right now is to provide the simple and powerful access to the large existing astronomical datasets in the VO compatible way. That is why we want present the results of the recent development of Sternberg Astronomical Institute Catalogue Access Services (SAI CAS) project – the first and the only project in Russia, which provides on-line access to the major astronomical catalogues and different services on top of them. It is developed by a group of astronomers in a framework of SAI Astronet project, funded by Russian Foundation for Basic Research. SAI CAS is an open-source implementation of the general Catalogue access service (influenced by SDSS CASjobs & OpenSkyQuery projects), based on original algorithms and open-source software. We decided to build our own system providing an effective access to the major astronomical catalogues and different services including cone-searches and cross-matching of user data with hosted catalogues.

We use only open-source software with good support and stable history of development – Apache Tomcat, Axis, PostgreSQL and Java. SAI CAS consists of the database server running PostgreSQL 8.1+, which stores the catalogues and all their metadata, the application server with set of web-services and a pool of JDBC connections to database server, and the frontend for the interaction with users and web-interface.

Clients could be web browsers or any programs communicating with SAI CAS via HTTP and SOAP protocols and retrieving the data in different formats (VOTable, CSV, etc.). The spatial queries (cross-matches and cone-searches) in the database are operated using the developed by us Q3C sky indexing scheme and corresponding package for PostgreSQL v.8.1+ which is freely available. It provides the very fast access to the data with spherical attributes and combines advantages of well-known HTM and HEALPIX indexing schemes. Currently we provide the ConeSearch service (registered in the VO registry) for USNO-A2/B1, 2MASS, UCAC-2, NOMAD, DENIS, Tycho-2, 2XMM (SDSS DR5 will be loaded very soon). These catalogues are also available for the crossmatches. It is possible to perform the crossmatches of the supplied by user VOTable with one of the hosted catalogues.

All the SAI CAS system can be easily operated with the set of web services, with which the data can be uploaded to the system, the data and the metadata can be queried and updated. The formal Skynode interface of querying our catalogues is being built. The SAI CAS system is mainly targeted for science problem solving, and currently it is already used by several science projects in Russia. The web-site of the SAI CAS project is: `<http://vo.astronet.ru>`.

Highlights of Astronomy, Volume 14
IAU XXVI General Assembly, 14-25 August 2006
Karel A. van der Hucht, ed.

doi:10.1017/S1743921307011854

ANCHORS: an Archive of *Chandra* Observations of Regions of Star formation

Scott J. Wolk and Bradley D. Spitzbart

High-Energy Division, Harvard-Smithsonian Center for Astrophysics,
60 Garden Street, Cambridge MA 02138, USA
email: swolk,bspitzbart@head.cfa.harvard.edu

ANCHORS is a web based archive of all the point sources observed during *Chandra* observations of regions of star formation. It is designed to aid both the X-ray astronomer with a desire to compare X-ray datasets and the star formation astronomer wishing to compare stars across the spectrum. For some 50 *Chandra* fields, yielding 10 000+ sources, the database contains X-ray source properties including position, net count rates, flux, hardness ratios, lightcurve statistics and plots.

Spectra are fit using several models, with final parameters and plots recorded in the archive. Multi-wavelength images and data are cross-linked to other archives such as 2MASS and SIMBAD. The pipeline processing ensures consistent analysis techniques for direct comparisons among clusters. Results are presented on-line with sorting, searching, and download functions HTML/XML interface. We are hoping to add linkage to the VO. We will solicit users' feedback.

doi:10.1017/S1743921307012409

End-to-end science from major facilities: does the VO have a role?

Gerard F. Gilmore

Institute of Astronomy, University of Cambridge, Madingley Road, Cambridge, CB3 0HA, UK
email: gil@ast.cam.ac.uk

The Virtual Observatory provides a natural solution to the existence problem in communications: how can one ask a question of another unless you know the other exists? Many think e-mail from apparent strangers, e-blogs, etc., suggest there is no shortage of possible such solutions. In that context, is the Virtual Observatory in fact the necessary and desirable part of the solution? Specifically, does the VO necessarily play a critical role in delivering end-to-end facility science, from ideas, through proposals, resources/ facilities, to distributed, reviewed, knowledge? If not, what else needs to be added?

Highlights of Astronomy, Volume 14
IAU XXVI General Assembly, 14-25 August 2006
Karel A. van der Hucht, ed.

doi:10.1017/S1743921307011878

'Retooling' data centre infrastructure to support the Virtual Observatory

Séverin Gaudet
Canadian Astronomy Data Centre, Herzberg Institute of Astrophysics,
National Research Council, Canada
email: severin.gaudet@nrc.gc.ca

The Canadian Astronomy Data Centre manages a heterogeneous collection of data from the following ground and space-based telescopes: CFHT, DRAO, *FUSE*, Gemini, *HST*, JCMT, and *MOST*. The archive data models implemented for these data collections are ten years old and pre-date two important developments: the Virtual Observatory and the systematic generation and management of data products. Three years ago, we began the process of supporting access to processed data products through IVOA protocols such as SIA by building a layer over the archive data models. Today, we now realise that this approach of layering VO models on archive models is not sufficient and that every archive must be re-tooled to properly support the VO – from the storage model through to the query, processing and access models. The CADC has begun an ambitious software development effort to implement a new infrastructure to serve both telescope archive and Virtual Observatory needs.

At the core of this new infrastructure is the Common Archive Data Model. This data model, to be implemented in each archive, will standardize the way observations are characterized and relationships are expressed. The design of this model is inspired from IVOA standards under discussion (Observation, Characterization, Simple Image Access, Simple Spectral Access) and leverages the CADC's archive, data modeling and data engineering experience. The observation characterisation is based on FITS WCS papers I, II and III. The data model will become the sole metadata interface between the CADC's archives and the CADC's data warehouse upon which the VO services are built. The data model will also allow the CADC to adapt to the evolving VO standards.

The CAOM is only but one element of the retooling of the CADC to support the VO. The storage model is being modified to support caching of data products for synchronous retrieval, the generation of cutouts, and the decompression and recompression of files in streaming modes of access. The processing model is also being significantly re-designed to use the CAOM and to persist processing metadata and the relationships between complex multi-observation products and their simpler inputs. The retrieval model is being changed to support additional programmatic interfaces and the retrieval of complex data packages. An access control model is being developed to support authenticated access to proprietary data through VO interfaces.

These changes represent fundamental changes to a mature data centre infrastructure that, in 2005, archived more that 61 terabytes of data and distributed 38 terabytes to more than 2,500 distinct IP address. So what path has the CADC chosen? The design process began in the fall of 2005. This was followed by a prototype implementation for the JCMT archive with its new instrumentation scheduled for released at the end of 2006. This will be followed by an evaluation of the prototype (lessons learned) and planning for the conversion of the remaining CADC archives. The target completion date is set for the end of 2007.

Highlights of Astronomy, Volume 14
IAU XXVI General Assembly, 14-25 August 2006
Karel A. van der Hucht, ed.

doi:10.1017/S174392130701188X

Prototype development for a *Hubble* Legacy Archive

Bradley C. Whitmore, Helmut Jenkner, Warren Miller and Anton M. Koekemoer

Space Telescope Science Institute, 3700 San Martin Dr., Baltimore, MD 21218, USA
email: whitmore,jenkner,wmiller,koekemoe@stsci.edu

The Space Telescope Science Institute (STScI), in collaboration with the European Coordinating Facility (ECF), and the Canadian Astronomy Data Centre (CADC), is studying the development of an enhanced archive for the *Hubble Space Telescope* (*HST*), with the goal of improving the scientific value of the data. The primary enhancements would be:

(*a*) making HST data VO compatible;
(*b*) improving the science products for the legacy instruments, e.g., providing CR-rejected multi-drizzled images for the *HST*-WFPC2;
(*c*) providing 'real time' access to the data;
(*d*) adding a cutout service for super-fast access;
(*e*) providing more extensive 'composite images', e.g., stacked images, mosaics, ...;
(*f*) improving absolute astrometry by a factor of $\simeq 10$, i.e., from $1'' - 2''$ to $< 0''.2$;
(*g*) adding a footprint service to make it easier to browse and download images;
(*h*) developing source catalogs for many datasets.

The current *HST* archive is organized as a collection of datasets as observed by the telescope. The *Hubble* Legacy Archive will use a sky atlas approach; combining the various datasets into higher level products in various ways. At the top level will be mosaics that combine all data sets of a given target to achieve the widest possible field of view. Below that will be stacked images that provide the deepest possible images of a single instrument field of view. The next level will be single-epoch stacked images that allow time-resolution. The final level will be single exposures similar to what is available from the current *HST* archive.

The project will use existing methodologies as much as possible due to the limited manpower available at this time. In addition, products and capabilities will be introduced in phases beginning with beta testing several of the prototypes in the fall of 2006. The goal is to provide a significant fraction of the current *HST* archives starting with the most popular instruments (e.g., ACS, WFPC2, NICMOS, STIS). We will not attempt completenes (e.g., inclusion of the first generation instruments). See Jenkner *et al.* (2006, ASP-CS 351, 406) for a more detailed description of the project.

Acknowledgements

We would like to acknowledge several key members of the *Hubble* Legacy Archive team. At STScI: Niall Gaffney (software engineering), Bob Hanisch (interface), Steve Lubow and Gretchen Greene (footprints), Brian Mclean (astrometry), and Rick White (cutouts). At ECF: Rudi Albrecht, Bob Fosbury, Richard Hook, and Alberto Micol. At CADC: David Schade and Daniel Durand. While each person is generally only contributing a small fraction of their time to this effort, the expertise they bring to the project will be instrumental in its success.

Highlights of Astronomy, Volume 14
IAU XXVI General Assembly, 14-25 August 2006
Karel A. van der Hucht, ed.

doi:10.1017/S1743921307011891

New generation wide-fast-deep optical surveys: petabytes from the sky

J. Anthony Tyson

Physics Department, University of California, Davis, CA 95616, USA
email: tyson@physics.ucdavis.edu

Many fundamental problems in optical astronomy – from planetary science, galactic structure, optical transients, to large-scale structure and cosmology – could be addressed though the same data set with millions of exposures in superb seeing, in multiple pass-bands, to very faint magnitudes over a large area of sky. This capability is largely driven by technology. In a logical progression towards this scientific capability, several increasingly ambitious wide-field optical surveys are planned in the next few years. A uniform high quality database covering all these science areas would be an ideal match to the VO. The above utopian goal of simultaneous pursuit of parallel surveys is achievable, but it relies on the ability to image a wide field quickly and deeply, and it is a non-linear function of the camera + telescope *étendue.*

A figure of merit for sky surveys is étendue, the product of aperture area in square meters and field of view in square degrees. The power of a survey facility to execute a given survey down to some limiting flux in selected wavelength bands over some area on the sky is proportional to its étendue. In practice one designs the survey strategy based on the étendue, in the context of the science driver(s). How does one decide on a overall survey strategy for a given facility? The science goals are often so different that they cannot be pursued simultaneously. Current surveys (as well as the upcoming PanSTARRS-1) optimize their survey strategy by pursuing different surveys mostly in series, giving them appropriate names like 'deep survey' 'wide survey' 'galactic plane survey' 'near-Earth object survey', etc., each with its own set of filters and observing cadence. However, there is in principle a thrshold in tendue above which is becomes increasingly possible to pursue multiple surveys in parallel. This multiplex capability occurs above about étendue $150\,\mathrm{m}^2\mathrm{deg}^2$ because the most demanding flux limit may be reached in a short time through standard filters. Breakthroughs in three areas of technology (large aspherical optics fabrication, microelectronics, and software) permit this major leap in capability. For the first time it will be possible to survey wide, fast, and deep simultaneously.

Data from a single active optics telescope with sufficient tendue can therefore address many scientific missions simultaneously. This is because short exposures also go deep, and with a large field of view the entire visible sky can be rapidly covered in multiple filters. The Large Synoptic Survey Telescope (LSST) with an tendue of 350 will provide unprecedented sky coverage, cadence, and depth. The 8.4 meter LSST and its 3.2 Gpixel camera and data system can attack high-priority scientific questions that are far beyond the reach of any existing or planned facility. The 20 – 30 terabytes of data obtained each night will open a new window on the deep optical universe – the time domain – enabling the study of variability both in position and time. This enables control of systematics required for precision probes of dark energy. Rarely observed events will become commonplace, new and unanticipated events will be discovered, and the combination of

LSST with contemporary space-based and ground-based missions at other wavelengths will provide powerful synergies.

The LSST is a public-private consortium and the software and data will be open – a necessary condition for the effective operation of the VO links and access. It will be sited on Cerro Pachon in northern Chile, and commissioning is scheduled to begin in 2013. Up to 30 000 square degrees of sky will be covered to 27.8 AB mag ($5\,\sigma$). Six filter bands are planned: ugrizy covering the near UV to the near IR. This will enable photometric redshifts of 4 billion galaxies. The short 15 s exposures will enable unprecedented time domain studies. The unprecedented data volume must be processed immediately for both quality assessment and the science programs. Alerts on flaring or moving objects will be issued within seconds, and the database will be accessible through a data access center. Massively parallel astrophysics will result.

In each of these areas of astronomy, the full-hemisphere, high-time-resolution coverage of LSST will increase sample sizes by the largest factor ever achieved in optical astronomy, creating challenges and opportunities for the Virtual Observatory.

Highlights of Astronomy, Volume 14
IAU XXVI General Assembly, 14-25 August 2006
Karel A. van der Hucht, ed.

doi:10.1017/S1743921307011908

Galaxy formation and evolution using multi-wavelength, multi-resolution imaging data in the Virtual Observatory

Paresh Prema, Nicholas A. Walton and Richard G. McMahon

Institute of Astronomy, University of Cambridge Madingley Road, Cambridge CB3 0HA, UK
email: pprema,naw,rgm@ast.cam.ac.uk

Observational astronomy is entering an exciting new era with large surveys delivering deep multi-wavelength data over a wide range of the electromagnetic spectrum. The last ten years has seen a growth in the study of high redshift galaxies discovered with the method pioneered by Steidel *et al.* (1995) used to identify galaxies above $z > 1$. The technique is designed to take advantage of the multi-wavelength data now available for astronomers that can extend from X-rays to radio wavelength. The technique is fast becoming a useful way to study large samples of objects at these high redshifts and we are currently designing and implementing an automated technique to study these samples of objects. However, large surveys produce large data sets that have now reached terabytes (e.g. for the Sloan Digital Sky Survey, `<http://www.sdss.org>`) in size and petabytes over the next 10 yr (e.g., LSST, `<http://www.lsst.org>`). The Virtual Observatory is now providing a means to deal with this issue and users are now able to access many data sets in a quicker more useful form.

We describe our development of a Spectral Energy Distribution (SED) matching technique that characterises objects at high redshift detected in the ultraviolet to infrared passbands. The observational SEDs are then matched to model SEDs that yield physical parameters of each object such as star formation rates (SFR), star formation histories (SFH), ages, stellar masses and colours. The technique uses model spectral synthesis codes that include those from Bruzual and Charlot (Galaxev), PEGASE and Starburst99. The technique can be broken down into the following sections. Data discovery, source extraction, cross match catalogues, photometric redshift creation, sample selection, model generation, model fitting and ouputs. However, each section has many implementation issues that need to be dealt with in order to produce scientifically viable results. For example, one important step in the source extraction step would be to include upper limits on flux measurements during the fitting process. The outputs would include plots of the best fit models to the data along with tabular data representing the best fit model and the closest matches, a standard error analysis on the various physical parameters and finally a set of image cutouts of the object.

The technique would be implemented as a workflow within the AstroGrid system. The workflow incorporates the step by step procedure that is required to complete the technique from retrieving the raw imaging data to the final outputs listed above. The execution of jobs within a workflow through various applications and services is done in an asynchronous manner as jobs are run can in different locations depending on which applications are used. Thus, the technique is ideally suited for this type of implementation.

We describe the technique, and how this is being developed as an application available through standard Virtual Observatory interfaces, specifically AstroGrid's Common Execution Architecture (CEA, `<http://www.astrogrid.org>`).

Highlights of Astronomy, Volume 14
IAU XXVI General Assembly, 14-25 August 2006
Karel A. van der Hucht, ed.

doi:10.1017/S174392130701191X

NVO study of Super Star Clusters in nearby galaxies

Bradley C. Whitmore

Space Telescope Science Institute, 3700 San Martin Dr., Baltimore, MD, 21218, USA
email: whitmore@stsci.edu

We are using NVO tools to obtain a large (N $\simeq$ 100), uniform (SDSS images, cross checked with *HST* images; analyzed using WESIX) database of super star clusters in nearby star-forming galaxies in order to address two fundamental astronomical questions: (*i*) Is the initial luminosity function of star clusters *universal*?; and (*ii*) What fraction of super star clusters is 'missing' in optical studies (i.e., are hidden by dust)?

It is now well established that large numbers of young massive star clusters (i.e., 'Super Star Clusters'; hereafter SSCs) form in gas-rich mergers and starburst galaxies (e.g., see review by Whitmore 2003, in *A Decade of HST Science*, eds. M. Livio, *et al.*, Cambridge: CUP, 153). Larsen & Richtler (2000, A&A, 354, 836) have shown that compact young clusters form in normal spiral galaxies as well, with the same luminosity function found for the clusters in merging galaxies. The difference is that these clusters are generally much fainter and less massive than their counterparts in merging systems. These studies have sparked an on-going discussion concerning the question of whether star formation in violent environments (i.e., in mergers and starbursts) is fundamentally different from star formation in quiescent settings (i.e., in spirals).

Whitmore (2003) addressed this question by making a simple plot of the brightest cluster versus the number of clusters in a galaxy, which suggested a universal rather than bimodal physical process. Our program is designed to readdress this issue using a more uniform database and analysis procedure.

Another important aspect of this project is to quantify the fraction of SSCs which are 'missing' (i.e, seen in the near infrared by 2MASS but not in optical studies; see Whitmore & Zhang; 2002, AJ, 124, 1418). This will also allow us to address the question of whether the fraction of obscured SSCs differs for mergers and starbursts when compared with normal spirals. We are also interested in determining whether *HST* preview images can be used to do serious science (i.e., what is the agreement between magnitudes determined from SDSS, *HST* observations, and *HST* preview images)? At present, this looks quite promising, with uncertainties of $\simeq$ 10 %. While this is large by normal photometric standards, for the purposes of this study this might be quite sufficient.

We plan to use two independent approaches for sample selection: (1) a volume limited sample, and (2) galaxies selected to provide a roughly uniform distribution in log(N). The primary NVO tools that we use for the project are: WESIX, Aladin, VOTool, and OpenSKYQuery.

Acknowledgements

I would like to acknowledge my co-investigators on this project: Kevin Lindsay, Rupali Chandar, Chris Hanley, and Ben Chan. This project is supported by a NVO Research Initiative grant.

Highlights of Astronomy, Volume 14
IAU XXVI General Assembly, 14-25 August 2006
Karel A. van der Hucht, ed.

doi:10.1017/S1743921307011921

Science projects with the Armenian Virtual Observatory (ArVO)

Areg M. Mickaelian

Byurakan Astrophysical Observatory, Byurakan 378433, Aragatzotn province, Armenia
email: aregmick@apaven.am

The main goal of the Armenian Virtual Observatory is to develop efficient methods for science projects based on the digitized famous Markarian survey (Digitized First Byurakan Survey, DFBS) and other large astronomical databases, both Armenian and international. Two groups of projects are especially productive: search for new interesting objects of definite types by low-dispersion template spectra, and optical identifications of new gamma, X-ray, IR and radio sources. The first one is based on modeling of spectra for a number of types of objects: QSOs, Seyfert galaxies, white dwarfs, subdwarfs, cataclysmic variables, planetary nebulae, late-type stars (K-M, S, carbon), etc. Each kind of object appears in the DFBS with its typical SED and spectral lines (for objects having broad lines only, like white dwarfs and subdwarfs, quasars and Seyferts, etc.), however affected also by its brightness, so that each template works for definite range of magnitudes. The search criteria define how many objects will be found for further study, and may restrict these numbers leaving with the best candidates. Optical identifications have been proven to be rather efficient for IR sources from *IRAS*-PSC and -FSC. Tests have been carried out for X-ray and radio sources as well. Special emphasis is put on search for bright QSOs missed by the Sloan Digital Sky Survey (SDSS), rather important for making up their complete sample, studies of the properties of the Local Universe, a comparison of their X-ray, IR, and radio properties and making up their multiwavelength SEDs, as well as for a refinement of the AGN classification. A project for search for new bright QSOs using the DFBS has been started in the region with $\delta > 0°$ and $|b| > 20°$. The Byurakan 2.6 m telescope is being used for the spectral identification of the candidates. The first test resulted in 145 objects found, 81 being known QSOs/Sys, and 64 new candidates (including 23 NVSS and FIRST radio sources).

A number of sub-projects have been tested concerning the asteroids search on the DFBS plates. All known bright ($< 15^{\rm m} - 16^{\rm m}$) asteroids have been grouped into fast and slow ones with a division parameter, estimated as the motion of $3''$ during 20 min (the typical exposure time of a DFBS plate). Extraction of the spectra of asteroids found in DFBS by SkyBote has been started, and they were grouped into extended (fast asteroids) and star-like (slow asteroids). The further plans include modeling of a template spectra of asteroids by means of the star-like spectra, search for new candidate asteroids by similar spectra and comparison with DSS1/DSS2 fields for elimination of the stars, spectral analysis of the asteroid spectra to get some physical parameters, etc.

Another project is the identification of the newly found IR sources from Spitzer Space Telescope (SST). 73 unidentified sources in the Bootes region have been found and classified on the DFBS plates. All available additional data from DSS1/DSS2, other optical and multiwavelength catalogs were used to clarify the nature of these objects. 51 were found to be known objects from existing catalogs, including 1 QSO, 28 galaxies, and 22 stars. The 22 new objects were classified as five candidate QSOs, ten galaxies (four AGN candidates and six interacting systems), and seven stars (six G-M type, and one carbon). Eight of the known stars not having spectral classification, were classified too.

Highlights of Astronomy, Volume 14
IAU XXVI General Assembly, 14-25 August 2006
Karel A. van der Hucht, ed.

doi:10.1017/S1743921307011933

Mapping Galactic spiral arm structure: the IPHAS survey and Virtual Observatory access

Nicholas A. Walton[1], Janet E. Drew[2], Eduardo A. Gonzalez-Solares[1], Robert Greimel[3], Ella C. Hopewell[2] and Mike J. Irwin[1]

[1]Institute of Astronomy, University of Cambridge, Madingley Road, Cambridge CB3 0HA, UK
email: naw,eglez@ast.cam.ac.uk

[2]Astrophysics Group, Imperial College London, Prince Consort Road, London SW7 2BZ, UK
email: j.drew,e.hopewell@imperial.ac.uk

[3]Isaac Newton Group of Telescopes,
Apartado de corras 321, E-38700 Santa Cruz de La Palma, Tenerife, Spain
email: greimel@ing.iac.es

There is now considerable interest in how stellar streams in the Milky Way can be used to probe how the earlier merger history of our galaxy, which in turn can be related to hierarchical models of galaxy evolution.

One example stream is the Sagittarius stream, caused by the disruption of the Sagittarius Dwarf galaxy. Related to this is the ability to accurately map the major spiral arms in our galaxy. There is much uncertainty, with Russeil (2003) providing a recent spiral arm model. Early type emission line stars are an easily identified class of stars which can be (and have been) utilised effectively to map spiral arm structure.

We report on the use of IPHAS (<http://www.iphas.org>) (the INT/WFC Photometric Hα Survey of the Northern Galactic Plane (Drew *et al.* 2005) carried out in the Hα, r' and i' bands to a depth of ~ 20 mag in r') as a rich source of these emission line stars. This paper includes a pilot mapping of Galactic spiral arm structure at $l \simeq 250°$, based on the older Schmidt Hα survey where the candidate tracer stars have been selected via their colours – and then confirmed via multi-object spectroscopy. Analysis of these data underline how the increased numbers of fainter tracer stars available from IPHAS, have the potential to pin down spiral arm structure in the outer Galaxy.

We describe how the AstroGrid Virtual Observatory system has been utilised to provide access to the IPHAS data products (see <http://www.astrogrid.org>). This is supporting the analysis of the large pipeline processed data products, being analysed by members of the IPHAS consortium. We note both the deployment of Virtual Observatory components in making the data available through the VO, but also how the AstroGrid system is being used by the astronomer, for instance in the use of workflows to automate certain routine analysis operations, such as generating lists of candidate emission line stars, based on database queries of the IPHAS catalogue data.

References

Drew, J. E., Greimel, R., Irwin, M. J., *et al.* 2005, *MNRAS*, 362, 753
Russeil, D. 2003, *A&A*, 397, 133

Highlights of Astronomy, Volume 14
IAU XXVI General Assembly, 14-25 August 2006
Karel A. van der Hucht, ed.

doi:10.1017/S1743921307011945

Large surveys and determination of interstellar extinction

Oleg Yu. Malkov[1,2] and Erkin Karimov[2]

[1]Center for Astronomical Data, Institute of Astronomy, Russian Academy of Sciences (INASAN), Pyatnitskaya ul 48, RU-119017 Moscow, Russian Federation
email: malkov@inasan.ru

[2]Sternberg Astronomical Institute, Moscow State University, Universitetskij pr. 13, RU-119992 Moscow, Russian Federation
email: youju@mail.ru

The study of the spatial distribution of interstellar extinction, A_V, is important for many investigations of galactic and extragalactic objects. Three-dimensional (3D) extinction models have been produced using spectral and photometric stellar data, open cluster data, star counts, the Galactic dust distribution model.

The standard approach to construct a 3D extinction model has been to parcel out the sky in angular cells, each defined by boundaries in Galactic coordinates (l, b). From the stars in each cell, the visual extinction $A_V(l, b)$ can then be obtained as a function of distance $A_V(l, b, r)$. The angular size of the cells has varied from study to study, although each cell was generally chosen to be large enough to contain a statistically significant number of calibration stars at different distances. Published 3D models, using spectral and photometric data, are based on 104 – 105 stars. Modern large surveys contain photometric (3 to 5 bands) data for 107 – 109 stars. But to make that data useful for a 3D extinction model construction one needs to run a correct cross-identification of objects between surveys. Another problem is a lack of spectral data in photometric surveys.

Identification of objects requires the federation of multiple surveys obtained at different wavelengths and with different observational techniques. Such cross-matching of catalogs is currently laborious and time consuming. But using VO data access and cross-correlation technologies a search for counterparts in a subset of different catalogues can be carried out in a few minutes. Particularly, information on interstellar extinction may be obtained from modern large photometric surveys data.

The goal of our paper is to design a procedure for construction of 3D interstellar extinction model, based on data from large surveys. To test the procedure we have selected a two-arc-minute area on the sky with $l = 323$, $b = +6$. For further analysis the following multicolor surveys were chosen (photometric bands are given in brackets): DENIS (J, K'), 2MASS (J, H, Ks), USNO-B (SERC-J). Our two-arc-minute test area contains 134 objects cross-identified in all three surveys. For 36 of them all required photometry is available. We approximate our result by the relation: $A_V = 0.01|cosec6^o|[1 - exp(-0.008r|sin6^o|)]$. The uncertainty of A_V is about $0.^{m}1$ depending primarily on the uncertainties of intrinsic colors. The relative error of the distance is about 25 %, depending primarily on the uncertainties of absolute magnitudes.

The proposed method has a number of advantages. One does not need for spectral type data and trigonometric parallaxes for calibration stars. One uses 104 - 106 times more stars than in 'traditional' models (it allows to choose angular cells on the sky small enough so that individual interstellar clouds can be resolved). 'On-line' model can be constructed to calculate $A_V(l, b, r)$ based on available data for a user defined area on the sky. When available, other multi-wavelength surveys like DPOSS, SDSS, UKIDSS can be incorporated using VO techniques.

Highlights of Astronomy, Volume 14
IAU XXVI General Assembly, 14-25 August 2006
Karel A. van der Hucht, ed.

doi:10.1017/S1743921307011957

Brown dwarfs and star forming regions in the framework of the Spanish Virtual Observatory

Enrique Solano[1], Eduardo L. Martín[2] and José A. Caballero[3]

[1]INTA – Laboratorio de Astrofísica Espacial y Física Fundamental,
P.O. Box 50727, E-28080 Madrid, Spain
email: esm@laeff.inta.es

[2]Instituto de Astrofísica de Canarias,
C/ Vía Láctea s/n, E-38200, La Laguna, Tenerife, Spain
email: ege@iac.es

[2]Max-Planck-Institut für Astronomie, Königstuhl 17, D-69117 Heidelberg, Germany
email: caballero@mpia.de

In this paper we describe three lines of work that are being developed in the framework of the Spanish Virtual Observatory.

• *Discovery of new ultra-cool brown dwarfs*

Building a statistically significant census of substellar objects implies the use of queries that combine attributes available from different archives, an approach out of the scope of the 'classical' methodology but that perfectly fits into the Virtual Observatory. Using 2MASS, DENIS and SDSS and Aladin as VO-tool, we are focusing on the nearby ($d \leqslant 10$ pc) population of brown dwarfs with T spectral type with the aim of shedding light on the problem of the form of the stellar mass function at the lower end, an issue that has been identified as a key VO-Science case both by AstroGrid and EURO-VO. A similar methodology will be apply for the exploitation of UKIDSS, a near-IR survey three magnitudes deeper than 2MASS and whose first Data Release took place in July 2006. With UKIDSS it will be possible to discover faint T-type dwarfs beyond the 2MASS limiting magnitude and to proof the existence of the theoretically predicted Y-type brown dwarfs, a class of objects cooler than the T-type dwarfs and whose discovery constitutes one of the key science drivers for the project.

• *Characterization of star forming regions*

Taking advantage of VO tools we have built the mass function of the σ Orionis cluster from 20 to 1 $M_\odot$. Membership criteria based on proper motions and colour-magnitude diagrams have permitted the identification of four new cluster members. A study of the 1 – 0.05 $M_\odot$ regime, down to the brown dwarf domain, by cross-matching 2MASS and DENIS is presently ongoing. In the short-term our aim is to repeat the same type of analysis for all the clusters of the Orion OB1b association.

• *Formation of brown dwarfs*

The way how brown dwarfs are formed is still a matter of debate. Among the competing theories, Reipurth & Clarke (2001, AJ, 122, 432) proposed that brown dwarfs are stellar embryos ejected from the formation region due to gravitational interactions before their hydrostatic cores could build up enough mass to eventually start hydrogen burning. Our goal is to check this model by cross-correlating IPHAS (Drew et al. 2005, MNRAS, 362, 753) and 2MASS to search, for the first time, young brown dwarfs over large areas of the sky on the basis of their Hα emission and IR colors. Preliminary analysis of the follow-up observations of some of the candidates seems to confirm the existence of, at least, one young brown dwarf among them.

Highlights of Astronomy, Volume 14
IAU XXVI General Assembly, 14-25 August 2006
Karel A. van der Hucht, ed.

doi:10.1017/S1743921307011969

Criteria for spectral classification of cool stars using high-resolution spectra

David Montes[1], Raquel M. Martínez-Arnáiz[1], Jesus Maldonado[1,2], Juan Roa-Llamazares[1], Javier López-Santiago[1,3], Inés Crespo-Chacón[1] and Enrique Solano[4]

[1]Departamento de Astrofísica, Facultad de Ciencias Físicas, Universidad Complutense de Madrid, E-28040 Madrid, Spain
email: dmg@astrax.fis.ucm.es

[2]Departamento de Física Teórica, C-XI, Facultad de Ciencias, Universidad Autónoma de Madrid, Cantoblanco, Spain

[3]INAF – Osservatorio Astronomico di Palermo Giuseppe S. Vaiana, Piazza Parlamento 1, I-90134 , Palermo, Italy
email: jlopez@astropa.unipa.it

[4]INTA – Laboratorio de Astrofísica Espacial y Física Fundamental, P.O. Box 50727, E-28080 Madrid, Spain
email: esm@laeff.inta.es

We have compiled a large number of optical spectra of cool stars taken with different high-resolution echelle spectrographs ($R \simeq 40\,000$). Many of those are available as spectral libraries (Montes *et al.* 1997, 1998, 1999, `<http://www.ucm.es/info/Astrof/invest/actividad/spectra.html>`.

We intend to include all these spectra in the Virtual Observatory (VO) following the standards of the International Virtual Observatory Alliance (IVOA).

The many VO tools that are or will be ready for the astrophysical community will make easier the use of these spectra in many areas, such as the study of chromospheric activity, spectral classification, determination of atmospheric parameters (T_{eff}, $\log g$, [Fe/H]), modeling stellar atmospheres, spectral synthesis applied to composite systems, and spectral synthesis of stellar population of galaxies.

In this contribution, as an example of the potential use of these spectra, we describe different spectral classification criteria for the cool stars (F5 to M5) based on equivalent width and equivalent width ratios of several photospheric lines, which are sensitive to effective temperature and luminosity class. To calibrate these relationships we have used a large number of optical spectra of spectral type standard stars (Morgan and Keenan (MK) standards) taken with different echelle spectrographs of similar spectral resolution. In addition, we have tested the behaviour of the photospheric lines with temperature by using synthetic spectra of main sequence stars ($T_{eff} = 6500 - 3500$ K) with solar abundance computed using the ATLAS9 code by Kurucz (Kurucz, 1993), adapted to work under linux platform by Sbordone *et al.* (2004) and Sbordone (2005).

We describe in detail the behaviour of the equivalent width of photospheric lines like Fe I λ6430, Ca I λ6439, and Ti I λ5866. Some equivalent width ratios as Fe I λ6430/Fe II λ6432, and Fe I λ4071/Sr II λ4077 and broad indices like MI6 (λ5125–5245)/(λ5245–5290) are specially useful to determine effective temperature and discriminate between dwarf and giant stars.

The use of additional high-resolution spectra of MK standard stars throughout the VO will be very useful to improve these calibrations and obtain good criteria for spectral classification.

Highlights of Astronomy, Volume 14
IAU XXVI General Assembly, 14-25 August 2006
Karel A. van der Hucht, ed.

doi:10.1017/S1743921307011970

Extracting parameters for stellar populations of SDSS galaxy spectra using evolution strategies

Juan C. Gomez and Olac Fuentes

Computer Science Department, Instituto Nacional de Astrofísica, Óptica y Electrónica, Luis Enrique Erro 1, Tonantzintla, Puebla 72840, México
email: jcgc,fuentes@inaoep.mx

Current surveys from modern observatories contain a huge amount of information; in particular, the Sloan Digital Sky Survey (SDSS) has reached the order of terabytes of data in images and spectra. Such amount of information needs to be exploited by sophisticated algorithms that automatically analyze the data in order to extract useful knowledge from the mega databases.

In this work we employ Evolution Strategies (ES), a stochastic optimization method used in other astronomy areas with good results, to automatically extract a set of physical parameters for stellar populations studies (ages, metallicities and reddening) and their relative contributions from a sample of galaxy spectra taken from SDSS. Such parameters are useful in cosmological studies and for understanding galaxy formation and evolution.

Extraction of parameters is treated as an optimization problem and then it is solved by using ES. The idea is to reconstruct each galaxy spectrum from the sample by means of a linear combination of three simple theoretical models of stellar population synthesis (young, intermediate and old). The goal of the fitting task with ES is to find, efficiently and automatically, the best combination for the redenning (r_1, r_2, r_3), metallicities (m_1, m_2, m_3) ages (a_1, a_2, a_3) and relative contributions (c_1, c_2, c_3) to produce a model that matches the original spectrum. Model is contructed using the equation: $\mathbf{g}_\lambda = \sum_{i=1}^{3} c_i s_i(a_i, m_i, \lambda)(10^{-4r_i k})$, where $s_i(a_i, m_i, \lambda)$ represents a model for specific age and specific metallicity and $10^{-4r_i k}$ is the redenning term.

The process starts with an observed spectrum $\mathbf{o}_\lambda$ that is passed to a preprocessing module to be adjust by rest frame, redenning correction, cut, re-bin and normalization; afterwards we begin an iterative process, which starts with a set of randomly created population synthesis parameters that is passed to a model creator program. This module produce a modeled spectrum $(\mathbf{g}_\lambda)$ that is compared with the original spectrum. If we obtain a good fit we stop the process otherwise we modify the parameters using ES operators (mutation an cross-over) and we return to create another model.

Our implementation uses a set of 50 models taken from Bressan *et al.* (1994) while all the spectra used were cut in order to cover the 3800 Å - 8000 Å range. Bressan's models cover the following ages: young=$[10^6, 10^{6.3}, 10^{6.6}, 10^7, 10^{7.3}]$ yr, intermediate=$[10^{7.6}, 10^8, 10^{8.3}]$ yr, old=$[10^9, 10^{10.2}]$ yr and the following metallicities: [0.0004, 0.004, 0.008, 0.02, 0.05] $Z_\odot$.

Until now we have obtained results for a set of 50 spectra from SDSS Data Release 2 as a first approach to the use of this method, proving ES are very well suited to extract stellar population parameters from galaxy spectra.

Highlights of Astronomy, Volume 14
IAU XXVI General Assembly, 14-25 August 2006
Karel A. van der Hucht, ed.

doi:10.1017/S1743921307011982

Virtual Observatory as a tool for stellar spectroscopy

Petr Škoda

Astronomical Institute of the Academy of Sciences of the Czech Republic,
Fričova 298, 251 65 Ondřejov
email: skoda@sunstel.asu.cas.cz

The Virtual Observatory (VO) is often presented as an all-purpose tool for the astronomer to do all his current work in a single web browser, from which he can control all data searches, reductions and mainly the full data analysis. This idea seems to be already fulfilling in certain fields of astronomy like is the search of unknown correlations between various parameters in large scale surveys, galaxy statistics or cross-identification of objects. The current VO tools are, however, lacking support for important capabilities needed in stellar spectroscopy. Especially the high-resolution spectroscopy of stars with rapid line profile variations or of those with complicated emission profiles (like Be or symbiotic stars) benefits from a number of specific methods implemented today only in stand-alone legacy packages working with local data.

We concentrate on such methods and give recommendations for their implementation as a VO services using both remote and local spectra in an unified manner as well as the distributed computing power or nice graphics available in current VO infrastructure.

Among the common tasks required for analysis of stellar spectra belong mainly: overplotting many spectra of the same object in time or folded by phase, measurement of radial velocities or red-shift (using fitting of profiles by various functions, profile mirroring around given laboratory wavelength or cross-correlation with reasonable templates), bisector and moment analysis of line profiles, complex period analysis (mostly using PDM or CLEAN methods), dynamic spectra (quotient or differential, showing line profile changes in time or in dependence on orbital phase) and for multiple stars there is a very powerful method of spectra disentangling (in wavelength domain or in Fourier space).

Implementing these tasks as VO services would benefit from the already existing VO infrastructure like is unified data format (VO-Table, UCD semantics of variables), transparent data conversion (including the physical unit transformation already available in SSAP), on-the-fly rescaling as well as powerful display and presentation tools capable of seamless work with remote data.

The precise analysis of optical stellar spectra requires in addition the introduction of certain capabilities into the current VO-enabled spectral tools. For example the flexible curve fitting using more complicated functions than simple polynomials or splines (e.g., INTEP or AKIMA methods) is necessary for good normalisation of the continuum or the measurement of radial velocities sometimes needs different fits for the core and line wings.

Therefore, we also give a review of several VO-compatible spectral tools (SpecView, SPLAT and VOSpec) emphasising their deficiencies and advantages for these tasks.

We hope our review will help the VO developers to identify the most important features to concentrate on in further improvements of VO-enabled spectral tools to allow the full-fledged analysis of stellar spectra to be done in VO domain.

Highlights of Astronomy, Volume 14
IAU XXVI General Assembly, 14-25 August 2006
Karel A. van der Hucht, ed.

doi:10.1017/S1743921307011994

Interconnecting the Virtual Observatory with computational grid infrastructures

Fabio Pasian, Giuliano Taffoni and Claudio Vuerli

INAF – Osservatorio Astronomico di Trieste, Via G.B. Tiepolo 11, I-34143 Trieste, Italy
email: pasian@oats.inaf.it

The term 'grid', in the Virtual Observatory (VO) context, has mainly been used to indicate a set of interoperable services, allowing transparent access to a set of geographically distributed and heterogeneous archives and catalogues, data exchange and analysis, etc. The design of the VO has been however mainly geared at allowing users to access *registered services.*

This is rather different from the approach other scientific communities are taking, mainly based on using the grid for computational tasks (e.g., EGEE). The Grid concept is to have a highly controlled 'coordinated resource sharing': individuals and/or institutions defined by the sharing rules form a 'virtual organization'. If the user belongs to one, he/she can run his/her *own applications.*

Within this framework, it appears as extremely important to be able to interconnect the VO and the computational grid infrastructures. This is particularly relevant in the case of massive computational problems connected to the production of theoretical and simulated data. In some countries, there is a well-established and growing community of 'Grid Astronomers'.

To implement this interconnection, on one side, mechanisms are being designed and implemented aimed at allowing VO users to exploit (through registered applications available at the VO data centres) the processing capabilities offered by the computational Grid. This fulfills a 'data centre' type of scenario: a VO user connects to a theory database, asking for a model by specifying a set of parameters: the model is provided if available, a local computation allows to interpolate between available models, while Grid computing is used to build a full-fledged new model.

In another 'user-centered' type of scenario, a user extracts information from the VO, performs heavy batch computations with own code on the Grid, then re-connects to the VO to compare results. If such a scenario is to be supported, high priority needs to be given to the access of VO-compliant archives and databases with the computational Grid using the proper interfaces and standards.

A two-way infrastructural approach is therefore needed for linking Grid-enabled computations with data, and viceversa, within the VO. This approach will allow users not only to run registered VO applications on computational Grids but, if they belong to a Grid Virtual Organization, also their own. Technically, on one side the Common Execution Architecture (CEA) is expected to be expanded to allow access to distributed computing architectures. On the other side, prototype systems are being built, which make use of all relevant VO standards and EGEE Grid middleware, with some extensions to allow VO-compliant access to data. As an example, a Grid middleware component (QE) has been developed by INAF with EGEE/LCG staff at INFN/CNAF, working under Globus Toolkit 2 and interoperable with OGSA-DAI.

The analysis of VO-Grid interoperability will be carried out as a work package of the VO-DCA project, funded by the EU Sixth Framework Programme.

Highlights of Astronomy, Volume 14
IAU XXVI General Assembly, 14-25 August 2006
Karel A. van der Hucht, ed.

doi:10.1017/S1743921307012008

Subject mediation approach for scientific problem solving in Virtual Observatories

Leonid Kalinichenko

Institute of Informatics Problems, Russian Academy of Sciences,
Vavilov street 44, 2, Moscow, 119333, Russian Federation
email: leonidk@synth.ipi.ac.ru

There exist two principally different approaches to the organization of problem solving in VO: (*i*) information resources driven approach (choice and integrated definition of resources are made independently of the problem specification); and (*ii*) scientific problem driven approach (a specification of a problem domain is created, the relevant to the problem resources are identified and semantically mapped into the domain). Intrinsic difficulties of the first approach: semantic gap between resources and the problem, instability of global schema w.r.t. a set of resources, inability of automatic identification of resources for the problem. To implement the second approach a mediation technology is required. On the consolidation phase of the mediator the efforts of the scientific community are focused on the problem definition by specifying the mediator. During the operational phase relevant information resources are identified and expressed in terms of the mediator. Advantages of the mediator approach include truly semantic integration of heterogeneous resources due to their semantic mapping into the mediator; multiple subjects can be semantically integrated applying recursive structure of the mediators.

Basic methods supporting the mediation technology include: (*i*) identification of relevant resources by metadata, ontologies, information structure, behaviour while registering in mediator resource classes in terms of the mediator classes; (*ii*) rewriting queries (rules) expressed in terms of a mediator into rules over relevant data and services; and (*iii*) running workflows with rules as tasks over distributed heterogeneous resources.

In May 2005 the RVO information infrastructure (RVOII) project report has been published (joint effort of SAO RAS, INASAN, IPI RAS). Basic principles of RVOII are defined as: representing infrastructure as a network of interoperating web (Grid) services, moving processing to data, encouraging code reuse and composition in SOA, emphasizing subject mediators to support various subject domains in astronomy. A Community centre in Moscow for support of scientific astronomical problem solving over distributed repositories of astronomical information has been created. Currently the Centre includes two AstroGrid installations, one of them at the Joint Supercomputer Center of RAS to make possible call supercomputer tasks from workflows. First experience shows that usage of AstroGrid as the RVOII core is suitable according to the RVOII principles.

The mediator approach shows a way for considering IVOA artifacts as well conceptualized and consolidated facilities (not as separate entities – data models, DAL protocols, database schemas, query languages, problem formulation models, applying different modeling facilities). Thus, the early IVOA document "A unified domain model for astronomy" has been attempted actually as a mediator definition for the whole astronomy. In IVOA such sort of models is not supported by suitable methodological idea. IVOA specific data types and classes of problems (e.g., Spectra, Quantity, Passband, Simulation) look conceptually better as specifications of parts of respective mediators. IVOA standards for uniform access to heterogeneous VO data (e.g., image, table, spectrum, etc.) look more consistent as the mediator layer.

Highlights of Astronomy, Volume 14
IAU XXVI General Assembly, 14-25 August 2006
Karel A. van der Hucht, ed.

doi:10.1017/S174392130701201X

An ontology of sstronomical object types for the Virtual Observatory

Sebastian Derriere[1], André Richard[1,2] and Andrea Preite-Martinez[1,2]

[1]CDS, Observatoire astronomique de Strasbourg, UMR 7550,
11 rue de l'Université, F-67000 Strasbourg, France
email: derriere@astro.u-strasbg.fr

[2]INAF – Istituto di Astrofisica Spaziale e Fisica Cosmica,
Via del Fosso del Cavaliere 100, I-00133 Roma, Italy

The Semantic Web and ontologies are emerging technologies that enable advanced knowledge management and sharing. Their application to Astronomy can offer new ways of sharing information between astronomers, but also between machines or software components and allow inference engines to perform reasoning on an astronomical knowledge base. The first examples of astronomy-related ontologies are being developed in the european VOTech project.

We present the current status of an ontology describing knowledge about astronomical object types, originally based on the standardization of object types used in the SIMBAD database. We discuss the strategies that have been adopted for structuring this ontology, the problems that have been solved during the construction, and the links with current works in the Semantics working group of the IVOA.

We present applications of this ontology on several use cases:

• Support advanced resource queries in the VO Registries, by using datasets correspondence with ontology concepts. Figure 1 presents snapshots of this use case: the astronomer selects a starting concept from a scroll-list; the concept hierarchy is then recursively explored to find all more specific concepts having an associated registry 'Subject' keyword; the registry can then be queried based on these keywords.

• Validate multiple object classification in SIMBAD by checking the consistency of the types associated to the object's identifiers.

• Refine object classification in SIMBAD when adding new measurements or identifiers.

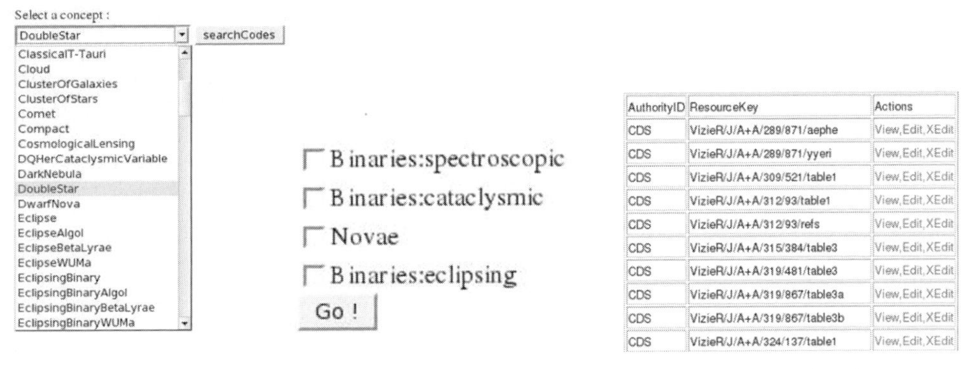

Figure 1. List of concepts related to astronomical objects in the ontology (left). Keywords corresponding to the selected concepts, and all more specific concepts (middle). List of relevant resources in the registry (right).

Highlights of Astronomy, Volume 14
IAU XXVI General Assembly, 14-25 August 2006
Karel A. van der Hucht, ed.

doi:10.1017/S1743921307012021

The Virtual Observatory paradigm and national projects

Ajit K. Kembhavi

Inter-University Centre for Astronomy and Astrophysics (IUCAA),
Post Bag 4, Ganeshkhind, Pune University Campus, Pune 411 007, India
email: akk@iucaa.ernet.in

A Virtual Observatory (VO) seeks to provide access to very large volumes of astronomical data over the internet to every interested user, along with software for the data anlysis, and tools for data visualisation, statistics and any other applications which are necessary for mining science from the data. Handling large volumes of data requires extensive computer resources, which may not be available to the end user; transferring data from the data centre to the user needs high bandwidth, which also may not be available. A VO therefore seeks to provide computing resources as well, which could be spread over a grid. The user can locate data of interest through registries, then access the data and analyse it using the computing resources, all through simple user interfaces.

The Virtual Observatory programme, which is now several years old, is being implemented through various national virtual observatory projects, which are federated under the International Virtual Observatory Alliance (IVOA). The IVOA provides a platform for international coordination and collaboration which are necessary to develop standards and tools which can work in a highly interoperable manner. Some of the national projects are large, with a significant number of people working on the development of standards, formats, registries, data models, query languages etc, and also manage immense data bases. Other national projects involve a smaller number of people, who work on the the development of specialised or generic tools compatible with VO standards, which can be accessed either from a specific website, or through other data services.

Some examples of the smaller VO projects, on can mention: (*i*) The Hungarian VO, which has successfully developed VO spectrum and filter services, created the first dynamical synthetic spectrum service, and has released photometric redshifts for more than 100 million objects. (*ii*) The Russian VO provides Russian astronomers effective access to international resources and conversely integrates Russian resources into the international VO structure. (*iii*) The Chinese VO focuses on the development of applications like VOFilter and SkyMouse, and infrastructure deployment through a VO Data Access Service, which will provide VO compliant uniform access interfaces for various astronomical resources. (*iv*) The German Virtual Observatory which serves *ROSAT* data, and has provided cone searches and other tools for the *ROSAT* All Sky Survey. This provides large databases developed through cosmological and hydrodynamical simulations.

Virtual Observatory India (VOI) is the result of a collaboration between the Inter-University centre for Astronomy and Astrophysics in Pune, and Persistence Systems Pvt. Ltd., based in the same city. The VOI is partially supported by a grant from the Ministry of Communications and Information Technology of the Government of India. This unique collaboration between the academia and industry has produced many useful tools for data visualisation, data management, statistical analysis and so forth, often in collaboration with other VO projects in the world.

Highlights of Astronomy, Volume 14
IAU XXVI General Assembly, 14-25 August 2006
Karel A. van der Hucht, ed.

doi:10.1017/S1743921307012033

Connecting the literature with on-line data

Guenther Eichhorn, Alberto Accomazzi, Carolyn S. Grant, Edwin A. Henneken, Michael J. Kurtz, Donna M. Thompson and Stephen S. Murray

Harvard Smithsonian Center for Astrophysics,
60 Garden Street, MS-67, Cambridge, MA 02138, USA
email: gei@cfa.harvard.edu

Over the last few years there has been considerable progress in linking the published scholarly literature with on-line data. This will greatly help with data discovery and aid the efforts to the VO.

In an initial effort to provide the means for such linking, the Astrophysics Data Center Executive Committee (ADEC), a collaboration of NASA data centers in the USA, has worked with the American Astronomical Society (AAS) and the University of Chicago Press, the publisher of the journals of the AAS, to establish a system that allows authors to specify data sets that they used in an article. This information is then used by the journal publisher to link to the on-line data. It is also forwarded to the Astrophysics Data System (ADS) and to the data centers to provide similar links between the literature and the data in these other systems.

We have developed the software infrastructure to handle the all aspects of this system from registering data center's data holdings, through automatic verification of the data set identifiers, to persistent linking from the journal articles to the on-line data.

The ADEC agreed on a format for the data set identifiers that is compatible with current VO identifier structures. Data set identifiers have the form:
`ADS/FacilityId#PrivateId` .

The AuthorityId string 'ADS' has been specified. This simply recognizes the current role of ADS in managing the namespace used for these identifiers, in the absence of a community-wide namespace granting authority. It does not suggest nor imply that ADS controls or manages the dataset itself.

The ResourceId token will be interpreted as a Facility. An ever-growing list of facilities is maintained by ADS. Data centers should contact ADS should they need to register new entries.

The PrivateId string can be anything that the data center desires, with the provision that the identifiers string as a whole should abide by the general syntax of a URI, as required by the IVOA identifiers specification.

Data centers who wish to participate in this effort, should register with the ADS. While it is expected that the appropriate metadata will one day be made available by a public VO registry, its format and access methods are at this time not available. As an intermediate solution to the problem, we require that the data centers maintain a simple profile which will provide ADS with the necessary metadata.

The data center profile is simple XML document that lists the data center name and description, the name and email address of the person responsible for the maintenance of the profile, the URL of the web service to be used for dataset verification, and the list of facilities that the datacenter has data for.

Highlights of Astronomy, Volume 14
IAU XXVI General Assembly, 14-25 August 2006
Karel A. van der Hucht, ed.

doi:10.1017/S1743921307012045

Astrophysical integrated research environment

Jianfeng Zhou[1] and Yang Yang[2]

[1]Department of Engineering Physics, Center for Astrophysics, Tsinghua University, Beijing, 100084, China
email: zhoujf@tsinghua.edu.cn

[2]Department of Physics, Center for Astrophysics, Tsinghua University, Beijing, 100084, China
email: yangx@tsinghua.edu.cn

Abstract. Astrophysical Integrated Research Environment (AIRE), aims to integrate astrophysical data, analysis software and astrophysical knowledge into an easy-to-use Internet based environment. Therefore, astrophysicists from different institutes can constitute virtual research groups which are favorable to study some complex multi-band astrophysical phenomena. The AIRE was put into use in Center for Astrophysics, Tsinghua university in 2003. Up to now, there are 219 advanced users in this environment. Several astrophysical researches base on AIRE have generated some important published results.

Keywords. astronomical data bases: miscellaneous

1. Basic components

The objective of AIRE is to provide an Internet based collaborative research environment for astrophysicists. AIRE has two important components to fulfill the above requirement. One is the Data Processing Center (DPC), Another is the Collaborative Astrophysical Research Project System (CARPS).

Data Processing Center a Linux based working environment which can be accessed by any java supported web browser . DPC has two notable advantages : Firstly, from traditional point of view, it is a Linux system. Almost all astrophysicists are familiar with Linux, and can do anything about astrophysical research in it. Secondly, from VO point of view, it is web-based, open and free. A user can not only obtain hard resources, like CPU time, memory and hard disk area etc, but also get the support from other users.

The CARPS provides a simple but effective way to do collaborative research. It has a HTML interface and a DPC interface. In HTML interface, a user can update part or whole documentation tree to get the latest progress of the project, or commit his/her newest work to other members of the project. All ASCII files of the project can be edited in HTML interface. The DPC interface provides more powerful functions for collaborative research, such as compile and run programs, read references, process data, model fitting, write journal papers etc..

2. Current status and future plans

The AIRE has 219 advanced users who can use both DPC and CARPS since 2003 when it was put into use. Many individual and collaborative researches were carrying out in this environment, and several of them have generated published results.

In the next step, we are going to develop Astrophysical Research Protocol (ARP) and relevant tools which is the third important component for AIRE.

Highlights of Astronomy, Volume 14
IAU XXVI General Assembly, 14-25 August 2006
Karel A. van der Hucht, ed.

doi:10.1017/S1743921307012057

The Astro-Wise system: a federated information accumulator for astronomy

Edwin A. Valentijn and Gijs Verdoes Keijn

Kapteyn Astronomical Institute, University of Groningen,
P.O. Box 800, NL-9700AV Groningen, the Netherlands
email: verdoes@astro.rug.nl

The Astro-Wise consortium has designed a new paradigm, 'Awe', and implemented a fully scalable and distributed information system to overcome the huge information avalanche in wide-field astronomical imaging Surveys.

In contrast to a TIER node setting, Awe allows the end-user to trace the data product, following all its dependencies up to the raw observational data and, if necessary, to re-derive the result with better calibration data and/or improved methods:'Awe target processing'. This is achieved by:
(*i*) emphasis on project management and documentation;
(*ii*) translating the data model to an object model stored in a database;
(*iii*) having all I/O reside in a distributed database, containing all metadata of bulk data, parameter values and catalogues;
(*iv*) connecting to the database a federated file server that stores 100s hundreds of Terabytes of bulk data;
(*v*) an own Awe compute-GRID which sends jobs to parallel clusters, which then request data from the database.

The database architecture allows tools for rapid trend analysis, complex queries and fast hunting in Terabyte-sized catalogues. Thus, the system provides the user with transparent access to all stages of the data processing and thereby allows the data to be re-processed and add knowledge to the system, the Awe paradigm.

All system components are distributed over Europe, enabling research groups to collaborate on shared projects. The web portal includes data viewing, quality labelling and compute-services. The Astro-Wise system will be connected to the Euro-VO, and publishing is effectively done by raising a flag by the Awe administrator. Currently, researchers use Awe with 10's Tbyte of image data.

Hundreds of Tbytes of data will start entering the system when the OmegaCAM camera starts operations in Chile. Several large surveys plan to use Awe: e.g., the 1500 Square Degree KIDS Survey (Kuijken *et al.*), the Vesuvio Survey of nearby superclusters, the OmegaWhite white dwarf binary survey, and OmegaTrans search for variables.

Highlights of Astronomy, Volume 14
IAU XXVI General Assembly, 14-25 August 2006
Karel A. van der Hucht, ed.

doi:10.1017/S1743921307012069

Wide field optical surveys

Naoki Yasuda
National Astronomical Observatory, Tokyo, Japan
email: naoki.yasuda@nao.ac.jp

The products of optical wide-field survey are very valuable for their own purposes like the studies of large scale structure of galaxies, evolution of galaxies, Galactic structure and so on. At the same time optical view of sky will provide basic reference for the observation at other wavelength ranges. For this reason Palomar all sky survey has been used for various astronomical studies over 50 years. Now in the era of electronic devices, digital archives, and powerful computer systems, modern observation will replace the Palomar all sky survey.

One of the pioneering work in this field is the Sloan Digital Sky Survey (SDSS) which has been carried out during 2000–2005. SDSS gives us the five band images over 8,000 square degrees of the sky, nearly 200 milliion celestial objects detected, and spectra of more than 675 000 galaxies, 90 000 quasars, and 185 000 stars. These huge and well-calibrated datasets of SDSS has been a good material for the development of Virtual Observatory (VO) and the demonstration of the effectiveness of VO. We will introduce the scientific results which were obtained by the large SDSS database. From these scientific studies we can learn about how VOs should be and what functionality VOs need.

We also discuss about the different types of optical surveys, i.e., deep surveys and time transient surveys. In order to reveal the structure and evolution of galaxies at the early Universe, a number of deep surveys have been carried out. As an example, we will discuss the results of deep surveys done by Subaru telescope.

On the other hand, time transient surveys give us a different aspect of astronomy. Many distant supernovae are discovered recently from time transient surveys. These supernovae are used for the investigation of the nature of dark energy. These different kinds of surveys also give us some insight for VOs.

Highlights of Astronomy, Volume 14
IAU XXVI General Assembly, 14-25 August 2006
Karel A. van der Hucht, ed.

doi:10.1017/S1743921307012070

Photometric identification of quasars from the Sloan Survey

Rameshwar P. Sinha[1], Ninan S. Philip[2], Ajit K. Kembahvi[1] and Ashish A. Mahabal[3]

[1]Inter-University Centre for Astronomy and Astrophysics (IUCAA),
Post Bag 4, Ganeshkhind, Pune University Campus, Pune 411 007, India
email: rsinha,akk@iucaa.ernet.in

[2]St. Thomas College, Kozhencheri, Kerala, India

[3]Astronomy Department, California Institute of Technology, Pasadena, CA 91125, USA
email: milan@astro.caltech.edu

We have developed a neural network based technique for identifying quasars from the photometric database of the Sloan Digital Sky Survey (SDSS). We have queried the SDSS data release 5 (DR5) to produce a dataset of spectroscopically identified samples of unresolved objects consisting of quasars and stars which forms the training set.

Using a representative subsample of this dataset we train a Difference Boosting Neural Network (DBNN) that looks for differences in the colours of the objects to classify them. The four colours, namely $u-g$, $g-r$, $r-i$, $i-z$ and the spectral classes, i.e., star and quasar are used as the inputs for the network which returns a predicted spectral class of each object, assigning a confidence level to the prediction.

We use a few subsamples of 10 000 randomly selected objects from the full dataset and train the network. Subsequently, we divide the whole colour space into different sections each having a different density of quasars and again train the DBNN. After a few rounds of training with each subsample set we get the best trained neural net based on adaptive data selection.

The quality of training data is adjudged only on the basis of the accuracy achieved when the trained data is tested on unseen data. We use our best trained network to find quasars from the test dataset which consists of spectroscopically confirmed quasars and stars. From the results we estimate the level of completeness and the degree of contamination to the quasars identified using the DBNN. We report our best trained network which has a completeness of 98 % and a contamination of 2 % from stars. The trained network can be used to identify quasars from samples of unresolved objects selected from DR5.

We have made use of VO tools such as VOPlot and VOPlot-3D. In particular, VOPlot-3D has helped in visually demonstrating the separation between stars and quasars in a multi-dimensional colour space as well as highlighting the hypersurface created by the DBNN for each of the two classes. We also make use of few statistical tools such as box-plot, histogram and k-mean clustering from VOStat to substantiate our results further and facilitate our investigation. In the near future, we plan to make a web-based tool with a user-friendly interface for making selections using DBNN. The technique can have a wide application.

Highlights of Astronomy, Volume 14
IAU XXVI General Assembly, 14-25 August 2006
Karel A. van der Hucht, ed.

doi:10.1017/S1743921307012082

A multiwavelength study of a sample of Texas Radio Survey steep spectrum sources

Ray A. Lucas[1], Neal Miller[2], Anton M. Koekemoer[1], Jeffrey Van Duyne[3] and Kenneth C. Chambers[4]

[1]Space Telescope Science Institute, 3700 San Martin Drive, Baltimore, MD 21218, USA
email: lucas,koekemoe@stsci.edu

[2]National Radio Astronomy Observatory and Johns Hopkins University Department of Physics and Astronomy, 3400 N. Charles Street, Baltimore, MD 21218, USA
email: nmiller@pha.jhu.edu

[3]Yale University, Astronomy Department, P.O. Box 208101, New Haven, CT 06520-8101, USA
email: vanduyne@astro.yale.edu

[4]Institute for Astronomy, University of Hawaii, 2680 Woodlawn Road, Honolulu, HI 96822, USA
email: chambers@ifa.hawaii.edu

VLA A-array snapshots were obtained of a complete sample of steep-spectrum radio sources from the Texas Radio Interferometer survey. Though similar in sensitivity to the FIRST Survey, our A-array snapshots have better resolution, and are complementary to FIRST. All initial A-array maps are made, and we are comparing them to FIRST.

We are using multicolour Sloan Sky Survey data which covers our survey area for optical detections and identifications, in addition to DSS images. Though likely not deep enough to put useful constraints on high-z objects in our sample, 2MASS survey images may be used for brighter infrared detections, and we will search for any other multiwavelength data readily available via VO resources. We aim to thus perform a multiwavelength characterization of these objects. The selection criteria have picked up many FR2-type sources (ones with bright twin lobes of radio emission.) This may be expected, since at higher fluxes, the radio source counts are dominated by powerful AGN such as FR2 galaxies, and the spectral index of their radio lobes is usually steep, thus they are bright at the low frequency used in the Texas survey observations.

There are also AGN unification models for FR2 galaxies and quasars which can be explored using this sample. Radio morphologies from our 71 VLA A-array snapshots: ~ 43 double-lobed, 3–4 of the 43 show distinct core; ~ 21 of 43 show diffuse emission; 9 resolved; 9 point sources; 7 apparent non-detections (now being checked versus FIRST); 4 spurious (from 1989 prepublication Texas list.) Using multiwavelength data which is available for these sources, in conjunction with our radio data, we will attempt to build a database of our sample and to better understand the nature of these objects, and how they fit into various galaxy formation and evolution scenarios, and AGN unification models, etc.

Acknowledgements

We acknowledge support from the NSF, USNVO, and NRAO Jansky Fellowships, VLA staff, and 2004 NVO Summer School colleagues David Rohde and Takayuki Tamura.

Reference

Douglas, J. N. *et al.* 1996, *AJ*, 111, 1945

Highlights of Astronomy, Volume 14
IAU XXVI General Assembly, 14-25 August 2006
Karel A. van der Hucht, ed.

doi:10.1017/S1743921307012094

Near-IR properties of *Spitzer* sources

Eduardo A. Gonzalez-Solares[1], Nicholas A. Walton[1], Anita M. S. Richards[2], Jonathan A. Tedds[3] and the AstroGrid Collaboration

[1]Institute of Astronomy, University of Cambridge, Madingley Road, Cambridge CB3 0HA, UK
email: eglez,naw@ast.cam.ac.uk

[2]Jodrell Bank Observatory, University of Manchester, Macclesfield, Cheshire SK11 9Dl, UK

[3]Department of Physics and Astronomy, University of Leicester,
University Road, Leicester LE1 7RH, UK
email: jat@star.le.ac.uk

We present an analysis of the near-IR properties of galaxies and quasars detected in the mid-IR by the Spitzer Space Telescope. Using optical photometry from the Wide Field Survey and near-IR fluxes from the First Data Release of the UKIDSS (Dye *et al.* 2006) survey we carry out a preliminary characterization of the surface density of different samples of galaxies and their clustering properties.

We also show how the near-IR photometry and morphology information allows an even better selection of AGNs than the mid-IR alone or optical alone. The combination of all the multiwavelenth photometry results in a powerful tool to perform AGN selection and study the fraction of obscured AGNs missing from the optical surveys.

We describe how Astrogrid the UK Virtual Observatory (`<http://www.astrogrid.org>`) has been used to perform the data discovery, query the relevant catalogues and perform their cross match and different sample selections. Using the different tools available we carry out the determination of photometric redshifts and clustering distribution. Individual research of outlier sources is significantly easy within the system.

Using the Data Set Access (DSA) component, AstroGrid provides access to the public *Spitzer*-SWIRE (Lonsdale *et al.* 2003) catalogues, the SDSS DR5 (Adelman-McCarthy *et al.* 2006) catalogues as well as authenticated access to the UKIDSS catalogues. All of them are queried using Astronomical Data Query Language (ADQL; based on standard SQL). Astrogrid provides tools, integrated within the system, to perform cross match of the different catalogues and produce interactive plots. Outliers are then studied using AstroScope, a powerful tool for finding images, catalogues and spectra over a region of sky.

References

Dye, S., Warren, S. J., Hambly, N. C., *et al.* 2006, *MNRAS*, 372, 1227
Lonsdale, C. J., Smith, H. E., Rowan-Robinson, M., *et al.* 2003, *PASP*, 115, 897
Adelman-McCarthy, J. K., Agüeros, M. A., Allam, S. S., *et al.* 2006, *ApJS*, 162, 38

Highlights of Astronomy, Volume 14
IAU XXVI General Assembly, 14-25 August 2006
Karel A. van der Hucht, ed.

doi:10.1017/S1743921307012100

Solar physics with the Virtual Solar Observatory

Frank Hill

National Solar Observatory, NOAO, 950 N Cherry Ave, Tucson, AZ 85719-4933, USA
email: fhill@nso.edu

The Virtual Solar Observatory (VSO) is a lightweight web service unifying twelve major solar data archives. With the VSO, users can simultaneously search for data from 50 space- and ground-based instruments covering the time period from 1915 to the present.

In addition to a web-based interface, an API allows users to directly connect other software systems, such as SolarSoft or AstroGrid, to the VSO. Results can be returned in VOTable format, allowing the use of VO technology to process VSO results.

With the availability of the VSO, solar physicists can now tackle a variety of scientific questions that were previously unfeasible. Some examples are: linking the solar wind properties to subsurface flows; detailed studies of the life cycle of active regions; development of space weather predictive techniques; the statistics of granulation over the activity cycle; and mechanisms of energy blocking by sunspots.

Highlights of Astronomy, Volume 14
IAU XXVI General Assembly, 14-25 August 2006
Karel A. van der Hucht, ed.

doi:10.1017/S1743921307012112

Mexican Virtual Solar Observatory project

Alfredo J. Santillán[1], Liliana Hernández[2], Guillermo Salas[1], Antonio Sánchez[3], Alejandro González[4] and José Franco[2]

[1]Cómputo Aplicado-DGSCA, UNAM, Mexico City, DF 04510 Mexico
email: alfredo@astroscu.unam.mx

[2]Instituto de Astronomía, UNAM, Mexico City, DF 04510 Mexico

[3]Area de Astronomía, DIFUS, Universidad de Sonora, Mexico

[4]Instituto de Ecología, UNAM, Mexico City, DF 04510 Mexico

The Virtual Solar Observatory (VSO) concept outlines a software environment for searching, obtaining and analyzing data from archives of solar data that are distributed at many different observatories around the world (Hill 2006, in this volume). The VSO, however, not only provides fast and reliable access to the existing data of Solar Active Regions, but also represents a powerful and unique tool to perform numerical simulations of the evolution and present state of solar phenomena. Two centers at UNAM, the Institute of Astronomy (IA) and the Supercomputer Center (DGSCA), along with the Sonora University, are working together to create the Mexican Virtual Solar Observatory (MVSO) that will be part of a wider national effort.

Here, we present a general description of the MVSO project, as well as the most recent advances. This project has three principal areas: Computational, Observational, and Educational. In the Computational area, we have developed a Portal that allows users to run Numerical Simulations Remotely. This Portal consist of three layers: the Graphics User Interface (GUI), the Numerical Simulation, and the Data Archives. The GUI is programmed in *ASP* and *.NET*, and we are using *MONO* & *XSP* so that it runs in Linux Systems. For the Numerical Simulations, we are using the MHD ZEUS-3D code, though this is not the only option, and a similar scheme be applied to use of any other numerical code.

Finally, the results produced by the numerical simulations are stored in a set of Data Archives, that are written in HDF format. They contain the following physical variables: density, internal energy, velocity, and magnetic fields. The three layers are related, which allows the user to make numerical simulations or searches within the data base of the MVSO, all through the GUI. We have applied this tool to the *Evolution of Coronal Mass Ejections in the Interplanetary Medium*. Regarding the observational data, the University of Sonora has made Solar Observations during the last decade with a set of filters (Ca II and Hα) and have made a Catalog of Solar Active Regions. In order search the Catalog, we have developed a GUI that allows the user to access the information via *date*, *filters* or *active region*.

Regarding the observational data in radio frequencies, the Geophysical Institute of UNAM is constructing the Mexican Array Radiotelescope (MEXART) that uses the technique of Interplanetary Scintillation (IPS). With MEXART, it will be possible to obtain information on the large-scale shape and velocity of solar wind disturbances within a range in the interplanetary medium for which no other technique exists. They will contribute to better understanding solar storm propagation and to space weather predictions. All observational data produced by this radiotelescope will be available for the astronomical community via MVSO.

Highlights of Astronomy, Volume 14
IAU XXVI General Assembly, 14-25 August 2006
Karel A. van der Hucht, ed.

doi:10.1017/S1743921307012124

Solar active region emergence and flare productivity

Silvia C. Dalla[1], Lyndsay Fletcher[2] and Nicholas A. Walton[3]

[1]School of Physics and Astronomy, University of Manchester,
P.O. Box 88, Manchester M60 1QD, UK
email: s.dalla@manchester.ac.uk

[2]Department of Physics and Astronomy, University of Glasgow,
Kelvin Building, Glasgow G12 8QQ, UK
email: lyndsay@astro.gla.ac.uk

[3]Institute of Astronomy, University of Cambridge, Madingley Road, Cambridge CB3 0HA, UK
email: naw@ast.cam.ac.uk

Abstract. We use the workflow capabilities of the AstroGrid Virtual Observatory system (<http://www.astrogrid.org>) to analyse the relation between flare productivity and location of Active Region (AR) emergence on the Sun. Specifically, we investigate whether emergence of a new region near existing ones results in increased productivity of the new and/or pre-existing AR. To address this question, we build a series of workflows that perform queries to catalogues of regions and flares, and operations on the results of the queries. There is a strong East-West asymmetry in the location of emergence of new regions. We do not find a significant difference between the flaring rate of paired and isolated regions, when we choose a value of 12° as the cutoff between the two populations.

Keywords. Sunspots, Sun: magnetic fields, Sun: flares

Emergence of magnetic flux through the solar photosphere creates new Active Regions, identified from sunspot, magnetogram and Hα observations. New flux emergence has been recognised as an important trigger of solar activity, including solar flares.

We analyse a large sample of regions, by means of AstroGrid workflows that retrieve information from separate catalogues of regions and flares, identify newly emerged regions, establish whether they emerged paired or isolated, and evaluate flare productivity. We use the USAF/Mt Wilson catalogue of sunspot regions (24 years of data, starting in 1981) and the NOAA Solar Region Summary (from 1986 onwards, available via the EGSO Solar Event Catalogue). Results of workflows are visualised with Topcat.

Using the Mt Wilson catalogue, we identify 3212 new regions that emerged in view from Earth, with strong asymmetry in their location of emergence: 825 regions emerged in the longitude bin E60–E40, and 177 in the bin W40–W60. This finding may be related to previously reported asymmetries in the total number of sunspots (Maunder 1907).

We divide our sample of newly emerged regions into paired and isolated, by calculating the distances of the new region to all pre-existing ones: we call a region paired if it emerged within 12° of another one. We evaluate the flare productivity of each new region using the GOES Soft-X-Ray flare catalogue (via EGSO/SEC). Here flare productivity is defined as the number of medium to large flares (class $>$ C1.0) produced in the first 4 days since emergence. We find no clear indication that being 'paired' makes a region or its companion more flare productive.

Reference

Maunder, A. S. D. 1907, *MNRAS*, 67, 451

Highlights of Astronomy, Volume 14
IAU XXVI General Assembly, 14-25 August 2006
Karel A. van der Hucht, ed.

doi:10.1017/S1743921307012136

Solar system bodies 'Observations in the Past' with the plate archive of the Main Astronomical Observatory of the Ukrainian National Academy of Sciences

Tetyana P. Sergeeva[1] **and Aleksandre V. Sergeev**[2]

[1]Department of Astrometry, Main Astronomical Observatory,
Zabolotnogo str., 27, Kiev, 03680 Ukraine

[2]Centre of Astronomical and Medico-Ecological Researches, Kiev, Ukraine
email: sergeev@mao.kiev.ua

The improvement of the dynamical models of solar system bodies' motions will be very useful for the future space astrometry mission *Gaia* for a fast identification of objects, to discriminate between the well-known objects and the new ones. 'Observations in the Past' with plate archives allow realising it.

The plate archive of the Main Astronomical Observatory of the National Academy of Sciences of Ukraine contains more then hundred thousand images of minor planets with magnitude up to 16.7 mag. About 10 % of minor planets, which may be found on our archive plates, were firstly discovered after time when plates have been taken.

So we can re-discovery them by so called 'observation in the past' and obtain their positions. Now we intend to find, measure and define of orbital elements for: 'rediscover' asteroids, asteroids with out-of-truth orbits and Near Earth Objects.

Other solar system bodies for which we try to get those 'observation in the past' are external planets' satellites. The objects choose criteria, methods of it search, identification and determination of position are discussed. The results of asteroids and external planets' satellites search in MAO plate archive will be presented.

Highlights of Astronomy, Volume 14
IAU XXVI General Assembly, 14-25 August 2006
Karel A. van der Hucht, ed.

doi:10.1017/S1743921307012148

Massive physical and dynamical characterization of asteroids

William Thuillot[1], Jerôme Berthier[1], Armand Sarkissian[2], Areg M. Mickaelian[3], Lena A. Sargsyan[3], Jesus Iglesias[1], Valery Lainey[1], Mirel Birlan[1] and Guy Simon[4]

[1]IMCCE, Paris Observatory, 77 avenue Denfert Rochereau, Paris, France
email: thuillot,berthier@imcce.fr

[2]Service d'aéronomie, Institut Pierre Simon Laplace, France

[3]Byurakan Astrophysical Observatory, Armenia

[4]GEPI, Observatoire Paris-Site de Meudon, 5 Pl J. Janssen, F-92195 Meudon Cedex, France
email: guy.simon@obspm.fr

Numerous new Solar System objects, mainly asteroids, are still detected nowadays but their physical and dynamical characteristics remain not accurately determined, until dedicated observations and analysis are made. However, the Virtual Observatory is a perfect framework to search for this characterization by data mining. We are developing two studies for this goal by exploring the DENIS infrared and DFBS spectroscopic surveys. In order to scan the survey catalogues and to search for Solar System objects, we have developed a VO compliant workflow founded on the use of a recent VO tool labelled SkyBoT (Sky Bodies Tracker) (Berthier *et al.* 2005) and on the VizieR service provided by CDS. A public version of SkyBoT is implemented in the sky atlas Aladin since January 2006. It allows us to quickly get the precise coordinates of the asteroids, planets and natural satellites in any star field, provided we know its center, its size and the precise date of the exposure (between 1949 and 2009). Our data mining workflow performs cone search queries on SkyBoT and VizieR to extract all the known astronomical objects observed by the survey. Then the sources which are detected in the survey are correlated with the known objects to determine the matched and unmatched objects.

The DENIS survey has been performed from 1995 to 2000 on a 1m telescope at La Silla, ESO. It leads to astrometric and I, J, K' photometric measurements. A previous work carried out by Baudrand *et al.* (2004) had already led to the detection of 2190 associations and the identification of 1931 asteroids by comparison to the ephemerides of only 9000 objects. We are carrying out a new exploration of this wide database by comparing now to the ephemerides of more than 340,000 asteroids. The DFBS (Digitized First Byurakan Spectroscopic Survey) has been performed from 1965 to 1980 on the 1m Schmidt telescope of the Byurakan Observatory equipped with a $1^\circ.5$ prism objective (Mickaelian *et al.* 2005). It was mainly dedicated to galaxies study. More than 15,000 spectra of low resolution (50 Å) in the spectral range 3400–6900 Å are present on each one of the 2180 plates (4° FOV). This led to about 40 millions of spectra in which we will identify those of Solar System objects. This is the first work dedicated to the extraction of Solar System data from this data base.

References

Baudrand, A., Bec Borsenberger, A., & Borsenberger, J. 2004, *A&A*, 423, 381
Berthier J., Vachier F., Thuillot W., *et al.* 2005, in: Proc. ADASS XV, *ASP-CS*, 351, 367
Mickaelian, A. M., Hagen, H.-J., Sargsyan, L. A., & Mikayelyan, G. A 2005, in: CDS VizieR On-line Data Catalog: VI/116

Highlights of Astronomy, Volume 14
IAU XXVI General Assembly, 14-25 August 2006
Karel A. van der Hucht, ed.

doi:10.1017/S174392130701215X

Future directions of the Virtual Observatory

Alexander S. Szalay

Physics and Astronomy Department, Johns Hopkins University,
Charles and 34th St, Baltimore, MD 21218-2686, USA
email: szalay@jhu.edu

The Virtual Observatory effort has undergone a substantial evolution over the last four years. In the talk we will discuss the directions of the current effort, and consider the direction along which the Virtual Observatory might evolve along with various other large astronomy projects. We will also discuss similar efforts in other disciplines and the relation of the VO to Grid Computing.

Highlights of Astronomy, Volume 14
IAU XXVI General Assembly, 14-25 August 2006
Karel A. van der Hucht, ed.

doi:10.1017/S1743921307012161

Image processing and scientific workflows in the Virtual Observatory context

Eric Slezak[1], André Schaaff[2] and Jean-Julien Claudon[2]

[1]Observatoire de la Côte d'Azur, BP 4229, F-06304 Nice cedex 4, France
email: slezak@obs-nice.fr

[2]Observatoire de Strasbourg, 11 rue de l'Université, F-67000 Strasbourg, France
email: schaaff@astro.u-strasbg.fr

Since 2005, the workflow working group of OV-France gathers astronomers from five french institutes with common objectives: (*i*) defining use cases of general interest; (*ii*) identifying the simplest workflow structure allowing portability; and (*iii*) suggesting solutions for designing and exploiting easily such workflows (cf. Astrogrid developments). Our workflow definition is the following: a sequence of tasks executed within a controlled context by an architecture taking into account VO standards.

The analysis of multivariate images is one of these uses cases requiring advanced algorithms. These data are indeed intrinsically complex due to potential non-gaussian behaviours, to their inherent sparsity and to hidden correlations. Basically, three main reduction tasks have to be carefully performed to properly handle them. First, one has to reduce the dimensionality of the data space by selecting or building variables carrying the relevant information. Then, after the choice of a metric and a model allowing a sparse representation, data are segmented and, finally, primitives are extracted and classification achieved. Let us consider for instance the multisource data fusion problem. The goal is to combine in an optimal way all the data into a single model while preserving all the information. To do so, different strategies are possible considering either a set of 2D images, a set of spectra or a data cube, as well as different tactics for each of them like for instance a chi-square sum or a bayesian approach for co-adding the images. Therefore we are developping new algorithms (MDA project) relevant to such data sets as well as related VO compliant web services.

To achieve a particular goal, several of these tools must in addition be organised in a rigorous plan including detailed knowledge about the use of each program and about the overall task. Such workflows encode a know-how information for solving complex tasks which has also to be made available to the end-user. For this purpose, we are developping the AIDA framework for graphically building and executing analysis systems from elementary processing blocks. Since the AIDA architecture itself is described in the next paper of this special session, let us just summarize two of its features from the user's point-of-view. First, the integration of each (local or distant) program in the AIDA repository is straightforward since it only requires to upload a file describing the number and type of the I/Os and another one storing the values of the variables needed by the algorithm; each code is then automatically wrapped to become accessible through CGI and Web Services. Second, to build any workflow, one has only to chain programs by (i) dragging them from the repository to the board using the mouse for selecting them and (ii) interconnecting the inputs/outputs at the edges of their graphical icons with a line (compatibility checks are automatically performed). A genuine action plan is built in this (graphical) way, which can be executed on the local computer running the AIDA workflow engine.

Highlights of Astronomy, Volume 14
IAU XXVI General Assembly, 14-25 August 2006
Karel A. van der Hucht, ed.

doi:10.1017/S1743921307012173

Implementing astronomical image analysis pipelines using VO standards

Mireille Y. Louys[1], François Bonnarel[2], André Schaaff[2], Jean-Julien Claudon[2] and C. Pestel[2]

[1]LSIIT, Université Louis Pasteur, 4 rue Blaise Pascal, F-67070 Strasbourg, France
email: louys@astro.u-strasbg.fr

[2]CDS, Observatoire de Strasbourg, 11 rue de l'Université, F-67000 Strasbourg, France

We consider here image analysis pipelines and examine how data and processes could be described in the context of the VO. The tasks chain is considered as a workflow, not only in terms of computing and resource allocation as in the Grid community, but in terms of data analysis know-how. Such pipelines may be published as coarse grain tool boxes and provide reference examples to the user.

We have designed a prototype named AIIDA (Astronomical Image processing Distribution Architecture) which allows encapsulating image processing programs developed in any language such as C, C++, FORTRAN, and MATLAB. The AIIDA client allows to sketch out a chain of processing steps using a graphical tool (JLOW library) encodes it into an XML description and passes it to a workflow engine which in turn interprets the language and orchestrates the execution of the workflow. The server part executes the workflow via CGI and Web Services interfaces. From this simple project has been an interesting collaborative tool between astronomers and signal processing developers inside and outside our laboratory. It helped as a test bed to understand how data analysis workflows could be represented and documented.

What is to be described, lays into two categories. On one hand: *tools and data*, i.e, the scientific purpose of each tool or program, with input and output parameters, and the physical content of the data file consumed by each processing tool. On the other hand, *the workflow execution*, i.e, the sequence of steps as a graph or a list, the data flow within the graph, the allocation of computing resources and the execution status of each step.

There are emerging VO standards matching these requirements: Processing blocks can be described, using the VOApplication Model, with parameters described using the Common Execution Architecture (CEA) data model or via a hierarchical structure as proposed for numerical simulation codes. Many parameters in data analysis workflow are related to observations files whose physical content could be checked before launching a complex workflow. The Spectrum data model for 1D spectra as well the Characterisation data model for the higher dimensions allow to describe for input files the physical axes and the data properties such as coverage or resolution, and check for the compliance with the signal expected in the processing block.

The requirements for workflow description are partly covered by the Astrogrid Workflow System, which provides a workflow scripting language (Groovy), a workflow engine and a user interface for scripting the task chain. It supports interfaces to VO applications via CEA and relies on a distributed storage (MySpace). The description of task allocation and execution on GRID installations is more in the hands of the grid community. Some further work is needed to bridge the gap, between VO workflow descriptions and grid execution logs in order to give feed back to the user, in terms of VO procedures.

Highlights of Astronomy, Volume 14
IAU XXVI General Assembly, 14-25 August 2006
Karel A. van der Hucht, ed.

doi:10.1017/S1743921307012185

VOMegaPlot: efficient plotting of large VOTable datasets

Nilesh Urunkar[1], Ajit K. Kembhavi[2], Ameya Navelkar[1], Jagruti Pandya[1], Vivekananda Moosani[1], Prameela Nair[1] and Mohasin Shaikh[2]

[1]Persistent Systems Pvt. Ltd., Bhageerath 402, Senapati Bapat Road, Pune 411016, India
email: urunkar,amey,jagruti_pandya,vivekananda_moosani,prameela_nair, shaikh@persistent.co.in

[2]Inter-University Centre for Astronomy and Astrophysics (IUCAA),
Post Bag 4, Ganeshkhind, Pune University Campus, Pune 411 007, India
email: akk@iucaa.ernet.in

Most plotting tools tend to load all the data to be plotted into main memory and then use the in-memory data for the actual operations such as plotting. Our Analysis shows that in case of such interactive applications, as the memory usage grows, the response time increases in a significant fashion resulting in poor user interactivity. Further, in cases where the data size exceeds the limits imposed by the available physical memory, the entire data cannot be loaded in memory for performing any kind of operation.

This paper introduces a way of effectively managing large quantity of VOTable data (of the order of millions of data points) for the purpose of plotting and analysis. The approach is to pre-process the data to create intermediate data structures, which are stored on the disk. These intermediate data structures are then used for plotting and/or analysis by the application. This technique has been successfully used for VOMegaPlot, the Java based plotting tool developed under the VO-India initiative. The technique is fairly generic and can be easily extended to astronomical data formats other than VOTable.

Pre-processing operation divides the input data into a number of blocks. For plotting and/or analysis, individual blocks are loaded as and when required which leads to lesser memory consumption and greater degree of user interactivity. Storing data in this fashion also helps in handling subsets of data as only those data blocks containing the subset can be loaded and used which results in faster operations.

VOMegaPlot supports various types of plots like scatter plot, density plot, projection plot and histogram. Following are the results of pre-processing and plotting operation for datasets, Tycho$^+$, Tycho-2$^+$ and UCAC2*. Results were obtained on a single CPU Pentium 4 (2.66 GHz) machine on Redhat Linux 9.0 platform.

	Tycho	Tycho-2	UCAC2
Data Size	1 million rows, 56 columns	2.5 million rows, 32 columns	48.3 million rows, 9 columns
Pre-processing time	18 minutes	30 minutes	3 hours and 26 minutes
Plotting time for scatter plot	9 seconds	22 seconds	5 minutes and 46 seconds

Notes:
$^+$: Tycho and Tycho-2 datasets were pre-processed by keeping the memory size at 64 MB.
*: UCAC2 dataset was pre-processed with 256 MB memory size.

Highlights of Astronomy, Volume 14
IAU XXVI General Assembly, 14-25 August 2006
Karel A. van der Hucht, ed.

doi:10.1017/S1743921307012197

Science applications of the Montage image mosaic engine

G. Bruce Berriman[1], Anastasia C. Laity[1], John C. Good[1], Daniel S. Katz[2], Joseph C. Jacob[2], Ewa Deelman[3], Gurmeet Singh[3], Mei-Hu Su[3] and Thomas A. Prince[4]

[1]Infrared Processing and Analysis Center, Caltech, MS 100-22, Pasadena, CA 91125, USA
email: gbb,laity,jeg@ipac.caltech.edu

[2]Jet Propulsion Laboratory, 4800 Oak Grove Drive, Pasadena, CA 91109, USA

[3]USC Information Sciences Institute, 4676 Admiralty Way, Marina del Rey, CA 90292, USA

[4]Division of Physics, Mathematics and Astronomy, Caltech, Pasadena, CA 91125, USA

Montage is a portable, scaleable toolkit that runs on end-users desktops, clusters, and computing grids. It generates astronomical image mosaics that preserve the calibration and photometry in the input FITS files. The code is available for download at the project website at ¡http://montage.ipac.caltech.edu¿, and includes independent modules for image discovery (including support for VO image access protocols), image reprojection, rectification of the sky background to a common level and image co-addition.

Montage is in active use to support product generation, and in generation of products for use in on-line data access services. Customers are taking advantage of the modular design, and are in some cases using Montage as a reprojection engine, or as a background rectification engine. Three *Spitzer Space Telescope* Legacy teams, the Galactic Legacy Infrared Mid Plane Survey Extraordinaire (GLIMPSE), the *Spitzer* Wide-Area Infrared Extragalactic Survey (SWIRE) and Surveying the Agents of a Galaxy's Evolution (SAGE), have integrated Montage into their pipelines, where it supports the generation of science products for public release. The INT/WFC Photometric Hα Survey (IPHAS) of the Northern Galactic Plane are making similar use of Montage. The Visible and Infrared Survey Telescope for Astronomy (VISTA) is evaluating Montage to stitch together image 'pawprints' to produce a fully-sampled tile under evaluation. The COSMOS *Hubble* Treasury Program used Montage to validate Advanced Camera for Surveys (*HST*-ACS) image mosaics generated at the *Hubble Space Telescope* Science Institute. The Multi-Mission Archive at Space Telescope (MAST) and the Stellar Archive and Retrieval System (StARS) (unreleased) have used Montage to generate 2MASS postage stamp images centered on their target sources. The NASA/IPAC Infra Red Science Archive is using Montage to create 3-color browse products and support image cutout services. The *Spitzer Space Telescope* Outreach program is generating mosaics in uncommon projections, often the best ones for E/PO products. Montage components have been used by the NVO Quick Sky Statistics and Sky Coverage Service to Montage components used to determine sky coverage, and to generate single-survey and composite-survey coverage maps.

An on-request image mosaic service is under evaluation by astronomers. This service has already been used by Beaton06 to confirm the presence of a bar in M 31. The rectification of the background revealed the 'boxy' signature characteristic of a bar.

Reference

Bearman, R., *et al.* 2006, *ApJ* (Letters), in press

Highlights of Astronomy, Volume 14
IAU XXVI General Assembly, 14-25 August 2006
Karel A. van der Hucht, ed.

doi:10.1017/S1743921307012203

VisIVO: an interoperable visualisation tool for Virtual Observatory data

Ugo Becciani[1], Marco Comparato[1], Alessandro Costa[1], Claudio Gheller[2], Bjorn Larsson[1], Fabio Pasian[3] and Riccardo Smareglia[3]

[1]INAF – Osservatorio Astrofisico di Catania, Via S. Sofia 78, I-95123 Catania, Italy
email: ube,marco.comparato,acosta@oact.inaf.it

[2]CINECA – Consorzio Interuniversitario, Casalecchio di Reno (Bologna), I-40033, Italy

[3]INAF – Osservatorio Astronomico di Trieste, Via G.B. Tiepolo 11, I-34131 Trieste, Italy
email: pasian,smareglia@ts.astro.it

We present VisIVO a software for the visualisation and analysis of astrophysical data which can be retrieved from the Virtual Observatory framework and for cosmological simulations. VisIVO is VO standards compliant and supports the most important astronomical data formats such as FITS, HDF5 and VOTables. Data can be retrieved directly connecting to an available VO service (i.e., VizieR WS), loaded in the local computer memory where they can be further selected, visualised and manipulated.

VisIVO can deal with both observational and simulated data and it is particularly effective in handling multidimensional datasets (e.g. catalogues, computational meshes, etc.). It is open source and a pre-release can be downloaded from `<http://visivo.cineca.it>`. VisIVO is also able to interoperate with the other astronomical VO compliant tools through PLASTIC (PLatform for AStronomical Tool InterConnection). This feature allows VisIVO to share data with many other astronomical softwares in order to obtain further information on the data loaded. VisIVO is an interoperable tool: it can automatically use data in the VO framework by using the VizieR CDS archive, Aladin, Topcat and so on. Moreover important applications can be done with pipelines of dedicated and specific tools in the VO frameworks. These can be designed to obtain a scientific use case that allows users to discover new properties and new features in the dataset that could not be easily recognised. The Institute for Astronomy of Edinburgh is developing the idea of a new tool called PLASTIC (PLatform for AStronomy Tool Inter Connection). PLASTIC is a platform to enable heterogeneous visualisation tools to interoperate on a user's desktop. VisIVO uses PLASTIC to communicate to other astronomical tools, such as Aladin, Topcat, etc., in a standard and extendable way, and share with them the same data. The interoperable tools must be connected with the PLASTIC software hub that allow them to broadcast a VOTable loaded in one of them: all the applications manages the same data. Using VisIVO (or other connected tools) the user can select some data points and can send them, like a new object, to all other tools.

The multiple usage of tools with the same data allow the researcher to enhance properties of some data points that cannot be easily discovered. Many data sets contain data in the form of vectors, such as the velocities of the objects. These vectors can be visualised using the vector viewer in VisIVO. Furthermore these vectors can be scaled and coloured according to the magnitude of the vector, aiding at increasing the knowledge about the data set. At present the software Windows XP release is already ready and the Linux version (Debian Sarge) will be ready at the end of 2006 VisIVO is integrated in the VO framework and it is supported as Italian contribution in the VO-TECH project. The European Virtual Observatory – VO Technology Centre (VO-TECH) is a Design Study implemented as Specific Support Action funded by EU in the FP6 Program.

Highlights of Astronomy, Volume 14
IAU XXVI General Assembly, 14-25 August 2006
Karel A. van der Hucht, ed.

doi:10.1017/S1743921307012215

Summary of Special Session 3

Françoise Genova

CDS, Observatoire Astronomique, 11 rue de l'Université, F-67000 Strasbourg, France
email: genova@astro.u-strasbg.fr

The Virtual Observatory is one of the very few truly global endeavours of astronomy, and IAU General Assemblies are the natural places to present the VO to the community. The three-year time scale of IAU is also well suited to measure progress. At the IAU XXV GA in Sydney, 2003, half of the two-day Joint Discussion 8 on *Large Telescopes & Virtual Observatories: Visions for the Future* was devoted to VO talks: the main goal was to review what was expected from the VO in different disciplines of astronomy; many projects were also presented in the poster sessions. This time the three-day VO Special Session has given an excellent overview of present status of the VO, showing both tremendous progress and the work which remains to be done. Several scientific communities (astronomy, solar and heliospheric physics) are working on VO development, national communities have been organizing themselves in VO projects. Implementation by data centres has begun. Additional technical development is required, and there are still technical challenges. There is also a huge work ahead for providing high quality data and services, but the VO development already has a very positive influence on astronomers' work environment, and more is to come.

VO projects are very diverse, depending on local data holdings, technical and scientific expertise, and also of course on the specific demands of the different funding agencies. IVOA provides a common framework to this increasingly diverse community, in particular interoperability standards, which have to be useful and usable. In the coming years, IVOA will have to adjust its organization to support VO uptake by data centres and user communities, and to gather their feedback from implementation and scientific usage. IVOA also has a role to play in the definition of metadata allowing to use information from data pipelines properly. Knowing the origin, limits, strengths, errors, systematics of data is required, not only for the dissemination of data products, but also for data analysis, complex queries, quality control. Proper metadata also has to be defined for theoretical services. Being able to compare data from different surveys, or modelling results with observations, is another aspect of interoperability.

The VO endeavour is science driven and must remain so, but it faces technical challenges, relevant to interdisciplinary work with the IT community. This requires time and effort, and we have to continue to aim at win-win collaborations, in which the VO provides IT teams with excellent research topics and test-beds, and expects operational solutions from the IT developments. When it comes to implementation, adoption of new technologies has to find a proper balance between risk and sustainability – in other words, new technologies have to be implemented not too early, not too late. Of particular importance is the relation with the computational GRID. The VO is a grid of data and services, which fits well with the 'knowledge grid' concept. Many VO aspects do not require the use of the computational GRID, but some important ones do, e.g. for massive data analysis, or 'Virtual data' creation and storage from theoretical modelling. There is a growing community of GRID astronomers, and willingness to build operational services. A bridge has thus to be built between the VO and the GRID. Because of the world-wide nature of the VO, topics such as Single Sign On and interoperability between the different GRID implementations are especially relevant.

The diversity of the data and service provider community appeared well in the presentations. Large organizations, motivated in particular by the VO development, move towards provision of 'science-ready' data and enhanced data archives, even for complex data such as interferometry, and in some cases size the opportunity to implement deeper changes to improve data and archive management – this is one of the very positive effects of the VO. New facilities and large projects include the VO concept in their plans. Networking of expertise around large facilities, including instrument teams and PIs of key programs, is certainly worth exploring. Several emerging national projects plan to implement digitized archives of older data, with science usage in mind and thus with a strong emphasis on proper data description, astrometric registration, calibration, etc. Many smaller teams are willing to share their knowledge with the community and to provide specific data and expertise to the VO. New services are emerging, such as theoretical services providing modelling results, or matching models with observations; software suites; data analysis services and algorithms; services dedicated to help solving a well defined science question; ambitious, full data analysis and research environments. The VO is not dictating how to manage archives and services – a thin layer on top of the services translates local parameters into standard ones, with a small overhead on data and service providers. But there is a cost in producing and maintaining quality services, and data centres have to be sufficiently funded. A career path to attract, keep and properly reward scientists working in these fields is also needed. Community support is required (e.g., in advisory committees) for funding and careers. Quality assessment is another open question: 'private' publication without quality assessment by some authority is felt as a risk for VO credibility. IVOA has no vocation at present to be the VO 'quality police', and national projects certainly have a role to play here.

The VO user community also showed up during the meeting. All disciplines are concerned, from users of large surveys to '(not so) old-fashioned stellar spectroscopists' (as one of them showed in an excellent, provocative talk). The VO will help, e.g., for multi-wavelength, multi-instrument astronomy, for integration of heterogeneous data, but also for comparizon between observations and modelling results, data analysis, statistical analysis – 'high fidelity statistics', search for diamonds in haystacks, etc. The different user communities have requirements on VO tools, which have to be taken into account by the projects, and are welcome to provide their own VO-enabled tools – the VO projects will not write all the tools. There will be a diversity of VO portals, because of the diversity of science needs. It is likely that most VO science usage will not be VO-enabled end-to-end, but rather that VO-enabled data and tools will be used at many stages of research work. Management of simple and complex workflows, and the possibility to keep track of and 'publish' data analysis paths, are important issues.

Widely used tools have already benefited from the VO development, so many astronomers use the VO without being aware of it. Interoperability between tools, collaboration between projects, are already producing innovative and powerful functionnalities. One important role of the VO projects is to find efficient ways of helping users, in particular on advanced usage of tools – actions targetted on specific domains, and on students and post-docs, would certainly be useful. VO does not mean, in spite of some stereotypes, ready-made, press-button astronomy, and astronomers' skills, knowledge, and critical eye on results, will continue to be the key for high quality science!

The VO has not been starting from scratch, there are expertise and lessons learnt from the on-line archive and service providers, and the community is already accustomed to use on-line resources. But the VO is a change in scale, providing a sustainable framework for service publication and seamless science usage, aiming at becoming part of astronomers' everyday life.

Highlights of Astronomy, Volume 14
IAU XXVI General Assembly, 14-25 August 2006
Karel A. van der Hucht, ed.

doi:10.1017/S1743921307012227

Special Session 3 Poster Abstracts

Nicholas A. Walton[1], Andrew Lawrence[2] and Roy Williams[3] (eds.)

[1]Institute of Astronomy, University of Cambridge, Madingley Road, Cambridge CB3 0HA, UK
email: naw@ast.cam.ac.uk

[2]University of Edinburgh, Royal Observatory, Blackford Hill, Edinburgh EH9 3HJ, UK
email: al@roe.ac.uk

[3]Center for Advanced Computer Research, California Institute of Technology, MC 158-79, 1200 E. California Boulevard, Pasadena, CA 91125, USA
email: roy@cacr.caltech.edu

An all-sky 2MASS mosaic constructed on the TeraGrid: processing steps for generation of a 20-terabyte 2MASS all-sky mosaic

G.B. Berriman[1], J.C. Good[1], A.C. Laity[1], D.S. Katz[2], J.C. Jacob[2], L. Brieger[3], R.W. Moore[3], R. Williams[4], E. Deelman[5], G. Singh[5] and M.-H. Su[5]

[1]*Infrared Processing and Analysis Center, Caltech, Pasadena, CA 91125, USA*
[2]*Jet Propulsion Laboratory, Pasadena, CA 91109, USA*
[3]*San Diego Supercomputer Center, La Jolla, CA 92093, USA*
[4]*Division of Physics, Mathematics and Astronomy, Caltech, Pasadena, CA 91125, USA*
[5]*Information Sciences Institute, Marina del Rey, CA 90292, USA*

Abstract. The Montage mosaic engine supplies on-request image mosaic services for the NVO astronomical community. A companion paper describes scientific applications of Montage. This paper describes one application in detail: the generation at SDSC of a mosaic of the 2MASS All-sky Image Atlas on the NSF TeraGrid. The goals of the project are: to provide a 'value-added' 2MASS product that combines overlapping images to improve sensitivity; to demonstrate applicability of computing at-scale to astronomical missions and surveys, especially projects such as LSST; and to demonstrate the utility of the NVO Hyperatlas format. The numerical processing of an 8-TB 32-bit survey to produce a 64-bit 20-TB output atlas presented multiple scalability and operational challenges. An MPI Python module, MYMPI, was used to manage the alternately sequential and parallel steps of the Montage process. This allowed us to parallelize all steps of the mosaic process: that of many, sequential steps executing simultaneously for independent mosaics and that of a single MPI parallel job executing on many CPUs for a single mosaic. The Storage Resource Broker (SRB) developed at SDSC has been used to archive the output results in the Hyperatlas. The 2MASS mosaics are now being assessed for scientific quality. The input images consist of 4,121,440 files, each 2 MB in size. The input files that fall on mosaic boundaries are opened, read, and used multiple times in the processing of adjacent mosaics, so that a total of 14 TB in 6,275,494 files are actually opened and read in the creation of mosaics across the entire survey. Around 130,000 CPU-hours were used to complete the mosaics. The output consists of 1734 6-degree plates for each of 3 bands. Each of the 5202 mosaics is roughly 4 GB in size, and each has been tiled into a 12×12 array of 26-MB files for ease of handling. The total size is about 20 TB in 750 000 tiles.

Aladin, a portal for the Virtual Observatory

T. Boch, P. Fernique, F. Bonnarel, M.G. Allen, O. Bienaymé and S. Derrière

CDS, Observatoire de Strasbourg, 11 rue de l'Université, F-67000 Strasbourg, France
email: boch@astro.u-strasbg.fr

Abstract. Created nine years ago, the Aladin sky atlas tool has become a widely-used VO (Virtual Observatory) portal capable of locating data of interest, accessing and exploring distributing datasets, and visualizing multi-wavelength data. Its compliance with existing or emerging VO standards, interconnection with other visualization or analysis tools, and its ability to easily compare heterogeneous data are key features that make Aladin a powerful data exploration and integration tool as well as a science enabler. Latest developments include simultaneous access and query to data servers discovered via the IVOA registry, overlays of user-defined instrument fields of view, access to solar system database (SkyBOT) in collaboration with IMCCE, extraction of postage stamps around regions of interest, (re)calibration of images/catalogues and compatibility with the PLASTIC (Platform for AStronomical Tools InterCommunication) protocol to facilitate easy connexion between VO tools. Some developments have been performed in the frame of the european *VO-Tech* project.

The Aladin software is available from the website <http://aladin.u-strasbg.fr/>

BRAVO (Brazilian Astrophysical Virtual Observatory): data mining development

R.R.de Carvalho, H.V. Capelato and H.C. Velho

INPE, São José dos Campos, Brazil
email: reinaldo@das.inpe.br

Abstract. The primary goal of the BRAVO project is to generate investment in information technology, with particular emphasis on datamining and statistical analysis. From a scientific standpoint, the participants assembled to date are engaged in several scientific projects in various fields of cosmology, astrophysics, and data analysis, with significant contributions from international partners. These scientists conduct research on clusters of galaxies, small groups of galaxies, elliptical galaxies, population synthesis, N-body simulations, and a variety of studies in stellar astrophysics. One of the main aspects of this project is the incorporation of these disparate areas of astrophysical research within the context of the coherent development of database technology.Observational cosmology is one of the branches of science experiencing the largest growth in the past few decades. large photometric and spectroscopic surveys have been carried out in both hemispheres. As a result, an extraordinary amount of data in all portions of the electromagnetic spectrum exists, but without standard techniques for storage and distribution. This project will utilize several specific astronomical databases, created to store data generated by several instruments (including SOAR, Gemini, BDA, etc), uniting them within a common framework and with standard interfaces. We are inviting members of the entire Brazilian astronomical community to partake in this effort. This will certainly impact both education and outreach efforts, as well as the future development of astrophysical research. Finally, this project will provide a constant investment in human resources. First, it will do so by stimulating ongoing short technical visits to Johns Hopkins University and Caltech. These will allow us to bring software technology and expertise in datamining back to Brazil. Second, we will organize the Summer School on Software Technology in Astrophysics, which will be designed to ensure that the Brazilian scientific community can take full advantage of the benefits offered by the VO project

---

Status report of Virtual Observatory at the National Central University of China Taipei

C.K. Chang[1], C.M. Ko,[2] and D. Kinoshita[1]

[1] *Institute of Astronomy, National Central University, Jhongli, China Taipei*
[2] *Department of Physics and Center for Complex System, National Central University, Jhongli, China Taipei*
email: rex@astro.ncu.edu.tw

Abstract. The idea of virtual observatory (VO) has started to get some attention in Taiwan. However, in order to join the VO community with minimum resources, we identify ourselves as a VO user instead of a developer. We implement the JVO skynode package to create a 2MASS star count map. According to level 6 Hierarchical Triangular Mesh (HTM), the sky was divided into 32768 leaves with approximately 1 square degree resolution. We can study Galactic structure and initial mass function with this map. Our next goal is to publish a rather unique light curve data. National Central University is part of Taiwan-America Occultation Survey (TAOS) team. TAOS project will conduct a census of the number of Kuiper belt objects down to a few km size by monitoring chance stellar occultations by these cometary nuclei. Owing to a special observation strategy, called Zipped mode observation, TAOS can have sub-second photometry. We would like to publish this rather unique light curve database to the VO world.

Theoretical Virtual Observatory and Grid Web services: VisIVO and new capabilities

A. Costa[1], U. Becciani[1], C. Gheller[2], M. Comparato[1] and B. Larsson[1]

[1] *INAF OACT, Catania, Italy;* [2] *CINECA, Bologna, Italy*
email: acosta@oact.inaf.it

Abstract. VisIVO is a tool for 3D visualization, it provides an effective and intuitive way of managing, visualizing and analysing the large amount of data produced by observations and numerical simulations. The software is specifically designed to deal with multidimensional data. Catalogues and numerical simulations represent the basic target of VisIVO. The package is written in C++. This poster describes VIsIVO's Grid Web Service (VWS) and its counterpart client side developed in VisIVO. VisIVO's Grid Web Service is developed as a part of the Italian Virtual Observatory, it allows to run applications in grid using the web service technology. The VWS has been designed to work within a Virtual Observatory environment so that the interface for the current application, shown in this poster, and the interfaces for other applications are described by a small and constant piece of WSDL code. Our first application is HOP: an algorithm for finding groups of particles based on the one developed and coded by Daniel Eisenstein & Piet Hut, Institute for Advanced Study, Princeton, NJ. We have developed VWS using Java AXIS libraries for the server side and C++ AXIS libraries for the client side. The access to the computational resources and storage areas is based on grid services in the INFN Production Grid and from this the VWS inherits asynchronous features, scheduling and matching algorithms. HOP was distributed as RPM Package and was installed in the Worker Node Elements of the INFN Production Grid. The idea is to use external tools through the grid avoiding the integration in our application. We can focus on the I/O management of the jobs and on the standardization of the access methods for the different analysis tools improving the scalability of our solution.

SkyMouse, a smart on-line astronomical information collector

C. Cui, H. Sun and Y. Zhao

National Astronomical Observatories, Chinese Academy of Sciences, Beijing 100012, China
email: ccz@bao.ac.cn

Abstract. With the progress of network technologies and astronomical observation technologies, as an example of cyber-infrastructure based sciences, Virtual Observatory is initiated and spreading quickly. More and more on-line accessible database systems and all kinds of services are available. Although astronomers have been aware the importance of 'interoperability', integrated access to the on-line information is still difficult. SkyMouse is a smart system developed by Chinese Virtual Observatory project to let you access different online resource systems easily then ever. Not like some VO efforts on uniformed access systems, for example, NVO DataScope, SkyMouse tries to show a comprehensive overview for a specific object, but not to snatch as much data as possible. Stimulated by a simple 'Mouse Over' on an interested object name, various VO-compliant and traditional databases, i.e., SIMBAD, NED, VizieR, DSS, ADS, are queried by the SkyMouse. An overview for the given object, including basic information, image, observation and references, is displayed in user's default web browser. In the poster, current status and trends of on-line astronomical services are introduced; architecture, topology and key technologies of the SkyMouse system are described; current functions and future plans are listed. The system is accessible at : <http://skymouse.china-vo.org>.

The Virtual Solar-Terrestrial Observatory: interdisciplinary data-driven science

P.A. Fox[1], D.L. McGuinness[2], D. Middleton[3], L. Cinuini[3], J. Garcia[1], P. West[1], J.A. Darnell[1] and J. Benedict[2]

[1] *HAO/ESSL/NCAR, Boulder, CO, USA*
[2] *McGuinness Associates, Stanford, CA, USA*
[3] *AC/SCD/CISL/NCAR, Boulder, CO, USA*
email: pfox@hao.ucar.edu

Abstract. Virtual Observatories can provide access to vast stores of scientific data: observations and models. As these electronic stores become widely used, there is potential to improve the efficiency, interoperability, collaborative potential, and impact of a wide range of interdisciplinary scientific research. In order to realize this potential, technical challenges need to be addressed concerning (at least) representations and interoperability of data, access, and usability. In the Virtual Solar Terrestrial Observatory (VSTO) project, we are providing an electronic repository of observational data spanning the solar-terrestrial physics domain. We are also implementing semantic web tools and infrastructure for accessing and using the data. Our main contributions include the repository, infrastructure, and tools for the particular solar terrestrial physics as well as the design and infrastructure that may be broadened to cover more diverse science areas and communities of use. In this presentation, we describe the goals, design, current and planned prototypes, and technical infrastructure. We present what we have learned about the processes involved in developing VSTO and the required semantics, how they affect the framework architecture, choice of technologies and service interfaces. VSTO is an NSF-funded joint effort between the High Altitude Observatory and the Scientific Computing Division at the National Center for Atmospheric Research (NCAR) and McGuinness Associates Consulting.

Tools and services from the French VO

F. Genova and the French VO Teams

CDS, Observatoire Astronomique, UMR CNRS/ULP 7550,
11 rue de l'Université, F-67000 Strasbourg, France
email: genova@astro.u-strasbg.fr

Abstract. The French VO (*Action Spécifique Observatoires Virtuels France – ASOV*), a joint action of INSU and CNES, coordinates French participation in the VO for astronomy, solar, heliospheric and space plasma physics, and the study of the planets. It supports teams which develops VO services to uptake VO standards and methods, in particular by organizing tutorials for developers, provides forums for discussion, and funds travel of French participants to IVOA meetings. INSU, on behalf of the French VO, is a member of the IVOA and of Euro-VO, and coordinates the Euro-VO Data Center Alliance, which will be funded by the European Commission as a Coordination Action of the Infrastructure, Communication Network Development program from September 1, 2006. ASOV was created in 2004. Within a few years it has succeeded in creating a national community of VO developers. Several teams participate very actively in the IVOA Interoperability standard development, and collaborate with IT laboratories on VO-related subjects. Most French astronomy laboratories are involved in the development of VO-oriented services and tools, with a wide diversity of actions (observation archives/'science ready' data, value-added data bases and services; tools for visualisation, image analysis, ...; software suites; numerical simulations/theoretical astronomy services, thematic services). A recent census received more than 40 answers describing projects at different scales, some operational, some in development. Details on French VO teams and ASOV actions are available from the French VO Web site: <http://www.france-vo.org/>

Three steps to CIELO

M. Guainazzi and S. Bianchi

European Space Astronomy Center, Madrid, Spain
email: Matteo.Guainazzi@sciops.esa.int

Abstract. The origin of the soft X-ray emission in obscured Active Galactic Nuclei (AGN) is still largely unknown, despite important progress made possible by recent measurements with *Chandra* and *XMM-Newton*. Our understanding of the evolution of accretion onto supermassive black holes, and of its interaction with gas andstars in the dense nuclear environment would receive a dramatic burst by thesolution of this mystery. In this paper, we will: (*a*) show why high-resolution X-ray spectroscopy is crucial to the solution of this issue; (*b*) present CIELO, the first catalogue of soft X-ray emission lines in obscured AGN (~ 80 sources), built from observations of the Reflection Grating Spectrometer (RGS) on-board *XMM-Newton*; and (*c*) discuss the implementation of the IVOA Line Data Model in VO tools (such as the SED builder VOSpec), and its application to CIELO. The combination of the unprecedented RGS sensitivity in the soft X-ray regime, and of the VO protocol power leads us to be closer than ever to unveiling the nature of soft X-ray emission in obscured AGN.

Graphics interfaces and numerical simulations: Mexican Virtual Solar Observatory

L. Hernández[1], A. González[2], G. Salas[3] and A. Santillán[3]

[1] *Instituto de Astronomía, UNAM, Mexico City, DF-04510 Mexico*
email: liliana@astroscu.unam.mx
[2] *Instituto de Ecología, UNAM, Mexico City, DF-04510 Mexico*
[3] *Cómputo Aplicado-DGSCA, UNAM, Mexico City, DF-04510 Mexico*

Abstract. Preliminary results associated to the computational development and creation of the Mexican Virtual Solar Observatory (MVSO) are presented. Basically, the MVSO prototype consists of two parts: the first, related to observations that have been made during the past ten years at the Solar Observation Station (EOS) and at the Carl Sagan Observatory (OCS) of the Universidad de Sonora in Mexico. The second part is associated to the creation and manipulation of a database produced by numerical simulations related to solar phenomena, we are using the MHD ZEUS-3D code. The development of this prototype was made using mysql, apache, java and VSO 1.2. based GNU and 'open source philosophy'. A graphic user interface (GUI) was created in order to make web-based, remote numerical simulations. For this purpose, Mono was used, because it is provides the necessary software to develop and run .NET client and server applications on Linux. Although this project is still under development, we hope to have access, by means of this portal, to other virtual solar observatories and to be able to count on a database created through numerical simulations or, given the case, perform simulations associated to solar phenomena.

Integration between solar and space science data for space weather forecast using web services

S. Kato

Hyogo College of Medicine, Nishinomiya, Japan

Abstract. As the technology develops, the opportunity that the human beings behave in space, and it is still understood that the solar activities (especially the solar flare) influence the air-lines communication, the ship communication and the power generator of the electric power company, etc. Forecasting the effects of the solar activities is becoming very important because there is such a background. Our goal is that constructs the detailed model from the Sun to the magnetosphere of the earth and simulates the solar activities and the effects. We try to integrate the existing observational data including the ground observational data and satellite observational data using by web service technology as a base to construct the model. We introduce our activity to combine the solar and space science data in Japan. Methods Generally, it is difficult to develop the virtual common database, but web service makes interconnection among different databases comparatively easy. We try to connect some databases in the portal site. Each different data objects is aggregated to a common data object. We can develop more complex services. We use RELAX NG in order to develop these applications easily. We begin the trial of the interconnection among the solar and space science data in Japan. In the case of solar observational data, we find the activity such as VO, for example, VSO and EGSO, but space science data seems to be very complex. In addition to this, there is time lag that solar activity has an effect on the magnetosphere of the Earth. We discuss these characteristic in the data analysis between the solar and space data. This work was supported by the Grant-in-Aid for Creative Scientific Research 'The Basic Study of Space Weather Prediction' (17GS0208) from the Ministry of Education, Science, Sports, Technology, and Culture of Japan

---

Access to high-energy astrophysics through Virtual Observatory: scientific motivation and status of INTEGRAL prototype

P. Kubanek and R. Hudec

INTEGRAL Science Data Center, Versoix, Switzerland
email: Petr.Kubanek@obs.unige.ch

Abstract. We discuss motivations to create Virtual Observatory access to high-energy data. On example of our *INTEGRAL* data access we discuss the challenges which we have attacked on creating it, and the current status of the development the VO enabled interface to the public *INTEGRAL* data.

Web Services for public cosmological surveys: the VVDS-CDFS application

L. Paioro[1], B. Garilli[1], V. Le Brun[2], P. Franzetti[1], M. Fumana[1] and M. Scodeggio[1]

[1]*INAF, IASF Milano 'G. Occhialini', via Bassini 15, I-20133 Milano, Italy*
email: luigi@lambrate.inaf.it
[2]*Laboratoire d'Astrophysique de Marseille, Traverse du Siphon, F-13376 Marseille, France*

Abstract. Cosmological surveys (like VVDS, GOODS, DEEP2, COSMOS, etc.) aim at providing a complete census of the universe over a broad redshift range. Often different information are gathered with different instruments (e.g., spectrographs, *HST*, X-ray telescopes, etc.) and it is only by correctly assembling and easily manipulating such wide sets of data that astronomers can attempt to describe the universe; many different scientific goals can be tackled grouping and filtering the different data sets. When dealing with the huge databases resulting from public cosmological surveys , what is needed is: (*a*) a versatile system of queries, to allow searches by different parameters (like redshifts, magnitude, colors, etc.) according to the specific scientific goal to be tackled; (*b*) a cross-matching system to verify or redefine the identification of the sources; and (*c*) a data products retrieving system to download data related images and spectra. The Virtual Observatory Alliance defines a set of services which can satisfy the needs described above, exploiting Web Services technology. Having in mind the exploitation of cosmological surveys, we have implemented what we consider the most fundamental VO Web Services for our scientific interests: Conesearch (retrieves physical data values from a cone centered on one point in the sky – the simplest query), SkyNode (allows to filter on the physical quantities in the database in order to select a well defined data subset), SIAP (retrieves all the images contained in a sky region of interest), SSAP (retrieves 1D spectra). Our testing bench is the VVDS-CDFS data set, made public in 2004, which contains photometric and spectroscopic information for 1599 sources (Le Fèrve *et al.*, 2004, A&A, 428, 1043, see <http://cencosw.oamp.fr>). On this data set, we have implemented and published on US NVO registry the first three services mentioned above, to demonstrate the viability of this approach and its usefulness to the astronomical community. Implementation of SSAP service for spectra retrieval will be the next step.

The Golosiiv on-line plate archive database, management and maintenance

L. Pakuliak and T. Sergeeva

Department of Astrometry, Main Astronomical Observatory,
Zabolotnogo str.,27, Kiev, 03680 Ukraine
email: pakuliak,sergeeva@mao.kiev.ua

Abstract. We intend to create online version of the database of the MAO NASU plate archive as VO-compatible structures in accordance with principles, developed by the International Virtual Observatory Alliance in order to make them available for world astronomical community. The online version of the log-book database is constructed by means of MySQL+PHP. Data management system provides a user with user interface, gives a capability of detailed traditional form-filling radial search of plates, obtaining some auxiliary sampling, the listing of each collection and permits to browse the detail descriptions of collections. The administrative tool allows database administrator the data correction, enhancement with new data sets and control of the integrity and consistence of the database as a whole. The VO-compatible database is currently constructing under the demands and in the accordance with principles of international data archives and has to be strongly generalized in order to provide a possibility of data mining by means of standard interfaces and to be the best fitted to the demands of WFPDB Group for databases of the plate catalogues. On-going enhancements of database toward the WFPDB bring the problem of the verification of data to the forefront, as it demands the high degree of data reliability. The process of data verification is practically endless and inseparable from data management owing to a diversity of data errors nature, that means to a variety of ploys of their identification and fixing. The current status of MAO NASU glass archive forces the activity in both directions simultaneously: the enhancement of log-book database with new sets of observational data as well as generalized database creation and the cross-identification between them. The VO-compatible version of the database is supplying with digitized data of plates obtained with MicroTek ScanMaker 9800 XL TMA. The scanning procedure is not total but is conducted selectively in the frames of special projects.

VObs.it – the Italian Virtual Observatory

F. Pasian[1], U. Becciani[2], S. Cassisi[3], A. Fontana[4], B. Garilli[5], C. Gheller[6], P. Giommi[7], G. Longo[8], A. Preite Martinez[9], R. Smareglia[1] and A. Volpicelli[10]

[1] *INAF - O.A. Trieste, Via G. B.Tiepolo 11, I-34143 Trieste, Italy*
[2] *INAF - O.A. Catania, Via S. Sofia 78, I-95123 Catania, Italy*
[3] *INAF - O.A. Teramo, Via Mentore Maggini, I-64100 Teramo, Italy*
[4] *INAF - O.A. Roma, Via di Frascati 33, I-00040 Monte Porzio Catone, Italy*
[5] *INAF - IASF Milano, Via E. Bassini 15, I-20133 Milano, Italy*
[6] *CINECA, Via Magnanelli 6/3, I-40033 Casalecchio di Reno (Bologna), Italy*
[7] *ASI - Unità 'Osservazione dell'Universo', Viale Liegi 26, I-00198 Roma, Italy*
[8] *Università di Napoli 'Federico II', via Cinthia, I-80126 Napoli, Italy*
[9] *INAF - IASF Roma, Via del Fosso del Cavaliere 100, I-00133 Roma, Italy*
[10] *INAF - O.A. Torino, Via Osservatorio 20, I- 10025 Pino Torinese (TO), Italy*
email: pasian@oats.inaf.it

Abstract. The participation of Italy in VO activities has been initially delegated to a Grid project (i.e., DRACO). After the MoU between INAF and Euro-VO was signed, the VObs.it project was established. The aim of the VObs.it project is to coordinate within a unified approach the archives and databases developed by the Italian community. The first steps in this direction are to foster the adoption of IVOA standards, to provide Grid-aware VO applications and to build a national registry containing the list of VO-compliant services available to the international community. The Italian participation in EU-funded international projects is targeted to the development of tools to be subsequently used within the international VO. This allows our developers to be at the forefront of technology in the field. Equally important are participation in the IVOA Executive Committee and in the IVOA working groups defining standard protocols and formats. The activities forming the initial core of the VObs.it are the following: the Long-Term Archive of the TNG available at the INAF centre for Astronomical Archives (IA2), the archive of the VIMOS-VLT Deep Survey, DSS-II and GSC-II (both the databases and the all-sky uncompressed images), BaSTI a widely-used database of theoretical stellar evolution predictions specifically suited for population synthesis analysis, and the Italian Theoretical Virtual Observatory (ITVO) a distributed database of simulated data. Extension to other data and future projects (e.g., the raw LBT and science LBC data, VST) is foreseen. From an initial core of activities carried out by INAF, other groups and organizations are joining the collaboration and expanding VObs.it scientific and technological capabilities. CINECA is active in ITVO and in the definition of IVOA standards, while at the ASI Science Data Center VO-related work has started within the VObs.it and Euro-VO frameworks.

Interoperability and integration of theoretical data in the Virtual Observatory

F. Pasian[1], S. Ameglio[2], U. Becciani[3], S. Borgani[2], C. Gheller[4], V. Manna[1], P. Manzato[1], L. Marseglia[1], R. Smareglia[1] and G. Taffoni[1]

[1] *INAF – O.A. Trieste, Via G.B.Tiepolo 11, I- 34143 Trieste, Italy*
[2] *Università di Trieste, Dip. Astonomia, Via G.B. Tiepolo 11, I-34143 Trieste, Italy*
[3] *INAF - O.A. Catania, Via S.Sofia 78, I-95123 Catania, Italy*
[4] *CINECA, Via Magnanelli 6/3, I-40033 Casalecchio di Reno (Bologna), Italy*
email: pasian@oats.inaf.it

Abstract. The aim of the Virtual Observatory has recently expanded from seeking interoperability among astronomical catalogue and archive systems to including also access to analysis tools, computational services and numerical simulations. As a matter of fact, beside the observational data, there is also a huge amount of theoretical data generated by computer simulations that can be useful if published in Virtual Observatory compatible form. Therefore, considerable interest has been shown in including products of theoretical research. A data model for theoretical data is being designed. At the same time, an interim Simple Numerical Access Protocol (SNAP) as part of the Data Access Layer provides, through negotiation between the client and the theoretical dataset service, a standardized access mechanism to distributed theoretical data objects. We present the first integration within the Virtual Observatory of a set of theoretical data structured with a prototype of the SNAP data access protocol. Our resource provides not only access to simulation data stored in a dedicated archive but also, interfacing web services, a visualization service and the possibility to extract a number of astronomical observables. We focussed our work on a set of numerical simulations of galaxy clusters identified at redshift 0 and produced with the GADGET2 code. We show the possibility of computing the temperature and density profiles, of visualizing theoretical results with VO-enabled astronomical tools, of comparing the results with astronomical observations. The activity is being carried out as part of VO-Tech/DS4, ITVO and VObs.it projects.

Hera - The HEASARC's new data analysis service

W.D. Pence and P. Chai

NASA Goddard Space Flight Center, Greenbelt, MD 20771, USA
email: William.D.Pence@nasa.gov

Abstract. Hera is the new data analysis service provided by the HEASARC at the NASA Goddard Space Flight Center that enables qualified student and professional astronomical researchers to immediately begin analyzing scientific data from astrophysics missions. All the necessary resources needed to do the data analysis are freely provided by Hera, including (*i*) the latest version of the hundreds of scientific analysis programs all installed and ready to run; (*ii*) high-speed access to the terabytes of data in the HEASARC's data archive, (*iii*) a cluster of fast Linux workstations to run the software; and (*iv*) ample local disk space to temporarily store the data and results. Full information about how to use Hera is available from <http://heasarc.gsfc.nasa.gov/hera>. Users only need to download the small Fv FITS file viewer program, which serves as the portal to the Hera services. Users can then access Hera either by using the Hera Graphical User Interface to control an interactive data analysis session on the Hera server machines, or by executing individual data analysis commands from a command window on the user's local machine. In the latter mode, any input data files on the user's local machine are automatically copied up to the Hera server machine where the software task is executed. Any output files from the task are then copied back to the user's machine. The Hera services have also been integrated into the HEASARC's data archive web pages, so researchers can simply click on a link to immediately begin analyzing the corresponding data set without having to first download the data or install any software other than Fv.

The Belgrade Plate Archive Database : current status and scientific tasks

V. Protitch-Benishek[1], A. Mihajlov[2], T. Jakshich[3], and V. Benishek[1]

[1] *Astronomical Observatory, Belgrade, Serbia*
[2] *Faculty of Physics, Belgrade, Serbia*
[3] *Department of Astronomy, Faculty of Mathematics, Belgrade, Serbia*
email: vprotic@aob.bg.ac.yu

Abstract. The plate archives of the Belgrade Astronomical Observatory contain more than 15000 glass photographic plates from the period 1936–1996. In addition to the other equipment the Observatory disposed of four instruments devoted especially to astrophotographic observations: Zeiss Refractor 65/1055 cm with special camera, Zeiss Astrograph 16/80 cm, Zeiss Refractor 20/302 cm with two photographic cameras 16/80 cm and Askania Equatorial refractor 13.5/100 cm. Scientific observations were performed in the framework of the programs like: minor planet follow-up, search for the new objects (33 new minor planets were discovered from BAO), comet investigation, systematic observations of the Sun, Moon, giant planets, natural and artifical satellites, variable stars, double and multiple stars, stellar clusters, etc. Rare phenomena, such as passages of Mercury and Venus across the solar disc, lunar occultations of stars and planets and special objects have been observed too. The current status of Belgrade Astrophtopgraphic Plate Archive (BAPA) Database is reported and a brief description of all phases of such a large Project is given. The preliminary computer-readable catalogue of relevant data from the period 1936–1966 is finished as a representative sample only. The Catalogue BAPA is included into WFPDB (<http://www.skyarchive.org>) as one of the basic sources. A couple of statistical distributions as an example of the kind of informations which will be possible to extract from the database is given.

Space-time coordinated metadata for the Virtual Observatory

A.H. Rots

Harvard-Smithsonian Center for Astrophysics, 60 Garden Street, Cambridge, MA 02138, USA
email: arots@head-cfa.harvard.edu

Abstract. Space-time coordinate metadata are at the very core of understanding astronomical data and information. This aspect of data description requires very careful consideration. The design needs to be sufficiently general that it can adequately represent the many coordinate systems and conventions that are in use in the community. On the other hand the most basic requirement is that the space-time metadata for queries, for resource descriptions, and for data be complete and self-consistent. It is important to keep in mind that space, time, redshift, and spectrum are strongly intertwined coordinates: time has little meaning without knowing the location, and *vice-versa*; redshift and spectral data require position and velocity for correct interpretation. The design of the metadata structure has been completed at this time and will support most, if not all, coordinate systems and transformations between them for the Virtual Observatory, either immediately or through extensions. This work has been supported by NASA under contract NAS 8-03060 to the Smithsonian Astrophysical Observatory for operation of the *Chandra* X-ray Center.

The Golosyiv plate archive digitisation

T.P. Sergeeva[1], A.V. Sergeev[2], L.K. Pakuliak[1] and A.I. Yatsenko[1]

[1] *Department of Astrometry, Main Astronomical Observatory, Kiev, 03680 Ukraine*
[2] *Centre of Astronomical and Medico-Ecological Researches, Kiev, 03680 Ukraine*
email: sergeev@mao.kiev.ua

Abstract. The plate archive of the Main Astronomical Observatory of the National Academy of Sciences of Ukraine (Golosyiv, Kyiv) includes about 85 000 plates which have been taken in various observational projects during 1950–2005. Among them are about 25 000 of direct northern sky area plates and more than 600 000 plates containing stellar, planetary and active solar formations spectra. Direct plates have a limiting magnitude of 14.0–16.0 mag. Since 2002 we have been organising the storage, safeguarding, cataloguing and digitization of the plate archive. The very initial task was to create the automated system for detection of astronomical objects and phenomena, search of optical counterparts in the directions of gamma-ray bursts, research of long period, flare and other variable stars, search and rediscovery of asteroids, comets and other Solar System bodies to improve the elements of their orbits, informational support of CCD observations and space projects, etc. To provide higher efficiency of this work we have prepared computer readable catalogues and database for 250 000 direct wide field plates. Now the catalogues have been adapted to Wide Field Plate Database (WFPDB) format and integrated into this world database. The next step will be adaptation of our catalogues, database and images to standards of the IVOA. Some magnitude and positional accuracy estimations for Golosyiv archive plates have been done. The photometric characteristics of the images of NGC 6913 cluster stars on two plates of the Golosyiv's double wide angle astrograph have been determined. Very good conformity of the photometric characteristics obtained with external accuracies of 0.13 and 0.15 mag. has been found. The investigation of positional accuracy have been made with A3± format fixed bed scanner (Microtek ScanMaker 9800XL TMA). It shows that the scanner has non-detectable systematic errors on the X-axis, and errors of $\pm 15\,\mu$m on the Y-axis. The final positional errors are about $\pm 2\,\mu$m ($\pm 0''.2$). have been obtained after corrections for systematic errors of the scanner and averaging four scans. So we may conclude that astrometric and photometric investigations may be done with precise commercial scanners. It will be necessary to scan plates at a minimum of two positions. We plan to scan the plate archive according to the priority of scientific tasks. Scanning will be done with an optical resolution of 1200×1200 dpi (pixel size $20\,\mu$m), and with maximum amplitude resolution. The plate archive of MAO NASU is a unique well equipped instrument for conducting a range of astronomical investigations with a time scale of more than 50 yr.

VOSED: a tool for the characterization of developing planetary systems

E. Solano[1], R. Gutiérrez[1], A. Delgado[1], L.M. Sarro[2] and B. Merín[3]

[1] *SVO-LAEFF. P.O. Box 50727, E-28080 Madrid, Spain*
[2] *SVO-UNED. Dpto. Inteligencia Artificial, C/ Juan del Rosal 16, E-28040 Madrid, Spain*
[3] *Leiden Observatory, P.O. Box 9513, NL-2300RA, Leiden, the Netherlands*
email: esm,raul,Arancha.Delgado@laeff.inta.es, lsb@dia.uned.es, merin@strw.leidenuniv.nl

Abstract. The transition phase from optically thick disks around young pre-main sequence stars to optically thin debris disks around Vega type stars is not well understood and plays an important role in the theory of planet formation. One of the most promising methods to characterize this process is the fitting of the observed SED with disk models. However, despite its potential, this technique is affected by two major problems if a non-VO methodology is used: on the one hand, SEDs building requires accessing to a variety of astronomical services which provide, in most of the cases, heterogeneous information. On the other hand, model fitting demands a tremendous amount of work and time which makes it very inefficient even for a modest dataset. This is an important issue considering the large volume of data that missions like Spitzer is producing. In the framework of the Spanish Virtual Observatory (SVO) we have developed VOSED <http://sdc.laeff.inta.es/vosed/> an application that permits to characterize the protoplanetary disks around young stars taking advantage of the already existing VO standards and tools. The application allows the user to gather photometric and spectroscopic information from a number of VO services, trace the SED, and fit the photospheric contribution with a stellar model and the IR excess with a disk model. The Kurucz models described in Castelli *et al.* (1997, A&A, 318, 841) are used to reproduce the photospheric contribution whereas the grid of models of accretion disks irradiated by their central stars developed by D'Alessio *et al.* (2005, <http://cfa-www.harvard.edu/youngstars/dalessio>) is used for the disk contribution. In both cases, the models are retrieved from the SVO Theoretical Model Web Server using the TSAP protocol. As pointed out before, model fitting constitutes a fundamental step in the analysis process. VOSED includes a tool to estimate the model parameters (both stellar and disk) based on bayesian inference. The main aim of the tool is to quantitatively analyse the data in terms of the evidence of models of different complexity, evaluate what other alternative models can compete with the most *a posteriori* probable one and what are the most discriminant observations to discard alternatives.

The 2XMM pre-release catalogue: a test case for VO cross correlation of large archives

J.A. Tedds[1], **D. Law-Green**[1], **M.G. Watson**[1], **K.T. Noddle**[1], **D. Morris**[2] **and N.A. Walton**[2]

[1] *Dept. of Physics & Astronomy, University of Leicester, Leicester LE1 7RH, UK*
email: jat@star.le.ac.uk
[2] *Institute of Astronomy, Madingley Road, Cambridge CB3 0HA, UK*

Abstract. The *XMM-Newton* Survey Science Centre has made a pre-release of the second serendipitous source catalogue, 2XMM. This is the largest X-ray source catalogue ever made, including $\sim$150 K objects derived from $\sim$2500 *XMM-Newton* observations since launch and covering a total sky area of approximately 400 sq.deg. The typical flux limit of the survey is $\sim 10^{-14}$ cgs, well matched to the dominant source population of the cosmic X-ray background emission. 2XMM will be a unique database to carry out evolutionary studies of different X-ray source populations. In order to maximise the scientific potential of 2XMM it is crucial to cross correlate with multi-wavelength archives and hence characterise the various source samples in detail. One can then go on to identify rare and unique sources as well as addressing crucial science questions using the large, statistically significant samples identified. These serve as the essential training sets for statistical identifications and VO data mining tools. We describe the work undertaken to make the catalogue fully VO compliant via the LEDAS high energy database and identify strategies to address the technical challenges of cross correlating a large catalogue such as this with other prime archives such as SDSS DR4 and UKIDSS. We will illustrate how 2XMM is being used as a test case within the VO community by the UK AstroGrid project to address the complex requirements of astronomers. This includes the handling of multiple matches to other catalogues and the adoption of a figure of merit to grade quality of matches that is not based on variable radius with error alone but can assign probabilities based on sky density and quality criteria from within each archive.

A VO 'container' for astronomical optical/UV spectra

R. Thompson[1], D. Durand[2], I. Kamp[3], K. Levay[1], A. Micol[4] and M.A. Smith[1]

[1] *STScI/CSC, Baltimore, USA*
[2] *CADC, Victoria, Canada*
[3] *ESA/STScI, Baltimore, USA*
[4] *ESA/ST-ECF, Garching, Germany*
email: msmith@stsci.edu

Abstract. FITS formats have provided a convenient means to access and interpret spectroscopic data for many years. However, each mission has been free to choose its FITS flavor and provide necessarily instrument specific keywords during this time. Over time, and with new missions developed with unique instrumental configurations, this has created a challenge for spectroscopic applications, which must be written in a complex fashion to recognize, read, and decipher the idiosyncrasies relating to each instrument. Moreover, some HST heritage instruments (GHRS, FOS) have stored different vectors in separate files, requiring their assembly before they can be used. With the advent of the VO and the first generation Simple Spectral Access Protocol (SSAP), it is possible to design a Spectral Container that addresses these issues by serving as a translation layer between the standardized VO protocols and the current FITS file formats. We have constructed SSAP services that point to a secondary data archive holding Spectral Container-packaged files for several MAST (Multi-Mission Archive at Space Telescope) missions. To date, these missions include those for which single-order observations are available: *GHRS*, *FOS*, *EUVE*, *HUT*, *WUPPE*, *IUE*, and *STIS*. In this poster we discuss the current status and examples of the Container using the Specview and VOSpec applications. We also discuss the need for second-generation VO protocols that will provide for multiple spectra (echelle multi-orders, time-series) within a single Container file.

An alternative catalogue collection and a system to exploit it: CATS

O. Verkhodanov[1], S. Trushkin[1], H. Andernach[2] and N. Chernenkov[1]

[1] *Special Astrophysical Observatory, RAS, 369167 Karachaevo-Cherkesia, Russian Federation*
[2] *Departamento de Astronomía, Universidad de Guanajuato, Guanajuato CP 36000, Mexico*
email: vo@sao.ru

Abstract. Conceived in 1996 to support observations with the RATAN-600 radio telescope, CATS is now a publicly accessible database allowing to search through ~ 400 catalogs, all of them accessible by anonymous ftp, with a total of ~ 109 records (<http://cats.sao.ru>). CATS' content overlaps only partly with that of VizieR at CDS and is more complete for radio sources due to contributions from the catalog collection of one of us (HA, see Norris *et al.*, this volume, for a summary on SpS 6). Similar to existing databases CATS offers data extraction both interactively and via batch requests to ¡cats@sao.ru>, with object selection either from a subset or from all available catalogues, by coordinates, flux, frequency, and other parameters. CATS offers a larger variety of radio source lists but only partial inter-catalog cross-identifications. It also provides interactive creation of radio spectra of certain source samples as well as over 1300 radio images of Galactic SNRs.

AstroGrid Virtual Observatory release 2006.3

N.A. Walton and the Astrogrid Consortium

Institute of Astronomy, University of Cambridge, Madingley Road, Cambridge, CB3 0HA, UK
email: naw@ast.cam.ac.uk

Abstract. AstroGrid (<http://www.astrogrid.org>), the UK's Virtual Observatory system, will be releasing its latest fully operational release system, 2006.3, prior to the IAU XXVI General Assembly. This poster describes the increased functionalities offered to the end user in the 2006.3 release. The key data resources and applications accessible through the AstroGrid workbench will be itemised, Details of how astronomers can gain access to the system will be given. We note the use of technologies and interoperability standards, and how use of AstroGrid gives end user access truly global resources, including large repositories of data across the UK, Europe, the USA and elsewhere. We show how the astronomer can easily discover and visualise data from any major data centre by use of the 'AstroScope'. How then data from this can be stored in 'MySpace' and processed using applications either on the uses local machine or on remote servers. Specific example science use cases are noted, ranging from the mining of deep field survey data to investigate the clustering of galaxies at high redshifts, to studies of active regions on our Sun. AstroGrid is ready to try out, and to use – just go to <http://www.astrogrid.org/launch> to find out how. Without registration you'll be able to use AstroScope to discover and locally visualise data. With a simple registration procedure you'll gain access to the full features of the AstroGrid Workbench – including being able to save your data and results to your MySpace virtual storage. See <http://software.astrogrid.org> for further details.

Highlights of Astronomy, Volume 14
IAU XXVI General Assembly, 14-25 August 2006
Karel A. van der Hucht, ed.

doi:10.1017/S1743921307012239

Special Session 5
Astronomy for the developing world

John B. Hearnshaw[1] and Peter Martinez[2] (eds.)

[1]Department of Physics and Astronomy, University of Canterbury,
Private Bag 4800, Christchurch 8020, New Zealand
email: john.hearnshaw@canterbury.ac.nz

[2]South African Astronomical Observatory, P.O. Box 9, Observatory Road,
7935 Observatory, Cape Town, South Africa
email: peter@saao.ac.za

Preface

The International Astronomical Union has a strong commitment to the development of astronomical education and research throughout the world, especially in those countries developing economically. This commitment is in part through the work of IAU Commission 46 for astronomy education and development. Within that commission, the Program Group for the *World-wide Development of Astronomy* (PGWWDA) coordinates many of these activities, promoting the development of astronomy in developing countries.

Six years ago, at the time of the IAU XXIV General Assembly in Manchester, Alan Batten, who was then chair of the PGWWDA, organized a special session on 'Astronomy for developing countries' (A. Batten. ed., 2001, *Astronomy for Developing Countries*, Proc. IAU XXIV GA Special Session (San Francisco: ASP). The success of that meeting has led Commission 46 to propose another Special Session, this time at the IAU XXVI General Assembly, in Prague, 2006. These pages present highlights from that two-day session, known as Special Session 5 on *Astronomy for the Developing World.*

A key theme proposed for SpS 5 was a survey of the development of astronomy in different geographical regions of the world, such as Latin America, eastern Europe, Africa, central Asia and the Far East. There are contributions here from all these places. In addition SpS 5 strived to bring together several other programmes promoting astronomy and space science in the developing world, from agencies outside the IAU. These include the United Nations Office for Outer Space Affairs (UNOOSA), the International Heliophysical Year (IHY) program for 2007, and the Committee on Space Research (COSPAR), amongst others.

Moreover, SpS 5 had as one of its aims to promote the concept of establishing a Third-world Astronomy Institute or Network (TWAI/TWAN) – an idea championed by Professor Jayant Narlikar and presented in the introductory session. Moves are already underway to give the Inter-University Centre for Astronomy and Astrophysics in Pune, India, an international dimension. This will be a step towards realizing this dream, to do for astronomy what the International Centre for Theoretical Physics in Trieste is already doing for physics.

The interest in SpS 5 'Astronomy for the developing world' was much greater than expected. In these pages, the abstracts of 61 papers are presented. Sixteen of these were invited papers, 26 were contributed oral talks and the rest were poster papers. The first authors came from 37 different countries. What is more, about 280 astronomers from 61 different countries registered their interest in participating in the SpS 5 session; it was a truly multinational gathering.

The following pages no more than summarize the papers presented at SpS 5. The full text of these papers is published as by J.B. Hearnshaw & P. Martinez, P. (eds., 2007, *Astronomy for the Developing World*, Proc. IAU XXVI GA Special Session No. 5, Cambridge: CUP).

Scientific Organizing Committee

Abdul A. Alsabti (UK/Iraq), Julieta Fierro (Mexico), Michele Gerbaldi (France), Hans J. Haubold (Germany), John B. Hearnshaw (New Zealand, chair), Barrie W. Jones (UK), Ajit K. Kembhavi (India), Hugo Levato (Argentina), Peter Martinez (South Africa), Jayant V. Narlikar (India), Jay M. Pasachoff (USA), John R. Percy (Canada), Boonrucksar Soonthornthum (Thailand), A. Peter Willmore (UK), and James C. White (USA).

John Hearnshaw and Peter Martinez,
co-editors of Special Session 5,
Christchurch, NZ, and Cape Town, SA, November, 2006

1. Overview to astronomy in the developing world

1.1. *Introduction*

This introductory section to the Special Session on 'Astronomy for the developing world' takes a general overview of this theme. There are just three papers: that by Rajesh Kochhar takes a look at the historical and cultural antecedents to modern astronomical science, and the different perceptions of the origin of science seen in the East and the West. He makes an interesting closing comment on the gifting of scientific facilities to developing nations.

John Hearnshaw's paper is an analysis of the present state of astronomical activity in the world. He uses statistical data on author affiliations from ADS to measure activity in different countries by the number of papers published, and correlates this with other parameters, such as GDP per capita and number of IAU members.

Finally Jayant Narlikar makes a bold proposal for a Third-World Astronomy Institute or Network, in the hope that instituions can be established where astronomers from the developing world can visit and do research work.

1.2. *TWAN: a way of networking third-world astronomers*

Jayant V. Narlikar, Inter-University Centre for Astronomy and Astrophysics, Ganeshkind, Pune, India, email: jvn@iucaa.ernet.in

This talk will elaborate upon the concept of Third World Astronomy Network (TWAN) discussed at an earlier IAU meeting. TWAN is suggested as a way of improving the level of teaching, research and development in Astronomy and Astrophysics (A&A) in developing nations. Networking of astronomers in these countries with fast e-mail/internet connection and a few selected institutions serving as nodes are proposed. A range of activities may be carried out within the network. These are briefly outlined and the budgetary aspects described.

1.3. *A survey of published astronomical outputs of countries from 1976 to 2005 and the dependence of output on population, number of IAU members and gross domestic product*

John B. Hearnshaw, Department of Physics and Astronomy, University of Canterbury, Christchurch, New Zealand, email: john.hearnshaw@canterbury.ac.nz

In this paper I report the results of a survey of the astronomical outputs of all 63 IAU member countries as well as several non-member countries, based on an analysis of the affiliations of the authors given for nearly 900 thousand astronomical papers appearing in ADS between the years 1976 and 2005. The results show a roughly three-fold increase in the number of published papers per year over this 30-year interval. This increase is seen both in developed and also in most developing countries. The number of publications per IAU member correlates strongly with gross domestic product per capita. It is over 2 papers per IAU member per year in the countries with the strongest economies but less than 0.5 in the countries with low GDP per capita.

Since 2000 there has been a dramatic increase in the number of multi-author multinational papers published. This increase is especially noticeable for authors in developing countries, indicating that astronomers in these countries are increasingly participating in international collaborations for their research activities.

1.4. *Promoting astronomy in developing countries: an historical perspective*

Rajesh Kochhar, NISTADS, New Delhi, India, email: rkochhar2000@yahoo.com

Any international effort to promote astronomy world wide today must necessarily take into account its cultural and historical component. The past few decades have ushered in an age, which we may call the Age of Cultural Copernicanism. In analogy with the cosmological principle that the universe has no preferred location or direction, Cultural Copernicanism would imply that no cultural or geographical area, or ethnic or social group, can be deemed to constitute a superior entity or a benchmark for judging or evaluating others.

In this framework, astronomy (as well as science in general) is perceived as a multi-stage civilizational cumulus where each stage builds on the knowledge gained in the previous stages and in turn leads to the next. This framework however is a recent development. The 19th century historiography consciously projected modern science as a characteristic product of the Western civilization decoupled from and superior to its antecedents, with the implication that all material and ideological benefits arising from modern science were reserved for the West.

As a reaction to this, the orientalized East has often tended to view modern science as 'their' science, distance itself from its intellectual aspects, and seek to defend, protect and reinvent 'our' science and the alleged (anti-science) Eastern mode of thought. This defensive mind-set works against the propagation of modern astronomy in most of the non-Western countries. There is thus a need to construct a history of world astronomy that is truly universal and unselfconscious.

Similarly, the planetarium programs, for use the world over, should be culturally sensitive. The IAU can help produce cultural-specific modules. Equipped with this paradigmatic background, we can now address the question of actual means to be adopted for the task at hand. Astronomical activity requires a certain minimum level of industrial activity support. Long-term maintenance of astronomical equipment is not a trivial task. There are any number of examples of an expensive facility falling victim to AIDS: Astronomical Instrument Deficiency Syndrome. The facilities planned in different parts of the world should be commensurate with the absorbing power of the acceptor rather than the level of the gifter.

2. Astronomy in Latin America and in the Caribbean

2.1. *Introduction*

There were many participants at the Special Session 5 from Latin America and the Caribbean. Fourteen papers are summarized here, while others from this part of the world appear in section 7 on astronomy education.

Hugo Levato's paper provided an authoritative overview of astronomy in Latin America, and divided countries into three groups, depending on the level of astronomical education in each. The strong group (Mexico, Argentina, Chile, Brazil) all have well developed astronomical infrastructure. The weakest group have little or no astronomical activity.

Julieta Fierro and Patricia Rosenzweig discuss various aspects of astronomy education in respectively Mexico and Venezuela. Overviews of astronomy in Trinidad, Cuba and Colombia are given by Shirin Haque, Ramón Rodríguez and William Cepeda. Important public outreach initiatives in Chile and Cuba are reviewed by Antonieta García, G. Argandoña (both in Chile) and Oscar Álvarez. Summaries of some research facilities are given by José Ishitsuka and Erick Vidal (Peru) and Gonzalo Tancredi (Uruguay). The work of the Ibero-American Astronomy League (LIADA) is discussed by Paulo Bretones (Brazil).

There is an astonishing range of astronomical activity in Latin America, at all levels of development, and this diversity is reflected in the contributions within this section.

2.2. *Formal education in astronomy in Latin America*

Hugo Levato, CASLEO, Complejo Astronómico el Leoncito, San Juan, Argentina,
email: hlevato@casleo.gov.ar

I will make a summary of the formal programs for education in astronomy in different countries of Latin America. I will provide a list of three different groups of countries, one with those that have very well developed careers in astronomy, another group of countries that will need some efforts to consolidate incipient careers, and a third group that will need strong efforts to develop formal education in astronomy.

2.3. *Astronomy for teachers in Mexico*

Julieta Fierro, Institute of Astronomy, UNAM, Mexico City, Mexico,
email: fierroju@astroscu.unam.ac.mx

Mexico has added five more years of compulsory education to its national education system. In the past it only included six years of elementary (grammar) school. Now three years of pre-school (kindergarten) and three years of middle school are being implemented. At present an optional course on astronomy is offered in high school (pre-college). During my presentation I shall discuss problems concerning education in Mexico; these are mainly the lack of continuity in different levels of education, the lack of teacher training in science in general and the small number of topics in astronomy that are addressed. I shall mention support for teacher training and public education, which includes books, lectures and videos.

2.4. *Astronomy – the Caribbean view from the ground up*

Shirin Haque, Department of Physics, University of the West Indies, St Augustine,
Trinidad, email: shirin@tstt.net.tt

This presentation reviews the historical development of astronomy in the Caribbean within its cultural and environmental framework. The present status of astronomy in education, research and at the popular level will be presented as well. The focus will be on its development in the island of Trinidad and Tobago in particular. The presentation will review what works in small developing islands versus larger developed countries and the peculiar trials and tribulations of our circumstances as well as the rewards of such efforts. The critical role of students and volunteer effort will be highlighted. The psychological and cultural aspect of the human response to its development in the Caribbean will also be examined in the paper. Based on an examination of the impacting variables on its development, a proposal will be presented for the next 10-year development of this area in the Caribbean, with consideration of the importance of the development of the Third World Institute of Astronomy and the possibility of it being located in this region of the world.

2.5. *Encounters with science at ULA, Venezuela: an incentive for learning*

Patricia Rosenzweig, Universidad de Los Andes, Facultad de Ciencias, Mérida,
Venezuela, email: patricia@ula.ve

In the School of Science of the Universidad de Los Andes (ULA), in Mérida, Venezuela, a very successful event focused on high school students and primary school students, was founded in 2000. The name of this event is 'Encounters with Physics, Chemistry, Mathematics, and Biology' (hereinafter 'Encounters with Science'), and it integrates these disciplines as well as astronomy.

Its main purpose is that young minds can become familiar with the methods of science inquiry and reasoning, and can understand the concepts and processes of the sciences through thoroughly prepared experiences. This flourishing program is continuing to grow and to become strong. As a matter of fact, in its sixth edition (2005), the number of high and elementary school students coming from all over the country, has reached the outstanding number of nine thousand.

Among all the experiences that the students could be engaged in were many involving astronomy. These experiences were prepared by professors, together with graduate and undergraduate students, who are pursuing their degrees in all branches of science including astronomy. Although there is this incredible team of faculty members and graduate and undergraduate students working together, the target is the students of the high and elementary schools.

We certainly focus on engaging and encouraging students to experience scientific work first hand. Additionally, our professors have prepared excellent didactic material that can, together with hour-long lectures, prepare high school and elementary school students for a better understanding of science, particularly, helping in this way for a better education in astronomy.

The main event of the Encounters lasts five days in the School of Science of ULA, but subsidiary events are spread all over the year and around the country. As a successful program, it can be interesting to see if other countries can adopt this method to recruit or to trigger the interest of students to pursue their studies in the sciences.

2.6. *Developing astronomy in Cuba* *Ramón Rodríguez Taboada, Instituto de Geofísica y Astronomía, La Habana, Cuba, email:* ramone@infomed.sld.cu

Beginning from a brief historical introduction the present-day situation of astronomy in Cuba is presented and the topics relevant for astronomy development are analyzed from the view point of a person actually working in astrophysics. Arising from national needs, astronomical calculations is the only 'native-born' branch of astronomy in Cuba. Cuba was an observational platform capable to provide the Soviet Union with a 24-hour solar patrol, needed by its Space Agency System to protect men in orbit. This was the beginning of a very fruitful development of solar research in Cuba. Russia installed the instruments, trained the people to operate them, and gave the academic environment to develop the scientific work in solar physics, space weather, and related topics.

What about stellar astronomy? The Cuban astro-climate is not good to develop an observational base. We are trying to develop stellar astronomy in collaboration with institutions capable to provide both the academic and technical environment; but to continue developing stellar astronomy we need to influence public opinion and convince people they need groups working in astronomy. How do we do that? By publishing, giving conferences, talking about OUR work, not only like spectators of science. Showing science is the culture of modern times. Showing projects in astronomy can be cheap. This is very important! Astronomy is not a luxury.

As a real possibility, I consider the Virtual Observatory concept as the more appropriate in the near future, but it is necessary to have an internet connectivity level that is not commonly provided in Cuba, and to train the people.

Concluding remarks: From my experience 'engagement' is the key word for astronomy development in developing countries. Astronomy can not be developed without an appropriate academic environment, and we do not have not this at present. It is not 'only' about financial resources, it is also about 'real collaboration' with a mature partner and common research goals.

2.7. *Planetario Habana: a cultural centre for science and technology in a developing nation*

Oscar Álvarez, Ministry of Science, Technology and the Environment (CITMA), Havana, Cuba, email: oscar@citma.cu

Astronomical education in Cuba is not widespread in the educational system; nevertheless the public interest in sciences in general but particularly in astronomy issues is very high, as it has become reflected by the attention paid to educational and scientific program broadcasts in the national television channels. The 'Planetario Habana' Cultural Centre for Science and Technology, which is under construction, is aimed at guiding the interest towards basic sciences and astronomical formation of the people, in the most populated and frequented area of the country. A key objective of this project shall be serving as an instructive motivation and entertainment for the casual or habitual visitors to these facilities, offering them the possibility to enjoy vivid representations, play with interactive amusement equipment and listen to instructive presentations on astronomy and related sciences, all guided by qualified specialists.

Another fundamental purpose shall be the establishment of a plan for complementary education in coordination with schools, in order to allow children and young people to participate in activities enabling them to get into the fascinating world of astronomy, exploration of outer space and life as a cosmic phenomenon.

The setting up of the Planetario Habana Cultural Centre for Science and Technology is under the general administration of the Office of the Historian of the City of Havana, and methodologically is being led by the Ministry of Science, Technology and the Environment, and will show in operation the GOTO Planetarium G Cuba custom, obtained under a Japanese Cultural Grant Aid. It will develop into a an unparalleled centre in the national environment for scientific outreach and education of these sciences.

Surrounded by the attractiveness of the colonial 'ambience', it shall become a centre for dissemination of information about new discoveries and scientific programs developed at national and international level. Here we present a general view of the project, and its present and future development.

2.8. *Astronomy in Colombia*

William Cepeda-Peña, Observatorio Astronómico Nacional, Universidad Nacional, Bogotá, Colombia, email: wecepedap@unal.edu.co

Astronomy in Colombia has been done since the beginning of the nineteenth century, when in 1803 one of the oldest (or may be the oldest) astronomical observatories of America was built. This is a very beautiful, historical and ancient building. A small dome with a small telescope is also on the university campus.

The observatory has led, since then, the development of astronomy in Colombia as a professional science. At the present time a Master's program and a specialization program are successfully carried out with a good number of students. The observatory has a staff of eleven professors, all with a master's degree in the sciences; two of them also have a PhD, and in a couple of years, five staff members will have a PhD in physics. With some international collaboration, they will undertake in a few years a doctoral astronomical program.

There are several research lines, mainly in the fields of astrometry, galactic and extragalactic astronomy, cosmology, astro-statistics and astro-biology. Three research groups have got recognition from the governmental institution that supports research in the sciences, COLCIENCIAS. Several papers have been published in national and international journals. Besides the professional activities in astronomy, the observatory also sponsors several non-professional Colombian astronomical groups that work enthusiastically in the field of astronomy.

2.9. *Astronomy education and popularization facilities at Guanajuato University in Mexico*

Hector Bravo-Alfaro, K.-P. Schroeder, and L. Ramirez, Universidad de Guanajuato, Guanajuato, Mexico, email: hector@astro.ugto.mx

At the Astronomy Department of Universidad de Guanajuato, 400 km NW of Mexico City, nine professional astronomers do research and teaching at both graduate and undergraduate level. In addition, in the last few years, this group has carried out astronomy popularization activities at three different sites. First, there is a rudimentary observatory named 'La Azotea' (the roof) on the top of the main building of the University (at Guanajuato centre), which includes a 16-cm refractor in a dome, a couple of XIXth century astronomical instruments, and a classroom with capacity for 50 people. The refractor was out of use for about twelve years, but will be fully operational before summer 2006.

Second, the 'Observatorio de La Luz', 20 km away from Guanajuato centre, includes a professional 0.6 m Cassegrain and a 2 m radio telescope, with a 21 cm receiver. Finally, on the roof of the Astronomy Department headquarters, an optical 0.4 m Dobsonian is available. We also have internet connections everywhere and six portable 8-inch telescopes (two at each site), devoted to regular astronomical observations for the general public, specially for scholars. Numerous repair works are currently carried out on the building of 'La Azotea', and recently a project to establish there a Centre for Popularization of Astronomy has been approved by the Regional Science Council.

The main activities, some of them currently developed at these sites are: (1) a permanent program of astronomical observations for a wide audience; (2) training in observational astronomy for physics undergraduate students; (3) regular talks on astronomy and other science domains; (4) summer schools in astronomy for elementary and high-school teachers; and (5) in the near future, the foundation of an amateur society of astronomy.

2.10. *Implementing an education and outreach program for the Gemini Observatory in Chile*

M. Antonieta Garcia, Gemini Observatory, La Serena, Chile, email: agarcia@gemini.edu

Beginning in 2001, the Gemini Observatory began the development of an innovative and aggressive education and outreach program at its Southern Hemisphere site in northern Chile. A principal focus of this effort is centered on local education and outreach to communities surrounding the observatory and its base facility in La Serena Chile. Programs are now established with local schools using two portable StarLab planetaria, an internet-based teacher exchange called StarTeachers and multiple partnerships with local educational institutions. Other elements include a CD-ROM-based virtual tour that allows students, teachers and the public to experience the observatory's sites in Chile and Hawaii. This virtual environment allows interaction using a variety of immersive scenarios such as a simulated observation using real data from Gemini. Pilot projects like 'Live from Gemini' are currently being developed which use internet video-conferencing technologies to bring the observatory's facilities into classrooms at universities and remote institutions. Lessons learned from the implementation of these and other programs will be introduced and the challenges of developing educational programming in a developing country will be shared.

2.11. *ESO strategy to promote astronomy and science culture in Chile*

G. Argandoña, European Southern Observatory, Santiago, Chile,
email: gargando@eso.org

With three astronomical sites operating in Chile (La Silla, Paranal and Chajnantor), the European Organization for Astronomical Research in the Southern Hemisphere, ESO, has developed multiple approaches to foster astronomy and science culture in the country, implemented both nationally and locally. At the national level, an annual fund has been established to provide grants for individual Chilean scientists, research infrastructures, scientific congresses, workshops for science teachers and astronomy outreach programmes for the public. This has been complemented by multiple partnerships and formal collaborations with relevant bodies, like the Chilean Ministry of Education, the National Science and Technology Commission (CONICYT), science museums and the national mass media.

At the local level, the education and outreach program includes traditional public visits to the observatories, support to science teaching at local schools, promotion of astronomy clubs, organization of a mobile observatory, among other activities. An overview of these national and local projects will be given, along with a review of the development of Chilean astronomy in recent years, including the latest statistics on the number of professional astronomers, science productivity and the percentage of access to international observatories by the Chilean astronomical community.

2.12. *Projects of the Ibero-American League for Astronomy (LIADA) teaching and popularization section*

Paulo S. Bretones and V.C. Oliveira, Instituto de Geociências, Universidade Estadual de Campinas, Campinas, Brazil; and Instituto Superior de Ciências Aplicadas,
email: bretones@mpc.com.br

The goal of this work is to present an analysis of the projects developed by the Teaching and Popularization Section of the Liga Ibero-Americana de Astronomía (LIADA). We first present a brief outline of the LIADA and its objectives, with emphasis in the attempts to organize, conduct and stimulate the collaboration between the professional and amateur astronomers in Latin-America. The Section is based in Brazil and counts with the support of 16 coordinators from most Latin-American countries. Presently several projects are being developed, such as astronomy in schools, oppositions, conjuctions, eclipses and transits.

The home page (<http://www.iscafaculdades.com.br/liada>) of the section communicates the observational projects and this aims to attract the attention of the general public, teachers and students to encourage the observation. The participants gather astronomical data at the home page, supplemented by the aid of local coordinators and spontaneous collaborators. The strategy is to circulate important support material and open a discussion forum about each of the observed phenomena, so as to enhance their public consideration and visibility.

We have analyzed the records and present an evaluation of the projects executed jointly with other institutions and individuals, their importance for scientific education, the nomination and relationship with the coordinators, the difficulties with written reports, the need for a dynamical maintenance of the home page, the question of the language, the establishment of a useful communications network and the visibility of the LIADA activities.

We conclude with a critical assessment of these activities, their strengths and weaknesses, as observed by us, and future projects of astronomy education.

2.13. *Activities of the Observatorio Astronómico Los Molinos, Uruguay*

Gonzalo Tancredi, S. Roland, R. Salvo, F. Benitez, A. Ceretta, and E. Acosta, Observatorio Astronómico Los Molinos, Montevideo, Uruguay,
email: gonzalo@fisica.edu.uy

The Observatorio Astronómico Los Molinos (OALM) is the only professional observatory in Uruguay and among the few observatories in the Southern Hemisphere mainly dedicated to the observations of asteroids and comets. At present we have the following observing programmes: (1) Confirmation and follow-up of recently discovered Near-Earth Objects (NEOs); (2) Photometric and astrometric follow-up of Comets to determine the perihelion light-curve; (3) Search for NEOs in the direction of the radiants of their orbits during twilight. The surveys are performed in regions of the sky where we expect a higher probability to find NEOs; and (4) Observations of asteroids in cometary orbits (ACOs) to detect the existence of a possible residual activity.

Though we have very modest equipment (a wide field 46 cm telescope), we have been able to make a relevant contribution in this field of research. The number of astrometric and photometric reports to the IAU-Minor Planet Center has increased a lot in the last year. We will present some results of this successful strategy to concentrate our research in a field where we can make a contribution at an international level, even with modest equipment.

In addition to these research programmes, we are conducting a very intense outreach activity that will be described.

2.14. *A new astronomical facility for Peru: transforming a telecommunications 32-metre parabolic antenna into a radio-telescope*

José Ishitsuka[1], *M. Ishitsuka*[2], *M. Inoue*[1], *N.Kaifu*[1], *S. Miyama*[1], *M. Tsuboi*[3], *M. Ohishi*[1], *K. Fujisawa*[4], *T. Kasuga*[5], *T. Kondo*[6], *S. Horiuchi*[7], *T. Umemoto*[3], *M. Miyoshi*[1], *K. Miyazawa*[1], *T. Bushimata*[1], *and E.D. Vidal*[2],
[1]*National Astronomical Observatory of Japan, Tokyo, Japan;* [2]*Instituto de Geofisico del Peru, Lima, Per;* [3]*Nobeyama Radio Observatory, Nobeyama, Japan;* [4]*Yamaguchi University, Yamaguchi, Japan;* [5]*Hosei University, Tokyo, Japan;* [6]*National Institute of Information and Communications Technology, Kashima, Japan;* [7]*Swinburne University of Technology, Victoria, Australia, email:* pepe@hotaka.mtk.nao.ac.jp

In 1984 Nippon Electric Company constructed an INTELSAT antenna at 3 370 meters above the sea level on the Peruvian Andes. Entel Peru, the Peruvian telecommunications company, managed the antenna station until 1993. This year the government transferred the station to a private telecommunications company, Telefónica del Peru. Since the satellite communications were rapidly replaced by transoceanic fiber optics, the beautiful 32-metre parabolic antenna has been unused since 2002. In cooperation with the National Astronomical Observatory of Japan we began to convert the antenna into a radio-telescope.

Because researchers on the interstellar medium around Young Stellar Objects (YSO) will be able to observe the methanol masers that emit at 6.7 GHz, initially we will monitor the 6.7 GHz methanol masers and survey the southern sky. An ambient temperature receiver with $T_{\rm rx} = 60\,$K was developed at Nobeyama Radio Observatory and is ready to be installed. The antenna control system is the Field System FS9 software installed in a Linux PC. An interface between the antenna and the PC was developed at Kashima Space Research Center in Japan. In the near future we plan to install 2 GHz, 8 GHz, 12 GHz and 22 GHz receivers.

The unique location and altitude of the Peruvian Radio Observatory will be useful for VLBI observations in collaboration with global arrays such as the VLBA array for astronomical observation and geodetic measurements. For Peru, where few or almost no astronomical observational instruments are available for research, the implementation of the first radio observatory is a big and challenging step, and it will foster science at graduate and postgraduate levels in our universities.

Worldwide, telecommunications antennas are possibly unused, and with relatively modest investments they could be transformed into useful observational instruments.

2.15. *Application of Field System-FS9 and a PC to the Antenna Control Unit interface in Radio Astronomy in Peru*

Erick Vidal[1], *J. Ishitsuka*[2] *and K.Y. Koyama*[3],
[1]*Instituto Geofisico del Peru, Lima, Peru;* [2]*National Astronomical Observatory of Japan, Tokyo, Japan;* [3]*National Institute of Information and Communications Technology, Kashima, Japan, email:* evidal@axil.igp.gob.pe

We are in the process to transform a 32-m antenna in Peru, used for telecommunications, into a radio-telescope to perform radio-astronomy in Peru. The 32-m antenna of Peru constructed by NEC was used for telecommunications with communications satellites at 6 GHz for transmission, and 4 GHz for reception. In collaboration with the National Institute of Information and Communications Technology (NICT) Japan, and National Observatory of Japan we have developed an antenna control system for the 32-m antenna in Peru. It is based on the Field System FS9, software released by NASA for VLBI station, and an interface to link a PC with FS9 software (PC-FS9) and the Antenna Control Unit (ACU) of the 32-meter antenna. The PC-FS9 controls the antenna, commands are translated by the interface into control signals compatible with the ACU using an I/O digital card with two 20-bit ports to read azimuth and elevation angles, one 16-bit port for reading the status of the ACU, one 24-bit port to send pulses to start or stop operations of the antenna, while two channels are analogue outputs to drive the azimuth and elevation motors of the antenna, an LCD display to show the status of the interface and error messages, and one serial port is used for communications with PC-FS9.

The first experiment of the control system was made with the 11-m parabolic antenna of Kashima Space Research Center (NICT), where we tested the correct working of the routines implemented for the FS9 software, and simulations were made with looped data between output and input of the interface. Both tests were done successfully. With this scientific instrument we will be able to contribute to research in astrophysics. We expect in the near future to work at 6.7 GHz, to study methanol masers, and higher frequencies with some improvements of the surface of the dish.

3. Astronomy in Africa

3.1. *Introduction*

By any measure that one cares to define, Africa lags significantly behind the rest of the world in the development of astronomy. There are currently 143 members of the IAU based in the 52 countries that comprise Africa and the Independent Island States. Thus it is important that a section of the Special Session on ‘Astronomy for the Developing World’ is devoted to the African continent.

Six papers are presented in this section. The first is by Peter Martinez, who presents an overview of the status of professional astronomy on the continent. He notes that the advent of large-scale facilities for ground-based astronomy in southern Africa presents a unique opportunity to promote astronomy throughout the continent. He also presents some lessons learnt from various capacity-building activities.

Paul Baki argues that astronomy can be used to teach basic skills of scientific reasoning. He suggests that by incorporating traditional African beliefs about the night skies into astronomy curricula, astronomy may be perceived less as a 'foreign' subject by the people of Africa. Baki also discusses how to create an enabling environment for astronomy in Africa. The theme of astronomy education is further developed by Hassane Darhmaoui and K. Loudiyi, who discuss astronomy education in Morocco. They discuss their programmes, successes and challenges.

Africa has some of the best astronomical sites in the world. The last three papers in this section focus on site selection issues. The Moroccan Atlas Mountains have been identified as one of five potential locations for the Extremely Large Telescope (ELT). Zouhair Benkhaldoun describes the ELT site prospecting programme in Morocco. The determination of astronomical extinction and aerosol content at two potential ELT sites in Morocco is discussed by E.A. Siher, A Bounhir and Z. Benkhaldoun.

3.2. *Capacity building for astronomy education and research in Africa*

Peter Martinez, South African Astronomical Observatory, Cape Town, South Africa,
email: peter@saao.ac.za

About 1.5 % of the world's professional astronomers are based Africa, yet in terms of research output, African astronomers produce less than 1 % of the world's astronomical research. The advent of new large-scale facilities such as SALT and HESS provides African astronomers with tools to pursue their research on the continent. Such facilities also provide unprecedented training opportunities for the next generation of African astronomers. This paper discusses recent efforts to develop astronomy education and research capacity on the continent. Various capacity-building initiatives are discussed, as well as the lessons learnt from those initiatives.

3.3. *Spreading astronomy education throughout Africa*

Paul Baki, Department of Physics, University of Nairobi, Kenya,
email: pbaki@uonbi.ac.ke

Although Astronomy has been an important vehicle for effectively passing on a wide range of scientific knowledge, for teaching the basic skills of scientific reasoning and for communicating the excitement of science to the public, its inclusion in the teaching curricula of most institutions of higher learning in Africa is rare. This is partly due to the fact that astronomy appears to be only good at fascinating people but not at providing paid jobs and also due to the lack of trained instructors, teaching materials and a clear vision of the role of astronomy and basic space science within the broader context of education in the physical and applied sciences. In this paper we survey some of the problems impeding the spread of astronomy in Africa and discuss some interdisciplinary traditional weather indicators. These indicators have been used over many years to monitor the appearance of constellations, such as Orion, and are closely intertwined with the indigenous cultures of some African societies. They could be incorporated into the standard astronomy curriculum as a way of making the subject more 'home grown' and to be able to reach out to the wider populace in popularizing astronomy and basic sciences. We also discuss some of the other measures that ought to be taken so as to create an enabling environment for sustainable teaching and the spread of astronomy throughout Africa.

3.4. *Astronomy education in Morocco – a new project for implementing astronomy in high schools*

Hassane Darhmaoui and K. Loudiyi, Al Akhawayn University in Ifrane, School of Science and Engineering, Ifrane, Morocco, email: H.Darhmaoui@aui.ma

Astronomy education in Morocco, like in many developing countries, is not well developed and lacks the very basics in terms of resources, facilities and research. In 2004, the International Astronomical Union (IAU) signed an agreement of collaboration with Al Akhawayn University in Ifrane to support the continued, long-term development of astronomy and astrophysics in Morocco. This is within the IAU program 'Teaching for Astronomy Development' (TAD). The initial focus of the program concentrated exclusively on the University's Bachelor of Science degree program. Within this program, and during two years, we were successful in providing adequate astronomy training to our physics faculty and a few of our engineering students. We also offered our students and community general astronomy background through courses, invited talks and extra-curricular activities. The project is now evolving towards a wider scope and seeks promoting astronomy education at the high-school level. It is based on modules from the Hands on Universe (HOU) interactive astronomy programme. Moroccan students will engage in doing observational astronomy from their PCs. They will have access to a world-wide network of telescopes and will interact with their peers abroad. Through implementing astronomy education at this lower age, we foresee an increasing interest among our youth, not only in astronomy but also in physics, mathematics, and technology. The limited astronomy resources, the lack of teachers' experience in the field and the language barrier are amongst the difficulties that we will be facing in achieving the objectives of this new programme.

3.5. *ELT site prospecting in the Moroccan Atlas Mountains*

Zouhair Benkhaldoun, University Cadi Ayyad, Laboratoire de Physiques des Hautes Energies et Astrophysique, Marrakech, Morocco, email: zouhair@ucam.ac.ma

The Extremely Large Telescope site testing working group has selected Morocco's mountains, as one of five locations over the world, to test for this European project. For that purpose, we first of all carried out a selection of two sites basing on their location relative to the dominant wind flow, the cloud cover and the circulation of the Saharan aerosols. We will detail in the communication which we present here, the methodology followed and results obtained. We also present the localizations of both sites with a cartographic, and geological study and some seismic information. The first measurements of the seeing will be also presented.

3.6. *Astronomical extinction over the ELT Moroccan sites from aerosol satellite data*

El Arbi Siher[1], *Z. Benkhaldoun*[2], *and A. Bounhir*[2], [1]*Faculté des Sciences et Techniques, Beni Mellal, Morocco;* [2]*Laboratoire de Physiques des Hautes Energies et Astrophysique, Faculté des Sciences, Semlalia, Marrakech, Morocco; email:* siher@ucam.ac.ma

Two Moroccan sites have been selected to be characterized as ELT candidate sites. These sites are in the Atlas Mountains, between Oukaimeden (where the national observatory is located) and the Canary Islands. For a preliminary study, we will use the TOMS/Nimbus7 aerosol index (AI), threshold 0.7, to extract the astronomical extinction (AE), threshold 0.2 mag/airmass. In fact, on the one hand, one previous work showed the link between these parameters over the Canary Islands (ORM Observatory). On the other hand, many studies proposed the dust characterization for any future extremely large telescope as a mandatory qualification.

3.7. *Effect of altitude on aerosol optical properties*

Aziza Bounhir and Zouhair Benkhaldoun, UCAM University, Marrakech, Morocco email: bounhir@fstg-marrakech.ac.ma

The ELT project is currently underway in Europe and North America. Astronomical sites critically depend on sky transparency and then on aerosol loadings. A quantitative survey of aerosol optical properties at candidate sites is an essential part of the site selection process. There are basically two methods to scan and characterize aerosol properties: ground based measurements and satellite measurements. In this paper we will establish a full climatology of two sites very close to each other, but with a difference of 2300 m in altitude. They are Izaña and Santa-Cruz, located in the Canary Islands. Both have sun photometers from the AERONET Network. AERONET provides a set of aerosol optical properties: atmospheric optical thickness, aerosol optical thickness, angstrom parameter, aerosol size distribution, aerosol refractive index, single scattering albedo, water vapour content, phase function, direct sun radiance and sky radiance. We also use satellite data from TOMS to determine the aerosol index.

The aim of this work is to see how these properties change with altitude. We establish a correlation between the TOMS index and the aerosol optical thickness at both sites. Aerosol optical properties show very good correlation between Izaña and Santa-Cruz. As a result, we establish a set of relationships helpful to characterize sites at high altitude from the data of a neighbouring site. In the ELT site selection and evaluation process, a preliminary study from satellite measurements and from AERONET neighbouring sites is very important.

4. Astronomy in eastern Asia and the Pacific

4.1. *Introduction*

This section presents several papers from eastern Asia and the Pacific. Boonrucksar Soonthornthum from Thailand reviews astronomy in his region. In fact two other papers, by Busaba Kramer, are also from Thailand, which reflects the strong development of astronomy in that country at the present time. (Thailand has just joined the IAU and founded the National Astronomical Research Institute of Thailand, of which Boonrucksar is the inaugural director.)

There are also papers by Nguyen Quynh Lan on astronomy in Vietnam, Osamu Hashimoto on the collaboration between ITB Bandung Indonesia and Gunma in Japan, and Sergei Gulyaev on the nascent development of radio astronomy in New Zealand.

4.2. *Astronomy in Asia*

Boonrucksar Soonthornthum, National Astronomy Research Institute of Thailand, email: boonraks@chiangmai.ac.th

Astronomy in Asia has continuously developed. Local wisdom in many Asian countries reflects their interest in astronomy since the historical period. However, the astronomical development in each country is different which depends on their cultures, politics and economics. Astronomy in some Asian developing countries such as China and India are well-developed, while some other countries, especially in south-east Asia, with support such as new telescopes, training, experts etc., from developed countries, are trying to promote relevant research in astronomy as well as to use astronomy as a tool to promote scientific awareness and understanding for the public. Recently, a new national research institute in astronomy with a 2.4-metre reflecting telescope has been established in Thailand. One of the major objectives of this research-emphasis institute would to aim at a collaborative network among south-east Asian countries, so as to be able to contribute new knowledge and research to the astronomical community.

4.3. *Astronomy in Thailand*

Busaba Hutawarakorn Kramer, National Astronomical Research Institute, Chiang Mai, Thailand, email: busaba@nari.or.th

During the last few years, Thailand has seen a significant change in the way astronomical research and education are pursued in the country. The government has approved the establishment of the National Astronomical Research Institute (NARI) which aims to develop not only astronomical research but also astronomy education at all levels, both in formal and informal education. A framework of national key projects exists which includes national facilities, national collaborative research networks, teacher training and public outreach programmes. Examples of these programmes will be presented in this talk.

4.4. *Astronomy development in Thailand: Roles of NARI*

Busaba Hutawarakorn Kramer, Boonrucksar Soonthornthum, and S. Poshyachinda, National Astronomical Research Institute, Chiang Mai, Thailand, email: busaba@nari.or.th

The development of astronomy in Thailand has improved significantly during the last few years. The government has approved the establishment of the National Astronomical Research Institute (NARI). The roles of NARI in the development of astronomical research and astronomy education in Thailand include a national framework, national facilities, collaborative research networks, teacher training and public outreach programmes. The new 2.4-metre reflecting telescope will serve not only the astronomical community in Thailand, but also in Southeast Asia.

4.5. *Astronomy in Vietnam*

Nguyen Quynh Lan, Hanoi University of Education, Hanoi, Vietnam, email: nquynhlan@dhsphn.edu.vn

We overview in this paper the development of astronomical education in Vietnam. We also discuss proposals to advance the development of astronomy in Vietnam, with support and assistance from the International Astronomical Union.

4.6. *Collaboration and development of radio-astronomy in Australasia and the South-Pacific region: New Zealand perspectives*

Sergei Gulyaev and Tim Natusch, Centre for Radiophysics and Space Research, Auckland University of technology, New Zealand, email: sergei.gulyaev@aut.ac.nz

Radio telescopes in the Asia-Pacific region form a natural network for VLBI observations, similar to the very successful networks in North America (Network Users Group) and Europe (European VLBI Network). New Zealand's VLBI facility, which we have been developing since 2005, has the potential to strengthen the Asian-Pacific VLBI network and its role in astronomy, geodesy and geoscience. It will positively influence regional and international activities in geoscience and geodesy that advance New Zealand's national interests.

A self-contained radio astronomy system for VLBI, including a 1.658 GHz (centre frequency), 16 MHz bandwidth RF system (feed and down-conversion system locked to a rubidium maser and GPS clock), an 8-bit sampler/digitization system, and a disk-based recording system built around a commodity PC, was developed in New Zealand at the Centre for Radiophysics and Space Research. This was designed as a portable system for use on various radio telescopes. A number of trans-Tasman tests has been conducted in

2005 – 2006 between the CRSR system installed on a 6-metre dish located in Auckland and the Australia Telescope Compact Array in Narrabri, Australia. This work has been successful, with fringes located from the recorded data and a high resolution image of the quasar PKS1921−231 was obtained.

Experiments were recently conducted with Japan and new tests are planned with Korea and Fiji. Plans have been made to build a new 16.5-m antenna in New Zealand's North Island and to upgrade an 11-m dish in the South Island. A possible future of New Zealand's participation in the SKA is being discussed.

4.7. *Mutual collaboration between the Institute of Technology Bandung (ITB), Indonesia and the Gunma Astronomical Observatory (GAO), Japan*

Osamu Hashimoto[1], *Hakim Malasan*[2], *H. Taguchi*[1], *K. Kinugasa*[1], *B. Dermawan*[2], *B. Indradjaja*[2], *and Y. Kozai*[1]

[1]*Gunma Astronomical Observatory, Gunma, Japan;* [2]*Institute of Technology Bandung, Bandung, Indonesia, email;* osamu@astron.pref.gunma.jp

The Institute of Technology Bandung (ITB), Indonesia, and the Gunma Astronomical Observatory (GAO), Japan, have been proceeding with several programs of mutual collaboration in the fields of astronomical research and education since 2002. ITB with Bosschca observatory has a great interest in education of astronomy for the general public, as well as in university education and research of their own. GAO is a public observatory operated by Gunma Prefecture (local government) and is equipped with a 150-cm reflector and some smaller telescopes, which are capable of scientific research of high grade.

We will report some of our cooperative activities, including the remote accessing of the telescopes of each observatory by each other, which can provide opportunities for astronomical experiences in the opposite hemisphere for various people of each country. Some scientific collaboration programs, such as common instruments and data analysis systems, which have been developed on both sites, will be also presented.

5. Astronomy in the Middle East and central Asia

5.1. *Introduction*

This section on astronomy in central Asia was one of the most compelling and interesting sessions of the SpS 5 meeting. Once again it is a very diverse region, stretching from the Ukraine to Mongolia, and including several countries of the former Soviet Union, such as Uzbekistan and Armenia.

Especially pertinent to the present political situation was the account by Athem Alsabti of the struggle to develop astronomy again in Iraq after two major conflicts in recent years. The initiative to rebuild the Mt Korek Observatory in the northern Kurdish region of Iraq is an initiative he is leading. The talk by Yousef Sobouti on astronomy in Iran and the proposed founding of a new astronomical observatory there was of great interest. Unfortunately Professor Sobouti was unable to come to Prague and Edward Guinan (Villanova University, USA) gave this talk for him. Sona Hosseini's paper gives information on site-testing for the Iranian observatory.

There are papers on astronomy in Uzbekistan (theoretical astrophysics from Bobomurat Ahmedov) and the new Suffa radio-telescope project from G.I. Shanin. Batmunkh Damdin discusses astronomy in Mongolia, a country that has just joined the IAU. The outstanding meterological conditions there make it a prime country for the future development of optical astronomy.

Finally interesting papers by Nikolai Bochkarev (Russia), Svetlana Kolomiyets (Ukraine) and Hayk Harutyunian (Armenia) discuss astronomy in the former Soviet states. Bochkarev gives a wide-ranging review of astronomy in many former Soviet territories, not just Russia.

5.2. *Astronomy in Iraq*

Athem Alsabti, University College London, UK, email: a.alsabti@ucl.ac.uk

The history of modern Iraqi astronomy is reviewed. During the early 1970s Iraqi astronomy witnessed significant growth through the introduction of the subject at university level and extensively within the school curriculum. In addition, astronomy was popularized in the media, a large planetarium was built in Baghdad, plus a smaller one in Basra. Late 1970 witnessed the construction of the Iraqi National Observatory at Mount Korek in Iraqi Kurdistan. The core facilities of the Observatory included 3.5-metre and 1.25-metre optical telescopes, and a 30-metre radio telescope for millimetre wavelength astronomy. The Iraqi Astronomical Society was founded and Iraq joined the IAU in 1976.

During the regime of Saddam Hussain in the 1980s, the Observatory was attacked by Iranian artillery during the Iraq-Iran war, and then again during the second Gulf war by the US air force. Years of sanctions during the 1990s left Iraq cut off from the rest of the international scientific community. Subscriptions to astronomical journals were halted and travel to conferences abroad was virtually non-existent. Most senior astronomers left the country for one reason or another. Support from expatriate Iraqi astronomers existed (and still exists); however, this is not sufficient. Recent changes in Iraq, and the fall of Saddam's regime, have meant that scientific communication with the outside world has resumed to a limited degree.

The Ministry of Higher Education in Baghdad, Baghdad University and the Iraqi National Academy of Science, have all played active roles in re-establishing Iraqi astronomy and re-building the damaged Observatory at Mount Korek. More importantly, the University of Sallahudin in Erbil, capital of Iraqi Kurdistan, has taken particular interest in astronomy and the observatory. Organized visits to the universities, and also to the observatory, have given us a first-hand assessment of the scale of the damage to the observatory, as well as the needs of astronomy teaching and research. Joint supervision for postgraduate level research was organized between local and expatriate Iraqi astronomers. The IAU was among the first international organizations to offer assistance. Many observatories worldwide have also given support. Plans will be proposed for re-building the observatory, supporting teaching and research, and establishing an institute for astronomy in Erbil, together with further suggestions on how the international astronomical community can assist Iraqi astronomers.

5.3. *Astronomy in Iran* *Yousef Sobouti, Institute for Advanced Studies in Basic Sciences, Zanjan, Iran, email:* sobouti@iasbs.ac.ir

In spite of her renowned pivotal role in the advancement of astronomy on the world scale during 9th to 15th centuries, Iran's rekindled interest in modern astronomy is a recent happening. Serious attempts to introduce astronomy into university curricula and to develop it into a respectable and worthwhile field of research began in the mid 1960s. The pioneer was Shiraz University. It should be credited for the first few dozens of astronomy- and astrophysics-related research papers in international journals, for training the first half a dozen of professional astronomers and for creating the Biruni Observatory. Here, I take this opportunity to acknowledge the valuable advice of Bob Koch and Ed

Guinan, then of the University of Pennsylvania, in the course of the establishment of this observatory.

At present the astronomical community of Iran consists of about 65 professionals, half university faculty members and half MS and PhD students. The yearly scientific contribution of its members has, in the past three years, averaged about 15 papers in reputable international journals, and presently has a healthy growth rate. Among the existing observational facilities, Biruni Observatory with its 51-cm Cassegrain, CCD cameras, photometers and other smaller educational telescopes, is by far the most active place. Tusi Observatory of Tabriz University has 60- and 40-cm Cassegrains, and a small solar telescope. A number of smaller observing facilities exist in Meshed, Zanjan, Tehran, Babol and other places.

The Astronomical Society of Iran (ASI), though some 30 years old, has expanded and institutionalized its activities since the early 1990s. ASI sets up seasonal schools for novices, organizes annual colloquia and seminars for professionals and supports a huge body of amateur astronomers from among high school and university students. Over twenty ASI members are also members of the IAU and take an active part in its events.

In the past five years, astronomers of Iran have staged an intensive campaign to have a National Observatory of their own (NOI). Initial planning is for one 2-m telescope and appropriate measuring devices. The project is approved and will be funded by the government in the course of five years. The site selection for NOI, however, is already in its third year and has been and is being generously funded by the government.

Last, but not least, Nojum, the only astronomical monthly magazine of the Middle East, is presently in its fifteenth year. It has a good readership among both professionals and amateurs of Farsi speaking communities within the country and abroad.

5.4. *Measurement of the light pollution at the Iranian National Observatory*

S. Sona Hosseini[1] *and S. Nasiri*[2]

[1]*Zanjan University, Zanjan, Iran;* [2]*Institute of Advanced Studies for Basic Sciences, Zanjan, Iran; email:* s.sona.h@gmail.com

The problem of light pollution became important mainly since 1960, by the growth of urban development and using more artificial lights and lamps at nighttime. Optical telescopes share the same range of wavelengths as are used to provide illumination of roadways, buildings and automobiles. The light glow that emanates from man-made pollution will scatter off the atmosphere and affects the images taken by the observatory instruments. A method of estimating the night sky brightness produced by a city of known population and distance is useful in site testing of new observatories, as well as in studying the likely future deterioration of existing sites.

Now under planning, the Iranian National Observatory will house a 2-metre telescope and the project of site selection is underway. Hence studying the light pollution is being carried out in Iran. Thus, we need a site with the least light pollution, beside other parameters, i.e. seeing, meteorological, geophysical and local parameters. The seeing parameter is being measured in our four preliminary selected sites at Qom, Kashan, Kerman and Birjand, since two years ago, using an out of focus Differential Image Motion Monitor. These sites are selected among 33 candidate sites by studying the meteorological data obtained from the local synoptic stations and the Meteosat. We use Walker's law to estimate the sky glow of these sites having the population and the distances of the nearby regions. The results are corrected by the methods introduced by Treanor and Berry using the atmospheric extinction coefficients. The data obtained using an 11-inch telescope with an ST7 CCD camera for the above sites are consistent with the estimated values of the light pollution.

5.5. *Astronomy in the former Soviet territory: fifteen years after the USSR disintegration*

Nikolai G. Bochkarev, Sternberg Astronomical Institute, Moscow, Russia, email: boch@sai.msu.ru

During the post-Soviet period, the main infrastructure of astronomy over the territory of FSU was kept intact, in spite of a dramatic decrease of financial support. The overall situation in FSU astronomy is stable. In Latvia, the 32-m radio-dish has been put into working order and this allows its joining VLBI programs. It has been handed over to the Venspils University. In Russia, all the three 32-metre radio dishes of the QUASAR VLBI system have been put into operation, as well as the 2-m telescope with a high-resolution spectrograph (up to resolution $R \simeq 500\,000$) and the horizontal solar telescope ($R = 320\,000$) of the Russian-Ukrainian Observatory on Peak Terskol (Caucasus, altitude 3100 m). But the situation with the observatory is worrying, because of the regional authorities' attempt to privatize its infrastructure. The process of equipping a number of CIS (including Russian) observatories with CCD-cameras is in progress.

To solve the staff problems, Kazakhstan, Tajikistan and Uzbekistan have begun to prepare national specialists in astronomy and the Baltic States, Armenia, Azerbaijan, Georgia, Russia, and the Ukraine all continue to train astronomers.

The teaching of astronomy at schools is obligatory only in the Ukraine and the Baltic countries. To maintain a 'common astronomical space' the Eurasian Astronomical Society (EAAS) continues the program of reduced-price subscription to Russian-language astronomical journals and magazines over the territory of the FSU, and the organization of international conferences and Olympiads for school students, lectures for school teachers and planetarium lecturers, etc.

5.6. *Relativistic astrophysics and cosmology in Uzbekistan*

Bobomurat Ahmedov[1,2], *R.M. Zalaletdinov*[1], *and Z. Ya. Turakulov*[1,2]
[1]*Institute of Nuclear Physics, Tashkent, Uzbekistan;* [2]*Ulugh Beg Astronomical Institute, Tashkent, Uzbekistan; email:* ahmedov@astrin.uzsci.net

Theoretical Astrophysics is the subject which has got an essential development in Uzbekistan during the last decade, especially through the newly established collaborations with western and eastern institutions. Our regional collaboration is supported by the ICTP (Trieste, Italy), TWAS (Trieste, Italy) and IUCAA (Pune, India) in the framework of BIPTUN (Bangladesh - India - Pakistan - Turkey - Uzbekistan) Network on Relativistic Astrophysics and Cosmology. Other important scientific collaboration is with the western partners, mainly at SISSA (Trieste, Italy), ICRA (Pescara, Italy), Dalhousie University (Halifax, Canada) and at the ICTP. Local scientific activity in theoretical astrophysics in Uzbekistan is partly supported through the Affiliation Scheme (ICAC-83) of the ICTP. These years some financial support towards the research in theoretical astrophysics was through NATO grants.

The theoretical results obtained in Uzbekistan in the field of relativistic astrophysics and cosmology are presented. In particular electrostatic plasma modes along the open field lines of a rotating neutron star and Goldreich-Julian charge density in general relativity are analyzed for the rotating and oscillating neutron stars. The impact that stellar oscillations of different type (radial, toroidal and spheroidal) have on electric and magnetic fields external to a relativistic magnetized star has been investigated.

A study of the dynamical evolution and the number of stellar encounters in globular clusters with a central black hole is presented.

In a cosmological setting the theory of macroscopic gravity is a large-distance scale generalization of general relativity. Exact cosmological solutions to the equations of macroscopic gravity for a flat spatially homogeneous, isotropic space-time are found. The gravitational correlation terms in the averaged Einstein equations have the form of spatial curvature, dark matter and dark energy (cosmological constant) with particular equations of state for each correlation regime. Interpretation of these cosmological models to explain the observed large-scale structure of the accelerating Universe with a significant amount of the nonluminous (dark) matter is discussed.

5.7. *The astronomical observatory 'Khurel Togoot' in Mongolia*

Batmunkh Damdin, Research Centre of Astronomy and Geophysics, Mongolian Academy of Sciences, Ulaanbaatar, Mongolia, email: btmnh_d@yahoo.com

In my presentation the basic researches, telescopes and devices of our astronomical observatory, which was founded during the International Geophysical Year, are briefly described. Our astronomical observatory is located on Bogd Mountain near the capital city Ulaanbaatar. Almost 50 years of scientific works have been carried out there. In particular, astrometric researches, GPS, solar researches and observations of minor planets are conducted. Now these scientific researches basically are kept and extended, with the introduction of modern technology. As an example of the data received by our solar telescope 'Coronograph', some solar images will be shown. Recently we equipped this telescope with a CCD camera. Because of the transformation of the economy in Mongolia, there are at present difficulties with the preparation of young professional astronomers and with the purchase of new astronomical equipment.

5.8. *Astronomical education in Armenia*

Hayk A. Harutyunian, Byurakan Astrophysical Observatory, Yerevan, Armenia, email: hhayk@bao.sci.am

Astronomy pupils in Armenia get their first ideas on astronomy at elementary schools. Astronomy as a distinct subject is taught at all secondary schools in the country. Teaching is conducted according to a unified program elaborated jointly by professional astronomers and astronomy teachers. Unfortunately only one hour per week is allotted for teaching astronomy, which obviously is not enough workload to hire specialized astronomy teachers at every school, and at many schools this subject is tutored by non-specialists. Many schools partly compensate this lack of teachers by organizing visits to the Byurakan Observatory (BAO) for pupils, where they also attend short lectures on astronomy. In some schools optional training in astronomy is organized by amateurs, for the purpose of a deeper understanding in astronomy.

During recent years annual competitions for revealing gifted pupils in astronomy have been organized. These competitions have three rounds, namely, in schools, in districts and the final round is, as a rule, held at BAO. The national winners successfully participate in and win prestigious prizes at international astronomical Olympiads as well.

At Yerevan State University (YSU) there is a department for astrophysics, which was set up in 1946 and is operating to date. This department trains specialists for a career in astrophysics. Only one or two students graduate from this department yearly at present, while in the 1980s a dozen specialists were trained every year. BAO serves as the scientific base for the students of YSU as well, and a number of staff members from BAO conduct special courses for YSU students. YSU provides a Master's degree in astrophysics, and BAO is granting a Doctor's (PhD) degree since the 1970s.

5.9. *On progress on the Suffa large radio-telescope project*

G.I. Shanin[1], *A.S. Hojaev*[1], *and Yu.N. Artyomenko*[2]
[1] *Center for Space Research, Uzbek Academy of Sciences, Tashkent, Uzbekistan;*
[2] *Astro-Space Center, Lebedev Phyical Insitute, RAS, Moscow, Russia;*
email: ash@astrin.uzsci.net

The large-scale radio astronomy facility complex (analogous to the GBT at NRAO) is being created not far from Samarkand (Uzbekistan) on the Suffa plateau at 230 m (Trimble, 2001, A Year of Discovery: Astronomy Highlights of 2000, S&T, 101, 51). Originally it was designed as a basic part of the Earth-Space VLBI system (Kardashev *et al.*, 1995, Acta Astron., 37, 271), URL <http://www.asc.rssi.ru/suffa/>) and contains the radio telescope for 0.8-60 mm bandwith 70-m main reflector, two removable sub-reflectors; satellite communication station; data receiving and processing system and other necessary infrastructure. The adaptive optics principle will be used for control of the surface of the main mirror, consisting of 1200 trapezoidal panels.

The site location provides good seeing conditions for the cm–mm range. Averaged annual atmospheric transmission coefficients at the zenith were derived as 0.90 - 0.98 for 3.1 mm and 5.8 mm wavelengths and about 0.60 for 1.36 mm (Hojaev & Shanin, 1996, JKAS, 29, S411).

The project started as far back as the period when the Soviet Union was stalled, since its disintegration. Quite recently the firm decision on completing the project has been endorsed by our Governments, and Russia will invest for this; therefore the project's layouts have been considerably modernized and updated in order to build up a state-of-the-art instrument. It should be operational in 2009.

Now we are arranging for a scientific consortium further to explore the Suffa site more deeply and to learn the main 'radio-astro-climate' parameters by means of new technology ('radio-seeing', radio transparency in different sub-mm, mm and cm bands, PWV, their intercorrelation and correlation with meteo-parameters) for the atmospheric modelling at the site, and to try to forecast the 'radio-weather' for reliably planning the scientific schedule of the future telescope.

5.10. *IHY: Meteor astronomy and the New Independent States (NIS) of the Former Soviet Union*

Svetlana Kolomiyets, Kharkiv National University of Radioelectronics, Kharkiv, Ukraine, email: s.kolomiyets@gmail.com

The purpose of this paper is to emphasize, that there are some specific features of the development of science in the New Independent States (NIS) of the Former Soviet Union. These features demand enhanced attention of the organizers of the IHY. The creation of effective mechanisms for the stimulation of connections to world science is necessary. This is because there exists a dormant sector of fundamental scientific knowledge in these countries, which has been saved up for the fifty years since the IGY in 1957. Probably, the IHY is the last opportunity for rescuing the dormant part of this knowledge from full oblivion.

The method used in this paper is to display the general tendencies in individual displays. The features and history of the development of meteor astronomy during the existence of the Soviet Union and the subsequent period give a key to an understanding of the problem. Meteor astronomy can be assumed to be a young science. It is an example of a cross-disciplinary science. It is also an example of a science having a sharp rise, due to the project of the IGY and to subsequent geophysical projects. Meteor astronomy is a science directly connected with the launching of the first space satellite of the Earth and with the resolution of problems of meteoroid danger to space missions.

Commission 22 (Division III) of the IAU coordinated the development of meteor astronomy during the IGY. The well-known Soviet researcher of meteors, V. Fedynskiy, has headed this Commission during four years since 1958. In the USSR, numerous meteor centres were created and activated. The general management was concentrated in Moscow. Despite a close interaction under global projects of the Soviet Union with other countries, there existed a language barrier. The language barrier, together with other reasons, led to the creation in the USSR of a powerful meteor science only in the Russian language. After the disintegration of the Soviet Union, the meteor centers have remained, but without the normal central management. The scientific outputs have therefore remained as isolated, and inaccessible, although the science is published in English.

A reunification of the scientific achievements of the last few years with international science should become the task of the IHY in the NIS. Revival of the activity of some of the centres will be useful.

6. Astronomy in eastern Europe

6.1. *Introduction*

These three papers give an overview of different aspects of astronomy in the former Yugoslavia. Olga Atanackovic-Vukmanovic gives a review of astronomy in Serbia and in Montenegro (the two countries dissolved their federation just before the IAU XXVI General Assembly). Aleksandra Andic discussed astronomy education in Bosnia and Herzegovina. Finally Davor Krajnovic presents information on a new program of distance education in astronomy in Split, Croatia. All these regions of the former Yugoslavia have undergone very different political and economic developments, so evidently the support for astronomy differs widely across this relatively compact region, as is reflected in these papers.

6.2. *Astronomy in Serbia and in Montenegro*

Olga Atanackovic-Vukmanovic, Faculty of Mathematics, University of Belgrade, Belgrade, Serbia, Yugoslavia, email: olga@matf.bg.ac.yu

A review of professional and amateur astronomy in Serbia and in Montenegro is given. After a brief historical survey of the foundations and development of astronomy education in Serbia and in Montenegro, special attention is given to a new curriculum that is being prepared for all educational levels.

6.3. *Astronomy and astrophysics in Bosnia and Herzegovina*

Aleksandra Andic, Prirodno Matematicki Fakultet, Banja Luka, Bosnia and Herzegovina, and Queen's University, Belfast, UK, email: a.andic@qub.ac.uk

In Bosnia and Herzegovina, astronomy teaching is almost non-existent. There are only several courses in universities and they are usually taught by physicists who often had only elementary courses in astrophysics. On the other hand, there is a huge interest for astrophysics in the student population. When it comes to educational outreach, the situation is even more grim. There are several solutions and possibilities which I will present, together with the main obstacles which need to be overcome.

6.4. *Study astrophysics in Split!*

Davor Krajnovic, University of Oxford, Oxford, UK, email: dxk@astro.ox.ac.uk

Beginning in autumn 2008 the first generation of astronomy master's students will start a two-year course in astrophysics offered by the Physics Department of the University of

Split, Croatia (<http://fizika.pmfst.hr/astro/english/index.html>). This unique master's course in south-eastern Europe, following the Bologna convention and given by astronomers from international institutions, offers a series of comprehensive lectures designed greatly to enhance students' knowledge and skills in astrophysics, and to prepare them for a scientific career. An equally important aim of the course is to recognize the areas in which astronomy and astrophysics can serve as a national asset, and to use them to prepare young people for real life challenges, enabling graduates to enter modern society as a skilled and attractive work-force.

I will present this new programme as an example of a successful organization of international astrophysics studies in a developing country, which aims to become a leading graduate program in astrophysics in the broader region. I will focus on the goals of the project, showing why and in what way astronomy can be interesting for third-world countries, what are the benefits for the individual students, nation and region, and also for research, science and the astronomical community in general.

7. Astronomy education in developing countries

7.1. *Introduction*

Astronomy education provides the foundation upon which any astronomical community rests. This foundation provides the human capital that is necessary for developing the infrastructure of astronomy in a country. The eleven papers presented in this section fall into three broad themes: University-level education, the use of small telescopes for teaching and research, and popularisation of astronomy.

Many developing countries suffer from the isolation of astronomers and their graduate students, working in very small groups, with little contact with the mainstream astronomical community. For such a group, an infusion of international expertise can be tremendously beneficial, not only in terms of scientific contacts, but also in terms of raised levels of support for astronomy in the country. For nearly fourty years the IAU has conducted the International Schools for Young Astronomers (ISYA). Michele Gerbaldi describes the ISYA programmes, of which she has been a key driver for many years.

John Percy discusses undergraduate and graduate programmes in astronomy for developing countries. He highlights the importance of imparting not only subject knowledge, but also practical skills that can find application outside of astronomy, as well as exposing the students to a culture of research and learning. Where attendance at a formal university course is not possible, distance education is a possibility. Barrie Jones of the Open University discusses his experiences gained from teaching astronomy via distance education for the past 30 years. Paulo Bretones and his colleagues discuss the pedagogic aspects of astronomy education. He presents the Latin-American Journal of Astronomy Education, which addresses the need for teaching resources in Spanish and Portuguese, as well as general pedagogical aspects of the teaching of astronomy.

Small telescopes have a huge role to play in the teaching of astronomy and in the development of observing skills by astronomy students. Shiva Pandey presents examples of the use of small, commercially available telescopes equipped with photometers or CCD cameras to perform laboratory exercises in astronomy and small, publishable research projects. Eder Martioli and J. Jablonski describe how they use small telescopes and commercial CCDs to search for planetary transits from a university campus, where the seeing conditions may not be excellent, but where accessibility for students is. John Baruch and his colleagues take the concept of small telescopes for teaching into the realm of the internet by presenting the potentials of small robotic telescopes to support basic astronomy education through the internet by servicing thousands of users spread across the globe for free. Taking this theme yet one step further removed from the telescope, Hayk Harutyunian and M. Mickaelian discuss the potentials of Virtual Observatories as teaching tools in astronomy.

The final group of papers in this section focusses on the role of public outreach and the amateur community in developing astronomy. Veteran eclipse chaser Jay Pasachoff discusses the immense potentials of solar eclipses to capture the imagination of the public and policy makers alike for the development of astronomy. A.P. Sule and colleagues discuss the role of amateur astronomy organisations in astronomy popularisation, pointing to examples of the Indian experience. The funding of capacity building acitivities is always a challenge. The final paper in this section is a proposal by Meelis Kaldalu that the IAU consider the marketing potentials associated with the naming of celestial objects as a possible source of revenue for astronomy development.

7.2. *International Schools for Young Astronomers (ISYA): their new horizon*

Michèle Gerbaldi, Institut d'Astrophysique, Paris, France, email: `gerbaldi@iap.fr`

This talk outlines the main features of this programme developed by the International Astronomical Union (IAU) since 1967 and its perspective at the time of the development of virtual observatories. The main goal of this programme is to support astronomy in developing countries by organizing a school over three weeks for students typically with an M.Sc. degree.

The context in which the ISYA were developed changed drastically in the past ten years. From a time when the access to any large telescope was difficult, and mainly organized on a national basis, nowadays the archives are developed at the same time that any major telescope is planned, whether ground-based or in space, and these archives are accessible from everywhere. The concept of the virtual observatory reinforces this access. However, the technological development of telecommunications and of world-wide internet connections do not remove all the difficulties, among which is the problem of the isolation of the scientist working in a small institution. In this context, the role of the ISYA will be addressed.

7.3. *Undergraduate and graduate programs in astronomy for developing countries*

John Percy, University of Toronto, Toronto, Canada, email: `jpercy@utm.toronto.ca`

This presentation will discuss some aspects of the design of undergraduate and graduate astronomy curricula, broadly defined, for developing countries. A fundamental requirement is to develop students' ability and desire to learn, both in university and beyond. I will then discuss several topics of the curriculum:

(*i*) The program of course work in astronomy and related topics, such as physics and mathematics;

(*ii*) The associated practical and project work to develop skills as well as knowledge;

(*iii*) linking the course work, effectively, to various aspects of research;

(*iv*) Development of general academic and professional skills such as oral and written communication, teaching, planning and management, and the ability to function as part of an interdisciplinary team; and

(*v*) Orientation to the culture of the university and to the science and profession of astronomy.

To accomplish all of the goals may seem daunting, especially as many of them are not achieved in the most affluent universities. But much can be achieved by recognizing that there are well-established 'best practices' in education, achieved through research, reflection and experience. Simple resources, effectively used, can be superior to the highest technology, used without careful thought. It is often best to do a few things well; 'less can do more'. And effective partnership, both within the local university and with the outside astronomical community, can also contribute to success.

7.4. *Distance education at university level – how useful for developing countries?*

Barrie Jones, The Open University, Milton Keynes, UK,
email: b.w.jones@open.ac.uk

Many countries now have institutions devoted to distance education at university level. Not all of these are in the developed world. In this talk I will outline those in developing countries. I will then describe the main features of distance education as presently practised by The Open University in the UK. It first admitted students in 1970, and since then we have learned what is really necessary for distance education to be successful, not just in the UK, but beyond. Distance education in other countries can, and does, differ from that in the UK, and a few examples of the different challenges and practises will be given. But the focus of my presentation will be the time I hope to allow for comments and questions. It will be invaluable to have input from those present with experience of other open learning institutions, and from those eager to begin to explore the potential of distance learning in their own countries.

7.5. *The first two years of the Latin-American Journal of Astronomy Education (RELEA)*

Paulo S. Bretones[1,2], *L.C. Jafelice*[3], *and J.E. Horvath*[4]
[1]*Instituto de Geociências, Universidade Estadual de Campinas, Campinas, Brazil;* [2]*Instituto Superior de Ciências Aplicadas;* [3]*Departamento de Fisica, Universidade Federal do Rio Grande do Norte, Natal, Brazil;* [4]*Instituto de Astronomia, Geofisica e Ciências Atmosféricas, Universidade de São Paulo, São Paulo, Brazil;*
email: bretones@mpc.com.br

We present and discuss in this work the motivations, goals and strategies adopted for its creation and launch of the e-journal *Latin-American Journal of Astronomy Education* (RELEA). The RELEA 'first light' was in August, 2004 with the appearance of No. 1, and it is now completing two years of existence. The creation of the new journal was prompted by: (*a*) the noteworthy absence of a specific publication in the field in Latin-America; (*b*) the lack of classroom material in Spanish or Portuguese that could be directly used without too many adaptations; and (*c*) the need for a regional forum to discuss and suggest public policies concerning the teaching of sciences in general and astronomy in particular.

We identify and present the difficulties encountered for the achievement of the proposed objectives and operational issues in this period, together with the adopted solutions (refereeing procedure, periodicity, etc.). Finally, we attempt to evaluate the long-run impact of such initiatives on scientific education as a tool for effective citizenship decision making, so critical for third-world countries.

7.6. *Teaching and research in astronomy using small aperture optical telescopes*

Shiva K. Pandey, Pt Ravishankar Shukla University, Raipur, India,
email: skp@iucaa.ernet.in

Small aperture (<1 m, typically 20 - 50 cm) optical telescopes with adequate back-end instrumentation (photometer, CCD camera and CCD spectrograph, etc.) can be used for spreading the joy and excitement of observational astronomy among postgraduate and research students in colleges. On the basis of over a decade's experience in observing with small optical telescopes it has been amply demonstrated that such a facility, which any university department can hope to procure and maintain, can be effectively used for teaching as well quality research.

The Physics Department of Pt Ravishankar Shukla University at Raipur, India offers astronomy and astrophysics as one of the specializations as a part of the M.Sc. program in physics. A set of observational exercises has been incorporated with a view to provide training in observations, analysis and interpretation of the astronomical data to the students. Observing facilities available in the department include $8-14$ inch aperture telescopes (CGE series from Celestron) equipped with the new-state-of-the-art back-end instrumentation-like photometer, CCD camera and also a CCD spectrograph.

An observing facility of this kind is ideally suited for continuous monitoring of a variety of variable stars, and thus can provide valuable data for understanding the physics of stellar variability. This is especially true for a class of variable stars known as chromospherically active stars. The stars belonging to this class have variable light curves, and the most puzzling feature is that their light curves change year after year in a rather strange way. A large fraction of these active stars are bright and, hence, the importance of a small aperture telescope for collecting the much needed photometric data.

For over a decade the research activity using a 14-inch optical telescope has been focused on photometric monitoring of well known as well suspected active stars. This, together with spectroscopic data using the observing facility at Indian observatories has led to identification of new chromospherically active stars. The talk is aimed at sharing our experiences, quoting examples with professional colleagues on the usage of small optical telescopes for teaching and research in colleges and universities.

7.7. *How to look for planetary transits using small telescopes and commercial CCDs in developing countries*

Eder Martioli and F. Jablonski, Divisão de Astrofísica, Instituto Nacional de Pesquisas Espaciais, São José dos Campos, Brasil, email: `eder,chico@das.inpe.br`

The main goal of this work is to have a better understanding of the problems and characteristics of photometric surveys with small-sized affordable equipment, like the one available at the Astrophysics Division/INPE, in São José dos Campos, Brazil. The use of low-cost instruments also has an appeal in the context of the detection of Extrasolar Planets (ESP), in the sense that many observers are available for survey and follow-up programs. It could also make possible the inclusion of many developing countries in the search for planetary transits. We describe the data collection and analysis procedure for differential photometry of the transit of HD209458 b, using a small telescope and a commercial CCD camera. According to the HST observations of Brown *et al.* (2001), the transit produces a box-shaped light-curve with 2 % depth and 184 min duration. The orbital period is $\sim$3.5 days. The equipment consists of a f/10, 11-in Schmidt-Cassegrain Celestron telescope equipped with a SBIG ST7E CCD camera. Since the seeing at the campus is quite poor, we used a focal reducer to produce an effective focal ratio of about f/5, still keeping a good sampling of the PSF but with a larger field-of-view. The larger field-of-view allows the simultaneous observation of a relatively bright nearby star, suitable for differential photometry. We discuss the IRAF reduction procedures for the large number of images collected and present the results obtained in the transit of September 8, 2004.

7.8. *Basic astronomy as part of a general higher education in the developing world*

John E.F. Baruch[1], *D.G. Hedges*[1], *J. Machell*[1], *K. Norris*[2], *and C.J. Tallon*[1]
[1] *University of Bradford, Bradford, UK;* [2] *Bradford College, Bradford, UK*
email: john@telescope.org

This paper describes a new initiative in support of the aim of Commission 46 of the IAU to develop and improve astronomy education at all levels throughout the world. This paper discusses the ideal specification of a facility to support basic astronomy within education programmes which are delivered to students who have access to the internet. The available robotic telescopes are discussed against this specification and it is argued that the Bradford Robotic Telescope, uniquely, can support many thousands of users in the area of basic astronomy education, and the resource is free. Access to the internet is growing in the developing world and this is true in the education programmes.

This paper discusses the serious problems of delivering to large numbers of students a web-based astronomy education programme supported by a robotic telescope as part of a general education. It examines the problems of this form of teaching for teachers who have little experience of working with IT and little knowledge of basic astronomy and proposes how such teachers can be supported. The current system (<http://www.telescope.org/>) delivers astronomy education in the language, culture and traditions of England. The paper discusses the need to extend this to other languages, cultures and traditions, although for trainee teachers and undergraduates, it is argued that the current system provides a unique and valuable resource.

7.9. *Modern facilities in astronomy education*

Hayk A. Harutyunian and M.A. Mickaelian
Byurakan Astrophysical Observatory, Yerevan, Armenia, email: hhayk@bao.sci.am

Astronomical education in Armenia enters a new stage of organization and development. Though the economic difficulties restricted the former interest in astronomy for a decade or so, the present young generation gradually finds more and more attraction in space sciences. Knowledge of computers and the internet is their typical difference from the previous students. Thus, astronomy education requires heavy use of computer facilities and the internet, as it is in the case of modern astronomical research. The students need powerful computers, and also computing methods, the internet (including grid technologies), usage of large astronomical databases, virtual observatories (VOs), etc.

The Armenian astronomy has a unique database of the famous Markarian survey (Digitized First Byurakan Survey, DFBS), as well as the newly created Armenian Virtual Observatory (ArVO). Since 2005, we have introduced a new (for the first time in the world!) subject for the Yerevan State University graduate students called 'Astronomical surveys, databases and virtual observatories', which is connected directly with the modern understanding and treatment of large multi-wavelength data volumes. The new requirements suggest also training of new kind of (so-called) astronomy-computing specialists, who could heavily push the modern research and make it much more efficient.

7.10. *Observing solar eclipses in the developing world*

Jay M. Pasachoff, Hopkins Observatory, Williams College, Williamstown, MA, USA,
email: jay.m.pasachoff@williams.edu

The paths of totality of total solar eclipses cross the world, with each spot receiving such a view about every 300 years. The areas of the world from which partial eclipses are visible are much wider. For the few days prior to a total eclipse, the attention of a given country is often drawn toward the eclipse, providing a teachable moment that we can use to bring astronomy to the public's attention. Also, it is important to describe how to observe the partial phases of the eclipse safely. Further, it is important to describe to those people in the zone of totality that it is not only safe but also interesting to view totality. Those who are misled by false warnings that overstate the hazards of viewing the eclipse, or that fail to distinguish between safe and unsafe times for naked-eye viewing, may well be skeptical when other health warnings – perhaps about AIDS or malaria prevention or polio inoculations – come from the authorities, meaning that the penalties for misunderstanding the astronomical event can be severe. Through the International Astronomical Union's Working Group on Solar Eclipses and through the IAU's Program Group on Public Education at the Times of Eclipses, part of the Commission on Education and Development, we make available information to national authorities, to colleagues in the relevant countries, and to others, through our websites at http://www.eclipses.info and `<http://www.totalsolareclipse.net` and through personal communication. Among our successes at the 29 March 2006 total solar eclipse was the distribution through a colleague in Nigeria of 400 000 eye-protection filters.

7.11. *Role of voluntary organizations in astronomy popularization: a case study of 'Khagol Mandal', Mumbai, India*

A.P. Sule[1], *S. Joshi*[2], *A. Deshpande*[3], *H. Joglekar*[3], *and Y. Soman*[3]

[1]*Astrophysikalisches Institut Potsdam, Potsdam, Germany;* [2]*Jodrell Bank Observatory, Manchester, UK;* [3]*Khagol Mandal, Mumbai, India; email:* `asule@aip.de`

In India, astronomy research institutions are few and far spaced as compared to the population density. Further, the public outreach activities of research institutes cannot cover most of the academic institutes in their area as they way out-number public outreach resource potential of any institute. The organizations of amateur astronomy enthusiasts do come handy in this scenario. We here present a case study of 'Khagol Mandal', a voluntary organization primarily based in Mumbai, India's economic capital. In 20 years since its inception in 1985 - 86, Khagol Mandal has given more than 1000 public outreach programmes in various schools, undergraduate colleges, famous city hangouts, apart from their regular overnight programmes in Vangani, a sleepy village on the outskirts of the city. Study tours on special occasions like TSE'95 and TSE'99 as well as regular study tours to meteor crater at Lonar, Maharashtra facilitate their volunteers with glimpses of real research work in astronomy. These have inspired a number of students to take professional astronomy careers. With a volunteer force, probably the largest in India or even South Asia, Khagol Mandal is well poised to take advantage of the newest tools like the Virtual Observatory and make the use of existing goodwill to take these tools to the layman. With little guidance from senior researchers, organizations like these can provide a solution to the ever-increasing need of man-power for secondary data analysis.

7.12. *A packet of proposals for collaboration between the IAU and the developing world*

Meelis Kaldalu, University of Tartu, Tartu, Estonia, email: `ifoundu@hot.ee`

The purpose of the presentation is to outline several opportunities on how astronomical society can contribute benefit to the developing world. The package of solutions described includes methods of peer-to-peer networking, collaboration with the international business community and a widespread dissemination campaign about the results of astronomical research. Preliminary realistic calculations estimate that at least 400 000 euro, or more, can be raised for the realization of the proposed plan of action. The presentation will be followed by a discussion about possible future activities, and ad hoc brainstorming for new solutions to popularize astronomy all over the world.

8. Promoting astronomy development through the UN, the International Heliophysical Year and COSPAR

8.1. *Introduction*

International scientific unions, such as the IAU and COSPAR, together with inter-governmental organisations, such as the UN and UNESCO, have done much to promote astronomy in the developing world. The United Nations Office for Outer Space Affairs has been instrumental in harnessing the prestige of the United Nations to organise a series of UN/ESA Workshops on Basic Space Science in developing countries in all regions of the world since 1991. Hans Haubold describes the philosphy behind this workshop series and their future thematic emphasis on the International Heliophysical Year. Keith Arnaud and Peter Willmore describe the COSPAR capacity-building workshops organised during 2000 – 2007. These are hands-on workshops intended to familiarise participants with the content of space mission data archives and the tools to use them.

The next group of papers focus on the capacity-building opportunities associated with the International Heliophysical Year (IHY) 2007. Nat Gopalswamy and his colleagues discuss the United Nations Basic Space Science Initiative for IHY 2007 in terms of which there is a programme to deploy arrays of small, inexpensive instruments around the globe. This is an exciting opportunity to initiate space science research at interested institutions that can afford very little in the way of capital outlay, but who have interested and committed scientists to operate these insturements and to participate in the scientific work. David Webb and Nat Gopalswamy discuss the IAU's role in the IHY. Within the IAU, the IHY activities are conducted under Division II, Sun and Heliosphere.

The IHY presents an excellent opportunity to promote education in basic space science world-wide. Indeed, this is a major thrust of the IHY. Maria Cristina Rabello-Soares and colleagues discuss the IHY Education and Outreach Programme, which is coordinating the education and outreach activities of all the IHY partners. Developing countries may access teaching resources and educational particpation opportunities through this programme. One such 'hands-on' project is the Space Weather Monitor Project, presented by Deborah Scherrer and her colleagues. This project allows students around the world to track solar-induced changes to the ionosphere.

8.2. *The United Nations Basic Space Science Initiative*

Hans Haubold, United Nations Office for Outer Space Affairs, Vienna, Austria,
email: Hans.Haubold@unvienna.org

Pursuant to recommendations of the United Nations Conference on the Exploration and Peaceful Uses of Outer Space (UNISPACE III) and deliberations of the United Nations Committee on the Peaceful Uses of Outer Space (UNCOPUOS), annual UN/European Space Agency workshops on basic space science have been held around the world since 1991. These workshops contribute to the development of astrophysics and space science, particularly in developing nations. Following a process of prioritization, the workshops identified the following elements as particularly important for international cooperation in the field: (*i*) operation of astronomical telescope facilities implementing TRIPOD; (*ii*) virtual observatories; (*iii*) astrophysical data systems; (*iv*) concurrent design capabilities for the development of international space missions; and (*v*) theoretical astrophysics such as applications of non-extensive statistical mechanics.

Beginning in 2005, the workshops focus on preparations for the International Heliophysical Year 2007 (IHY2007). The workshops continue to facilitate the establishment of astronomical telescope facilities as pursued by Japan and the development of low-cost, ground-based, world-wide instrument arrays as lead by the IHY secretariat.

Further reading:
Wamsteker, W., Albrecht, R. and Haubold, H.J. (eds.), 2004, *Developing Basic Space Science World-Wide: A Decade of UN/ESA Workshops* Dordrecht: Kluwer
<http://ihy2007.org>
<http://www.unoosa.org/oosa/en/SAP/bss/ihy2007/index.html>
<http://www.cbpf.br/GrupPesq/StatisticalPhys/biblio.htm>

8.3. *The COSPAR Capacity-Building Workshop programme, 2000 – 2007*

Keith Arnaud[1] *and Peter Willmore*[2],
[1] *University of Maryland, College Park, MD, USA;* [2] *University of Birmingham, UK,*
email: apw@star.sr.bham.ac.uk

The objectives and mode of operation of the COSPAR Capacity-Building Workshop Programme will be described, together with the activities carried out and the results achieved in the first seven years of its existence. The policies in place to embed the skills imparted into the host scientific communities and to the methods used to monitor the effectiveness of the programme, will be discussed.

8.4. *The United Nations Basic Space Science Initiative for IHY 2007*

Nat Gopalswamy[1], *J.M. Davila*[1], *B.J. Thompson*[1], *and Hans Haubold*[2],
[1] *NASA Goddard Space Flight Center, Greenbelt, MD, USA;* [2] *United Nations Office for Outer Space Affairs, Vienna, Austria; email:* gopals@ssedmail.gsfc.nasa.gov

The United Nations, in cooperation with national and international space-related agencies and organizations, has been organizing annual workshops since 1990 on basic space science, particularly for the benefit of scientists and engineers from developing nations. The United Nations Office for Outer Space Affairs, through the IHY Secretariat and the United Nations Basic Space Science Initiative (UNBSSI) will assist scientists and engineers from all over the world in participating in the International Heliophysical Year (IHY) 2007.

A major thrust of the IHY/UNBSSI program is to deploy arrays of small, inexpensive instruments such as magnetometers, radio telescopes, GPS receivers, all-sky cameras, etc. around the world to provide global measurements of ionospheric and heliospheric phenomena. The small instrument program is envisioned as a partnership between instrument providers, and instrument hosts in developing countries. The lead scientist will provide the instruments (or fabrication plans for instruments) in the array; the host country will provide manpower, facilities, and operational support to obtain data with the instrument typically at a local university.

Funds are not available through the IHY to build the instruments; these must be obtained through the normal proposal channels. However, all instrument operational support for local scientists, facilities, data acquisition, etc will be provided by the host nation. It is our hope that the IHY/UNBSSI program can facilitate the deployment of several of these networks world wide. Existing data bases and relevant software tools that can be used will be identified to promote space science activities in developing countries. Extensive data on space science have been accumulated by a number of space missions. Similarly, long-term data bases are available from ground based observations. These data can be utilized in ways different from originally intended for understanding the heliophysical processes. This paper provides an overview of the IHY/UNBSS program, its achievements and future plans.

8.5. *The IHY program and associated IAU activities*

David Webb[1] *and Nat Gopalswamy*[2],
[1]*Boston College, Chestnut Hill, MA, USA;* [2]*NASA Goddard Space Flight Center, Greenbelt, MD, USA; email:* david.webb@hanscom.af.mil

The International Heliophysical Year is an international program of scientific collaboration planned for the time period starting next year, the 50th anniversary of the International Geophysical Year. The physical realm of the IHY encompasses all of the solar system out to the interstellar medium, representing a direct connection between in-situ and remote observations. The IHY is of great interest to the IAU because of this broad astronomical coverage as well as its emphasis on international cooperation and developing nations. The IHY program is promoting worldwide participation in its activities that include dispersing networks of inexpensive instrumentation to achieve its scientific goals. Within the IAU the IHY program is organized under Division II, which covers the Sun and Heliosphere. Nat Gopalswamy is the IHY International Coordinator and Chair of the IHY subgroup within the IAU's Working Group on International Collaboration on Space Weather. David Webb is the IAU representative for the IHY and the outgoing President of Division II. The United Nations IHY effort is led by Hans Haubold under the UNBSS program and will be discussed next by Dr. Gopalswamy. Under this program the IAU is supporting the annual IHY Workshops and is facilitating the communications between scientists in developed and developing countries.

8.6. *Globalizing space and Earth science – the International Heliophysical Year education and outreach program*

M. Cristina Rabello-Soares[1], *C. Morrow*[2], *and B.J. Thompson*[3],
[1]*Stanford University, Stanford, CA, USA,* [2]*Space Science Institute, Boulder, CO, USA,* [3]*NASA Goddard Space Flight Center, Greenbelt, MD, USA,*
email: csoares@sun.stanford.edu

The International Heliophysical Year (IHY) in 2007 and 2008 will celebrate the 50th anniversary of the International Geophysical Year (IGY) and, following its tradition of international research collaboration, will focus on the cross-disciplinary studies of universal processes in the heliosphere. The main goal of IHY Education and Outreach Program is to create more global access to exemplary resources in space and earth science education and public outreach. By taking advantage of the IHY organization with representatives in every nation and in the partnership with the United Nations Basic Space Science Initiative (UNBSSI), we aim to promote new international partnerships. Our goal is to assist in increasing the visibility and accessibility of exemplary programs and in the identification of formal or informal educational products that would be beneficial to improve the space and earth science knowledge in a given country; leaving a legacy of enhanced global access to resources and of world-wide connectivity between those engaged in education and public outreach efforts that are related to IHY science. Here we describe how to participate in the IHY Education and Outreach Program and the benefits in doing so. Emphasis will be given to the role played by developing countries; not only in selecting useful resources and helping in their translation and adaptation, but also in providing different approaches and techniques in teaching.

8.7. *The Space Weather Monitor project: bringing hands-on science to students of the developing world for the IHY2007*

Deborah K. Scherrer[1], *M.C. Rabello-Soares*[1], *and C. Morrow*[2]
[1]*Stanford University, Stanford, CA, USA;* [2]*Space Science Institute, Boulder, CO, USA,*
email: dscherrer@solar.stanford.edu

Stanford's Solar Center, Electrical Engineering Department, and local educators have developed inexpensive Space Weather Monitors that students around the world can use to track solar-induced changes to the Earth's ionosphere. Through the United Nations Basic Space Science Initiative (UNBSSI) and the IHY Education and Public Outreach Program, our Monitors are being deployed to 191 countries for the International Heliophysical Year, 2007. In partnership with Chabot Space and Science Center, we are designing and developing classroom and educator support materials to accompany the distribution. Materials will be culturally sensitive and will be translated into the six official languages of the United Nations (Arabic, Chinese, English, French, Russian, and Spanish). Monitors will be provided free of charge to developing nations and can be set up anywhere there is access to power.

9. The virtual observatory and developing countries

9.1. *Introduction*

Virtual Observatories, and the tools to use them, are being developed in all the astronomically developed countries. For astronomers in the developing countries Virtual Observatories have the potential to level the playing fields in terms of access to data and processing tools, *provided that the users are appropriately trained, and know what they are doing.* Ajit Kembhavi discusses these potentials and illustrates some of the facilities that are already available. Virtual Observatories are one area of astronomy where the instruments or tools may be contributed by people in developing countries, where there is much software development potential. Ganghu Lin and colleagues discuss VO software that they have developed to do common data processing tasks and the methods they have devised for data management and data sharing.

In order to realise their potential use in developing countries, the existing capacity building programmes of the IAU, COSPAR and the UN will have to create opportunities, such as hands-on workshops, to allow the users to interact with VOs under the guidance of experienced astronomers.

9.2. *Developing countries and the virtual observatory*

Ajit Kembhavi, Inter-University Centre for Astronomy and Astrophysics, Pune, India,
email: akk@iucaa.ernet.in

A Virtual Observatory is a platform for launching astronomical investigations: it provides access to huge data banks, software systems with user friendly interfaces for data visualization and analysis, and even access to computers on which the analysis can be carried out. Virtual observatories the world over are seamlessly networked, and their resources can be accessed over the internet by astronomers regardless of their location, expertise and the level of access to their own advanced computing facilities. Due to their nature, virtual observatories can make an immense impact on the way astronomy is done in the developing world. I will consider in my talk some of the facilities that virtual observatories provide, discuss their possible use by astronomers, and also how even small groups in the developing world can contribute to the setting up of virtual observatories.

9.3. *Exploiting software towards easier use and higher efficiency*

Ganghu Lin, J.T. Su, and Y.Y. Deng, National Astronomical Observatories, Chinese Academy of Sciences, Beijing, China, email: `lgh@sun10.bao.ac.cn`

In developing countries, using data based on instruments made by themselves to the maximum extent is very important. It is not only related to maximizing science returns upon prophase investment, deep accumulations in every aspects but also science output. Based on the idea, we are exploiting a software (called THDP: Tool of Huairou Data Processing). It is used for processing a series of issues, which is necessary in processing data.

This paper discusses its design, purpose, functions, method and specialities. The primary vehicle for general data interpretation is through various techniques of data visualization, techniques which are interactive. In the software, we employed an Object-Oriented approach. It is appropriate to the vehicle. It is imperative that the approach provide not only function, but do so in as convenient a fashion as possible. As result of implementing the software, it is not only easier to learn data processing for a beginner but also more convenient. For experienced researchers TDHP has increased greatly the efficiency in every phase, including analysis, parameter adjusting, and the display of results. Under the framework of the virtual observatory, for developing countries, we should study more and newer related technologies, which can advance the ability and efficiency in scientific research, like the software we are developing.

9.4. *A series of technologies exploiting data sharing*

Ganghu Lin, National Astronomical Observatories, Chinese Academy of Sciences, Beijing, China, email: `lgh@sun10.bao.ac.cn`

One of the purposes of the Virtual Observatory is providing data sharing. The Solar Multi-Channel Telescope in Huairou Solar Observing Station, Beijing, China is not only used for science research but also for solar activity prediction and space environment prediction. For providing these services, we have been carrying through a series of technologies that enable data sharing. In this article, we will discuss the exploiting of this technology. The exploiting includes setting up a WWW server, network, network safety facility, data processing software and designing international unified meta data for our speciality, etc.

So far, as result of this work, the initial needs have been reached. We still have further work to implement, such as uploading data in real time, setting up a database with a query function, and continuous improvement of the software, etc.

Highlights of Astronomy, Volume 14
IAU XXVI General Assembly, 14-25 August 2006
Karel A. van der Hucht, ed.

doi:10.1017/S1743921307012240

Special Session 6
Astronomical data management

Raymond P. Norris[1] (ed.), Heinz J. Andernach[2], Güenther Eichhorn[3], Françoise Genova[4], R. Elizabeth Griffin[5], Robert J. Hanisch[6], Ajit K. Kembhavi[7], Robert C. Kennicutt[8] and Anita M.S. Richards[9]

[1]CSIRO ATNF, PO Box 76, Epping, NSW 1710, Australia
email: Ray.Norris@csiro.au

[2]Departamento de Astronomía, Universidad de Guanajuato, Guanajuato CP 36000, Mexico
email: heinz@astro.ugto.mx

[3]Astrophysics Data System, Smithsonian Astrophysical Observatory,
60 Garden St., MS-67, Cambridge, MA 02138, USA
email: gei@cfa.harvard.edu

[4]CDS, UMR CNRS/ULP 7550, Observatoire Astronomique,
11 rue de l'Université, F-67000 Strasbourg, France
email: genova@astro.u-strasbg.fr

[5]NRC Herzberg Institute of Astrophysics,
5071 West Saanich Road, Victoria, BC V9E 2E7, Canada
email: Elizabeth.Griffin@hia-iha.nrc-cnrc.gc.ca

[6]Space Telescope Science Institute, 3700 San Martin Drive, Baltimore, MD 21218, USA
email: hanisch@stsci.edu

[7]Inter-University Centre for Astronomy and Astrophysics, Pune, India
email: akk@iucaa.ernet.in

[8]Institute of Astronomy, University of Cambridge, Madingley Road, Cambridge, CB3 0HA, UK
email: robk@ast.cam.ac.uk

[9]MERLIN/VLBI National Facility, University of Manchester, Jodrell Bank Observatory,
Macclesfield, Cheshire SK11 9DL, UK
email:amsr@jb.man.ac.uk

Abstract. We present a summary of the major contributions to the Special Session on *Astronomical Data Management* held at the IAU XXVI General Assembly in Prague in 2006. While recent years have seen enormous improvements in access to astronomical data, and the Virtual Observatory aims to provide astronomers with seamless access to on-line resources, more attention needs to be paid to ensuring the quality and completeness of those resources. For example, data produced by telescopes are not always made available to the astronomical community, and new instruments are sometimes designed and built with insufficient planning for data management, while older but valuable legacy data often remain undigitised. Data and results published in journals do not always appear in the data centres, and astronomers in developing countries sometimes have inadequate access to on-line resources. To address these issues, an 'Astronomers Data Manifesto' has been formulated with the aim of initiating a discussion that will lead to the development of a 'code of best practice' in astronomical data management.

Keywords. astronomical data bases: miscellaneous, atlases, catalogs, surveys , instrumentation: miscellaneous, techniques: miscellaneous

1. Introduction

The last few years have seen a revolution in the way astronomers use data. Data centres such as ADS (Astrophysics Data System, `<http://adswww.harvard.edu/>`), CDS

(Centre de Données astronomiques de Strasbourg, <http://cdsweb.u-strasbg.fr/>), and NED (NASA/IPAC Extragalactic Database, <http://nedwww.ipac.caltech.edu/>) have transformed the way we access the literature and carry out our research. Archival research with space observatory data is becoming the dominant driver of new research and publications, while major ground-based mission archives are not far behind. Access to electronic data and publications has brought front-line research capabilities to all corners of the developed world, and a growing number of archives from major telescopes are being placed in the public domain. Other success stories include the vigorous international development of the Virtual Observatory (VO), the revolutionary public data releases from individual astronomical projects, and the rapid dissemination of results made possible by forward-thinking journals, the ADS, and the astro-ph preprint server. All of these position astronomy as a role model to other sciences for how technology can be used to accelerate the quality and effectiveness of science.

On the other hand, our management of astronomical data is still inadequate, to the detriment of our science. For example, there exist international pressures to surround our open-access databases in a morass of legal red tape, and we are poorly prepared to resist them (Norris 2005). There remains a bottleneck between journals and data centres, so that a significant fraction of important data published in the major journals never appears in the data centres, and there is little provision for the preservation of the digital data underlying the results published in peer-reviewed journals. New instruments are still being built with little planning or budgeting for data management, so that while the instrument may technically perform well, the quality of the delivered science fails to meet expectations or capacity. While astronomers in the developed world revel in instant access to data and journals, their developing-world colleagues still rely on photocopied preprints. Valuable legacy data which might prove crucial to the understanding of the next supernova lie undigitised and inaccessible in some remote storage room, at the mercy of natural hazards and human ignorance.

In many astronomical institutions, data management as a discipline is not yet taken seriously. For example, an astronomer making a large database publicly available is not given the recognition that is given to the author of a paper, even though the database may effectively attract far more 'citations' than the paper.

In an effort to raise awareness of these issues, and to work towards a policy of 'best-practice' astronomical data management, a Special Session on 'Data Management' was held at the IAU XXVI General Assembly in Prague in 2006. In addition, a lively and productive electronic discussion <http://www.ivoa.net/twiki/bin/view/Astrodata> took place over several months preceding the IAU General Assembly. This paper presents a summary of the contributions to this Session, including not only the oral and some poster presentations, but also key points from the e-discussion. It is co-authored by the main contributors to the oral sessions and to the preceding e-discussion, and represents a consensus view. Inevitably, it falls short of conveying the spirited discussion that livened the meetings.

2. The Virtual Observatory

2.1. *Overview of the Virtual Observatory*

The VO aims to provide astronomers with seamless access to on-line resources. A good overview of its present status is provided by the proceedings of Special Session 3 in these *Highlights of Astronomy.* Although there are many national VO projects, each working in a specific context with its own goals, strongly dependent on the local funding agencies, the

VO is a world-wide, global endeavour, and all projects work together in the International Virtual Observatory Alliance (IVOA), which was formed in 2001. Among its main tasks, the IVOA defines the VO interoperability standards, including the standards for resource registries, query language, data access layer, content description (semantics), and data models. Most of the essential VO standards are now ready or nearly ready, and the VO is now in transition from R&D infrastructure definition to implementation by data centres. Many participants in the groups in charge of defining the standards, formed by staff from national projects, are knowledgeable about data archives and on-line services, and care about providing useful and usable standards, with the aim that data and services can be published in the VO through a thin interface layer.

Because the VO has successfully developed these standards, together with a growing range of tools and services, it has attracted a huge visibility, interest, and respect in the IT (information technology) community. And yet perceptions in the broader astronomical community are mixed, ranging from those who enthusiastically use VO tools to generate science, to those who consider that its value and relevance have yet to be demonstrated. VO proponents acknowledge that some critical challenges remain (e.g., long-term curation, quality control, certification, intellectual property guidelines, version control) and a number of key tasks remain unfinished, but the momentum of VO development is steadily meeting this challenge.

2.2. *Data centres in the VO*

The definition of a data centre can range from ‘a place distributing observational data’ to a service such as the CDS or NED which also distributes information, tools, and value-added services. In the VO context, new types of data and service providers emerge, and it is more appropriate to define a ‘data centre’ in terms of attributes such as *service to the community*, *added-value linked to expertise*, *sustainability*, and *quality*.

Many teams are willing to provide VO-compliant data and services in their domains of expertise. Key participants in the VO include the ‘classical’ data centres such as observatory archives, discipline-specific data centres, and data centres like CDS and NED which provide reference services and tools. In addition, a growing number of scientific teams are willing to participate by providing specific value-added services and tools in their domains of expertise. For instance, when the French VO performed a census in 2004, more than 40 teams planned to participate in VO-related actions, and most of these confirmed their activity in a census update two years later, indicating real commitment.

The community of VO data and service providers is therefore diverse both in the size of the teams, and in the context in which they work, and range from large national or international agencies to small teams working in scientific laboratories. Many types of services are being implemented, such as:

- observation archives, with a strong emphasis on ‘science ready’ data,
- value-added services and tools, with compilations, including additional data required for data interpretation, such as data on atomic and molecular lines,
- theoretical services, with on-demand services, or sets of pre-cooked modelling results,
- software suites, in particular for data analysis,
- specific services, to help solving specific science questions, and
- full research environments.

One key objective for the VO projects in the coming years is to create a community of VO service providers, who will help data centres to use the VO framework, and gather their feedback from implementation. This is a very important role for the national VO projects, and IVOA has to take into account the implementation feedback. The functionality of the VO in this increasingly operational phase is illustrated by the schema of the Euro-VO, which has three facets interacting together: a Data Centre Alliance, a distributed Technology Centre in charge of infrastructure definition, and a Facility Centre (ESA, ESO and national projects) which provides general information and supports users.

2.3. *The future of the VO*

Since the advent of the Internet, astronomy has been at the forefront for provision and networking of on-line data and services. This has already produced a revolution in the way astronomers work, even if they do not always realize it and simply use the tools. The VO is the next step, providing new resources and seamless access to them. New data and tools are already here and will be continue to be added.

The VO development provides a strong incentive to observatories and scientific teams to make their data and services available to the whole community, so that many teams want to become VO data centres. This is excellent news, since it will increase the sharing of information and knowledge among the community. Data centres have certain requirements, including

- a *critical mass* adapted to the aims;
- *medium term sustainability*, which requires strong support from the funding agencies;
- *national/international scientific niche* to gain community support.

Data centres also have significant responsibilities, including *curating data*, which comes with a large overhead for selecting, homogenizing, describing, and distributing data, and data centres must be sufficiently funded to perform these tasks. One has to keep in mind that data centres can be terminated and that it is critical for them to define a long term strategy, and to adjust it to the scientific evolution of astronomy, to technical evolution, and to the evolution of context, such as the development of the VO.

Why should data providers join the VO? They will have to care more about data quality and metadata, which means more work, but they will improve their service, and will have more occasions to collaborate with colleagues and build synergies between their services, and their visibility and usage statistics will increase significantly. The most difficult task will probably be to provide and maintain the service and to ensure quality, not to implement the VO framework!

Success of an operational VO network will ultimately be measured by customer participation and satisfaction, where the 'customers' include both users and data providers. And while many VO elements, such as format standardization and user tools, are already in place, others are still being addressed. These remaining challenges include long-term data access, data quality and curator certification, version control, histories, and intellectual property standards. These require a shift in focus away from technological tools towards a suite of data management processes.

3. Open access and observatory archives

3.1. *Open access*

Because the advance of astronomy frequently depends on the comparison and merging of disparate data, it is important that astronomers have access to all available data on the objects or phenomena that they are studying. Astronomical data have therefore always enjoyed a tradition of open access, best exemplified by the astronomical data centres, which provide access to data for all astronomers at no charge. There exist a number of exceptions to this open access tradition, some of which are widely supported, such as the initial protection of observers data by major facilities.

At the 2003 IAU XXV General Assembly a resolution was adopted which says, broadly, that publicly-funded archive data should be made available to all astronomers. This is aligned with ICSU (International Council for Science) and OECD (Organisation for Economic Cooperation and Development) recommendations, and may be regarded as a first step towards articulating the principles by which the astronomical community would like to see its data managed. Since then, a number of observatories, notably the European Southern Observatory (ESO), have embraced an open-access policy, but there remain a number of observatories that have not yet made their archival data publicly available, typically because of resource constraints. There also remain a few observatories (primarily privately-funded) which allow data archive access only to affiliated scientists, while still benefiting from the open access policies of other institutions.

The adoption of an open-access policy is not just for the public good. Roughly three times as many papers (and citations) result from data retrieved from the *Hubble* archive as those based on the original data (Beckwith 2004). In the parallel case of *IUE* (*International Ultraviolet Explorer*) spectra, five times as many publications resulted from archive data (Wamsteker & Griffin 1995). So, in principle, observatories might quadruple their science by making their archive data public. Since the funding for most major observatories depends on performance indicators such as publications and citations, it may be an expensive decision for an observatory not to adopt an open-access policy.

3.2. *Data needs for new telescopes*

Part of the success of modern astronomy can be attributed to astronomers who continue to strive for bigger and better instruments. But as plans are developed for a new telescope, data processing and management are sometimes neglected. However, half the cost of a modern ground-based telescope is typically in the software and data processing. These need to be planned and developed at the same time as the hardware, rather than leaving it to graduate students or support staff to figure out when the data arrive. This may seem obvious, especially to those major projects that already routinely follow this practice. However, some projects have not shown such foresight, resulting in instruments which perform well technically, but which have not delivered the expected scientific impact. To avoid this, it is vital to think about these issues before, rather than after, the telescope is funded and built.

4. Journals and data

4.1. *The changing face of astronomical publications*

`arXive/astro-ph` is now the primary channel for disseminating new research results, and ADS is now the primary channel for accessing published papers. Electronic editions have become the main journals of record, and the days of paper journals are numbered. The primary journals (*A&A*, *AJ*, *ApJ*, *MNRAS*) are adapting to this new publishing paradigm, but the future of commercial and small journals is unclear. Meanwhile, astronomical monographs and conference proceedings generally remain locked in the old paradigm, and consequently their impact is declining. Other components of 'grey literature', the observatory reports and technical papers, are locked out of the new paradigm, and are being lost.

A further consequence of this changing paradigm is that the current business model for astronomical publications is being challenged. Most astronomers accept the need for high-quality peer-reviewed journals, while searching for ways in which they can be improved, and made cheaper. But there is a growing demand for open access, or free, journals, although it has yet to be demonstrated how an open-access journal can afford to maintain the quality that we have come to expect from our mainstream journals. Thus, the future business model for peer-reviewed publication is unclear.

In addition to accessing the journal article itself, astronomers are demanding better links between publications and data, where the term 'data' is taken to include primary observational data, published results based on those data, and graphical representations of results. Astronomers requirements vary from field to field, and include:

- tables published in a journal should be accessible by catalogue browsers such as VizieR;
- results published in the journal should appear in object-searchable or position-searchable databases such as NED or SIMBAD;
- readers should be able to click on an object in a journal to obtain more information about that object from a database such as NED or SIMBAD;
- users of NED or SIMBAD should be able to trace a link back to a refereed publication which validates and authenticates the data; and
- links should be given in a publication to an archive containing its source data.

Both the journals and the data centres are actively addressing these issues. For example, a collaboration between the American Astronomical Society (AAS) and the Astronomical Data Center Executive Committee (ADEC) has put in place a system that allows authors to specify data that they used in their articles. This information is then processed by the publisher and used to link from articles to on-line data, both in the journals, and the ADS. It is hoped that other publishers and data centres world-wide will participate in this system to provide such links world-wide.

Although the journals are embracing opportunities engendered by the new technologies such as active links from electronic journals to the data centres, the metadata (and for that matter, the quality of error-checking in the tables themselves) provided by authors are not currently sufficient to enable completely automated transfers of the results from journal to data centre. Consequently, data centres often have to go through published tables by hand. Their capacity to do so is strongly limited by available resources, and so a significant fraction of published results never appears in a data centre, or does so only after a period of some years.

For example, Andernach (2006) has conducted a case study of over 2000 published articles, for which he collected or restored (via OCR) the electronic tables they contain, and finds that typically only about 50 % of results published in journals ever appear in the data centres, and lists some surprising and significant omissions. Strangely, this fraction did not appear to change significantly as journals changed from print-only to electronic formats.

One solution to this data bottleneck would presumably be to increase funding for the data centres to enable them to employ more staff to transcribe and interpret the journal data, but the finite resources available make this option unlikely. An alternative option is to define formats and metadata that are author-friendly, journal-friendly, and data centre-friendly, and define the data sufficiently well. Then, if an author chooses to supply these metadata, and certifies that he has checked the data using appropriate tools (many of which are already available), they could be imported automatically into the data centres. This effectively redistributes the transcription workload from the data centres to the authors, and necessarily entails more work for authors. However, the authors will benefit from the greater scientific impact and the higher citation rate that will result from their data being in the data centres. In many cases, the paper itself will benefit from this further level of checking, which will remove the errors that are still too common in published papers, which therefore require checking by data centre staff before their data are accepted by the data centres.

There is no widespread agreement whether such a system can ever be made to work reliably without reducing the quality of the data in the centres. Some astronomers are concerned that using such tools would allow more errors to remain undetected when data are deposited in data centres. Others argue that this concern is outweighed by the advantages of easier access to data. Data centres experienced in handling author-provided data and metadata have expressed doubt that quality can be insured without a final stage of checks by data centre or journal experts. A pilot study seems justified to explore the feasibility and examine whether results based on data published in this way are in fact less reliable.

4.2. *Digital data preservation for astronomy journals*

Astronomers are producing and analysing data at ever more prodigious rates. NASA's Great Observatories, ground-based national observatories, and major survey projects have archive and data distribution systems in place to manage their standard data products, and these are now interlinked through the protocols and metadata standards agreed upon in the Virtual Observatory. However, the digital data associated with peer-reviewed publications are only rarely archived. Most often, astronomers publish graphical representations of their data but not the data themselves. Other astronomers cannot readily inspect the data to either confirm the interpretation presented in a paper or extend the analysis. Highly processed data sets reside on departmental servers and the personal computers of astronomers, and may or may not be available a few years hence.

A project led by Hanisch is investigating ways to preserve and curate the digital data associated with peer-reviewed journals in astronomy. The technology and standards of the VO provide one component of the necessary technology. A variety of underlying systems can be used to physically host a data repository, and indeed this repository need not be centralized. The repository, however, must be managed and data must be documented through high quality, curated metadata. Multiple access portals must be available: the original journal, the host data centre, the Virtual Observatory, or any number of topic-oriented data services utilizing VO-standard access mechanisms.

The near-term goal of this project is to implement an end-to-end prototype digital data preservation facility using astronomy scholarly publications as a test-bed. Astronomy is an ideal discipline to start with, as most data are available in a single standard format (FITS), the community is small and highly aware of e-publishing, and there are few restrictions on data access and exchange. The prototype will be implemented using commodity open-source technologies and will utilise the infrastructure already being developed by the VO in order to minimize development costs and maximize flexibility. Specific development tasks include metadata definition, evaluation and selection of a content management tool (Fedora, DSpace, etc.), deployment of storage applications and layered storage management software (VOSpace), and adapting the publication process for data capture. By implementing a prototype, it is hoped to understand operational costs and thus be able to develop a long-term business plan for the preservation of peer-reviewed journal content and the associated supporting data. The availability of such facilities for digital data preservation will undoubtedly lead to changes in policies affecting data access. Peer pressure may initially encourage researchers to contribute their data to the repository, but eventually such contributions might become mandatory. It seems likely that published papers having digital data available will be more heavily used, and thus more heavily cited, than papers lacking such data. In the longer term it will be important to evaluate the impact on scientific productivity through citation analyses and community feedback.

5. Challenging the Digital Divide

The 'Digital Divide' refers to the widening gulf between those who have high-bandwidth access to information, data, and web services, and those who do not. For those who do not, their lack of access results in even more disadvantages, making it even less likely that they will gain access in the future.

Astronomers in developing countries are better positioned than their colleagues in some other disciplines, because the astronomical data centres maintain immense databases, and electronic archives of scientific periodicals, while the latest research is available through preprints on astro-ph. Furthermore, some of the leading astronomical journals provide free or cheap access to astronomers in developing countries. In the future, the situation is set to improve further, as powerful Virtual Observatory tools will provide even better access to astronomical data. Meanwhile, facilities such as SALT and GMRT are already demonstrating the feasibility of building leading-edge facilities, complete with well-managed data archives, in developing countries. However, many astronomers in these countries lack the bandwidth, expertise and the environment to make use of these riches. Further obstacles include resistance to the use of new concepts and tools, and reservations about exposing hard-won data to international access.

For example, India possesses several research institutes with state-of-the-art facilities, including access to high bandwidth, databases, literature and computing facilities. The Indian software industry is one of the most successful in the world, and yet very few Indian astronomers make extensive use of archival data for large scientific projects, and little attention is given to software aspects of large astronomical projects. As a result, the Information Technology prowess of India in the business domain has not been exploited by the scientific community, astronomical data from Indian observatories have not been archived and made available to the community, and India remains on the wrong side of the digital divide.

The situation in Africa, which does not have the technological advantages available to India, is even worse. Most institutions do not have good internet bandwidth, and ADS access statistics show that although African ADS usage is increasing, African astronomers are not yet taking full advantage of the available digital information.

However, because of a number of initiatives, the situation seems to have improved slightly in the past few years. For example, associates and their students from all over India are funded to spend a few months every year at the Inter-University Centre for Astronomy and Astrophysics, where they develop their own research programs and set up collaborations. The resulting technology and expertise are transferred to the universities, helped by the decreasing cost of personal computers.

This shows that such efforts are producing results and need to be continued and supported as much as possible. In particular, the Indian experience could be replicated elsewhere in the developing world. Although the digital divide problem extends over all disciplines, astronomy is well-positioned to lead the charge to challenge this divide. Astronomers in the developing world could help build archives, develop software and provide much needed human resources, using a platform provided by the Third World Astronomy Network.

6. Safeguarding data

There are many reasons, both scientific and economic, why a properly-managed data archive is an essential facility in astronomy. Most modern astronomers agree with the principle of archiving data for the wider community benefit, but in practice our achievements are patchy, particularly in the case of the preservation and accessibility of historic data. While the VO is currently focussing on modern space- and ground-based data that were recorded digitally, much less attention is being paid to astronomy's rich legacy of photographic observations, some of which date back to the late 19th century.

The value of such data to modern science has been demonstrated repeatedly, through studies of very long-period variability (something that is predicted, but scarcely figures even today in the astrophysics landscape), identifying and measuring non-recurrent events such as spectrum changes in AGB stars, refining small-body orbits (including those crucial near-earth objects), studying the pre-outburst phases of a supernova (such as was very fortunately possible for SN 1987A), resolving important ambiguities or anomalies through more precise re-measurements of historic data, or in inter-disciplinary science such as measuring the Earth's ozone concentrations as extracted from historic stellar spectra. There are also many data sets on magnetic tape that were abandoned as incompatible technology moved ahead without them. Although the physical longevity of photographic data far outmatches that of tapes, resources for safeguarding them are necessarily in competition with those required to generate new data. It is therefore important to determine, as far as we can, what value to place on the historical archives, and to determine a workable solution for their long-term storage and digitization before we lose the opportunity to make that decision.

The migration of present-day digital data is now well orchestrated in data centres, so that as technology moves on, data are migrated seamlessly to new media or new formats. Outside the data centres, however, the problem remains. Astronomers and small observatories keep magnetic tapes, including DAT and Exabytes tapes, well beyond their recommended life. Few now have the technology to read a round magnetic tape or a $5\frac{1}{4}$-inch floppy. How long before an Exabyte, or even a CDROM or a DVD, becomes unreadable?

Astronomy needs to take a broader view of safe-guarding its data. The cost of recovering historic observations as a common-user resource is small compared to new installations or space missions.

7. The astronomers data manifesto

In an attempt to raise awareness of these issues, and define the goal, the IAU Working Group for *Astronomical Data* proposed the following manifesto. This is intended not as a rigid declaration, but as a stimulus for discussion. It is hoped that such discussion will converge to a consensus on how the astronomical community would like to see its data managed.

"We, the global community of astronomy, aspire to the following guidelines for managing astronomical data, believing that they would maximise the rate and cost-effectiveness of scientific discovery. We do not underestimate the challenge, but believe that these goals are achievable if astronomers, observatories, journals, data centres, and the Virtual Observatory Alliance work together to overcome the hurdles.

(*a*) All significant tables, images, and spectra published in journals should appear in astronomical data centres.

(*b*) All data obtained with publicly-funded observatories should, after appropriate proprietary periods, be placed in the public domain.

(*c*) In any new major astronomical construction project, the data processing, storage, migration, and management requirements should be built in at an early stage of the project plan, and costed along with other parts of the project.

(*d*) Astronomers in all countries should have the same access to astronomical data and information.

(*e*) Legacy astronomical data can be valuable, and high-priority legacy data should be preserved and stored in digital form in the data centres.

(*f*) The IAU should work with other international organisations to achieve our common goals and learn from our colleagues in other fields."

8. Conclusion

The revolution in the way that astronomy manages its data has already resulted in enormous scientific advances. The potential for further advances in the future is even greater, but we need to have a clear vision and a clear goal if we are to succeed in realising that potential. In particular, the astronomical community needs to have a clear picture of what represents 'best practice' in astronomical data management. Astronomy does not have any strategic data framework to provide policies or guidelines for astronomical data management, and is not therefore able to represent the interests of astronomical data management to external parties. For example, data quality, long-term accessibility, and provenance carry real costs but are critical requirements for success.

The Virtual Observatory is a powerful tool that will enable us to make even more effective use of our data, but we should not regard it as a cure-all for our current deficiencies in data management. Whilst the VO is attempting to make major databases accessible to all astronomers, it cannot do so unless those databases are properly constructed and managed. Now that the infrastructure is in place we need to focus on building a user base and bringing in all key archives and collections.

These advances also have the potential to overcome the Digital Divide, but only if further initiatives enable open access to these facilities by astronomers in developing countries. Such initiatives are likely to be cost-effective, as the VO, electronic publication, and effective archives will enable science to tap an enormous intellectual base, with fresh ideas and approaches, which will benefit all of us.

Scientific Organizing Committee

Giuseppina Fabbiano (USA), Françoise Genova (France), Robert J. Hanisch (USA), Ajit K. Kembhavi (India), Andrew Lawrence (UK), Oleg Yu. Malkov (Russia), Atilla Meszaros (Czech Republic), Raymond P. Norris (Australia, chair), Masatoshi Ohishi (Japan), Peter Quinn (Germany), Isabelle F. Scholl (France), Enrique Solano (Spain), Alexander Szalay (USA), Nicholas A. Walton (UK), Roy Williams (USA), and Yongheng Zhao (China).

References

Andernach, H. 2006, <http://adsabs.harvard.edu/abs/
Beckwith, S. V. W. 2004, private communication
Norris, R. P. 2005, *IAU Information Bulletin*, No. 96, p. 16
Wamsteker, W., & Griffin, R. E. M. 1995, *Ap&SS*, 288, 383

Highlights of Astronomy, Volume 14
IAU XXVI General Assembly, 14-25 August 2006
Karel A. van der Hucht, ed.

doi:10.1017/S1743921307012252

Special Session 7
Astronomy in Antartica

Michael G. Burton[1,2] (ed.)

[1]School of Physics, University of New South Wales, Sydney, NSW 2052, Australia
email: M.Burton@unsw.edu.au

[2]Armagh Observatory, Armagh, BT61 9DG, Northern Ireland, UK

Abstract. The high, dry and stable climatic conditions on top of the Antarctic plateau offer exceptional conditions for a wide range of observational astronomy, from optical to millimetre wavelengths. This is principally on account of the greatly reduced thermal backgrounds, the improved atmospheric transmission and the supurb seeing, in comparison with conditions at temperate latitude sites. The polar plateaus in the Arctic may also offer excellent conditions for astronomy, though these have yet to be quantified. We briefly review the history of astronomy in Antarctica and outline some of the activities now taking place on the polar plateaus, and plans for the future.

Keywords. Antarctica, site testing, infrared, optical, millimetre, neutrinos, meteorites

---

1. A brief history of antarctic astronomy

Antarctica is the highest, driest and coldest continent, facets which all offer gains for observational astronomy. The field goes back as far as 1912 with the discovery of the 'Adelie Land Meteorite' during Douglas Mawson's Australasian Antarctic Expedition (see Indermuehle, Burton & Maddison 2005). It was to be another 60 years, however, before this investigation advanced further, with the discovery in Antarctica, by Japanese scientists, of a number of different meteorites in close proximity – an event which could not have happenned by chance (Nagata 1975). More meteorites have since been discovered on the continent than the rest of the world put together, a result of the favourable conditions for their collection.

Cosmic ray detectors were installed in the Australian base of Mawson in 1955 and the US base of McMurdo during the International Geophysical Year of 1957, and are now widespread around the continent, taking advantage of the high geomagnetic latitude to detect lower energy particles than reach the ground at mid-latitudes.

The modern era in Antarctic astronomy began in 1979 with measurement of solar oscillations using an 8 cm optical telescope at the South Pole by a team led by Martin Pomerantz (see Grec *et al.* 1980). The field began in earnest with the establishment in 1991 of the Center for Astrophysical Research in Antarctica (CARA) at the South Pole, led by Al Harper. Experiments were begun in infrared, sub-millimetre and CMBR astronomy, together with an extensive site testing program to quantify the conditions for observation. Particularly successful were the sub-mm observations pursued with the 1.7 m AST/RO telescope (resulting in nearly 50 refereed publications; see Stark *et al.* 2001), and a series of increasingly precise measurements of fluctuations in the cosmic microwave background (e.g., the first measurement of its polarization – Kovac *et al.* 2002).

An extensive bibliography of publications relating to Astronomy in Antarctica can be found at the JACARA website `<www.phys.unsw.edu.au/jacara>`. Papers by Burton

(2004), Ashley *et al.* (2004), and Storey (2005) provide further background on conducting astronomy in Antarctica and some of the current activities taking place.

2. The South Pole

The South Pole suffers from a turbulent boundary layer, approximately 200 m thick, driven by the katabatic wind flowing off the summit of the Antarctic plateau (i.e., from Dome A). The seeing at ice level is thus modest, and so optical and infrared observations have not been pursued there beyond using modest-sized telescopes (e.g., the 60 cm SPIREX; see Rathborne & Burton 2005 for a review), where the diffraction limit is comparable to the seeing. Nevertheless, extremely deep thermal IR images have been obtained. For instance, it was only in 2004 that the 8 m VLT achieved deeper (ground-based) images at 3.5 μm than those obtained with SPIREX in 1998 of the 30 Doradus star forming complex in the LMC (see Maercker & Burton 2005).

Conditions for millimetre astronomy are, however, supurb, as they are for neutrino detection (making use of the vast quantities of pure ice to track the Cerenkov radiation following extremely rare interactions with nuclei). Astronomy at the South Pole is today dominated by two experiments being built to exploit these conditions. These are the 10 m South Pole Telescope (SPT, Ruhl *et al.* 2004), designed to measure the SZ-efect towards galaxy clusters in the sub-millimetre, and IceCube, a 1 km^3 neutrino *telescope*, designed to locate neutrino sources in the northern skies (Ahrens *et al.* 2004). A new station has been constructed at Pole by the US in order to meet the required infrastructure needs.

3. Dome C

Concordia Station, at the 3,200 m Dome C (one of the 'summits' along the ridge of the Antarctic plateau), is the newest scientific station on the continent. Run jointly by France and Italy, the Station's first 'winter' occurred in 2005. Already the median visual seeing, above an $\sim$30 m thick boundary layer, has been shown to be $\sim 0.''25$, and to fall below $0''.1$ on occassion (Lawrence *et al.* 2004; Agabi *et al.* 2006). Amongst the gains that have been quantified for optical and infrared astronomy are an isoplanatic angle 2–3 times smaller than at good temperate sites, coherence times 2–3 times longer, IR sky backgrounds 20–100 times lower, image sizes 2–4 times smaller (when using the same size telescope), and scintillation noise 3–4 times smaller.

An 80 cm prototype mid-infrared telescope (IRAIT) is shortly to be commissioned at Dome C (Tosti *et al.* 2006), in order to provide a first demonstration of the science potential, as well as to conduct a range of science programs. This may be followed by the 2.4 m optical/IR PILOT (Storey 2006; Burton *et al.* 2005). PILOT would be large enough to have comparable IR sensitivities as temperate latitude 8 m-class telescopes (on account of the low background), but be able to obtain high angular resolution (on account of the supurb seeing) over wide fields of view.

Beyond this, there are several possible options for telescopes under active discussion. These include optical/IR interferometers (API, Swain *et al.* 2003; and KEOPS, Vakili *et al.* 2005), as well as large (8 m++) telescopes (e.g., LAPCAT, Storey *et al.* 2006; and GMTA, Angel *et al.* 2004). Whether these are built at Dome C, or eleswhere, will depend on when other high plateau sites are opened for astronomy.

4. Dome A and Dome F

There are other sites along the summit ridge of the Antarctic plateau that will provide comparable conditions to Dome C, and possibly be superior in some respects. These are the highest point, the 4,200 m Dome A, and the 'northern' end of the summit ridge, the 3,800 m Dome F. Dome F is already the site of a Japanese ice-core drilling station ('Fuji') and has wintered over, though no measurements have yet been made of the astronomical site conditions. Dome A ('Argus') was visited for the first time by humans in January 2005 on a Chinese traverse, and is the subject of expeditions for the International Polar Year of 2007–08. Compared to Dome C, the most significant gain may be the lower water vapour content of the atmosphere, opening up windows in the terahertz regime, and possibly a slightly lower IR background (colder) and thinner boundary layer (even lower katabatic winds).

5. The Arctic

The polar plateaus of the Arctic also offer some promise for observational astronomy, though no site testing has yet been carried out. In particular, the summit ridge of the Greenland icecap (including the 3,200 m Summit Station) and northern Ellesmere Island in the North-West territories of Canada (reaching 2,600 m) warrant investigation. While not as cold or as dry as the Antarctic plateau, conditions should still be favourable for infrared and sub-millimetre astronomy. However it is possible that higher winds (storms?) and greater cloud cover may limit their use for astronomy?

Acknowledgements

The Antarctic astronomy program at UNSW could not have been developed without the support of a great many colleagues from around the world. I particularly wish to thank John Storey, Michael Ashley & Jon Lawrence at UNSW, without whose trojan efforts over the past decade the program would be but a shadow of what is has now become.

References

Agabi, A., Aristidi, E., Azouit, M., *et al.* 2006, *PASP*, 118, 344

Ahrens, J., Bahcall, J. N., Bai, X., *et al.* 2004, *New Astron. Revs.*, 48, 519

Angel, R., Lawrence, J.S., & Storey, J. W. V. 2004, *SPIE*, 5382, 76

Ashley, M. C. B., Burton, M.G., Lawrence, J. S., & Storey, J. W. V. 2004, *AN*, 325, 619

Burton, M. G. 2004, in: A. Heck (ed.), *Organisations and Strategies in Astronomy*, Vol. 5 (Dordrecht: Kluwer), p. 11

Burton, M. G., Lawrence, J. S., Ashley, M. C. B., *et al.* 2005, *PASA*, 22, 199

Grec, G., Fossat, E., & Pomerantz, M. 1980, *Nature*, 288, 541

Indermuehle, B. T, Burton, M. G., & Maddison, S. T 2005, *PASA*, 22, 73

Kovac, J., Leitch, E. M., Pryke, C., *et al.* 2002, *Nature*, 420, 772

Lawrence, J. S., Ashley, M. C. B., Tokovinin, A., & Travouillon, T. 2004, *Nature*, 431, 278

Maercker, M., & Burton, M. G. 2005, *A&A*, 438, 663

Nagata, T. (ed.) 1975, *Mem. Nat. Ins. Polar Res.*, 5

Rathborne, J., & Burton, M. G. 2005, in: O. Engvold (ed.), *Highlights of Astronomy*, Vol. 13, as presented at the XXVth General Assembly of the IAU (San Francisco: ASP), p. 937

Ruhl, J., Ade, P. A. R., Carlstrom, J. E., *et al.* 2004, *SPIE*, 5498, 11

Stark, A., Bally, J., Balm, S. P., *et al.* 2001, *PASP*, 113, 567

Storey, J. W. V. 2005, *Antarctic Science*, 17, 555

Storey, J. W. V. 2006, *Acta Ast. Sin. Sup.*, 47, 407

Storey, J., Angel, R., Lawrence, J., Hinz, P., Ashley, M., & Burton, M. 2006b, *SPIE*, 6267, 45

Swain, M. R., Coude du Foresto, V., Fossat, E., & Vakili, F. 2003, *Mem. S.A.It. Suppl.*, 2, 207

Tosti, G., Busso, M., Nucciarelli, G., *et al.* 2006, *SPIE*, 6267, 47

Vakili, F., Aristidi, E., Schmider, F.X., *et al.* 2005, *EAS-PS*, 14, 211

Highlights of Astronomy, Volume 14
IAU XXVI General Assembly, 14-25 August 2006
Karel A. van der Hucht, ed.

doi:10.1017/S1743921307012264

Ten years from the Antarctic Sub-millimeter Telescope and Remote Observatory

Christopher L. Martin, on behalf of the AST/RO Team
Department of Physics and Astronomy, Oberlin College, Oberlin, OH 44074, USA
email: Chris.Martin@oberlin.edu

1. The telescope

Beginning with the winter season of 1995 and for the next ten years, the Antarctic Sub-mm Telescope and Remote Observatory (AST/RO, Stark *et al.* 1997; Stark *et al.* 2001; `<http://www.cfa.harvard.edu/ASTRO/>`), a 1.7 m diameter, offset Gregorian telescope located at an altitude of 2847 m at the Amundsen-Scott South Pole Station collected sub-mm and Terahertz data in the 1.3 mm to 200 μm wavelength bands. From its location just a few hundred meters away from the geophysical South Pole, AST/RO was the first sub-mm telescope to over-winter on the polar plateau, a location uniquely suited to high quality sub-mm observations due its very low humidity, high atmospheric stability and thin troposphere (Chamberlin *et al.* 1997).

While there are a number of scientists around the world still making use of AST/RO data, the telescope itself was regrettably decommissioned in December 2005. This was due in part to the slow drifting of snow at the South Pole which caused the snow level surrounding the building to slowly rise over time. In Fig. 1, you can compare two photos of the telescope building taken a few years after construction and then again from a similar vantage point in late 2005. In the background can also be seen one of the other grand changes to the landscape during this ten year time-span, the construction of the new Amundsen-Scott South Pole Station to replace the iconic dome.

2. Observing highlights

Over the past ten years scientific results using AST/RO data have appeared in nearly fifty peer-reviewed articles, seven Ph.D. theses, and numerous conference proceedings. For the full list see the AST/RO web site. While the full details of all of AST/RO's results are far too numerous to list here, I would like to highlight just a few.

- First detection of C I emission in the Magellanic Clouds (Bolatto *et al.* 2000a)
- First detection of C I in absorption (Staguhn *et al.* 1997)
- Surveys of C I and CO $J = 7 \rightarrow 6$, CO $J = 4 \rightarrow 3$, and CO $J = 2 \rightarrow 1$ emission from
  - The H II region/molecular cloud complex NGC 6334 (Kim & Narayanan 2006)
  - The inner few degrees of the Milky Way (Martin *et al.* 2004)
  - Nine strips across the Galactic plane (Lane, in prep.)
  - The Carina spiral arm region around η Carinae (Zhang *et al.* 2001)
  - N 159 / N 160 region in the LMC (Bolatto *et al.* 2000b)
  - Lupus Clouds as part of the Spitzer Legacy Program 'Cores to Disks' (Tothill, in prep.)
- Maps in the [N II] 205 μm line of G287.57−0.59 using SPIFI
- Numerous PhD theses
  - Staguhn (1996), U. Cologne – Galactic Center
  - Ingalls (1999), Boston U. – High Latitude Clouds
  - Bolatto (2001), Boston U. – Magellanic Clouds

Figure 1. On the left is a picture taken of AST/RO in the late 1990's, a few years after its construction. For contrast, another picture of AST/RO taken in late 2005 during decommissioning is shown on the right. You can clearly see how the snow level has risen around the building over the course of the decade due to drifting in the Dark Sector at the South Pole.

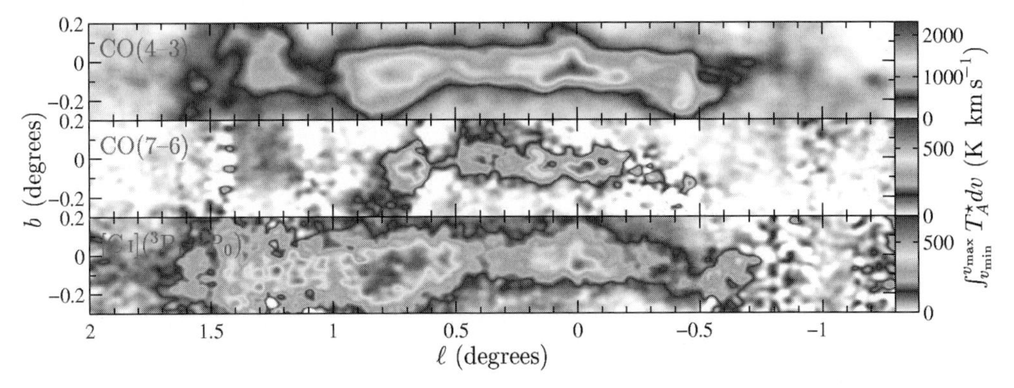

Figure 2. The three panels show spatial-spatial (ℓ, b) integrated intensity maps for the three transitions observed with AST/RO in the Galactic Center region. Transitions are identified at left on each panel. The emission is integrated over all velocities where data are available. All have been smoothed to the same $2'$ resolution.

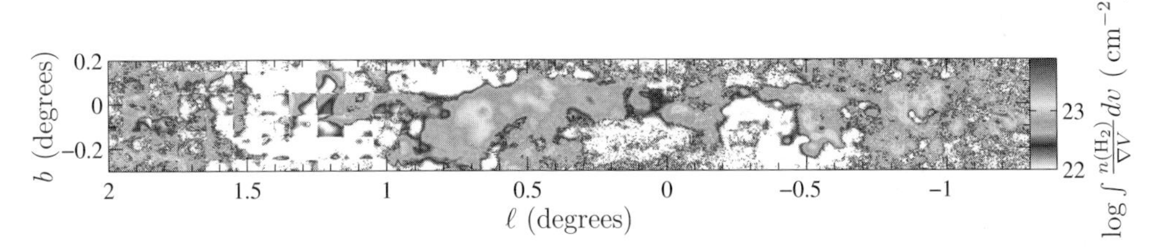

Figure 3. This figure shows one of the most exciting results of AST/RO's Galactic Center Survey. We used a Large Velocity Gradient (LVG) analysis to calculate quantities like the column density (shown here) for every point in our (ℓ, b) map. This independent estimate of column density is then available for a wide range of scientific studies. For electronic versions of results from this region as published in Martin *et al.* (2004), contact the author.

- Huang (2001), Boston U. – Southern H II Regions
- Kulesa (2002), U. Arizona – Dark Interstellar Clouds
- Groppi (2003), U. Arizona – Star Forming Regions

3. Galactic Center survey

While each of the highlights listed above is equally impressive, I would like to focus in on one to serve as an example for what AST/RO has been capable of.

To understand the strongly excited gas near the center of our own galaxy, detailed surveys in a variety of higher excitation states are required. To aid in this effort, AST/RO completed a fully sampled survey of CO(7-6), CO(4-3), [CI]($^3P_2$-$^3P_1$), and [CI]($^3P_1$-$^3P_0$) in a three square degree region around the Galactic Center (Martin *et al.* 2004) as shown in Fig. 2. In addition to this inner region, AST/RO has recently completed a survey area around Clump 1 and 2, thus covering the bulk of strongly excited gas near the center of the galaxy.

To collect this dataset required nearly a million distinct telescope pointings over many square degrees of the sky. To handle a sub-mm dataset of this size required the development of new automated observational methodologies, reduction techniques, and visualizations not to mention a substantial amount of observing time. Fortunately AST/RO was designed from the start as a survey instrument with a beam-size of 103–109″ at 461–92 GHz and 58″ at 807 GHz. So while covering a few square degrees was still a sizable proposition, we could reasonably contemplate making the multiple passes necessary to acquire data with the uniform signal-to-noise ratio required.

One of the interesting features of this data set is that by using the wide range of emission lines available to AST/RO, we can accurately estimate the kinetic temperature and density over a wide region of the survey using a Large Velocity Gradient (LVG) technique as shown in Fig. 3.

Acknowledgements

While there are many people to whom thanks is due (see the AST/RO website for the full list), none of the work described here would have been possible without the tireless efforts of the winter-over scientists who spent a full year stationed at the South Pole with the telescope taking the data and maintaining the instrument. AST/RO is supported by NSF grant number ANT-0126090

References

Bolatto, A. D. 2001, *The Interstellar Medium in Low Metallicity Environments*, PhD Thesis, Boston University

Bolatto, A. D., Alberto D., Jackson, J. M., Israel, F. P., *et al.* 2000a, *ApJ*, 545, 234

Bolatto, A. D., Jackson, J. M., Kraemer, K. E., & Zhang, X. 2000b, *ApJ* (Letters), 541, L17

Chamberlin, R. A., Lane, A. P., & Stark, A. A. 1997, *ApJ*, 476, 428

Groppi, C. E. 2003, *Submillimeter Heterodyne Spectroscopy of Star Forming Regions*, PhD Thesis, University of Arizona

Huang, M. 2001, *Interstellar Carbon Under the Influence of H II Regions*, PhD Thesis, Boston University

Ingalls, J. G. 1999, *Carbon Gas in High Galactic Latitude Molecular Clouds*, PhD Thesis, Boston University

Kim, S. and Narayanan, D. 2006, *PASJ*, 58, 753

Kulesa, C.A. 2002, *Molecular Hydrogen and Its Ions in Dark Interstellar Clouds and Star-Forming Regions*, PhD Thesis, University of Arizona

Lane, A. P. 1998, in: G. Novack & R. H. Landsberg (eds.), *Astrophysics From Antarctica, ASP-CS* 141, 289

Martin, C. L., Walsh, W. M., Xiao, K.-C., *et al.* 2004, *ApJS*, 150, 239

Staguhn, J. 1996, *Observations Towards the Sgr C Region Near the Center of our Galaxy*, PhD Thesis, University of Cologne

Staguhn, J., Stutzki, J., Chamberlin, R. A., *et al.* 1997, *ApJ*, 491, 191

Stark, A. A., Bally, J., Balm, S. P., *et al.* 2001 *PASP*, 113, 567

Stark, A. A., *et al.* 1997, *Rev. Sci. Instr.*, 68, 2200

Zhang, X., Lee, Y., Bolatto, A., and Stark, A. A. 2001, *ApJ*, 553, 274

Highlights of Astronomy, Volume 14
IAU XXVI General Assembly, 14-25 August 2006
Karel A. van der Hucht, ed.

doi:10.1017/S1743921307012276

The IceCube neutrino observatory: latest results on the search for point sources and status of IceCube construction

Thierry P.A. Castermans[1] and Albrecht Karle[2], for the IceCube collaboration†

[1]Université de Mons-Hainaut, B-7000 Mons, Belgium
email: Thierry.Castermans@umh.ac.be

[2]Department of Physics, University of Wisconsin, Madison, WI 53706, USA
email: jauffenb@uni-wuppertal.de

Abstract. The AMANDA neutrino telescope, prototype instrument of the IceCube neutrino observatory at South Pole, has collected data since 2000 in its final configuration. A period of 1001 days of livetime between 2000 and 2004 has been analysed in order to find evidence of a neutrino signal coming from point-like sources such as *microquasars*, *active galactic nuclei*, *supernovae remnants* or *gamma ray bursts*. A sensitivity to fluxes of $\nu_\mu + \bar{\nu}_\mu + \nu_\tau + \bar{\nu}_\tau$ of $d\Phi/dE = 1.0 \cdot 10^{-10} (E/\mathrm{TeV})^{-2} \cdot \mathrm{TeV}^{-1} \mathrm{cm}^{-2} \mathrm{s}^{-1}$ was reached in the energy range between 1.6 TeV and 1.6 PeV. No significant excess over the background has been found so far. Flux upper limits infered from this study can constrain certain neutrino emission models of X-ray binaries. IceCube will have a substantially higher sensitivity. Currently at 10 % of its final extension, it will comprise 4800 optical sensors deployed along 80 strings by early 2011, instrumenting one cubic kilometre volume of ice and 1 km^2 at the surface.

Keywords. Neutrinos, cosmic rays, point sources, gamma rays

1. Introduction

High energy neutrinos constitute highly valuable astronomical messengers. Unlike photons or protons, they can travel cosmic distances without being absorbed or deflected from their initial direction of propagation and deliver unaltered information related to the site of their emission. The Universe being transparent to photons only up to modest energies of order 1 TeV, neutrinos thus can be the indispensable partners of "conventional" astronomy to probe the most violent astrophysical objects.

Another major potential of neutrino astronomy resides in the fact that neutrinos could help us to understand the origin of cosmic rays. Indeed, these are thought to be accelerated in the expanding shocks of supernovae remnants, active galactic nuclei, gamma ray bursts or microquasars. In the vicinity of these extremely energetic sources, cosmic rays have the opportunity to interact with local hadronic matter or radiation fields giving rise to a flux of neutrinos and gamma rays (respectively by decay of charged or uncharged pions):

$$\begin{aligned} p + (p \; or \, \gamma) &\rightarrow \pi^0 \;\; \rightarrow \gamma\gamma \\ &\rightarrow \pi^\pm \rightarrow \nu_e \, \nu_\mu \rightarrow \nu_e \, \nu_\mu \, \nu_\tau \text{ (after oscillations)} \end{aligned}$$

The discovery of neutrino point sources would thus unambiguously reveal the sites of cosmic ray acceleration in the Universe.

† <http://icecube.wisc.edu/science/publications/vulcano2006.html>

Table 1. Flux upper limits for the sources in the catalog of potential neutrino emitters. From left to right are given the source name, its sky position, the number of observed and expected events, the upper limit on the contribution from signal events at 90 % confidence level μ_{90} and the expected number of events from muon neutrino s_{ν_μ} and tau neutrino interaction s_{ν_τ} for a differential flux of $d\Phi/dE = 10^{-11}\ \mathrm{TeV}^{-1}\mathrm{cm}^{-2}\mathrm{s}^{-1}(E/\mathrm{TeV})^{-2}$. In the last three columns, upper limits are presented on the differential flux $d\Phi/dE = \Phi_0^\nu(E/\mathrm{TeV})^{-2}$ of muon neutrinos, tau neutrinos and both channels combined in units of $10^{-11}\ \mathrm{TeV}^{-1}\mathrm{cm}^{-2}\mathrm{s}^{-1}$.

Source name	RA[h]	Dec[°]	N_{obs}/N_{bg}	μ_{90}	s_{ν_μ} / s_{ν_τ}	$\mathbf{\Phi_0^{\nu_\mu}}$	$\Phi_0^{\nu_\tau}$	$\mathbf{\Phi_0^{\nu_\mu+\nu_\tau}}$
			TeV Blazars					
Markarian 421	11.1	38.2	6 / 7.37	4.1	0.97 / 0.15	**4.2**	27.8	**7.4**
1ES 1426+428	14.5	42.7	5 / 5.52	4.8	0.90 / 0.13	**5.4**	36.6	**9.4**
Markarian 501	16.9	39.8	8 / 6.39	7.9	0.93 / 0.14	**8.5**	57.2	**14.7**
1ES 1959+650	20.0	65.1	5 / 4.77	5.6	0.71 / 0.11	**7.8**	52.2	**13.5**
1ES 2344+514	23.8	51.7	4 / 6.18	3.1	0.89 / 0.15	**3.5**	20.9	**5.9**
			GeV Blazars					
QSO 0219+428	2.4	42.9	5 / 5.52	4.9	0.89 / 0.13	**5.5**	37.6	**9.6**
QSO 0528+134	5.5	13.4	4 / 6.08	3.2	1.06 / 0.14	**3.0**	22.8	**5.3**
QSO 0954+556	9.9	55.0	2 / 6.26	1.4	0.91 / 0.15	**1.6**	9.2	**2.7**
3C273	12.5	2.1	8 / 4.72	9.6	0.96 / 0.10	**10.0**	94.3	**18.0**
			other AGN					
NGC 1275	3.3	41.5	4 / 6.75	2.7	0.95 / 0.14	**2.9**	19.7	**5.0**
M87	12.5	12.4	6 / 6.08	5.3	1.07 / 0.14	**4.9**	38.6	**8.7**
			Microquasars & Neutron star binaries					
LSI +61 303	2.7	61.2	5 / 4.81	5.6	0.75 / 0.13	**7.4**	44.0	**12.6**
SS433	19.2	5.0	4 / 6.14	3.1	1.16 / 0.13	**2.7**	23.6	**4.8**
Cygnus X-1	20.0	35.2	8 / 7.01	7.3	0.95 / 0.15	**7.7**	48.4	**13.2**
Cygnus X-3	20.5	41.0	7 / 6.48	6.4	0.95 / 0.14	**6.8**	46.7	**11.8**
			Supernova Remnants & Pulsars					
Crab Nebula	5.6	22.0	10 / 6.74	10.1	0.98 / 0.15	**10.2**	68.9	**17.8**
Geminga	6.6	17.9	3 / 6.23	2.0	1.01 / 0.14	**2.0**	14.0	**3.5**
PSR 1951+32	19.9	3.3	4 / 6.72	2.7	0.94 / 0.14	**2.9**	19.0	**5.0**
Cassiopeia A	23.4	58.8	5 / 6.00	4.4	0.86 / 0.13	**5.1**	33.2	**8.9**
			Unidentified high energy gamma-ray sources					
3EG J0450+1105	4.8	11.4	8 / 5.94	8.4	1.08 / 0.14	**7.8**	61.6	**13.8**
TeV J2032+4131	20.5	41.5	7 / 6.75	6.1	0.95 / 0.14	**6.4**	43.8	**11.2**

2. Different strategies to search for extraterrestrial neutrinos

Completed in 2000, AMANDA has demonstrated the capability to detect neutrinos and determine their direction of origin (Andres *et al.* 2000). Five years of data have been accumulated between 2000 and 2004, corresponding to a livetime of 1001 days. After applying the reduction procedure aimed to reject the atmospheric muon background (Ahrens *et al.* 2004a), a sample of 4282 candidate neutrinos was selected in the northern hemisphere.

In a first step, a search for neutrinos was done by looking for excesses of events with respect to background coming from selected potential sources. Table 1 shows the different candidate sources reviewed as well as the number of observed and expected events. All the observations are compatible with the atmospheric neutrino background hypothesis. The highest excess found corresponds to the direction of 3C 273 with eight observed events compared to an average of 4.72 expected background (1.2 σ).

A complete survey of the northern hemisphere neutrino sky was then achieved by means of a grid search (Ackermann 2006). The highest significance (3.74 σ) obtained is located at $\alpha = 12.6$ h and $\delta = 4°$. However, the probability to observe this or a higher

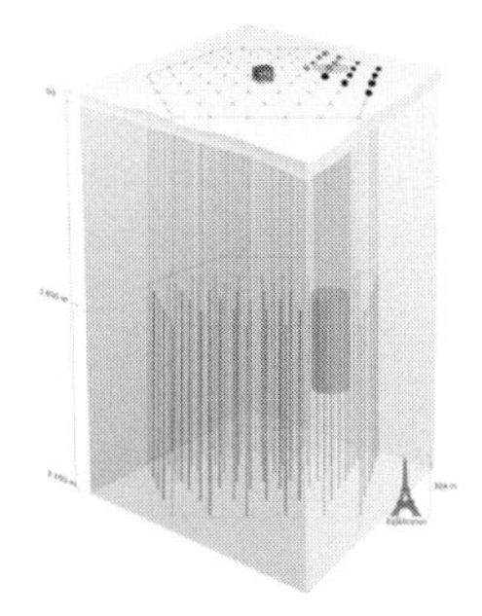

Figure 1. Representation of the IceCube neutrino observatory in its final extension. AMANDA is shown as a darker cylinder.

excess from a random fluctuation of the background, taking into account the trial factor, is 69 %. There is thus no claim of discovery.

As no evidence of a neutrino signal was found in the previous analyses, upper limits on the neutrino flux were determined (cf. Table 1) and compared to specific theoretical predictions. In the case of SS 433, the limit infered in this analysis excludes 0.4 times the corresponding flux predicted in Distefano *et al.* (2002). The upper limit for Cygnus X-3 and the predicted flux in Bednarek (2005) are of the same order of magnitude. However, the upper limits obtained for pulsar wind nebulae and AGN's are at least one order of magnitude above predictions, except for the most optimistic case of neutrino production in the jets of *EGRET* blazars predicted by Neronov & Semikoz (2002) and Neronov *et al.* (2002).

Finally, in order to further increase the sensitivity to particular AGN classes, a stacking analysis was developed: the cumulative signal coming from several AGN's of the same class was evaluated and compared to the corresponding background level (Achterberg *et al.* 2006a). No particular excess was found but the limits could be significantly improved.

3. IceCube status and prospects

The deployment of the IceCube neutrino observatory has started at South Pole during the austral summer 2004–05 (Achterberg *et al.* 2006b). IceCube is currently composed of 9 strings deployed in the ice and 16 IceTop surface cosmic ray air shower detector units (cf. Fig. 1). The strings consist of 60 optical sensors each deployed at a depth between 1450 m and 2450 m. Once completed in early 2011, one cubic kilometre volume of ice will be instrumented with 80 strings (4800 optical modules) and an array of one square kilometre will be covered by IceTop stations at the surface (320 optical modules). The 604 deployed sensors consist of 25 cm diameter photomultipliers and associated electronics, housed in a transparent pressure vessel. The first data taken with the 9 string array (about 140 events/second) are consistent with expectations based on detailed computer simulations.

According to initial potential performance studies (Ahrens *et al.* 2004b), the angular resolution of IceCube will be better than 1° for muon energies above 1 TeV and the effective area for muon detection will exceed 1 km^2 above 10 TeV. It is expected that the limit on an E^{-2} flux of diffuse neutrinos will be about thirty times smaller than the limit reached during a similar period of observation with AMANDA.

First tests show that the deployed hardware meets its performance goals (Achterberg *et al.* 2006b). IceCube is already producing physics data and will rapidly reach an unprecedented sensitivity to sources of extra-terrestrial neutrinos in the next few years.

References

IceCube : <http://icecube.wisc.edu>
Achterberg, A., Ackermann, M., Adams, J., *et al.* 2006a, *Astroparticle Phys.*, 26, 129
Achterberg, A., Ackermann, M., Adams, J., *et al.* 2006b, *Astroparticle Phys.*, 26, 282
Ackermann, M. 2006, Ph.D. thesis, Humboldt-Universität, Berlin
Ahrens, J., Bai, X., Bay, R., *et al.* 2004a, *Nucl. Instrum. Meth. in Phys. Res.* A, 524, 169
Ahrens, J., Bahcall, J. N., Bai, X., *et al.* 2004b, *Astroparticle Phys.*, 20, 507
Andres, E., Askebjer, P., Barwick, S. W., *et al.* 2000, *Astroparticle Phys.*, 13, 1
Bednarek, W. 2005, *ApJ*, 631, 466
Distefano, C., Guetta, D., Waxman, E., & Levinson, A. 2002, *ApJ*, 575, 378
Neronov, A., & Semikoz, D. 2002, *Phys. Rev.* D, 66(123003)
Neronov, A., Semikoz, D., Aharonian, F., & Kalashev, O. 2002, *Phys. Rev. Lett.*, 89, 051101

Highlights of Astronomy, Volume 14
IAU XXVI General Assembly, 14-25 August 2006
Karel A. van der Hucht, ed.

doi:10.1017/S1743921307012288

Site testing at Dome C: history and present status

Jean Vernin, Abdelkrim Agabi, Eric Aristidi, Max Azouit, Merieme Chadid, Eric Fossat, Tatiana Sadibekova, Hervé Trinquet and Aziz Ziad

LUAN, UMR 6525, Université de Nice – Sophia Antipolis, F-06108 Nice Cedex 1, France
email: vernin@unice.fr

1. Introduction

The idea of starting an astronomical site testing in Antarctica began during a congress organized by French Académie des Sciences, in 1992, and entitled 'Recherches polaires-Une Stratégie pour l'an 2000'. At this time, one of us (Vernin 1994) gave a proposal for an astronomical site testing in Antarctica. This proposal was rapidly followed by a meeting between Al Harper (from 'Center for Astrophysical Research in Antarctica', Chicago), Peter Gillingham (from the Anglo Australian Observatory, Australia) and Jean Vernin (from Nice University) at Lake Geneva, Wisconsin, in 1993. It was decided to investigate what was the astronomical quality of South Pole station, each institute bringing its own participation: CARA, the South Pole infrastructure, University of New South Wales, a PhD student and Nice University its expertise and instruments.

On September 7th, Mike Dopita and John Storey presented a proposal to Roger Gendrin (former head of Institut Polaire, Michel Glass (former IFRTP head) and Jean Vernin. Later, following the project of a French-Italian base to be setup at Dôme C, we presented the first France-Italian-Australian proposal for Astrophysics at Dôme C, Paris, November 11th 1994.

2. South Pole site testing 1994–1995

The first astronomical site testing at South Pole took place from April to August 1994. We investigated the surface layer, attaching many sets of microthermal probes at various altitude on a mast (Marks *et al.* 1996) and next year we launched 15 balloons instrumented with microthermal probes to assess the optical turbulence vertical profile (Marks *et al.* 1999). From this two year campaign, it appeared clearly that almost all the optical turbulence was concentrated within the first 200 m of the surface layer. The overall seeing was $1''.86$, but only $0''.37$ when excluding those first 200 m.

From Fig. 1 (*middle*) one can see the huge injection of kinetic energy in the surface layer. This kinetic energy is mixing parcels of air with very large variation in potential temperature (refractive index) giving rise to huge optical turbulence (*bottom*).

It was thus clear that katabatic winds were inducing this important wind shear and that one had better to investigate antarctic sites where no katabatic winds were expected. Indeed, at South Pole, even if the terrain seems quite flat, there remains still some slope from Dôme A and Dôme C, which triggers such a flow.

3. Summer-winter Dôme C site testing

Site testing operations began at Dôme C in 1995 during the stay of one of us (JV). Operations started again during summers 2000 to 2004 during which many DIMM measurements were performed (Aristidi *et al.* 2003, 2005). One of the main conclusion is that

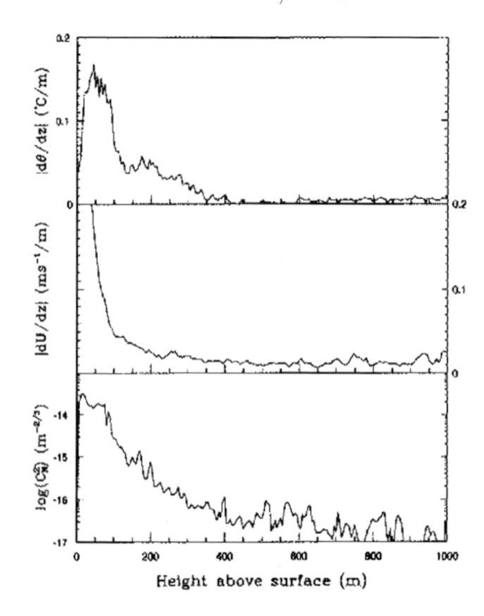

Figure 1. From *top* to *bottom*: vertical profile of the vertical gradient of the potential tempearature, the gradient of the wind velocity and the $C_N^2(h)$ within the first km above ice, as observed at South Pole. The same behavior is observed at Dôme C with less intensity and over a 30 m thick surface layer.

the median seeing is 0″.54 and that every day during a 4 hr period the seeing is better that 0″.5 arcsec. For solar as well as IR astronomy (in bands were sky background from the sun is not annoying) Dôme C seems one of the best place in the world.

In 2005, AstroConcordia station was first open for winterover. One of us (AA), setup two DIMMs and launched successfully about 40 balloons (Azouit & Vernin 2005) instrumented for $C_N^2(h)$ and **V**(h) profiles. Again, it became obvious that most of the turbulence was generated within the surface layer (Agabi *et al.* 2006), as it was found at South Pole, but, the depth of the SL is 30 m instead of 200 m and the optical turbulence is much less. This site is much better since almost all the large telescope have their mirror at such an altitude. For smaller telescopes, stiff platform might be envisaged in order to operate above the SL.

In 2006, a second winterover took place with the installation of two new instruments: the Generalized Seeing Monitor (GSM) and the Single Star Scidar (SSS). From the first instrument the outer scale of the wavefront of the light seems to be smaller than everywhere in the world, i.e., 10 m. The SSS worked all along the whole polar night during about 400 hours, giving thousands of $C_N^2(h)$ and **V**(h) profiles.

References

Agabi, A., Aristidi, E., Azouit, M., *et al.* 2006, *PASP*, 118, 344
Aristidi, E., Agabi, A., Vernin, J., *et al.* 2003, *A&A* (Letters), 406, L19
Aristidi, E., Agabi, A., Fossat, E., *et al.* 2005, *A&A*, 444, 651
Azouit, M., & Vernin, J. 2005, *PASP*, 117, 536
Marks, R. D., Vernin, J., Azouit, M., *et al.* 1996, *A&AS*, 118, 1
Marks, R. D., Vernin, J., Azouit, M., Manigault, J.F., & Clevelin, C. 1999, *A&AS*, 134, 161
Vernin, J. 1994, *Recherches polaires – Une stratégie pour l'an 2000, Paris, 16-17 Décembre 1992*

Highlights of Astronomy, Volume 14
IAU XXVI General Assembly, 14-25 August 2006
Karel A. van der Hucht, ed.

doi:10.1017/S174392130701229X

Single Star Scidar first light from Dôme C

Jean Vernin, Merieme Chadid, Eric Aristidi, Max Azouit, Tatiana Sadibekova and Hervé Trinquet

LUAN, UMR 6525, Université de Nice - Sophia Antipolis, F-06108 Nice Cedex 1, France
email: vernin@unice.fr

1. Introduction

In the recent years, a lot of instruments have been put into operation during the polar summer at Dôme C., Then, during the first polar night when the Astro-Concordia sation was open for the first time during winter, about 40 balloons (Azouit & Vernin (2005)) instrumented to measure optical turbulence profiles and 2 Differencial Image Motion Monitors (DIMM) were setup. The main results from this first important campaign are found in Agabi *et al.* (2006). It appears from this first night time observations that almost all the optical turbulence was concentrated in the first 30 m above the ice. At an elevation of 8.5 m above the ice the seeing is about $1''.4$, while above an elevation of 30 m the seeing drops down to $0''.36$. This last figure is coherent with the estimation from Lawrence *et al.* (2004) if one takes into account that they were not sensitive to the first 30 m., which corresponds to the turbulent surface layer.

For the second winter, we decided to implement the so-called Single Star Scidar (for SSS see Habib *et al.* 2005, 2006) in order to assess continuously the vertical profiles of both the optical turbulence and the wind speed. Indeed, a balloon gives a cut of the atmosphere with a very good vertical resolution but it traverses optical turbulent layers in few seconds. The SSS technique is able to retreive both $C_N^2(h)$ and $\mathbf{V}$(h) vertical profiles from the groud up to 25–30 km each 15 s, during hours. At Dôme C, and tracking Canopus bright star, it becomes possible to monitor $C_N^2(h)$ and $\mathbf{V}$(h) during days almost continuously.

2. First light

During spring 2005 began the construction of the Antarctic SSS which was sent to Dôme C during the fall of the same year. Then the instrument was setup on top of a 8.5 m high plateform (see Fig. 1) by one of the authors (MC) with the precious help of E. Aristidi, and, on February 4th, we got the first light from this 40 cm telescope (see <http://www-luan.unice.fr/CHADID/chadid-aristidi.htm>).

Then, night time measurements where performed by E. Aristidi during the whole winter, from March to August. Thousands of profiles were obtained during almost 400 hr of observations. From this huge set of measurements, only a small part have been processed yet in our laboratory, since only few minutes of observations can be sent by e-mail per day.

3. First profiles

To imagine what will be the installation of a large telescope at Dôme C, and what will be the consequences of an interaction between such a building and the optical turbulence concentrated in the surface layer, it was of major importance to have the detailed structure of both $C_N^2(h)$ and $\mathbf{V}$(h) profiles. But the vertical resolution of the SSS is around one km. Thus we decided to leave the Simulated Annealing method to reconstruct four

Figure 1. Installation of the Single Star Scidar on a 8.5 m high plateform at Dôme C during summer 2005.

arbitrary layers at ice level, and the other layers distributed every 1 km. Of course, SSS is not able to distinguish between the altitude of the first four layers, but we assumed that the wind speed is increasing from the bottom to the top of the surface layer, and thus it became possible to sort the four first layers with increasing speed.

As we already know that most of the optical turbulence is concentrated within the first 30 m, and that we are very interested in the $C_N^2(h)$ and $\mathbf{V}$(h) profiles within those 30 m, we left the SA algorithm to reconstruct four layers within the SL. In Fig. 2 one can see the rapid decrease of the optical turbulence intensity and the rapid increase of the wind speed. The seeing deduced from the optical turbulence profile is 0″.56, very close to the 0″.6-0″.7 measured by the DIMM at the same time.

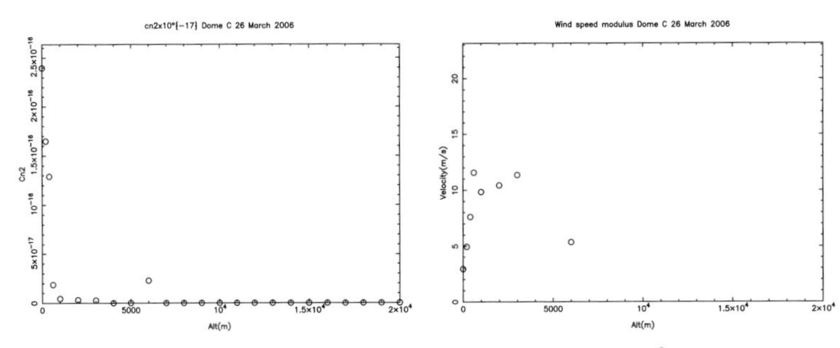

Figure 2. *Left*: Optical turbulence profile. Rapid decrease of the C_N^2 in the surface layer and a layer at 6 km. *Right*: Wind speed rapidly increases from 3 to 13 ms^{-1} in the SL.

4. Conclusion

The Antarctica Single Star Scidar was installed at Dôme C during January–February 2006 and ran almost continuously from March to August giving thousands $C_N^2(h)$ and $\mathbf{V}$(h) profiles. This will help to have a better knowldge of the optical turbulence within the surface layer as well as the free atmosphere.

References

Agabi, A., Aristidi, E., Azouit, M., *et al.* 2006, *PASP*, 118, 344
Aristidi, E., Agabi, A., Vernin, J., *et al.* 2003, *A&A* (Letters), 406, L19
Azouit, M., & Vernin, J. 2005, *PASP*, 117, 536
Habib, A. H., *et al.* 2005, *Comptes Rendus Acad. Sci. Paris*, 3, 385
Habib, A., Vernin, J., Benkhaldoun, Z., & Lanteri, H. 2006, *MNRAS*, 368, 1456
Lawrence, J. S., Ashley, M. C. B., Tokovinin, A., & Travouillon, T. 2004, *Nature*, 431, 278

Highlights of Astronomy, Volume 14
IAU XXVI General Assembly, 14-25 August 2006
Karel A. van der Hucht, ed.

doi:10.1017/S1743921307012306

Optical sky brightness at Dome C, Antarctica

Suzanne L. Kenyon and John W.V. Storey

School of Physics, University of New South Wales, Sydney, NSW 2052, Australia
email: suzanne@phys.unsw.edu.au , j.storey@unsw.edu.au

1. Introduction

Dome C, Antarctica is a prime site for astronomical observations in terms of climate, wind speeds, turbulence, and infrared and terahertz sky backgrounds (for example, see Aristidi *et al.* 2005; Storey *et al.* 2005). However, at present little is known about the optical sky brightness and atmospheric extinction. Using a variety of modelling techniques, together with data from the South Pole, the brightness of the night sky at Dome C is estimated in Kenyon & Storey (2006) including the contributions from scattered sunlight, moonlight, aurorae, airglow, zodiacal light, integrated starlight, diffuse Galactic light and artificial sources. The results are compared to Mauna Kea, Hawaii. We summarise the main conclusions.

2. Discussion

The high latitude of Dome C has an impact on the number of formal astronomical dark hours that the site experiences. Although Dome C has less total dark time than sites closer to the equator, when cloud-cover is taken into account Dome C may have a comparable number of cloud-free dark hours to Mauna Kea. The atmosphere at Dome C is very clear and this should lead to reduced sky brightness contributions from scattered sunlight and moonlight, and should reduce the atmospheric extinction in the optical. At Dome C the Moon never rises higher than between about 33° and 43°, depending on the 18 yr lunar nodal cycle. Modelling shows that moonlight is expected to contribute less at Dome C than at Mauna Kea because of the lower elevation angles. Dome C is close to the centre of the annular auroral region in the southern hemisphere. Aurora will generally be no higher than 7° above the horizon and further than about 1,200 km away. Aurorae are expected to have a minor impact in the optical. Zodiacal light is expected to be less at Dome C than at Mauna Kea because the ecliptic plane is always close to the horizon. Airglow emissions at Dome C are thought to be about the same brightness as those at temperate sites. Integrated starlight is anticipated to be negligible because of the excellent seeing and low atmospheric extinction at Dome C. Diffuse Galactic light may be brighter at Dome C than Mauna Kea because the Galactic plane is always close to the zenith, however this contribution is not large when compared to other sources of sky brightness. Sky brightening by artificial light sources should be non-existent at Dome C, if proper planning is put into place.

We conclude that Dome C is a very promising site not only for infrared and terahertz astronomy, but for optical astronomy as well.

References

Aristidi, E., Agabi, A., Fossat, E., *et al.* 2005, *A&A*, 444, 651

Kenyon, S. L., & Storey, J.W.V. 2006, *PASP*, 118, 489

Storey, J. W. V., Ashley, M. C. B., Burton, M. G., & Lawrence, J. S. 2005, in: M. Giard, F. Casoli & F. Paletou (eds.), Proc. Dome C Astronomy and Astrophysics Meeting, *EAS-PS*, 14, 7

Highlights of Astronomy, Volume 14
IAU XXVI General Assembly, 14-25 August 2006
Karel A. van der Hucht, ed.

doi:10.1017/S1743921307012318

Atmospheric scintillation at Dome C, Antarctica

Suzanne L. Kenyon[1], Jon S. Lawrence[1], Michael C. B. Ashley[1], John W. V. Storey[1], Andrei Tokovinin[2] and Eric Fossat[3]

[1]School of Physics, University of New South Wales, Sydney, NSW 2052, Australia
email: suzanne@phys.unsw.edu.au

[2]Laboratoire Universitaire d'Astrophysique de Nice, UMR 6525,
Université de Nice – Sophia Antipolis, F-06108 Nice Cedex 1, France

[3]Cerro Tololo Inter-American Observatory, Casilla 603, La Serena, Chile

1. Introduction

Dome C, Antarctica is one of the most promising astronomical sites in the world (Fossat & Candidi 2003, and references therein). Dome C boasts low wind speeds, very cold temperatures and little precipitation. The atmospheric turbulence is very weak compared to temperate sites, leading to sub-arcsecond seeing conditions (Lawrence *et al.* 2004; Agabi *et al.* 2006).

A Multi-Aperture Scintillation Sensor (MASS) was operated at Dome C (123°21″ E, 75°06″ S, 3260 m) during the first two months of the 2004 Antarctic winter season. The MASS instrument measures the scintillation of a single star; from this information the vertical distribution of atmospheric turbulence is derived. These data have been analysed in terms of seeing (Lawrence *et al.* 2004) and scintillation (Kenyon *et al.* 2006), here we summarise the main conclusions of the second paper and look at the implications for photometry and astrometry. The results are compared to similar data from Cerro Tololo (70°48″ W, 30°09″ S, 2215 m) and Cerro Pachon (70°44″ W, 30°14″ S, 2738 m) in Chile.

2. Results

A comparison of the turbulence profiles measured above Dome C with those of the two Chilean sites shows that Dome C has significantly less turbulence in all layers except the lowest layer. The most striking result is that the turbulence measured in the highest layer above Dome C is negligible compared to the Chilean sites. It is this high-altitude turbulence that has the largest influence on the astrometric and photometric precision achievable at a particular site. Using average wind speed profiles, we assess the photometric noise produced by scintillation, and the atmospheric contribution to the error budget in narrow angle differential astrometry.

2.1. *Photometry*

High-precision photometry is important, for example, for the detection of extra-solar planets and for observations of objects with very fast intensity changes (e.g., asteroseismology). The calculation of the photometric precision from atmospheric turbulence profiles depends on the length of the integration time and the size of the telescope. For the case of a long integration time on a large diameter telescope the photometric precision is expressed as

$$\sigma_I = \left[10.7 \int h^2 C_n^2(h) V^{-1}(h) \mathrm{d}h\right]^{1/2} D^{-2/3} t^{-1/2}, \tag{2.1}$$

where h is the height above the site, C_n^2 is the refractive index structure constant, V is the wind speed, D is the telescope diameter and t is the integration time. Fig. 1 (*left*) shows the median photometric precision at each site as a function of telescope diameter

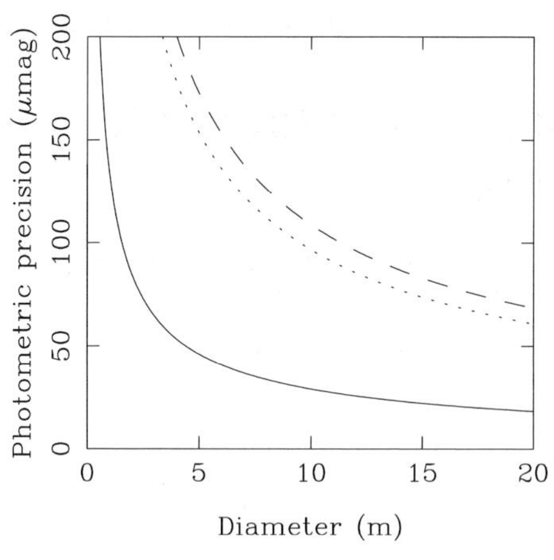

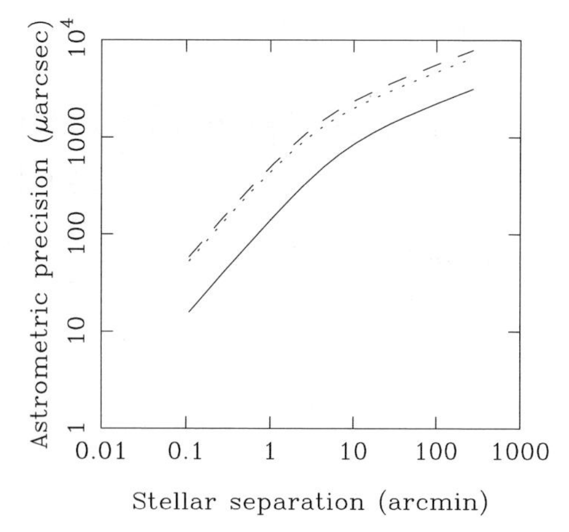

Figure 1. *Left*: the median photometric precision at Dome C (*solid line*), Cerro Tololo (*dashed line*) and Cerro Pachon (*dotted line*) as a function of telescope diameter, for a 60 s integration time. *Right*: the median astrometric precision at each site as a function of angular stellar separation, for a 1 hr integration time and 10 m baseline.

for a 60 s integration. Dome C offers a gain of about 3.6 in photometric precision over the two Chilean sites.

2.2. *Astrometry*

Long baseline interferometry can be used to achieve high precision very narrow angle differential astrometry, benefiting science programs such as extrasolar planet searches and the study of close multiple star systems. Uncertainties in astrometric precision arise from instrumental effects and atmospheric effects. The astrometric uncertainty caused by the atmosphere can be calculated for two regimes, narrow-angle and very narrow angle, using the equations from Shao & Colavita (1992).

$$\sigma_{atm} = \begin{cases} \theta^{1/3} t^{-1/2} \left[5.25 \int h^{2/3} C_n^2(h) V^{-1}(h) \mathrm{d}h\right]^{1/2} & \text{narrow angle, } \theta\bar{h} \gg B \\ \theta B^{-2/3} t^{-1/2} \left[5.25 \int h^2 C_n^2(h) V^{-1}(h) \mathrm{d}h\right]^{1/2} & \text{very narrow angle, } \theta\bar{h} \ll B \end{cases} \tag{2.2}$$

where, B is the baseline length, θ is the angular separation between the two celestial objects and $\bar{h}$ is the turbulence weighted atmospheric height. Figure 1 (*right*) shows the median astrometric precision at each site as a function of angular separation for a 1 hr integration and a 10 m baseline. Dome C offers a significant advantage in achievable astrometric precision.

3. Conclusions

Although the data from Dome C cover a fairly limited time frame, they lend strong support to expectations that Dome C will offer significant advantages for photometric and astrometric studies. Dome C offers a potential gain of about 3.6 in both long integration photometric precision and narrow-angle astrometry precision when compared to two mid-latitude sites in Chile.

References

Agabi, A., Aristidi, E., Azouit, M., *et al.* 2006, *PASP*, 118, 344
Fossat, E., & Candidi, M. (eds.) 2003, *Mem. S.A.It. Supp.*, 2, 3
Kenyon, S. L., Lawrence, J. S., Ashley, M. C. B., *et al.* 2006, *PASP*, 118, 924
Lawrence, J. S., Ashley, M. C. B., Tokovinin, A., & Travouillon, T. 2004, *Nature*, 431, 278
Shao, M., & Colavita, M. M. 1992, *A&A*, 262, 353

Highlights of Astronomy, Volume 14
IAU XXVI General Assembly, 14-25 August 2006
Karel A. van der Hucht, ed.

doi:10.1017/S174392130701232X

AMICA – the infrared eye at Dome C

Alberto Riva[1], Mauro Dolci[2], Oscar Straniero[2], Filippo Maria Zerbi[1], Emilio Molinari[1], Paolo Conconi[1], Vincenzo De Caprio[1], Gaetano Valentini[2], Gianluca Di Rico[2], Maurizio Ragni[2], Danilo Pelusi[2], Igor Di Varano[2], Croce Giuliani[2], Amico Di Cianno[2], Angelo Valentini[2], Favio Bortoletto[3], Maurizio D'Alessandro[3], Carlotta Bonoli[3], Enrico Giro[3], Daniela Fantinel[3], Demetrio Magrin[3], Leonardo Corcione[4], Maurizio Busso[5], Gino Tosti[5], Giuliano Nucciarelli[5], Fabio Roncella[5], Carlos Abia[6] and the IRAIT Team

[1]INAF Osservatorio Astronomico di Brera – Merate, Merate, Italy
[2]INAF Osservatorio Astronomico Collurania Teramo, Teramo, Italy
[3]INAF Osservatorio Astronomico di Padova, Padova, Italy
[4]INAF Osservatorio Astronomico di Torino, Torino, Italy
[5]University of Perugia, Perugia, Italy
[6]University of Granada, Granada, Spain

1. Introduction

AMICA (Antarctic Multiband Infrared Camera) is a dual-channel Infrared Imager (2-28 μm), that will be located at the Nasmyth focus of the IRAIT telescope at Dome C. Dome C base, on Antarctic plateau offers an unique chance for infrared astronomy. It has several advantages like temperature, pressure and site environment. Temperature, around $-60°$C (mean) allows a good atmospheric stability (good seeing and good windows transparency) a low atmospheric background and the reduction of instrumental background. Pressure (equivalent of 4000 m a.s.l.), implies low content of water vapors; this means higher transmission, broader and new astronomical windows. The site offers the possibility of very long observations (about 6 months winter night).

2. Science

Most of the scientific targets of the IRAIT-AMICA project have been selected to take profit of the environmental conditions described above. Indeed the first goal is the site testing and characterization of the Dome C sky, in the bands between 2 to 28 μm, followed by a survey in these bands. Once the above general goals will be achieved a variety of cool IR targest could be observed with IRAIT-AMICA. Among these: AGB stars in the Milky Way, star formation regions in our and nearby galaxies, Solar System bodies, supernova remnants, extragalactic point-like sources, etc. Last, but not least, the IRAIT-AMICA collaboration will allow the development of the first permanent astronomical observatory in the Dome C base, with the possibility of find and solve a lot of environmental problems of a quasi-space situation.

3. The instrument

AMICA is a near-mid infrared camera, with a wavelength coverage between 2–28 μm. The camera has an all-reflective design (in order to reduce aberrations) and is composed by two off axis parabolas and two gold-coated plain mirrors. The wavelength coverage is done with two detectors: the SWA (Short Wavelength Array) that covers from 1 to 5.5 μm (InSb, 258 × 258 pixels, Raytheon), and the LWA (Long Wavelength Array) that covers from 7 to 28 μm (SiAs, 128×128 pixels, DRS Technologies). The instrument toggles between the LWA and the SWA by means of a folding mirror, before the detectors. The focal reduction of AMICA is 1:1.47 and the best sampling is at 3.42 μm for SWA and 8.54 μm for LWA. Plate scales are $0''.538$/pixel (SWA) and $1''.345$/pixel (LWA). The field of view is $2'.29 \times 2'.29$ for SWA and $2'.87 \times 2'.87$ for LWA. The entrance window is made of CdTe that guarantees and high transmission for all the 1–25 μm microns band. The first-light set of filters will be a standard set in order to calibrate the camera with existing standards (K, L, M, N_1, N_2, Q_1, and Q_2).

4. The cryostat

The internal temperature of the cryostat will be around 35 K except for the LWA detector that will be at 7 K. All internal components (mirrors, optical bench, mountings, filterwheel, etc.) will be in aluminum, in order to have an homotetic contraction during the cooling-down process. This allows to align the optical components at room temperature and maintain the alignment at any temperature, drastically reducing the number of regulations and the necessity of human operations during the functioning time. The cryocooler will be an ARS 2-stage with a power consumption of 3.5 kW. The choice of this element was a critical point: the power supply from the base is limited, and the low density of the atmosphere, causing limited convection, requires an increased capability of heat removal.

5. Electronic and housekeeping

Due to atmospheric conditions (temperature, pressure, etc.) we decided to keep only few elements at external temperature: the cryocooler head and the 'remote' unit. All the other subsystems will be placed in a temperature-controlled rack. Also for the electronic system we decided to divide the readout and control electronic in two subsections: the first is a 'remote unit' next to the telescope, with essential controls, the second is a 'local unit', inside the base. The software development has been designed as a modular structure, in order to drive both the telescope and the camera.

6. Testing subsystems

The AMICA team has designed and built ANTARES (ANTARctic Environment Simulator), a climatic chamber specifically designed to test each component before the delivery at Dome C of AMICA. ANTARES allows the test of small and medium size components with a pressure of 410 mbar, a temperature of -60°C and relative humidity of 6 %.

References

Dolci, M., Straniero, O., Valentini, G., *et al.* 2006, *SPIE*, 6267, 48
Tosti, G., Busso, M., Nucciarelli, G., *et al.* 2006, *SPIE*, 6267, 47
Valentini, G., Magrin, D., Riva, A., *et al.* 2006, *SPIE*, 6267, 40
Bailey, J. 1996, *PASA*, 13, 7
Burton, M. G., Storey, J. W., & Ashley, M. C. 2000, *SPIE*, 4005, 326B
Burton, M. G., Lawrence, J. S., Ashley, M. C. B., *et al.* 2005, *PASA 22*, 199
Chamberlain, M. A., Ashley, M. C. B., Burton, M. G., *et al.* 2000, *ApJ*, 535, 501
Harper, D. A. 1989, *AIP-PC*, 198, 123
Lawrence, J. S. 2004, *PASP*, 116, 482
Pel, J.-W., Glazenborg-Kluttig, A. W., de Haas, J. C., *et al.* 2000, *SPIE*, 4006, 164
Storey, J. W. V., Ashley, M. C. B., Lawrence, J. S., *et al.* 2003, *Mem. S.A.It. Suppl.*, 2, 13
Walden, V. P., Town, M. S., Halter, B., & Storey, J. W. V. 2005, *PASP*, 117, 300

Highlights of Astronomy, Volume 14
IAU XXVI General Assembly, 14-25 August 2006
Karel A. van der Hucht, ed.

doi:10.1017/S1743921307012331

Design and construction of the moving optical systems of IRAIT

Josep Colomé[1], Carlos Abia[2], Inma Domínguez[2], Jordi Isern[1,3], Gino Tosti[4], Maurizio Busso[4], Giuliano Nucciarelli[4], Fabio Roncella[4], Oscar Straniero[5] and Mauro Dolci[5], for the IRAIT Collaboration

[1]Institut d'Estudis Espacials de Catalunya (IEEC), E-08034 Barcelona, Spain
email: colome@ieec.cat

[2]Dpto.Física Teórica y del Cosmos, Universidad de Granada, E-18071 Granada, Spain

[3]Institut de Ciències de l'Espai (CSIC), E-08034 Barcelona, Spain

[4]Dipartimento di Fisica, Universitá di Perugia, via A. Pascoli, I-06100 Perugia, Italy

[5]INAF-Osservatorio di Teramo, Via Maggini, I-64100 Teramo, Italy

1. Introduction

The IRAIT (International Robotic Antarctic Infrared Telescope) project (Tosti *et al.* 2006) is based on a 80 cm aperture telescope to observe in the infrared range. It is due to start operations in spring 2008, several months after installation in Dome C (Antarctica). We describe the contributions made to such project by the Institute for Space Studies of Catalonia (IEEC) and the University of Granada, whose participation has been mainly focused in developing the moving optical system for the secondary (M2) and tertiary (M3) mirrors of the telescope. Moving parts of the optical system provide focusing and chopping capabilities, implemented in M2, and a rotation mechanism, implemented in M3, allow observation in either Nasmyth foci. The work package includes the design and construction of both mirrors, the mechanical supports, the electronics and the control software, all prepared to work at the low temperatures at Antarctica. A Spanish company, NTE, was contracted to carry out the design and manufacture. Tests at low temperature and integration in the telescope were finished during summer 2006, before sending the telescope to Antarctica, scheduled by the end of the same year.

2. Secondary mirror driver subsystem

The M2 Drive Subsystem is the system in charge of providing the following movements to the mirror: (1) Focus: movement of the mirror in Uz axis, (2) Chopping: tip-tilt mirror movement. Three main parts compose the assembly: (*a*) fixed subassembly; (*b*) mobile subassembly or focuser; and (*c*) chopper and M2 mirror subassembly. A number of parts are made of stainless steel in order to prevent corrosion or important dimensional variations caused by the coefficient of thermal expansion that could block the system, leaving it unusable at certain temperatures. Other materials are used: anodized aluminium, carbon fiber composite and Teflon.

(*a*) Fixed subassembly: holds the cryogenic motors and also contains the guiding system, the safety switches and the backward and forward actuator.

(*b*) Focuser: the focusing mechanism is realized by means of a linear actuator, manufactured by INA, which include a motor and a reduction stage (Pythron VSS32.200.1.2.UHVC and VPGL 32 i-50 UHVC units). The stroke of the actuator is 100 mm, and the screw pitch is 2 mm per revolution. Two limit switches prevent the actuator from getting over the limits. These guiding systems have been considered as the most appropriate due to its compactness, high load capacity and accuracy on corrosion resistance. The cryogenic motor supplies the mechanical power to move this assembly over a temperature range that goes from $-270°$C up to $30°$C. The rotational movement of the motor is converted into a lineal movement through a roller screw. The roller screw is manufactured on a G1 ISO quality corrosion proof that assures the accuracy, repeatability and steadiness of the advancing movement. The kinematics mechanism has been dimensioned for not being back drivable. All mobile parts are lubricated with cryogenic grease with melting point of $-90°$C in order to avoid crystallization that would lead to a halt of the system.

(*c*) Chopper and mirror assembly: the chopper assembly is the mechanism that provides the tilting movement on the XY plane to the M2 mirror. The mechanism is designed to provide an angular oscillation from 0 to ± 4.6 mrad (equivalent to $5' \times 5'$ in the sky) over the specified temperature range. This device allows us to obtain a maximum chopping frequency of 25 Hz, compatible with the lower resonance frequency of the telescope top ring which was estimated to be about 80 Hz. Within the mobile parts there are the piezoactuators, manufactured by PiezoMechanik technology, that provide the displacement to activate the chopping movement. They have cryogenic capabilities and are close loop servo actuated. The sensors that supply the feedback are Eddy current. The reason of this choice instead of a capacitance sensor is that they are more suitable for adverse environment conditions, such as moisture presence or possible ice deposition on the sensor surface. The chopper subassembly is composed for the following parts: radial spring, piezo stacks, mirror support, M2 mirror, sensor position, tilting adjusting system.

3. Tertiary mirror driver subsystem

The M3 Drive Subsystem disposes of the following performances: (*i*) Locates the mirror in proper place to receive the light from M2 and send it to the cameras located at the Nasmyth focus. To obtain a correct position, two adjusting mechanisms have been implemented: Tilt Correction Mechanism, in base, and Mirror Position Mechanism, in mirror base; (*ii*) Allows mirror rotation in *z*-*z* axis from 0 to 180° with high accuracy; (*iii*) Allows fixation of the whole subsystem in the M3 Interface Area.

This subassembly can be divided in the next main parts and mechanisms:

(*a*) M3 Mirror: main element of the subsystem. It has an oval shape with a rear square interface to place and fix the element.

(*b*) Mirror position mechanism: mechanism in charge of adjusting the mirror orientation once the M3 Drive subsystem is assembled. It consists on a rear flexure, used to join the mirror and the tube, and one adjustment screw, used to modify the angle orientation.

(*c*) Tube: element in charge of providing stiffness to the subsystem. It is made of stainless steel and disposes of a lower flange to fix it to the rotating actuator and an upper flange to accommodate the mirror position mechanism. This upper flange is 45° inclined to direct the M2 light to the Nasmyth focus.

(*d*) Rotating actuator: it is used to allow z-z rotation and *z*-*z* adjustment of the subsystem is based on a Micos PRS-110 precision rotation stage modified for operation in a low-temperature environment (down to $-80°$C) The unit incorporates two electrical limit switches that will be set at the 0° and 180° positions. The repeatability of these switches (5 μm) provides an angular repeatability of $0°.008$.

(e) Axial alignment mechanism: mechanism located below the rotating actuator and it is able to rectify misalignment in the mirror due to mechanical tolerances. This mechanism can tilt the tube in two axis thanks to a platform driven by three screws.

(f) Base: element in charge of the following issues: support all the subsystem with enough stiffness, fix all the assembly in the interface area, allow to plug and unplug the Rotating Actuator wires and connectors, allow to actuate the tilt correction mechanism, allow inner access for cleaning purposes. An Hex tool accommodation is placed in the base centre. It permits to drive the adjustment screw of the Mirror Position Mechanism in order to tilt the mirror.

Acknowledgements

This project is supported by the Spanish Ministry of Education and Science. IRAIT is a project approved by the ARENA consortium, network funded by the European Community's 6th Framework Programme. We acknowledge NTE S.A. and Marcon Telescopes for the outwork.

Reference

Tosti, G., Busso, M., Nucciarelli, G., *et al.*, 2006, *SPIE*, 6267, 47

Highlights of Astronomy, Volume 14
IAU XXVI General Assembly, 14-25 August 2006
Karel A. van der Hucht, ed.

doi:10.1017/S1743921307012343

Multi-aperture interferometry at Concordia

Eric Fossat[1], Farrokh Vakili[1], Eric Aristidi[1], Bruno Lopez[2], François-Xavier Schmider[1], Karim Agabi[1], Jean-Baptiste Daban[1], Fatmé Allouche[1], Adrian Belu[1], Pierre-Marie Gori[1], Géraldine Guerri[1] and Bruno Valat[1]

[1]LUAN, UMR 6525, Université de Nice – Sophia Antipolis, Parc Valrose, F-06108 Nice, France
email: eric.fossat@unice.fr

[2]Observatoire de la cote d'Azur, BP 4229, F-06304 Nice, France

1. Introduction

The next generation (post-VLTI) of multi-telescope interferometric arrays operated in optical/infrared wavelengths should be kilometric, from 1 to 10 km. The Concordia station offers a unique opportunity to set such an interferometer in the best atmospheric conditions presently known on Earth.

2. KEOPS – the concept

The Dome C site astronomical qualities begin to be well bracketed. After several summers and now almost two winter-over site testing campaigns, it is clear that it is, for many astronomical parameters, the best, or one of the very best sites on Earth. Some of these parameters still demand additional investigation or more statistics. But the global quality has been proved to be enough out of range for attracting an ever increasing scientific community. French and Italian funding has started to be raised, so that beyond the site testing, real astronomical programmes are expected to be operated in 2008 (IRAIT from Italy and A-STEP from France).

On the longer term range, medium and far infrared imaging on one hand, and on the other hand Extremely High Resolution Imaging even in visible light are among the favorite targets in the prospect studies. Some are thinking of an Antarctic ELT, to be set above the 30-m turbulent surface boundary layer, others would prefer an multi mirror interferometer. Such an interferometer, that could be called an optical equivalent to the VLA in radio waves or ALMA in millimetric, can possibly be regarded as the next generation, post-VLTI, of large size optical interferometry.

Of course, optical long baseline imaging interferometry is extremely difficult, as the technical challenges go more or less as the inverse of the wavelength, and that means a factor 100 to 1000 for optical or near-IR as compared to the millimetric case of ALMA. However and to some extent, it can now be regarded as a mature observing technique. Several optical arrays are able to provide 2-D maps: NPOI in Arizona, COAST at Cambridge, UK, CHARA at Mount Wilson, California and of course the VLTI at Paranal, Chile. At the 2004 Liège International Astrophysical Colloquium devoted to the Science case for next generation optical/infrared interferometric facilities (the post-VLTI era), it was recognised by Pierre Lena that on one hand *"the next interferometer generation should operate at least from 1 to 12 μm and have kilometric baselines (1 to 10 km at most)"*, and on the other hand *"the Dome C site characteristics, as far as they are known today, appear to be of an entirely different class than any other ground-based site: in fact, this site classifies as an intermediate one between space and conventional ground.*

They appear especially favorable for interferometry (transparency, isoplanetism, stability of the atmosphere, area)".

The concept of KEOPS results from these statements. It emerges as an imaging array of optical diffraction limited telescopes of 1.5 to 2-m diameter in Dome C conditions. These telescopes are spread over three concentric rings of 200, 348 and 676 meter radii. Six or seven telescopes on the first ring, 12 or 13 on the second, 18 or 19 on the outer one. These numbers offer optimized *u-v* coverage to achieve a 1 mas resolution at 10 μm in order to resolve the angular distance between a star and its exo-Earth at a one kpc distance. KEOPS is an implicitly co-phased array operated in the so-called hypertelescope mode (Labeyrie *et al.* 2003), but using a more efficient nulling design named IRAN (Vakili *et al.* 2004). KEOPS has an equivalent collecting surface comparable to the Keck interferometer, but located in extreme cold, dry and excellent seeing conditions of the Antarctica plateau. It will challenge a 30 m-class ELT, and the number of available square kilometers on the polar plateau is essentially unlimited!

3. KEOPS – the science rationale

The bottom line of an interferometric instantaneous field of view is the Airy disc of individual telescopes. Considering the 1.5 to 2-m diameter proposed for KEOPS, one may expect a sub-mas resolution across a 1 arcsec field of view. Thus, unlike classical wide field telescopes, KEOPS offers Ultra High Spatial Resolution imaging with a reasonable wide field of 2000×2000 resolution elements (resels). The inefficient filling factor (less than 10^{-3}) could be compensated by Earth rotation synthesis for imaging compact objects which benefit from the long polar night of Antarctica. Beyond the search for exo-Earths, it could bring significant breakthroughs in the study of galactic and extra-galactic objects from the visible to the thermal infrared wavelenths inaccessible from any other ground based site. That could be stellar surface imaging, the central engines of YSO's, the cores of AGN, or even the ballet of stars rotating around the central black hole of galaxies as far as a few millions light-years. In fact the number of possible scientific scoops with such an instrument is nearly infinite.

References

Labeyrie, A., Le Coroller, H., Dejonghe, J., *et al.* 2003, *SPIE*, 4852, 236
Vakili, F., Aristidi, E., Abe, L., & Lopez, B. 2004, *A&A*, 421, 147

Highlights of Astronomy, Volume 14
IAU XXVI General Assembly, 14-25 August 2006
Karel A. van der Hucht, ed.

doi:10.1017/S1743921307012355

Antarctica – a case for 3D-spectroscopy

Andreas Kelz
Astrophysikalisches Institut Potsdam, An der Sternwarte 16, D-14482 Potsdam, Germany
email: akelz@aip.de

1. Advantages of 3D-spectroscopy

DS or Integral-Field Spectroscopy (IFS) provides multiple spectra for each point of a 2-D field, rather than along a narrow, 1-D spectrograph slit only. Therefore, IFS does not require very accurate telescope pointing, nor do pre-assumptions about slit or aperture sizes have to be made. It avoids any 'slit-losses' due to seeing or atmospheric dispersion, which eliminates the need for any parallactic alignment or a dispersion compensator (see Fig. 1).

Integral-field units (IFUs) with 100 % fill factor (e.g., PMAS, Roth *et al.* 2005) can be used for accurate spectrophotometry (Kelz & Roth 2006). As all the information is gathered at the same time, 3D-spectroscopy is more efficient than any scanning technique and insensitive to variable instrumental and atmospheric conditions. The resulting data-cube (with coordinates in RA, Dec, and lambda) allows both a PSF-optimized extraction of single and combined spectra, as well as the re-construction of narrow- and broad-band images, without the need for filters. As the sky background around the target is recorded with better coverage than with slits, an improved background subtraction, in particular in crowded fields, is possible (Becker *et al.* 2004). Additional results from post-processing, such as differential images, abundance ratio maps, or velocity fields can be extracted with little effort from the data cube. Obviously, spectroscopy of any complex structures such as galaxies, mergers, nebulae, winds, or jets benefits from the 2-dimensional field-of-view. The various advantages of 3DS are discussed in Roth *et al.* (2004).

Certain IFUs, such as the PPak fiber bundle (Kelz *et al.* 2006), provide very high instrumental grasp, i.e., light collecting power. The availability of 2-D information allows spatial binning of spectra to improve the signal-to-noise, in particular for low surface brightness objects, even further. For projects where flux collection, rather than spatial resolution is an issue, binning the IFU spaxels has the same effect as increasing the aperture size of a telescope. In case the spatial position of the target is not known well enough (e.g., optical counterparts of X-ray sources, γ-ray bursts, or because the target is too faint to be visible at the acquisition system), the integral-field provides an increased error circle to ensure that the target is not missed altogether. If the location of spectral features is uncertain (e.g., because the redshift is unknown a priori), 3DS is the only technique that can reliable detect these. For extra-galactic or cosmological applications, the 3D-data cube corresponds to a volume in space, which otherwise can only be recorded with time-consuming scanning techniques using tunable filters (Bland-Hawthorn 2006).

2. Relevance for Antarctica

While the above advantages of 3DS are of general nature, some of them are particularly important at a remote location such as in Antarctica, where highly autonomous or robotic telescopes are required (Ashley *et al.* 2004). The case stated here is applicable to the optical/near-IR domain, i.e. to future spectroscopic instrumentation and related science cases as proposed for a PILOT-like telescope (Burton *et al.* 2005).

Given the environmental conditions in Antarctica (Storey *et al.* 2005), it is desirable to reduce the amount of movable components as a potential source of failure. 3DS completely

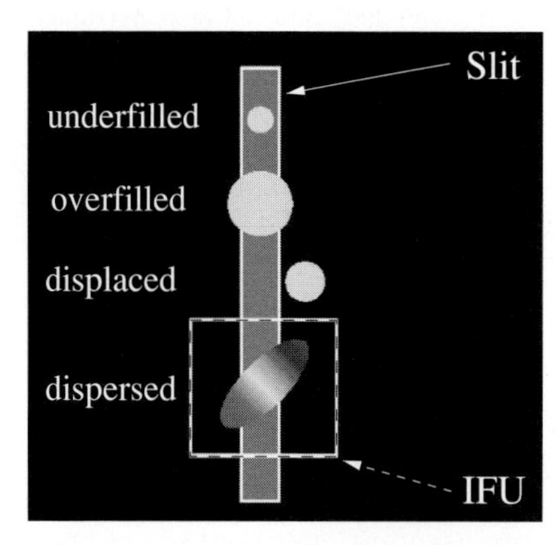

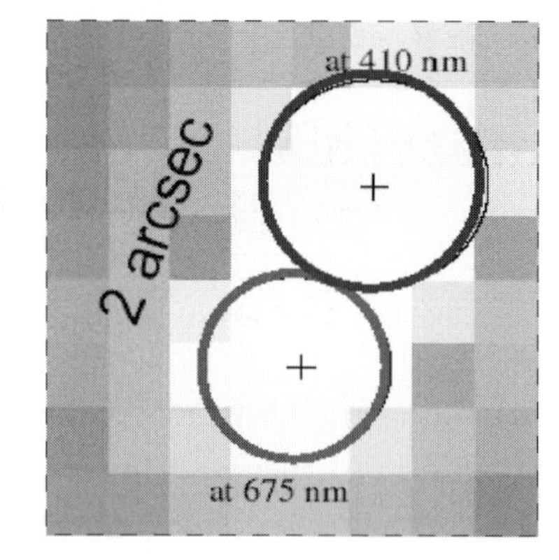

Figure 1. *Left*: Sketch of the common problems present in classical slit-spectroscopy. From top to bottom: under-filling and over-filling of the slit, mispointing, atmospheric dispersion and parallactic misalignment. *Right*: A re-constructed image of a star, observed with an integral–field-unit (IFU) at an air mass of 1.7. Despite a dispersion of $2''$ between 410 nm and 675 nm, the IFU records the entire flux, avoiding any slit-losses or chromatic errors.

avoids the need for a (rotatable) filter wheel, any slit width or angle adjustments or an ADC. If the IFU is fiber-coupled, the subsequent instrumentation can be mounted remotely from the telescope in a stable and climatized environment. This would imply that the telescope and fiber-link needs to be adapted to the Antarctic conditions, but not the spectrograph as such. The background subtraction, in particular for the OH-bands in the NIR, is improved by IFS. Furthermore, IFS may be operated with a nod-&-shuffle mode (Roth *et al.* 2002) or fiber Bragg gratings (Bland-Hawthorn 2006) may be used for future fiber-coupled instruments. The precision requirements for telescope pointing, target acquisition, guiding and tracking are less stringent for IFUs, which greatly relaxes the demands on the accuracy of drives, gears and motors for the telescope and reduces frequent re-calibrations due to any ice-drift.

In summary, the use of innovative IFUs eliminates much of the complexity, present in classical spectroscopy (Kelz 2004). It relaxes acquisition requirements and removes critical, movable parts from the system. This simplifies the instrumental design and minimizes potential sources of failure. 3DS allows a fast and reliable 'point-and-expose' observational approach, which is ideally suited for remote or robotic observations. At the same time, it offers multiplex and time-saving advantages for a broad range of scientific projects, ranging from stellar population studies to cosmology, that are proposed for a large telescope at Antarctica.

References

Ashley, M. C. B., Burton, M. G., Lawrence, J. S., & Storey, J. W. V. 2004, *AN*, 325, 619
Becker, T., Fabrika, S., & Roth, M. M. 2004, *AN*, 325, 155
Bland-Hawthorn, J. 2006, *New Astron. Revs.*, 50, 237
Burton, M. G., Lawrence, J. S., Ashley, M. C. B., *et al.* 2005, *PASA*, 22, 199
Kelz, A., & Roth, M. M. 2006, *New Astron. Revs.*, 50, 355
Kelz, A., Verheijen, M., Roth, M. M., *et al.* 2006, *PASP*, 118, 129
Kelz, A. 2004, *AN*, 325, 673
Roth, M. M., Kelz, A., Fechner, T. *et al.* 2005, *PASP*, 117, 620
Roth, M. M., Becker, T., Kelz, A., & Schmoll, J. 2004, *ApJ*, 603, 531
Roth, M. M., Fechner, T., Wolter, D., *et al.* 2002, *Exp. Astron.*, 14, 99
Storey, J. W. V., Ashley, M. C. B., Burton, M. G., & Lawrence, J. 2005, *EAS-PS*, 14, 7

Highlights of Astronomy, Volume 14
IAU XXVI General Assembly, 14-25 August 2006
Karel A. van der Hucht, ed.

doi:10.1017/S1743921307012367

CAMISTIC: THz/submm astronomy at Dome C in Antarctica

Vincent Minier, Gilles Durand, Pierre-Olivier Lagage and M. Talvard

Service d'Astrophysique/DAPNIA/DSM/CEA Saclay, F-91191 Gif-sur-Yvette, France
email: vincent.minier@cea.fr

1. Project context

Submillimetre (submm) astronomy is the prime technique to unveil the birth and early evolution of a broad range of astrophysical objects. It is a relatively new branch of observational astrophysics which focuses on studies of the cold Universe, i.e., objects radiating a significant – if not dominant – fraction of their energy at wavelengths ranging from $\sim 100\,\mu$m to ~ 1 mm. Submm continuum observations are particularly powerful to measure the luminosities, temperatures and masses of cold dust emitting objects. Examples of such objects include star-forming clouds in our Galaxy, prestellar cores and deeply embedded protostars, protoplanetary disks around young stars, as well as nearby starburst galaxies and dust-enshrouded high-redshift galaxies in the early Universe.

A major obstacle to carry out submm observations from ground is the atmosphere. Astronomical observations in the submm spectral bands can only be achieved from extremely cold, dry and stable sites (e.g., high altitude plateau, Antarctica) or from space (e.g., the *Herschel Space Observatory*) to overcome the atmosphere opacity and instability that are mainly due to water vapour absorption and fluctuation in the low atmosphere. Chile currently offers the best accessible (all-year long) sites on Earth, where the precipitable water vapour (PWV) content is often less than 1 mm. Chile hosts the best astronomical facilities such as ESO VLT, APEX and Chajnantor plateau will be the ALMA site.

At longer term, and particularly if global warming severely restricts the 200 – 350 – 450 μm windows on ESO sites, Antarctica conditions with less than 0.2 mm PWV, could offer an exciting alternative for THz/submm astronomy (Fig. 1). This is an attractive opportunity for the 200 μm windows, especially, which are normally explored with space telescopes (e.g., *Herschel*).

Observations of submm continuum emission are usually carried out with bolometer detectors. Recently, two Research Departments at CEA (DSM/DAPNIA/SAp and DRT/LETI/LIR) developped filled bolometer arrays for the PACS submm/far-infrared imager on the *Herschel Space Observatory*, to be launched by ESA in 2007. The R&D was based on a unique and innovating technology that combines all silicon technology (resistive thermometers, absorbing grids, multiplexing) and monolithic fabrication. The bolometers are assembled on a mosaic 'CCD-like' array that provides full sampling of the focal plane with $\sim 2{,}000$ pixels that are arranged in units of 256 pixels. They are cooled down to 300 mK to optimise the sensitivity down to the physical limit imposed by the photon background noise. The PACS bolometer arrays have passed all the qualification tests (Billot *et al.* 2006). The newly started ArTéMiS project at CEA Saclay capitalises on this achievement by developing submm (200 – 450 μm) bolometer arrays with $\sim 4{,}000$ pixels for ground-based telescopes. A prototype camera operating in the 450 μm

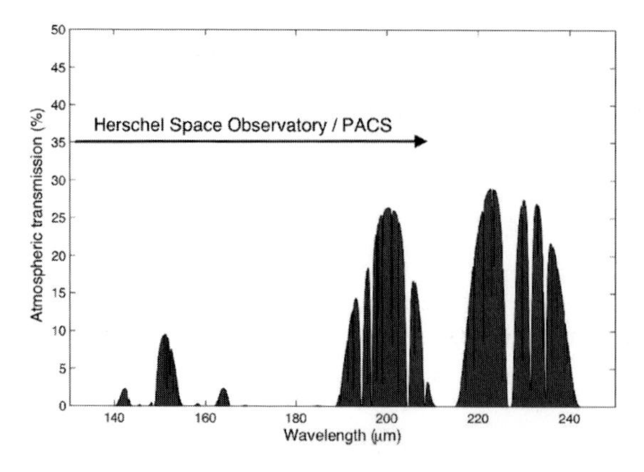

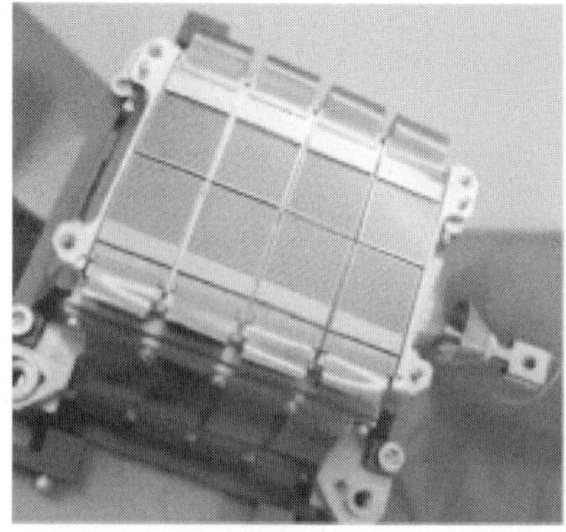

Figure 1. *Left*: Atmospheric transmission between 130 and 250 μm assuming a PWV = 0.2 mm. *Right*: CEA bolometer array for *Herschel Space Observatory*-PACS.

atmospheric window has successfully been tested in March 2006 on the KOSMA telescope (Talvard *et al.* 2006).

In the future, placed on a 12-m single-dish telescope at Dome C, a bolometer camera with $\sim$10,000 pixels at 200 – 450 μm will be particularly powerful to undertake wide field surveys of star-forming complexes in our Galaxy as well as deep field surveys of dust-enshrouded high-redshift galaxies in the early Universe.

2. CAMISTIC objective

The CAMISTIC project aims to install a filled bolometer-array camera with 16×16 pixels on a small telescope (e.g., IRAIT) at Dom C and explore the 200 μm (i.e., THz) windows for ground-based observations. Many windows between 150 and 250 μm are reachable if PWV is below 0.2 mm (Fig. 1), which is an expected value at Dome C (cf. Vernin, this SpS 7). Opening these windows would be an important achievement as this part of the electromagnetic spectrum is usually observed by the mean of space telescopes. Ground-based submm telescopes will have the advantages to be potentially larger than space telescope and, therefore, allow observations with higher angular resolution.

CAMISTIC will be located at about 500 m from the base, with very reduced access. Autonomous and automated cryogenic devices specifically designed for the harsh conditions in Antarctica will therefore be needed. We plan to demonstrate the reliability of a novel cryogenic system with all static parts placed next to cryostat at outer temperature conditions and a warmed cabinet for compressors, motors and valves. Extensive tests in wintering condition will be performed before expedition.

CAMISTIC will be equipped with novel bolometer technology. The filled bolometer array with a monolithic grid of 256 pixels was designed by CEA for the far-IR/submm imager *Herschel Space Observatory*-PACS. It can operate in the 150 – 250 μm range with an adequate filter for each specific window.

References

Billot, N., Agnèse, P., Auguères, J.-L., Béguin, A., *et al.* 2006, *SPIE*, 6265, 9
Talvard, M., André, P., Rodriguez, L., Minier, V. *et al.* 2006, *SPIE*, 6275, 2

Highlights of Astronomy, Volume 14
IAU XXVI General Assembly, 14-25 August 2006
Karel A. van der Hucht, ed.

doi:10.1017/S1743921307012379

A test for the detection of vegetation on extrasolar planets: detection of vegetation in Earthshine spectrum and its diurnal variation

Danielle Briot[1], Karim Agabi[2], Eric Aristidi[2], Luc Arnold[3], Patrick François[4,1], Pierre Riaud[5], Patrick Rocher[1] and Jean Schneider[5]

[1]Observatoire de Paris-Meudon, 61 avenue de l'Observatoire, F-75014 Paris, France
email: Danielle.Briot@obspm.fr

[2]Université de Nice – Sophia Antipolis, 28 avenue Valrose, F-06108 Nice cedex 2, France

[3]Observatoire de Haute-Provence-CNRS, F-04870 Saint Michel l'Observatoire, France

[4]European Southern Observatory (ESO), 3107 Av. de Cordova, Vitacura, Santiago 19, Chile

[5]Observatoire de Paris-Meudon, 5 place Jules Janssen, F-92195 Meudon, France

Abstract. The search for life in extraterrestrial planets is to be tested first with the only planet known to shelter life. If the planet Earth is used as an example to search for a signature of life, the vegetation is one of its possible detectable signature, using the Vegetation Red Edge due to chlorophyll in the near infrared ($0.725\,\mu$m). We focus on the test of the detectability of vegetation in the spectrum of Earth seen as a simple dot, using the reflection of the global Earth on the lunar surface, i.e., Earthshine. On the Antartic, the Earthshine can be seen during several hours in a day (not possible at our latitudes) and so variations due to different parts of Earth, that is to say oceans and continents facing the Moon could be detected.

Keywords. Earth, Moon, astrobiology, techniques : spectroscopy, vegetation red edge, biomarker, Earthshine, exolife

1. Introduction

It could be hoped that in few years (e.g., beyond 2015) we will search for detection of life in terrestrial extrasolar planets. Life on extrasolar planets will probably present unusual and unknown forms. However, as we know nothing about these forms of life, we look for indices of presence of life similar to the one we know on Earth. Firstly, we explore classical biosignatures like H_2O, CO_2, O_3 and O_2, but it is also interesting in visible wavelengths to search how vegetation can be distinguished on a planet seen from space.

2. Detection of vegetation

Vegetation spectrum presents an increase at $0.5\,\mu$m in the green range, which implies that plants are seen as green, but mostly a very sharp rise at $0.725\,\mu$m, known as the Vegetation Red Edge (VRE, Arnold *et al.* 2002), the signature of photosynthetic plants. The Vegetation Red Edge can be much more easily detected than the bump at $0.5\,\mu$m, and this signature corresponds hardly to other elements than chlorophyll. The search for vegetation on exoplanets should be tested with the only planet known to shelter life. Vegetation can be detected on the planet Earth from a spacecraft as done by Sagan *et al.* (1993) using the *Galileo* spacecraft but in that case, vegetation has been detected vertically, and obliquity, limb effects, nor cloud cover have been considered. Earth has to be observed as a whole like we see a extrasolar planet, i.e., as a point source. Presently, no distant spacecraft has the cability to take a spectrum of the whole Earth.

3. Earthshine

Another possibility is to use the Moon as a giant reflector and to observe ashen light or Earthshine. Earthshine can be seen on the dark part of the Moon during the first or the last days of the lunar cycle. This corresponds to a Earth light on the Moon. The light of the Sun arrives on Earth, is reflected by Earth, arrives on the Moon, is reflected by the Moon and comes back on Earth. The light coming from the different parts on Earth is blended and thus, as in the case of an exoplanet, seen integrated. Then: $[Earthshine\, Spectrum] = [Solar\, Spectrum] \times [Earth\, Albedo] \times [Moon\, Albedo]$ and transmitted three times through Earth atmosphere; and: $[Moonligh\, spectrum] = [Solar\, Spectrum] \times [Moon\, Albedo]$ and transmitted once through Earth atmosphere.

Arcichovsky V.M. (1912) suggested to look for chlorophyll absorption in the Earthshine spectrum, with the aim to calibrate chlorophyll in the spectrum of other planets, but at these times, Earthshine observations did not have sufficient spectral resolution (Tikhoff 1914; Danjon 1928). Earthshine shows Rayleigh scattering in the Earth atmosphere and allowes to predict that from space Earth is seen as blue. The red side of the Earth reflectance spectrum shows the presence of O_2 and H_2O absorption bands, while the blue side clearly shows the Huggins and Chappuis ozone (O_3) absorption bands.

4. Results obtained

The first detections of vegetation from the Earthshine spectrum were obtained by Arnold *et al.* (2002) at Haute-Provence, and by Woolf *et al.* (2002) at the Tucson. Observations made at ESO NTT (Hamdani *et al.* 2006) obtain a VRE lower than previous studies, which were near 8 – 10% when Africa and Europe light the Moon (Arnold *et al.* 2002). The present results are from 3 to 4 % when Africa faces the Moon and 1.3% when the Pacific faces the Moon. Even with these lower values, VRE differs over Pacific Ocean *vs.* Africa, thus allowing detection of vegetation on Earth. These observations also show significant variations in Rayleigh scattering depending on cloud cover, implying that Earth as 'pale blue dot' can be almost white.

5. Importance of observations from Dome C, Antarctica

Observations of Earthshine can be done during the first and last days of the lunar cycle. From intermediate Earth's latitudes, observations of the waxing Moon are possible in the evening, and of the waning Moon in the morning, in both cases twilight observations. Only at high latitudes it is possible to observe the Moon in the first or the last days of the cycle during several hours, sometimes even all the day long. This happens at Dome C (75°06′S, 123°21′E), about six times per year. During one observing run, continents and oceans successively face the Moon and the variations of the VRE corresponding to successive 'landscapes' of the planet Earth can be detected. A small telescope and low resolution spectrograph can be used to detect VRE in Earthshine spectra. Observations are carried out by one of us (E.A.) since March 2006.

References

Arcichovsky, V. M. 1912, *Don Cesarevitch Alexis a Novotcherkassk*, Vol. 1, no. 17, 195
Arnold, L., Gillet, S., Lardière, O., Riaud, P., & Schneider, J. 2002, *A&A*, 352, 231
Danjon, A. 1928, *Ann. Obs. Strasbourg*, 2, 165
Hamdani, S., Arnold, L., Foellmi, C., Berthier, J., Billeres, M., Briot, D., François, P., Riaud, P., & Schneider, J. 2006, *A&A*, 460, 617
Tikhoff, G.A. 1914, *Mitteillungen der Nikolai-Haupstrenwarte zu Pulkovo*, no. 62, Band VI2, 15
Sagan, C., Thompson, W. R., Carlson, R., Gurnett, D., & Hord, C. 1993, *Nature*, 365, 715
Woolf, N. J., Smith, P. S., Traub, W. A., & Jacks, K. W. 2002, *ApJ*, 574, 430

Author Index